VEGETATION SCIENCE IN FORESTRY

Handbook of Vegetation Science

FOUNDED BY R. TÜXEN

H. LIETH, EDITOR IN CHIEF

Volume 12/1

Vegetation Science in Forestry

Global Perspective Based on Forest Ecosystems of East and Southeast Asia

Papers from four symposia from the International Congress of Ecology, Yokohama 1990

Edited by

E.O. BOX
Institute of Industrial Science, University of Tokyo, Japan

with
R.K. PEET, T. MASUZAWA, I. YAMADA, K. FUJIWARA
and P.F. MAYCOCK

Kluwer Academic Publishers
Dordrecht / Boston / London

Library of Congress Cataloging-in-Publication Data

Vegetation science in forestry : global perspective based on forest ecosystems of East and Southeast Asia / edited by E.O. Box ... [et al.].
p. cm. -- (Handbook of vegetation science ; pt. 12/1)
Includes bibliographical references and index.

1. Forest ecology. 2. Forests and forestry. 3. Forest ecology--Asia. 4. Forests and forestry--Asia. I. Box, Elgene Owen. II. Series.
QK911.H3 pt. 12/1
[QK938.F6]
581 s--dc20
[581.5'2642'095] 94-683

ISBN 978-90-481-4361-0

Published by Kluwer Academic Publishers,
P.O. Box 17, 3300 AA Dordrecht, The Netherlands.

Kluwer Academic Publishers incorporates
the publishing programmes of
D. Reidel, Martinus Nijhoff, Dr W. Junk and MTP Press.

Sold and distributed in the U.S.A. and Canada
by Kluwer Academic Publishers,
101 Philip Drive, Norwell, MA 02061, U.S.A.

In all other countries, sold and distributed
by Kluwer Academic Publishers Group,
P.O. Box 322, 3300 AH Dordrecht, The Netherlands.

Printed on acid-free paper

Table of Contents

Foreword

More than a decade has passed since Prof. Gisela Jahn completed Vol. 12 of the Handbook of Vegetation Science series dealing with the application of vegetation science in forestry, mostly European forestry. The volume was well received by the critics with the exception that they wanted a more diversified demonstration of forestry related vegetation science work and a wide representation of forest types from different continents.

During the INTECOL Congress in Yokohama Prof. Fujiwara was charged with the organization of a series of symposia which originally was meant to deal primarily with East and Southeast Asian forest vegetation. While this symposium developed, more and more vegetation scientists from other parts of the world offered contributions. Vol. 12 was frequently mentioned since it was at hand in the book exhibits, and the request for a broader coverage of the topic was voiced again.

Drs. Fujiwara, Peet, Box and myself got together and discussed the matter. The contributions were too diverse to justify one particular title but on the other hand so rich in information on the application of vegetation science to forestry that it appeared to me that these symposia provided exactly what the critics had asked for, i.e. an extension of Vol. 12 of the Handbook of Vegetation Science.

Therefore, instead of simply publishing the contributions as a "proceedings" we offered this idea to the organizers of the symposium. Of course this was a new aim for the collection of papers, and the authors had to be asked to respond to this goal.

It was to the merit of Prof. Box that this aspect of the new aim took shape in the form that appears now. He took most of the burden of editing and served as a link between the various editors and symposium organizers. Both Prof. Fujiwara and myself owe many thanks to him.

As it stands now the new volume 12.1 of the Handbook of Vegetation Science is a substantial improvement on the coverage of vegetation science in forestry. The topics covered in this volume widen the scope of vegetation science work in forestry over the phytosociological work, which was originally the main focus in the Handbook (as perceived by Dr. Tüxen but which gradually changed under my editorship).

It is always difficult to place the topics in the Handbooks in particular categories, especially if we are dealing with ecological approaches to vegetation analysis. This might require a future volume (vol. 12.2). As time goes on, the demands of scientific fields change and consequently the research carried out in this particular field changes.

We offer this new volume to the field workers. It contains many aspects for new research to be done in many areas of the world. We hope that the book will be accepted as such. On the other hand, it solidifies the initial objective of Dr. Jahn, who in her volume tried to demonstrate and generalize the application of vegetation analyses for forestry.

Again, both Prof. Fujiwara and I thank all co-organizers of the Symposia and the editorial board and especially Prof. Dr Elgene Box for their work on behalf of this volume, which will substantially add to the completeness of this handbook for the application of vegetation science in forestry.

Helmut Lieth
Osnabrück and Yokohama
October 20, 1993

1. Forest ecosystems of East and South-East Asia in a global perspective*

TATUO KIRA
Lake Biwa Research Institute, Otsu, Shiga, 520 Japan

Abstract

Owing to the influence of the Tibeto-Himalayan Highland, the east- and southeast-coastal zone of the Asian continent lacks the subtropical dry belt, which elsewhere in the world separates temperate forest areas from the humid tropics. Thus forest climates with sufficient rainfall cover coastal areas of the Western Pacific from subarctic eastern Siberia to equatorial SE Asia, giving rise to a sequence of five forest formations. The warm-temperate evergreen (lucidophyll) forest formation is peculiar to this region and represents the east-coast counterpart of Mediterranean sclerophyll forest, which develops under the west-coast climate with a different pattern of seasonal rainfall distribution.

The characteristics of the five formations and their latitudinal distribution were briefly described as being related to climatic environments. Certain ecosystem properties such as floristic diversity and the rate of organic matter decomposition were found to be very sensitive to thermal gradients, whereas the responses of plant biomass, primary productivity, etc. to changes in thermal climate were less apparent.

Key words: East Asia, ecosystem functions, forest zones, thermal climate, Western Pacific

* First appeared in Ecological Research 1991; Vol. 6: pp. 185–200. Reproduced with permission from The Ecological Society of Japan.

E.O. Box et al. (eds.), Vegetation Science in Forestry, 1–21.

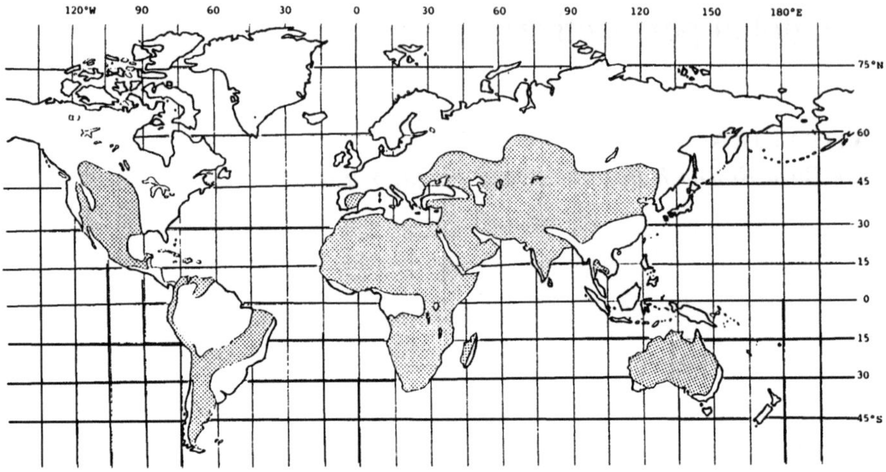

Fig. 1. Approximate distribution of arid areas (shaded), including desert, thorn scrub, steppe, savanna and open woodland ecosystems, throughout the world.

Introduction

The forest vegetation of the Western Pacific, or the eastern and southeastern coasts of Eurasia with their attached continental islands, is unique from the viewpoint of global phytogeography in the following aspects.

(1) This region lacks the subtropical dry belt characterized by the year-round prevalence of a dry subtropical air mass, which elsewhere in the world separates the temperate forest area from the humid tropics (Fig. 1). Closed forest vegetation covers this area continuously from subarctic eastern Siberia to tropical Southeast Asia.

(2) Therefore, the Western Pacific coast offers a unique opportunity to observe the full differentiation of forest zones along a latitudinal thermal gradient and associated changes in the structure and function of forest ecosystems.

Climatological background

According to the classical diagram (Fig. 2) of global climatic zones by Alisov (1954), four zones with more or less stable air masses, i.e. equatorial (tropical), subtropical, mid-latitude (temperate) and polar, encircle the globe at different latitudes. In each of the zones indicated by odd numbers, 1, 3, 5 and 7, a single air mass prevails throughout the year. In the areas between

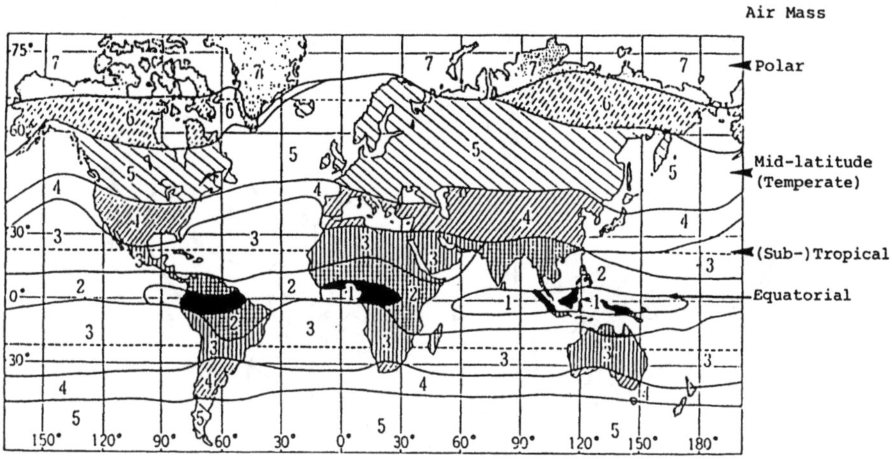

Fig. 2. Climatic zones of the world (Alissow, 1954).

these stable zones, two adjacently located air masses alternate in the summer and winter halves of the year due to their seasonal shift in a north/south direction. This results in seasonal reversal of the prevailing wind direction, so that the zones with even numbers, 2, 4 and 6, are called monsoon climate zones.

The subtropical air mass is by far the most stable of the four, and gives rise to the subtropical dry high-pressure belt in which most desert areas of the world lie. It is, however, noteworthy that the continuity of subtropical air mass Zone 3 is broken only in South and Southeast Asia, south of the Himalayan Range.

The existence of the Tibeto-Himalayan Highland affects the climate of the Western Pacific profoundly. In winter, it serves as a barrier against the southward outflow of cold polar air, which thus accumulates to develop a powerful continental high-pressure system in the inland area of eastern Siberia. This leads to acceleration of the NW winter monsoon over the Western Pacific, causing it to extend more southward than in other parts of the northern hemisphere. The NW monsoon reaches as far south as the southern tip of Taiwan and the central mountain range of Hainan Island (Fig. 3, upper).

With the beginning of the hot season, the warming-up of this vast highland causes a sudden translocation of the subtropical jet stream from its winter course along the southern fringe of the Himalayas to its summer route along the northern border of the highland toward central Japan (Fig. 3, lower). This jump of the subtropical jet stream, which normally takes place within

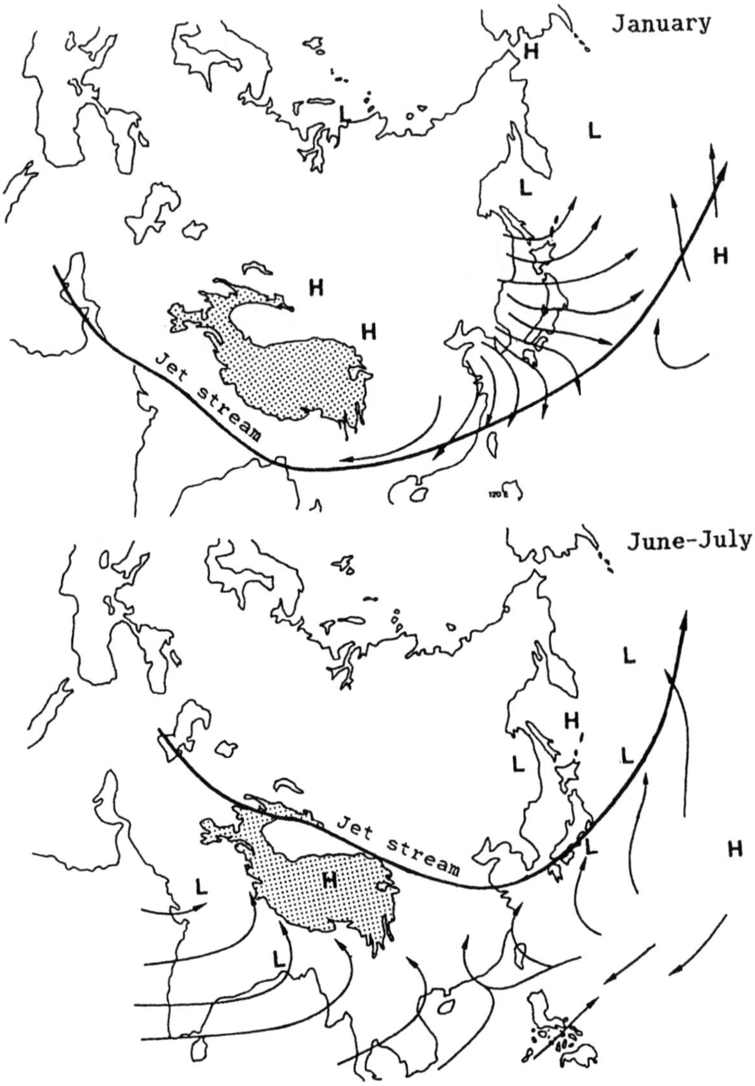

Fig. 3. Winter and summer courses of subtropical jet stream over Asia, with prevailing wind systems (fine arrows). Shaded area indicates the Tibeto-Himalayan Highland over 3000 m in altitude. H and L refer to high- and low-pressure areas, respectively. Redrawn and simplified from the original figure by Ogawa and Matsumoto (1987).

several days in May–June, allows the influence of the equatorial air mass as the SW monsoon to extend almost to the northern end of Japan.

The extensive development of summer and winter monsoon winds brings abundant rainfall over a wide latitudinal range of temperate to subtropical coasts of eastern Asia, thereby preventing the year-round prevalence of the ever-dry subtropical air mass. As a result, forest vegetation covers the coasts and islands of the Western Pacific continuously from the equator to the arctic forest limit without interruption at middle latitudes.

The climate of this region is remarkably continental, as expected from the above conditions. The Okhotsk Sea is known for its extensive freezing in winter in spite of its relatively low latitude. The winter monsoon brings heavy snowfall over the northwestern shores of the Japanese Archipelago between 35°30′ N and 45°30′ N, while the summer in the southern half of Japan is as hot as in Southeast Asia. Certain properties of the East-Asian forest vegetation are related to this climatic continentality.

The north/south continuity of humid climate along the Western Pacific coasts seems to have persisted through the Pleistocene glacial periods, when no continental ice sheet developed in this part of the Eurasian continent. The northward migration route has always been open for southern floral elements, making them abundant in temperate forests of Eastern Asia, particularly Japan. For example, the subarctic spruce/fir forests and cool-temperate beech forests of the Japanese Archipelago are characterized by the dominance of *Sasa* and other dwarf bamboos as undergrowth. The physiognomy of warm-temperate evergreen forests in SW Japan and south-central China is often 'tropical' with abundant woody lianas, epiphytes, and an undergrowth of ferns, zingibers, etc.

Latitudinal differentiation of forest formations

Subarctic evergreen conifer forest formation

Subarctic forests of the Maritime Province (Primorye) of Siberia and the northernmost part of Japan are generally dominated by *Picea jezoensis* and *Abies* spp., and are therefore similar in floristic composition to those in other parts of the Holarctic Region.

The *Picea*/*Abies* forests of Sakhalin and Hokkaido are different from those on the continental side in that the former are often characterized by a dense undergrowth of dwarf bamboo, as mentioned above. The species of *Abies* are also different, *A. sachalinensis* being found in Hokkaido and Sakhalin, and *A. nephrolepis* in northern Korea, northeastern China and eastern Siberia. Such a difference in corresponding forest types between the chain of islands from Sakhalin to Taiwan and the adjacent coasts of continental Eurasia may be attributed at least partly to the difference in climatic continentality, and help to understand the environmental background of forest distribution in the Western Pacific as discussed later.

Cool-temperate deciduous broadleaf forest formation

The main part of Japanese cool-temperate forest vegetation is dominated by Japanese beech, *Fagus crenata*. Beech forest is found in the northernmost part of Honshu (main island) as well as on its backbone mountain ranges. It also extends northward to the southernmost part of Hokkaido (Fig. 4). However, the entire lowlands of Hokkaido except its northeastern end (Shiretoko Peninsula) still belong to the cool-temperate zone, where the original vegetation is believed to have been a different type of forest consisting of a deciduous oak (*Quercus mongolica* var. *grosseserrata*), other deciduous hardwood species such as *Tilia*, *Acer* and *Ulmus*, and also *Abies sachalinensis*.

This cool-temperate mixed forest has several dominant hardwood species in common with its continental counterpart in northeastern China, Korea and Ussuri, and is apparently a derivative of the latter, which also lacks *Fagus* and is characterized by an abundance of *Pinus koraiensis* instead of *Abies* (Yim, 1977; Editorial Committee for the Vegetation of China, 1980; Zhao, 1980). Although the mixed hardwood/conifer forest of Hokkaido is located to the north of the beech forest, the distribution of the two forest types cannot be ascribed to a thermal gradient, because the beech forest covers the whole cool-temperate zone in the altitudinal sequence of forest vegetation on mountains of Honshu. It seems to result from the northward increase in climatic continentality along the Japanese island chain.

The distribution of potential vegetation types in Hokkaido can be reasonably interpreted in terms of the *warmth index* (Kira, 1945, 1949, 1977a and b), as shown in Fig. 4. Apparently, the distribution of subarctic conifer forest and that of cool-temperate mixed forest are limited by the critical warmth index values of 15°C month and 45°C month, respectively.

Warm-temperate lucidophyll forest formation

This is the forest formation peculiar to the warm-temperate zone of the Far East, and is dominated by evergreen oaks of the genera *Quercus* (*Cyclobalanopsis*), *Castanopsis* and *Lithocarpus* with associated rich evergreen tree species such as *Persea*, *Cinnamomum*, *Ilex*, *Symplocos* and *Camellia*.

It is the east-coast counterpart of the Sclerophyll or Mediterranean Forest formation on the western side of Eurasia and other continents under climates with a winter concentration of rainfall, and differs from the latter in having

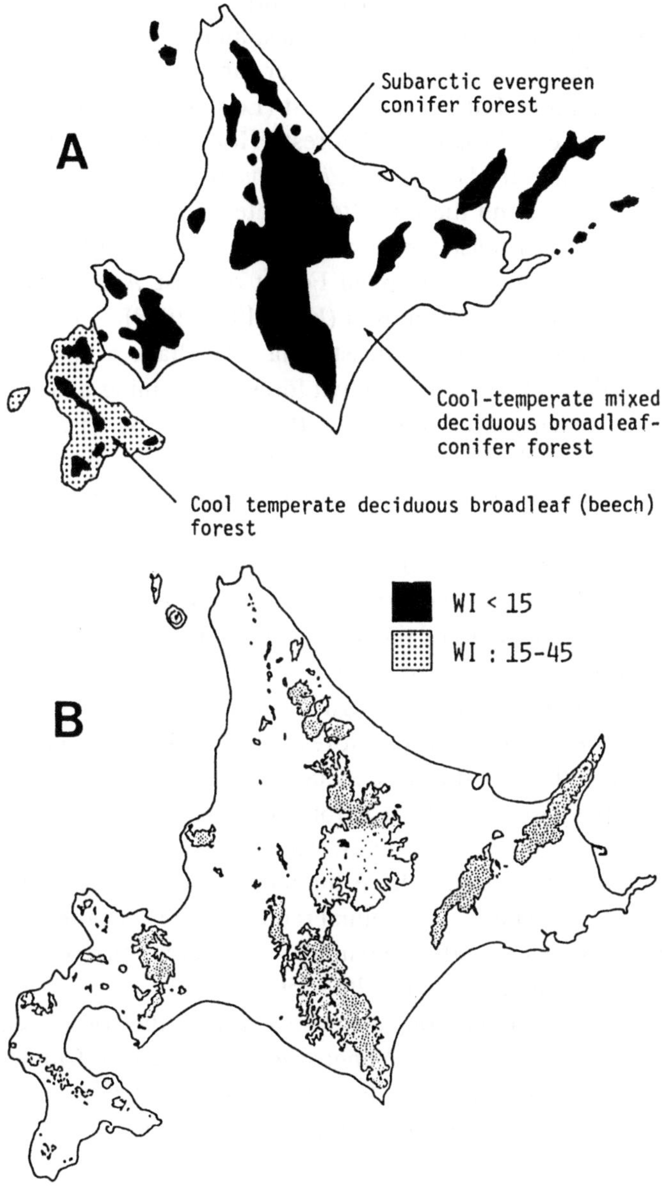

Fig. 4. Potential areas of main forest types (upper: after Miyawaki and Itow, 1966 and Yoshioka, 1973) and the distribution of warmth index (WI) (lower: °C month) in Hokkaido. The warmth index is the summation of monthly mean temperature (t°C) with 5°C as the threshold; i.e. WI = $\Sigma(t-5)$ where the summation is made for the months in which $t > 5$.

less xeromorphic characteristics such as larger shiny leaves, a richer flora including many epiphytes and woody lianas, and larger tree size.

This forest type has been described under various names such as temperate rain forest (Schimper, 1903), laurel forest (Lorbeerwald) (Rübel, 1930) and subtropical evergreen broadleaved forest (Hou et al., 1956). However, it does not contain laurel (*Laurus nobilis*), it is not located in the subtropical zone, as stated later, and the term 'temperate rain forest' should preferably be reserved for supermoist temperate forests in the southern hemisphere. The new nomenclature, Lucidophyll Forest or East Asian Evergreen Oak Forest, has therefore been proposed (Kira, 1977; 1978).

Montane rain forests of Southeast Asia often have a floral composition somewhat similar to that of lucidophyll forest, containing a large number of evergreen oak species. Most probably, these two forest types are genealogically related to each other, but they can hardly be identified with the same forest formation because of the wide difference in associated flora and physiognomy.

The warm-temperate zone of the southeastern United States used to be included in the Deciduous Forest Climax by U.S. plant ecologists (Weaver & Clements, 1929) in spite of the similarity of climate to the corresponding zone of the Western Pacific, particularly in the seasonal distribution of rainfall (concentrated in summer). Although some authors have emphasized the significance of relic evergreen oak forests in Florida and Louisiana (Box & Fujiwara, 1988), there is still little evidence that the lucidophyll forest once spread extensively over this area before the original vegetation changed to the fire subclimax of southern pines. The shortage of rainfall as compared with the Western Pacific may have limited the development of lucidophyll forest in the United States. The annual rainfall in the warm-temperate region of the United States is 1000–1400 mm yr^{-1}, whereas that in the Asian lucidophyll forest area amounts to 1300–2500 mm yr^{-1}.

The same rainfall limitation may also apply to the forest vegetation on both sides of the Chang Jiang (Yangtze River) in central China, where the original vegetation is presumed to have been mixed deciduous/evergreen forest instead of typical lucidophyll forest (Fig. 5) (Hou et al., 1956; Wang, 1961; Song, 1983, etc.).

The borderline between cool-temperate beech forest and warm-temperate lucidophyll forest corresponds approximately to the warmth index isotherm of 85°C month under a maritime climate. However, the northward or upward (in altitudinal sequence) distribution of broadleaf evergreens actually tends to be limited by the duration of low temperature in winter instead of the 'warmth' during the growing season, while the summer temperature is responsible for the southward or downward limits of beech (*Fagus crenata*) forest. As a result, there are areas in the inland districts of Honshu where

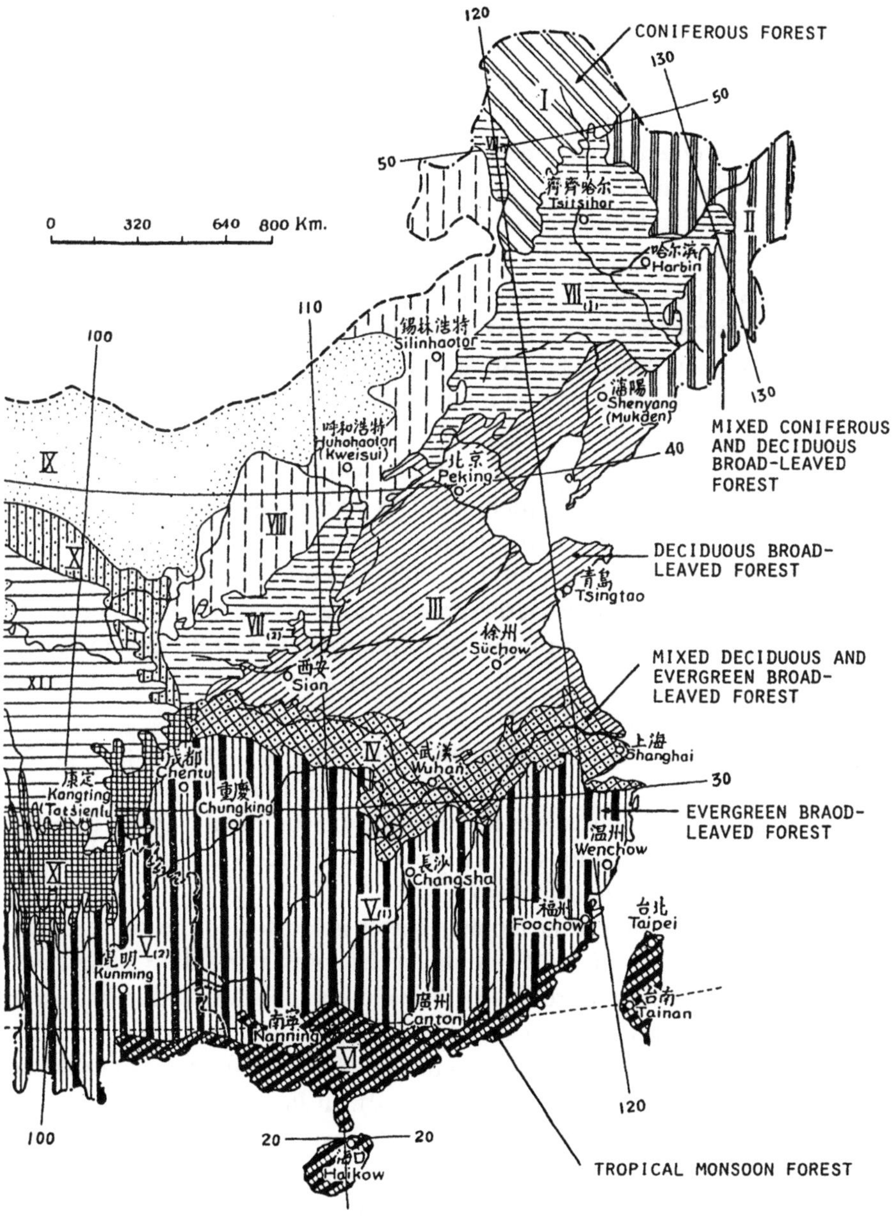

Fig. 5. Potential vegetation map of China (Hou et al., 1956).

the growing season warmth is still within the range of the warm-temperate zone but cannot support the growth of broadleaf evergreen tree species due to increased winter coldness. Kira (1949, 1977) identified a type of deciduous forest, lacking both lucidophyll species and *Fagus crenata* and characterized by trees such as *Quercus serrata*, *Carpinus* spp., *Fagus japonica* and sometimes *Abies firma*, and called it the Warm-temperate Deciduous Broadleaf Forest as distinguished from the cool-temperate deciduous forest types. This term may also be applied to the deciduous forest types of the southeastern United States and central China mentioned above, in which the critical climatic factor is probably the shortage of rainfall rather than winter coldness. The critical winter coldness limiting the distribution of warm-temperate broadleaf evergreen forest can be approximated by a critical *coldness index* value of −10°C month (Kira, 1949, 1977a and b).

In the northeastern coastal area of continental Asia, warm-temperate deciduous broadleaf forest covers most of its warm-temperate zone, i.e. the central and southern parts of the Korean Peninsula, the southern part of Liaoning Province in China, etc. The only exception is a narrow strip of lucidophyll forest along the southernmost shore of the Korean Peninsula (Fig. 6). Such distribution of the two warm-temperate forest types in Korea is also subject to the same climatic limitations as in Japan as stated above (Yim and Kira, 1975; Yim, 1977).

The occurrence of deciduous broadleaf forest in the northern part of the warm-temperate zone thus offers further evidence that a slight increase in climate continentality may allow the invasion of continental types into the sequence of forest formations which developed under the mild maritime climate of the Japanese Archipelago.

The vast area of Chinese lucidophyll forest is located to the south of the warmth index isotherm of 140°C month, which runs along the Chang Jiang (Figs. 5 and 6). Local differences in floristic composition within this area have not yet been described in detail, but one of the most representative components is apparently a large number of *Castanopsis* species (Editorial Committee for the Vegetation of China, 1980). In Japan, on the other hand, a greater part of the lucidophyll forest zone lies within a warmth index range of 100–140°C month, and *Castanopsis* is represented by a single small-leaved species, *C. cuspidata*. The floristic composition of Japanese lucidophyll forest is obviously much less diverse than that of its Chinese counterpart. It may therefore be reasonable to divide the warm-temperate lucidophyll forest zone into two subzones, northern and southern, by the 140°C month isotherm. The warm-temperate zone on the southern slope of the central Nepal Himalayas is also divided into an upper zone (*Quercus* (*Cyclobalanopsis*) zone) and a lower zone (*Schima*/*Castanopsis* zone) (e.g. Kawakita, 1956).

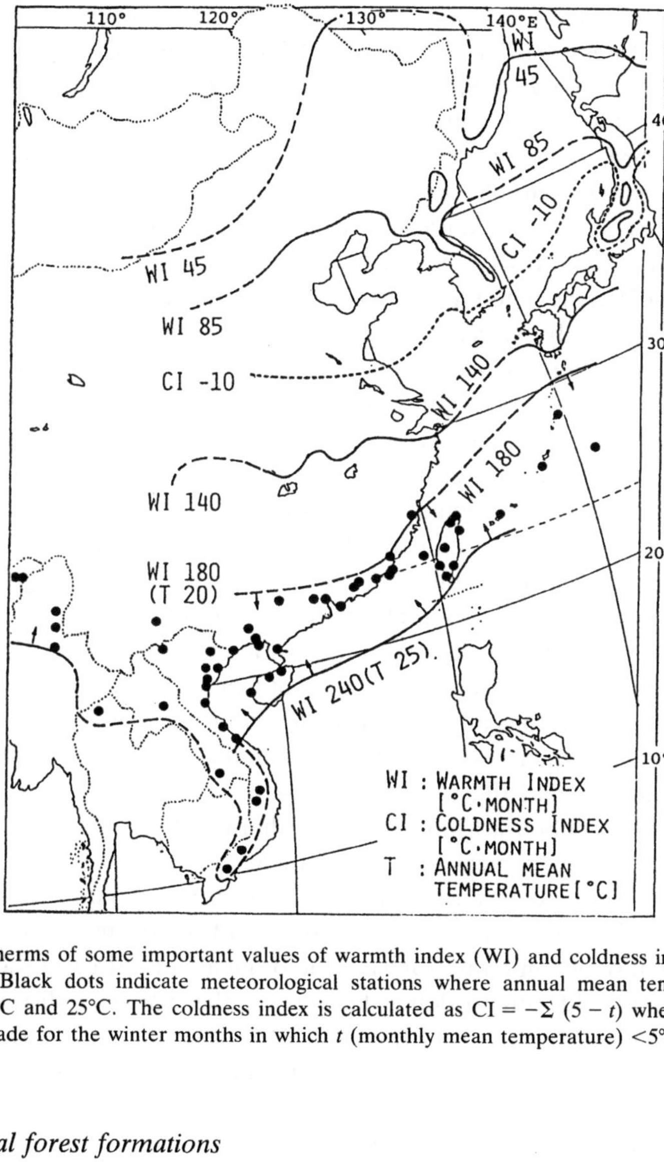

Fig. 6. Isotherms of some important values of warmth index (WI) and coldness index (CI) in East Asia. Black dots indicate meteorological stations where annual mean temperature is between 20°C and 25°C. The coldness index is calculated as $CI = -\Sigma\ (5 - t)$ where the summation is made for the winter months in which t (monthly mean temperature) <5°C.

Subtropical forest formations

The lucidophyll forest of Central China stretches southward across the main ridge of the Nanling Range to its foot-slopes in Guangdong Province. This borderline approximately corresponds to the warmth index isotherm of 180°C month, which is identical with the annual mean temperature isotherm of 20°C, since the monthly mean temperature at this low latitude is always higher than the threshold of 5°C (Fig. 6). The original vegetation south of this line is considered to be a mixed deciduous/evergreen forest containing both tropical and warm-temperate floral elements. Chinese ecologists call this type of forest the *Tropical Seasonal Forest*, which is sometimes intermingled with the *Tropical Rain Forest* where precipitation is locally abundant (Guangdong Botanical Institute, 1976).

With regard to floristic composition, however, these are not typically *tropical* forests comparable with those of Southeast Asia. Dipterocarpaceae, the dominant family of Southeast Asian tropical forests, are very poorly represented in the forests of South China, which are therefore considered to be a kind of intermediate type between warm-temperate and tropical forests and should properly be called *subtropical*. Only on the southern slope of Hainan Island, where the mean annual temperature reaches 25°C (equivalent to a warmth index of 240°C month), the remaining natural forests are more closely related to Southeast Asian rain forests both physiognomically and floristically, containing dipterocarp trees such as *Vatica* and *Hopea*.

The nomenclature of climate/vegetation zones adopted by Chinese ecologists seems to have been influenced by the European and Soviet systems, in which the beech/deciduous oak forest zone is usually assigned to the temperate zone while the boreal conifer forest zone is often called cool-temperate. According to this system, the Mediterranean sclerophyll forest and the East Asian lucidophyll forest, both of which lie south of the beech/oak forest, are logically placed in the subtropical zone. However, the Western Pacific sequence of thermal vegetation zones described so far leads to the conclusion that the term *subtropical* would better be retained for a zone ranging from South China to the northern half of Vietnam as discriminated from the real tropical zone of Southeast Asia.

The subtropical zone thus defined includes also the chain of islands from Okinawa to Taiwan, where the climate is sufficiently moist to allow the development of genuine rain forests. The original natural vegetation of lowland Taiwan is difficult to define because of the extensive disturbance by human activities, but there is little evidence to suggest that it is much different from that of South China with regard to its floristic composition.

On the other hand, the native forests on the Okinawa Islands between 30° N and 24° N are closely related to Japanese lucidophyll forests in having *Castanopsis cuspidata*, *Quercus* (*Cyclobalanopsis*) spp. and other dominant species in common. However, it is notable that many tropical elements, of which the seeds or spores are disseminated by wind, sea current and probably birds, are abundantly represented in the Islands' flora, e.g. tree ferns, mangrove species and other littoral plants, strangler *Ficus*, *Pandanus*, etc. The origin of this unusual composition of Okinawa forests might be explained by the geological history of the islands, which were isolated from the Asian continent quite early in the Pleistocene, probably 1–1.5 million yr ago (Kizaki, 1980). The dominant oak species may be relics of an age when a cooler climate prevailed and the islands were connected either with the continent or with the main islands of Japan, and have survived in coexistence with the tropical flora that arrived later from across the sea. Based on this hypothesis, the vegetation of the Okinawa Islands is here regarded as a type of subtropical rain forest.

Tropical rain forest formation

Kano (1936–1937, 1941) pointed out a remarkable discontinuity in fauna and flora between the main island of Taiwan and an islet, Lan Hsu (Botel Tobago), located some 70 km to the east of the southern tip of Taiwan, and emphasized the biogeographical importance of the narrow intervening strait. He suggested further that the native forest of Lan Hsu was more closely related to the tropical rain forest of the Philippines than to the subtropical forest of Taiwan, not only floristically but also in its structure and physiognomy.

There is a difference of about 2°C in annual mean temperature between the two sides of the strait, through which the annual mean temperature isotherm of 25°C is drawn (Fig. 6), showing that the winter monsoon from the north rarely reaches Lan Hsu across the strait. Since the whole Southeast Asian tropics are thermally quite homogeneous, with an annual mean temperature generally ranging between 25°C and 26°C, the thermal limit of distribution of tropical rain forest seems to lie approximately at an annual mean temperature of 25°C or a warmth index value of 240°C month.

The 25°C isotherm extends further southwestward across the South China Sea, barely including the southern coast of Hainan Island, and reaches the central part of Vietnam to divide the country into the subtropical north and the tropical south (Fig. 6). The distribution of *tropical* forest on Hainan Island mentioned above is consistent with this temperature distribution.

Summarized thermal zonation of forest formations

The sequence of forest formations along a thermal gradient under sufficiently moist climates of the Western Pacific coasts is summarized in Table 1.

The boundaries between the formations are usually more or less obscured by the presence of transitional zones. It is, however, obvious that at least five zones of forest formations have to be discriminated, even if transitional zones are disregarded. This probably represents the maximum number of forest formations which can develop along a latitudinal thermal gradient without interruption by the dry climate at middle latitudes, and may serve as a criterion for the geographical classification and ordination of terrestrial vegetation types.

Table 1. Thermal forest zones in the Western Pacific coasts and islands with sufficiently moist climates

Zone and formation	Warmth index (°C mo)	Coldness index (°C mo)	Annual mean temp. (°C)
(1 Arctic zone)	0–15		
2 Subarctic zone			
*Evergreen conifer forest	15–45		
3 Cool-temperate zone			
*Mixed broadleaf/conifer forest (continental type)	45<		
*Deciduous broadleaf forest (beech type)	45–85		
4 Warm-temperate zone			
*Deciduous broadleaf forest (continental type)	85<	<−10	
*Lucidophyll forest	85–180	>−10	<20
5 Subtropical zone			
*Subtropical rain forest (Okinawa type)	180–240		20–25
6 Tropical zone			
*Tropical rain forest	240<		25<

Latitudinal changes in forest structure and functions

Although the continuous distribution of forest vegetation along the Western Pacific coasts, from the equator to the arctic tree line, offers an ideal research opportunity, only limited quantitative information is yet available on changes in the structure and function of forest ecosystems corresponding to the latitudinal/thermal gradient. Furthermore, differences in sample size and field survey methods make it difficult to collate existing data for a fair comparison. The following accounts are therefore only preliminary.

Forest structure and floristic diversity

Some structural properties of the forest community are known to be affected by thermal climate to some extent; e.g. a tendency for increased aboveground tree height, increased plant biomass, and a greater degree of forest canopy multi-layering toward warmer regions. However, the trend of these changes is rarely monotonous and is sometimes not so obvious.

The aboveground biomass of undisturbed tropical rain forests in Southeast Asia is very large, amounting to 450–500 t ha^{-1} dry weight (Kato et al., 1974; Kira, 1987; Yamakura et al., 1986), but similarly large biomass has also been reported for temperate conifer forests dominated by *Abies firma* or *Tsuga sieboldii* in the transitional zone between the warm- and cool-

Table 2. Comparison of species diversity in three forest types of the Western Pacific under perhumid climates

	Tropical lowland rain forest (Pasoh, Malaysia)	Warm-temperate lucidophyll forest (Nara, Japan)	Cool-temperate beech forest (Nagano, Japan)
Sample area [ha]	2.0	2.2	2.5
Total number of trees	400	455	427
Total number of species	139	28	4
Simpson's index of species diversity	62.8	6.06	1.13
Sources of data	Ashton (1971)	Unpublished data by K. Naka	Watanabe (1983)

temperate zones of Japan (Ando et al., 1977). Since the aboveground biomass of temperate broadleaf forests and subarctic spruce/fir forests in Japan may well reach the level of 350–400 t ha^{-1} under locally favorable conditions (Kira and Shidei, 1977; Kira et al., 1978), the latitudinal gradient of change in forest biomass is by no means steep, although it does exists. Also the leaf area index of a whole forest community including undergrowth layers does not differ widely between broadleaf forest communities in different thermal zones, usually ranging from 6 ha ha^{-1} to 8ha ha^{-1} (Kira, 1976).

On the other hand, the diversity of component tree flora is extremely sensitive to changes in thermal climate. Table 2 compares the diversity of tree flora in three mature forest stands in the Western Pacific calculated for the same range of tree size and sample.plots of similar sizes. Fully developed stands of Japanese beech forest are generally poor in tree flora, while the cool-temperate mixed forest of Hokkaido described before has considerably richer flora. Nevertheless, it is quite obvious that the diversity of tree flora decreases sharply from the tropics toward higher latitudes.

Primary productivity

Table 3 shows the ranges of annual net primary production in different forest types of East and Southeast Asia. Apparently the net productivity tends to be higher under warmer climates, but the difference between tropical and boreal forests is again not as large as would be expected. The rate of net production appears to be more or less proportional to the length of the growing period. It is noteworthy, however, that cool-temperate deciduous broadleaf forests are less productive than evergreen conifer forests of temperate zones, and even less so than some subarctic conifer forests. This may result from the fact that deciduous trees have to use a considerable proportion of the thermally favorable period for unfolding new leaves and for leaf

Table 3. Summary of aboveground net primary productivity in different forest types of East and Southeast Asia (Kira, 1976)

Forest type	No. of stands surveyed	Rate of net production (ton d.w. ha^{-1} yr^{-1}) Mean	Variance
Subarctic evergreen conifer forest	88	11.15	3.75
Cool-temperate deciduous broadleaf forest	55	8.74	3.47
Temperate conifer forest[a,b]	96	14.25	5.78
Temperate pine forest[a]	44	13.64	5.00
Warm-temperate evergreen broadleaf forest[c]	33	20.65	7.21
Tropical evergreen broadleaf forest	2	24.6	–

[a] Both cool- and warm-temperate.
[b] Mostly plantations of *Cryptomeria japonica*. Others are plantations of *Chamaecyparis obtusa* and natural forests of *Abies firma* and *Tsuga sieboldii*.
[c] Lucidophyll forests and a few *Acacia* plantations.

shedding, whereas evergreen conifer leaves are physiologically active throughout the whole period.

The net primary productivity is, however, not always a reasonable measure of the potential productivity of forests, since the ratio of net production (the amount of plant bodies actually produced) to gross production (the total amount of assimilated organic matter) varies greatly depending on the developmental stage or age of forest stands. The respiratory consumption, which is the difference between gross and net production,.is strongly tree-size dependent (Kira and Shidei, 1967; Kira, 1975).

The gross production may thus be a better parameter for assessing the effect of climatic conditions on forest productivity. Figure 7 presents the result of an attempt to correlate gross productivity of forest communities in the Western Pacific with their structure and thermal environment. The product of the leaf area index of a forest stand and the warmth index at its site was found to be significantly correlated with its gross productivity.

Rate of organic matter decomposition

As compared with the rate of organic matter production, that of decomposition processes is much more sensitive to changes in thermal climate. The accumulation and decomposition of dead organic matter in and on the soils of three undisturbed forests in the cool-temperate, warm-temperate and tropical zones of the Western Pacific are compared in Fig. 8 based on the

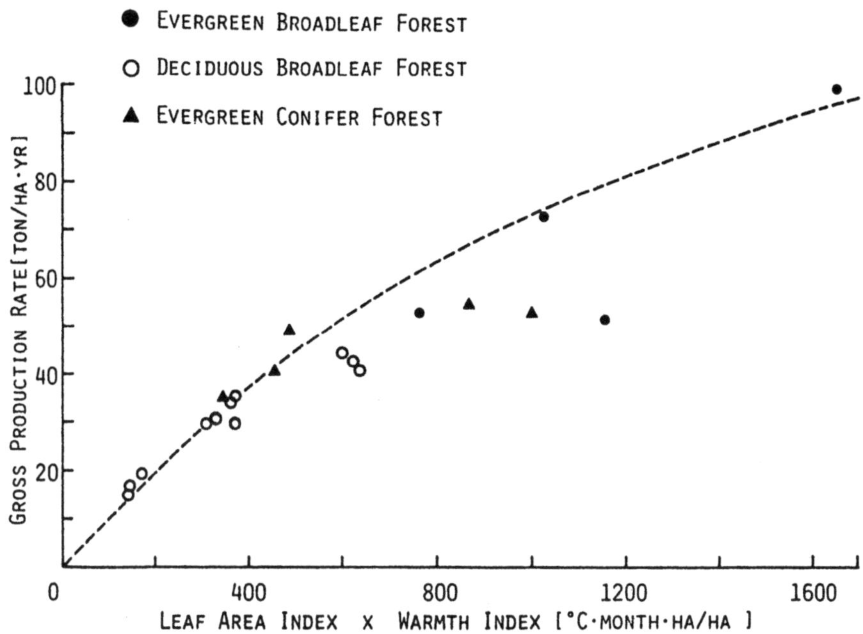

Fig. 7. Correlation between gross production rate and the product, (leaf area index) × (warmth index), in the Western Pacific forests.

carbon dynamics model proposed by Nakane (1975, 1978a and b) to illustrate the temperature-dependency of these processes.

Apparently the flow of carbon between the four carbon pools, ground-surface litter (A_0) layer, humus in mineral soil, dead root and live root biomass, through litterfall, humification of litter, decomposition and root respiration, grows larger with rising ambient temperature. The rates of the flows and soil respiration are almost three times as large in tropical rain forest as in cool-temperate beech/fir forest. This may be logically expected, since the rate of soil respiration should balance that of net primary production in more or less homeostatic mature forest stands. The estimates of soil respiration in Fig. 3 correspond approximately to the rates of net production given in Table 3.

On the other hand, the accumulation of dead organic matter as the A_0–layer and soil humus decreases toward warmer habitats. The size of the ground-surface litter pool is particularly sensitive to temperature. The time necessary for 95% disappearance (due to decomposition and downward transport into mineral soil) of A_0-layer litter calculated from Fig. 8 is 1.1, 3.3

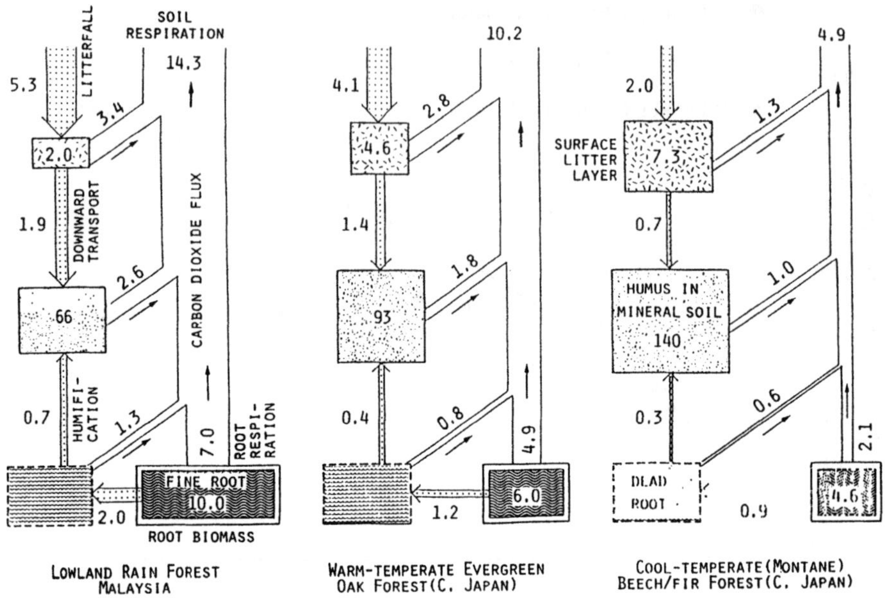

Fig. 8. Dynamics of carbon storage and flow in soil-vegetation systems in tropical, warm-temperate and cool-temperate zones of the Western Pacific. Redrawn after Nakane (1980).

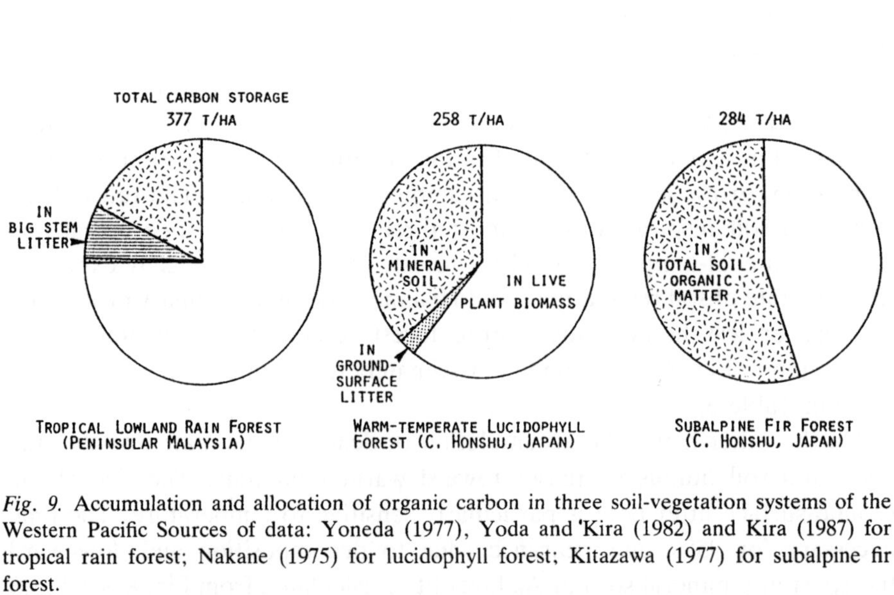

Fig. 9. Accumulation and allocation of organic carbon in three soil-vegetation systems of the Western Pacific Sources of data: Yoneda (1977), Yoda and 'Kira (1982) and Kira (1987) for tropical rain forest; Nakane (1975) for lucidophyll forest; Kitazawa (1977) for subalpine fir forest.

and 11.0 yr, respectively, in tropical rain forest, warm-temperate lucidophyll forest and cool-temperate beech/fir forest, while that for decomposition of soil humus amounts to 76, 155 and 420 yr, respectively. The amount of surface litter accumulation is also sensitive to factors other than temperature that affect the rate of organic matter decomposition, and thus serves as a good indicator of the suitability of habitat conditions for forest growth.

As a result, more than 50% of the total amount of organic carbon held by a forest-soil system exists in and on the soil as dead organic matter in the boreal zone, whereas nearly 75% is contained in the live biomass of tropical rain forests under favorable conditions (Fig. 9). This should be borne in mind if we expect to increase forest area as a carbon pool to control the increase of atmospheric carbon dioxide and resulting global warming.

References

Alisov, B. P. 1954. Die Klimate der Erde. 277 pp. Berlin.

Ando, T., Chiba, K., Nishimura, T. & Tanimoto, T. 1977. Temperate fir and hemlock forests in Shikoku. Kira et al. (eds.), pp. 213–245.

Ashton, P. S. 1971. IBP Malayan project; Pasoh forest reserve vegetation survey (second report). (Mimeographed, unpublished material).

Box E. O. & Fujiwara, K. 1988. Evergreen broadleaved forests of the southeastern United States: preliminary description. Bull. Inst. Environ. Sci. Technol. Yokohama Nat. Univ. 15:71–93.

Editorial Committee for the Vegetation of China (1980) The vegetation of China. 1375 pp. Scientific Publications, Beijing (In Chinese).

Guangdong Botanical Institute 1976. Vegetation of Guangdong. 341 pp. Scientific Publications, Beijing (In Chinese).

Hou, H.-Y., Chen, C.-T. & Wang, H.-P. 1956. The vegetation of China with special reference to the main soil types: Report for the 6th International Congress of Soil Science. 26 pp. Soil Science Society of China. Peking.

Kano, T. 1935–1936. Some problems concerning the biogeography of Kotosho, near Formosa. Geogr. Rev. Jpn. 11:950–959, 1027–1055; 12:33–46, 154–177, 911–935, 997–1020, 1107–1133 (In Japanese).

Kano, T. 1941. Biogeography of the island of Kôtôsho (Botel Tobago) with special reference to the New-Wallace Line. "Dai-nanyô (Greater South Seas)" Taiheiyô Kyôkai (ed.), (The Institute of the Pacific), 219–323. Kawade Shobô, Tokyo (In Japanese with English summary).

Kato, R., Tadaki, Y. & Ogawa, H. 1978. Plant biomass and growth increment studies in Pasoh Forest. Malay. Nat. J. 30:211–224.

Kawakita, J. 1956. Land and crops of Nepal Himalaya: Scientific results of the Japanese expeditions to Nepal Himalaya 1952–1953, Vol. II. in: Kihara, H. (ed.), Vegetation, pp. 1–65. Fauna & Flora Research Society, Kyoto University, Kyoto.

Kira, T. 1945. A new classification of climate in eastern Asia as the basis for agricultural geography. 23 pp. Horticultural Institute, Kyoto Univ., Kyoto (In Japanese).

Kira, T. (1949) Forest zones of Japan. 41pp. Ringyô Gizyutu Kyôkai, Tokyo (In Japanese).

Kira, T. 1976. Terrestrial ecosystems – an introduction (Handbook of ecology, Vol. 2). 166 pp. Kyôritsu Shuppan, Tokyo (In Japanese).

Kira, T. 1977a. A climatological interpretation of Japanese vegetation zones. Vegetation science and environmental protection Miyawaki, A. & Tüxen, R. (eds.), pp. 21–30. Maruzen Co. Ltd., Tokyo.

Kira, T. 1977b. Forest vegetation of Japan. Shidei & Kira (1977), 1–9. Univ. Tokyo Press, Tokyo.

Kira, T. 1978. Introduction. Kira et al. 1–7.

Kira, T. 1987. Tree crop physiology. in: Sethuraj, M.R. & Raghavendra, A.S. (eds.), Primary production and carbon cycling in a primeval lowland rainforest of Peninsular Malaysia. 99–119. Elsevier, Amsterdam.

Kira, T., Ono, Y. & Hosokawa, T. (ed.) 1978. Biological production in a warm-temperate evergreen oak forest of Japan (JIBP synthesis, Vol. 18). 288 pp. Univ. Tokyo Press, Tokyo.

Kitazara, Y. (ed). 1977. Ecosystem analysis of the subalpine coniferous forest of the Shigayama IBP Area, central Japan (JIBP Synthesis Vol. 15). 199 pp. Univ. Tokyo Press, Tokyo.

Kizaki K. (ed.) 1980. Natural history of the Ryukyus. 282 pp. Tsukiji Shokan, Tokyo (In Japanese).

Miyawaki, A. & Itow, S. 1966. Phytosociological approach to the conservation of nature and natural resources in Japan. Divisional Meeting of Conservation, 11th Pacific Science Congr., Tokyo.

Nakane, K. 1975. Dynamics of soil organic matter in different parts on a slope under evergreen oak forest. Jpn. J. Ecol. 25:206–216 (In Japanese with English summary).

Nakane, K. 1978a. A mathematical model of the behavior and vertical distribution of organic carbon in forest soils. Jpn. J. Ecol. 28:111–122.

Nakane, K. 1978b. A mathematical model of the behavior and vertical distribution of organic carbon in forest soils. II. A revised model taking the supply of root litter into consideration. Jpn. J. Ecol. 28:169–177.

Nakane, K. 1980. Comparative studies of cycling of soil organic carbon in three primeval moist forests. Jpn. J. Ecol. 30:155–172 (In Japanese with English summary).

Ogawa, H. & Matsumoto, A. 1987. The climate and vegetation of Japan (Nippon no Fûdo). in: Matsui, T. & Ogawa, H. (ed.), What characterizes Japanese climate. Japan and its nature, Vol. 2, pp. 41–54 Heibonsha, Tokyo (In Japanese).

Rübel, E. 1930. Pflanzengesellschaftern der Erde. Bern.

Schimper, A. F. W. 1903. Plant geography upon a physiological basis. Clarendon Press, Oxford.

Shidei, T. & Kira, T. (ed.) 1977. Primary productivity of Japanese forests (JIBP Synthesis Vol. 6). 289 pp. Univ. Tokyo Press, Tokyo.

Song, Y.-C. 1983. Die räumliche Ordnung der Vegetation Chinas. Tuexenia, N.S. 3:131–157.

Wang, C. W. 1961. The forests of China with a survey of grassland and desert vegetation. Maria M. Cabot Foundation Publication No. 5. 313 pp. Harvard Univ. Press, Cambridge, Mass.

Watanabe, R. 1983. Studies on the primeval beech forest in Kayanodaira, I, DBH census in the scientific forest reserve in Kayanodaira. Bull. Inst. Nat. Educ. Shiga Heights, Shinshu Univ. 21:7–18 (In Japanese with English summary).

Weaver, J. E. & Clement, F. E. 1938. Plant ecology. McGraw-Hill, New York.

Yamakura, T., Hagihara, A., Sukardjo, S. & Ogawa, H. 1986. Aboveground biomass of tropical rain forest stands in Indonesian Borneo. Vegetatio 68:71–82.

Yim, Y.-J. 1977. Distribution of forest vegetation and climate in the Korean Peninsula, IV, Zonal distribution of forest vegetation in relation to thermal climate. Jpn. J. Ecol. 27:269–278.

Yim, Y.-J. & Kira, T. 1975. Distribution of forest vegetation and climate in the Korean Peninsula, I, Distribution of some indices of thermal climate. Jpn. J. Ecol. 25:77–88.

Yoda, K. & Kira, T. 1982. Accumulation of organic matter, carbon, nitrogen and other nutrient elements in the soils of a lowland rainforest at Pasoh, Peninsular Malaysia. Jpn. J. Ecol. 32:275–291.

Yoneda, T. 1977. Accumulation and decomposition of big wood litter in Pasoh Forest, West Malaysia. Jpn. J. Ecol. 27:53–60.

Yoshioka, K. 1973. Plant geography (Handbook of ecology, 12). 84 pp. Kyôritsu Shuppan, Tokyo (In Japanese).

Zhao, D.-C. 1980. The altitudinal distribution belts of vegetation on Changbai Mountain. Research of Forest Ecosystem 1:65–70. Changbai Mountain Research Station of Forest Ecosystem, Academia Sinica, Shenyang (In Chinese).

Note

The author is responsible for any bias resulting from the omission of the courses of other jet streams, etc. from Fig. 2 against the original paper's intention.

2. Climatic relations of the forests of East and South-East Asia

ELGENE O. BOX
Institute of Industrial Science, University of Tokyo, Minato-ku, Roppongi 7-22-1, Tokyo 106, Japan; Permanent affiliation: University of Georgia, Geography Department and Institute of Ecology, Athens, Georgia 30602, USA

Abstract

The forest region of East and Southeast Asia has four main subregions: perhumid Japan, continental China plus Korea, humid-subtropical Southeast Asia, and Yunnan, the world's largest area of subtropical-warm temperate uplands and mountains. In addition to latitude, elevation, and the land-sea configuration, the Asian monsoon system is also a major control on climates, generating some of the unique features of the region. Climate-based systems of zonal forest and other vegetation types have been developed and describe the regional relationships fairly well. Interesting, more local climatic effects on East Asian vegetation are also apparent, however, some with parallels in other parts of the world. The prospect of future climate change poses special problems for East Asia, due to the many obstacles to northward migration.

Key words: climate controls, climatic regions, cool-temperate, maritime climate, monsoon climate, perhumid climate, seasons, subalpine belt, subtropical, warm-temperate, water balance, typhoon

Introduction

East Asia, as treated here, is the region of potentially wooded landscapes extending from boreal eastern Siberia to Southeast Asia, including Japan, Korea, roughly the eastern half of China (i.e. west to Xinjiang and Tibet), Taiwan, and southward to Thailand and Malaysia. It is essentially a region of forest climates, except for the drier, more continental influences penetrating from the northwest, across northern China. East Asia, thus defined, can be divided quasi-latitudinally into five broad climatic zones:

1. tropical Southeast Asia and adjacent islands (plus coastal areas extending further north);

E.O. Box et al. (eds.), Vegetation Science in Forestry, 23–55.

2. a humid subtropical region including much of southern China plus Taiwan, the Okinawa (Nansei) and Ogasawara islands;
3. a warm-temperate region, covering parts of eastern and southern China, the southern half of Japan (main islands), and the southern tip of Korea;
4. a typical/cool-temperate region in northeastern China, most of Korea and the northern half of Japan; and
5. a subalpine/subarctic region in eastern Siberia, northeasternmost China, small parts of Hokkaido, and in mountains further south.

These zones (see also Table 1) correspond fairly readily to those found normally in the mid-latitudes on continental east sides, as described in global climate classification systems such as those of Walter (1977, 1985, Walter & Box 1976), Troll & Paffen (1964), Köppen (1931), and Bailey (1990).

East and Southeast Asia are dissected by many mountain ranges, as shown schematically in Fig. 1. In addition to the horizontal climatic zonation one can also recognize vertical bioclimatic belts, including small alpine belts occurring on the highest and most wind-exposed summits of Japan, northeastern China, and Korea. Climatic patterns throughout the region are dominated by the Asian monsoon system, with dry continental winds in winter and moist, oceanic winds in summer. The combination of strong seasonal winds and numerous mountain ranges results in many local climatic effects, from seasonally alternating windward and leeward slopes and rain shadows to the unusual, persistently cloudy climate of the Sichuan Basin. The relationships become more complex and the expected wet-dry seasonality often reversed when water bodies intervene between land masses (e.g. the Japan Sea side of Honshu).

As in other countries, climate study in East Asia has been closely related to the study of vegetation and environmental relationships (e.g. Kira 1945, Shidei 1974, Kira et al. 1976, Yim & Kira 1975–1976, 1977, Yoshino 1978, Yazawa 1980, Environment Agency 1982). Similarities between East Asia and other parts of the Northern Hemisphere, both in vegetation and climate, were recognized early (e.g. Gray 1846, *cf.* Boufford & Spongberg 1983). The purpose of this chapter is to give a relatively brief overview of such relationships, as a background for the more detailed ecological studies which follow (see Chapter 1). More complete accounts of East Asian climates, in English, are given by Fukui (1977) and Box (in press) for Japan; by Zhao (1986) for China; and by Yoshino (1973b) for monsoon Asia in general.

Climate controls

Climates in general are controlled by latitude (zonal position), the local/regional land-sea configuration, by ocean currents in coastal areas, by the

Table 1. Bioclimatic zones and corresponding vegetation on continental east sides

Climate zone	Walter symbol	Climatic conditions		Zonal vegetation	Main Locations
		Summer	Winter		
Equatorial	I	warm/hot and perhumid (ITC dominant)		Tropical rainforests	Malay Peninsula
Tropical wet-dry	II	Warm & wet (ITC)	Very dry (STHP)	Tropical deciduous forests/scrub	Thailand-SW China, western Hainan
Humid subtropical	I-V	Hot, much rainfall	Cooler, less rain	Evergreen broad-leaved forests (subtropical)	S China, Taiwan, montane SE Asia
Warm-temperate	V	Warm, much rainfall	Cool/cold, less rain	Evergreen broad-leaved forests (warm-temperate)	SE-E China, S Japan, S-most Korea
Typical temperate (nemoral)	VI	Warm, with rainfall	Cold, with frost	Summergreen deciduous forests	NE China, N Japan, most of Korea
Boreal/Subarctic	VIII	Cool, with rainfall	Long, very cold	Boreal coniferous forests	Siberia, montane subalpine NE China
(Alpine belts)	IX	Short, cool	Very long, very cold	Tundra, heaths, meadows	Alpine belts (only)

The following four transitional zones can also be recognized: 1. a sub-equatorial zone (I-II) with short dry season and seasonal evergreen or semi-evergreen forests (as in various parts of southern China and SE Asia); 2. a more continental warm-temperate zone (V-VI) with warm-temperate deciduous forest, especially in south-central China; 3. a sub-boreal zone (VI-VIII) with mixed forest involving broad-leaved deciduous trees and boreal conifers; 4. a subalpine or treeline belt (*cf.* lowland VIII-IX) at the upper limit of tree growth, with conifer or sometimes *Betula* krummholz. Abbreviations: ITC = intertropical convergence zone, STHP = subtropical high-pressure belt.

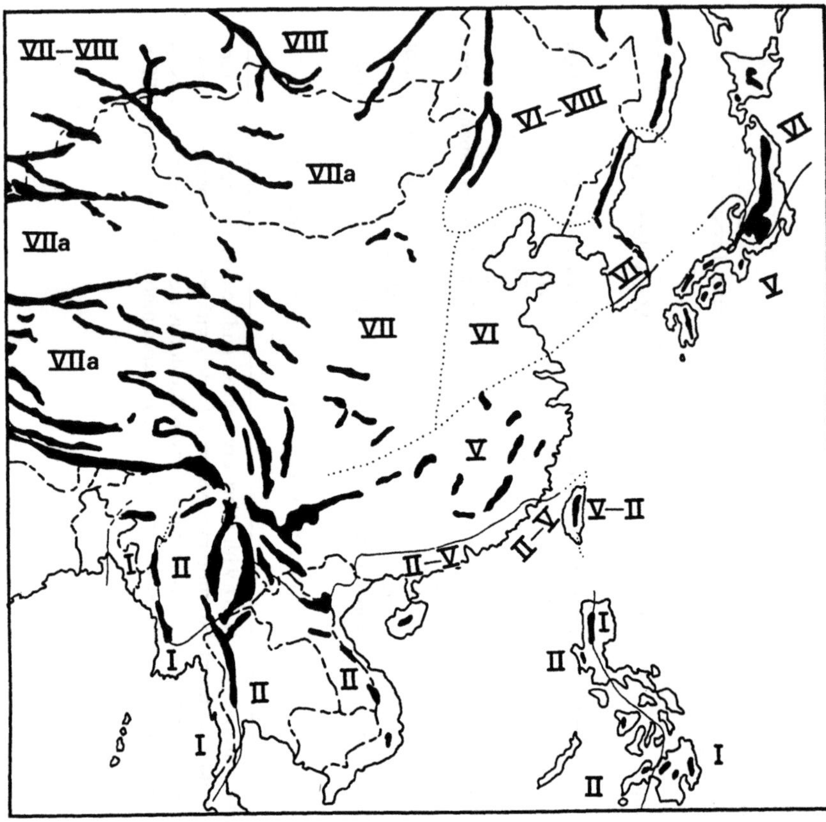

Fig. 1. Main mountain ranges and climatic regions of East and Southeast Asia. The climatic regions are redrawn from maps by Walter and Lieth (1960–67), with the modification that type V (warm-temperate climate) is shown here more as a maritime climate with less inland extent. The climate types are: I Equatorial, II Tropical wet-dry, V Warm-temperate, VI Typical temperate, VII Temperate continental, VIIa Temperate arid, VIII Boreal. The coastal humid-subtropical to humid-tropical climates, with much ameliorated dry seasons, are indicated as transitional climates I-V and II-V.

height and orientation of mountain ranges, and by relative geographic position for particular places. An additional factor in much of Asia is the strong monsoon system, which results from the seasonal heating and cooling of the large Asian land-mass and the consequent development of strong high (winter) and low (summer) pressure centers over the mid-latitude interior. The effect of the resulting monsoon wind systems is a stronger effect than the zonal westerly winds over much of the year. The monsoon winds blow generally from the Pacific and Indian Oceans toward the continent in summer and outward from Mongolia in winter. The summer winds are weaker and more variable since the monsoonal low pressure cell is weaker than the winter high pressure. The shifts from one monsoonal pattern to the other occur near the equinoxes. These transitions bring with them early-summer and early-autumn rain fronts, which represent boundaries between wet oceanic and drier continental air masses.

The air masses influencing East Asia at different times of year were explained in detail by Yoshino (1977a, *cf.* Takahashi 1955).

1. Cold, dry continental ("Siberian") air masses dominate continental East Asia in winter and keep winter temperatures low throughout the region, even in maritime Japan (Yoshino 1977b). This dominance of cold, dry air begins in October and lingers through springtime in most of northern and eastern China.
2. Warm, dry air masses occur over China in spring and early autumn; these may also bring generally dry, anticyclonic circulation to Japan, both before and after the summer wet season.
3. Cool maritime air masses are brought by the early-summer rain front from the Pacific and Sea of Okhotsk north and northeast of Japan; this front moves gradually across Japan and on into Korea and China in early summer, bringing the more or less extended, cool rain season known as "tsuyu" in Japan or "plum rains" in China (*cf.* Yoshino 1963, 1965–66, 1966).
4. Tropical maritime air masses from the warm Pacific follow the rain front and dominate the hot, humid East Asian summer (mainly July-August). These humid air masses may not be strongly unstable themselves but are readily activited by the hotter land surface to produce heavy rainfall.

In addition to these four main types of air mass, Yoshino also lists equatorial air masses (which may bring rain in summer) and cold, stable upper-air masses (which act as a ceiling on upward convection but may also help trigger thunderstorms in mid-summer, when acting in combination with strong surface heating and the large Pacific water source).

The situation in Southeast Asia is less complex, being dominated by the north-south migration, over the course of the year, of the Intertropical Convergence Zone (rainy season) and the subtropical high-pressure belts (dry season), plus the intervening trade winds (*cf.* Yoshino 1969). The contrasting wet-dry seasonality is, however, in phase with and thus strengthened by the monsoonal seasonality. This results in even heavier rainfall during the summer wet season (3 m or more on windward coasts and slopes) and cooler, drier conditions in winter – except along the extensive coastlines, where the sea provides humidity and some off-season rainfall.

Within the above framework, ocean currents and typhoons provide additional, local effects in coastal areas. Warm currents flow northward along both sides of Japan, bringing milder winter conditions to mid-Honshu (see Fukui 1977, p. 4). On the Pacific side of northern Japan, on the other hand, the cold Oya-shio current brings cooler conditions in summer. The meeting of opposing warm and cold ocean currents, off both southeastern and northernmost Hokkaido, off Cape Inubozaki (eastern Honshu), and off the southeast coast of Korea, produces persistent local fog in summer.

The warm, humid Pacific air masses of summer also provide ideal conditions for the development of typhoons, the same type of tropical storm as the "hurricanes" of the tropical North Atlantic and the "cyclones" of the Indian Ocean. After drifting from east to west in the tropical Pacific, typhoons during their main season (July-October) drift northward over land and eventually back to the northeast if caught in the westerlies flow. This means that they may hit southern Japan and Korea from the south or even southwest, as well as from the southeast, and may bring large amounts of rain even to northern Japan. Typhoons may also occur in winter but are then restricted to the tropical region, mainly affecting the northern Philippines and southeastern China (*cf.* map by Takahashi 1977). The main climatic effect of the typhoons may be the large amounts of rainfall which they bring, but they are also important factors in forest dynamics, preventing the development of stable, mature forests in the most typhoon-exposed areas.

Strong winds are common in East Asia, not only during typhoons but also in coastal areas in general (especially in winter), in mountain areas (*cf.* Yoshino 1973a, Okitsu 1984), in some basins, and over the large flatter, deforested areas especially of eastern China. On the Japan Sea side of Honshu, the dominance of deciduous broad-leaved trees in what would appear to be the montane conifer belt has been explained as related to unusually high winds (Makita & Ishikawa 1990), though heavy snow may also be involved.

The seasons in most of temperate East Asia are relatively well-defined but include the two rainy seasons in addition to the four normal temperate-zone seasons. The second rainy season, at least in Japan ("shurin" or "akisame") and Korea, occurs in late summer (September), associated with the retreat of the polar front from Asia. This and the late-summer typhoons combine to produce heavier rain than during the early-summer rainy season. True summer is hot and very humid but is relatively short (July-August) in Japan and Korea (*cf.* Matsumoto 1984). Autumn, because of the strong oceanic influence, is long in more coastal areas and can extend on into December, with temperatures falling only gradually to winter levels. The coldest month is usually January, but February can be almost as cold. Spring comes gradually, especially in maritime Japan. In Southeast Asia there are essentially two main seasons, wet and dry, but with hottest conditions ("summer" – best developed in India) occurring in some areas during the month or two before the arrival of the monsoonal-ITC rain season.

Climatic regionalization

For lowland East Asia the global bioclimatic zones described in Table 1 correspond readily to the zones recognized by Kira and other East Asian ecologists (e.g. Numata et al. 1972, Numata 1974, Miyawaki 1980–89, Wu 1980, Hou 1983). Walter's global climate system did not have a humid subtropical climate type as such, but this type, which occurs extensively throughout East Asia, can be represented as a transition between the Walter equatorial (I) and warm-temperate (V) types. The tropical wet-dry climate (Köppen Aw, Walter II), which normally comes between types I and V, dominates much of tropical Asia and extends, in somewhat moderated form, well into southern China. Its dry season is significantly moderated in coastal areas (e.g. southeastern China) and is almost not recognizable on some offshore islands (e.g. Okinawa).

Climatographic zonation systems for East Asia include those of Kira (1945), T. Suzuki (1953), Hämet-Ahti et al. (1974), and Wolfe (1979), as well as the adaptations of Thornthwaite by Fukui (1958) and systems for particular countries (e.g. Sekiguchi 1959, H. Suzuki 1962, Yoshino 1980, China Nat. Geogr. Edt. Comm. 1984, etc.). The zonal approach is elegantly represented by the Warmth Index (WI) and Coldness Index (CI) of Kira (1945), which provide a quantitative (thermal) mapping basis and correspond to potential forest zones, as shown in Table 2.

Table 2. Major forest zones and corresponding ranges of the Kira warmth and coldness indices. Kira (1949) proposed a division of the Japanese Islands into four forest zones based on his warmth and coldness indices. The warmth index is defined as the annual sum of all positive differences between monthly means and + 5°C, the coldness index as the corresponding negative sum for monthly means below +5°C

	Warmth index (°C)	Coldness index (°C)
Subtropical evergreen forest	>180	
Warm-temperate evergreen forest	85–180	>−10
(transitional deciduous forest)	85–180	<−10
Cool-temperate deciduous forest	45–85	
Subarctic/subalpine coniferous forest	15–45	

Within this zonal framework, though, there are four rather distinct regions, which provide another perspective on East and Southeast Asia in general. The following sections are organized around these four particular situations, each of which may include one or more latitudinal zones:

1. essentially perhumid (temperate) Japan, with moderated extreme temperatures, frequent precipitation, no dry season, and rarely even any short-term drought;
2. continental (temperate) China plus Korea and southeastern Siberia, with

more extreme temperatures and essentially dry conditions, especially toward the north and interior, in autumn and spring as well as winter;

3. humid tropical and subtropical Southeast Asia, with moderated dry seasons in coastal areas; and
4. the large region of subtropical uplands and mountains, unique in the Northern Hemisphere, mainly in Yunnan but also in smaller adjacent areas as well as southestern China, Taiwan and Okinawa.

Each of these represents a somewhat unique situation of regional aberration within the global climatic framework.

Maritime Japan

Japan (23°–46° N) includes the four main islands, the Nansei (=southwest) or Ryukyu (Okinawa) Islands extending toward Taiwan, and the small Ogasawara (Bonin) and Iwo-Jima (Volcano) Islands in the Pacific Ocean. Its climates range from cool-temperate to subtropical, but with no truly tropical climate except perhaps in immediate coastal areas of southern Okinawa. By virtue of its small size relative to the large surrounding sea, Japan has high precipitation and humidity throughout, and even short drought periods are rare. The position of Japan, as islands off the east side of the large Eurasian land mass results, however, in the unusual combination of greatly moderated absolute (and perhaps diurnal) temperate extremes but mean temperature patterns which are almost continental, especially toward the north.

The Asian monsoon system brings wet air masses from the Pacific in summer and continental air masses from Asia in the winter. The Japan Alps and other mountains, though, lying perpendicular to the monsoon winds, divide Japan into distinct Pacific and Japan Sea "sides", which are alternately windward or leeward with the seasonal wind shift. In winter, the leeward Pacific side of Japan has relatively dry weather, with clear skies common and no snowfall at all in some years. The windward Japan Sea side, however, receives the monsoonal air masses from the continent, which have taken up much water while crossing the warmer (Tsushima current) Japan Sea. This water is then dumped as snow on the colder Japanese coastal plain and windward slopes, mainly of Honshu, producing average snowfall totals of several meters and extreme (average) accumulations of over ten meters in some locations, perhaps the highest in the world. During summer the situation is reversed, with windward Pacific Japan receiving heavy rainfall and the leeward Japan Sea side receiving less.

Japan has the six recognizable seasons described above, except for Hokkaido, which has no rainy seasons. Seasonal patterns of the main climatic elements are suggested by the climate diagrams in Fig. 2, most of which

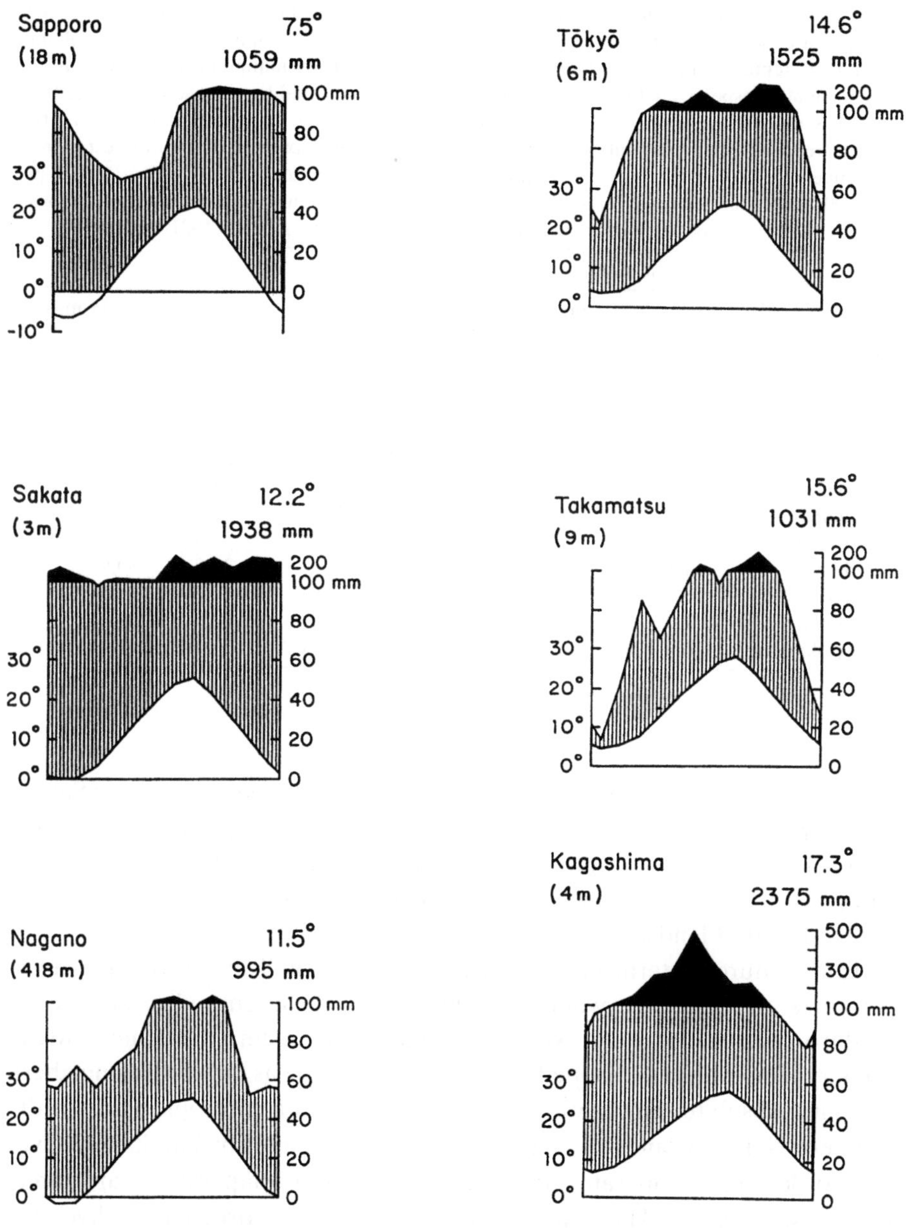

Fig. 2. Climate diagrams for selected sites in Japan. Sapporo shows the colder winters of Hokkaido. Sakata shows the heavy autumn and winter precipitation characteristic of the winter-windward Sea of Japan coast. Tokyo shows the drier winters of the Pacific side, and Nagano, at the same latitude, the generally drier climate of the inner valleys of central Honshu. Takamatsu shows the rain-shadow effects of the Seto Inland Sea, the driest region in Japan. Kagoshima, at the southern tip of Kyushu, shows the warm-perhumid climate characteristic of many coastal areas.

Table 3. Temperature means and extremes for selected cities in Japan. The cities are arranged from north to south. All cities are coastal unless indicated by elevation. The first two columns of numbers represent the mean temperature and the mean daily minimum temperature in the coldest month, respectively (all January unless February, indicated by superscript 2). The third column is the coldest temperature ever recorded, with the year in parentheses. The last two columns represent mean monthly temperature and mean daily maximum of the warmest month, respectively, all August except Naha. Source: Tokyo Astronomical Observatory (1987), pp. 198, 200, 214. This source is used because it includes older data, sometimes showing lower minima, than do many other data compilations (e.g. Japan Met. Agency 1981–83, 1984)

	Coldest month			Warmest month	
	mean	mean min.	abs. min.	mean	mean max.
Asahikawa (112 m)	−8.5	−13.4	−41.0 (1902)	20.4	25.8
Kushiro	−6.4	12.5	−28.3 (1922)	17.6	21.3
Sakata	1.0^{2}	-2.5^{2}	−12.8 (1978)	24.0	28.2
Sendai	0.9	−2.7	−17.7 (1945)	23.9	28.0
Nagano (418 m)	−1.2	−5.0	−17.0 (1934)	24.5	30.2
Toyama	2.1	-0.8^{2}	−11.9 (1947)	25.8	30.2
Yokohama	4.9	0.6	−8.2 (1927)	26.2	30.3
Kyoto (41 m)	3.9	0.1	−9.4 (1917)	27.5	32.8
Hiroshima	4.3	0.6	−8.6 (1917)	26.8	30.9
Nagasaki	6.4	3.0	−4.3 (1977)	27.5	31.2
Kagoshima	7.0	2.4	−6.7 (1923)	27.7	32.2
Naha (Okinawa)	16.0	13.7	6.6 (1967)	28.1^{7}	31.0^{7}

show a distinct summer precipitation maximum. Only the diagram for Sakata (38° N) really represents the Japan Sea side, with its heavy snowfall in winter. All stations show significant seasonal variation in temperature, even in Okinawa. All stations also show large climatic moisture surpluses and the potential for forested landscapes.

Temperature patterns in Japan are summarized in Table 3. Summer temperatures are high everywhere, except at high elevation and around foggy Kushiro (17.6°C) near the Nemuro Peninsula of eastern Hokkaido. Mean January temperatures (mapped in Fig. 3), exceed 0°C as far north as northern Honshu. Absolute temperature minima range from a subtropical 7°C in the southern Nansei Islands to nearly boreal levels in interior Hokkaido. Over most of lowland temperate Japan, however, winter temperatures are rarely far below freezing. The lowest temperature ever recorded in Tokyo, for example, was only −9.2°C in 1876 (Tokyo Astron. Obsv. 1987) and since 1931 only −7.9°C (Jap. Met. Agency 1984). Absolute minima are only −4°C to −8°C for most of coastal southern Japan, and mean nighttime January minima are above freezing. In coastal northern Japan (north of the Tokyo area), absolute minima are still not extreme (e.g. only −12.8°C as far north as Sakata at 38° N), and mean minima are only −2° to −5°. As a result, evergreen broad-leaved forest remnants extend as far north as 38° N, and

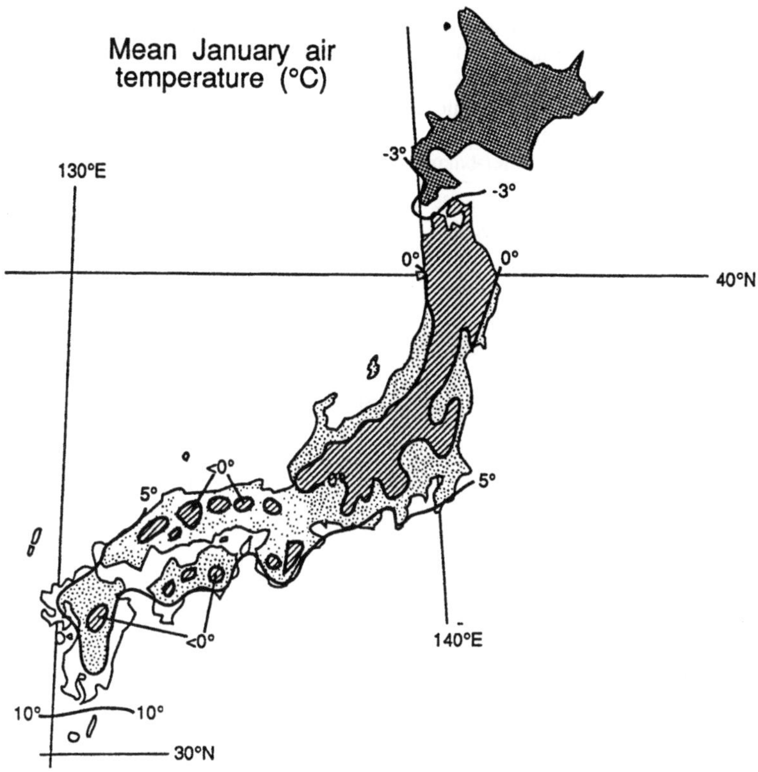

Fig. 3. Mean January air temperature in Japan. The map shows clearly the ameliorated temperatures in coastal areas of Japan. A similar pattern exists for mean minimum and absolute minimum temperatures, permitting evergreen broad-leaved forests to extend to 38° N on both sides of Honshu.

some evergreen broad-leaved tree species extend to Aomori (40° N) at the northern tip of Honshu.

Average annual precipitation ranges from less than 1200 mm to over 4000 mm in some parts of Pacific-facing southern Japan. Some windward middle elevations on Yakushima island, off southern Kyushu, may receive as much as 10 meters. Climatic water-balance conditions are shown in Table 4 for selected sites in Japan. As one can see from these data, all parts of Japan have humid if not perhumid forest climates, with significant climatic water surpluses. Four regions of Japan are somewhat drier: northeastern Hokkaido (Rubeshibe), interior basins of central Honshu (Nagano), the rain shadow around Japan's Inland Sea (Takamatsu, on northern Shikoku), and the Daito and other small, low-lying subtropical islands. Seasonal climatic water deficits (mostly in August) are significant only in these four drier areas,

Table 4. Water-balance data for selected Japanese cities. The sites represent different climatic regions in Japan and are arranged by latitude from north to south. Potential evapotranspiration (PET) is the average of the Thornthwaite (and Mather 1957) and Holdridge (1959) estimates, found to be more reliable for geographic comparisons. MI is the annual moisture index (precip./PET). AET is actual evapotranspiration. Climatic water surplus and deficit represent annual totals of seasonal values. All water-balance values (except MI) are in millimeters.

	Mean temp.	Annual precip.	PET	MI	AET	Surplus	Deficit
Rubeshibe (Hok.)	5.9	650	516	1.26	488	162	28
Sapporo (Hok.)	7.5	1059	549	1.93	538	521	10
Sakata	12.2	1938	710	2.73	710	1228	0
Sendai	11.4	1222	670	1.82	670	552	0
Nagano (418 m)	11.5	995	690	1.44	682	313	8
Tokyo	14.6	1525	821	1.86	813	712	8
Kyoto	15.0	1537	848	1.81	847	688	1
Hiroshima	15.0	1525	840	1.82	833	692	7
Takamatsu (Shik.)	15.6	1031	875	1.18	833	198	43
Kagoshima (Kyushu)	17.3	2375	956	2.48	956	1419	0
Naha (Okinawa)	22.0	2144	1208	1.77	1208	936	0
Daito (Daito Isls.)	22.4	1436	1231	1.17	1149	287	82

although very slight deficits may occur also in some other areas. Southern Kyushu (Kagoshima), Pacific Kyushu, windward southern Shikoku, and the Kii Peninsula (southern Honshu), on the other hand, are in the path of both summer monsoonal rains and late-summer typhoons, and show no summer water deficits.

The distinctly different seasonal patterns of the Pacific and Japan Sea sides of Japan have some subtle ecological consequences. The deep snow accumulations over much of the Japan Sea side provide protection for shorter plants in winter and ensure water supplies throughout the summer, resulting in distinctly larger leaf sizes in some species and the occurrence of more evergreen species in the deciduous forest understoreys. The deep snow also, however, results in a shorter snow-free growing season and lower springtime temperatures, with consequences for rice crops as well as natural vegetation.

In Köppen's (1931) climate system Japan is mostly a Cfa climate, except for mostly Dfb in interior central and northern Honshu plus all of Hokkaido and ET in high mountains (*cf.* Fukui 1933). Using the 0°C isotherm to separate C and D climates (instead of Köppen's original − 3°C isotherm), Sekiguchi (1949) was able to distinguish significant Dfa as well as Cfb and Dfc areas more clearly and felt that a Cw climate occurs in some smaller Pacific areas (e.g. parts of the Kanto, Kochi, and Miyazaki Plains). He also felt that the Cfa/Dfa boundary coincided closely with the boundary between the evergreen and deciduous broad-leaved forest regions. The relationship between climate and potential forest zones was explained more completely by Kira's (1945, 1977) Warmth and Coldness Indices, presented already in Table 2. The well known regionalization by Sekiguchi, with map, is explained in English by Mizukoshi (1977).

Continental China and Korea

The climate of the large East-Asian mainland, as opposed to Japan, is dominated by the full effect of the dry winter monsoon for much of the year.

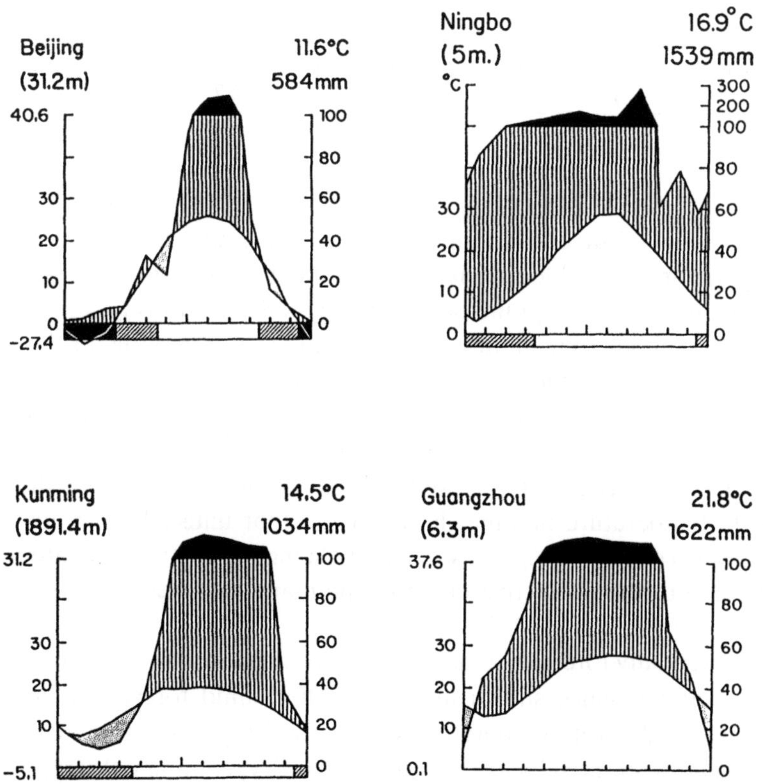

Fig. 4. Climate diagrams for selected sites in China. All sites show the clear monsoonal seasonality characteristic of eastern China. Beijing, in addition, shows the extreme seasonality of the subhumid areas of northern China, which extend from Mongolia to quite near the east coast. There is very little precipitation in spring and autumn as well as winter. Ningbo, south of Shanghai, shows the more moderated, humid transitional climate (warm-temperate to subtropical) of much of eastern China. Guangzhou shows the similar climate to subtropical southeastern China, with greater dryness in winter. Kunming represents the somewhat unique upland-subtropical climate of the Yunnan Plateau in southwestern China.

Table 5. Temperature summary for selected cities on the East Asian mainland. The cities are arranged from north to south. All cities are at relatively low elevation and nearly coastal unless indicated by elevation. The first two columns of numbers represent the mean temperature and the mean daily minimum temperature in the coldest month, respectively (all January). The third column is the coldest temperature ever recorded (year not available). The last two columns represent mean monthly temperature and mean daily maximum of the warmest month, respectively, all July (as opposed to August in Japan). Sources: Müller (1982), Wu (1980).

	Coldest month			Warmest month	
	mean	mean min.	abs. min.	mean	mean max.
Blagoveshchensk (50°)	−24.2	−28.3	−42	21.5	27.2
Haerbin (45°)	−20.1	−24.4	−41.4	23.3	27.8
Beijing (40°)	−4.7	−10.0	−22.8	26.1	32.0
Seoul (38°)	−4.9	−9.4	−22.2	25.4	30.6
Xian (410 m) (34°)	−0.5	−5.0	−19.1	27.9	33.9
Shanghai (31°)	3.4	0.6	−12.1	27.2	32.2
Chengdu (500 m) (31°)	6.2	3.3	−4.0	26.5	29.4
Kunming (1900 m) (25°)	9.6	2.8	−5.4	20.2	25.0
Guangzhou (23°)	13.6	9.4	0.0	28.8	32.8
Jinghong (550 m) (22°)	15.5	5.7	2.7	25.6	–

Not only winter but also spring and autumn are unusually dry, except in southern, more coastal locations. The continental conditions are clearly suggested by the climate diagrams shown in Fig. 4. Summer brings heavy rainfall but only for about four months in the north (to 6–8 months in the south). Temperatures show wide annual ranges and lower winter extremes.

Temperature patterns for representative sites in continental East Asia are summarized in Table 5. Mean winter temperatures are relatively low, due to the pervasive effect of the cold, dry monsoonal winds from the north. Absolute temperature minima, however, are not unusually low for a large mid-latitude continent (e.g. only −4°C at Chengdu, at 500 m elevation). This may be due to the protective effect of the many east-west mountain ranges in China.

The potentially forested areas of China are suggested by the map of annual climatic water balance shown in Fig. 5. The humid forest climates of the subtropical and warm-temperate southeast and typical-temperate to boreal northeast are separated by an extension of the dry, continental steppe climate from Mongolia into the East China Plain. This large region, from Beijing to Shanghai, has been settled for a very long time, has very little remaining natural vegetation, and represents a major gap in the understanding of vegetation zonation in East Asia.

Climatic zonation and regionalization have been a major focus of research in China since World War II. This was to provide a more rational basis for agriculture, forestry, and other aspects of economic development, and there is a very large literature of such work (e.g. Phys. Regionaliz. Comm. 1958–59, Han 1981, Inst. Botany 1982, Hou 1983, China Nat. Geogr. 1983, 1984, Agric. Nat. Res. Comm. 1984). There is an excellent treatment of the vegetation of China (Wu 1980), and many provinces have their own detailed vegetation descriptions (e.g. Guangdong Plant Res. Inst. 1976, Yunnan Veg. Ed. Comm. 1987). Some of this work is summarized briefly in English by Zhao (1986), but most remains untranslated.

In the course of their regionalization work, the Chinese relied considerably on various Russian approaches, notably that of Alisov (*cf.* 1954). Viewing the world from a more northerly perspective, Alisov expanded the concept of the subtropics, consequently shifting the term warm-temperate northward to apply mainly to the summergreen deciduous forest region (see Table 6). These differences in zonal terminology have been well explained by Song (1988) but are still confusing and must be kept well in mind when reading the Chinese literature.

The effect of the drier climate in continental East Asia is especially important for the potential for forest regrowth after cutting. On the East Asian mainland, in the few places where forests may remain at lower elevations, such forests occur at lower water-balance values, often near one (*cf.* Fig. 5). Lush "laurel forests" may remain only because their own microclimates are more humid than the macroclimate outside. If cut, these forest areas would be subject to the drier macroclimate, and forests may not be able to re-grow readily. This may be a significant factor in the reforestation of the wide, totally deforested gap in the driest part of eastern China, between the more humid areas to the north and south (in which some forest patches do remain).

Subtropical mountains

The Yunnan Plateau (about 1800–2000 m average elevation) represents the only large occurrence of uplands in the humid warm-temperate to subtropical latitudes of the Northern Hemisphere. Various other mountain ranges, mainly in subtropical China but extending also into warm-temperate areas, show similar conditions in smaller areas. The area involved is not large but shows some unique climatic effects and vegetation relationships, including:

- the "perpetual springtime" climate of Kunming (see Fig. 4) and the Yunnan Plateau in general;
- the unusual, persistently cloudy climate of the Sichuan Basin;
- the occurrence of hygrophilic evergreen "laurel forest" (*Castanopsis* and other Fagaceae) in climates with a true dry season (precipitation near zero in the driest winter month); and

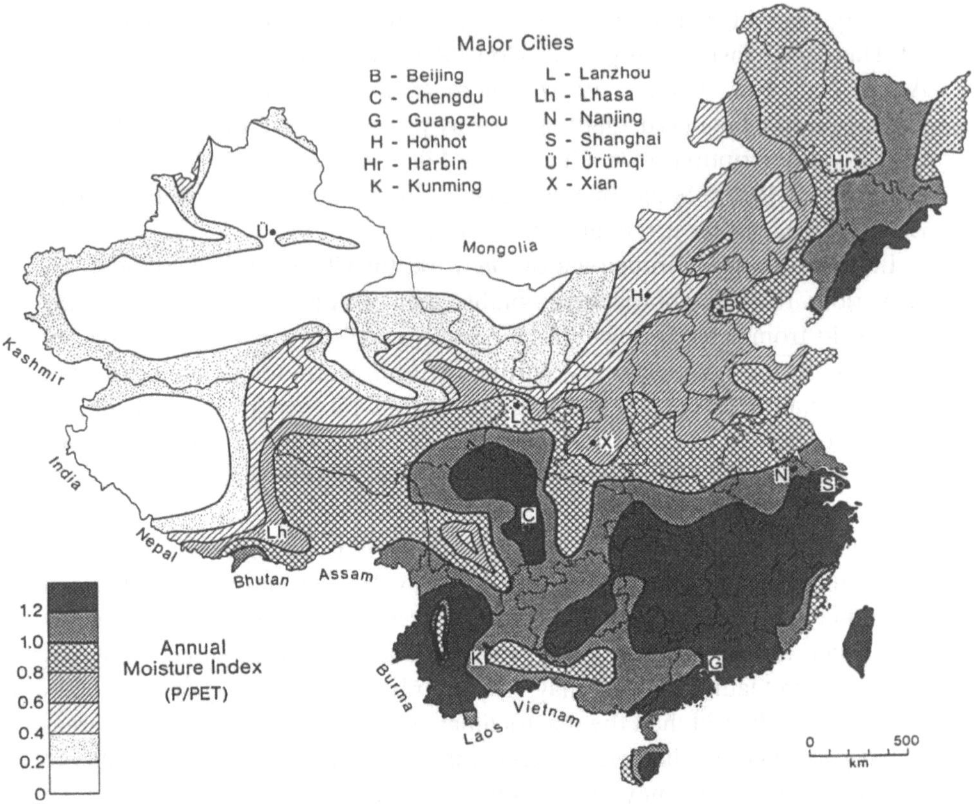

Fig. 5. Annual climatic water balance ("Moisture Index") in China. Areas with annual climatic water balance over 1.0 (i.e. precipitation in excess of potential evapotranspiration) are potential forest climates. These occur throughout southeast China as well as in less continental parts of Dongbei (northeastern China). Climates with annual water balance between 1.0 and about 0.8 may support grassland to open woodland or more closed low forest. These areas occur through central China, all the way to the east coast in the area of the East China Plain. Such areas are today mostly in agricultural use. Areas with annual water balance below 0.8 generally do not support woody vegetation.

- an unusual abundance of non-boreal conifer genera (*Torreya*, *Keteeleria*, *Sciadopitys*, etc.) and of non-boreal species in otherwise largely boreal conifer genera (e.g. *Abies*).

Differences in the species composition of the evergreen broad-leaved forests of southeastern and winter-dry southwestern China (i.e. Yunnan and Sichuan), with numerous cases of apparent vicariant species pairs, were noted by Wang (1961), who also provided some climatic data. The composition of these forests is described in more detail by Wu (1980). It has been suggested that these conifer taxa (see list by Box, in press) are relicts and that their abundance is the result of less accessible locations where primitive taxa are

Table 6. Climatic zonation terminology on continental east sides in the Northern Hemisphere. The main difference between Chinese and more global terminology involves the concept of the subtropical and warm-temperate zones. In Japan (e.g. Kira 1977, Miyawaki 1984, Numata 1974) and in more global perspective (e.g. Walter 1985, *cf.* Woodward 1987), climates which have frequent frost are not called subtropical. Source: Box (in press, based on Song 1988). EG = evergreen

China	Vegetation structure	Japan and east N America
Warm-temperate	Summergreen deciduous forest (deciduous understoreys) Summergreen canopy with EG/semi-EG understoreys	Temperate (typical)
Northern subtropical	Semi-evergreen broad-leaved forest (mixed) Summergreen-evergreen forest mosaics (unmixed stands)	Temperate/warm-temperate
Middle subtropical	Evergreen broad-leaved forest (temper. & subtrop. taxa; secondary forests deciduous; lianas/buttresses not large)	Warm-temperate
Southern subtropical	Evergreen broad-leaved forest (mainly tropical taxa; secondary forests evergreen; larger lianas/buttresses)	Subtropical
Tropical	Evergreen seasonal (rain)forest (only tropical taxa; some trees drought-deciduous; large lianas/buttresses)	Tropical/subtropical
Tropical	Tropical rainforest (only tropical taxa; many epiphytes; cauliflory; large lianas/buttresses)	Tropical/equatorial

somewhat protected from competition with the full range of faster-growing Angiosperms. The locations of these conifers can also be separated climatically, though, and constitute combinations of climatic value-ranges which correspond to some perhaps similar (but less diverse) conifer areas on other continents, e.g. southern South America, eastern Australia (Box 1981 and in press).

Humid-tropical Southeast Asia

The climate of Southeast Asia, including Thailand and Myanmar (Burma), tropical southernmost China, the Malay Peninsula, and the East Indies, varies from aseasonal equatorial in the south to humid-subtropical at its northern limit. Between the equatorial belt and the humid warm-temperate regions there is normally a zone of tropical summer-rain climate (Walter

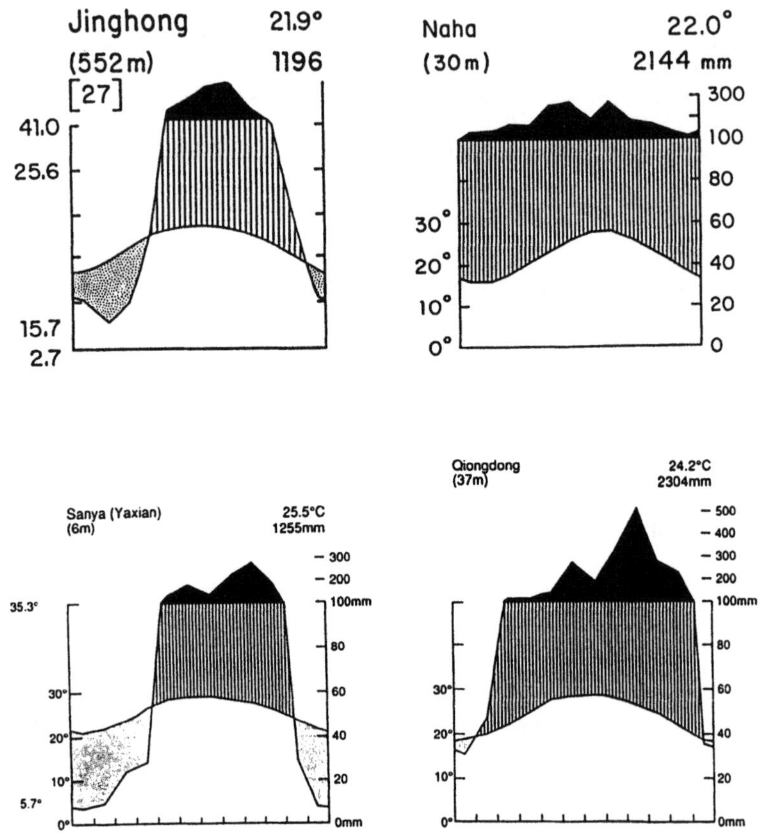

Fig. 6. Climate diagrams for selected sites in subtropical Southeast Asia. Naha (Okinawa) and Qiongdong (eastern Hainan) represent the humid-subtropical to tropical climates of coastal and island locations in Southeast Asia. These continue, with decreasing temperature seasonality, southward to Indonesia. Jinghong (southern Yunnan) and Sanya (rain-shadow of southwestern Hainan) represent the more zonal wet-dry climates of inland areas and suggest the climatic conditions of much of mainland Southeast Asia.

1973) or Aw climate (Köppen 1931), with wet season associated with the intertropical convergence zone (ITC) and (winter) dry season associated with the subtropical high-pressure belt. This climate also occurs across Southeast Asia, in typical latitudes (roughly 10° to 25° N), but the dry season is not completely rainless and the humidity not so extremely low (as in Africa and India, for example), due to the land-sea configuration and local winds. The resulting climate can be called humid-subtropical or humid wet-dry. The availability of some soil water even in the dry season permits the widespread occurrence of evergreen and semi-evergreen forests in areas with a distinct dry season (e.g. the "dry evergreen" dipterocarp forests of Thailand). Some of these climatic aspects are illustrated by the climate diagrams in Fig. 6.

Table 7. Water-balance amelioration with increasing elevation in mountains (example: Mt. Jianfengling, Hainan). The annual moisture index increases from that of a dry woodland (Dongfang) or at best a mesic but deciduous tropical forest (Changjiang) near sealevel to that of a humid "laurel" forest at only 890 m. There is still a dry season, but it is shorter and less severe. (From: Box et al. 1989.)

	Dongfang	Changjiang	Jianfengling (890 m)
Mean annual temperature	24.7°C	24.2°C	19.7°C
Mean July temperature	29°C	28°C	24°C
Mean January temperature	19°C	19°C	14°C
Average annual precipitation	993 mm	1677 mm	2500 mm
Annual pot. evapotrans. (PET)	1456 mm	1426 mm	1160 mm
Annual moisture index (P/PET)	0.68	1.18	~2.1
Precipitation of driest month (January)	7 mm	10 mm	~15 mm
Estimated January PET	91 mm	93 mm	~70 mm
Estimated moisture index for January	0.08	0.11	~0.21

The natural forests of tropical lowlands with a dry season are either deciduous or at least subject to seasonal water stress while retaining more xeromorphic evergreen leaves (e.g. the "dry evergreen forests" of Thailand). As one ascends mountains in these areas, however, one finds a rapidly changing water balance, as precipitation (and humidity) increase and temperature (and potential evapotranspiration) decrease. The degree of change, in less than 1000 m, is illustrated by a projected water balance in western Hainan (Table 7), where there is a relatively strong dry season in the lowlands (Box et al. 1989). In addition to the improved water balance, temperatures decrease (roughly 6°C per 1000 m) such that near-freezing conditions may occasionally (or more regularly) occur in winter at elevations as low as 1000–1500 m (depending also on the latitude). Under such conditions, truly tropical trees and other species are excluded, and the forest may change very rapidly from a "tropical" forest to the East Asian "laurel" type forest dominated by evergreen Lauraceae and Fagaceae, etc. (*cf.* Whitmore 1984, Fujiwara 1985, Ohsawa 1983). The abruptness of this change is illustrated (Box et al. 1991a) by a situation in Xishuangbanna in which two forests separated by about 350 m of elevation, on the same mountain, contained no tree species in common.

Zonal vegetation

The natural forest and other vegetation regions of East Asia, and less clearly Southeast Asia, have been related to climatic zones by many regional and global ecologists (e.g. Kira 1945, Hämet-Ahti et al. 1974, Wolfe 1979, Hou 1983, Tuhkanen 1984, Box 1986). As a result, natural lowland forest regions, with corresponding vertical sequences of vegetation belts, can be readily identified, as summarized in Table 8. This variety of forest and related vegetation regions occurs probably more completely in East and Southeast Asia than on other continents due to the larger size of the Asian land mass. The geographic occurrence of these forest types is shown on maps by Schmithüsen (1976), Miyawaki (1979), and Inst. of Botany (1982).

In practice, these forest types can be compressed into the four more general regions shown as headings in Table 8. These also represent the main divisions of this book:

1. tropical forests
2. evergreen broad-leaved forests (i.e. subtropical plus warm-temperate "laurel" forests)
3. summergreen broad-leaved forests
4. boreal/montane coniferous forests (including related subalpine and alpine vegetation).

The upward temperature decrease and improved water balance results in an easily recognizable vertical sequence of vegetation belts which can be seen on many mountains throughout tropical and subtropical East Asia (*cf.* Table 8). This sequence is also well described for tropical Southeast Asia by Whitmore (1984).

The four main forest regions were related to temperature by means of Kira's Warmth Index (WI) and Coldness Index (CI), as shown in Table 2. The Kira system of vegetation zones and climatic indices has been applied to China by Fang (1988) and to Korea by Yim & Kira (1975–76) and by Yim (1977). A particular forest type which does not immediately follow simpler bioclimatic zonations involves the occurrence, in both China and Japan, of deciduous forests, apparently as the potential natural vegetation, in transitional areas which appear to be warm enough to be called warm-temperate. These "warm-temperate deciduous forests" (*Quercus-Zelkova* forests in interior western Honshu, deciduous and semi-deciduous *Quercus* forests in central China, *cf.* Wang 1961) occur in more continental transitional areas, from the temperate deciduous to the warm-temperate evergreen forest region, and were explained by Kira (1949) as belonging to the warm-temperate zone in terms of summer warmth but with winters still too cold (CI 0°C to −10°C) for evergreen broad-leaved species.

An attempt to summarize the climatic ranges and some other relationships of the main forest types of East and Southeast Asia is given in Table 9. Forests generally occur only where the annual (potential) climatic water balance does not show a significant deficit (MI $>$ about 0.9), but there are two exceptions:

Table 8. Characteristic natural forest regions and related sequences of mountain vegetation on continental east sides (Northern Hemisphere)

Lowland	Characteristics	Montane belt	Subalpine belt
Tropical			
Tropical rainforests	Tall trees, lauraceous to coriaceous leaves, high diversity & productivity	Evergreen broad-leaved or mixed forests	Cloud/ mossy forest
Tropical seasonal evergreen or semi-evergreen forests	Forests with coriaceous to sclerophyll leaves, short dry season synchronizing plant activity patterns		
Tropical deciduous (raingreen) forest/woodlands	Significant, very dry season; most trees completely deciduous	Semi-evergreen broad-leaved forests	Cloud/ mossy forest
Subtropical			
Subtropical "laurel" forests	Glossy (lauraceous) evergreen leaves; permanently moist; many tropical affinities	Evergreen mixed forests	Mixed or conifer forest
Warm-temperate "laurel" forests	Glossy evergreen leaves; permanently moist but cooler; many temperate-zone affinities	Summergreen or semi-evergreen forest	Conifer forest
Temperate (incl. cool-temperate)			
Summergreen deciduous forests	High summer productivity but complete winter dormancy	Conifer forests with some deciduous elements	Boreal-like conifer forest
Boreal/Subarctic			
Boreal/subarctic and similar montane/ subalpine conifer forests	Few species of evergreen conifers (mainly *Abies* and *Picea*); complete winter dormancy, moist in summer	Subalpine krummholz/montane tundra	
Ultra-continental boreal *Larix* forests/woodlands	Forests to open woodlands, dominance by single species; dry summer ameliorated by slowly melting soil water		

References: 1. global: Troll (1948, 1961), Walter (1968, 1973), Walter & Box (1976), Wolfe (1979), Tuhkanen (1984). 2. regional: Miranda & Sharp (1950), Whittaker (1956), Troll (1972), Walter (1974), Numata (1974), Numata et al. (1972), Kira et al. (1976), Yoshino (1978), Gomez-Pompa (1977), Wu (1980), Greller (1980, 1988, 1989), Fujiwara (1981–86), Hou (1983), Ohsawa (1983), Miyawaki (1984), Chen (1986), Barbour & Billings (1988).

Table 9. Climatic ranges of main forest types of East and Southeast Asia

	BT	T_{max}	T_{min}	DTy	MI	Dry months
Tropical rainforests (lowlands only)	>25	>26	>20	<5	>1.0	none
Dry evergreen forests (lowlands only)	>23	>26	>17 (19)	<12	0.8–1.3	>1
Tropical deciduous forests (lowlands)	>20	>23	>16	<14	>0.6	>2
Evergreen broad-leaved forests	>12.7 (to 26)	>17.8	>3.0	>8	>0.9 (1.3)	<2 (winter only)
Temperate deciduous forests	7–21	>14	<9.0 (to −25)	>11	>0.95	(winter only)
Boreal/montane (EG) conifer forests	3.5–7	12–20	<−3.0 (to −30)	>17	>0.9	(winter only)
Boreal larch taiga (ultra-continental)	3.8–5 (to 8)	16–19	<−30	>45	>0.34	(winter only)

BT = annual biotemperature (sum of positive mean monthly temperatures, divided by 12); T_{max} = mean temperature of the warmest month; T_{min} = mean temperature of the coldest month; DTy = annual range of monthly mean temperatures; MI = annual moisture index (precipitation divided by potential evapotranspiration); dry months = months with essentially no precipitation (average); EG = evergreen.

1. the drier tropical deciduous low forests and, more commonly, open woodlands, in which winters are not cold (so the MI ratio is greatly reduced); and
2. the ultra-continental *Larix* taiga of the coldest parts of eastern Siberia, mainly open woodlands, in which the modest amount of soil water available is efficiently rationed over the course of the climatically rather dry summer by the slow thawing of the soil.

Broad-leaved forests are potentially evergreen (water not limiting) where mean winter temperatures are above freezing and extremes not below about −15°C (*cf.* Woodward 1987). This means that evergreen "laurel" forests can extend as far north in maritime Japan as northern Honshu (only along the coast), to January mean temperatures of only +0.5°C (!) (Miyawaki 1980–89, vol. 8). In eastern China, subtropical evergreen broad-leaved tree species can still be found near Ningbo (south across the estuary from Shanghai), where January mean temperature is around 4°C (*cf.* Forests of Jiangxi Comm. 1986, Box et al. 1991b). Where there is a dry season (SE Asia to southwestern China), evergreen "laurel" forest, dry evergreen forest, and tropical seasonal evergreen forest can generally tolerate 1–2 months with no rainfall but disappear if the extreme dry season exceeds two months. True tropical rainforests usually do not occur if average monthly rainfall drops to zero.

Global perspective

Comparing East and Southeast Asia with other parts of the world, one can make the following observations.

1. The general environment for plant growth and biological diversity over most of East and Southeast Asia is unusually favorable, due to abundant rainfall/humidity, diverse topography, etc. Primary productivity is generally high, and many plant species are distinctly mesomorphic in structure, except in localized harsh situations such as the windy coastlines, highest mountains, and the continental interior. Estimates of primary productivity and standing biomass based on field measurements in Japan as well as tropical Asia are summarized by Shidei & Kira (1977). These values have been generalized and compared globally by Uchijima & Seino (1985), among others (*cf.* Box et al. 1989).

2. In general, East and Southeast Asia correspond to global vegetation-climate relationships. This was seen in Table 8 and, more rigorously, in quantitative comparisons of vegetation zones (Wolfe 1979) and predictions of plant types (Box 1981). The largest exceptions to the general relationships seem to involve minimum temperature (see below) and/or the unusual northward extension of evergreen broad-leaved forests in maritime Japan, to unusually low mean winter temperatures (*cf.* Box 1988).

3. The temperature relationships of East Asia in general are somewhat different from world patterns. As a result, quantitative extrapolations of Asian climatic zonations, such as that of Kira, to other continents do not always fit the local conditions, perhaps especially in North America. Mean, mean minimum, and absolute minimum temperatures for selected coastal areas in Japan, eastern China, and corresponding latitudes in eastern North America are shown in Table 10. Although winter temperature extremes are much lower in China and Korea than at comparable latitudes in Japan, the absolute extremes are not lower than in eastern North America. Eastern North America, with no east-west mountains to block colder northern air, but a source of warm moist air to the south (i.e. the Gulf of Mexico), has higher winter mean temperatures than would be expected but also infrequent, quite low absolute minima (e.g. − 1°C at Miami). In East Asia, on the other hand, the strong winter monsoon winds depress both the minima and the monthly and daily means. Absolute minimum temperatures for various cities in the world are plotted against mean cold-month temperature in Fig. 7. East Asian sites generally show smaller differences between cold-month mean and absolute minimum than in other, otherwise comparable temperate-zone areas.

4. The region elsewhere in the world which is most comparable to East Asia is clearly eastern North America at comparable latitudes (about 48° N

Table 10. Mean and absolute temperature minima at coastal sites in Pacific Japan, eastern China, and eastern North America (1930–1980). Data periods vary, due to the less ready availability of the values for mean and absolute minima. Values represent the following sites. Eastern N. America: Washington (39° N), New Holland (N. Car., 35.5° N), Charleston (33° N), Jacksonville (30.4° N). China: Lüda (39° N), Tianjin (39° N), Qingdao (36° N), Shanghai (31° N). Japan: Sendai (38° N), Tokyo (36° N), Hiroshima (34° N), Kagoshina (31° N). Colder minima were recorded in Japan before 1930 (Tokyo Astron. Observ. 1980) and in eastern N. America since 1985 (Box 1988)

	Japan	China	Eastern N. Amer.
Mean of coldest month			
38–39° N	0.1		2.7
35–36° N	3.7	−1.1	6.8
33–34° N	4.2	–	10.2
30–31° N	6.6	3.4	13.3
Mean daily minimum			
38–39° N	−7.1	−8.9	−2.8
35–36° N	−1.2	−4.4	1.1
33–34° N	−0.6	–	6.1
30–31° N	2.8	0.6	8.3
Absolute minimum			
38–39° N	−11.7	−20.2	−25.6
35–36° N	−9.2	−16.4	−12.2
33–34° N	−8.3	–	−12.2
30–31° N	−5.7	−12.1	−9.4

Sources: Japan Environmental Agency (1982, 1984), Müller (1982), National Climatic Center (1978).

to 20° N) (*cf.* Gray 1846, Graham 1972, Wolfe 1979, Boufford & Spongberg 1983). The main vegetation regions of East Asia also occur in eastern North America, but in the latter the deciduous forest region is by far the largest, extending even onto the warm-temperate southeastern coastal plain of the USA. The forests of the southeastern USA, where northern species are replaced by southern but still deciduous species (e.g. *Qu. rubra* by *Qu. falcata*, *cf.* Daubenmire 1978), have been called "southern mixed hardwoods" (Quarterman & Keever 1962, *cf.* Braun 1950) and seem to be explained well by Kira's (1949) concept of warm-temperate deciduous forest, even though the mean temperature levels involved are considerably higher. The extensive evergreen broad-leaved forest zones of East Asia have their North American counterparts only on mesic to moist sites in relatively narrow bands along the Gulf of Mexico and the southern Atlantic coast (Greller 1980, Box & Fujiwara 1988, Christensen 1988) and in the small subtropical zone of south Florida. Here the floristic composition of the small subtropical "hammock" forests involves mainly Neotropical taxa and a few remaining warm-temperate species (see for example Myers & Ewel 1990), but there are definite similarities to subtropical East Asia at the level of families and

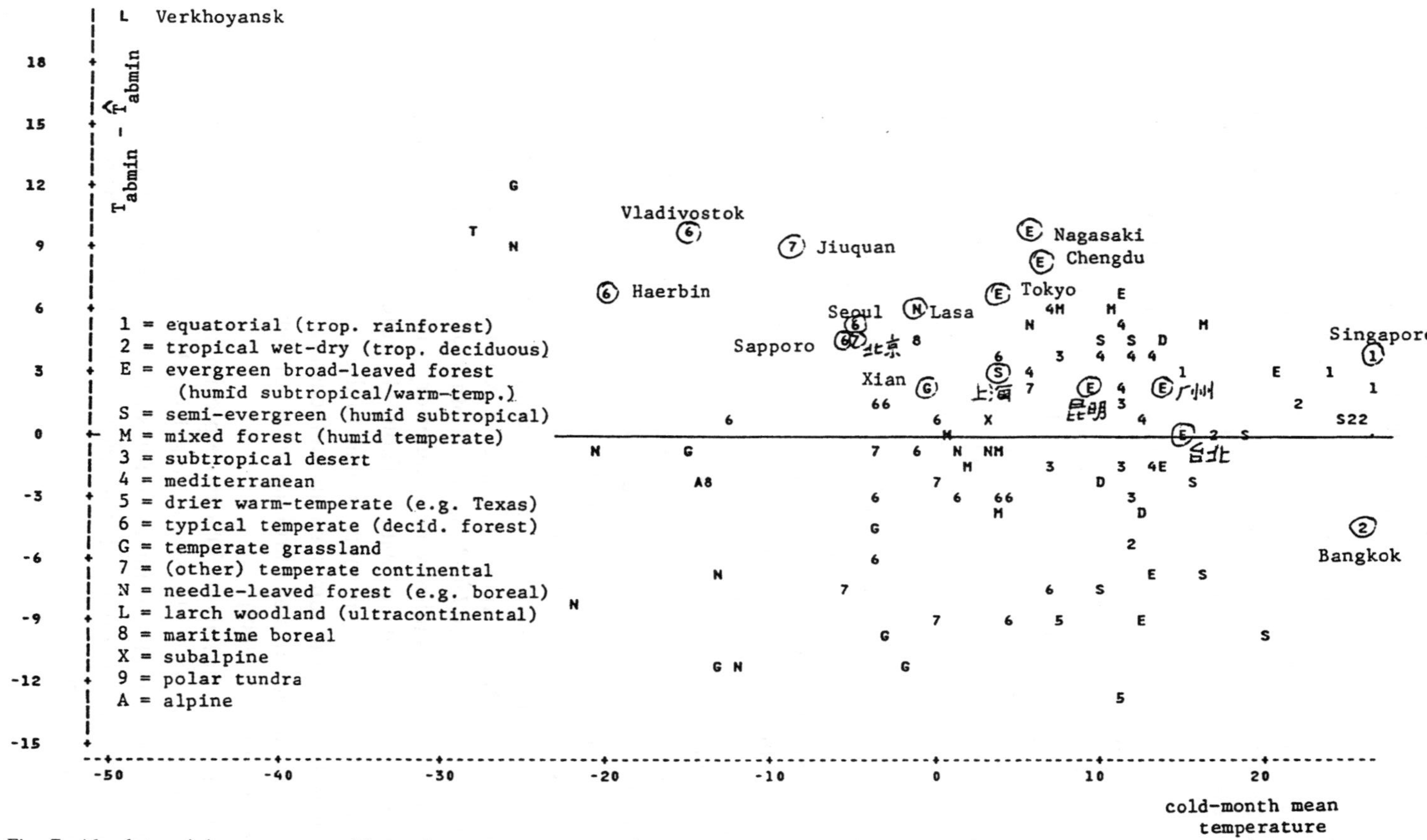

Fig. 7. Absolute minimum versus cold-month mean temperatures for representative mid-latitude sites in the Northern Hemisphere. A global trend pattern in absolute minimum temperature (coldest temperature ever measured) was obtained by linear regression ($n = 103$, $r = 0.4$) of absolute minimum temperature against the mean temperature of the coldest month (horizontal axis). The deviation of absolute minimum temperature from this global trend line (horizontal line) is shown on the vertical scale and represents an indication of an interesting aspect of continentality. Most sites in the forest regions of East and Southeast Asia (circled) fall above the trend line, i.e. they have absolute minimum temperatures not as far below their winter mean temperatures as might be expected based on global relationships. Sites with absolute minima lower than expected from global relationships are found below the trend line and come mainly from North America, especially the southeastern USA. Sites with cold-month means below − 10°C are mainly from interior Eurasia and occur both above and below the trend line. The symbols for the individual sites represent landscape (i.e. general climatic) types, e.g. E = evergreen

general physiognomy. Fagaceae are not important as far south in the Caribbean area as in Asia, but warm-temperate *Quercus virginiana* still occurs in subtropical hammocks of south Florida and in Cuba. Many islands of the northern Caribbean are limestone and carry dense, mainly evergreen low subtropical forests reminiscent of the Ryukyu Islands. Small mountain areas in subtropical northeastern Mexico (Sierra Madre Oriental) may also show similarities to East Asia (see Gomez-Pompa 1977, Miranda & Sharp 1950, *cf.* Peters, this volume).

5. Other forest regions similar to those of East and Southeast Asia include at least the following:

- evergreen broad-leaved forests (extra-tropical) of eastern Mexico, the south slope of the Himilaya, southern Brazil, Australia (mainly eastern, see Beard, this volume), and southern and montane eastern Africa, as well as in cool-temperate western Patagonia (see Hildebrand-Vogel, this volume), New Zealand (see Meurk, this volume), and southernmost Australia;
- summergreen deciduous forests of cooler temperate regions (*cf.* northern Japan), such as in northwestern Europe (*cf.* Klötzli & Zarzycki, this volume), northwestern North America, and southern Patagonia;
- montane and boreal coniferous forests of Europe (but probably not of much more continental Siberia); and
- tropical and subtropical forests in many places but perhaps especially some sclerophyll forests in Australia (*cf.* dry evergreen forests of Thailand), the cerrado of Brazil, the lowland to montane forest sequences in eastern and central Africa, and the Amazonian rainforest.

6. The closest equivalent to the drier regions of East Asia (e.g. Japan's Inland Sea area, various rain shadows in interior China, much of Korea) may also be in eastern North America. The East Asian drier areas, including areas with *Quercus phillyraeoides* (Japan) and other sclerophyllous *Quercus* species, are not similar climatically to summer-dry Mediterranean Europe, as is shown by the water-balance data in Table 11. Instead, the East Asian climate and *Quercus* scrub seem more similar to the sclerophyllous *Quercus* forests and woodlands of the summer-rain southeastern USA, particularly in the drier area from the coastal Carolinas into northern (but not western) Florida (Box & Fujiwara 1988).

7. In general, Japan has about 50% more precipitation per year, over its full range of latitudes, than do comparable latitudes in China or eastern North America. As a result, the Japanese forests seem especially well established, with perhaps a higher potential for successful recovery and/or reestablishment after disturbance.

Table 11. Annual water balance in coastal warm-temperate to subtropical areas of East Asia, Europe, and eastern North America. No part of Japan and only subtropical dry, deep valleys of southwestern China (not shown) have water-balance ratios as low as in mediterranean climates. Dry areas of East Asia, and their vegetations, are more comparable to drier parts of the southeastern USA (or perhaps eastern Australia)

	Precip./PET	Annual deficit	Natural vegetation
Japan and Korea			
S Honshu, Kyushu, southernmost Korea	>1.4	0–10	Laurel forest
Inland Sea rain-shadow	0.9–1.2	10–50	Moderately sclerophyllous *Quercus* forest
China			
Shanghai area	0.9–1.2	5–130	(semi-evergreen?)
Zhejiang and inland Fujian	1.3–1.6	0–100	Laurel forest
Fujian coastal rain-shadow	0.8–1.0	100–230	(evergreen scrub/woods?)
Guangdong, most of Taiwan	1.2–1.5	10–60	Subtropical laurel forest
Southeast USA			
N Gulf coast	1.2–1.3	25–50	EG *Quercus*, *Fagus-Magnolia*, deciduous hardwood forests
S Atlantic coast (to central Fla.)	1.0–1.2	10–100	EG *Quercus* and deciduous hardwood forests
Mediterranean Europe			
S France to central Italy	0.7–0.9	100–375	EG *Quercus* maquis, some EG and decid. *Quercus* forest sclerophyll maquis
S Spain to Sicily	0.3–0.9	300–700	

Climate change and East Asia

Several global climate models (GCM's) of the earth's atmosphere have now been developed and used to provide initial predictions of changes in temperature and precipitation which might be expected from "greenhouse" warming of the earth. Results from three of these models were compared by Schlesinger & Mitchell (1985). Estimated increases in winter mean temperature range from 3°C in Southeast Asia to as much as 7°C in Japan and northeastern China. Precipitation changes are much more complex, but the greatest variability in estimates occurs in southern China and Southeast Asia (including predicted increases as well as decreases, mainly in summer). The general picture of a more humid, less continental world, though, as expected from GCM simulations, would bring great changes to East and Southeast Asia, subject as they are to the monsoonal seasonality.

Conclusion

In conclusion it can be seen that the climatic relations of the forests and forest regions of East and Southeast Asia are similar to those elsewhere, but with their own particular character:

- Japan is maritime (humid, with moderated temperature extremes) – but not its mean temperature patterns.
- China is continental (drier, monsoonal, with a greater range of temperatures) – but not its extreme minimum temperatures (as compared at least with Siberia to the north).
- Southeast Asia is tropical wet-dry – but not always raingreen deciduous.
- Yunnan is unique.

The zonal forests show their own corresponding characteristics, for example especially lush "laurel" (evergreen) and deciduous forests in more maritime areas as well as other areas where moderate temperatures permit.

Evergreenness in broad-leaved extra-tropical forests seems to be limited (poleward) by mean minimum temperatures, perhaps in combination with absolute minima. This permits such forests to occur at quite low overall mean (monthly) temperatures (e.g. coastal northern Japan, *cf.* New Zealand) if minima are not too low. Tropical and subtropical evergreen trees seem to be limited in much the same way, extending in eastern China north to areas with mean January temperature as low as around 5°C but with minima not far below freezing, as in higher tropical mountains.

Deciduousness may be an appropriate response to cold winters where (moist) summers are warm (i.e. in the temperate zone), as well as in boreal areas with moderated winters and thus perhaps longer growing seasons (e.g. Kamchatka, Hokkaido). Short-summer taxa such as *Betula* and *Populus* seem to do well in monsoonal areas with warm summers but only short rainy seasons, as in northern China. Evergreenness (conifers) reappears in more continental boreal areas (with shorter albeit perhaps warmer summers) – but then disappears again in the most ultra-continental areas with the most severe winters (eastern Siberia).

Summergreen deciduous forests as well as boreal/montane coniferous forests (and secondary pine forests) regenerate relatively well after even severe disturbances, throughout the Northern Hemisphere. Evergreen "laurel" forests also seem to regenerate relatively well, at least in favorably humid maritime areas such as Japan. There is then a rather drastic discontinuity, equatorward, between the resilience of the temperate-zone types just mentioned, on the one hand, and the tropical and even subtropical forest types on the other hand. The latter generally do not recover so easily, and frequent cutting may lead to their permanent replacement by grassland, as has happened over enormous areas throughout southern and southeast Asia. The reasons for this sharp discontinuity in resilience may involve climate (or microclimate), soil, and/or nutrient dynamics, perhaps especially in the rhizosphere. Clarification of this discontinuity in forest resilience is an important task for present-day ecologists.

It has been fashionable in some circles to contradict ecological generalizations almost automatically, in order to show that a particular region (perhaps one's own) is somehow "unique." This is relatively easy to do, since every place becomes unique at some level of detail. Generalizing is more difficult but provides insights at higher levels. Some areas, such as the Yunnan Plateau, do seem to be unique even by rather general standards. One such consideration behind the organization of this series of four INTECOL 1990 symposia on "Forest Ecosystems of East and Southeast Asia, in Global Perspective," involved the suggestion that the world's extra-tropical evergreen broad-leaved forests have something in common, even though they have rarely been studied comparatively (*cf.* Ovington 1983, Box, in press). It is hoped that a comparative approach, studying climate and other environmental conditions as well as vegetation composition, history, and dynamics, will help to clarify the ecological position and potentials of the Asian forests and their counterparts elsewhere.

References

Agricultural Division of China (Committee) 1984. (A Scheme of Natural Divisions for China.) Beijing: Science Press. 165 pp. (in Chinese).

Alisov, B. P. 1954. Die Klimate der Erde (ohne das Gebiet der UdSSR). German translation of Russian original. 277 pp.

Bailey, R. G. 1990. Ecoregions map of the continents (world map with explanatory supplement). Envl. Conserv. 16(4): 307–309, plus map.

Barbour, M. G. & Billings, W. D. (eds.) 1988. North American Terrestrial Vegetation. Cambridge University Press. 434 pp.

Beard, J. S. (this volume) Historical and ecological development of the evergreen broad-leaved forests of Australia.

Boufford, D. E. & Spongberg, S. A. 1983. Eastern Asian-Eastern North American phytogeographical relationships: a history from the time of Linnaeus to the twentieth century. Annals Missouri Bot. Garden 70(3):423–439.

Box, E. O. 1981. Macroclimate and Plant Forms: An Introduction to Predictive Modeling in Phytogeography. Tasks for Vegetation Science, Vol. 1. Dr. W. Junk BV, Publishers, The Hague. 258 pp.

Box, E. O. 1986. Comparing the Mountain Vegetation of East Asia and eastern North America. In: Chen L.-Zh. and Chang Hs.-Sh. (eds.) Proceedings of Internat. Symp. on Mountain Vegetation, Beijing, September 1986. 8 pp.

Box, E. O. 1988. Some similarities in the climates and vegetation of central Honshu and central eastern North America. Veröff. Geobot. Inst. Rübel 98:141–168.

Box, E. O. (in press) The climate of Japan. In: A. Miyawaki et al. (eds.) Vegetation and Environments of Japan. Shibundo, Tokyo.

Box, E. O. (in press) Japanese vegetation in global perspective. In: A. Miyawaki et al. (eds.) Vegetation and Environments of Japan. Shibundo, Tokyo.

Box, E. O. & Fujiwara, K. 1988. Evergreen broad-leaved forests of the southeastern United States: preliminary description. Bull. Inst. Env. Sci. Techn., Yokohama Nat. Univ. 15:71–93.

Box, E. O., Fujiwara, K., Hao Y.-L., Zhong Yi, Fu Qi-Hao & Xiao B.-Sh. 1989. A tropical montane evergreen forest and other vegetation on Hainan Island, southern China. Bull, Inst. Envl. Sci. Techn., Yokohama Nat. Univ. 16:75–94.

Box, E. O., Holben, B. N. & Kalb, V. 1989. Accuracy of the AVHRR Vegetation Index as a predictor of biomass, primary productivity and net CO_2 flux. Vegetatio 80:71–89.

Box, E. O., Fujiwara, K. & Qiu X.-Zh. 1991a. Diversity and dissimilarity of three forest types in Xishuangbanna, tropical southern China. Bull. Inst. Env. Sci. Techn. Yokohama Nat. Univ. 17:85–105.

Box, E. O., Song Y.-Ch., Miyawaki, A. & Fujiwara, K. 1991b. An evergreen broad-leaved forest in transitional eastern China. Bull. Inst. Env. Sci. Techn. Yokohama Nat. Univ. 17:63–84.

Braun, E. L. 1950. Deciduous Forests of Eastern North America. Blakiston Co, Philadelphia. 596 pp.

Chen L.-Zh. 1986. The oak forest in North China. In: Chen L.-Zh. & Chang Hs.-Sh. (eds.) Proceedings of International Symposium on Mountain Vegetation, pp. 203–208. Botanical Society of China.

China Natural Geography Editing Committee (eds.) 1983. (China Natural Geography: Plant Geography.) Science Press, Beijing. 129 pp. (in Chinese).

China Natural Geography Editing Committee (eds.) 1984. (China Natural Geography: Climate.) Science Press, Beijing. 161 pp. (in Chinese).

Christensen, N. L. 1988. Vegetation of the southeastern coastal plain. In: Barbour, M. G. & Billings, W. D. (eds.) North American Terrestrial Vegetation, pp. 317–363. Cambridge University Press.

Daubenmire, R. F. 1978. Plant Geography with Special Reference to North America. Academic Press, New York.

Environment Agency 1982. The Natural Environment of Japan. Tokyo. 249 pp. (in Japanese and English).

Fang J.-Y. 1988. Climate and Distribution of Vegetation in China. Osaka City University. 172 pp.

Forests of Jiangxi Editorial Committee 1986. (Forests of Jiangxi.) Jiangxi Publ. House of Sci. and Technology, Nanchang. 890 pp. (in Chinese).

Fujiwara, K. 1981–86. (Phytosociological investigation of the evergreen broad-leaved forests of Japan.) Bull. Inst. Env. Sci. Technol., Yokohama Nat. Univ. 7:67–133, 8:121–150, 9:139–160, 13:99–149 (in Japanese).

Fujiwara, K. 1985. Comparison of the subtropical evergreen broad-leaved forests of Japan and Thailand. In: Yang H.-X. (ed.) Proceedings of Internat. Symp. on Ecology of Development of Tropical and Subtropical Mountain Areas, pp. 19–21. China Academic Publishers, Beijing.

Fujiwara, K. (in press) Evergreen broad-leaved forest region. In: A. Miyawaki et al. (eds.) Vegetation and Environments of Japan. Shibundo, Tokyo.

Fukui, E. 1933. (Climatic regions in Japan.) Geogr. Rev. 9:1–17, 109–127, 195–219, 271–300 (in Japanese).

Fukui, E. 1958. (The second Thornthwaite classification of Asia.) Tokyo Geogr. Papers II:47–64 (in Japanese).

Fukui, E. (ed.) 1977. The Climate of Japan. Kodansha, Tokyo. 317 pp.

Gomez-Pompa, A. 1977. Ecología de la Vegetación del Estado de Veracruz. Mexico: Cia. Editorial Continental, S. A. de C. V. 91 pp.

Graham, A. (ed.) 1972. Floristics and Paleofloristics of Asia and Eastern North America. Elsevier, Amsterdam. 278 pp.

Gray, A. 1846. Analogy between the flora of Japan and that of the United States. Amer. J. Sci. Arts II(2):135–136 (reprinted also in Graham 1972a, see above).

Greller, A. M. 1980. Correlation of some climate statistics with distribution of broad-leaved forest zones in Florida, USA. Bull. Torrey Botan. Club 107:189–219.

Greller, A. M. 1988. Deciduous forest. In: Barbour, M. G. & Billings, D. W. (eds.) North American Terrestrial Vegetation, pp. 287–326. Cambridge University Press.
Greller, A. M. 1989. Correlation of warmth and temperateness with the distributional limits of zonal forests in eastern North America. Bull. Torrey Botan. Club 116:145–163.
Guangdong Plant Research Institute 1976. Guangdong Zhibei (Guangdong Vegetation.) Science Press, Beijing. 341 pp. (in Chinese).
Hämet-Ahti, L., Ahti, T. & Konen, T. 1974. A scheme of vegetation zones for Japan and adjacent regions. Ann. Bot. Fenn. 11:59–88.
Han Y.-L. 1981. (A discussion on the northern boundary of the subtropical evergreen broadleaf forest zone in Anhui Province.) Acta Phytoecol. Geobotan. Sinica 5:54–57 (in Chinese).
Hildebrand-Vogel, R. & Vogel, A. (this volume) Evergreen broad-leaved forests of Southern South America.
Holdridge, L. R. 1959. A simple method for determining potential evapotranspiration from temperature data. Science 130:572.
Hou X-Y. 1983. Vegetation of China with reference to its geographical distribution. Ann. Missouri Botan. Garden 70:509–548.
Huo Y.-Q. (no year) A brief account of vegetational regionalization and zonal types in Guangdong, China (unpublished manuscript).
Institute of Botany 1982. ("Vegetation Map of China.") 2nd edition. Academia Sinica, Inst. of Botany. Cartographic Publishing House, Beijing (in Chinese, with legend booklet in Chinese and English).
Japan Meteorological Agency 1981–83. Climatic Tables of Japan, 1951–1980. 5 parts. Jap. Meteorol. Agency, Tokyo (in Japanese, with English key).
Japan Meteorological Agency 1984. Climatic Atlas of Japan. Tokyo. 188 pp. (in Japanese and English).
Kira, T. 1945. (A new classification of climate in eastern Asia, as the basis for agricultural geography.) Horticultural Inst., Kyoto University, Kyoto. 23 pp. (in Japanese).
Kira, T. 1949. (Forest Zones in Japan.) Ringyo Kaisetsu, no. 17: 105–141. Nippon Ringyo Gijutsu Kyokai, Tokyo (in Japanese).
Kira, T. 1977. A climatological interpretation of Japanese vegetation zones. In: Miyawaki, A. (ed.) Vegetation Science and Environmental Protection, pp. 21–30. Maruzen, Tokyo.
Kira, T. 1994. Forest ecosystems of East and South-East Asia in a global perspective. In: Box, E. O. et al. (eds.) Vegetation Science in Forestry, pp. 1–21.
Kira, T., Shidei, T., Numata, M. & Yoda, K. 1976. (Vegetation of Japan – its ordination in the global system of vegetation.) Kagaku (Science) 46(4):235–247 (in Japanese).
Köppen, W. 1931. Grundriss der Klimakunde. Walter de Gruyter, Berlin.
Makita, H. & Ichikawa, T. 1990. Summergreen Broad-Leaved Forests of Japan: Nature and Human Life in the Past and Present. Presentation in Symposium S3–4–02, IVth Internat. Congress of Ecology (INTECOL), Yokohama, August 1990.
Matsumoto, J. 1984. Summertime circulation patterns over eastern Asia. Jap. Progress Climatol. 1984:14–22.
Meurk, C. D. (this volume) Evergreen broad-leaved forests of New Zealand and their bioclimatic definition.
Miranda, F. & Sharp, A. J. 1950. Characteristics of the vegetation in certain temperate regions of eastern Mexico. Ecology 31:313–333.
Miyawaki, A. 1979. Vegetation und Vegetationskarten auf den japanischen Inseln. In: Miyawaki, A. & Okuda, Sh. (eds.) Vegetation und Landschaft Japans, pp. 49–70. Bull. Yok. Phytosoc. Soc. 16, Yokohama.
Miyawaki, A. (ed.) 1980–1989. (Japan Vegetation Record.) 10 vols., each 400–600 pp., with color veg. maps and tables. Shibundo, Tokyo (in Japanese, with German summary).
Miyawaki, A. 1984. A vegetation-ecological view of the Japanese Archipelago. Bull. Inst. Env. Sci. Technol., Yokohama Nat. Univ. 11:85–101.
Mizukoshi, M. 1977. Climatic division and climatography. In: Fukui, E. (ed.) The Climate of Japan, pp. 225–270. Kodansha, Tokyo.
Müller, M. J. 1982. Selected Climatic Data for a Global Set of Standard Stations for Vegetation Science. Tasks for Veg. Sci., vol. 5. Dr. W. Junk bv Publ., The Hague (now Kluwer).

Myers, R. L. & Ewel, J. J. (eds.) 1990. Ecosystems of Florida. University of Central Florida Press. Orlando. 765 pp.
National Climate Center 1978. Climatic Data of the States. US Department of Commerce, Asheville, North Carolina.
Numata, M. (ed.) 1974. The Flora and Vegetation of Japan. Kodansha, Tokyo; Elsevier, Amsterdam. 294 pp.
Numata, M., Miyawaki, A. & Itow, D. 1972. Natural and semi-natural vegetation in Japan. Blumea 20:435–481 (with 26 photos, veg. maps).
Ohsawa, M. 1983. (Comparative study on zonation of vegetation in East Asia.) In: Ikushima, I. et al. (eds.) (Contemporary Ecology of Japan) pp. 206–216. Kyoritsu Shuppan, Tokyo (in Japanese)
Okitsu, S. 1984. Control factors of the forest limit on Mt. Taisetsu, central Hokkaido, Japan. Jap. J. Ecol. 34:439–444.
Ovington, J. D. (ed.) 1983. Temperate Broad-leaved Evergreen Forests. Ecosystems of the World, vol. 10. Elsevier, Amsterdam.
Peters, R. (this volume) Architecture and development of Mexican beech forest.
Physical Regionalization Committee 1958–59. (General Natural Divisions of China.) Science Press, Beijing (in Chinese).
Quarterman, E. & Keever, Ch. 1962. Southern mixed hardwood forest: climax in the southeastern coastal plain, USA. Ecol. Monogr. 32:167–185.
Schlesinger, M. E. & Mitchell, J. F. B. 1985. Model Projections of the Equilibrium Climatic Response to Increased Carbon Dioxide. DOE/ER-0237, pp. 81–147. US Department of Energy, Washington.
Schmithüsen, J. 1976. Atlas zur Biogeographie. Meyers Grosser Physischer Weltatlas, vol. 3. Bibliographisches Institut, Mannheim/Wien/Zürich.
Sekiguchi, T. 1949. (The Köppen climatic classification of Japan.) Geogr. Social Life (15):12–16 (in Japanese).
Sekiguchi, T. 1959. Climatic classification of Japan. Tokyo Geography Papers 3:65–78.
Shidei, T. 1974. Forest vegetation zones. In: Numata, M. (ed.) Flora and Vegetation of Japan, pp. 87–124. Kodansha, Tokyo.
Shidei T. & Kira, T. (eds.) 1977. Primary Productivity of Japanese Forests. JIBP Synthesis, vol. 16. Univ. Tokyo Press. 289 pp.
Song Y.-Ch. 1988. Broad-leaved evergreen forests in central Japan in comparison with eastern China. Veröff. Geobot. Inst. Rübel 98:197- 224.
Suzuki, T. 1953. The forest climaxes of East Asia. Jap. J. Bot. 14:1–12.
Suzuki, H. 1962. (The classification of Japanese climates.) Geogr. Rev. Japan 35:205–211 (in Japanese).
Takahashi, K. 1955. (Dynamic Climatology.) Iwanami, Tokyo. 316 pp. (in Japanese).
Takahashi, K. 1977. Typhoons and Shurin (Autumn Rain). In: Fukui, E. (ed.) The Climate of Japan, pp. 113–134. Kodansha, Tokyo.
Thornthwaite, C. W. & Mather, J. R. 1957. Instructions and tables for computing potential evapotranspiration and the water balance. Publ. in Climatol. (Univ. Delaware) 10(3):185–311.
Tokyo Astronomical Observatory (eds.) 1987. Rika Nenpyo (Chronological Scientific Tables). Maruzen, Tokyo. 1007 pp. (in Japanese).
Troll, C. 1948. Der asymetrische Aufbau der Vegetationszonen und Vegetationsstufen auf der Nord- und Südhalbkugel. In: Jahrbuch Geobotan. Inst. Rübel f. 1947, pp. 46–83. Zürich.
Troll, C. 1961. Klima und Pflanzenkleid der Erde, in dreidimensionaler Sicht. Die Naturwiss. 48:332–348.
Troll, C. (ed.) 1972. Geoecology of the High-Mountain Regions of Eurasia. Franz-Steiner-Verlag, Wiesbaden.
Troll, C. & Paffen, K. H. 1964. Karte der Jahreszeitenklimate der Erde. Erdkunde, Arch. Wiss. Geogr. 18:5–28.
Tuhkanen, S. 1984. A circumboreal system of climatic-phytogeographical regions. Acta Botan. Fennica 127:1–50.

Uchijima, Z. & Seino, H. 1985. Distribution Maps of the Global Solar Radiation, Photosynthetically Active Radiation, Net Radiation, Net Primary Productivity and Solar Energy Efficiency. National Institute of Agro-Environmental Sciences. Kyushu Nat. Agric. Exper. Stn. Chikugo (Kyushu). 121 pp.

Walter, H. 1968. Die Vegetation der Erde in öko-physiologischer Betrachtung. Vol. II: Die gemäßigten und arktischen Zonen. VEB Gustav-Fischer-Verlag, Jena. 1002 pp.

Walter, H. 1973. Die Vegetation der Erde in öko-physiologischer Betrachtung. Vol. I: Die tropischen und subtropischen Zonen. Gustav-Fischer-Verlag, Stuttgart.

Walter, H. 1974. Die Vegetation Osteuropas, Nord- und Zentralasiens. Gustav-Fischer-Verlag, Stuttgart.

Walter, H. 1977. Vegetationszonen und Klima. 3rd ed. Eugen-Ulmer-Verlag, Stuttgart. 309 pp.

Walter, H. 1985. Vegetation of the Earth, 3rd ed. Springer-Verlag, Berlin/New York/Tokyo. 318 pp.

Walter, H. & Box, E. O. 1976. Global classification of natural terrestrial ecosystems. Vegetatio 32:75–81.

Walter, H. & Lieth, H. 1960–67. Klimadiagramm-Weltatlas. Gustav-Fischer-Verlag, Stuttgart.

Wang Chi-Wu 1961. The Forests of China. Cabot Foundation Publ. no. 5. Harvard University, Cambridge (Massachusetts). 313 pp.

Whitmore, T. C. 1984. Tropical Rainforests of the Far East, 2nd ed. Clarendon Press, Oxford.

Whittaker, R. H. 1956. Vegetation of the Great Smoky Mountains. Ecol. Monogr. 26:1–80.

Wolfe, J. A. 1979. Temperature parameters of humid to mesic forests of eastern Asia and relation to forests of other regions of the Northern Hemisphere and Australasia. U.S. Geological Survey, Professional Paper No. 1106, Washington. 37 pp.

Woodward, I. 1987. Climate and Plant Distribution. Cambridge University Press. 174 pp.

Wu Zh.-Y. (ed.) 1980. Zhongguo Zhibei (Vegetation of China). Science Press, Beijing. 1375 pp. (in Chinese).

Yazawa, T. 1980. Der Ablauf der Jahreszeiten in Japan aus der Sicht der Witterungsklimatologie und seine Bedeutung für das japanische Leben. Erdkunde 34:88–95.

Yim, Y.-J. 1977. Distribution of forest vegetation and climate in the Korean Peninsula. Parts 3 and 4. Jap. J. Ecol. 27:177–189, 269–278.

Yim, Y.-J. & Kira, T. 1975–76. Distribution of forest vegetation and climate in the Korean Peninsula. Parts 1 and 2. Jap. J. Ecol. 25:77–88, 26:157–164.

Yoshino, M. M. 1963. Rainfall frontal zone and jet stream in early summer over East Asia. Bonn Met. Abh. 3:1–127.

Yoshino, M. M. 1965–66. Four stages in the rainy season in early summer over East Asia (I, II). J. Met. Soc. Japan 43:231–245, 44:209–217.

Yoshino, M. M. 1966. Bai-u, the rainy season in early summer in Japan. Special Publ., Assn. Jap. Geogr. (1):66–71.

Yoshino, M. M. 1969. Climatological studies on the polar frontal zones and the Intertropical Convergence Zone over South, Southeast and East Asia. Climat. Notes, Hosei Univ. 1:1–71.

Yoshino, M. M. 1973a. Wind-shaped trees in the subalpine zone in Japan. Arctic Alpine Research 5:A115–A126.

Yoshino, M. M. (ed.) 1973b. (Water Resources of Monsoon Asia.) Kokon-Shoin, Tokyo (in Japanese).

Yoshino, M. M. 1977a. Dynamic and synoptic aspects of Japan's climate. In: Fukui, E. (ed.) The Climate of Japan, pp. 29–63. Kodansha, Tokyo.

Yoshino, M. M. 1977b. The Winter monsoon. In: Fukui, E. (ed.) The Climate of Japan, pp. 65–84. Kodansha, Tokyo.

Yoshino, M. M. 1978. Altitudinal vegetation belts of Japan with special reference to climatic conditions. Arctic Alpine Research 10:449–456.

Yoshino, M. M. 1980. The climatic regions of Japan. Erdkunde 34:81–87 (+ map supplement).

Yunnan Vegetation Editing Committee 1987. Yunnan Zhibei (Yunnan Vegetation). Science Press, Beijing. 1024 pp. (in Chinese).

Zhao S.-Q. 1986. Physical Geography of China. Science Press, Beijing (with John Wiley & Sons). 209 pp. + 64 color photos and 11 Landsat images.

3. Species diversity in East Asia in a global perspective

KUNIO IWATSUKI
Botanical Gardens, University of Tokyo, Tokyo 112, Japan

Key words: fern phytogeography, floristic manuals, Sino-Japanese floristic region, species diversity

Introduction

In the latter half of this century, very comprehensive progress in biology has occurred in molecular genetics, based on a single species *Escherichia coli*. A great contribution to understanding the nature of life was made through the study of this particular species. At this moment, though, we are faced with analyzing the other main aspect of biology, i.e. the diversity of the organisms, that phenomenon brought about by the evolution of life through more than three billion years. The techniques of the molecular through population levels of biological organization are now available to analyze the diversity of organisms. Topics in biodiversity vary, but the focus here will be on species diversity, introducing some aspects of floristic and phytogeographical studies, mostly in East Asia.

Compared with fundamental biological knowledge, information on biodiversity in a global perspectus is still insufficient, especially in the Far East. China and Malesia are the regions where species diversity is greatest on the earth. We have more advanced knowledge of the flora of Japan, although it is still far behind that of the other fields of the biological sciences.

To summarize knowledge of species diversity in East Asia, we begin with floristic regions. According to Good (1974), 37 floristic regions are recognized in the world, three of them in East Asia: the Sino-Japanese Region, Southeast Asian Region and Malesian Region. Floristic regions are different from vegetation divisions, since floristic regions are more strongly influenced by the geohistorical background than are vegetation divisions, which are mainly fixed in accordance with present environmental factors.

E.O. Box et al. (eds.), Vegetation Science in Forestry, 57–67.

Species diversity in the Sino-Japanese region

Among the 300,000 species of plants estimated to be on the earth, more than 30,000 are considered to occur in this Region, though even the fundamental enumeration is still incomplete in the most species-rich area, China.

Himalayan flora

Since the 1950's, Japanese botanists have explored the flora of the Himalaya very carefully, especially its eastern part. The results were combined with all other available information, and an enumeration of the flowering plants of Nepal was presented by Hara et al. (1978–82) in three volumes. The pteridophytes of Nepal were enumerated by Iwatsuki (1988). In Bhutan, floristic work has just started (Grierson & Long 1983), and fundamental collections are still much needed in this botanically interesting country.

Most parts of the Himalayas are densely inhabited by the hill tribes, and the primary forests have been destroyed to some extent. Artificial influences, especially pasturage, extend up to timberline, and it is difficult at present to describe the diversity of the natural flora in this area. This seems to be the case also in the European Alps.

Flora of China

Flora Republicae Popularis Sinicae is a large-scale national project to enumerate all the species native to China. More than half has already been published, but the remaining parts seem to be going rather slowly. Since China covers a very large area and our knowledge of its plants is still insufficient, it is difficult to enumerate the species completely, and it suffices initially to recognize some forms as distinct species. This often happens when a new fundamental enumeration of plants is begun, and this tendency to split species is evident in some cases in the studies by Chinese botanists. Thus, the number of species identified in the flora of China is sometime greater than that in other comparable areas. For example, the late Ching Reng-Cheng in volume 2 of the Flora (1959) recognized 81 species of Hymenophyllaceae. When I listed all the continental Asian species of this family (Iwatsuki 1985), I admitted only 43 species from China. I know that I often lump species, especially when the details of their structure are not elucidated. Even so, my enumerations of the pteridophytes of Nepal (1988), Thailand (1979–1989), and Japan (1992) are not very different from those of other botanists. Professor Ching (1982) could collaborate with me when he was still active in botany. This may be a particular case, but Chinese botanists generally split species based only on phenotypic features. Recently, however, some botanists there are more careful with their species concepts, and biosystematic analysis is being performed on some plant groups in order to elucidate the structure of the species. This may guide knowledge of Chinese species more soundly and lead to carefully revised floristic treatment. Actually, it is said that when the English version of the "Flora of China" is prepared, direct translation of the Chinese edition is not wanted but rather a simplified revision of the Chinese edition, including all information available.

The Chinese flora includes areas which belong to the Southeast Asian Region, but usually such floristic treatments are delimited by national boundaries, not by botanical boundaries. In the case of the Chinese flora, every province seems to expect to have its own floristic treatment. In many provinces the floristic treatments are completed or under preparation, and these are the important basis and/or represent additions to the general Flora of China.

Regional phytogeography was studied by Wu Cheng-yi and others (1980), and a phytogeographical subdivision of China has been proposed. A large-scale project on the flora and phytogeography of China is currently underway, under the leadership of Professor Wu, and we can expect to have a clearer scheme of the phytogeography of China shortly. Much of the material on the vegetation of China was edited by Wu (1980), and books on the vegetation of various provinces have been or are to be published, for instance Guangdong (1977), Sichuan (1980), Yunnan (1987), Xizang (Tibet) (1988), and so on.

Flora of Japan

The first "Flora of Japan" was published by Thunberg (1784). In 1890, Ryokichi Yatabe proposed that the Japanese flora should be described by Japanese botanists and described two new species. Since then, valuable contributions have been added; Ohwi compiled his Flora of Japan in 1953, with an English edition in 1965. Various of botanical research has elucidated the biology of plant diversity in Japan, and after a quarter of a century there is great progress in understanding the flora of Japan. There are many good books, often with beautifully coloured illustrations – though published mostly in Japanese, such as Satake et al., eds. (1981–89), Ono et al., eds. (1989), Kitamura et al. (1957–76) and so on.

The Japanese flora is rich for its small area. More than 5300 species are known, and nearly 40% of them are endemic to Japan. Unfortunately, about 900 species are extinct, endangered, or vulnerable.

As most of the Japanese Archipelago belongs phytogeographically to the Sino-Japanese Region, the flora of Japan is generally similar to that of China. It had been rather difficult, however, to make biosystematic comparisons between Japanese and Chinese plants until recently. Closer comparison of the plants in these two areas will greatly clarify the phytogeography of Sino-Japanese Region.

Compared with the other areas of East Asia, Japanese plants have been better analysed, by applying various techniques in biology. After enumerating all the species of vascular plants, the biological features throughout their life cycles have been observed in many plant groups. Comparative morphological, cytological, phytochemical, and ecological information has been accumulated, and molecular-biological information has also been applied recently to analyse the diversity of the flora. It is not intended here to introduce such examples in detail.

Relationships of fern phytogeography between Eastern North America and Japan

One of the conspicuous features of the flora of East Asia is its resemblance to that of eastern North America. Fern phytogeography is used here as an actual example.

Phytogeographic relationships between eastern North America and eastern Asia have been discussed by many botanists since Asa Gray and are comprehensively summarized by Boufford & Spongberg (1983). Changes in land and climate in the Northern Hemisphere since the Tertiary have given rise to replacement of boreal, temperate, or warm-climate floras in various areas and have enabled plants to migrate between North America and Asia through the Bering Strait and Aleutian chain. This migration probably occurred several times and in both eastward and westward directions.

The species migration of ferns is initiated primarily by spore dispersal. Based on a study of the fern flora of oceanic islands, Tryon (1970) noted the effectiveness of long dispersal of fern spores, estimating that "distances up to 300 miles are only a slight barrier to the migration of a fern flora and that a distance of 500 miles is not a significant barrier". Actually, most fern spores, except green ones such as *Osmunda*, *Matteuccia*, filmy ferns, Grammitidaceae and so on, are viable for several years or more, and fern spores are reported (Page 1979) to resist, to a considerable extent, physical and chemical environmental stress from low temperature, desiccation, UV- and X-rays, and other factors that operate during transport. Nevertheless, most fern species are restricted in distribution, and only a limited number have very wide ranges. The factors regulating fern distribution are still unknown. Many biotic and non-biotic ecological relations involved in fern migration were discussed by Page (1979). Klekowski (1972) considered that the long-distance establishment and future evolutionary development of fern popula-

Table 1. Fern genera native to North America but not found in Japan

Ophioglossaceae	*Cheiroglossa*
Schizaeaceae	*Anemia*
Parkeriaceae	*Aspidotis, Notholaena, Pellaea, Bommeria*
Dryopteridaceae	*Phenerophlebia*
Thelypteridaceae	*Meniscium*
Polypodiaceae	*Neurodium, Campyroneuron, Microgramma, Pleopeltis, Phlebodium*
Marsileaceae	*Pilularia*

tions depend on their mating system (intragametophytic or intergametophytic) and genetic variability, as well as geographic reproductive isolation.

We have as yet no sufficient understanding of the mechanism of fern spore dispersal and species establishment at a new locality, except for one preliminary study on *Osmunda* species (Watano 1986). Here the phytogeographic relationships between ferns of North America and Japan will be discussed mostly based on the presumed taxonomic affinity of the fern species concerned.

About 350 species are credited to the pteridophyte flora of North America (Mickel 1979; Lellinger 1985); the non-endemic species among them are referred (Tryon 1969) to four geographic groups: circumpolar, amphioceanic, tropical, and Mexican. In Japan some 650 species of pteridophytes have been identified (Iwatsuki 1992). The Japanese Archipelago is much smaller than North America, yet Japanese species are more complex taxonomically. The Japanese pteridophytes are to be arranged roughly into the following phytogeographic groups as: circumpolar, Eurasian, trans-Pacific, Sino-Himalayan, and tropical south Asian, in addition to endemic species, and they are regionally divided into four areas (Nishida 1972).

The phytogeographic relationships between North America and Japan are found primarily in the temperate to boreal species. Kato & Iwatsuki (1983) compared the pteridophytes of Hokkaido (80,000 km^2) to those of Europe and eastern North America. The flora of Hokkaido is temperate to sub-boreal and includes 122 pteridophyte species, comparable with 116 in eastern North America (Morton 1968) and 141 in Europe (Tutin et al. 1964). The number of common species among Hokkaido, northeastern North America, and Europe is shown in Kato & Iwatsuki (1983, Fig. 1).

There are 44 fern genera growing both in North America and Japan, including five fern allies. The genera native to North America but not found in Japan are shown in Table 1; there are 47 genera native to Japan but not known in North America, in addition to the genera presumed to have been introduced into North America very recently, such as *Lygodium* and *Cyrtomium* (Table 2).

These comparisons suggest differences in the fern floras of North America

Table 2. Fern genera native to Japan but not found in North America

Ophioglossaceae	*Helminthostachys*
Marattiaceae	*Angiopteris, Marattia*
Plagiogyriaceae	*Plagiogyria*
Gleicheniaceae	*Gleichenia, Dicranopteris*
Hymenophyllaceae	*Crepidomanes, Cephalomanes*
Cyatheaceae	*Cyathea*
Dicksoniaceae	*Dicksonia*
Dennstaedtiaceae	*Microlepia, Histiopteris, Monachosorum*
Lindsaeaceae	*Lindsaea, Tapeinidium*
Davalliaceae	*Davallia, Pachypleuria*
Oleandraceae	*Arthropteris*
Parkeriaceae	*Coniogramme, Onychium, Pleurosoriopsis*
Vittariaceae	*Antrophyum*
Lomariopsidaceae	*Bolbitis, Elaphoglossum*
Dryopteridaceae	*Acrophorus, Hemigramma*
Woodsiaceae	*Acystopteris, Cornopteris, Deparia, Dictyodroma, Hypodematium*
Dipteridaceae	*Dipteris*
Cheiropleuriaceae	*Cheiropleuria*
Polypodiaceae	*Pyrrosia, Lepisorus, Neocheiropteris, Lemmaphyllum, Drymotaenium, Microsorium, Leptochilus, Colysis, Pseudodrynaria, Loxogramme, Crypsinus*
Grammitidaceae	*Grammitis, Xiphopteris, Ctenopteris*

Table 3. Species occurring only (identically) in North America and Japan

Lycopodium chinense
L. obscurum
L. sitchense
Selaginella sibirica
Osmunda cinnamomea
O. claytoniana
Hymenophyllum wrightii
Adiantum pedatum
Onoclea sensibilis
Polypodium virginianum

and Japan, but phytogeographical relationships can only been found by comparing key species in these two areas. This relationship is suggested by apparently identical or vicarious species that are confined to these two areas. About 70 species are common to these two areas; those found only in the two areas are shown in Table 3.

Of these, two species of *Osmunda* were studied by Miller (1967) palaeobotanically. *Osmunda claytoniana* appeared to be derived from a mother stock, *O. nathorstii*, which occurred in the Pliocene (although *O. cinnamomea* had not been specifically different from Miocene except for some slight characters). Both *O. claytoniana* and *O. cinnamomea* are often separated into varieties based on distribution, though we have as yet no evidence to establish the infraspecific classification for these two species. The paleontological study was succeeded by a molecular systematic study by Stein et al. (1979) who compared chloroplast DNA among *Osmunda* species. Their results supported the conclusion of Miller (1967).

The onocleoid ferns were revised by Kato and Sahashi (1980), yielding two genera. According to fossil records, the *O. sensibilis* group was more diversified and more widely distributed in the Cretaceous and Tertiary than at present and was known from Alaska, Canada, the United States, central and southern Europe, and Japan. At present, this species is divided into two varieties indicated by a difference in the fertile pinnae: the sori are contiguous in the North American var. *sensibilis* and interrupted in the Japanese var. *interruptus*.

Adiantum pedatum and *Polypodium virginianum*, to the contrary, have been considered conspecific in the areas in question. A recent study by Paris et al. (1990), based on electrophoretic analysis, chloroplast DNA and chromosome counts, shows that there are three distinct genetic forms in *Adiantum pedatum*, each occurring in E. Asia, W. North America and E. North America. The first has n = 29 chromosomes and the latter two, n = 30. *Polypodium virginianum* was also observed electrophoretically by Haufler (pers. comm.), who showed that the Japanese form is so different from the American form as to warrant a new specific name. This does seem to be a vicariant species of the *Polypodium vulgare* complex.

In addition to the so-called identical species strictly confined to these two areas, there are also vicariant species between North America and Japan. *Asplenium rhizophyllum* and *A. ruprechtii* were assigned to a different genus, *Camptosorus*, based on their reticulate venation, simple fronds with elongate and gemmiferous tips, and other features. They are distinct from each other by the form of the leaf base, and there is no fossil evidence to illustrate the migration and speciation of these two species.

Deparia section *Lunathyrium* occurs in the two areas in question. The only North American representative is *D. acrostichoides*, which occurs in the temperate eastern region. This is close to the Japanese *D. pycnosora* and its relatives, differing by having a creeping rhizome with more widely spaced leaves.

Diplazium pycnocarpon is the only member of *Diplazium* in temperate eastern North America. This is related to the Asiatic *D. javanicum* group, which is usually segregated to form a distinct genus *Diplaziopsis*. Kato (1977) discussed the resemblance between *Diplaziopsis* and *Dictyodroma*, and he extended the similarity to tropical American *D. marginatum* or *Hemidictyum*. We have no sound evidence at the moment to show that *D. pycnocarpon* is derived from *Diplaziopsis*.

Table 4. Some floristic manuals from the Southeast Asian Floristic region

Ryukyu islands
Hatusima 1971. Flora of the Ryukyus (in Japanese).
Walker 1976. Flora of Okinawa and the Southern Ryukyu Islands.

Taiwan
Li et al. 1975–79. Flora of Taiwan, 6 volumes.

Hainan
Flora of Hainan, 4 volumes. 1964–77.

Indo-Chinese Peninsula
Lecomte 1907–1951. Flore Général de l'Indo-Chine. 7 volumes.
Flore du Cambodge, du Laos et du Vietnam. 1960–
Larsen & Smitinand 1970–. Flora of Thailand.

Burma
Hooker 1875–97. Flora of British India, 7 volumes.

Malesia
Van Steenis 1950–. Flora Malesiana.
Ridley 1922–25. Flora of the Malay Peninsula, 5 volumes.
Revised Flora of Malay. 1953–.
Merrill 1922–26. Enumeration of Philippine Flowering Plants, 4 volumes.
Backer & Bakhuizen van den Brink 1963–68. Flora of Java, 3 volumes.

Even with the above examples, the history of establishment of the flora in each area, or species diversification, is hardly elucidated as yet, and phytogeographical comparison between these two areas is greatly needed. The populations of all these taxa in Japan and the United States seem to have some genetic differences, and we are measuring the genetic distances more accurately. Shortly, we can enumerate the differences between the Japanese and US populations, and then our phytogeographical discussion will be based more on real evidence.

Southeast Asian region

The southern parts of mainland China (Fujian, Guangdong, Guangxi, Xishuangbanna in Yunnan), as well as Hainan and Taiwan, are included in the SE Asian region, together with the Indo-Chinese Peninsula. The Ryukyu Islands also belong to this region.

Ryukyu. As the "Flora of Japan" by Ohwi (1953, 1965) did not cover the Ryukyus, the Flora of the Ryukyus was compiled nearly at the same time but independently by Egbert Walker (1976) in English and in Japanese by Sumihiko Hatusima (1971). Since then, the detailed structure of the Ryukyu flora was observed. Recent information is incorporated in the "Wild Flowers of Japan", edited by Satake et al. (1981–89).

Taiwan. The "Flora of Taiwan" (Li et al., eds. 1975–79) was completed during 1975–79, in which 3577 species of vascular plants were enumerated. The island itself is not very large (46,000 km^2), but it has a peak of 3997 m (Mt. Yu-shan) and complicated topography with a varied edaphic basis. In addition, Kuroshio, or the Black Current, brings warm, humid air to make the island very favorable for plant life.

Botanical exploration has been active since the beginning of this century, and this is botanically one of the best known areas in the Asian tropics. A new revision of the "Flora of Taiwan" is now under preparation in a collaborative project between Taiwan and US botanists. Detailed information on their materials are expected, and all herbarium specimens are to be recorded in a data-base. It is a pity to note here, however, that little biosystematic research is underway in this interesting island with its rich tropical flora.

Laos, Vietnam and Cambodia. The "Flore Général de l'Indo-Chine" (Lecomte 1907–1951) was completed in 7 volumes with many installments from 1907 to 1951, taking nearly half a century. A new series, comprising revisions is on the way under the title "Flore du Cambodge, du Laos et du Vietnam"(1960-) Owing to the unhappy confusion in this area in the recent past, botanical research is based mostly on the herbarium specimens. Even so, information on the flora there was advanced greatly, partly based on specimens from the area and partly from new information in neighboring areas.

Thailand. The "Flora of Thailand" (Larsen & Smitinand, eds. 1970-) has been published only since 1970, but the "Flore Général de l'Indo-Chine" contained information from this country as much as possible. Field exploration in Thailand has been very active since the 1950s, when the Forest Herbarium, Royal Forest Department, began collaborating very much with foreign botanists interested in the Thai flora. This country includes some elements of the Sino-Japanese region in the north and some of the Malesian region in the peninsula. Therefore, the flora includes great variety, on which more botanical research is necessary. The revision of the pteridophytes was completed, but various problems remain unsolved, which are to be analyzed by applying modern techniques.

Malesian region

The "Flora Malesiana" (van Steenis, ed. 1950-) was founded by the late C. G. G. J. van Steenis in the middle of this century, but less than half of the whole flora has been published. It may simply be noted here that fundamental records of this tropical area are urgently important, especially since the flora there is greatly threatened at the moment.

Since the "Flora Malesiana" covers very wide areas, various parts of this area were already treated floristically, as by the "Flora of Malaya" (Ridley 1922–25, Holttum 1953-), "Flora of Java"(Backer and van den Brink 1963–68), and "Flora of the Philippines" (Merrill 1922–26), (the last of which is expected to be revised).

References

Backer, C. A. & Bakhuizen van den Brink, R. C. Jr. 1963–68. Flora of Java, 3 vols. Noordhoff, Djakarda.

Boufford, D. E. & Spongberg, S. A. 1983. Eastern Asian – Eastern North American phytogeographical relationships – a history from the time of Linnaeus to the twentieth century. Ann. Missouri Bot. Gard. 70: 423–439.

Ching, R.-C. 1959. Hymenophyllaceae. In: Flora Republicae Popularis Sinicae. 2:132–196 (in Chinese).

Flore du Cambodge, du Laos et du Vietnam 1960–. Mus. Nat. Hist. Nat., Paris.

Good, R. 1974. The Geography of the Flowering Plants, 4th ed. Longman, London.

Grierson, A. F. J. & Long, D. G. 1983–. Flora of Bhutan. Roy. Bot. Gard. Edinburgh.

Hara, H., Stearn, W. T. & Williams L. H. J. (eds.) 1978–1982. An Enumeration of the Flowering Plants of Nepal, 3 vols. Brit. Mus. Nat. Hist., London.

Hatusima, S. 1971. Flora of the Ryukyus. Naha.

Holttum, R. E. 1953–. Revised Flora of Malaya. Kuala Lumpur.

Iwatsuki, K. 1985. Hymenophyllaceae of Asia, excluding Malesia. J. Fac. Sci. Univ. Tokyo III, 13:501–551.

Iwatsuki, K. 1988. An Enumeration of the Pteridophytes of Nepal. Univ. Mus. Univ. Tokyo Bull. 31:231–339.

Iwatsuki, K. 1992. Coloured Illustrations of the Pteridophytes of Japan. Heibon-sha, Tokyo (in Japanese).

Kato, M. 1977. Classification of *Athyrium* and allied genera of Japan. Bot. Mag. Tokyo. 90:23–40.

Kato, M. & Iwatsuki, K. 1983. Phytogeographic relationship of pteridophytes between temperate North America and Japan. Ann. Missouri Bot. Gard. 70:724–733.

Kato, M. & Sahashi, N. 1980. Affinities in the onocleoid ferns. Acta Phytotax. Geobot. 31:127–138.

Kitamura, S., Murata, G., Hori, M. & Koyama, T. 1957–1964. Coloured Illustrations of Herbaceous Plants of Japan, 3 vols. Hoikusha, Osaka (in Japanese).

Kitamura, S. & Murata, G. 1971–76. Coloured Illustrations of Woody Plants of Japan, 2 vols. Hoikusha, Osaka (in Japanese).

Klekowski, E. J. 1972. Genetical features of ferns as contrasted to seed plants. Ann. Missouri Bot. Gard. 59:138–151.

Larsen, K. and T. Smitinand (eds.) 1970–. Flora of Thailand. For. Herb., Royal Forest Department, Bangkok.
Lacomte, H. (ed.) 1907–1951. Flore Général de l'Indo-Chine. 7 vols. Mus. Nat. Hist. Nat., Paris.
Lellinger, D. B. 1985. A Field Manual of the Ferns and Fern-Allies of the United States and Canada. Smithsonian Inst. Press, Washington.
Li, H.-L., Lin, T.-S., Huang, T.-C., Koyama, T. & DeVol, C. E. (eds.) 1975–1979. Flora of Taiwan, 6 vols. Epoch, Taipei.
Merrill, E. D. 1922–26. Enumeration of Philippine Flowering Plants, 4 vols.
Mickel, J. T. 1979. How to know the ferns and fern allies. Dubuque, Iowa.
Miller, C. N. 1967. Evolution of the fern genus *Osmunda*. Contr. Mus. Paleontol, Univ. Mich. 21:139–203.
Morton, C. V. 1968. The genera, subgenera, and sections of the Hymenophyllaceae. Contr. U.S. Nat. Herb. 38:153–214.
Nishida, M. 1972. An Outline of the Distribution of Japanese Ferns. In: Graham, A. (ed.) Floristic and Palaeofloristics of Asia and Eastern North America, pp. 101–105. Elsevier, New York.
Ohwi, J. 1953. Flora of Japan. Shibundo, Tokyo (in Japanese).
Ohwi, J. 1965. Flora of Japan. Smithsonian Inst., Washington.
Ono, M., Ohba, H. & Nishida, M. (eds.) 1989. Revised Makino's New Illustrated Flora of Japan. Hokuryukan, Tokyo (in Japanese).
Page, C. N. 1979. Experimental aspects of fern ecology. In: Dyer, A. F. (ed.) The Experimental Biology of Ferns. Academic Press, London.
Paris, C. A., Nakato, N. and Stein, D. B. 1990. Divergent speciation in the *Adiantum pedatum* complex: evidence from chloroplast DNA variation, cytology, and isozyme electrophoresis. Symp. International Association of Pteridology, Ann Arbor 1990.
Ridley, H. N. 1922–25. Flora of the Malay Peninsula. 5 vols.
Satake, Y., Ohwi, J., Kitamura, S., Watari, S. & Tominari, T. (eds.) 1981–82. Wild Flowers of Japan, Herbaceous Plants, 3 vols. Heibon-sha, Tokyo (in Japanese).
Satake, Y., Hara, H., Watari, S. & Tominari, T. (eds.) 1989. Wild Flowers of Japan, Woody Plants, 2 vols. Heibon-sha, Tokyo (in Japanese).
Stein, D. B., Thompson, W. F. & Belford, H. S. 1979. Studies on DNA sequences in the Osmundaceae. J. Mol. Evol. 13:215–232.
Tagawa, M. & Iwatsuki, K. 1979–1989. Pteridophytes. In: Larsen, K. & Smitinand, T. (eds.) Flora of Thailand, vol. III.
Thunberg, C. P. 1784. Flora Japonica (in Latin).
Tryon, R. 1969. Taxonomic problems in the geography of North American ferns. Bioscience 19:790–795.
Tryon, R. 1970. Development and evolution of fern floras of oceanic islands. Biotropica 2:76–84.
Tutin, T. G., Heywood, V. H., Burges, N. A., Valentine, D. H., Walters, S. M. & Webb, D. A. (eds.) 1964. Flora Europaea, Vol. 1: Lycopodiaceae to Plantanaceae. Cambridge University Press, Cambridge.
Van Steenis, C. G. G. J. (ed.) 1950–. Flora Malesiana.
Vegetation of Guangdong. 1977. Science Press (in Chinese).
Vegetation of Sichuan. 1980. Sichuan Science Press (in Chinese).
Vegetation of Xizang. 1988. Science Press (in Chinese).
Vegetation of Yunnan. 1987. Science Press (in Chinese).
Walker, E. 1976. Flora of Okinawa and the Southern Ryukyu Islands. Smithsonian Inst. Press, Washington.
Watano, Y. 1986. Genetic Life History of the Homosporous Fern *Osmunda lancea* Thunb. In: Iwatsuki, K. et al. (eds.) Modern Aspects of Species, pp. 211–219. Univ. Tokyo Press.
Wu, Cheng-yi (ed.). 1980. Vegetation of China. Science Press, Beijing (in Chinese).
Yatabe, R. 1890. A few words of explanation to European botanists. Bot. Mag. Tokyo 4:355–359.

4. On the global position of the evergreen broad-leaved forests of China

SONG YONGCHANG
East China Normal University, Department of Environmental Science, Shanghai 200062, China

Key words: evergreen broad-leaved forest, lauraceous forest, monsoon evergreen broad-leaved forest, sclerophyll forest, subtropical rainforest, temperate rainforest

Introduction

Evergreen broad-leaved forest is one of the main vegetation types of China. It occurs in the region from 24° to 32° north latitude and from 99° to 123° east longitude, which makes up about one quarter of the total area of China. This forest type has been put into different vegetation categories (formation class of Rübel 1930), formation type of Schimper (1898) and given different names in the literature dealing with the vegetation of the world.

Some authors (Schimper 1903, Dansereau 1957, Polunin 1960) put the evergreen broad-leaved forest of China together with the forest of central and southern Japan and named it "temperate rain forest" or "warm-temperate rain forest." Other authors (Warming 1909, Shabalinna 1960) grouped the evergreen broad-leaved forests of China with mesophilous evergreen forests of southern Japan, southeastern North America, southern Brazil, eastern Africa, and eastern Australia and Tasmania, and called them "subtropical rain forest."

Some authors (Rübel 1930, Alechin 1950, Schröder 1983, Hübl 1988) put this forest type together with the forests of the Canary Islands, Madeira, Portugal, the Pacific coast of North America, Florida, western Patagonia (southern Chile and adjacent Argentina), Japan, the southern slope of the Himalaya, as well as New Zealand and named them Lauraceous forest (Laurisilvae, Lorbeerwald). Additionally, some authors (Müsel et al. 1965, Hämet-Ahti et al. 1974) named it the evergreen broad-leaved forest of the Meridional zone.

In their "Tentative Physiognomic-Ecological Classification of main Plant

E.O. Box et al. (eds.), Vegetation Science in Forestry, 69–84.

Formations of the Earth," Ellenberg & Mueller-Dombois (1967) named this forest type the "Temperate evergreen seasonal broad-leaved forest," pointed out that it corresponds to the "lauraceous forest" of Rübel, and grades into subtropical or temperate ombrophilous forest, probably including subpolar types.

In the work of Walter (1968, 1973, 1979), the "Vegetation of the Earth," this vegetation type was incorporated into Zonobiome V of the warm-temperate moist climatic zone and was kept with the evergreen broad-leaved forest of Japan, Florida, eastern Brazil, northern Argentina, the southern coast of Africa, Tasmania, and New Zealand as the same type which belongs to a very humid subzonobiome with rain throughout the year or with the rainfall minimum in the winter.

Ovington (1983) in his book on "Temperate Broad-Leaved Evergreen Forest" (Ecosystems of the World series) called this forest type the "Temperate broad-leaved rain forest" and put it with "Temperate broad-leaved sclerophyll forest" and "Temperate evergreen seasonal forest," together named with "Temperate broad-leaved evergreen forest."

This kind of forest also occurs widely in Japan and is usually called "Warm-temperate broad-leaved evergreen forest" (Miyawaki 1967, 1980, 1984, Numata et al. 1972, Numata 1974, 1987, Shidei 1974, Satoo 1983, Kira et al. 1978, Fujiwara 1981–1986) or "Laurel forest" (Nakano 1930) or "Lucidophyllous forest" (Imanishi & Kira 1953).

In China most authors called this forest the "Subtropical evergreen broad-leaved forest" (Chien et al. 1956, Hou 1960, 1982, 1983, 1988, Lin Y. 1981, Chen et al. 1965, Wu et al. 1980, Song 1983, 1988a, 1988b, Zhang 1983, Zhong 1988), occasionally using the term "Evergreen oak forest" (Hou 1960) or "Evergreen *Cinnamomum-Castanopsis* forest" (Lin Y. 1965) or "South-temperate lauraceous forest" (He 1951), or "Warm-temperate and subtropical evergreen broad-leaved forest" (Wang, Ch.-W. 1961).

As can be concluded from the above situations, the evergreen broad-leaved forest of China is thrown into confusion not only in the recognition of its type but also in the terminology. There are several main questions:

- What is the meaning of the evergreen broad-leaved forest?
- Are there any differences among the evergreen broad-leaved forest, subtropical rain forest, temperate rain forest, lauraceous forest, and sclerophyllous forest? Or are these terms synonyms?
- Do the evergreen broad-leaved forests of China and the evergreen forest of central and southern Japan, southernmost Korea, southeastern and west-coast North America, southeastern Australia, Tasmania, New Zealand, southeastern Africa, and the Mediterranean area belong to one type?
- To which climatic zone, temperate or subtropical, do the evergreen broad-leaved forests belong?

– What is the global position of this vegetation type?
These problems will be discussed in this paper.

Basic characteristics of evergreen broad-leaved forest of China

In order to compare the evergreen broad-leaved forests (EBLF) of China with other, similar vegetation types, at first we should clarify the basic characteristics of these forests (*cf.* Hou 1960, 1988, Song 1988a, 1988b, Jin 1979, Zhong 1988). These can be summarized as follows:

1. The stratification of EBLF is quite distinct, commonly being divided into 4–5 layers. The overstory is generally 15 to 25 m high, very rarely over 30 m, with dark green, hemispherical crowns and more regular canopies.
2. The EBLF mainly consists of mesophilous angiospermous trees, usually mixed with thermophilous, moisture-loving coniferous and deciduous trees.
3. The leaves are usually leathery smooth mesophylls and hang with their flat surfaces perpendicular to the strongest light. The buds are covered with scales.
4. The trees commonly have rough, thick, dark-brown bark which peels off in strips.
5. Vines and epiphytes are of certain importance in EBLF, but thick lianas and epiphytic spermatophytes are very rare or absent.
6. Commonly the trees have no plank-buttresses, but occasionally some smaller buttresses appear. Cauliflory and drip-tips are practically absent.
7. The ground herbaceous layer is usually sparse and dominated by species of Pteridophyta and Cyperaceae.
8. Species of Fagaceae, Lauraceae, Theaceae, Magnoliaceae and Hamamelidaceae play a prominent role in the floristic composition of EBLF.
9. Evergreen broad-leaved forests occur in a monsoon climate, with hot summers and fairly cold winters. The mean annual temperature ranges from 15 to 20°C. The minimum monthly temperature (January) is 3–10°C, and maximum monthly temperature (July) is generally above 26°C. The frost period is short, commonly 64–100 days. The average annual rainfall is about 1000–2000 mm, being concentrated in summer (above 40% of total precipitation, only 15% in winter).
10. The soils under the EBLF of China commonly are red and yellow earths. Their profiles show strong acidity (pH 4.0 to 5.5) with a thin humus layer. The contents of calcium, manganese, nitrogen, sulphur and phosphorus are very low. At higher altitudes the soils under EBLF are mountain yellow-brown forest soils, with characteristics of both brown soils and yellow earths, and with pH values between 5.0 and 6.0. Ad-

Table 1. Comparison of features between evergreen broad-leaved forest, subtropical rain forest, temperate rain-forest and evergreen sclerophyl forest

	Evergreen broad-leaved forest (EBLF)	Subtropical rain-forest (SRF)	Warm-temperate rain-forest (WRF)	Evergreen sclerophyll forest (ESF)
	1. Stratification is quite distinct, commonly being divided into 4–5 layers.	1. Stratification is not quite distinct, structure is much more complex, seems to tropical rainforest	1. Stratification is relatively distinct. Structure is rather complicated.	1. Stratification is very distinct. Structure is simple, commonly being divided into 3–4 layers.
	2. Commonly trees have no plank buttresses, occasionally some smaller buttresses appear.	2. Trees of upper layer are taller and always have plank-buttresses.	2. Plank-buttresses occur only exceptionally.	2. Trees have no plank-buttresses.
	3. Crowns are hemispherical and more regular canopies.	3. The canopies involved are smaller, crowns are more uneven.	3. The crowns of trees are various patterns.	3. The crowns widely spread. Tranks are generally lower, massive, branches gnarled.
Physiognomy & structure	4. Bark of tree is usually rough, thick, dark brown, and peels off in forms of scale of strips.	4. The bark is neither smooth nor rough with light or light-grey color.	4. The bark is neither smooth nor rough, usually with light color.	4. Bark is uaually very thick and rough, with a well developed cork layer.
	5. Leaves are usually leathery, smooth, mesophylls and microphylls with entire margins, generally perpendicular to the strongest light with their flat surfaces. The buds are covered with scales.	5. Leaves are usually macrophyll and mesophyll, not hang perpendicular to the strongest light. The buds are partly without scale, partly with scales.	5. Leaves are usually mesophylls and microphylls, large foliage is rare. In most cases periodically foliaged trees occur.	5. Leaves are usually smaller, thick and leathery, with entire or sharply toothed margins, always covered with hair on both surfaces, seldom shiny, more frequently dull, not perpendicular to the strongest light, but assume an oblique position. The buds are without scale.
	6. Lianas and epiphytes are rather abundant, but thick lianas, epiphytic ferns and phanerogams are rare or absent.	6. Lianas and epiphytes are abundant, thick lianas, epiphytic ferns and phanerogams, epiphylls and strangle plants present.	6. Lianas and epiphytes are abundant. Ferns and phanerogams always deck the stems and branches. Epiphylls and cauliflory are unknown.	6. Epiphytes are lacking. Lianas occasionally appear, only herbaceous climbers.

	7. The understory is usually sparse, mainly consists of species of ferns, Cyperaceae and Liliaceae.	7. The understory is well developed, in which the woody ferns often being seen.	7. The understory is well developed. Mosses and ferns densely cover the ground. Tree ferns and palms often occur.	7. The herbaceous layer is generally sparse with many tuberous and bulbous plants.
Ecological feature	1. Essentially formed by thermophilous, mesophilous plants. 2. Developed under the monsoon climate with hot and moist summers and fairly cold winters. 3. Soils are generally red and yellow earth.	1. Essentially formed by hydrophilous plants with macrophylls. The means of protecting against transpiration is weakly developed. 2. Climate is much hotter and moist in whole year round. 3. Soils and laterite and red earth	1. Essentially formed by hydrophilous plants. The means of protecting against transpiration is weakly developed. 2. Climate is temperate & extremely oceanic, seasonal distribution of rainfalls and temperature are more even. 3. Soils are more humiferous	1. Essentially formed by xerophilous trees, with well-developed means of protecting against transpiration and deep root systems. 2. Developed under the mediterranean climate with hot, dry summer and warm, moist winter. 3. The rendzina soils hold relatively little organic matter.

ditionally, there are black earths and purple soils under the EBLF. Both of these soils are fertile. The N, P, K and organic-matter contents are high, and the pH values are between 6.5 and 8.5.

Based on the above characteristics of the EBLF, three main points should be kept in mind:

1. The EBLFs are constituted by thermophilous plants which have relatively low resistance to cold and freezing. They cannot tolerate temperature below 0°C over a long period of time. Thus, the low temperature in winter is the limiting factor for their growth and distribution.
2. The EBLFs are constituted by mesophilous plants which can tolerate neither excessive wetness nor drought. Within the distribution region of the EBLFs, they usually grow on well drained soils. On waterlogged soils (as opposed to water-oozing soils, e.g. on limestone), some important evergreen species cannot grow and are replaced by thermophilous, moisture-loving or drought-resistant deciduous species. Therefore, summer drought is also a limiting factor for their distribution.
3. The EBLFs have developed under a monsoon climate. The coincidence of abundant rainfall and high temperature in the same season is favorable for their growth.

Delimitation toward other, similar vegetation types

We have already expounded the basic characteristics of the EBLF of China, now we compare it with other, similar vegetation types, in terms of physiognomy and structure, ecological features, and geographical distribution (see Table 1 and Fig. 1).

EBLF versus subtropical rain forest

A definition of subtropical rain forest had been given by Schimper (1903): "The tropical rain-forest occasionally extends beyond the tropics, at first without losing its characteristic features but with a gradual reduction in the wealth of forms and a decrease in specifically tropical peculiarities, such as large leaves, plank-buttresses, woody lianas, woody epiphytes, epiphyllous epiphytes, cauliflory, and water-storing calyces. The presence of certain temperate forms adds a distinguishing feature of a positive nature to these last traces of the tropical rain forest, which may be termed "Subtropical rain-forest." Ellenberg & Mueller-Dombois (1967) further explain that "the subtropical ombrophilous forests are present only locally and in small fragmentary stands, because the subtropical climate is typically a climate with a dry season; where the subtropical ombrophilous forest occurs, e.g. Queensland and Taiwan, it usually grades rather inconspicuously into the tropical ombrophilous forest. Its trees are less vigorous and allow some shrubs to

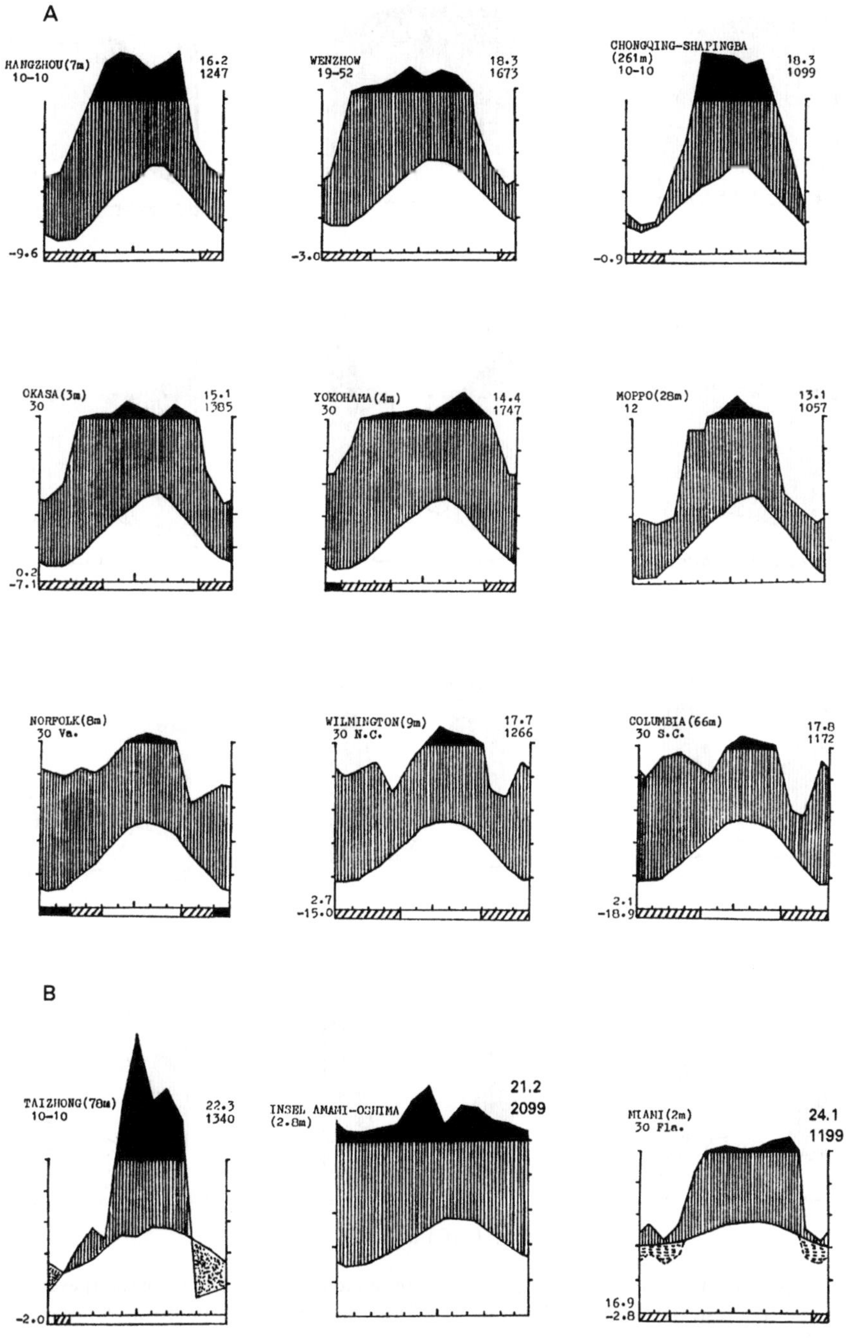
A
HANGZHOU(7m)
10-10
16.2
1247
-9.6
WENZHOW
19-52
18.3
1673
-3.0
CHONGQING-SHAPINGBA
(261m)
10-10
18.3
1099
-0.9
OKASA(3m)
30
15.1
1385
0.2
-7.1
YOKOHAMA(4m)
30
14.4
1747
MOPPO(28m)
12
13.1
1057
NORFOLK(8m)
30 Va.
WILMINGTON(9m)
30 N.C.
17.7
1266
2.7
-15.0
COLUMBIA(66m)
30 S.C.
17.8
1172
2.1
-18.9
B
TAIZHONG(78m)
10-10
22.3
1340
-2.0
INSEL AMAMI-OSHIMA
(2.8m)
21.2
2099
MIAMI(2m)
30 Fla.
24.1
1199
16.9
-2.8

Fig. 1. (A) and (B).

C

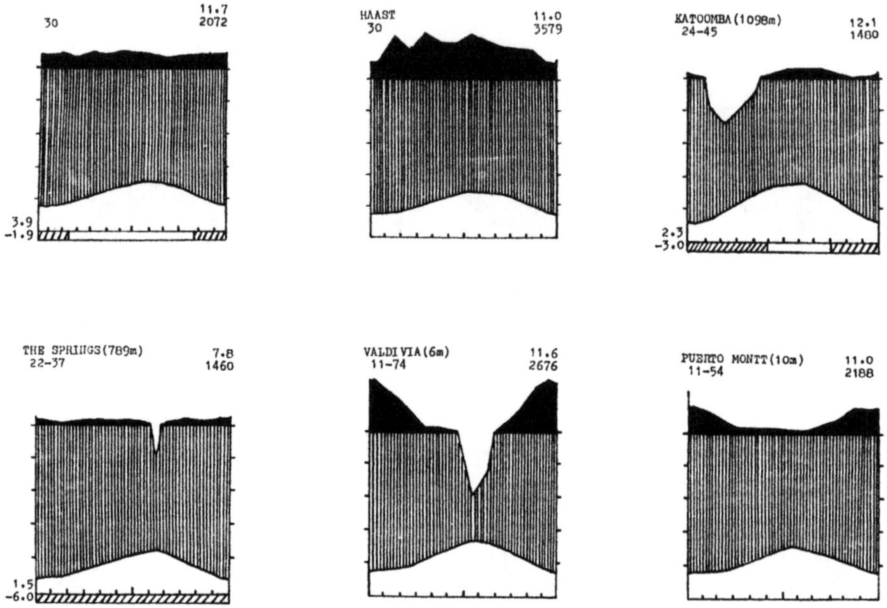

D

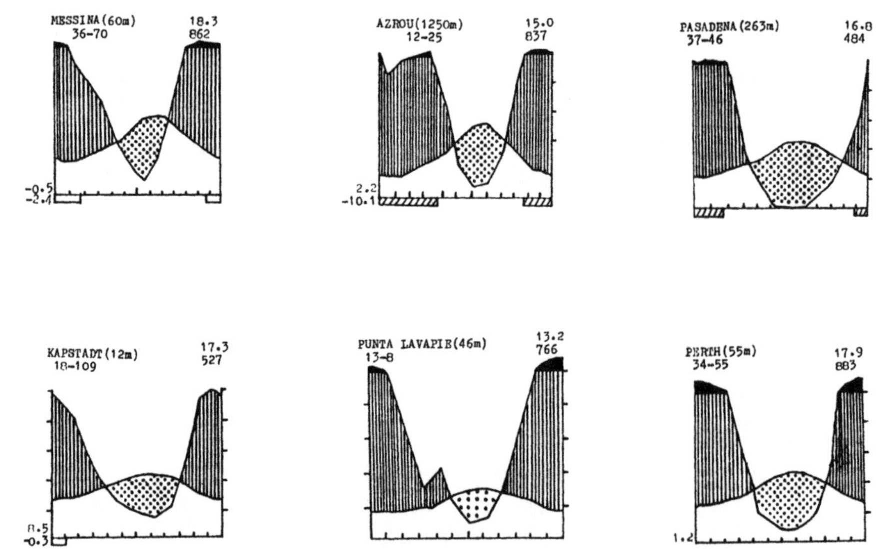

Fig. 1. (C) and (D).

Fig. 1. Comparison of typical climate diagrams between four biomes. (A) (Monsoon) evergreen broad-leaved forest. (B) Subtropical rain-forest. (C) Warm-temperate rain-forest. (C) Evergreen sclerophyll forest.

grow in the understory . . . . The subtropical ombrophilous forest is physiognomically more closely related to the tropical than to the temperate one." According to the above definitions, the main differences between subtropical rain forest (SRF) and EBLF can be summarized as follows:

1. The structure and physiognomy of SRF are much more complex than that of EBLF and more closely related to tropical rain-forest.
2. Subtropical rain-forests usually have plank-buttresses, thick lianas, epiphytic ferns and phanerogams, epiphyll and strangler plants, even if less than the tropical rain-forest.
3. There are many tropical elements in the SRF, palms and woody ferns often being seen.
4. In subtropical rain-forest the hydrophilous species with macrophylls are dominant. The means of protecting against transpiration is weakly developed. Occasional leaves have drip tips.
5. The climate of subtropical rain-forest is much hotter and more moist. The mean annual temperature is about 18–21°C. Minimum monthly temperature is usually above 9–13°C. The frost time is very short. The soils are commonly lateritic and red earths.
6. The distribution of SRF is restricted within narrow limits and is scattered locally. It occurs in southern China (Lin, P. 1961), especially Taiwan island, the southern Ryukyu Islands, the southern extremity of Florida, northern Mexico, the southernmost part of Brazil, northern Argentina, and northeastern Australia.

EBLF versus temperate rain-forest

Distinguishing these two biomes is sometimes difficult. They have more similar features and thus in the literature are usually treated as one type. As Klötzli (1988) wrote, though, the temperate rain-forests are more hydrophilous and present more biomass. The important differences are summarized as follows:

1. The temperate rain forest, like the subtropical rain-forests, is essentially formed by evergreen hydrophilous trees, for therein truly lie most of the essential features of rain forest. The means of protecting against transpiration is weakly developed, but the large leaf sizes of the subtropical woody plants are rare in temperate rain-forests. In most cases periodically-foliaged trees occur as subordinate components.
2. In the temperate as in the tropical and subtropical rain forests, the lianas and epiphytes are abundant. Ferns and phanerogams always densely deck the stems and branches of the trees. Plank-buttresses occur only exceptionally in temperate rain-forests, and epiphylls and cauliflory are unknown.
3. The herbaceous layer frequently appears as densely as in the tropics.

Mosses and ferns densely cover the ground. Tree ferns and palms often occur in the understory.

4. The distribution regions of temperate rain-forest have more evenly distributed rainfall, more even temperature, and have more or less no frost. The mean annual temperature is 10–14°C, the minimum monthly mean temperature is 4.5–8.5°C, maximum monthly mean temperature is 13.5–16.5°C, and average annual rainfall is above 1000 mm. The ratio of rainfall of wet months to that of dry months is only 1.2–1.7:1.
5. The occurrences of temperate rain-forests are quite isolated, occurring only in the extremely oceanic climates of the southern hemisphere. Sometimes these forests occur in mountain regions at high altitude. They occur mainly in New Zealand, southeastern Australia, western Tasmania, and southern Chile.

EBLF versus evergreen sclerophyllous forests

The difference between these two biomes is obvious:

1. The trees and shrubs constituting the sclerophyll forest are xerophilous sclerophyllous plants. The leaves are usually smaller, thick and leathery, with entire or sharply toothed margins. The leaves are always covered with hair on both surfaces, are seldom shiny, more frequently dull, perhaps owing to exudations of resins.
2. The trees are usually lower, their trunks generally massive and the branches gnarled, their crowns widely spread, the bark usually very thick and rough, with a well developed cork layer.
3. The community structure is simple and stratification very distinct, commonly consisting of 3–4 layers. Lianas are very rare, only herbaceous climbers. Epiphytes are absent (except for occasional herbaceous climbers).
4. The herbaceous layer is generally sparse, with many tuberous and bulbous plants.
5. The sclerophyllous plants usually have deep root systems which can absorb water from deep in the ground. When water is short, stomata close. When autumn rains begin, these plants grow immediately. Therefore, the sclerophyllous plants have greater competitive ability than the evergreen broad-leaved and deciduous broad-leaved species in the winter-rain regions.
6. The climate is relatively adverse. In summer it is hot and dry, the mean temperature of the hottest month being between 22 and 28°C, with average precipitation only 2–23 mm. In winter it is warm and moist. The mean temperature of the coldest month is 5–12°C. Most of the annual precipitation falls in winter.
7. The sclerophyll forest mainly occurs in the littoral countries of the Medit-

erranean Sea, the southwestern extremity of Africa, south and southwestern Australia, central Chile, and the greater part of coastal California.

The global position of evergreen broad-leaved forest

From the comparisons above it can be seen that there are not only similarities but also differences between the EBLF and subtropical rain-forest, temperate rain-forest, and sclerophyllous forest. Due to insufficient literature in the past and the different emphases of different authors, the above terms have been used with different meanings (see Table 2), causing some confusion. Since the EBLF developed under particular climatic conditions and has specific characteristics of structure, physiognomy, ecological adaptations, and geographical distribution, it seems reasonable to consider EBLF a specific type. We agree with Klötzli (1988), who defined the EBLF as a specific biome under the influence of the monsoon climate. In order to emphasize the specific climatic conditions and avoid the confusion of terms, we propose using the term "monsoon evergreen broad-leaved forest" or "summer-rain evergreen broad-leaved forest." This forest type occurs not only in southeastern China but also in central and southern Japan, southernmost Korea (Yim 1977), and southeastern North America (Olson 1983, Box & Fujiwara 1988). According to the ecological features and floristic composition, it can be subdivided into four subtypes:
1. moist monsoon EBLF of eastern China;
2. moist monsoon EBLF of Japan and Korea;
3. moist monsoon EBLF of southeastern North America; and
4. semi-moist monsoon EBLF of western China (see Table 3).

Here we must point out that the monsoon evergreen broad-leaved forest not only differentiates from the above mentioned three vegetation types but also from the laurineous forest (LF) of the Mediterranean area, which developed under the mediterranean-type climate, and the species composition is relatively poorer than that of the monsoon EBLF. There are many annual plants in the herbaceous layer of the LF. Its tree layer is dominated by *Laurus*, *Castanopsis* is absent. This type of vegetation can be named "winter-rain evergreen broad-leaved forest," which can be recognized as a synonym of laurineous forest.

In our opinion it is better to use the evergreen broad-leaved forest in a narrow sense which only indicates the monsoon EBLF. Only when the similar features of these four vegetation types (SRF, WRF, EBLF, and LF) are emphasized and they are put together as a higher rank should the term evergreen broad-leaved forest be used in a wider sense, which expresses all of the evergreen broad-leaved forests occurring in extra-tropical regions.

As for the EBLF (narrow sense) belonging to the temperate or subtropical

Table 2. Comparison of the terms used by different authors

<table>
<tr><td>1.Schimper (1903)</td><td>Subtropical rain-forest</td><td colspan="3">Temperate rain-forest</td><td>Temperate sclerophylous woodlands</td></tr>
<tr><td>2.Dansereau(1957)</td><td>Subtropical rain-forest</td><td colspan="3">Temperate rain-forest</td><td>Evergreen hardwood forest</td></tr>
<tr><td>3.Rubel (1933)</td><td colspan="2">Pluviisilvae</td><td colspan="2">Laurisilvae</td><td>Durisilvae</td></tr>
<tr><td>4.Hubel (1988)</td><td colspan="4">Lorbeerwalder</td><td>Hartlaubvegetation</td></tr>
<tr><td rowspan="2">5.Warming (1909)</td><td rowspan="2">Subtropical rain-forest</td><td rowspan="2">Antarctic forest</td><td rowspan="2">Subtropical rain forest</td><td colspan="2">Evergreen sclerophyllous forest</td></tr>
<tr><td>Evergreen broad-leaved forest</td><td>Sclerophyllous evergreen forest</td></tr>
<tr><td>6.Ellenberg & Muller-Dombois(1967)</td><td>Subtropical ombrophilous forest</td><td>Temperate evergreen ombrophilous forest</td><td colspan="2">Temperate evergreen seasonal broad-leaved forest</td><td>Winter-rain evergreen broad-leaved sclerophyllus forest</td></tr>
<tr><td rowspan="2">7.Ovington (1983)</td><td rowspan="2"></td><td colspan="2">Temperate broad-leaved rain forest</td><td>Temperate evergreen season forest</td><td>Temperate broad-leaved sclerophyllous forest</td></tr>
<tr><td colspan="4">Temperate evergreen broad - leaved forest</td></tr>
<tr><td>8.Schmithusen(1976)</td><td>Subtropical evergreen rain-forest</td><td>Temperate evergreen rain-forest</td><td>Temperate evergreen broad-leaved forest</td><td>Laurel forest</td><td>Sclerophyllous forest</td></tr>
<tr><td>9.Klotzli (1988)</td><td>Subtropical rain-forest</td><td>Temperate rain-forest</td><td colspan="2">Subtropical and temperate evergreen broad - leaved forest</td><td>Sclerophyllous forests (woodland)</td></tr>
</table>

Table 3. Four categories of monsoon evergreen broad-leaved forest

Eastern China moist MEBLF	Japan and Korea moist MEBLF	America moist MEBLF	Western China semimoist MEBLF
Climatic condition			
Typical monsoon climate with hot, moist summers and fairly cold winters	Typical monsoon climate, winters do not so drought and so cold, summers not so hot than that in Eastern China. The temperature range is less and rainfall is much more.	Similar monsoon climate, but the temperature range is much less than that of Asiatic monsoon climatic region. Distribution of rainfall is more even in a year.	Monsoon climate, in summer it is warm and moist, but in winter and spring warm and dry, so that wet and dry season alternate distinctly. Annual temperature range is smaller. The weather is semimost.
Floristic composition			
Dominated by Sino-Japanese elements: *Quercus glauca* *Quercus gracilis* *Quercus gilva* *Castanopsis carlesii* *Castanopsis sclerophylla* *Castanopsis eyrei* *Castanopsis fordii* *Lithocarpus glabra* *Schima superba* *Schima sinensis* *Camellia fraterna* *Neolitsea aurata* var. *chekiangensis* *Altingia chinensis* *Machilus thunbergii*	Dominated by Sino-Japanese elements: *Quercus glauca* *Quercus gilva* *Quercus myrsineafolia* *Quercus acuta* *Quercus salicina* *Castanopsis cuspidata* var. *sieboldii* *Camellia japonica* *Neolitsea sericea* *Acuba japonica* *Fatsia japonica* *Machilus thunbergii*	Dominated by North American elements: *Quercus virginiana* *Quercus laurifolia* *Quercus myrtifolia* *Magnolia virginiana* *Magnolia grandiflora* *Gordonia lasianthus* *Machilus borbonia*	Dominated by Sino-Himalayan elements: *Quercus glaucoides* *Quercus delavayi* *Castanopsis delavayi* *Castanopsis orthacantha* *Castanopsis ferox* *Schima wallichi* *Schima villosa* *Altingia yunnanensis*

climatic zone, there exist different viewpoints which have been discussed in another paper (Song 1988b). The different opinions come from the different understanding of the terms subtropic and temperate. According to the classification of climate proposed by Köppen (1931), except for the subtropical rain-forest, all of the above-mentioned types of vegetation are located in the warm, rainy climate zone (C), with minimum monthly temperature 0–18°C. Therefore, the EBLF can be recognized as a climatic climax of the warm-temperate climatic zone. Works of Kira (1949, 1954) support this viewpoint, who proposed the criterion for subtropical that the Warmth Index be above 180. This isotherm runs across the northern Ryukyu Islands, about 29° N in Japan, and the Nanling Mountains in China, at about 24° N (Fang 1988). If one accepts the classification system of Strahler (1978), on the other hand, the subtropics are recognized as an independent climatic zone which is controlled by the subtropical high pressure or influenced by its activity for a long period of the year. There are at least eight months with monthly temperature above 9.5°C, and on the equatorward sides the minimum monthly temperature is around 18.3°C. By this criterion, all of the above five vegetation types are in the subtropical zone and represent subtropical vegetation. Because of the divergence of opinion on climatic zones, it is difficult to agree on the bioclimatic zonation of evergreen broad-leaved forest. This problem remains to be solved in the future.

The distribution area of the evergreen broad-leaved forest on the earth is not as large as that of the other main biomes. The climatic conditions for EBLF are neither cold nor hot, neither wet nor dry, and their products are very abundant. These conditions are very suitable for human life, so the population density in EBLF regions is the highest in the world. Under long-term, uninterrupted anthropogenic interference, many stands of EBLF were transformed into farmland and economic plantations, or were degraded to secondary thickets or thicket-grassland. The remaining stands of natural forest are now few and small. This biome contains many species of plants, animals and microbes, so the resources are very abundant. We should strengthen our efforts toward their protection and study.

References

Alechin, B. B. 1950. (Plant Geography.) Chinese translation from Russian (1954). Chinese Book Press, Beijing.

Box, E. O. & Fujiwara, K. 1988. Evergreen broad-leaved forests of the southeastern United States: preliminary description. Bull. Inst. Env. Sci. Techn., Yokohama Nat. Univ. 15:71–93.

Chen Y.-Z., Song, Y.-C., Zhang, S., Feng, Z.-J. & Shi, J.-L. 1965. (The basic characteristics, classification and distribution of broad-leaved evergreen forest in Lu-Shan, Jiangxi Province.) J. E. China Normal Univ. (Science) 1:77–89 (in Chinese).

Chien, S.-S., Wu, Ch.-Y. & Chen, C.-T. 1956. (The vegetation types of China.) Acta Geographica Sinica 22(1):37–94 (in Chinese).

Dansereau, P. 1957. Biogeography, an Ecological Perspective. The Ronald Press, New York. 394 pp.

Ellenberg, H. & Mueller-Dombois, D. 1967. Tentative physiognomic-ecological classification of the main plant formations of the earth. Ber. Geobotan. Inst. ETH, Stiftung Rübel (Zürich) 37:21–55.

Fang, J.-Y. 1988. Climate and Vegetation in China. II: Distribution of main vegetation types and thermal climate. Ecol. Res. (Japan) 4:71–83.

Fujiwara, K. 1981–86. (Phytosociological investigation of the evergreen broad-leaved forests of Japan.) Bull. Inst. Env. Sci. Technol., Yokohama Nat. Univ. 7:67–133, 8:121–150, 9: 139–160, 13:99–149 (in Japanese).

Hämet-Ahti, L., Ahti, T. & Konen, T. 1974. A scheme of vegetation zones for Japan and adjacent regions. Ann. Bot. Fenn. 11:59–88.

He, J. 1951. Floristic regions and plant communities of Fujian. Scientia Sinica 2(2):193–213.

Hou, Hs.-Y. 1960. (Vegetation of China.) People's Education Publisher of China, Beijing. 201 pp. (in Chinese).

Hou, Hs.-Y. 1982. (Vegetation Geography of China and Chemical Composition of Dominants.) Science Press, Beijing. 418 pp. (in Chinese).

Hou, Hs.-Y. 1983. Vegetation of China with reference to its geographical distribution. Ann. Missouri Botan. Garden 70:509–548.

Hou, Hs.-Y. 1988. (Plant Geography of China.) Vol. 2 of "Vegetation Geography of China," Physical Geography series. Science Press. Beijing. 318 pp. (in Chinese).

Hübl, E. 1988. Lorbeerwälder und Hartlaubwälder (Ostasien, Mediterraneis und Makaronesien). Düsseldorfer Geobotan. Kolloq. 5:3–26.

Imanishi, K. & Kira, T. 1953. (Biogeography.) In: Fukui, E. (ed.) (Physical Geography II) pp. 235–313. Asakura Shoten, Tokyo (in Japanese).

Jin, Zh.-Zh. 1979. The types and characteristics of evergreen broad-leaf forests in Yunnan. Acta Botanica Yunnanica 1(1):90–105.

Kira, T. 1945. A new classification of climate in eastern Asia as the basis of agricultural geography. Horticultural Institute, Kyoto Univ., Kyoto (in Japanese).

Kira, T. 1949. (Forest Zones in Japan.) Ringyo Kaisetsu, 17: 105–141. Nippon Ringyo Gijutsu Kyokai, Tokyo (in Japanese).

Kira, T., Ono, Y. & Hosokawa, T. (eds.) 1978. Biological Production in a Warm-Temperate Evergreen Oak Forest of Japan. JIBP Synthesis, vol. 18. Univ. of Tokyo Press. 288 pp.

Klötzli, F. 1988. On the global position of the evergreen broad-leaved (non-ombrophilous) forest in the subtropical and temperate zones. Veröff. Geobotan. Inst. ETH, Stiftung Rübel (Zürich) 98:169–196.

Köppen, W. 1931. Grundriss der Klimakunde. Walter de Gruyter, Berlin.

Lin, P. 1961. On the type and distribution of subtropical rain forest in Fukien Province. Univ. Amoiensis, Acta Scientiarum Naturarium 8(1):24–34.

Lin, Y. 1965. On the characteristics of vegetation of the Nanling Mountains and its position in the vegetation zonation of China. Acta Phytoecol. et Geobotan. Sinica 3(1):50–74.

Lin, Y. et al. 1981. Vegetation of Jiulian-shan nature sanctuary, Jiangxi Province. Acta Phytoecol. et Geobotan. Sinica 5(2):110–120.

Miyawaki, A. (ed.) 1967. (Vegetation of Japan Compared with Other Regions of the World.) Encyclop. Sci. Technol., vol. 3. Tokyo. 535 pp. (in Japanese).

Miyawaki, A. 1980. Das System der Lorbeerwalder (Camellietea japonicae) Japans. In: Dierschke, H. (ed.) Syntaxonomie, pp. 589–599. Cramer, Vaduz.

Miyawaki, A. 1984. A vegetation-ecological view of the Japanese Archipelago. Bull. Inst. Env. Sci. Technol., Yokohama Nat. Univ. 11:85–101.

Muesel, H., Jaeger, E. & Weinert, E. 1965. Vergleichende Chorologie der zentraleuropaischen Flora. Fischer-Verlag, Jena. 538 pp.

Nakano, H. 1930. (Plant Communities and their Succession.) Chinese translation from Japanese original (1936). Commercial Press. 151 pp. (in Chinese).
Numata, M. (ed.) 1974. The Flora and Vegetation of Japan. Kodansha, Tokyo; Elsevier, Amsterdam. 294 pp.
Numata, M., Miyawaki, A. & Itow, D. 1972. Natural and semi-natural vegetation in Japan. Blumea 20:435–481 (with 26 photos, veg. maps).
Numata, M. 1987. Zonation and conservation of vegetation and related problems in Japan. In: Miyawaki, A. et al. (eds.) Vegetation Ecology and Creation of New Environments. Tokai Univ. Press.
Olson, D. F. jr. 1983. Temperate broad-leaved evergreen forest of Southeastern North America. In: Ovington, J. D. (ed.) Temperate Broad-Leaved Evergreen Forest, pp. 103–105. Elsevier, Amsterdam.
Ovington, J. D. (ed.) 1983. Temperate Broad-Leaved Evergreen Forest. Ecosystems of the World, vol. 10. Elsevier, Amsterdam. 241 pp.
Polunin, N. 1960. Introduction to Plant Geography and Some Related Sciences. Longmans, London.
Rübel, E. 1930. Pflanzengesellschaften der Erde. Verlag Hans Huber, Bern. 464 pp.
Satoo, T. 1983. Temperate broad-leaved evergreen forest in Japan. In: Ovington, J. D. (ed.) Temperate Broad-Leaved Evergreen Forest, pp. 169–189. Elsevier, Amsterdam.
Schabalinna, Y. A. 1960. (Plant Geography.) Shanghai Sci. Techn. Press. 774 pp. (in Chinese).
Schimper, A. F. W. 1898. Plant Geography upon a Physiological Basis. English translation of German original (1903). Oxford University Press. 839 pp.
Schmithüsen, J. 1976. Atlas zur Biogeographie. Meyers Grosser Physischer Weltatlas, vol. 3. Bibliographisches Institut, Mannheim/Wien/Zürich.
Schröder, F.-G. 1983. Die thermischen Zonen der Erde. Ein Beitrag zur Präzisierung der geobotanischen Terminologie, mit einer Vegetationskarte. Tuexenia 3:31–46.
Shidei, T. 1974. Forest Vegetation Zones. In: Numata, M. (ed.) Flora and Vegetation of Japan, pp. 87–124. Kodansha, Tokyo.
Song, Y.-C. 1983. Die räumliche Ordnung der Vegetation Chinas. Tuexenia 3:131–157.
Song, Y.-C. 1988a. The essential characteristics and main types of the evergreen broad-leaved forest in China. Phytocoenologia 16(1):105–123.
Song Y.-C. 1988b. Broad-leaved evergreen forests in central Japan in comparison with eastern China. Veroff. Geobot. Inst. Rubel 98:197–224.
Strahler, A. N. & Strahler, A. H. 1978. Modern Physical Geography, John Wiley & Sons.
Walter, H. 1968. Die Vegetation der Erde in ökologischer Betrachtung. Vol. II: Die gemäßigten und arktischen Zonen. VEB Gustav-Fischer-Verlag, Jena. 1002 pp.
Walter, H. 1973. Die Vegetation der Erde in ökologischer Betrachtung. Vol. I: Die tropischen und subtropischen Zonen. Gustav-Fischer-Verlag, Stuttgart.
Walter, H. 1979. Vegetation und Klimazonen. 4th ed. Eugen-Ulmer-Verlag, Stuttgart. 342 pp.
Wang Chi-Wu 1961. The Forests of China. Cabot Foundation Publ. no. 5. Harvard University. Cambridge (Massachusetts). 313 pp.
Warming, E. 1909. Oecology of Plants. An Introduction to the Study of Plant Communities. Oxford Univ. Press, London. 422 pp.
Wu Zh.-Y. & committee (eds.) 1980. Zhongguo Zhibei (Vegetation of China). Science Press, Beijing. 1375 pp. (in Chinese).
Yim, Y.-J. 1977. Distribution of forest vegation and climate in the Korean Peninsula. Part 4. Jap. J. Ecol. 27:269–278.
Zhang, H.-D. 1983. (Discussion on the characteristics of evergreen broad-leaved forest on the Nanling Mountains: the Monshan forest used as an example.) J. South China Agric. College. 3:1–17 (in Chinese).
Zhong, Z.-C. 1988. (Ecological Studies of Evergreen Broad-Leaved Forest.) Southwest China Normal Univ. Press. 696 pp. (in Chinese).

5. Pine forests in East Asia

NOBUKAZU NAKAGOSHI
Department of Environmental Studies, Faculty of Integrated Arts and Sciences, Hiroshima University, Kagamiyama, Higashi-Hiroshima 724, Japan

Abstract

In East Asia, including the Sino-Japanese floristic region, there are about 24 tall-tree species in genus *Pinus*. Generally *Haploxylon* pines are endemic and form the edaphic climax forests on rocky ridges or steep slopes. Among these pines, *P. koraiensis* is widespread in northern Korea and northeastern China and is one of the important canopy species of the temperate mixed forest. Only two microthermal pines, *P. sibirica* in northern East Asia and *P. sylvestris* var. *mongolica* in northeastern East Asia, occur as the dominant tree of climax boreal forests. The other pines are pioneer or seral species and develop even-aged secondary forests substituting for the broad-leaved forests in humid temperate, warm-temperate and subtropical regions where East Asian peoples have been settled for thousands of years. Typical secondary pines are *P. densiflora* and *P. thunbergii* in maritime East Asia, *P. luchuensis* in southern Japanese Islands, *P. tabulaeformis* in eastern China, *P. massoniana* and *P. taiwanensis* in southern China and Taiwan, and *P. yunnanensis* in southern China. In arid regions of East Asia, a few pines including *P. densata*, *P. griffithii* and *P. roxburghii*, grow in high-mountain areas. Pines are introduced and planted widely in East Asia. In South Korea, there are many man-made pine forests including those of two North American pines, *P. rigida* and a hybrid *P. rigi-taeda*, introduced for reforestation of denuded slopes after the Korean War.

Key words: chorology, climax forest, formation, historical disturbance, *Pinus*, secondary forest, Sino-Japanese floral region, species composition

Introduction

Pines (*Pinus*) are one of the common tree species in the Northern Hemisphere (Critchfield & Little 1966). Evolutionarily speaking, pines appeared

E.O. Box et al. (eds.), Vegetation Science in Forestry, 85–104.

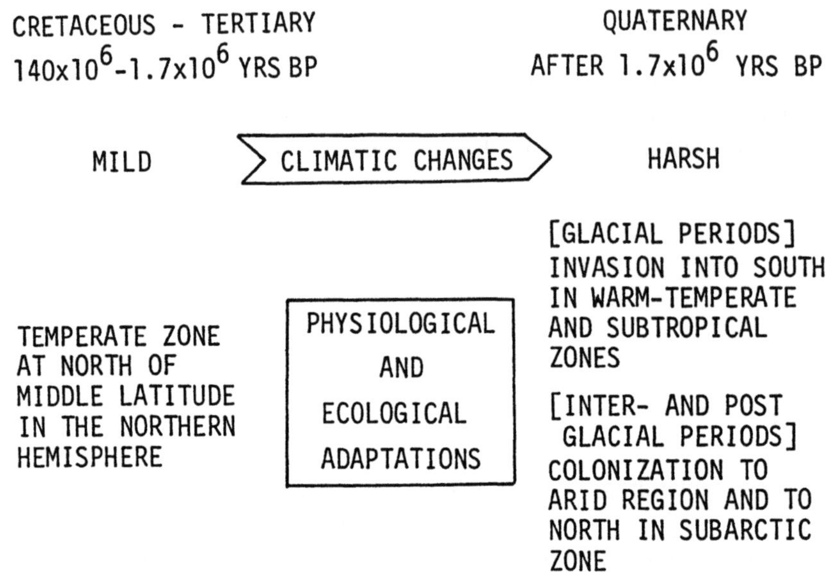

Fig. 1. The evolutionary development of pines.

in the Jurassic on our planet. Until the Tertiary, they grew in the temperate Northern Hemisphere. Thereafter, they invaded more southern zones during the harsh conditions in the Quaternary and also colonized north to boreal areas in the interglacial periods (Fig. 1). One reason for their success in migration was the high freezing resistance, developed during their evolution. Several pine species remain uninjured at less than −80°C (Oohata & Sakai 1982). Pines have varied thermal habitats. Also, they migrate rapidly by winged seeds, which prefer the bare ground surface for germination. The growth rate and biomass production of pines are quite large, and this fast growth and carbon fixation are due to mycorrhizae in their root systems in oligotrophic soils, such as the sandy soil originating from glaciers. Therefore, pines could predominant under the climatic conditions which are poor for forest development. These physiological and ecological characteristics are commonly recognized in the East Asian pines in the northern and southern groups.

In general, pine forests have a significantly stratified structure, namely with pine-canopy, shrub and field layers. The low shade tolerance of pines fosters their even-aged stands and distinct physiognomy. It is expected that understanding the floristic composition of pine forest associations would be easy. Intensive use of pine forests, however, lead to difficulties in ordination of pine forests, which lack their character species in many cases.

The pine forests have been investigated geobotanically since the 1950s in Japan, where several researchers educated by J. Braun-Blanquet and R. Tüxen of the Zürich-Montpellier school of phytosociology worked diligently. Korean phytosociologists started to investigate their pine forests in the 1970s (Rim et al. 1991). Until the 1980s, the pine forest distributions were not well known in the western part of China. Chinese botanist groups explored the pine forests in Xizang (Tibet) and Xinjiang Uygur, western China, a large and uncertain part of the pine territory (Chen 1988, 1991). In this Chinese literature, vegetation units are formations. It corresponds roughly to the alliance in the European system of phytosociology. In the Himalayas, the distributions of native montane pines were identified by their ecological niches only recently (Ohsawa et al. 1986). By synthesizing my original data set and these pine forest surveys, I attempted in this chapter to ordinate the pine forest formations in East Asia.

The area treated and its pine flora

East Asia as defined in this chapter is the region from about 18° to 53° N latitude and from 73° to 149° E longitude. This area includes Mongolia, China, Taiwan, Korea, Japan, Nepal, Bhutan, Primorye of the Russian Republic, and the Indian Himalayas. This geographical definition of East Asia coincides almost exactly with the East Asian cultural/historical region.

Plant-geographically, this also includes the Sino-Japanese floristic region. In this region, there are 24 native tall pine species, excluding a dwarf pine *Pinus pumila* in northeastern East Asia and the introduced pines from North America and other regions (Critchfield & Little 1966). Pines are ordered into two subgenera, *Strobus* and *Pinus*. The pines and their territories in East Asia are illustrated in Figs. 2 and 3, separately for the two subgenera.

Among the 12 species in subgenus *Strobus*, *P. sibirica* is isolated further north than the other pines. It grows on the Hangayn mountains in Mongolia, the Altai mountains in western China and northward to the Siberian boreal forest zone. *P. koraiensis* occurs in coastal East Asia, under a humid but cold climate. Its stands in southern Korea and the Japanese Alps are isolated. *P. parviflora* occurs as a montane pine in western Japan, including Tsushima Island, and on Ulreung Island of Korea, in the Japan Sea (Kim et al. 1986). Its variety *pentaphylla* occurs on mountains in eastern Japan. *P. bungeana* also grows on mountain slopes, in eastern China. This pine and *P. gerardiana* in the eastern Himalayas are sister species. *P. armandii* has a long history. Although the mother species occurs in inland southern China, more than 1000 km from the coast, it has two varieties, *masteriana* on Taiwanese mountains and *amamiana* on Yakushima and Tanegashima Islands, isolated from

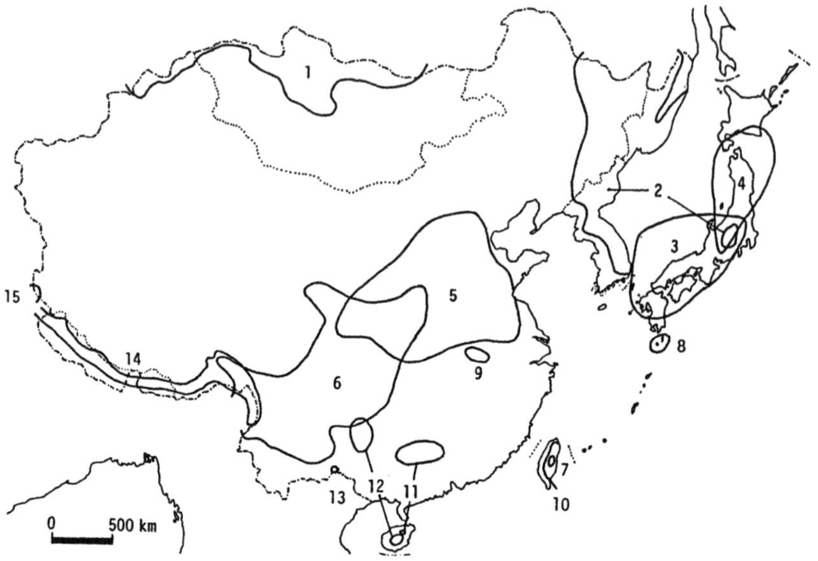

Fig. 2. The original distribution of pines belonging to subgenus *Strobus* in East Asia. 1. *Pinus sibirica*, 2. *P. koraiensis*, 3. *P. parviflora*, 4. *P. parviflora* var. *pentaphylla*, 5. *P. bungeana*, 6. *P. armandii*, 7. *P. armandii* var. *masteriana*, 8. *P. armandii* var. *amamiana*, 9. *P. dabeshanensis*, 10. *P. morrisonicola*, 11. *P. kwangtungensis*, 12. *P. fenzeliana*, 13. *P. wangii*, 14. *P. griffithii*, 15. *P. gerardiana*.

mainland Japan. *P. dabeshanensis* is an endemic pine on the Dabeishan mountains in China. *P. morrisonicola* is an endemic pine scattered in broad-leaved forests on mountain slopes in Taiwan (Flora of Taiwan Editorial Committee 1975). *P. kwangtungensis* occurs isolated in the Nanling mountains and on Hainan Island in the South China Sea. *P. fenzeliana* also occurs separately in Guizhou province and on Hainan Island. *P. wangii* is known only from limestone slopes in southern Yunnan, China. *P. griffithii* (=*P. wallichiana*) occurs on high mountains of southern China, and in the Himalayas, Kashmir and Hindukush.

Among all pines, *P. sylvestris* of subgenus *Pinus* is the most widespread species in the world. A variety *mongolica* occurs in the eastern part of this territory, in the northern Far East. Its southern limits are Selenge in Mongolia and the Amur river lowland. *P. tabulaeformis* occurs mainly in eastern China. This pine has several isolated localities and varieties, for example var. *umbraculifera* in Liaoning province of China and var. *mukdensis* in Liaoning and Korea. Under the humid temperate climate, *P. densiflora* is widespread in maritime China, Korea and Japan. *P. thunbergii* occurs at warm coastal regions in Korea and Japan. Only in intensively disturbed

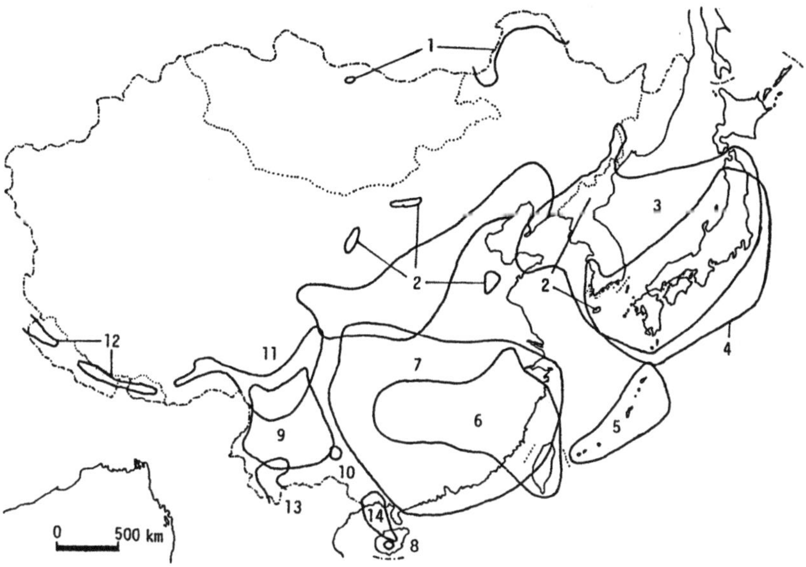

Fig. 3. The original distribution of pines belonging to subgenus *Pinus* in East Asia. 1. *Pinus sylvestris* var. *mongolica*, 2. *P. taburaeformis*, 3. *P. densiflora*, 4. *P. thunbergii*, 5. *P. luchuensis*, 6. *P. taiwanensis*, 7. *P. massoniana*, 8. *P. massoniana* var. *hainanensis*, 9. *P. yunnanensis*, 10. *P. yunnanensis* var. *tenuifolia*, 11. *P. densata*, 12. *P. roxburghii*, 13. *P. kesiya* var. *langbianensis*, 14. *P. latteri*.

places are there hybrid populations between *P. thunbergii* and *P. densiflora* in Japan (Kondo et al. 1985). *P. luchuensis* is limited to the Ryukyu Islands of southern Japan. *P. taiwanensis* occurs in southern China and Taiwan. *P. massoniana* including *P. henryi* grows at lower altitudes in southern China and Taiwan, as compared to *P. taiwanensis*. A variety *hainanensis* of *P. massoniana* is isolated on Hainan Island. *P. yunnanensis* mainly occurs in the Yunnan and southern Sichuan provinces of southern China. A variety *tenuifolia* of *P. yunnanensis* occurs in Guizhou province of southern China. *P. densata* occurs in Sichuan and Yunnan provinces and Xizang. *P. roxburghii* grows from the Himalayas to Kashmir.

Excluding *P. sibirica* and *P. sylvestris* var. *mongolica*, 22 species of East Asian pines occur within the Sino-Japanese floristic region and form a chain of pine habitats from Japan to Afghanistan. Three species of pines, *P. krempfii* in Vietnam, *P. insularis* and *P. merkusii* in Southeast Asia, occur on mountains far south from the Sino-Japanese floristic region (Critchfield & Little 1966). In addition, there is a large area lacking pines in western China and southern Mongolia, due to the extremely arid climate.

Table 1. Thermal conditions of the East Asian pines in native regions (Oohata et al. 1981, Oohata & Sakai 1982)

Species	Mean WI (°C month)	Mean temp. of coldest month (°C)	Difference of mean temp. of coldest and warmest months (°C)
Subgenus *Strobus*			
Pinus koraiensis	42.8	−14.8	38.6
P. parviflora	74.2	−1.4	23.0
P. parviflora var. *pentaphylla*	54.5	−5.3	25.8
P. griffithii	71.0	−2.2	–
P. bungeana	81.0	−7.2	29.1
P. armandii	96.3	4.0	23.0
Subgenus *Pinus*			
P. sylvestris	33.8	−12.7	30.8
P. tabulaeformis	66.8	−8.6	30.5
P. densiflora	86.4	−0.1	27.9
*P. rigida**	89.7	0.2	23.2
P. thunbergii	100.9	2.8	20.5
P. taiwanensis	104.7	8.2	12.8
P. massoniana	113.9	2.9	12.7
*P. taeda**	146.9	7.5	18.2
P. luchuensis	188.1	13.5	13.6
P. latteri	211.0	19.1	4.5

* North American pine.

Vegetation processes of pine forests in East Asia

The East Asian pines of the subgenera *Strobus* and *Pinus* have their own territories. The mean warmth indices (Kira 1945, 1977) of East Asian pines vary from 33.8 to 211 °C · month, representing subarctic to subtropical climates (Table 1). There are few areas of geographic overlap (Figs. 2 and 3). Even in the same geographical locations, each pine grows in a different habitat showing a different ecology. This means that East Asian pines have different courses of adaptation and evolutionary trends in each locality. In the subarctic zone, *P. sibirica* and *P. sylvestris* var. *mongolica* climax forests were never disturbed until the 20th century. Five montane pine (*P. bungeana*, *P. armandii*, *P. densata*, *P. roxburghii* and *P. griffithii*) forest types resisted human influences on high, steep mountains in China and the Himalayas.

The large areas of pine forest in temperate to subtropical zones of East Asia are identified as secondary forests influenced by human activities. Chinese vegetation has been disturbed by more than 5000 years of human history (Wen & Pimentel 1984). Korea (Lee 1986) and Japan (Tsukada 1981) have 3000 and 2000 years of agricultural history respectively. The primeval broad-leaved forests have been cut for cultivation, including slash-and-burn agricul-

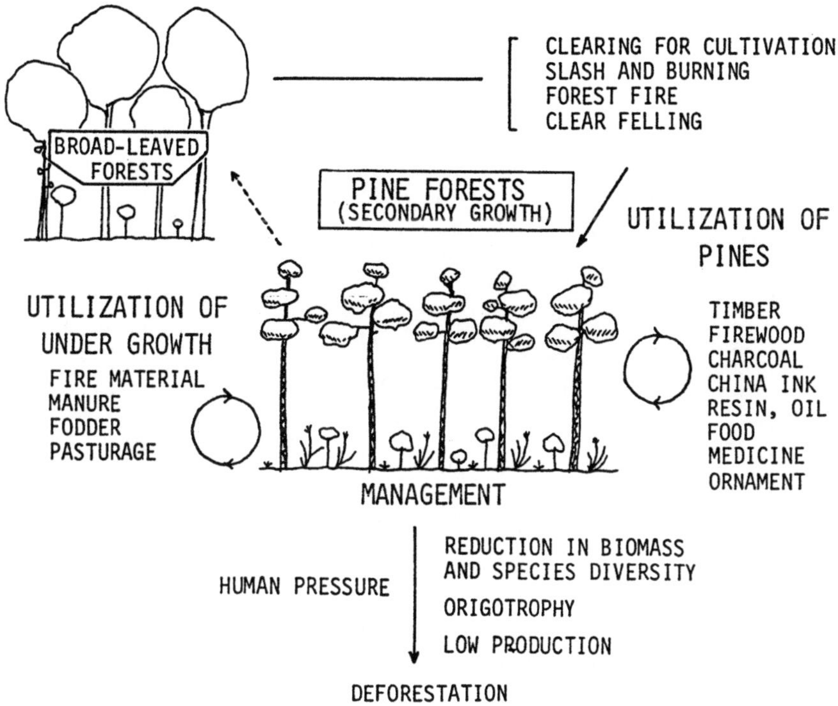

Fig. 4. Vegetation dynamics and the human use of rural pine forests in East Asia. The small circle shows the recycling use of pines and plant materials in pine forests.

ture, for producing timber and fuel, and so on. Sometimes, natural disturbances including forest fire might occur. These disturbances have converted the broad-leaved forests to secondary pine forests. Once this occurred, the particular manner of utilization of pine forests, including their undergrowth, was also established in rural areas. From pine itself, timber, firewood, charcoal, China ink, resin, oil, seeds as food, medicine, ornaments and so on, are produced. From the undergrowth of pine forest, fuel materials, manure for rice paddies, and fodder for livestock have been collected. Pine forests are also used for pasturage. These uses represent attempts at recycling management under the feudal regimes in East Asian countries (Fig. 4).

Even the successional pine species must have some original habitat. In general, it is estimated these origins were places such as steep mountain slopes. For example, the original habitats of *P. densiflora* and *P. parviflora* populations in Japan are on similar steep slopes and other harsh sites without hardwoods. Only the seral pine, *P. densiflora*, could colonize the disturbed sites, especially in the area ruled by human beings. On the contrary, edaphic-climax pine, *P. parviflora*, never forms secondary forest on gentle slopes and

Fig. 5. The mechanized collection of firewood, including pine branches, for "Ondol" from rural pine forests in Yangdong-myon, Kyonggi-do, Korea.

lowlands in Japan. The same is true for *P. densiflora* and *P. parviflora* var. *pentaphylla* in eastern Japan.

The secondary *P. densiflora* has been used intensively by humans, especially by farmers in Korea (Administration of Forestry 1989). The pine branches, needles and undergrowth were consumed as firewood for the heating and cooking system "Ondol" in rural Korea (Fig. 5). Forest management by cyclic, shifting utilization of rural *P. densiflora* forest in Japan and Korea was reported recently in detail (Kamada et al. 1991). In Japan before the 1960s, *P. densiflora* forests were used in various ways. Tall pine trees were felled for timber on about a 40 to 60-year cycle. Woody plants in the pine forests, especially broad-leaved trees, were cut on about a 20-year cycle to produce charcoal. Shrubs were cut for firewood on a shorter cycle. Herbaceous plants of the forest floor, *Miscanthus sinensis* in particular, were mowed to make compost and were used for cultivated fields. Sometimes pine trees were thinned to provide better light conditions for *M. sinensis*. This grass was also used for feed. For these uses, pine forests were kept in an early successional phase. Another significant disturbance of pine forest is man-made forest fire, both in Japan (Nakagoshi et al. 1987) and in Korea (Kim 1989). The denuded burned surfaces of post-fire stands give pines the chance to regenerate. By these means, not only the pine populations but also accompanying plant species in pine forests were limited by the continuous human impact, resulting in degraded species compositions.

In Yunnan province, a relatively well conserved area of southern China, most forest vegetation is pine forest and pine savanna. Closed forests of *P. yunnanensis* are scattered within a matrix of *P. yunnanensis* savannas (Vegetation of Yunnan Editorial Committee 1987). This shows typical overuse of forest vegetation, and only the pine forests can persist to sustain a forest structure influenced by the human impact in this region. The pine forests in southern and eastern China, where cultivated fields are predominant in the landscape with a long history of human disturbance, are the main forest type still remaining on hills and mountain slopes.

In the Himalayas, *P. griffithii* and *P. roxburghii* show the ecological character of both climax and successional trees (Ohsawa et al. 1986). Especially in the arid western Himalayas, these pines behave as dominant species in every stages of ecological succession, due to the lack of competitive, potentially climax trees.

The formations of pine forest in East Asia

The pine forest formations at the alliance level are described based on the original data sets and references. Useful reports come from Vegetation of China Editorial Committee (1980), Vegetation of Anhui Working Group (1981), Vegetation of Yunnan Editorial Committee (1987), Chen (1988, 1991) and Shimizu (1991) for China; Kamada et al. (1987), Kim et al. (1989), Lee & Lee (1989) and Rim et al. (1991) for Korea; Toyohara (1973), Itow & Kawasato (1978), Takeda (1981), Nakagoshi & Nehira (1982), Miyawaki et al. (1983) and Suzuki (1989) for Japan; and Ohsawa et al. (1986) and Ohsawa (1987) for the Himalayas. For the distribution of pine formations in China, four main regions are recognized because China has different climate zones (Song 1988, Fang & Yoda 1989) and biomes within the territory in this chapter (Fig. 6).

Pinus sibirica – Larix sibirica *forest*

P. sibirica – Larix sibirica forest is a boreal forest type occurring in northern Mongolia and northwestern China. It is a climatic climax forest and forms the subalpine forest on the Altai mountains. About 80% of the forest canopy is *P. sibirica*, with the other 20% being *L. sibirica*. The main shrub in the forest is *Vaccinium myrtillus*. Other shrubs are *Rosa suavis*, *Cotoneaster uniflorus* and *Lonicera angustifolia*. *Phalaris arundinacea*, *Poa compressa*, *Pyrola incarnata* and *Aquilegia glandulosa* provide only about 10% cover in

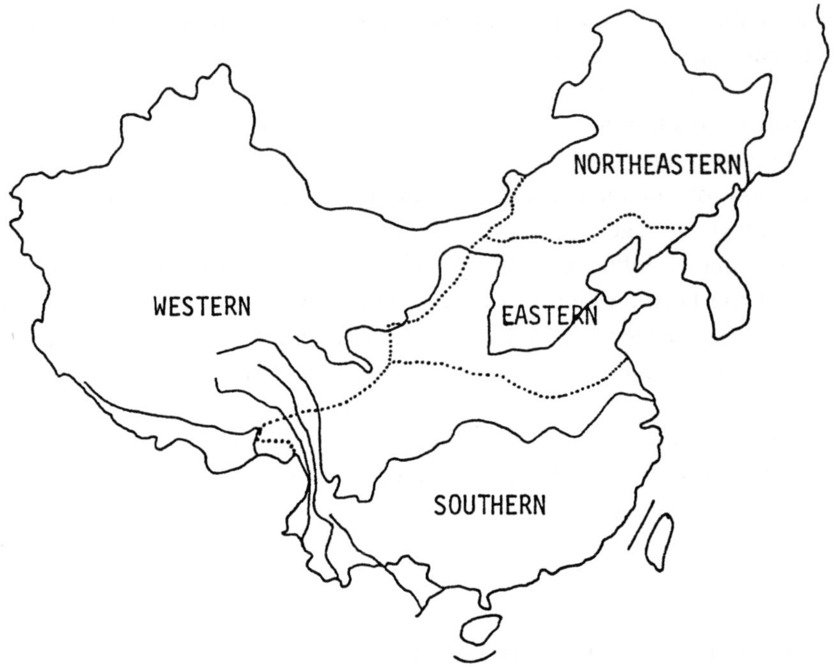

Fig. 6. Four biogeographic regions of China corresponding to the description in northeastern, eastern, southern and western regions correspond to cold-temperate, temperate, warm-temperate/subtropical zones and temperate highlands.

the field layer. On the other hand, the mosses are very rich in the ground layer.

Pinus sylvestris *var.* mongolica *forest*

This formation is also a climax boreal forest and occurs in the lowlands along the Amur river between China and Siberia and isolated in Mongolia. Its habitat is cold but humid in summer. The main shrubs are *Rhododendron dauricum*, *Ledum palustre* var. *angustrum*, *Spiraea sericea* and *Rosa suavis*. The common herbs are *Sanguisorba officinalis*, *Fragaria orientalis*, *Chamaenerion angustifolium*, *Convallaria keiskei* and *Equisetum sylvaticum*.

Pinus koraiensis *and hardwood mixed forest*

This forest type is a unique climax forest of conifers mixed with hardwoods (Fig. 7), occurring in the cold-temperate zone of Korea, northeastern China, and the Primorskii region of Siberia. This well conserved forest can be seen at the north foot of the Changbaishan mountains (Fig. 7). The main admixed hardwoods are *Quercus mongolica* (all oaks in this chapter are deciduous),

Fig. 7. A *Pinus koraiensis* and hardwood mixed forest in Jilin province, northeastern China.

Tilia amurensis and *Fraxinus mandshurica*. *Picea jezoensis*, *P. koraiensis*, *Abies nephrolepis* and *A. holophylla*, as other conifers, are often admixed in the canopy layer. *Corylus mandshurica*, *Philadelphus schrenkii* and *Maackia amurensis* are common in the shrub/arborescent layer, and *Carex pilosa*, *Brachybotrys paridiformis*, *Phryma leptostachya* and *Athyrium multidentatum* occur frequently in the field layer.

Pinus tabulaeformis *forest*

P. tabulaeformis forests occur mainly in eastern China, forming pure even-aged stands showing characteristics of a seral community. Sometimes *P. tabulaeformis* is mixed with *P. armandii*, *Juniperus rigida*, *Quercus*, *Populus*, *Betula*, *Celtis*, *Carpinus*, *Acer*, and *Tilia* over its wide territory. The resulting forests are temperate forests. In the *P. tabulaeformis* forest in eastern China, *Lespedeza bicolor*, *L. formosa*, *Spiraea cantoniensis* and *S. trilobata* are common shrubs. Graminoids *Themeda triandra* var. *japonica*, *Bothriochloa ischaemum*, *Arundinella hirta* and *Zoysia macrostachys* dominate in the field layer.

Pinus densiflora *forest*

P. densiflora forest is a secondary forest occupying rural areas in the temperate climate of Japan, Korea and northeastern China. The shrub flora in *P. densiflora* forest has a high variability from north to south, but *Vitex negundo*

Table 2. List of pine associations in Japan

***Pinus densiflora* forest:**
Fagetea crenatae MIYAWAKI, OHBA et MURASE 1964
Quercetalia serrato-grosseserratae MIYAWAKI et al. 1971
Pinion densiflorae SUZ.-TOK. 1966
Rhododendro-Pinetum densiflorae SUZ.-TOK. et USUI 1952
Rhododendro-Pinetum kinkianum SUZ.-TOK. 1966
Rhododendro nudipedis-Pinetum densiflorae TAKEDA 1981
Rhododendro reticulati-Pinetum densiflorae H. SUZUKI et TOYOHARA 1971
Rhododendro weyrichi-Pinetum densiflorae SUZ.-TOK. 1964
Rhododendro kiusiani-Pinetum densiflorae ODA et SUMATA 1962
***Pinus thunbergii* forest:**
Pinus thunbergii – Vitex rotundifolia community
Pinus thunbergii – Imperata cylindrica var. *koengii* community
Pinus thunbergii – Persea thunbergii community
Pinus thunbergii – Pittosporum tobira community
***Pinus luchuensis* forest:**
Pinion luchuensis K. SUZUKI 1978
Rhododendro amaoi-Pinetum 1uchuensis K. SUZUKI 1978
Carici brunneae-Pinetum luchuensis K. SUZUKI 1978
Pinus luchuensis – Melastoma candidum community
Pinus luchuensis – Ficus virgata community
Pinus luchuensis – Pandanus boninensis community

var. *heterophylla*, *Zizyphus jujuba*, *Indigofera macrostachys*, and *Forsythia suspensa* grow commonly throughout the *P. densiflora* territory in China.

On the other hand, only a few shrubs, including *Rhododendron mucronulatum*, *Lespedeza bicolor* and *L. cyrtobotrya*, grow sparsely in *P. densiflora* forest in Korea. The dominant species in the field layer are the graminoids *Arundinella hirta*, *Carex lanceolata*, *Miscanthus sinensis*, *Spodiopogon sibiricus*, as well as, *Artemisia japonica* and *Smilax china*. This species poverty is a result of intensive use of pine forests for a long time.

Japanese phytosociologists have been working on pine forests since the 1950s. The systematics of *P. densiflora* forests is almost completed. The other two secondary pine forest types in Japan, involving *P. thunbergii* and *P. luchuensis*, are being discussed now at the association level. As a standard unit for comparing the pine forests of the continent and those of Japan, the alliance level of the European tradition was acceptable. In Japan, *Rhododendron kaempferi*, *Lyonia ovalifolia* var. *elliptica*, *Clethra barbinervis* and *Ilex crenata* in the shrub layer, and *Miscanthus sinensis*, *Smilax china*, and *Pteridium aquilinum* var. *latiusculum* in the field layer are common throughout the Pinion densiflorae. Six associations in the Pinion densiflorae can be identified, based mainly on plantgeography, especially on the distribution of Ericaceae species, and on Japanese data (Table 2). The association Rhododendro reticulati-Pinetum densiflorae covers most of the slopes in western Honshu, especially in coastal regions and on islands in the Seto Inland Sea (Fig. 8).

Fig. 8. A *Pinus densiflora* forest (Rhododendro reticulati-Pinetum densiflorae) on mountain slopes in Kurahashi-cho, Hiroshima Prefecture, Japan.

Pinus thunbergii *forest*

Natural forests of *P. thunbergii* occur near the seacoast of Japan and southern Korea, in the warm-temperate zone. Four communities (association level) compose this formation (Table 2). They are *P. thunbergii – Vitex rotundifolia* community on coastal dunes in western Japan, *P. thunbergii – Imperata cylindrica* var. *koengii* community on dunes in northeastern Japan, *P. thunbergii – Persea thunbergii* community on deep sandy soils of western coasts in Japan and *P. thunbergii – Pittosporum tobira* community on coastal rocks in Japan and Korea. Every community shows successional phases and develops normally into broad-leaved forest. *P. thunbergii* needs warmth, but its habitats are oligotrophic, such as on rocks and dunes near the seashore. The common understorey species are *Eurya japonica* in the shrub layer and *Miscanthus sinensis* in the field layer.

Pinus luchuensis *forest*

The *P. luchuensis* forest in the humid subtropical climate of the Ryukyu Islands of southern Japan occurred originally as a secondary forest. Phytosociologically, this pine forest is the Pinion luchuensis, which contains two associations and three communities (Table 2). Common shrubs are *Rhododendron tashiroi*, *Vaccinium wrightii*, *Glochidion obovatum*, *Cinnamomum doederleinii*, and *Syzygium buxifolium*. Four phytocoenoses, Rhododendro amaoi-Pinetum luchensis, Carici brunneae-Pinetum luchensis, and the *P. luchuensis – Melastoma candidum* and *P. luchuensis – Ficus virgata* communities, can be seen in the Ryukyu Islands. The dominant herbs are *Dicranopteris dichotoma*, *Miscanthus sinensis*, *Oplismenus compositus*, *Smilax nervo-marginata*, and *S. china* var. *kuru*.

In the Bonin Islands in the Pacific Ocean, introduced *P. luchuensis* populations have spread to form pine forests (Shimizu & Tabata 1985). Excluding a few endemic shrubs such as *Schima mertensiana* and *Pandanus boninensis*, most shrubs (*Ardisia sieboldii*, *Planchonella obovata*, *Rhaphiolepis umbellata*, etc.) and herbs (*Gahnia tristis*, *Oplismenus compositus*, *Nephrolepis auriculata*, etc.) in this alien community are common to the Pinion luchuensis of the Ryukyu Islands. These alien pines are dying, however, due to attack by pine-wood nematodes (Shimizu 1986).

Pinus bungeana *forest*

P. bungeana forest is an edaphic-climax forest scattered on steep slopes in temperate eastern China. *Cotoneaster zabelli*, *Viburnum betulifolium*, and *Rhamnus utilis* are common in the shrub layer. *Anemone tomentosa*, *Patrinia scabiosaefolia*, *Bupleurum fulcatum*, and *Pulsatilla chinensis* grow frequently in the field layer.

Pinus armandii *forest*

The climax forest of *P. armandii* is a relict, but secondary forests occur widely in the temperate climate of eastern and southern China. In southern parts of this pine territory, trees such as *P. densata*, *Cupressus*, and *Quercus* are admixed in the canopy layer. Both in the shrub and field layers, species compositions vary by subregion, since the understorey floras vary from mountain to mountain. Two varieties isolated in the islands of Japan and Taiwan never form large forest stands.

Pinus massoniana *forest*

P. massoniana forest is a representative secondary forest in the warm-temperate and subtropical climates of southern China. The main shrubs are *Rhododendron simsii*, *Vaccinium bracteatum*, *Loropetalum chinense*, *Lespedeza formosa*, and *L. virgata*. The main herbs are *Imperata cylindrica* var. *major*, *Miscanthus floridulus*, *Themeda triandra* var. *japonica*, *Pteridium aquilinum* var. *latiusculum*, and *Dicranopteris dichotoma*.

Fig. 9. A *Pinus yunnanensis* forest in Yunnan province, southern China (Photo by Kazue Fujiwara).

Pinus taiwanensis *forest*

Most *P. taiwanensis* forests are seral communities in secondary succession. Their territory is on temperate mountain slopes in southern China and Taiwan. The species richness is large in the shrub layer. *Rhododendron oratum*, *Rh. mariesii*, *Rh. simsii*, *Conylopsis sinensis*, *Lindera obtusiloba*, *Symplocos chinensis* and *Wiegela japonica* are dominant shrubs. The forbs *Aster ageratoides*, *Solidago decurrens*, *Dendranthema indicum* and *Patrinia villosa* dominate in the sunny field layer. Climax *P. taiwanensis* populations occur on rocky ridges and steep slopes under harsh conditions (Flora of Taiwan Editorial Committee 1975, Zhang 1990).

Pinus yunnanensis *forest*

P. yunnanensis forest is a secondary forest in warm-temperate southern China (Fig. 9). The main shrubs are *Pyrus pashia*, *Rhododendron decorum*, *Rh. delavayi*, *Rh. microphyton*, *Rh. spiciferum*, *Rh. spenuliferum*, *Lyonia ovalifolia*, *L. villosa*, *Vaccinium fragile*, and *V. sprengellii*. Common herbs are *Eulalia quadrienervis*, *Eremopogon delavayi*, *Arundinella chenii*, *Themada triandra* var. *japonica* and *Pteridium aquilinum* var. *latiusculum*. The species composition and ecology of *P. yunnanensis* var. *tenuifolia* forest is little different from that of *P. yunnanensis* forest. Therefore, *P. yunnanensis* var. *tenuifolia* forest, whose territory is not large, is not treated as an independent formation in this chapter.

Pinus kesiya *var.* langbianensis *forest*

P. kesiya var. *langbianensis* forest is a subtropical, secondary pine formation in southern China. Its common shrubs are *Wendlandia scabra, W. paniculata, Aporosa yunnanensis* and *Phyllanthus emblica*. The main herbs are *Eupatorium adenophorum, Eulalia speciosa, Pogonatherum paniceum, Arundinella nepalensis*, and *Dicranopteris dichotoma*.

Pinus latteri *forest*

P. latteri forest is a typical secondary forest, influenced by human impact, in the warm-temperate climate of southern China. The coverage of the shrub layer is small. Common species in the field layer are *Andropogon chinensis, Hyparrhenia eberhardtii, Schizachyrium sanguineum* and *Heteropogon contortus*.

Pinus densata *forest*

This pine forest is a zonal montane pine formation in temperate-climate areas with dry season in southern and western China. *Rhododendron litangense, Rh. decorum, Rh. cephalantum, Lyonia vilosa, L. ovalifolia* and *Cotoneaster hebephyllum* are common shrubs. *Deyeuxia scabrescens, Drynaria baronii, Anaphalis gracilis* var. *ulophylla, Andropogon tristis, Arundinella yunnanensis*, and *Artemisia vulgaris* grow frequently in the field layer.

Pinus griffithii *forest*

P. griffithii forest is a climax formation in the arid western Himalayas. It also grows continuously eastward to the eastern Himalayas and southwestern China as a seral community. Its thermal range is wide, extending from the cold to warm-temperate zone. The main shrubs are *Rhododendron arboreum, Lyonia ovalifolia, Enkianthus deflexus* and *Gaultheria griffithiana*. In the field layer, *Arundinella chenii, Geranium refractum, G. donianum, Thalictrum rostellatum, Th. foliolosum, Th. saniculaeforme* and *Pteridium aquilinum* var. *latiusculum* are important. Several vines are also important, including *Hedera nepalensis* var. *sinensis, Schisandra grandiflora* and *Vitis semicordata*.

Pinus roxburghii *forest*

P. roxburghii forest is another climax forest in the dry western Himalayas, occurring also in patches on dry steep slopes in warm-temperate eastern Nepal, Bhutan and the Gyrong valley of western China. As a seral community it also invades disturbed natural areas. The main shrubs are *Rhododendron arboreum*, *Lyonia ovalifolia*, *Cornus capitata*, *Prunus padus*, *Rhus punjabensis* and *Elaeagnus umbellata*. Ground cover in the field layer is large. The main species are *Sinarundinaria falcata*, *Pteridium aquilinum* var. *latiusculum*, *Apluda mutica* and *Arundinella nepalensis*.

Pine plantations *and alien pines*

Pines are also useful trees for reforestation. Historically, *P. massoniana* in southern China and *P. thunbergii* in Shandong province of China and behind coastal dunes in Japan have performed well as successfully restored tree species in waste land. A *P. thunbergii* plantation in Anhui province, for example, contains shrubs such as *Glochidion puberum*, *Lespedeza virgata*, *L. cuneata* and *Albizzia macrophylla*, and herbs such as *Themeda triandra* var. *japonica*, *Potentilla discolor*, *Sanguisorba officinalis* and *Smilax china*. This species composition shows that the plantation site used to be heavily disturbed land.

After the Korean War, North American pines, especially *P. rigida*, were planted all over South Korea for restoration of the rural forests seriously damaged during the war. The area of *P. rigida* plantation is larger than those of other pine plantations (*P. rigi-taeda*, *P. koraiensis* and *P. thunbergii*) in South Korea. These plantations of *P. rigida* are commonly seen on foothills near residential area (Kamada et al. 1991). The community structure of *P. rigida* forest is similar to that of the native pine forests in Korea. There are also few differences between the species composition of alien pine forests and that of native pine forests.

Conclusion

In northern East Asia, climax pine forests compose the southern border of the boreal forests. On mountainous sites in China, pine forests still remain as edaphic-climax forests. On the other hand, human-influenced pine forests are predominant among the pine forests of eastern East Asia. From the viewpoint of area occupied, this type of pine forest is distinctly the major type. In addition, some man-made pine forests in East Asia are pine plantations.

The native pine forests recognized at large scale in East Asia were classified into 17 formations. In this classification, the unit of vegetation is the alliance level of phytosociology. In general, it is very difficult to identify formations, since East Asian pine forests have different climate and human influences even in forests of the same species. This was done in this paper in order to suggest further questions. Secondary pine forests seldom overlap semi-natural forests. Strong human impacts and different physiological backgrounds might cause this. Originally, these pine populations were limited to mountainous areas, from the Himalayas and southern China to Taiwan, Japan and Korea. At the subgenus level, these used to be isolated geographically and ecologically showing adaptive radiation. But they had not large genetical gaps in each other.

Seral pines forming secondary forests seem able to regenerate only on sites where continuous, cyclic disturbances occur. If the human impact is reduced, the pine forests in temperate East Asia might change to deciduous forests in which *Quercus*, *Carpinus*, *Castanea* and other summergreen tree genera are dominant, pines would and never occur due to the dark understorey environments in such well developed deciduous forests. Successional pine forests in the warm-temperate zone seem unable to maintain their populations when disturbance ceases. In these pine forests, broad-leaved trees grow up to the canopy layer, and finally the evergreen trees (such as *Cyclobalanopsis* (=evergreen *Quercus*, *Castanopsis* and *Schima*) dominate the canopy layer. There are few possibilities for this to occur, however, because man needs more plant materials, due to the overpopulation in present East Asia, especially in China and the Himalayas.

In conclusion, it seems that the most common forest in rural East Asia is a pine forest. These are also very important for East Asian people. After the 1960s, the energy source in Japan changed from plant materials to fossil energy. As a result, pine-forest landscapes are changing rapidly to broad-leaved forests (Kamada & Nakagoshi 1990). It may be difficult to find secondary pine forests in the latter half of the 21st century in Japan, if rural forests are abandoned. Because of harsh environmental pressures (industrial emissions, acid rain, insect damage, land overuse, desertification, etc.) on rural pine forests in recent years (Pimentel et al. 1986), discussion of sustainable uses of these forests is needed in each region of East Asia.

Acknowledgements

I heartily thank Elgene O. Box of the University of Georgia, USA, and Kazue Fujiwara of Yokohama National University, Japan, who kindly suggested this synthesis work. Thanks are also due to Rim Young-Deuk of

Inchon National University of Education, Kim Chul-Soo of Mokpo National University in Korea, Zhao Dachang of the Institute of Applied Ecology (Shenyang), and Chen Weilie of the Institute of Botany (Beijing) in China for their fruitful comments on this review.

References

Administration of Forestry 1989. Forest Statistics. Seoul (in Korean).

Chen, W.-L. 1988. Pine forests. In: Vegetation of Xizang. The Scientific Expedition to the Qinghai-Xizang Plateau, pp. 121–130. Science Press, Beijing (in Chinese).

Chen, W.-L. 1991. The pine forests in western China. In: Nakagoshi, N. & Golley, F. B. (eds.), Coniferous Forest Ecology from an International Perspective, pp 5–16. SPB Academic Publishing, The Hague.

Critchfield, W. B. & Little, E. L. Jr. 1966. Geographic Distribution of the Pines of the World. U.S. Dept. Agr. For. Serv., Washington, D.C. 97 pp.

Fang, J. & Yoda, K. 1989. Climate and vegetation in China. II. Distribution of main vegetation types and thermal climate. Ecol. Res. 4:71–83.

Flora of Taiwan Editorial Committee 1975. Flora of Taiwan, vol. 1. pp. 518–525. Epoch Publishing, Taipei.

Itow, S. & Kawasato, H. 1978. Distribution of secondary forests in Japan. In: Papers on Plant Ecology to the Memory of Dr. K. Yoshioka, pp. 281–284. Sendai (in Japanese).

Kamada, M. & Nakagoshi, N. 1990. Pattern and processes of secondary vegetation at a farm village in southwestern Japan after the 1960s. Jpn. J. Ecol. 40:137–150 (in Japanese with English synopsis).

Kamada, M., Nakagoshi, N. & Kim, C.-S. 1987. Human activities on the pine forests nearby farm villages in Chollanam-do, Korea. Bull. Biol. Soc. Hiroshima Univ. 53:3–10 (in Japanese with English summary).

Kamada, M., Nakagoshi, N. and Nehira, K. 1991. Pine forest ecology and landscape management: A comparative study in Japan and Korea. In: Nakagoshi, N. & Golley, F.B. (eds.), Coniferous Forest Ecology from an International Perspective, pp. 43–62. The Hague: SPB Academic Publishing.

Kim, C.-S., Park, Y.-W. & Nakagoshi, N. 1989. Phytosociological studies on the vegetation and flora of Pogil-do. Bull. Inst. Litt. Biota 6:65–95 (in Korean with English abstract).

Kim, S.-D., Kimura, M. & Yim, Y.-J. 1986. Phytosociological studies on the beech (*Fagus multinervis* Nakai) forest and the pine (*Pinus parviflora* S. et Z.) forest of Ulreung Island, Korea. Korean J. Bot. 29:53–65.

Kim, W. 1989. The secondary succession and species diversity at the burned area. Korean J. Ecol. 12:285–295 (in Korean with English summary).

Kira, T. 1945. A new classification of climate in eastern Asia as the basis for agricultural geography. Kyoto: Horticultural Institute, Kyoto Univ. 23 pp. (in Japanese).

Kira, T. 1977. A climatological interpretation of Japanese vegetation zones. In: Miyawaki, A. & Tüxen, R. (eds.), Vegetation Science and Environmental Protection, pp. 21–30. Tokyo: Maruzen.

Kondo, K., Segawa, M. & Yata, T. 1985. Hybridization of *Pinus densiflora* and *P. thunbergii* accelerated by human interference in natural pine-forest vegetation in Hiroshima, Japan. Фyton 45:45–54.

Lee, W.-T., & Lee, C.-H. 1989. Plant sociological studies on the *Pinus densiflora* forest in Korea. Korean J. Ecol. 12:257–285 (in Korean with English summary).

Lee, Y.-N. 1986. *Korean Coniferae*. Seoul: Ewha Women's Univ. Press. (In Korean).

Miyawaki, A., Okuda, S. & Mochizuki, R. 1983. Handbook of Japanese Vegetation, Rev. ed. Tokyo: Shibundo. 872 pp. (In Japanese).

Nakagoshi, N. & Nehira, K. 1982. Regeneration of vegetation in the burned pine forest in southern Hiroshima Prefecture, Japan. III. The control forests. Bull. Biol. Soc. Hiroshima Univ. 48:7–16 (in Japanese with English summary).

Nakagoshi, N., Nehira, K. & Takahashi, F. 1987. The role of fire in pine forests of Japan. In: Trabaud, L. (ed.), The Role of Fire in Ecological Systems, pp. 91–119. The Hague: SPB Academic Publishing.

Ohsawa, M. 1987. Vegetation zones in the Bhutan Himalaya. In: Ohsawa, M. (ed.), Life Zone Ecology of the Bhutan Himalayas, pp. 1–71. Chiba.

Ohsawa, M., Shakya, P. R. & Numata, M. 1986. Distribution and succession of west Himalayan forest types in the eastern part of the Nepal Himalaya. Mountain Research and Development, 6:143–157.

Oohata, S., Hasegawa, Y. & Sakai, A. 1981. Studies on freezing resistance and distribution of the genus *Pinus*. I. Seasonal changes in freezing resistance. Jpn. J. Ecol. 31:79–89 (in Japanese with English synopsis).

Oohata, S. & Sakai, A. 1982. Freezing resistance and thermal indices with reference to distribution of the genus *Pinus*. In: Li, P.-H. & Sakai, A. (eds.), Plant Cold Hardiness and Freezing Stress, vol. 2, pp. 437–446. New York: Academic Press.

Pimentel, D., Wen, D.-Z., Eigenbrode, S., Lang, H., Emerson, D. & Karasik, M. 1986. Deforestation: interdependency of fuelwood and agriculture. Oikos, 46:404–112.

Rim, Y.-D., Kang, H.-K. & Nakagoshi, N. 1991. Community ecology of *Pinus densiflora* forests in Korea. In: Nakagoshi, N. & Golley, F. B. (eds.), Coniferous Forest Ecology from an International Perspective, pp. 17–30. The Hague: SPB Academic Publishing.

Shimizu, Y. 1986. Serious damage to *Pinus lutchuensis* by the attack of the pine wood nematode at Chichijima in the Bonin (Ogasawara) Islands. J. Fac. Lett. Komazawa Univ. 44:169–178.

Shimizu, Y. 1991. Forest types and vegetation zones of Yunnan, China. J. Fac. Sci. Univ. Tokyo III, 15:1–71.

Shimizu, Y. & Tabata, H. 1985. Invasion of *Pinus lutchuensis* and its influence on the native forest on a Pacific Island. J. Biogeogr. 12:195–207.

Song, Y.-C. 1988. The essential characteristics and main types of the broad-leaved evergreen forest in China. Phytocoenologia, 16:105–123.

Suzuki, K. 1989. Secondary evergreen coniferous forest. In: Miyawaki, A. (ed.), Vegetation of Japan. Vol. 10. Okinawa and Ogasawara, pp. 393–396. Tokyo: Shibundo. (In Japanese).

Takeda, Y. 1981. A phytosociological study of the Rhododendro nudipedis-Pinetum densiflorae. Bull. Fac. Educ. Kobe Univ. 66:109–125 (in Japanese with English summary).

Toyohara, G. 1973. Pine forests. In: Sasaki, Y. (ed.), Phytosociology, pp. 48–53. Tokyo: Kyoritsushuppan. (In Japanese).

Tsukada, M. 1981. The last 12,000 years – the vegetation history of Japan. II. New pollen zones. Jpn. J. Ecol. 31:201–215 (in Japanese with English summary).

Vegetation of Anhui Working Group 1981. Vegetation of Anhui Hefei: Anhui. Sci. & Tech. Press. 321 pp. (In Chinese).

Vegetation of China Editorial Committee 1980. Vegetation of China. Beijing: Science Press. 1382 pp. (In Chinese).

Vegetation of Yunnan Editorial Committee 1987. Vegetation of Yunnan. Beijing: Science Press. 1024 pp. (In Chinese).

Wen D.-Z. & Pimentel, D. 1984. Energy flow through an organic agroecosystem in China. Agric. Ecosystems Environ. 11:145–160.

Zhang, L.-Q. 1990. Population structure and dynamics of *Pinus taiwanensis* Hayata at Songyang county, Zhejiang province, China. Vegetatio 86:119–129.

6. Comparison of the vegetation of subtropical China with that of the corresponding region of the USA

HUO YING-QIANG[1] & SHIRLEY A. HARTZLER[2]
[1]*Forest Research Institute of Guangdong, Guangzhou 510520 P.R. China*
[2]*The State Botanical Garden of Georgia, University of Georgia, Athens, Georgia 30605, USA*

Abstract

This paper initially addresses the confusing phytogeographical terminology of the subtropics, warm-temperate and temperate zones in China and the United States. Based upon the standards and data presented, the range of regions of the United States corresponding to China's subtropics is clarified. Climate, topography and soils as well as natural vegetation, artificial vegetation, and potential vegetation are compared in the two regions. Recommendations of appropriate plant species from China's subtropics are suggested for introduction into the southeastern United States.

Key words: China's subtropics, comparative phytogeography, evergreen broad-leaved forest, Southeastern USA, vegetation.

Introduction

Both the United States and China are vast countries and are situated over nearly the same range of latitudes of the northern hemisphere. Both possess a great diversity of ecological environments and include almost every kind of vegetation found in the northern hemisphere. It is seen that the vegetation of the two countries are comparable with each other as implied by phytogeography. Particularly in the subtropics of China and some southeastern states of the United States, there is a rich variety of vegetation, while parts of Africa, Southwest Asia and the central United States at the same latitude have a dry climate and are mostly deserts. Therefore, a comparison of the vegetation between the two regions has far-reaching significance both practically and theoretically. China has gained many economic benefits from the introduction and large-scale expansion of slash pine, *Pinus elliottii*, and

E.O. Box et al. (eds.), Vegetation Science in Forestry, 105–123.

Table 1. Subdivision of China's subtropics and the corresponding general climatic regions of the USA

Subdivision of China's subtropics (Fig. 1)	Climatic criteria	Corresponding general climatic regions of the USA
South subtropics (transitional tropics)	Basically no frost, minimum temperature 0°C	Subtropical region including south Florida
Typical or mid-subtropics	Light frost (−10°C) or near-frost every year	Warm-temperate region including north Florida, south Georgia and the coastal areas of the southeastern states (Fig. 2)
North subtropics (transitional subtropics)	Frost every year, generally significant, occasional temperatures below −15°C to −20°C except in maritime areas	Southern portion of typical temperate region, its northern boundary is nearly to the northernmost distribution limits of slash pine and loblolly pine (Fig. 3)

loblolly pine, *P. taeda*, in the subtropical region. On the other hand, there are more than 1,600 genera and 18,000 species of seed plants in China's subtropics, many of which are evergreen broad-leaved shrubs and trees (Wu 1980), providing great potential for introduction into the corresponding region of the southeastern U.S. for the purpose of enriching plant resources. From this point, comparative phytogeography should begin on a theoretical basis and develop further by the process of experiment and research.

1. Physical environmental factors of the two regions

The total area of China is 9.6 million square kilometers of which 22% is considered subtropical. China has the largest area of subtropics in the world. Due to the use of various terms in different countries, it is not always clear where boundaries lie, and the various terminology is often confusing. It is important to begin with a definition of the terminology and, based upon the same criteria, determine the corresponding region in the USA (Table 1; Figs. 1, 2, 3).

China's subtropical region is divided into three different zones (Fig. 1): (1) The south subtropics (transitional tropics) is situated from latitude 22° to 24° or 25° N and has an area of 0.40 million square kilometers; (2) The middle subtropics lie between 24° or 25° N and 31° or 32° N and include 1.5 million square kilometers; and (3) The north subtropics occur from 31° or 32° N to 34° or 35° N, and the area is about 0.30 million square kilometers.

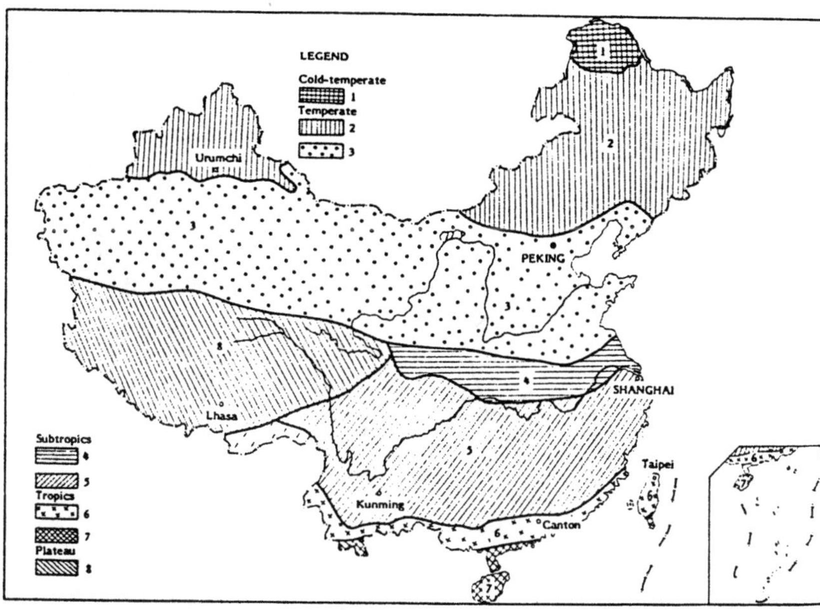

Fig. 1. China's subtropical region divided into three different zones. Map showing regions of atmospheric warmth of China: 1) Cold-temperate; 2) Temperate; 3) Warm-temperate; 4) Transitional subtropics (North subtropics); 5) Subtropics (Mid-subtropics); 6) Transitional tropics (South subtropics); 7) Tropics; 8) Cold high-mountain climate (Hou 1983). Reprinted by permission.

The northern boundary is the Chingling Mountains in the west and Huai River in the east. Climatic data for the three zones are shown in Table 2.

The total area of the USA is 9.3 million square kilometers. The region of the USA corresponding to China's subtropics is situated in the southeastern states. The evergreen broad-leaved forest, characteristic of humid warm temperate climates (equivalent to the term 'mid-subtropics' of China) is present in the southeastern USA, but only as small remnants in narrow strips along the Atlantic and Gulf coasts (Box and Fujiwara 1988). Box and Fujiwara suggest that a region of potential evergreen broad-leaved forest extends from southeasternmost Virginia to the eastern part of the 'hill country' of central Texas, in a coastal strip gradually widening to the south as shown very generally in Fig. 2. We propose that the region of the USA corresponding to China's subtropics should be at least as large as this region shown in Fig. 2, including also South Florida. This total region accounts for about 5% of the total area of the USA. The evergreen broad- leaved forest is not a zonal vegetation of the north subtropics in China, so the region of the USA should be larger than that shown in Fig. 2. According to data on introduced American slash and loblolly pines, their distribution extends

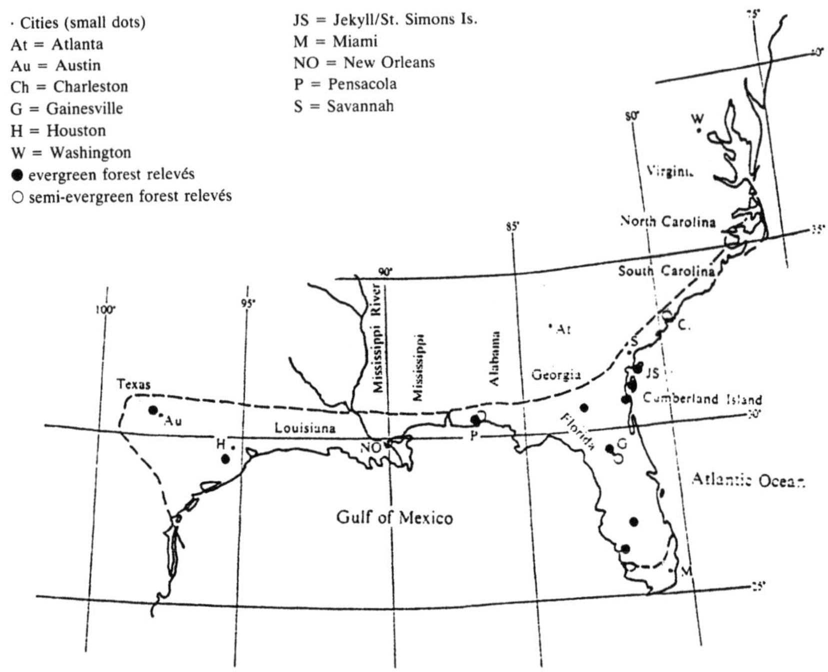

Fig. 2. Locations of evergreen broad-leaved and related forest relevés in the warm-temperate southeastern United States (Box and Fujiwara 1988).

throughout the whole subtropics of China (see Table 3). Therefore, we suggest that the corresponding region of the USA is roughly the same as the geographical distribution of slash pine and loblolly pine shown in Fig. 3, which amounts to about 10% of the total area of the USA (though still much less than that of China's subtropics).

1.1. Climate

China has a climate dominated by monsoonal winds rising out of the differences in the heat-absorbing capacity of the Asian Continent and the Pacific Ocean in the southeast as well as the Indian Ocean in the southwest (Chung 1978). The winter monsoons move southward causing a sharp drop in temperature. There are generally five or six strong cold waves every winter, and the greater part of China is therefore cold and dry in winter. The summer monsoons pass across the vast ocean surface, so the annual rainfall over the greater part of China is highly concentrated within a few summer months.

The advance and retreat of the monsoons has much to do with the timing of China's rainy season and the amount of rainfall in many parts of the country. When the monsoons are abnormal, drought or flooding occurs over

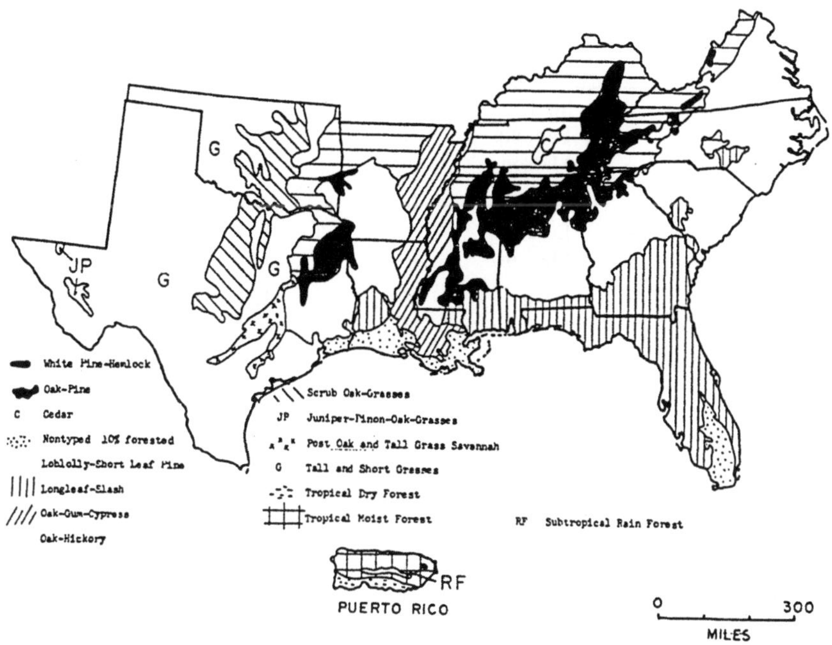

Fig. 3. Vegetation in the southeastern States. From U.S. Forest Service 196; and A. W. Kuchler 1964 (Buol 1973). Reprinted by permission.

Table 2. Various climatic indices in the three zones of China's subtropics

Zone	Annual rainfall (mm)	Annual mean temperature (°C)	Coldest-month mean temperature	Annual summation of temperature above 10°C
South subtropics	1400–2000	21–23	12–16	5000–6500
Middle subtropics	1000–2000	15–21	2–12	5000–6500
North subtropics	900–1000	14–15	0	5400–5000

large areas. In the tropics and subtropics of China, mainly because of the relatively low latitudes and the fact that the cold air from the northwest gradually becomes warmer as it moves southward, luxuriant vegetation thrives year-round. Influenced by the Atlantic Ocean and the Gulf of Mexico, the climate in the corresponding region of the USA is also warm and humid. It has six months or more of frost-free temperatures each year and abundant sunshine to promote vegetation growth. Mean annual temperatures are 16 to 24°C. Precipitation ranges from 1220 to 1625 mm per year, so the region has abundant rainfall. However, the total potential evaporation ranges from 1270 to 2030 mm per year, and considerable areas experience some water

Table 3. Growth of various pine species in different localities of China

Locality	Latitude (N)	Pine species	Age (yr)	Average ht. (m)	Average diam. (cm)
Zhanjiang, Guangdong	21°51′	Slash	6	3.5	5.2
		Loblolly	7	4.5	7.5
		Masson	6	2.5	2.8
Liuzhou, Guangxi	24°	Slash	16	10.8	16.2
		Loblolly	16	10.1	15.1
		Masson	16	11.4	12.2
Minhou, Fujian	26°	Slash	15	11.0	20.0
		Loblolly	15	11.5	27.0
		Sortleaf	15	6.5	19.0
Changsha, Hunan	28°	Slash	7	5.6	8.5
		Loblolly	7	4.1	8.1
		Masson	7	2.5	3.3
		Slash	20	11.3	24.0
		Loblolly	20	11.1	27.0
Fuyan, Zhejiang	30°	Slash	8	5.6	8.2
		Loblolly	8	5.1	8.8
		Masson	8	4.0	5.5
		Slash	25	12.3	18.8
		Loblolly	25	13.7	22.7
Maanshan, Anhui	21°32′	Loblolly	10	6.8	15.1
		Masson	10	5.8	13.3
		Loblolly	24	14.0	22.0
Laoshan, Jiangsu	32°	Slash	14	9.0	14.5
		Masson	14	8.5	8.8
		Slash	23	13.0	22.5
Damashan, Yunnan	25°12′	Slash	10	3.6	5.6
Meng Wang, Yunnan	21°52′	Slash	10	11.3	12.6
		Yunnan	10	1.8	2.1

shortage. In comparison with China's subtropics, the corresponding region in the USA is situated at a higher latitude and covers a much smaller area.

1.2. Topography

In China, the complex topography and the vast extent of territory add to the variety of climatic types. The mountain ranges extend in an east-west direction obstructing the cold air movement to a certain extent. Especially, the Chingling Mountains and the Nanling Mountains exert a great influence on the climate and vegetation in China's subtropics. Altitude and terrain both impose a strong effect on the climate. Pronounced dissimilarities in climate are found in basins and the areas outside, and on the windward and leeward sides of mountains. Because China has complex terrain which is cut by numerous mountain ranges, there is an extremely varied pattern of temperature and precipitation. China's complex topography and varied climate provide for a wealth of plant life. In the USA, the generally low-lying terrain with mountain ranges running in a south-north direction is not able to obstruct the cold air moving southward, and frost damage is a severe problem for vegetation. In fact, the distinctly continental climate of eastern North America, which can have relatively high mean temperatures in winter but then sudden nighttime minima as low as −15°C (absolute minima around −20°C) as far south as northern Georgia to northern Texas, would kill most evergreen broad-leaved plants. This is another limiting factor of the area in the USA corresponding to China's subtropics.

1.3. Soils

The soil types of both regions are similar. In the south subtropics of China, the zonal soil type is lateritic red soil which is equivalent to Typic Rhodudult in the USA. Broad areas of Spodosols, however, also occur on the Atlantic and Gulf Coastal Plain in Florida, and there are some types of Entisols and Histosols. In the mid-subtropics of China there are yellow-red soils, and there is a region of mainly red-yellow podzolic clays (a kind of Ultisols or Hydric Hapludult) in the USA, though sometimes only a thin layer over red clay or limestone. In the north subtropics of China are yellow-red, yellow-brown and yellow-cinnamon soils, and in the USA a kind of Ultisol or Leptic Dystrochrept which is roughly equivalent to the above types. Generally speaking, the soil types of both regions are palaeo-soils, because many parts of China's subtropics were not completely covered by the glaciers, and the south slope of the Appalachians was not covered by glaciers for sixty million years in the USA. However, the soil of the young coastal plains is mostly white sand (Buol 1963).

2. A comparison of natural vegetation

The present actual vegetation is luxuriant and flourishing because heat and water are plentiful in both regions (Walter 1984). However, in China because of long term disturbances, there is a longstanding dominance of much of the regions by successional and/or disclimax pine forest and secondary forests, and the full expression of natural zonal or climax vegetation is not permitted except in some remote areas, preserves and temple forests (Huo 1986).

2.1. South subtropical natural vegetation in China

The subtropical evergreen monsoon rain forest is a typical zonal vegetation and it is also a transitional type between the tropical monsoon rain forest and the mid-subtropical evergreen broad-leaved forest. There is a rich variety of plant species in the forest types, which are mainly composed of Lauraceae, Fagaceae, Myrtaceae, Theaceae, Euphorbiaceae, Rubiaceae, Hamamelidaceae, Fabaceae, Rutaceae, Sterculiaceae, Elaeocarpaceae, Myrsinaceae, Aquifoliaceae, Arecaceae, Annonaceae, Burseraceae, Moraceae, Meliaceae, Sapindaceae, and Symplocaceae. In comparison with the subtropical evergreen broad-leaved forest, the number of Fagaceae species decreases and that of Lauraceae species increases in the subtropical evergreen monsoon rain forest. The common species are *Castanopsis hystrix*, *C. tribuloides*, *C. fordii*, *Cyclobalanopsis glaucoides*, *Lithocarpus deabatus*, *Machilus yunnanensis*, *Cinnamomum burmannii*, *Cryptocarya chinensis*, *C. chingii*, *Schima wallichii*, *S. argentea*, *Altingia chinensis*, *Canarium album*, *C. uraiana*, *C. pimela*, *Exbucklandia tonkinensis*, *Toona ciliata*, *T. microcarpa*, *Dysoxylum binectariforum*, *D. hongkongene*, *Rhodoleia championi*, *Erythrophleum fordii*, *Syzygium hancei*, *Ficus* spp. and *Ormosia* spp., etc. An example is the *Castanopsis chinensis – Cryptocarya chinensis – Cryptocarya coccinna* community situated in Dinghushan, Gaoyao County, Guangdong Province. The canopy consists of three layers of trees with a cover of 80–90%. The height of the canopy top is 30 m and the thickness of this layer is about 10 m, d.b.h. (diameter at breast height) 60–80 cm, composed of trees with umbrella-shaped crowns, rough bark, straight trunk and well-developed buttress roots. Epiphytes are very abundant. The height of the second tree layer is about 10–18 m, and d.b.h. 15–30 cm, composed of trees with smooth bark and rather straight trunks. The height of the third tree layer is 4–10 m and d.b.h. 4–14 cm, composed of trees with irregular crown shape, narrow crown width, crooked trunk and rather smooth bark. The understory of shrubs and saplings is generally 1.5–2 m high, with a cover of 60–70%. The species composition is not diverse, of which the dominant ones are *Indocalamus congiaritus* and *Pseudosasa* sp. The herb understory (herbs and seedlings) is around 20 cm high with a cover of merely 3–5% mainly composed of shade-tolerant ferns. There are plenty of climbers including a certain amount of woody lianas in the community. In addition, mangroves occur discontinuously along the muddy beaches of the region. The communities of mangrove, with a cover of about 70–90% are usually very dense. The height of mangrove is generally 2–5 m and sometimes 1–2 m in a dense bushy form, but the height can reach 10–15 m in a few undisturbed mangroves. On coastal sandy land, the number of plant species is rather limited, i.e. about 400 species. There is a very close relationship between the flora of the local region and that of the south-central peninsula of southeast Asia.

The native vegetation of the Spodosols in Florida, USA, is forest. The dominant trees are *Pinus elliottii*, *P. taeda*, *P. caribaea*, and *P. palustris*, with an understory of saw palmetto (*Serenoa repens*), southern wax myrtle (*Myrica cerifera*), evergreen runner oak (*Quercus minima*), and wiregrass (*Aristida stricta*). The trees occur on nearly level to gently sloping landscapes, elevations of about 0.3 m to 40 m above mean sea level, with a fluctuating groundwater table at or near the surface during its seasonal high. The site index for slash pine ranges from 70 (21 m) to 85 (26 m) at 50 years of age. Loblolly and slash pine make good initial growth but the growth curve levels off at a relatively early age. Vegetation associated with Entisols is largely dependent upon water available to plants. The upland sandy Entisols in Florida are droughty and have native vegetation consisting of scrub oak (*Quercus ilicifolia*), turkey oak (*Q. laevis*), sand post oak (*Q. stellata* var. *margaretta*) some *Pinus palustris*, and other drought-tolerant hardwoods, shrubs, and native grasses. Natural vegetation in the wet area (psammaquents) consists of pond cypress (*Taxodium distichum* var. *nutans*), a scattering of pines, reeds, grasses, cabbage palmettos (*Sabal palmetto*) and other water-tolerant trees, shrubs and grasses. Degree of wetness and topography greatly influence the natural vegetation. However, the subtropical evergreen monsoon rain forest is rare in Florida because of the low-lying mosaic topography of white sand, smaller areas of silt or clay, plus limestone, much of which is too wet (too low) to permit full expression of a zonal vegetation. The growth of mangroves here seems better than that of mangroves in the south subtropics of China. Florida mangroves can reach a height of 18 m and would be taller, if not for the influence of hurricanes.

2.2. Mid-subtropical natural vegetation in China

The subtropical evergreen broad-leaved forest is a typical zonal vegetation or the climax forest community in the mid-subtropics. Fagaceae is the dominant family, especially in mountain regions. Dominant genera of the family are *Castanopsis*, *Lithocarpus*, *Cyclobalanopsis*, and *Quercus*. The other co-dominant families are *Lauraceae*, *Theaceae*, *Hamamelidaceae*, *Aquifoliaceae*, *Araliaceae*, *Styracaceae*, *Myrsinaceae*, and *Magnoliaceae*. The evergreen broad-leaved species of the north subtropics also occur in the mid-subtropics, with additional species of *Castanopsis eyrei*, *C. carlesii*, *C. fargesii*, *C. tibetana*, *C. jucunda*, *Cyclobalanopsis myrsinaefolia*, *C. stewardiana*, *C. nubium*, *Lithocarpus harlandii*, *Machilus leptophylla*, *Litsea coreana* var. *sinensis*, *L. cuboba*, *L. elongata*, *Manglietia fordiana*, *Michelia figo*, *M. maudiae*, *Schima superba*, *Camellia cuspidata*, *C. fraterna*, *Cleyera japonica*, *Daphniphyllum macropodum*, *Dendropanax chevabieri*, *Ilex latifolia*, *Prunus spinulosa*, *Symplocos stellaris*, and *Myrica rubra*. The most common and important deciduous broad-leaved species are *Castanea henryi*, *Quercus chenii*, *Fagus longipetiolata*, *Populus adenopoda*, *Betula luminifera*, *Cyclocarya paliurus*, *Nyssa sinensis*, *Alniphyllum fortunei*, *Pterostyrax corymbosa*, *Emmenopterys henryi*, *Tapiscia sinensis*, *Toxicodendron suceedaneum*, *Idesia polycarpa*, *Meliosma myriantha* and *M. flexuosa*. The structure of the plant community is generally divided into four stories – two of trees, one of shrubs and saplings, and one of herbs and seedlings, as in the *Castanopsis fargesii – Symingtonia tonkinensis – Altingia chinensis* community in the southern part. There are two stories of trees. The height of the first one is generally 14–18 m and the maximum 22 m, and d.b.h. 15–30 cm, with a cover of 60–70%. The second tree story is 7–10 m. The understory height of shrubs and saplings is 1–3 m. There are only a few species in the understory of herbs and seedlings, with a cover of 10% or so. *Loropetalum chinense* is the dominant species in understories of most communities in the mid-tropics. It is extensively distributed in hilly and mountainous areas, and is a marked characteristic which can be used as an index to identify the vegetation of mid-subtropics from that of south subtropics. In the central and northern parts of the zone, the average height of mature forests is generally 15–20 m. In the low mountains and hill areas below 500 m in altitude, however, the secondary *Castanopsis sclerophylla* and *Lithocarpus glaber* forests are common types, and the average height of these mature forests is usually 12–15 m. An example of this secondary young growth is the *Castanopsis sclerophylla – Lithocarpus glaber* community, a successional community following pioneer Masson pine (*Pinus massoniana*) forests and intolerant deciduous broad-leaved forests, such as the *Albizzia kalkora – Liquidambar formosana* community, as a result of conservation for vegetation recovery. Nevertheless, this community has high productivity and an effective ecological function. There are many seedlings and saplings of the dominant tree species under the canopy. The humus soil layer of the forest ecosystem is increasing gradually. Some evergreen broad-leaved forests at higher altitudes, for example *Cyclobalanopsis glauca* or *Castanopsis tibetana* forests, will gradually appear due to the process of vegetation succession. The successional sequence for plant communities in the mid-subtropics is shown in Fig. 4.

The corresponding region of the USA, or warm temperate zone, is characterized by evergreen vegetation. Evergreen broad-leaved trees, pines and palms are important components of this region. As proposed by Box & Fujiwara (1988), the evergreen broad-leaved forest is the zonal vegetation but is present now only in small patches along the Atlantic and Gulf coasts.

Of the evergreen broad-leaved forest, *Quercus* is the most important tree

(Deviation Climax)	(Normal Climax)	(Deviation Climax)
Bamboo Plantation ← →	Evergreen Broad-leaved Forest	→ ← Chinese fir Plantation
	⇅	
	Deciduous Broad-leaved Forest or Pine Mixed forest	
	⇅	
	Pine Forest	
	⇅	
	Scrub	
	⇅	
	Barren	

Fig. 4. The successional sequence of plant communities in China's mid-subtropical mountainous region.

genus, with six species and many hybrids, *Quercus virginiana*, *Q. geminata*, *Q. fusiformis*, *Q. laurifolia*, *Q. hemisphaerica*, and *Q. myrtifolia*. Other important species include *Persea* spp., *Magnolia virginiana*, *Gordonia lasianthus*, *Magnolia grandiflora*, *Ilex* spp., *Vaccinium arboreum*, *Lyonia ferruginea*, *Lyonia lucida*, *Myrica cerifera*, *Symplocos tinctoria*, *Bumelia lanuginosa*, *B. texax*, *Cliftonia monophylla*, and *Cyrilla racemiflora*. The pines of this region are mainly *Pinus elliottii*, *P. taeda*, *P. glabra*, *P. serotina*, *P. palustris*, and *P. echinata*. The palms are *Sabal palmetto*, *S. minor*, *S. etonia* and *Serenoa repens* (Box 1986, Box & Fujiwara 1988.)

2.3. Natural vegetation in the north subtropics of China

Mixed evergreen and deciduous broad-leaved forest is the natural zonal vegetation type on the yellow-brown soils of mountainous regions and yellow-cinnamon soils of hilly areas. It is a transitional vegetation type between the evergreen broad-leaved forest and the deciduous broad-leaved forest. Because of lower temperatures, the warmth-loving species of evergreen broad-leaved forests are replaced by cold resistant species mixed with deciduous broad-leaved trees. The trees of the mixed forests belong mainly to the *Fagaceae*, especially evergreen trees of genera such as *Castanopsis*, *Cyclobalanopsis*, and *Lithocarpus*, which occur in the northern subtropics. The dominant members of the eastern region include *Castanopsis sclerophylla*, *C. eyrei*, *Cyclobalanopsis glauca*, *C. gracilis*, *Lithocarpus glaber*, and are often intermingled with humidity-loving trees including *Fagus lucida*, *F. engleriana*, and *F. longipetiolata* in montane forests, which are not seen in temperate zones of China (Hou 1983, Fang 1988.) The other families are *Aquifoliaceae*, *Celastraceae*, *Oleaceae*, *Aceraceae*, *Betulaceae*, *Hamamelidaceae*, *Styraceae*,

Araliaceae, *Magnoliaceae*, *Rosaceae*, *Ulmaceae*, and *Tiliaceae*, *Lauraceae*, *Theaceae*, *Ericaceae*, *Symplocaceae*, and *Anacardiaceae*. The other evergreen broad-leaved species are *Phoebe sheareri*, *P. neurantha*, *P. hunanensis*, *P. chekiangensis*, *Machilus thunbergii*, *M. ichanpensis*, *Lindera megaphylla*, *Cinnamomum camphora*, *Ilex cornuta*, *Ligustrum lucidum*, *Photinia serrulata*, and *Osmanthus fragrans*. The deciduous woody component species are *Quercus acutissima*, *Q. variabilis*, *Q. aliena* var. *acuteserrata*, *Q. glandulifera*, *Q. fabri*, *Liquidambar formosana*, *Pistacia chinensis*, *Zelkova schneideriana*, *Loropetalum chinense*, *Pterocarya strobilaceae*, *Koelreuteria bipinnata*, *Albizia macrophylla*, *Dalbergia hupeana*, *Tilia henryana* var. *subglabra*, *Kalopanax pictus*, *Sassafras tsumni*, *Bothrocarvum controversum*, *Euonymus alatus*, and *Acer* spp. On limestone, the forests consist mainly of the species of *Ulmaceae* such as *Pteroceltis tatarinowii*, *Zelkova schneideriana*, *Celtis bungeana*, *Ulmus parvifolia*, and those of other genera such as *Dalbergia hupeana*, *Pistacia chinensis*, *Cudrania tricuspidata*, *Symplocos paniculata*, *Fortunearia sinensis* and *Diospyros rhombifolia*. It should be noted that the mixed montane forests embrace many characteristic species such as *Cercidiphyllum japonicum* var. *sinense*, *Tetracentron sinense*, *Bretschneidera sinensis*, *Davidia involucrata*, *Camptotheca acuminata*, *Nyssa sienesis*, *Aesculus chinensis*, *A. wilsonii*, *Liriodendron chinense*, and *Emmenopterys henryi*. The community structure of these forests generally consists of three to four layers, and the canopy height is about 15–20 m. It is obvious in Mt. Baimajian of Anhui Province, that the evergreen forest changed to deciduous forest via the mixed evergreen and deciduous forest with the change of humidity from wet to dry.

Native vegetation is predominantly forest on the Ultisols, the main soil type in the corresponding region of the USA. Species of deciduous *Quercus* are the most important in the forest. Other species of *Pinus* and *Carya* are also in the canopy of the forest. Species of genera such as *Cornus*, *Acer*, *Fraxinus*, *Gleditsia*, *Sassafras*, *Liquidambar*, *Diospyros*, *Liriodendron*, *Ulmus*, *Celtis*, *Juniperus*, and *Populus* are significant in the understory of the forest (Little 1983). There are three forest types as follows: subclimax forest of pine, with mixed hardwoods developing in undisturbed sites; oak-hickory-pine complex in well-drained soils derived from deposits of marine origin; and oak-hickory dominating the remainder of the Ultisols of the region. The structure of these forests generally consists of three to four layers and the canopy top is about 20 m. Essentially no evergreen broad-leaved species reach the canopy, and only a few are in the understory of the forest. Therefore, it is proposed there is the potential of introducing some cold-tolerant and drought-resistant evergreen broad-leaved tree species from the north subtropics of China into the corresponding region of the USA in order to enrich the forest resources.

3. Artificial vegetation of the corresponding regions in China and the USA

The local ecological environment is reflected in artificial vegetation or plantations and farming systems to a certain extent. Therefore, in addition to natural vegetation, a comparison of artificial vegetation is useful (Huo 1987).

3.1. South subtropics of China

Plantations mainly consist of rice fields and other economic crops. Rice can be harvested twice a year, sweet potato and wheat can be grown in winter, but the yield of the latter is very low. Sugar cane, peanut, jute, mulberry, tobacco and various kinds of fruit trees are in the majority of economic crops. The species and cultivated varieties of fruit trees are very abundant, of which large areas are covered by mainly tropical species such as *Litchi chinensis*, *Euphorbia longan*, *Poidium quajava*, *Musa paradisica* var. *sopientum*, *M. nana*, *Carica papaya*, *Ananas comosus*, *Clausena lansium*, *Canarium album*, *C. pimula*, *Dracontomelum dao*, *Syzygium jambos*, *Annona squamosa* and *Averhoa carambola*. The production of oranges also covers large areas. Deciduous fruit trees such as pears and plums are in considerable quantity, too. Economic forests such as *Livistona chinensis*, *Pseudosasa amabilis*, *Cinnamomum cassia* and *Illicium verum* are characteristic in the zone.

On Spodosols of Florida, USA, the economic vegetation is predominantly pine forests. Many stands of pines on these soils are managed for pulpwood. Spodosols with hyperthermic soil temperatures (south and central Florida) are used mostly for woodland grazing. Some areas of Spodosols are used for improved pasture and row crops. Adequate water control measures are necessary for successful crop production. Areas in southern Florida with hyperthermic soil temperature are used for citrus and winter vegetables, in addition to improved pastures.

3.2. Mid-subtropics of China

Rice is harvested twice a year. Other agricultural crops are sweet potato, corn, beans, winter wheat, tobacco, ramie, peanut, sugar cane, rape, *Polygonum fagopyrum*, *Canna edulis* and various kinds of winter sown green manure crops such as *Astragalus sinicus*, *Raphanus raphanistrum*, *Brassica* spp. and *Medicago* spp. The fruit trees of mid-subtropical species and varieties are mainly evergreen oranges (such as *Citrus reticulata*, *C. sinensis*, *Fortunella margarita*), *Myrica rubra* and *Eriobotrya japonica*. The deciduous fruit trees are peach, plum, Chinese chestnut and Chinese date. The species of timber forests, such as *Cunninghamia lanceolata*, *Pinus massoniana*, *Phyllostachys pubescens*, and forests of tung-oil tree (*Aleurites fordii*) and tea-oil tree (*Camellia oleifera*) are very important economically and cover a large area of the zone (Huo 1986). The seeds of tea-oil trees are pressed to provide a staple edible oil used by southerners. The products of tung-oil and lacquer trees (*Rhus verniciflua*) make excellent wood finishes. Tallow tree seeds (*Sapium sebiferum*) are used to make synthetic oils. In recent years, large areas have been seeded with pine seeds from the air, with good results. As a matter of fact, *Pinus massoniana*, *P. elliottii*, *P. taeda*, *Rhus verniciflua*, and *Sapium sebiferum* are widely distributed in the whole subtropics, especially the eastern part of China.

The long frost-free period, up to 260 days in the corresponding region of the USA, permits the production of a wide range of field, forage, truck, and horticultural crops. However, various pine and hardwood plantations have been established from abandoned crop and pasture lands of Ultisol soils. Bahia grass, coastal and bermuda grass are dominant forage and pasture species along the southern part of the Ultisol belt and on Alfisol soils such as Udalfs and Aqualfs. Bright-leaf tobacco and cigar wrapper tobacco are grown in north Florida and south Georgia. Sweet potato is a major crop in southwestern Alabama and pecans are grown extensively from south Georgia to east Texas. The dry sandy Entisol soils are used for citrus where the climate is favorable. Rice is a common crop on the level topography of Alfisol soils such as Aqualfs of southwestern Louisiana and southeastern Texas.

3.3. North subtropics of China

The artificial vegetation of this zone is a reflection of transitional characteristics. In the southernmost part, the common farming system is single or double rice, plus winter wheat, rapeseed or a green manure crop. There are plantations of ramie, mulberry, and orange. Masson pine forests occur widely in the southern areas of the zone. In the central part, the common farming system is summer rice plus winter wheat or rapeseed, two crops per year. There are plantations of tea, pomegranate, peach, pear, and loquat, an evergreen fruit tree. In the northern part, the common farming system is winter wheat plus corn, millet, soybeans or sweet potato, two crops per year. There are also plantations of deciduous fruit trees such as date, apple and pear.

The corresponding area of the USA is rather extensive, including the northern parts of Georgia, Alabama, and Mississippi, the whole area of South Carolina and North Carolina, and most of Virginia. Many of the areas now in cultivation have been in continuous use for row crops for more than 100 years and some along the Atlantic coast for 200 years or more. During the past 40 to 60 years, vast areas of the less productive as well as some of the most productive Ultisol soils have been abandoned for crops and pasture land use and have reverted to forest. Tree farms on which various pine and hardwood plantations have been established, are extensive in some areas, especially loblolly and slash pines. These farms are quite evident in the Piedmont and Coastal Plain. Tall fescue grass and white clover are used extensively from North Carolina through the northern part of South Carolina, Georgia, Alabama, and Mississippi, and a wide range of row crops is grown on the Ultisols. Corn, soybeans, small grains and grain sorghum are the dominant crops for the region as a whole. Apples are an important crop in the mountainous areas from north Georgia through Virginia, and peaches are important in central Georgia, South Carolina, and parts of Louisiana. High income crops such as vegetables are usually grown on soils suitable for irrigation.

Slash and loblolly pines, extensively distributed in the southeastern USA (Fig. 3), have been successfully introduced into the whole region particularly the eastern parts of China's subtropics. In the early 1930's, slash (*Pinus elliotii*), loblolly (*P. taeda*), shortleaf (*P. echinata*) and longleaf (*P. palustris*) pines were first introduced into Guangdong and Fujian Provinces. In 1947–48, slash and loblolly pines were extensively introduced to various provinces south of the Yangtze River. In the 1970's, introduction of slash pine from the USA was on the upsurge in the southeastern provinces of China. Up to 1988, the area of slash pine plantations in all of China was more than 500,000 ha. Performance of various pine species is shown in Table 3.

Slash and loblolly pines grow much better than local pines, i.e. *Pinus massoniana*, in southeastern China and *P. yunnanensis* in southwestern China. However, in comparison with data from the USA, productivity and growth performances of slash and loblolly pines in the USA are higher and better than those of China. For example, in Georgia, USA, 20 year old slash pine with a density of 1000 stems/ha on site index 60 have an average height of 16.2 m and diameter of 17.8 cm; 20 year old loblolly pine with a density of 1000 stems/ha on site index 65 have 19 m and 20 cm respectively.

4. Comparison of the potential vegetation of the two areas

The potential vegetation is the natural zonal or climax vegetation in a vegetational region. The clarifying of the distribution of the potential vegetation is important for land use and vegetation science. Generally speaking, the potential vegetation has greater ecological and economic benefits than the actual or successional one (Fang 1988).

4.1. The potential vegetation of China's subtropics

Many systems of vegetational division have been reported, but most of these have been based on the actual and agricultural vegetation in China. However, this kind of vegetation has been greatly impacted by high human population density and the exploitation from a long history of the country. The zonal or climax vegetation still exists in some undisturbed areas such as temple forests, preserves and remote localities. There is no doubt that the subtropical evergreen monsoon rain forest is the potential zonal vegetation type of the Chinese south subtropics, but there are other types for special sites. Evergreen broad-leaved forest in mid-subtropics and mixed evergreen and deciduous forest in north subtropics are the potential vegetation for these regions. Particularly, the mixed evergreen and deciduous broad-leaved forest region connects with the subtropical evergreen broad-leaved forest zone. The northern distribution of the evergreen broad-leaved forest is mainly limited by low temperatures in winter and climatic humidity/aridity. It seems reasonable to regard the mixed evergreen and deciduous broad-leaved forest as the zonal vegetation of the southern half and the deciduous broad-leaved forest as the zonal vegetation of the northern half with drier climatic conditions (Fang 1988). Furthermore, 50–60% of the forested area in the subtropics is occupied by successional pine forests and pine plantations such as *Pinus massoniana* or other subtropical pine species. Therefore, the potential vegetation map is an important theme in China.

4.2. The potential vegetation of the corresponding region in the USA

Most of the above mentioned forest types are well recognized. The problem is with the hypothesized upland evergreen oak forest, which may once have extended over the broad upland area between the Atlantic coast and the Gulf of Mexico but for which almost no representative stands remain. Although the area is presently covered mainly by deciduous forest, pine plantations and successional pine woods, theoretically potential natural vegetation is evergreen broad-leaved forest. Nevertheless, there are three basic types of successional forest which might have evergreen potential. A pine evergreen type occurs on sandy areas and contains mainly *Quercus virginiana*, *Q. geminata*, and sometimes *Persea borbonia* and even *Magnolia virginiana* (higher groundwater) in the understory, all of which are slow growing and may remain shorter than the pines on poorer sites. Such sites may develop into peinotrophic, mixed evergreen woodland (or forests) if left undisturbed. The second type is a pine deciduous successional type which generally occurs in better soils and typically develops, over 50–75 years, into a deciduous forest. The third type is the "southern mixed hardwood" type, which may have some evergreen broad-leaved canopy trees.

Succession of these types into an upland evergreen oak forest depends on various factors. In the southeastern coastal plain, with a somewhat irregular precipitation pattern and recurring droughts, the young sandy soils exacerbate climatic dryness. This factor, in combination with low lying topography and recurring disturbance by late summer hurricanes and/or winter ice storms, catenal and historical factors become especially important in the development of the upland evergreen oak forest. Evergreen broad-leaved forests should not be expected to occur in low lying, saturated soils or on the driest and/or most nutrient poor upland areas (Box & Fujiwara 1988). Even so, evergreen broad-leaved forests, especially with their draping of Spanish moss on spreading live oaks, were a distinctive feature of coastal landscapes, which are preserved in some areas in the form of parks and hunting tracts. Therefore, recurring disturbance has important implications for the concept of potential natural vegetation, which has largely ignored the role of disturbance.

4.3. Plant introduction and plant collections between the two regions

Based upon the analogous principle and analysis of similarity in geological history, ecological environment, flora and geographic distribution, phenological phenomena and growth performances, plant introduction and plant collections between the two regions should be successful. In fact, China has met with great success on introduction of slash and loblolly pines from the southeastern USA to its subtropics. On the other hand, the USA should introduce cold resistant, drought tolerant evergreen broad-leaved tree species such as *Castanopsis sclerophylla* and *Lithocarpus glaber* from the north subtropics and high altitude of mid-subtropics of China to enrich the plant resources and for landscape use.

Theoretically, a botanical garden is a laboratory of phytogeography for plant introduction and plant collections. The establishment of a plant exchange program between corresponding botanical gardens is a practical means of plant introduction between the two regions. For example, The State Botanical Garden of Georgia might exchange plant materials with Nanjing Botanical Garden in Jiangsu Province, Shanghai Botanical Garden, Wuhan Botanical Garden in Hubei Province, and Lushan Botanical Garden, located at a high altitude in Jiangxi Province. Further study on comparative phytogeography is necessary to determine the most promising area of China from which to introduce plant species that would be successful.

Conclusions

1. South Florida is the region corresponding to the south subtropics of China. The southern half of Georgia and northern two-thirds of Florida, plus coastal strips along the Atlantic and Gulf of Mexico is the region corresponding to China's mid-subtropics. The northern part of slash and loblolly pine distribution areas is the region corresponding to the north subtropics of China.
2. The climate and soils are similar between the two regions, but topography is quite different, and the latitudes in the corresponding region of USA are higher (Buol 1973, Chung 1978). Therefore, successful plant exchange between the two regions is highly probable but the plant introduction stations need to be carefully sited.
3. The subtropical evergreen monsoon rain forest, evergreen broad-leaved forest, and mixed evergreen and deciduous forest are the zonal vegetation types or climax forests of south, middle, and north subtropics in China, respectively. In the corresponding region of the USA, the evergreen broad-leaved forests are present, at least now, only as small remnants in narrow strips along the Atlantic and Gulf Coasts.
4. The artificial vegetation and farming systems are reflections of local conditions to a certain extent, therefore, the vegetational regionalization or division should be a combination of natural vegetation and artificial vegetation.
5. Actual vegetation has been greatly impacted by human activities and various disturbances, therefore, clarifying the distribution of potential vegetation is important for land use and for vegetation science because the potential vegetation has largely ignored the role of disturbances.
6. China has had great success with the introduction of slash and loblolly pines from the southeastern USA. The two southeastern U.S. pine species are now distributed in the whole region of China's subtropics. On the other hand, the USA should introduce the cold resistant and drought tolerant evergreen broad-leaved species from China's subtropics to enrich the plant diversity. The establishment of plant exchange programs between corresponding botanical gardens is an important mechanism for plant introduction between the two regions.

References

Box, E. O. 1986. Comparing the natural montane vegetation of eastern Asia and eastern North America. Internat. Symp. Mountain Vegetation, Beijing. Bot. So. China, Beijing:87–94.

Box, E. O. & Fujiwara, K. 1988. Evergreen broad-leaved forests of the southeastern United

States: preliminary description. Bulletin of the Inst. of Environmental Sci. and Tech. Yokohama National Univ. 15:71–93.
Buol, S.W. (ed.), 1973. Soils of the Southern States and Puerto Rico. Southern Cooperative Series Bulletin No. 174.
Chung, C. 1978. An Outline of Chinese Geography. Foreign Language Press, Beijing.
Fang, J. 1988. Climate and distribution of vegetation in China. Dissertation. Osaka City University, Osaka, Japan.
Hou, X. Y. 1983. Vegetation of China with reference to its geographical distribution. Ann. Missouri Bot. Garden. 70:509–548.
Huo, Y. Q. 1986. Studies on the productivity and water conservation capacity of various forest types in the mountainous region of East Hunan, China. Internat. Symp. Mountain Vegetation. Beijing. Bot. Soc. Beijing, China: 159–163.
Huo, Y. Q. 1987. China's two great green walls: A brief report on stabilizing drift sands along coastal areas and in deserts of China. Internat. Conf. on Sand Stabilization along Mediterranean Regions and in Developing Countries, Palermo, Italy. 11 pp.
Little, E. L. 1983. North American trees with relationships in eastern Asia. Ann. Missouri Bot. Garden 70:605–615.
Walter, H. 1979. Vegetation of the Earth in Relation to Climate and the Eco-physiological Conditions, 3rd Ed. Springer-Verlag, New York.
Wu, Z. Y. 1980. China's Vegetation. Academy Sinica Press, Beijing.

7. Evergreen broad-leaved forests of southern South America

RENATE HILDEBRAND-VOGEL[1] & ANDREAS VOGEL[2]
[1]*Bombeck 6, D-4425 Billerbeck, Germany*; [2]*Institut für Geographie, Universität Münster, Robert-Koch-Strasse 26, 44 Münster, Germany*

Abstract

The natural vegetation of western Patagonia (southern Chile and adjacent Argentina) is evergreen "laurel forest." In the north its distribution area borders the region of mediterranean-type sclerophyllous evergreen forest, which is not treated here. Within the laurel-forest area, two major vegetation zones are evident: that of the "Valdivian laurel forest" extending from 38 to about 43° S and the "North Patagonian-Subantarctic laurel forest," reaching as far as the archipelago of Tierra del Fuego at about 56° S.

Aside from the trees, the predominant life forms of these forests are shrubs and woody climbers. Stems are laden with thick coats of cryptogamic epiphytes (Hymenophyllaceae, lichens, mosses). Some forest types have tall-growing bamboo understoreys (*Chusquea*). The trees belong to various families, such as Monimiaceae, Myrtaceae, Cunoniaceae, Proteaceae, etc. as well as Fagaceae.

Species diversity decreases with temperature and length of the vegetative period as one proceeds from north to south.

The natural vegetation ("contact vegetation") adjacent to the laurel forests includes deciduous forests at higher elevations and to the east. (Under normal conditions laurel forest extends throughout the lowest elevations, to the surface of lakes, rivers or sea, and to the large mires in southernmost Patagonia.)

The influence of man has reached all types of laurel forest. Some communities of Valdivian rainforest are nearly gone, others are altered by grazing and introduction of European weeds. Vast areas of North-Patagonian laurel forest were destroyed in the 1930s and 1940s when the woods were burned for land-clearing. Nowadays they are threatened by industrial timber exploitation.

Key words: *Nothofagus*, Valvian laurel forest, western Patagonia

E.O. Box et al. (eds.), Vegetation Science in Forestry, 125–140.

Introduction

In parts of southwestern South America, evergreen broad-leaved forests make up the natural vegetation. These forests, known since Hauman (1916) as "Valdivian forest," can also be characterized as "laurel forests," i.e. extra-tropical mesomorphic evergreen forests with generally glossy broad leaves (see Oberdorfer 1960).

Their distribution area ranges from about 38° S to the southern part of the continent and from the Chilean coastal highland in the west to the eastern slopes of the Andean Cordillera in Argentina (Hueck & Seibert 1981). They occur in all altitudinal belts, from the coast to high montains. Above and to the east they are replaced by summergreen broad-leaved forests mostly of *Nothofagus pumilio* (Hildebrand-Vogel et al. 1990), while in the south they abut the Patagonian mires (see map). In addition to the zonal forest communities they also occur as azonal vegetation types in swamps and flood-plains.

Regional characteristics

Within their range, three main landscape types can be differentiated, occurring in a sequence of 60 km-wide strips:

1. in the west the coastal highland, with altitudes ranging up to 1000 m (actually a low mountain range), which reaches to south of the island Chiloé (43° S);
2. adjacent the so-called longitudinal depression; and
3. in the east the Andean Cordillera, the main ridge of which (about 2000 m high) is topped by numerous volcano peaks up to 3000 m high.

The perhumid, euoceanic climate of the rainforest region is characterized by high annual precipitation (2000–4000 mm) and a maximum during the winter. Episodic dry seasons during the summer are typical in the northern part of the region (see Fig. 1, showing a sequence of temperature-precipitation diagrams). The marked decrease of precipitation from west to east causes rapid vegetation changes: the zone of evergreen broad-leaved forests changes to the treeless Patagonian steppe within a distance of less than 50 km (Seibert 1977). Average annual temperatures range from 12°C in the north to 6°C in the south. The difference between summer and winter mean temperatures is only about 8°C in the north but drops to less than 5°C in the archipelago of southern Patagonia (Pisano 1977). While snow in the northern laurel forests usually does not remain long, it is of some significance in the southern types.

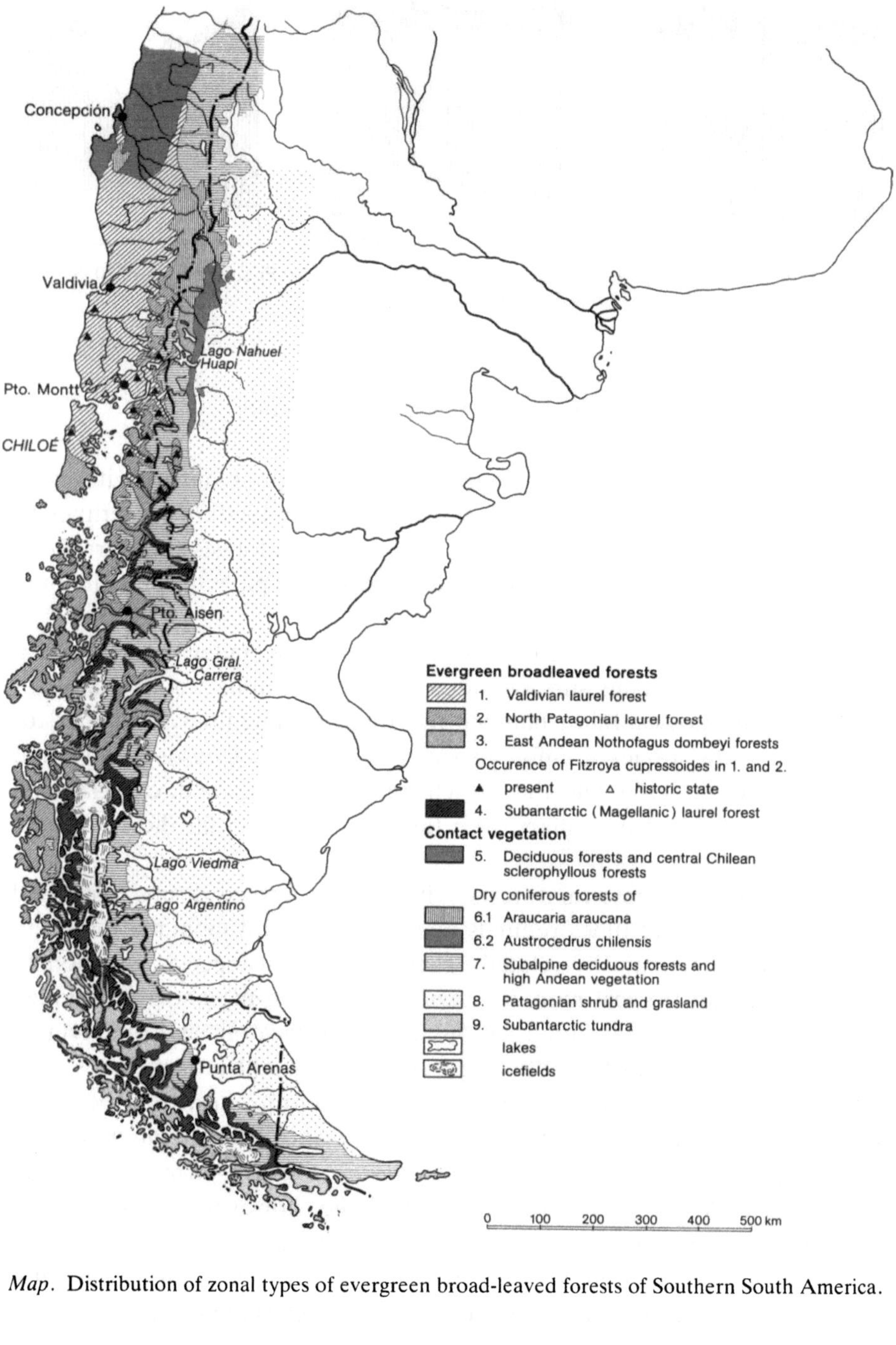

Map. Distribution of zonal types of evergreen broad-leaved forests of Southern South America.

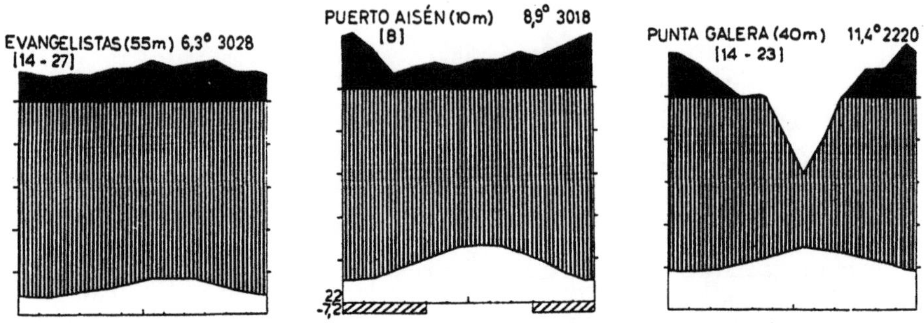

Fig. 1. Temperature-precipitation diagrams: Punta Galera (40° S), Puerto Aisen (45°20′ S), and Faro Evangelistas (52°24′ S). Source: Walter & Lieth (1967).

Flora and structure of the evergreen broad-leaved forests

In the laurel forests of southern Chile woody plants predominate with more than 60% in the spectrum of life forms (Oberdorfer 1960). The permanent woody biomass is of the order of that in tropical rainforests (Walter 1977). These forests also contain a remarkable number of epiphytes and climbing plants.

On the forest floor, which is mostly covered by fallen boles, only a few shade-tolerant plants survive, often only regenerating woody plants, in addition to the mosses. In the bamboo-rich forest communities, the understorey has very few species. Climbing species of the genus *Chusquea*, common in southern Chile, grow into the branchwork of the trees and, especially on steep slopes, form strong interwoven thickets which can fill the volume of the whole forest.

The structure of these rainforests, with their evergreen shrubs, ferns and stands of bamboo, at times reminds one of the tropical montane forests to which they are floristically related. There are various species common to Subantarctic and tropical montane regions, including the above-mentioned *Chusquea*, the large-leaved herbaceous *Gunnera tinctoria*, and the Hymenophyllaceae, as well as the following woody genera which, sometimes even with the same species, are represented in both floral zones: tree species of *Drimys*, *Embothrium*, *Nothofagus*, *Podocarpus* and *Weinmannia* as well as the shrubby *Dacrydium*, *Desfontainea*, *Fuchsia* and *Gaultheria*. The great number of endemic species from southern Chile, however, emphasizes the peculiarity and independence of these forest communities.

The present vegetation is, apart from recent environmental factors, also the result of floral displacements during the last glaciations. In this region the coastal highland and the island Chiloé show no traces of former glaciation and thus may have been refuges for the forest flora. In fact, through pollen analyses (Auer 1958, 1970; Heusser 1974, 1976) it could be shown that various species remigrated and spread eastward, including forest dominants and other major taxa such as *Aextoxicon*, *Eucryphia*, *Drimys* and *Weinmannia*, among others.

Based on plant-sociological studies, Oberdorfer (1960) and Villagran (1980), who also included the descriptions of Reiche (1907), Skottsberg (1916) and Kalela (1941), presented a rough classification of South-Chilean laurel forests, subdividing them into three broad units:

1. the northern "Valdivian laurel forest," Nothofago-Eucryphion, to be found between 38° and 43° S;
2. the "North Patagonian and Subantarctic laurel forest," Nothofago-Winterion, which prevails south of 42–43° S; and
3. the azonal myrtle woods of swamps and river marshes, the distribution area of which reaches to the central-Chilean sclerophyll region.

Within these units there is a broad spectrum of different communities, from which some zonal forest types are described.

Detailed description of the azonal moist forests involving various Myrtaceae is not given. Some details of these can be found in Oberdorfer (1960), Villagran (1980) and Ramirez et al. (1983). Debus (1987) studied the ecology of these forests.

The forests with the most species and life forms can be found within the alliance of the Valdivian laurel forests. The warmest, driest regional sites are characterized by Nothofago-Perseetum, the so-called Roble-Laurel-Lingue forest (*Nothofagus obliqua – Laurelia sempervirens – Persea lingue* forest). It is an evergreen forest with emergent summergreen *Nothofagus obliqua* (Photo 1).

Aextoxicon punctatum, the Tique, also requires heat but, in contrast to the roble, only occurs on sites with high atmospheric humidity and heavy precipitation (e.g. on the Pacific coast, as shown in Photo 2). Under good conditions this large-leaved evergreen species builds almost pure stands characterized by many climbing plants, ferns and epiphytes: *Hydrangea serratifolia* (Hydrangeaceae)[1] and the large-bloomed *Lapageria rosea* (Philesiaceae); the more than man-high *Lophosoria quadripinnata* and various smaller ground ferns as well as phanerogamic epiphytes *Luzuriaga radicans* (Philesiaceae), and *Mitraria coccinea* and *Sarmienta repens* (Gesneriaceae), which paint their red on the thick trunk cover of mosses, lichens and Hymenophyllaceae (for details of the Hymenophyllaceae flora of the Lapagerio-Aextoxiconetum see Ramirez et al. 1976). Tree-tops covered with Bromeliaceae (*Fascicularia bicolor*) and the life form represented by epiphytic chokeweed (*Pseudopanax laetevirens*, Araliaceae) remind one of tropical structures (Photo 3).

[1] Nomenclature according to Marticorena and Quezada (1985).

Photo 1. South-Chilean longitudinal valley, with park-like landscape instead of Valdivian laurel forest.

Photo 2. Pacific coast at 40.5° S, with *Aextoxicon* forest. Dense branching of the Tique specimen in the foreground is induced by salt spray.

Photo 3. Valdivian laurel forest with epiphytic *Fascicularia bicolor*.

In the south and above these forests, growing on warmer sites, we find the Coihue-Ulmo forest (Dombeyo-Eucryphietum), with the striking *Eucryphia cordifolia* (ulmo) among mighty trees of *Nothofagus dombeyi* (coihue,>40 m high, with widely spreading crowns which appear layered). During the summer *Eucryphia* (Eucryphiaceae) is covered with large blossoms of a shining white and thus marks the vertical dispersal area of its community (Photo 4). Even after fires it often remains as decorative individual trees (Photo 5). The structural composition of this forest corresponds to that of the Aextoxiconetum; instead of thermophilic elements like *Lapageria rosea* for example, which is already missing in the southernmost stands of the Tique forest (Villagran 1980), this forest includes many representatives of the North Patagonian forests, even the yew tree-like *Saxegothea conspicua* and *Podocarpus nubigena* (Podocarpaceae).

In the North Patagonian laurel forest (Nothofago-Winterion), the coihue remains predominant. In addition, there are the similar *Nothofagus nitida* as well as *Weinmannia trichosperma* (tineo, Cunoniaceae) and *Laurelia philippiana* (tepa, Monimiaceae), after which the Laurelio-Weinmannietum is named, the most significant community in terms of area covered (Photo 6). Of importance in the undergrowth, besides the above mentioned Podocarpaceae, are the canelo (*Drimys winteri*, Winteraceae) and several Myrtaceae. Typical elements of the shrub layer are the holly-leaved *Desfontainea spinosa* (Desfontainiaceae, Photo 7), *Maytenus magellanica* (Celastraceae) and

Photo 4. Dombeyo-Eucryphietum at Fjordo Comauo (42–42.5° S). Near its southern boundary, *Eucryphia* is restricted to a narrow belt just above sea level.

Blechnum magellanicum, which displays large rosettes (up to 2 m wide) of a shiny light-green at the end of a short trunk. Among the climbers and epiphytes in these woods are also magnificently blooming forms like *Asteranthera ovata* (Gesneriaceae), *Philesia magellanica* (Photo 8) and two species of *Luzuriaga* (Philesiaceae) and *Campsidium valdivianum* (Bignoniaceae). Here there are also many species and forms of cryptogamic epiphytes. (A table compiled from 14 relevés by Oberdorfer (1960) lists more than a dozen representatives of the Hymenophyllaceae.) Further research on these synusiae will show that, besides ubiquitous species, others also occur which are closely associated with specific forest communities (for the Pteridophyte flora, see Godoy & Figueroa 1989).

Ancient Cupressaceae occur on climatically and edaphically unfavorable

Photo 5. Eucryphia is often spared from clearing.

sites in the laurel forests. *Fitzroya cupressoides* (alerce) develops tall montane forests on steep slopes as well as on swampy flatlands. Some individuals reach 40 m high and are several thousand years old (Photo 9) (Schmithüsen 1960, Golte 1974). Our relevés from the national park "Alerce Andino" (41.5° S) illustrate, in the same way as those from the coastal Cordillera and Andean Vicente Pérez Rosales national park (40–41° S), by Oberdorfer (1960), Ramirez & Riveros (1975) and Villagran (1980), that between the mighty columnar trunks of the Alerce (dbh of up to 200 cm!) there grow broad-leaved forests of the Nothofago-Winterion. Our studies showed that the *Fitzroya* forest on steep slopes (30–40° S) occurs on organic soil with a peaty horizon about 1 m thick over the outcropping rock. Thus, the remarkably different sites, swampy flatlands on the one hand and steep slopes on the other, are modified greatly, and the clear floristic similarity of the Alerce forests of these two habitat types can be understood. A third type of laurel-

Photo 6. Laurelio-Weinmannietum at about 43° S. One can appreciate the size of the trees by comparing with the person at bottom, in front of the water-polished rock.

Photo 7. Desfontainia spinosa (Desfontainiaceae).

Photo 8. Philesia magellanica (Philesiaceae).

Photo 9. Millennial *Fitzroya cupressoides* forest at Alerce Andino National Park.

Photo 10. Pilgerodendron uvifera forest with *Drimys winteri* and Myrtaceae, at Golfo Elefantes (46.5° S).

conifer forest involving *Fitzroya*, recently no longer to be studied, occured on the wide lowland ñadis south of Lake Llanquihue (41.5° S). These were cut completely in the course of the colonization in the middle of the last century, and the devastated woodland remains today covered with undemanding shrubs (Hildebrand 1983, 1984).

Even more confined to moist soils than the Alerce is *Pilgerodendron uviferum*, which reaches northward to 41° S with some outposts but which occurs mainly south of 42° S, in the Patagonian lowland and archipelago. This species, which is mostly associated with *Drimys winteri* and Myrtaceae, is considerably smaller than the alerce; it forms coastal forests (as shown in Photo 10) and borders the mires of the Patagonian lowlands. Called "Ciprés de las Guaitecas," this species is common up to the southwestern part of the Tierra del Fuego archipelago. In the herb layer of these Ciprés forests there are species even more indicative of acidity and wet soil than in the Alercales.

On less moist soil in the Patagonian mountain country there are laurel forests consisting of *Nothofagus betuloides* (Coihue de Magallanes). The first signs of this "subantarctic laurel forest" are at 41° S; at about 43° S it already forms its own altitudinal belt at 600 m, reaching the upper forest limit under good conditions (Photo 11). Pure stands of *Nothofagus betuloides* in the submontane belt are to be found south of 48°S (Reiche 1907). In South

Photo 11. Nothofagus betuloides forest forming the upper altitudinal belt at Queulat National Park (46.5° S).

Patagonia, where the only evergreen beech is *Nothofagus betuloides*, laurel forests occur from the coast to the timberline (Pisano 1977), particularly on well drained slopes which are not influenced by cold air but only in areas of higher precipitation (the drier and/or colder sites are covered by decidious Lenga forests, *cf.* Hildebrand-Vogel et al. 1990). In comparison to northern types they are considerably impoverished: the number of species (which usually lies above 30) is reduced to fewer than 20. Even under optimum conditions the forest is not higher than 20 m, and often it is visibly influenced by the wind. On the islands south and west of the Isla Grande (de Tierra del Fuego), the laurel forest is finally reduced to a krummholz on wind-exposed sites (Dollenz 1981) and at timberline (about 600 m above sealevel, see Photo 12).

Photo 12. Nothofagus betuloides krummholz at Isla Navarino (55° S).

Anthropogenic transformation of the laurel forests

It has generally been shown that forests growing under favorable climatic conditions, in flat only slightly sloping terrain, are the first to be influenced by civilization. Thus, the Roble-Laurel-Lingue forests were nearly completely transformed into a park-like landscape. Numerous old forest trees suffer from dry crowns due to lack of the necessary, more humid microclimate of a closed forest. These still give historic evidence of the former natural forest vegetation on the current pasture areas and farmland. Especially in nearly inaccesible areas, as on the Pacific coast of Chiloé, beautiful Tique forests have survived. In those areas influenced by modern civilization, e.g. the shores of the pre-Andean lakes, traces of deforestation are visible without exception.

In modern times even the North Patagonian laurel forest has become the victim of an unsparing exploitation through clearing and deforestation, the consequences of road-building through the North Patagonian forest region during the last decade. Severely affected are, in particular, the Alerce forests, the exploitation of which continues despite the legal protection of *Fitzroya cupressoides*. Within the area of Aisén/Coihaique the forests fell victim to immense fires set by the colonists in the 1930s and 1940s under pressures for deforestation at that time.

Present industrial exploitation of the forests involves chipping the forestry production of entire mountain ranges and transporting it to wood-demanding industries overseas. This can be considered a further escalation of deforestation history.

Today there are already enough devastated tracts of land to give evidence that the Patagonian laurel forests are a fragile system and that their destruction leads to irreversible local transformations. Under the present precipitation regime, erosion of the delicate soils progresses faster than any possible natural reforestation.

References

Auer, V. 1958. The Pleistocene of Fuego-Patagonia. Part II: The history of the flora and vegetation. Ann. Acad. Sci. Fenn., Ser. A, III 50:1–239. Suomalaisen Tiedeakatemian Toimituksia, Helsinki.

Auer, V. 1970. The Pleistocene of Fuego-Patagonia. Part V: Quarternary problems of southern South America. Ann. Acad. Sci. Fenn., Ser. A, III 100:1–194.

Debus, R. 1987. Untersuchungen zum Wasserhaushalt von *Myrceugenia exsucca* und *Temu divaricatum* in Relation zur Morphologie und Anatomie der Wurzel an Überflutungsstandorten. Diss. Botanicae, 100:1–151. J. Cramer, Berlin/Stuttgart.

Dollenz, A. 1981. Estudios fitosociologicos en el Archipielago Cabo de Hornos. II. Relevamientos en la Isla Hornos. Anales del Instituto de la Patagonia 12:173–182. Instituto de la Patagonia, Punta Arenas.

Godoy, R. & Figueroa, H. 1989. Composition and distribution of the pteridophyte flora of continental and insular Chile. Nowa Hedwigia 48:437–453.

Golte, W. 1974. Ökophysiologische und physiogenetische Grundlagen der Verbreitung der Coniferen auf der Erde; dargestellt am Beispiel der Alerce (*Fitzroya cupressoides*) in den südlichen Anden. Erdkunde, Archiv für wissenschftl. Geographie (Bonn), 28(2).

Hauman, L. 1916. La Foret Valdivienne et ses Limites. Notes de Géographie Botanique. Trabajos del Instituto de Botanica y Farmacología. Buenos Aires. 34 pp.

Heusser, C. J. 1974. Vegetation and climate of southern Chilean Lake District during and since the last Interglaciation. Quaternary Research 4:290–315.

Heusser, C. J. 1976. Palynology and depositional environment of the Río Ignao nonglacial deposit, Province of Valdivia, Chile. Quaternary Research 6(2):273–279.

Hildebrand, R. 1983. Die Vegetation der Tieflandsgebüsche des süd-chilenischen Lorbeerwaldgebiets unter besonderer Berücksichtigung der Neophytenproblematik. Phytocoenologia 11(2):145–223.

Hildebrand-Vogel, R. 1984. Acerca de la vegetacion de los matorrales de las tierras bajas en la region del bosque laurifolio valdiviano en el sur de Chile. Phytocoenologia 12(2/3):251–259.

Hildebrand-Vogel, R., Godoy, R. & Vogel, A. 1990. Subantarctic-Andean *Nothofagus pumilio* forests. Vegetatio 89:55–68.

Hueck, K. & Seibert, P. 1981. Vegetationskarte von Süd-amerika. Vegetationsmonographien der einzelnen Großräume, vol. IIa. Gustav-Fischer-Verlag, Stuttgart. 90 pp.

Kalela, E. K. 1941. Über die Holzarten und die durch die klima-tischen Verhältnisse verursachten Holzartenwechsel in den Wäldern Ostpatagoniens. Ann. Acad. Sci. Fenn., Ser. A, IV 2:1–151. Suomalaisen Tiedeakatemian Toimituksia, Helsinki.

Marticorena, C. & Quezada, M. 1985. Catálogo de la flora vascular de Chile. Gayana Bot. 42(1–2):1–157.

Oberdorfer, E. 1960. Pflanzensoziologische Studien in Chile. Ein Vergleich mit Europa. Flora et Vegetatio Mundi 2:1–208.

Pisano, E. 1977. Fitogeografia de Fuego-Patagonia Chilena. I. Comunidades vegetales entre las latitudes 52 y 56° S. Anales del Instituto de la Patagonia, 8:121–250. Inst. de la Patagonia. Punta Arenas.

Ramirez, C., Ferriere, F. & Figueroa, H. 1983. Estudio fitosociológico de los bosques pantanosos templados del sur de Chile. Revista Chilena de Historia Natural 56:11–26.

Ramirez, C. & Riveros, M. 1975. Los alerzales de Cordillera Pelada: Flora y Fitosociología. Medio Ambiente 1(1):3–13.

Ramirez, C., Steubing, L. & Alberdi, M. 1976. La Familia Hymenophyllaceae (Pteridophyta) en el Fundo San Martín, Valdivia, Chile. I. Taxonomía y ecología. Medio Ambiente 2(1):21–28.

Reiche, K. 1907. Grundzüge der Pflanzenverbreitung in Chile. In: Engler, A. & Drude, O. (eds.), Vegetation der Erde, vol. 8, pp. 1–374. Verlag Wilhelm Engelmann, Leipzig.

Schmithüsen, J. 1960. Die Nadelhölzer in den Waldgesellschaften der südlichen Anden. Vegetatio 9:313–327.

Seibert, P. 1977. Klima- und Vegetationsgefälle in der argentinischen Südkordillere. In: Tüxen, R. (ed.), Ber. Internat. Symposien der Internat. Vereinigg. für Vegetationskunde: Vegetation und Klima, pp. 513–515.

Skottsberg, C. 1916. Die Vegetationsverhältnisse längs der Cordillera de los Andes S 41° S. Br. Svensk. Vet. Acad. Handl. 56(5):1–366.

Villagran, C. 1980. Vegetationsgeschichtliche und pflanzen-soziologische Untersuchungen im Vicente Perez Rosales Nationalpark (Chile). Diss. Botanicae 54:1–165. J. Cramer, Vaduz.

Walter, H. 1977. Vegetationszonen und Klima. Verlag Eugen Ulmer, Stuttgart. 309 pp.

Walter, H. & Lieth, H. 1967. Klimadiagramm-Weltatlas. Jena.

8. Historical and ecological development of the evergreen broad-leaved forests of Australia

J. S. BEARD
6 Fraser Road, Applecross, W.A., Australia

Abstract

Modern Australian forests are descended from an ancestral rain forest widespread at the end of the Cretaceous, in which southern conifers (Podocarpaceae, Araucariaceae), *Nothofagus* and other species of tropical and subtropical affinity were dominant. Climatic deterioration during the Tertiary led to retreat of this flora to refugial areas on the east coast of the continent. It was replaced by new forms which were adapted to seasonal aridity, poor exhausted soils and finally fire. Orthophyll forest was replaced by sclerophyll, at first dominated by *Casuarina* with only a small admixture of *Eucalyptus*. Study of fossil pollen profiles shows that at a particular stage *Eucalyptus* replaced *Casuarina* in dominance, accompanied by a marked increase in charcoal. Thus, initially, sclerophyll forests developed in response to nutrient deficiency. Later their flora was further selected by fire. There seems good reason to attribute this change to the activities of early man. Fire today maintains the boundaries between orthophyll and sclerophyll forests.

Key words: Australia, fire ecology, orthophyll forest, sclerophyll forest, southern conifers.

Historical aspects

Generally speaking, all Australian forests are evergreen and with very minor exceptions broad-leaved. The summergreen, winter-deciduous forests and needle-leaved forests to which everyone is accustomed in the Northern Hemisphere, do not exist in Australia and hardly at all in the Southern Hemisphere as a whole. This is due basically to the much higher proportion of ocean in the Southern Hemisphere, which has a moderating effect on temperature extremes and affects rainfall patterns. These influences are not only impor-

E.O. Box et al. (eds.), Vegetation Science in Forestry, 141–150.

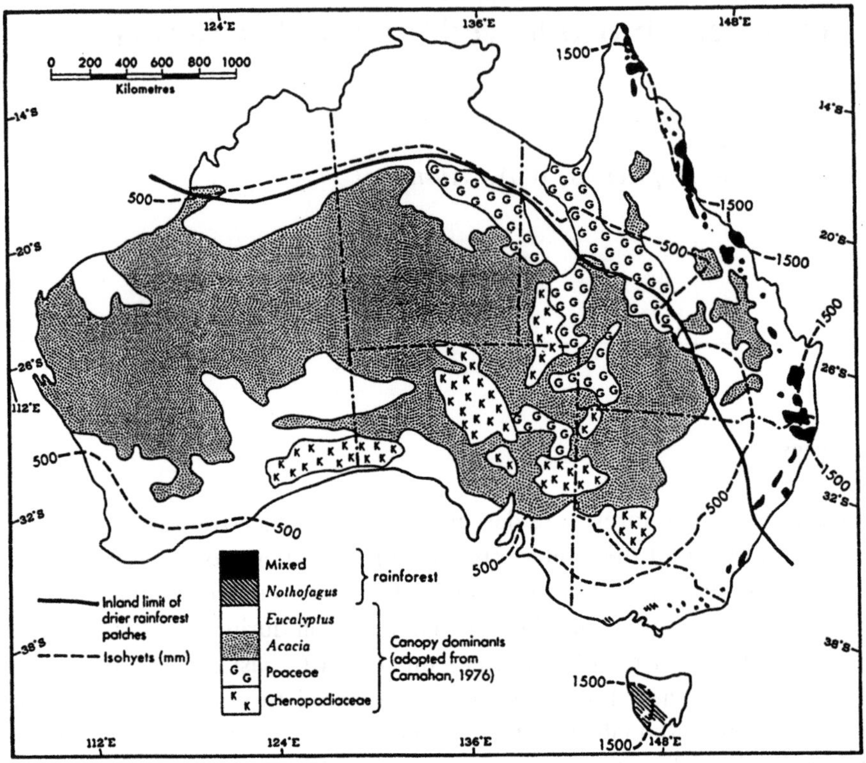

Fig. 1. Present day distribution of major vegetation types in Australia (from Sluiter & Kershaw 1982).

tant today but have been so during the evolution of modern vegetation in the Cenozoic.

Figure 1 shows a highly simplified vegetation map of Australia, from Sluiter & Kershaw 1982. Rain forests of two types – Mixed, coloured black, *Nothofagus* striped – are restricted to small disjunct patches along the eastern seaboard and in Tasmania. At the other extreme communities dominated by *Acacia*, Poaceae and Chenopodiaceae belong to the Arid Zone from which forests are absent. In the rest of the map, the areas left blank or uncoloured are *Eucalyptus*-dominated, but only a small proportion of these communities can be classed as forests. Across the whole of the north the eucalypts are dominants in savannas as scattered trees. These have no economic value as timber or other products and cannot be classed as forests. Eucalypts are generally present in the Arid Zone also as scattered small trees, but these also do not constitute forests. In the southern half of Australia eucalypts may form forests in both the southwest and southeast where rainfall exceeds about 800 mm per annum. Under drier conditions they occur in more open stands of shorter, more branchy trees which are of little economic value and are classed as woodlands rather than forests.

The restricted distribution of forests in Australia has not always been the general pattern. The fossil record for the Early Tertiary shows somewhat consistently throughout the continent a flora whose modern relatives constitute the temperate to sub-tropical relict rain forests in southeastern Australia as well as in New Zealand, New Caledonia and New Guinea. This Tertiary fossil flora is composed of mingled angiosperm and gymnosperm elements, angiosperms newly evolved during the Cretaceous associating with descendants of Mesozoic conifers. There is an abundance of southern beech (genus *Nothofagus*, whose pollen is mostly of the *N. brassii* type no longer found in Australia) and conifers of the families Podocarpaceae and Araucariaceae (i.e. *Dacrydium*, *Podocarpus*, and *Araucaria*, which seem to have been the forest dominants). It should be noted that these conifers are broad-leaved rain forest trees, a totally different group morphologically from the common conifers of the Northern Hemisphere. Much of the angiosperm pollen found in association with these is not referable to modern genera, but it is possible to recognise *Cupaniae* (Sapindaceae), *Anacolosa* (Olacaceae), *Santalum* and Proteaceae, which indicate a tropical to subtropical element.

At that time Australia lay in higher latitudes than now, attached to Antarctica from which it did not part company until the middle Eocene. The southern coast of Australia must have been situated in the vicinity of the mid-oceanic ridge, which has a modern latitude of 50° S. In spite of this, data on the temperature of the sea surrounding Australia in the Early Tertiary (Cockbain 1968) indicate that it was many degrees warmer than now, supporting the evidence on the nature of the flora. There is a general uniformity in the fossil record throughout Australia during the earlier stages of the Tertiary, showing dominance of these *Nothofagus*-podocarp warm-temperate rain forests and indicating a generally uniform humid and non-seasonal climate. This flora also remained in place for a remarkably long period, as it appears consistently at all depositional sites from the very beginning of the Tertiary until at least the middle Miocene, a period of 50 million years.

The uniformity of the fossil assemblages of the Tertiary led at first to the concept of a uniform rain-forest flora throughout the continent. It is increasingly recognised, however, that landscape diversity must have existed then as now, with different landscape components carrying different vegetation (Sluiter & Kershaw 1982). The largest rain forest area of the present world, the Amazon Basin, is not uniform; there are distinct differences between swamp forests and those of dry ground, between these and the vegetation of rocky outcrops, deep sands and seasonally waterlogged sites. It is assumed therefore that the ancestors of the modern Australian sclerophyll flora began to appear early on certain adverse sites, but have left little trace in the record because it is biased towards humid environments. Remains of both *Banksia* and *Casuarina* have been identified as early as the Eocene. A copious macrofossil flora has been found near Woomera in South Australia which includes a diversity of fruits similar to those of *Eucalyptus* and other types consistent with *Leptospermum*, *Calothamnus*, *Melaleuca-Callistemon* and *Angophora* (Ambrose et al. 1979). Unfortunately, the deposit cannot be precisely dated, and we have to accept an inferred age of Miocene. A more certain middle Miocene date is available for fruits and leaves of *Eucalyptus* at the Warumbungles in New South Wales (Holmes et al. 1982). The brown coal deposits of the Latrobe Valley in Victoria extend from the Eocene to the Middle Miocene and are biased in their fossil record towards vegetation of a lacustrine environment. None the less, an increase in sclerophylly from the early to the middle Miocene is detectable, with appearance of taxa such as *Banksia* and Casuarinaceae (Sluiter & Kershaw 1982).

Australia, in line with most parts of the world, experienced deteriorating climatic conditions from the Middle Miocene onwards (Bowler 1976, Kemp 1978). Factors contributing to lower temperatures and precipitation included the continuing northward movement of Australia into drier latitudes and the development of the Antarctic icesheet, the presence of which was instrumental in reducing surface ocean temperatures and intensifying atmospheric and oceanic circulation patterns. Maximum aridity for the Late Cenozoic may have been achieved by the Late Pliocene or Early Pleistocene (Bowler 1976, Singh et al. 1981), coinciding with the development of the Northern Hemisphere ice sheets (Shackleton & Kennett 1975). In response to these changes in climate, available evidence suggests that there was a gradual replacement of evergreen closed forest, which had dominated the landscape throughout much of the Tertiary, by more xerophytic vegetation (Sluiter & Kershaw 1982).

Several late-Tertiary sites illustrate the retreat of evergreen rain forest to coastal areas and its replacement by xerophytic communities, as effective rainfall decreased in the Late Miocene and Pliocene. In contrast to today, eucalypt communities had a very restricted distribution while drier rain forests were much more extensive, particularly in the northern and eastern parts of the continent. There is some evidence of Late Tertiary vegetation trends continuing into the Pleistocene with a further reduction in evergreen rain forest, but it may not have been until the Late Quaternary that there was any major disruption to the established Pliocene vegetation pattern. At this time there was a massive expansion of open eucalypt vegetation at the expense of "drier" rainforests and other sclerophyll communities, in response to an increase in burning, most likely as the result of the activities of aboriginal man. In recent years cores have become available from deep borings in a significant number of Australian lakes, providing long columns of palynological records. The most important of these include several crater lakes in Queensland, of which the record from Lynch's Crater is estimated to extend back for 200,000 years (Kershaw 1986), and Lake George in New South Wales, where the column is believed to cover 7 million years, of which deposits of the last 730,000 years have been intensively studied (Singh & Geissler 1985). The deepest levels at Lake George show the usual rain forest of *Nothofagus* and *Podocarpus* with Proteaceae and many other families. *Eucalyptus*-type pollen is present to the order of 1%. At 4 million years *Casuarina* pollen has become dominant and *Eucalyptus* has increased to 3.6% with 7 taxa, but rain forest elements are still present. Singh & Geissler (1985) wrote that "Casuarina appears to have dominated the vegetation and, together with taxa belonging to the Myrtaceae (including *Eucalyptus*) plus *Acacia* and *Haloragis*, is suggested to have formed an open forest in which many rain forest taxa lived side by side with sclerophyll taxa".

A continuous column of sediment was examined by Singh & Geissler (1985) for the past 730,000 years, divided into 17 zones numbered A to Q, representing periods of stability or constant change. From Zone Q, the oldest, up to Zone G, the flora is essentially of the above type, alternating with periods of herbaceous vegetation interpreted as colder, more alpine intervals. From Zone F to the present day we have a radical change. The rain forest taxa diminish markedly and eventually disappear. *Casuarina* loses its dominance. A sclerophyll forest appears in which seven taxa of *Eucalyptus* and numerous Myrtaceous shrubs are dominant, and the herbaceous component shows the forest to have been more open than before. Equally arresting, there is an extraordinary rise in the amount of charcoal particles in the sediment, which is maintained down to the present and appears to show that fire became more frequent. We know that *Eucalyptus* was in existence long before in small amounts. Then suddenly it became dominant, in association with signs of fire. The inference is that the eucalypt-dominated vegetation to which we are accustomed in Australia today is of no great antiquity and is dependent upon the fire regime. The inference also is that the intensification of the fire pattern was due to the arrival of primitive man.

The data from Lynch's Crater (Kershaw 1986) show the same state of affairs. Beginning about 38,000 years ago there is a marked increase in charcoal in the sediments, accompanied progressively by increase in both *Casuarina* and *Eucalyptus* and decline in rain forest elements, culminating at 26,000 years BP when sclerophyll vegetation had become totally dominant, remaining so until 9000 BP when increasing rainfall enabled rain forest to return. An important feature of the long-running record at Lynch's Crater is that it is dated with more certainty than that of Lake George and covers also the penultimate dry glacial period of 165,000 to 126,000 years BP, during which the vegetation was represented by drier types of rain forest and not taken over by sclerophylls, nor was charcoal noticeably abundant. In other words, the last dry glacial period was different. This seems to reinforce the conclusion from Lake George that primitive man and his fire stick succeeded in profoundly modifying the plant cover of Australia long before the arrival of European settlers. It is consistent with archaeological evidence for the existence of early man in Australia at least 38,000 years ago (Hallam 1983).

Ecological aspects

The sudden rise to dominance of *Eucalyptus* is clearly associated with its selection by fire, from a pre-existing sclerophyll flora, for its various fire-resisting, fire-surviving qualities. The dominance of *Eucalyptus* today is so complete that the book "Forest Trees of Australia" (Hall et al. 1970) lists – excluding rainforest taxa – 117 taxa of *Eucalyptus*, only 8 other angiosperms and 10 gymnosperms. It is also clear, however, that a pre-existing sclerophyll flora was already in place, which had become differentiated from "drier" types of rain forest.

Australian rain forests today comprise a number of forms, and especially in the north include a whole series along a moisture gradient from the most humid complex evergreen forests to low deciduous thickets (Webb 1959). Throughout the gradient the flora remains mixed, non-sclerophyll, orthophyll. Surviving patches at the "drier" end of the gradient tend to occur today on richer soils, usually derived from basalt (Webb 1963). Sclerophyll vegetation is typically associated with poor, siliceous soils. The terms sclerophylly and sclerophyll forest have been used freely throughout this paper owing to this very marked characteristic, not only of *Eucalyptus* but of its associated taxa in other genera and families. In the Myrtaceae and Proteaceae, leaves are stiff, harsh, brittle and often spiny. In more specialised cases the leaf blade is discarded in favour of a broadened petiole (phyllode), as in most Australian species of *Acacia*, or is suppressed altogether, its role taken over by leafless green stems (cladodes), as in *Casuarina*, some species of *Acacia* and some Fabaceae such as *Daviesia*. In these cases the new leaf-like structures also exhibit sclerophylly. While other floras than the Australian also exhibit this character, particularly in the other mediterranean-climate regions, it is never so strongly developed.

It was first thought that sclerophylly must be an adaptation to withstand drought, but this would not explain its predominance in Australia. Beadle (1954, 1966) first suggested that it could be a response to the low nutrient status of Australian soils, and this view has gradually received general acceptance. Due to their great age, Australian soils have lost most of their nutrients by leaching. Since the formation of soft, green tissue is nutrient-demanding, the plants adopt a strategy of forming leaves mainly of lignified tissue and withdrawing nutrients from them at an early stage. Sclerophyllous plants are more fire-prone than soft-leaved plants, so their development encouraged the formation of fire-adapted vegetation in Australia.

Before the intensification of burning, the sclerophyll and rainforest floras seem not to have been sharply differentiated. The fossil record often appears to show a mingling of both elements. Today, however, they are very sharply differentiated, with the fire factor holding the balance between the two. We have in fact two quite distinct biomes, the one orthophyll and fire-tender, the other sclerophyll and fire-resistant. As we have seen earlier, the sclerophyll forests are of several different types, of which the fire reponses and behaviour are also different.

So-called "wet sclerophyll" forests occur in the most humid parts of the continent, from southern Queensland to Tasmania and on the southwestern tip of Western Australia. Rainfall generally exceeds 1000 mm per annum, but the critical factor is of course the reliability of rainfall and the length of dry periods. "Dry" months, as shown in standard climate diagrams, should not exceed 3–4 per annum (Beard 1981). These are forests of giant trees reaching heights of 50 to 80 m, which rival the California redwoods as the world's tallest trees, though they do not compete in trunk diameter or longevity. Australian eucalypts may reach butt diameters of 2–3 m and ages of 300–400 years. Understories are variable but in general show a reduced degree of sclerophylly, frequently containing many rain-forest species. In areas where rain-forest relics are sufficiently extensive to provide seed sources, there is a constant tendency to invade the eucalypt forest, a process that has been well documented in Tasmania (Jackson 1968) as follows. A severe fire destroys an area of rain forest, which is then colonised by *Eucalyptus*. In the absence of further fire the eucalypts grow up into a dominant stand without recruitment, the lower layers being invaded by rain-forest species. In about 90 years, given continued absence of fire, the eucalypts will have a well-developed rain forest beneath them. In 300–400 years the eucalypts, still without recruitment, die off leaving a fully re-established rain forest. It will be seen, however, that a very long fire-free period is needed for the completion of the succession, which is rarely achieved. Normally fire occurs after a shorter interval, and a new stand of eucalypts is established. Surviving rain forest patches therefore have to occupy especially humid, sheltered or topographically protected places into which fire does not spread readily.

The life-cycle of *Eucalyptus*, on the other hand, requires fire at some time in the seed-bearing life of the trees, giving rise to the paradox that such forests produce vast amounts of fuel and generate high-intensity fires, yet consist themselves of fire-sensitive species (Ashton 1981). Wet sclerophyll eucalypts typically possess little fire resistance and are killed by severe fires, to be replaced by a new crop which springs up rapidly on the ash-bed left by the fire. Ashton (1975) estimated that 14 million seeds per hectare could be stored in the canopy of a *E. regnans* forest, to be released after fire, and that 2.5 million seedlings per ha could result (Beetham 1950)!

The ecology of "dry sclerophyll" forests is somewhat different. In the south of the continent any drier rain forest communities with which they may once have had a successional relationship no longer exist. The dry sclerophyll forests have inflammable sclerophyll shrub understories which are subject to frequent fires (within the range 3 to 20 years), and the canopy trees possess various fire-resisting adaptations. Regeneration may take place sporadically following ground fires, the seedlings then developing a lignotuber, a woody structure at ground level which bears dormant buds. Seedlings and saplings killed by succeeding ground fires resprout from the lignotuber until a sufficiently long fire interval enables some of them to grow large enough to develop a thick insulating bark. Adult stems are protected in this way and have the ability to produce epicormic shoots to replace twigs and branches that have been killed. Shrubs and small trees in the understory behave similarly, some being "resprouters" which can regenerate from the base of the plant and others "reseeders" which must regenerate from seed and are typically bradysporous, i.e. storing the seed on the plant in unopened capsules until fire causes them to open.

In the tropical north, eucalypts growing in savannas must possess similar fire-resistant adaptations if they are to withstand the fierce and usually annual burning of the grassy ground layer. One of the most common species, *E. tetrodonta*, has been shown to add root-suckering to its lignotuberous habit (Lacey & Whelan 1976). In the north such savannas may be contiguous with patches of drier rain forest vegetation *sensu lato*, i.e. semi-evergreen and deciduous vine thickets in Webb's (1959) classification. As in the south, these often enjoy some topographic or other protection against frequent burning. Under equilibrium conditions there is a sharp stable boundary between the savanna and the "rain forest" (Beard 1976, Hnatiuk & Kenneally 1981), but with intensification of the burning pattern the rain forest will recede, and conversely with reduction of burning, if the savanna is not under pastoral use or if aborigines have left the area, the rain forest species will begin to invade (Langkamp et al. 1981, Clayton-Greene & Beard 1985). A recent study (Russell-Smith 1986) has shown that the seeds of the rain forest species are disseminated widely by birds, giving them the ability to colonise vacant habitats and invade adjacent communities. The forest is not as fixed as might appear from the tiny patches which it occupies – in fact Russell-Smith (1986) described it as "vagile" (from the Latin *vagus*, "wandering") and entitled his thesis "The Forest in Motion". Therefore, just as in the south, there can be no doubt that the wet sclerophyll forests at least would be replaced by rain forest if the fire factor were in some way eliminated, so in the north the savanna would be replaced by vine forest and thickets.

Summary

To sum up, therefore, we have seen that deteriorating climate from the Miocene onwards led to partial replacement of rain forest by sclerophyll associations adapted not only to seasonal drought but to soil poverty and lack of nutrients as well. Finally, the arrival of early man in Australia about 40,000 years ago led to a marked intensification of burning and thereby a final recession of rain forests to their present refugial areas. The intensified fire pattern also selected the more fire-resistant elements of the sclerophyll flora and led to the almost universal dominance of the genus *Eucalyptus* in forests and woodlands and of the genera *Acacia* and *Triodia* in the Arid Zone. This was a unique evolutionary process, which has been responsible for the uniqueness of Australian vegetation. *Eucalyptus* in particular is an extraordinary phenomenon not paralleled on any other continent. In many respects it can be claimed that *Eucalyptus* is the Southern Hemisphere analogue of *Pinus* in the Northern Hemisphere, since both represent an evolutionary response to changing Tertiary environments: they tolerate poor soils, and fire plays an important role in their life cycle in communities. They are fast-growing pioneer species establishing rapidly after burning and requiring further fire for their regeneration. Among hardwoods, however, *Eucalyptus* is certainly unique. In consequence it has become popular for introduction into warm countries throughout the world, where it is esteemed for its ease of establishment even on poor soils, its rapid growth rate and its ability to regenerate from coppice. It has been of enormous benefit to poor Third World countries in desperate need of small timber and pulpwood. For all this, in the last resort, we have to thank the Australian aborigine, for it was he, thousands of years ago, who unwittingly selected the genus and promoted its speciation.

References

Ambrose, G. J., Callen, R. A., Flint, R. B. & Lange, R. T. 1979. *Eucalyptus* fruits in stratigraphic context in Australia. Nature 280:387–389.

Ashton, D. H. 1975. Studies of flowering behaviour in *Eucalyptus regnans* F. Muell. Aust. J. Bot., 23:399–411.

Ashton, D. H. 1981. Fire in tall open forests (wet sclerophyll forests). In: Gill, A. M., Groves, R. H. & Noble, I. R. (eds.) Fire and the Australian Biota. Aust. Acad. Sciences, Canberra.

Beadle, N. C. W. 1954. Soil phosphate and the delimitation of plant communities in eastern Australia. Ecology 25:370–374.
Beadle, N. C. W. 1966. Soil phosphate and its role in moulding segments of the Australian flora and vegetation, with special reference to xeromorphy and sclerophylly. Ecology 47:991–1007.
Beard, J. S. 1976. The monsoon forests of the Admiralty Gulf, Western Australia. Vegetatio 31:177–192.
Beard, J. S. 1981. The Vegetation of the Swan Area. Explanatory Notes to Vegetation Survey of Western Australia, 1:1 000 000 series, Sheet 7. Univ. W. Aust. Press, Nedlands.
Beetham, A. H. 1950. Aspects of forest practice in the regenerated areas of the upper Latrobe Valley. Dip. For. Thesis, Forests Comm. Victoria, Creswick.
Bowler, J. M. 1976. Aridity in Australia: age, origins and expression in aeolian landforms and sediments. Earth Sci. Rev. 12:279–310.
Clayton-Greene, K. A. & Beard, J. S. 1985. The fire factor in vine thicket and woodland vegetation of the Admiralty Gulf region, northwest Kimberley, Western Australia. In: Ridpath, M. G. & Corbett, L. K. (eds.) Ecology of the Wet-Dry Tropics. Procs. Ecol. Soc. Aust. 13:225–230.
Cockbain, A. E. 1968. The stratigraphy of the Plantagenet Group, Western Australia. Geol. Survey W. Aust., Annual Report 1967, pp. 62.
Hall, N., Johnston R. D. & Chippendale, G. M. 1970. Forest Trees of Australia. Aust. Govt. Publ. Service, Canberra.
Hallam, S. J. 1983. The peopling of the Australian continent. Indian Ocean Newsletter 4(2):11–14.
Hnatiuk, R. J. & Kenneally, K. F. 1981. A survey of the vegetation and flora of Mitchell Plateau, Kimberley, Western Australia. In: Biological Survey of Mitchell Plateau etc. Western Aust. Museum, Perth.
Jackson, W. D. 1968. Fire, air, water and earth – an elemental ecology of Tasmania. Proc. Ecol. Soc. Aust. 3:9–16.
Kemp, E. M. 1978. Tertiary climatic evolution and vegetation history in the southeast Indian Ocean region. Palaeogeog., Palaeoclim., Palaeoecol. 24:169–208.
Kershaw, A. P. 1986. Climatic change and aboriginal burning in northeast Australia during the last two glacial/interglacial cycles. Nature 322:47–49.
Lacey, C. L. & Whelan, P. I. 1976. Observations on the ecological significance of vegetative reproduction in the Katherine-Darwin region of the Northern Territory. Aust. Forestry 39:131–139.
Langkamp, P. J., Ashton, D. H. & Dalling, M. J. 1981. Ecological gradients in forest communities on Groote Eylandt, Northern Territory, Australia. Vegetatio 48:27–46.
Russell-Smith, J. J. 1986. The forest in motion: man, fire, and monsoon vine forest in western Arnhem Land. Ph.D. thesis, Aust. National Univ., Canberra.
Shackleton, N. J. & Kennett, J. P. 1975. Paleotemperature history of the Cenozoic and the initiation of Antarctic glaciation: oxygen and carbon isotope analyses in D.S.D.P. sites 277, 279, 281. In: Initial Reports Deep Sea Drilling Project, 29:743–755.
Singh, G., Kershaw, A. P. & Clark, R. 1981. Quaternary vegetation and fire history in Australia. In: Gill, A. M. et al. (eds.) Fire and the Australian Biota. Aust. Acad. Sciences, Canberra.
Singh, G. & Geissler, E. A. 1985. Late Cainozoic history of vegetation, fire, lake levels and climate, at Lake George, New South Wales, Australia. Phil. Trans. Roy. Soc. London 267B:467–501.
Sluiter, I. R. & Kershaw, A. P. 1982. The nature of Late Tertiary vegetation in Australia. Alcheringa 6:211–222.
Webb, L. J. 1959. A physiognomic classification of Australian rain forests. J. Ecol. 47:551–570.
Webb, L. J. 1963. The influence of soil parent materials on the nature and distribution of rain forests in south Queensland. Symp. Ecol. Res. Hum. Trop. Veg., Kuching, Sarawak, Malaysia.

9. Evergreen broadleaved forests of New Zealand and their bioclimatic definition

COLIN D. MEURK
Landcare Research New Zealand Ltd, Lincoln, New Zealand

Abstract

Specific biogeographic factors are discussed for New Zealand's forests. This is not a comprehensive treatment of forest phytosociology but focuses on some historical, evolutionary, structural and autecological considerations. Climatic envelopes are graphically presented for 33 representative forest species which, in aggregate, cover the range of forest conditions in New Zealand. A comparison is made between New Zealand's largely evergreen forests and those of other parts of the world. It is concluded that, in terms of climate, soils, ecosystem dynamics, growth forms and floristics, New Zealand forests are most similar to rain forests in South America, southeast Australia, the equatorial montane, and, to a lesser extent, oceanic parts of Europe and North America.

Key words: autecology, biogeography, broadleaved trees, climatic niches, ecosystem dynamics, evergreen forests, floristics, forest structure, hardwoods, New Zealand, oceanic climate, podocarps, soils

Introduction

Climatically, woody vegetation is believed to be capable of occupying virtually all terrain beneath New Zealand's Alpine zone. In pre-human times about 85% of the total land area was in forest or scrub; about 40% was lost to Polynesian fires – mainly in the drier eastern country – and a further 20–30% to European clearances.

Most of this indigenous forest can be defined as "evergreen, broadleaved rainforest" with only minor inclusions, generally less than a few hectares, of deciduous small trees. These latter may occupy seral sites on fresh, disturbed

E.O. Box et al. (eds.), Vegetation Science in Forestry, 151–197.

or dry soils, and include *Plagianthus*, *Aristotelia*, *Fuchsia*, *Hoheria*, *Sophora*, *Muehlenbeckia*, *Olearia* and *Discaria* (Haase 1987, Batcheler 1989). Leptophyllous woodlands form quasi-stable communities in the driest areas. Biomass and physiognomy of both seral and older mixed forests may be dominated by emergent podocarps, cedars and/or araucarians. The conifers are relatively broad-leaved or cupressoid; none have long needle leaves. Hardwoods, with evergreen nanophylls (1–2 cm, 6% of trees), microphylls (2–6 cm, 68%) to mesophylls (6–18 cm, 25%) (Dawson 1988), are invariably mixed with these other elements, dominating the canopy and lower tiers. Evergreen hardwoods can grow on any site that the conifers and deciduous species may dominate from time to time. It is therefore impractical, on a biogeographical scale, to treat the conifers, and deciduous and evergreen hardwoods as separate entities.

I describe some key properties of New Zealand's forests, their origins, evolutionary influences, contemporary factors, structural attributes, patterns and processes. Climatic definitions of the forests and principal species are illustrated, related to the segregation of forest types, and examined in a global context.

Sketch of the New Zealand biogeographical region

New Zealand is an archipelago of continental fragments and volcanic-oceanic crustal materials between 30° and 55° S, with the Norfolk Ridge (connecting to New Caledonia) and Lord Howe Rise extending to the northwest (Fig. 1). The main islands are more or less divided by the axial ranges of the Southern Alps which rise to 3760 m, about 1000 m above the permanent freezing level. Snowline is generally below this (Fig. 2) because of high precipitation.

Basal climate spans oceanic Subtropic, Warm Temperate, Cool Temperate, to Subantarctic (Figs. 1, 2, and Meurk 1984), with continentality scores of −5 to 17 on Conrad's (1946) 0–100 scale (Fig. 3, and Meurk & Blaschke 1990). Because of the prevailing, moist, westerly winds there is a strong foehn effect created by the north-south aligned topography, and rainfall declines sharply from 2000–5500 mm in the west to 300–800 mm in the east, over a distance of as little as 125 km (NZ Meteorological Service 1985). The climate is notably even (extreme temperature range for the country as a whole being −22°C to 42°C, NZ Meteorological Service 1983), but unseasonal events may be damaging to actively growing plants.

New Zealand has a moderate sized flora with about 2200 native vascular plants, of which 82% are endemic (Webb et al. 1990). The remainder are predominantly ferns (Brownsey & Smith-Dodsworth 1989), orchids and

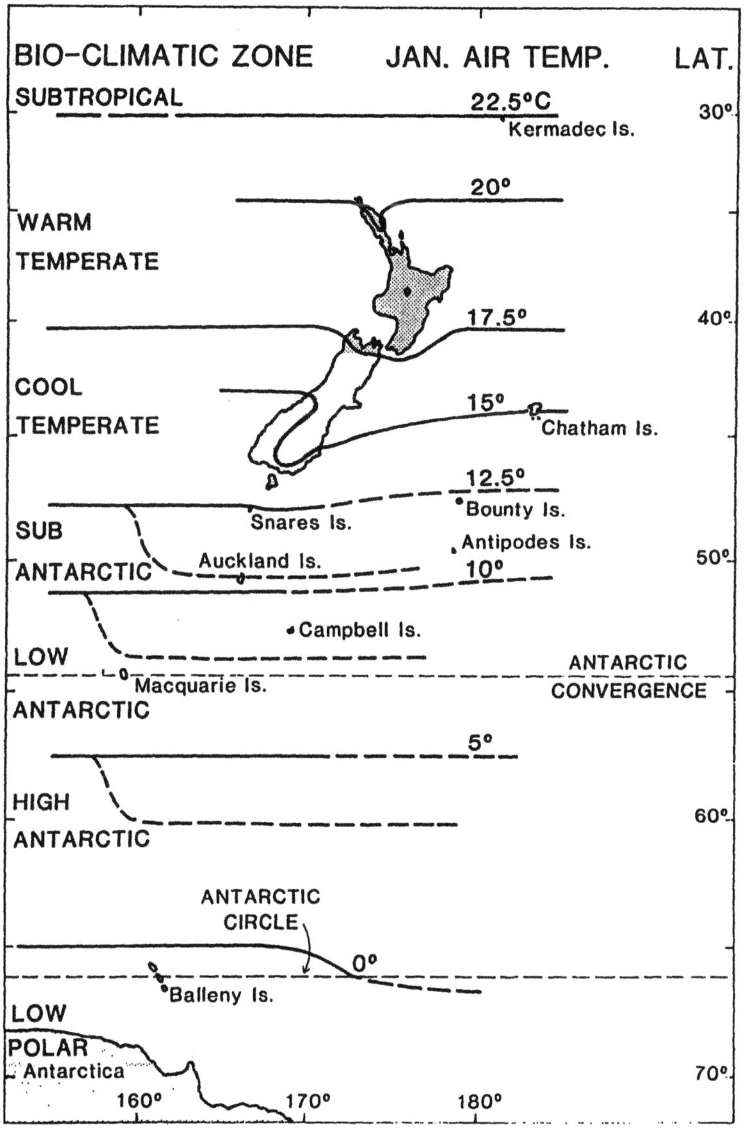

Fig. 1. New Zealand and outlying islands, with approximate positions of lowland mean January isotherms, marine convergences and sea-level (basal) bioclimatic zones. Short dashed lines curving below southern isotherms demonstrate the displacement of growth-form limits (bioclimatic zones) southwards resulting from isothermal climates with long growing seasons.

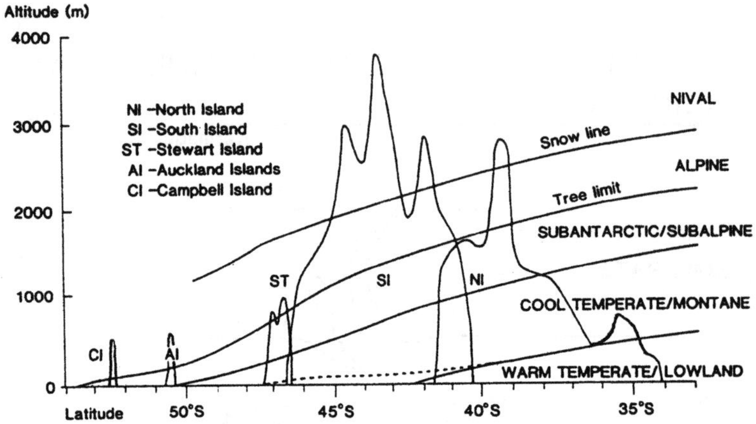

Fig. 2. Bioclimatic zones in New Zealand in relation to latitude and altitude. Irregularities in zonal patterns and a snow line lower than permanent freezing level are caused by oceanic influences.

water plants (Moore & Edgar 1970) conspecific with those in southeast Australia. New Zealand has almost 1% of the world's flowering plant species, whereas our 20 conifers make up 3.6%. Endemism is generally low at genus and family level. The indigenous flora is now matched in number by naturalised aliens from temperate Europe, Asia, the Americas, Africa and Australia (Webb et al. 1988). New Zealand has about 540 each of mosses and hepatics (A.J. Fife, pers. comm.), representing 4 and 7% of global floras respectively, and is one of the world centres of endemism for the latter. There are about 1500 lichens (>7% of world total), with a proliferation of southern foliose, fruticose and some crustose genera (Galloway 1985). About 70 families among flowering, coniferous and treefern plants occur in forests (Fig. 4).

Origins, sources and historical events

To appreciate the character of New Zealand's biota we must go back to the Late Cretaceous, over 80 million years ago. This was a crucial period in the biogeography of the southern hemisphere (Stevens et al. 1988). The splitting of the Gondwana supercontinent into several large and small fragments, the timing of individual separations and evolutionary events, the trajectories of the various pieces in relation to climatic zones and to other land masses, and eventually the availability of escape routes from glaciation all affected the biotas of the segregate continental fragments.

The derivatives shared an initial Gondwanic heritage which has been

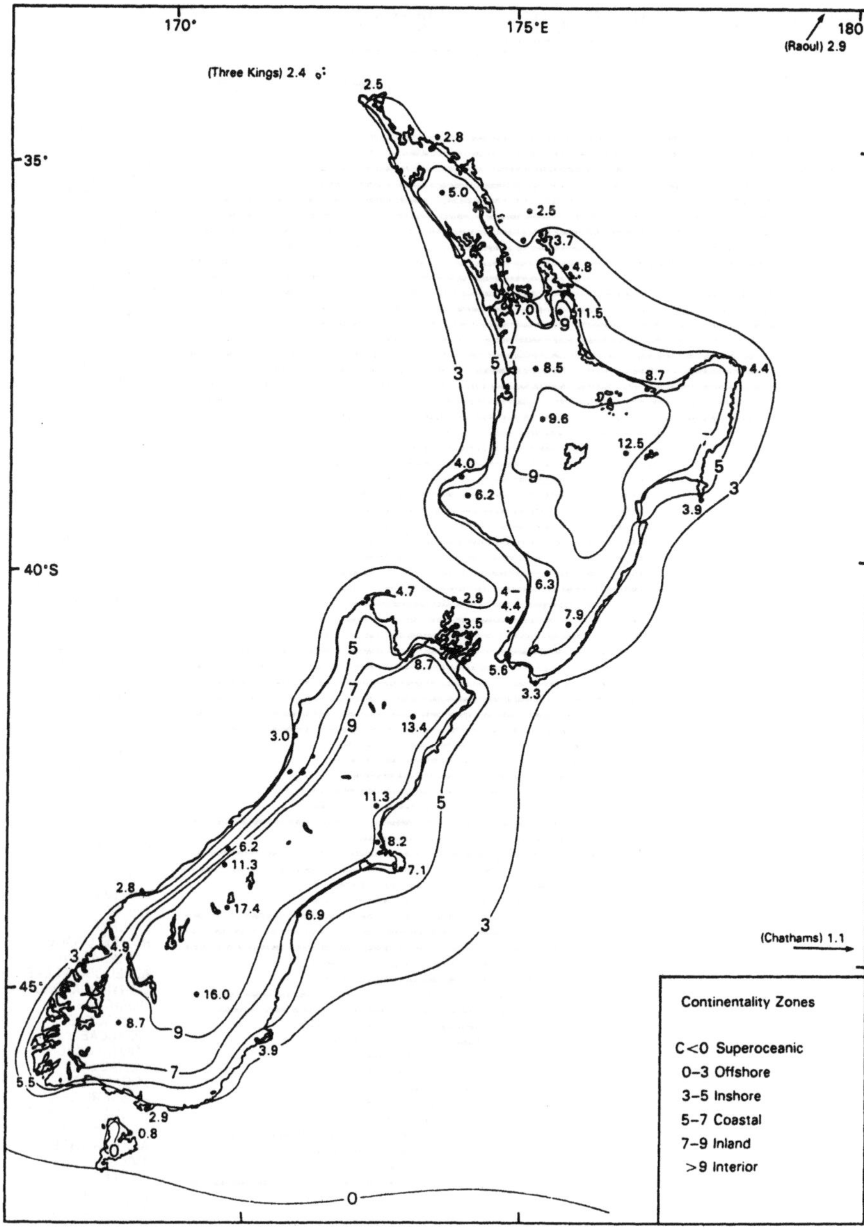

Fig. 3. Isolines of Conrad's Continentality Index (theoretical global range 0–100) in the New Zealand region, with representative calculated values shown. Values less than zero are classified as Super-oceanic and occur in the southern islands (not shown on the map).

Fig. 4. Geographical ranges of New Zealand forest families. Short cross bars denote the southernmost records of forest-inhabiting species within the families. A dot at the southern end of a bar indicates presence of genera on other Fuegian land masses but not necessarily Gondwanic relationships. All but Philesiaceae and Griseliniaceae have Malayo-Pacific or wider northern generic distributions. MI = Macquarie Island, CI = Campbell Island, AI = Auckland Islands, StI = Stewart Island, KI = Kermadec Islands.

progressively and variously diluted, whittled away or modified as the pieces rafted from their common ancestral home and/or fused with alien biotas. When New Zealand was cut adrift from its parent continent, it wandered into lonely obscurity in the vastness of the Pacific Ocean. Like some Noah's Ark, it carried a small parcel of the original Gondwanic life. New Zealand, perhaps with New Caledonia, is unique in having been separated for longer and more completely than any of the other Gondwanic derivatives. Consequently New Zealand's representation of the Mesozoic biota (at least at the generic and familial level) is the least diluted and modified. Wace (1965) estimated that 17% of New Zealand genera (29% of forest families excluding ferns) were of southern continental origin, as compared to 9% for Tasmania and 2–3.5% for all other Australasian and South American land masses.

Three factors may have helped preserve core Gondwanic elements. Large reptilian and mammalian herbivores failed to establish on the New Zealand subcontinent. Consequently a major shift to co-adapted plants, capable of responding vigorously to regular defoliation, did not eventuate. Atkinson & Greenwood (1989) and others, however, have speculated on adaptations in the flora to low-intensity bird browsing. Secondly, since "climax" forest species (with large propagules) tend to be filtered out by long distance dispersal, r-selected immigrants would largely have contributed to the weedy fringes of established ecosystems. Furthermore, the most tender or least adaptable of the Tertiary, **Malayo-Pacific** migrants were subsequently winnowed from the land by Pleistocene glaciation.

Ascribing Gondwanic qualities to the contemporary vegetation is nonetheless fraught with pitfalls. Despite accepted affinities at higher taxonomic levels it is sobering to note that, in the late-Tertiary macrofossil record, only *Ripogonum* and *Dacrycarpus* bear a close resemblance to modern-day representatives (Pole 1989). Prior to this, morphological relationships are frequently obscure.

Mesozoic elements

Based on fossil and biogeographical evidence (summarised by Fleming 1975, Mildenhall 1989, Pocknall 1989, Pole 1989, Stevens et al. 1988), it appears that much of the present forest flora has a continuous lineage back to Gondwana. The modern families which have contemporary connections to Fuegian lands are identified in Fig. 4 (Balgooy 1971). Of course there has been continual, circum-polar exchange of propagules between the southern lands. This either reinforces the relationships or perhaps suggests antiquity where it does not properly apply. The number of possibly Gondwanic or **Palaeoaustral** elements present on the definitely oceanic Hawaiian and Tahitian archipelagos (*Metrosideros*, Epacridaceae, *Weinmannia*, *Fuchsia*, treeferns, *Acacia*, araliads) give us reason for caution. Many archaic animal groups, relict in New Zealand, also have a Gondwanic pedigree.

Early Tertiary (Palaeocene, Eocene)

This period of increasing isolation, warmer climates and low relief was characterised by the steady dispersal of propagules (by wind, ocean currents and on birds) from first Australia and then increasingly from the Tropics through connections to Lord Howe Island and New Caledonia. The early Tertiary migrant animals produced a second order of animal endemism, which together with the Mesozoic biotas, make up the **Palaeoaustral** and **Endemic** elements of Fleming (1975).

Late Tertiary (late Eocene-Pliocene)

This period was characterised by a cooling trend, the establishment of the west wind drift, and the commencement of volcanic and tectonic mountain building. Several hundred taxa, originating in the west and north, make their first appearance in the fossil record. The **Australian** species included cold-adapated forms which colonised the rising mountains. Some Trans-Tasman bird migrants became heavier and flightless in response to weak predation and cold. On the other hand, many of the **Malayo-Pacific** elements stayed only a brief time as climates deteriorated.

Although the recorded assemblages (with Southern beech, podocarps, *Agathis*, *Casuarina*, *Ilex*) may be interpreted as reflecting moist temperate conditions, the presence of *Casuarina*, chenopods, *Sophora*, asterads, proteas, myrtles, acacias, Gyrostemonaceae (now prevalent in Australian sclerophyll woodlands and shrublands), Sapindaceae, eucalypts, grasses, and *Mallotus/Macaranga* (Euphorbiaceae) could equally suggest semi-arid conditions in the hinterlands of the swamps where these fossil remains have been found. Charcoals are found in the swamps; desiccation cracks have been observed in sediments but no evaporites. As in Australia, the growth forms implied above may have been adaptations to soil infertility as much as to aridity and fire.

Freeze and thaw of the Pleistocene-Holocene

During the Pleistocene the forests retreated and many of the more tropical elements died out. Tree limits on the mainland fell in the north to about 1000 m and in the south to 400 m (*cf.* 1500 m and 850–1200 m today, Fig. 2), representing an annual average temperature depression of about 4.5°C.

In post-glacial times woody vegetation expanded south and up from refugia that existed at least around the periphery of most of the mainland. The sequence was from grassland and shrubland of celery pine (*Phyllocladus alpinus*), *Halocarpus*, *Myrsine* and *Coprosma*, sometimes with patches of beech (*Nothofagus*) and mountain cedar (*Libocedrus bidwillii*), to matai (*Prumnopitys taxifolia*) and totara (*Podocarpus totara*) dominance, with kahikatea (*Dacrycarpus dacrydioides*) on less well drained terraces. Matai, totara, kanuka (*Kunzea*), kowhai (*Sophora*), *Discaria*, and black beech/mountain beech (*Nothofagus solandri*) persisted in dry basins and rainshadow plains up to the present. Elsewhere, rimu (*Dacrydium cupressinum*) became more prevalent as conditions became milder, moister and soils more leached. Associated were miro (*Prumnopitys ferruginea*), *Weinmannia* and rata (*Metrosideros*). Mountain cedar, mountain totara (*Podocarpus hallii*), celery pine, pink pine (*Halocarpus biformis*), shrubby daisies (*Olearia*, *Brachyglottis*) and heaths (*Dracophyllum*, *Cyathodes*) occupied higher elevations in the wet mountains above the limits of rimu and miro. Kauri (*Agathis australis*), tanekaha (*Phyllocladus trichomanoides*), rewarewa (*Knightia*) and *Libocedrus plumosa* expanded onto far northern hills and ridges. Presumably tawa (*Beilschmiedia tawa*) and other hardwoods also increased at the expense of podocarps. On the poorest montane soils, where there were handy seed sources, beech replaced podocarps and broadleaved trees (McGlone 1985, Stevens et al. 1988).

A cooling and possibly drying of the climate over the past 7000 years is implicated in the spread again of matai and totara (and beech) on formerly rimu-broadleaved forest sites. Kauri continued to expand in the north, however, as did rimu on Stewart Island. In the end, fires have had an increasing impact in the drier parts of the country from about 2500 years ago, particularly since the human advent (McGlone 1989).

The readjustments have certainly not been uniform across all taxa. In particular, the beeches are much slower to disperse than the fleshy-fruited podocarps and hardwoods. Consequently, the latter preferentially recolonised those areas that suffered piedmont glaciation or were rendered inhospitable to tree growth, creating the so-called "beech gaps" (Wardle 1988). McGlone (1985) has preferred to explain such disjunctions and centres of endemism on the basis of existing environments and lateral tectonism, although these viewpoints are not mutually exclusive.

Selection forces

The preconditions of the contemporary vegetation were:
- isolation from the mainstream of continental evolution (mammalian preda-

tors, herd grazers, browse resistant vegetation, pinaceous conifers, other vigorous, woody, seral species);
- a long history of humid, oceanic, equable, windy climates (but effective rainfall may have been limited because of low relief through most of the Tertiary and the prevailing Warm-Temperate to Subtropical latitudes);
- leaching, peat formation or laterisation over much of the landscape, and a predominance of sandstone geology, producing a generally low base fertility (except in lowland swamps, the source of most fossil evidence);
- impact of glaciation, combined with ineffectual escape routes, on the distribution and local extinctions of drought- and fire-tolerant sclerophylls and of tropically derived Tertiary species; and
- relative freedom from fire during the Holocene.

The biological response to this relatively unproductive, though stable and unthreatening environment was low rates of metabolism – manifested as large, long-lived, slowly growing and reproducing organisms, with probable losses of browse repellant chemistry and resistance to defoliation in some plants, and loss of predation defences in vertebrate animals. Among plants the giant kauri (*Agathis*) is a classic case. Ogden (1985) suggested its longevity was a strategy for retaining occupation of stable sites. Hardwoods and other more shade-tolerant species may supplant the kauri over many thousands of years, but catastrophic storms eventually rejuvenate some part of the landscape, stimulating fresh cohorts of kauri. Within the forest flora Ogden could identify an r-K spectrum on the basis of seed size and shade tolerance, and McSweeney (1982) contrasted the light-demanding, small-seeded totara of forest margins with the shade-tolerant, large-seeded matai capable of regenerating in dense stands. The distinction is subtle, however, and the indigenous r-strategists hardly match the growth rates of their exotic counterparts.

Climatic, tectonic and volcanic disturbances are a feature of New Zealand as a whole. The unpredictability of such events, however, and their often superficial or localised impact, impose weak selective pressures, especially in the absence of reinforcing factors such as herd grazing. This contrasts with the continental pines, oaks, eucalypts, wattles and proteas that have adapted vigorously to frequent and regular defoliation or devegetation from fire and grazing.

Fire has been a periodic, though minor influence in primordial New Zealand. Some shrubby plants (*Discaria*, *Cassinia*) can resprout after fire, and *Kunzea*/*Leptospermum* have fire-adapted capsules. Nevertheless, the continental perspective of Spurr & Barnes (1980) that "fire is the dominant fact of forest history" cannot be upheld here.

Apparently the duration and intensity of the glacial cold was insufficient for the development of boreal-type adaptations, i.e. deciduousness and tolerance of intense frost. Glacial landscapes in the south were treeless even where average temperatures were above the usual threshold for trees. With the larger land area, increased clear skies and katabatic air flows off the South Island neve, the interior would have been strongly continental or seasonal, with intensified summer drought and winter frost. Even today frost may settle over the outermost, coastal sand dunes in Canterbury (Fig. 3).

New Zealand's most frost-tolerant woody plants resist no less than about − 25°C and for most trees the limit is −15°C (Sakai & Wardle 1978, Bannister 1990). The outwash plains of the South and lower North Islands must have experienced winter temperatures beneath the tolerance of the local trees. True forest would have survived only on sunny aspects of steeply plunging coastal hills (Wardle 1988). McGlone & Webb (1981), however, have postulated that the exposed peri-glacial plains were the crucible in which certain "sub-tropical" elements evolved a distinctive divaricating habit, in lieu of "normal" treeline growth forms.

Whereas temperature extremes are currently modest, the climate is quite stressful in terms of unpredictable cold, warm or dry spells that can occur in or out of season. Some trees may take advantage of short bursts of suitable growing conditions at almost any time of year. Nevertheless, many species do not tolerate even moderate frosts as seedlings, but through a measure of shade tolerance may grow up under the shelter of other taller plants in frosty areas. The widespread lack of specialised resting buds indicates an ambivalent relationship to the seasons.

Many of New Zealand's dominant species could be regarded as generalists, i.e. capable of tolerating a wide range of soil fertility and climatic conditions (within the New Zealand context). Red tussock (*Chionochloa rubra*), bog pine, mountain totara and mountain beech are all "physiognomic" species capable of growing in quite dry, fertile environments as well as in acid bogs.

Overall, the indigenous biota has developed conservative or modest strategies to cope with the vagaries of a low pressure but unpredictable environment. It was poorly prepared for invasion by legions of continentally honed species, introduced with pioneering fervour and formidable technology. Nearly 100 million years of isolation were turned upside down in a few centuries (Wardle 1985b, McGlone 1989).

Forest-soil dynamics

In a temperate environment, with high rainfall, leaching saps the ecosystem. "Climax" condition is a temporary optimum gradually overtaken by inexorable deterioration of the edpahic environment (*cf.* Stevens & Walker 1970, Mokma et al. 1973). This is a different situation from that commonly encountered in northern continents where the concepts of succession and climax

Table 1. New Zealand soil types, their USDA equivalents, and their potential/"climax" vegetation in lowland sites. "!" in front of a vegetation element indicates presence or absence

NZ Soil Classes	USDA	Lowland "Climax"
Typical Chronosequence on Acid Rocks		
Recent	Entisol	Seral cryptogams, herbs or mats
Brown Grey Earth	Aridosol	Leptophyllous scrub-dwarf forest
Yellow Grey Earth	(Fragi) Alfisol-Inceptisol	Matai-totara low hardwood forest
Yellow Brown Earth	Inceptisol	Tall podocarp-hardwood forest !kauri
Podzolised Yellow Brown Earth	Inceptisol	Tall podocarp-hardwood forest !beech
Podzol	Spodosol	Ditto
Gley Podzol	Spodosol-Inceptisol	Stunted pod.-hardwood forest !beech
Gley	Aquept	Scrub-fern-sedges
Peat	Histosol	Ditto & cushions
Soils of Volcanic/Alkaline Rocks		
Brown Granular Loam on basalt	Inceptisol-Alfisol-Mollisol	Tall podocarp-hardwood forest !kauri
Yellow Brown Loam and pumice soils	Andisol	Ditto
Red Brown Loam on ancient tephra	Ultisol-Oxisol	Kauri-podocarp-hardwood forest
Rendzina on limestone	Rendoll	Tall podocarp hardwood forest

were promulgated (Burrows 1990). There, modest leaching of deep, zonal soils is balanced by rapid weathering and cycling of nutrients, or landscape rejuvenation by wild-fire. Exhaustion of soils is protracted – measured in geological time rather than millenia or centuries. For example, Spurr & Barnes (1980) report fine-textured profiles in North America which are little weathered after 3500–10000 years. On the other hand, 10% of New Zealand is alpine or coastal barrens, and *c.* 70% of the mantle is shallow, leached or developed in deficient or acid parent materials. Less than 20% of the (potentially forested) terrain is of richer basalt, andesite or limestone, and AB horizons seldom exceed 60 cm (N.Z. Soil Bureau 1968, Molloy 1988). Salient aspects of soil fertility are summarised in Tables 1 and 2. With this geological and climatological configuration, any discussion of New Zealand vegetation must incorporate a concept of "post climax" vegetation-soil systems.

Tedrow & Hill (1955) described such a situation in the Arctic and questioned the use of the term "zonal" for gleys, podzols or peats that dominated developmentally mature landscapes (in which humidity and wetness have

Table 2. Average pH and nutrient values for New Zealand soil classes developed on sedimentary or metamorphic, basaltic (BGran Loam), andesitic (YB Loam), and calcareous parent materials

Soil Type	pH (subsoil)	TEB% (subsoil)	Carbon% (topsoil)	Citric P (topsoil)
Recent	6.4	3.1	3.7	26
BGE	8.5	11.5	1.3	16
YGE	6	6.9	4.9	8
YBE	5.7	2.8	6.9	6.3
Podz YBE	5	1.7	18	1.5
Podzol	4.9	2.4	15	1
Gley Podz	4.9	0.5	31	5
Peat	4.6	23	48	7
BGran Loam	6.2	20	4.5	21
YB Loam	5.8	2.6	12	12
Rendzina	6.4	29	9.6	25

accelerated vegetation-soil processes). They reserved application of the "zonal" soil concept to "brown soils" (inceptisols) that had achieved their optimum, productive potential. The dominant soil of a landscape is a temporally arbitrary condition reflecting both its chronological age and climatically-driven rate of soil development. Since landscape dominance (by soils) is a transitory state, it is an unsatisfactory criterion for ascribing "zonality".

A classic chronosequence, in Westland, has been described for alluvium under 4000–5000 mm rainfall (Stevens & Walker 1970, Wardle 1980, 1991, Molloy 1988). Recent soils (entisols), with N-fixing *Carmichaelia* and *Coriaria*, develop into Yellow Brown Earths and Podzolised Yellow Brown Earths (the zonal inceptisols, supporting a climax podocarp-hardwood forest at about 5000 years); then to Podzols or spodosols (post-climax after 12 000 years, with available nutrients halved from initial levels); and finally to Gley Podzols (aquepts) and acid Peats (histosols) up to 22,000 years old (Table 2, Fig. 5), with stunted cupressoid and ericoid shrubs. Even in superhumid environments, a terminal steady-state may be attained only after 40–70,000 years have elapsed (Adams 1978). Clarkson et al. (1988) have inferred a similar sequence for the volcanic plateau of the North Island, inspired by Gibbs' (1983) downward (nutrient supply) spiral theory. Gibbs proposed a general model of pioneering podocarps (kahikatea, matai, totara) establishing on fresh tephra, a rise of hardwoods with increasing nutrient supply (*cf.* andesitic-basaltic loams and yellow brown earths), a transition to tolerant podocarps (rimu, miro, mountain totara) as podzolisation sets in (podzolised yellow brown earths), and replacement by beech and/or kauri with further deterioration. Thus, in the absence of major disturbance, in net leaching conditions, we have to view the forest ecosystems as points in space and time along a trajectory of first building then counterbalancing decline of subcanopy light and nutrient status.

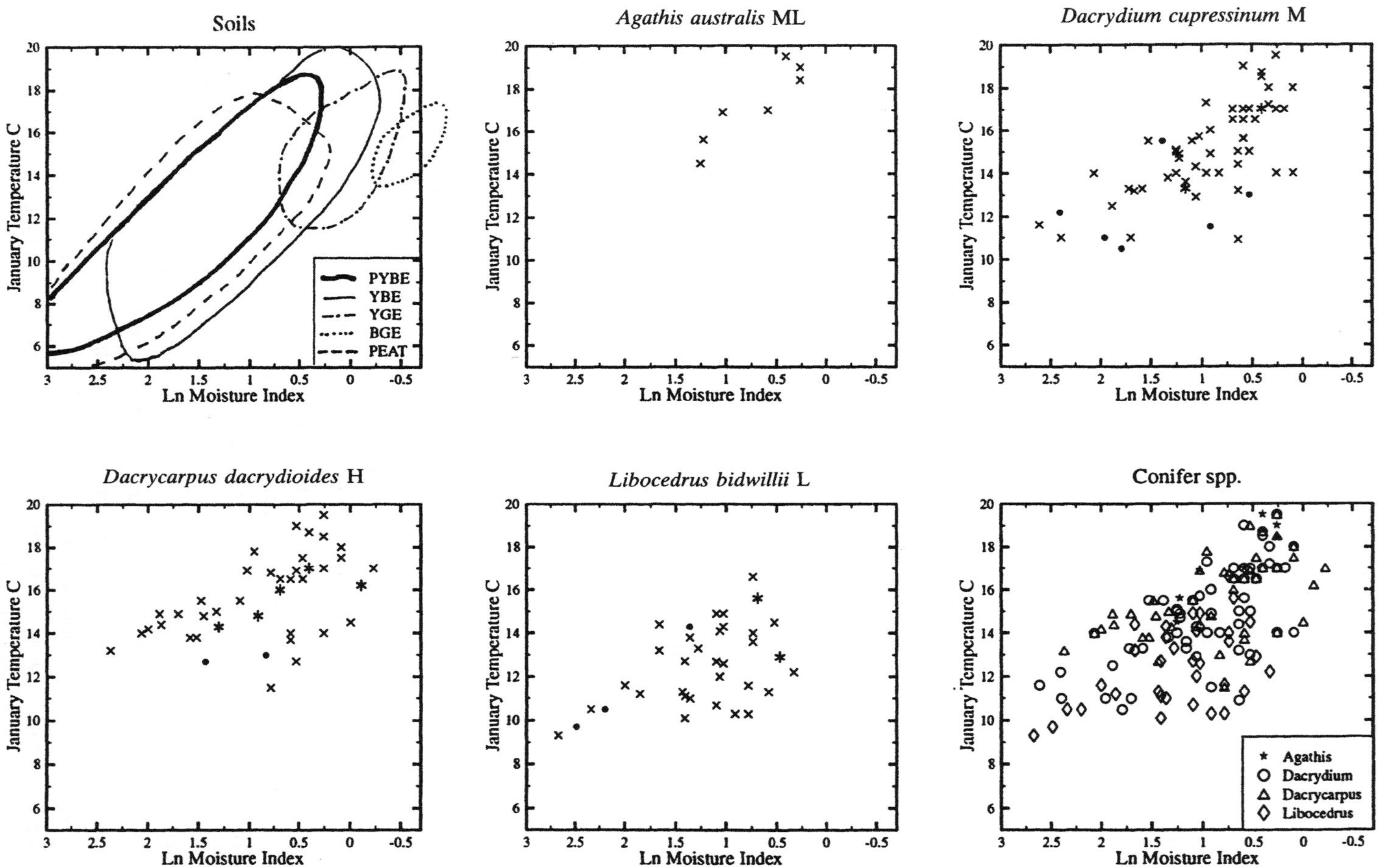

Fig. 5. Scattergrams of forest conifers within a climatic space defined by temperature (mean January air) and moisture (Ln Moisture Index). Dots denote Conrad Continentality of <4, "X" 4–9, and "*" >9 respectively. The soil diagram indicates the environmental envelopes for the standard New Zealand Soil Classes (see Table 1). The top-bottom, left-right orientation of the diagram can be equated with geographical north-south (low-high), west-east (high-low) transects through New Zealand.

Some northern kauri forests on soils ranging from deep inceptisols to ultisol or marginally oxisol status may have a pedigree predating the glaciation. Tight cycling of nutrients through the litter layer perhaps sustains these ancient systems. Burning the forest, however, results in degradation to a virtually sterile "gumland", for an indefinite period (Wardle 1991). The difference from the apparently perpetual tropical rainforest, on equivalently weathered soils, is that the warmth, high evaporative potential and less acid substrate permits complete decomposition and release of nutrients from the litter, which is efficiently recycled by the forest trees. In temperate, waterlogged, acid soils, on the other hand, some part of the nutrient pool is being continually sequestered into the unavailable organic fraction.

Superimposed on these regressive developments is cyclical rejuvenation. In New Zealand the causes are erosion and deposition associated with cyclonic storms, earthquakes, floods and slips, characteristic of tectonically active, rugged landscapes (Grant 1989), and vulcanism in the North Island. Gap-phase dynamics (Adams & Norton 1991) and cataclysmic disturbance such as vulcanism will result in small to large-scale mosaics, the latter on a very long rotation.

Total or partial destruction of vegetation under blankets of volcanic ash or lava is the typical starting point for forest in northern and central North Island. The andisol and basalt weathering sequence differs from that on acid sandstone in that initial base fertility is higher (Table 2). Clarkson (1990) has reviewed the evidence and cites examples which demonstrate a sequence of moss, scrub and herbs establishing after 10–50 years, N-fixing species by 35–100 years (montane), and forests with *Metrosideros*, *Weinmannia*, *Beilshmiedia*, *Phyllocladus*, *Melicytus*, *Libocedrus* and rimu, depending on elevation, in 100–450 years. Rejuvenation of North Island landscapes has maintained a preeminence of nutrient-demanding podocarps such as matai, totara and kahikatea. *Libocedrus bidwillii* is also stimulated to form dense stands on high-altitude andesitic tephra (Clayton-Greene 1977).

A gap-phase event is described for North Island podocarp-hardwood forests by Beveridge (1973): seral hardwoods and tree ferns establish in fallen podocarp gaps, kamahi establishes on tree ferns, podocarps establish by 50 years and are emergent at 100 years, podocarps decline or fall and dense angiosperm dominance is reestablished. Cameron (1954) also speculates on forest mosaics created by cycles of hardwoods (shade-tolerant, fast growing, mull formers) and podocarps (relatively shade-intolerant, slow growing mor formers).

Whereas kauri and podocarps have been observed regenerating under beech (Ecroyd 1982, Wardle 1988), evidence for beech invasion of podocarp-hardwood forests (at average spread rates of 5–8 m/century) has been cited by Holloway (1954), Wardle (1980), Allen (1988) and Rogers (1989). Soils under kauri become deeply podzolised, but subsequent angiosperm dominance may replenish the soil (*Leptospermum* → *Kunzea* → broadleaved spp. → kauri). Ecroyd (1982) reports that *Weinmannia* and beech can be "nurse" species for kauri, and there is monitored, site-specific evidence that podocarps and kauri are surging ahead of hardwood competitors in secondary forest on soils of low to medium nutrient status in the Coromandel Peninsula (Burns & Smale 1990). The evidence for some of these pathways (see also Ogden 1985) is often indirect and at times contradictory. Kauri, rimu, miro, *Halocarpus* and *Lepidothamnus*, however, are conifers adapted to impoverished soils and so enter the sequence later and retain possession of a site indefinitely.

Monocultural beech stands (Wardle 1984) support an understorey of "bonsai" beech seedlings which sit until the canopy is broken. A rush for the light ensues, with the weaker saplings being thinned out. The semi-deciduous red beech (*Nothofagus fusca*) is a mull former preferring mild, mesic, fertile conditions. In the mountains it is often confined to a midslope thermal zone with mountain (*N. solandri cliffortioides*) or silver (*N. menziesii*) beech above and below. The small-leaved beeches are mor-forming, and classical soil profiles, derived from Subalpine mountain beech under moderately high rainfall (about 1200 mm y^{-1}) have ginger subsoils and egg-cup podzols (Molloy 1964). Deforestation of these montane slopes (by induced fires) has accelerated erosion, accretion and colluviation. The rejuvenated surfaces have been reclothed, over a period of 700 years, by first tussock and more recently *Dracophyllum* shrubland, with much slower spread of beech.

In the past, Cockayne (1928) and Robbins (1962) believed they were witnessing the evolutionary subjugation of primitive conifers by modern hardwoods. Such replacements are now regarded as localised to particular landscape states, however, and are understood in terms of a cyclical paradigm and not a grand evolutionary struggle. Likewise, earlier ideas of climate-induced regeneration gaps in podocarp populations (Holloway 1954; Wardle 1963) have been superceded by a view of episodic rejuvenation of landscapes or patches, resulting in explosive colonisation by podocarps that come to dominate as a cohort for up to a millenium. Gap-phase disturbances gradually erode the even-aged nature of the forest. Lack of synchronous "regeneration gaps" across the country, and the examples of vigorous podocarp and kauri regeneration (especially when browsing mammals are excluded, *cf.* McSweeney 1982), argue against purely climatic interpretations of forest demographics.

We can piece together various fragments of New Zealand's forest history: waves of immigration, the quietly or sometimes violently restless land, glacial disruption and impoverishment of the flora, the generalist nature of the survivors (adapted to mediocrity), the variable dispersability and longevity of the trees, the consequent uneven rates of advance and retreat, the soil processes, and finally the human-facilitated influx of many specialists and specialist generalists drawn from the global genetic storehouse. Overall, the seemingly contradictory dynamics reported for New Zealand forests can be interpreted in terms of the building and degrading scenario proposed by Gibbs (1983) and Meurk (1982).

Contemporary forest pattern

Discrepancies exist between physiological limits and actual distributions of forest species. There are few that cannot be grown throughout the length of the country (Wardle 1988). Plants more or less endemic to islands off the northern tip of New Zealand (*Meryta*, *Tecomanthe*) are cultivated in Dunedin; kauri can grow on Stewart Island and in frost- and drought-prone Canterbury. Of course kauri survived in Auckland during glacial times under conditions similar to present-day Dunedin, therefore there seems little reason why it should not survive and compete in these southern latitudes now (Fig. 5), notwithstanding the more modest potentials estimated by Mitchell (1991) and the need to distinguish between the environment in which a tree establishes and that in which it later finds itself. Likewise there is no ecophysiological reason for the absence of podocarps and many other hardwoods from the northern Subantarctic islands (Fig. 6).

Given the broad amplitude of the flora, the relatively small size of the country, and the disequilibrium of the vegetation, it would be perilous to make assumptions about direct relationships between contemporary environment, species limits and their potential distribution. This point is implicit in Holloway's (1954) signal work describing the laboured migration of beech.

Within this potpourri of history, environment and biology, forest ecologists have recognised classes of forest, but they are very broad and must admit a great deal of variation, flexibility and intergradation. As Ogden (1985) has illustrated, surviving bits of a preexisting forest stand alongside mature trees of a later community and the first generations of a future forest. Each cohort has its own history.

The broad floristic potentials and patterns are outlined in Table 3 (Wardle 1984). The regional forest floras decline from north to south and from west (wet) to east (dry) (Fig. 4). This is a genuine relativity, even if the potential numbers have not yet been realised in all places.

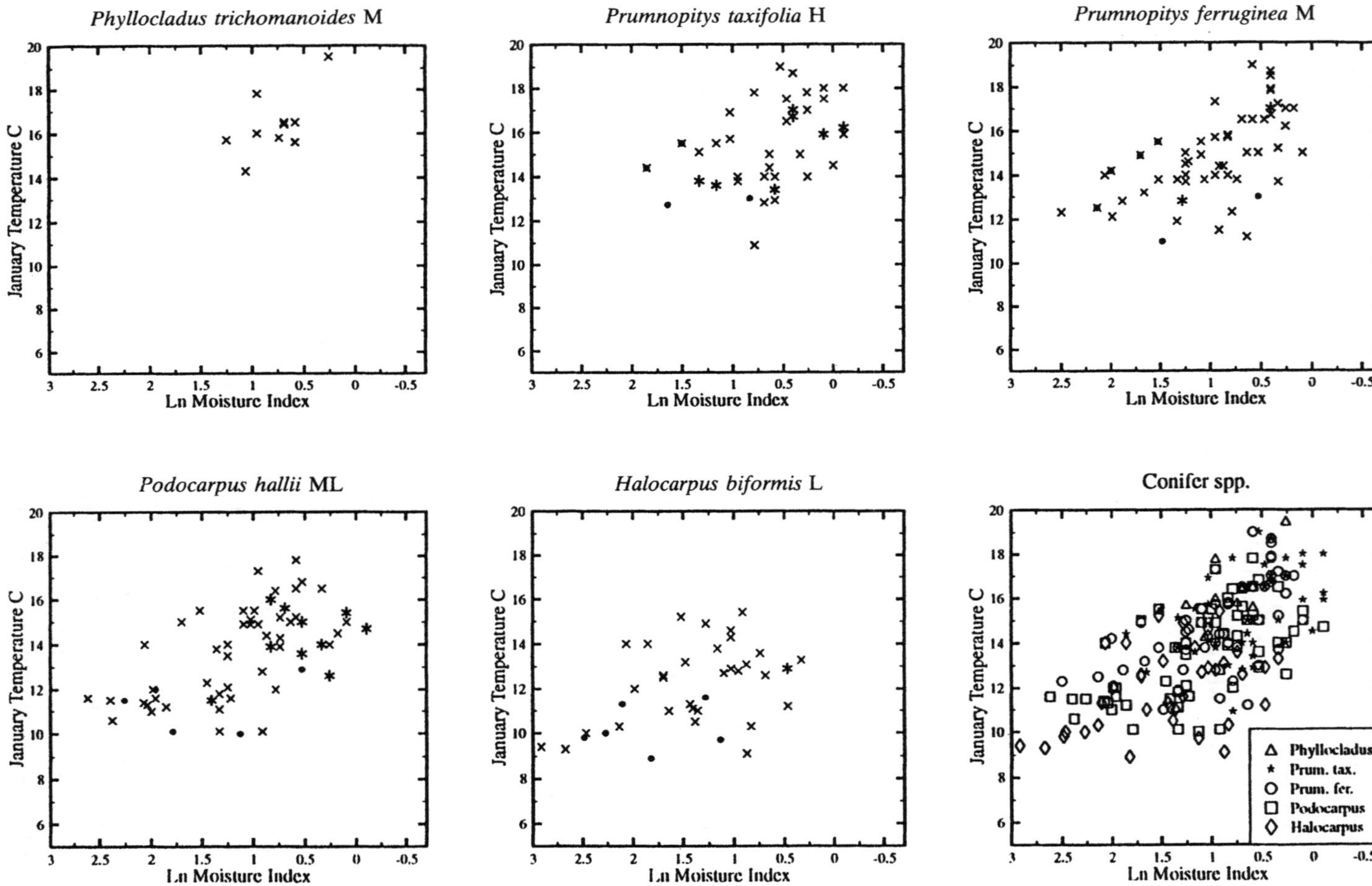

Fig. 6. Scattergrams of forest conifers (as in Fig. 5).

Table 3. Trends of species richness in New Zealand forests along gradients of latitude, altitude and annual rainfall (from Wardle 1984)

Gradient	Forest families	No. tree species	Vasc. spp. per stand
Top North I	69	>36	*c.* 25
Top South I	60	18	29
Stewart I	39	9	27
Auckland I	11	6	10
Campbell I	7	3	8
Middle New Zealand			
Sea Level		9	27
Footslopes		12	34
Timberline		2	14
South Island			
Rain 5000 mm			23
Rain < 1000 mm			<6

Forest formations

The following outline of largely unmodified forest formations is based on the physiognomic/timber categories developed by Nicholls (1976, 1983) and McKelvey (1984). It is a classification based on both subjective and statistical analysis. The main segregation centres on the dominance of kauri in northern North Island; *Beilschmiedia* spp. in the North Island and northern South Island; tall podocarps throughout the three main islands; southern beech species on poor or cold soils of the North and South Islands; low podocarps and hardwoods on impoverished and cold soils of the three main islands; and low hardwoods in coastal, seral and subantarctic island habitats. Forest-soil associations are indicated in Table 1. The class characterisations given below are the accumulated top four tree species (in constant, not abundance order) for all the subtypes (mapping units) of each class that these authors list. "N" and "S" suffixes indicate the species occurs only in the North or South Islands respectively. These categories have been the basis for forest mapping by the N.Z. Forest Research Institute. Other national or regional maps, with forest, have been compiled by Atkinson (1981), Clarkson (1986), Newsome (1987), Wilson (1987), Leathwick et al. (1988), and Meurk & Given (1990).

Kauri forest

A. Kauri
- [abundant, giant kauri with diverse understory; northern North Island]
- kauri (N), rimu, miro, mountain totara, tanekaha, rewarewa.

B. Kauri-Podocarp-Hardwood
- [large scattered or logged kauri; northern North Island]

- kauri (N), rimu, miro, mountain totara, *Halocarpus kirkii* (N), *Lepidothamnus intermedius*, toatoa (*Phyllocladus glaucus*)(N), tawa, rata (*Metrosideros robusta/umbellata*), *Weinmannia*, rewarewa, pohutukawa (*Metrosideros excelsa*)(N), kohekohe (*Dysoxylum*).

C. Kauri-Podocarp-Hardwood-Beech
- [small, perhaps secondary kauri and beech; northern North Island]
- kauri (N), rimu, miro, mountain totara, tanekaha, toatoa (N), hard beech (*N. truncata*), silver beech (*N. menziesii*), tawa, rata, hinau (*Elaeocarpus dentatus*), *Weinmannia*, *Ixerba brexioides* (N).

Podocarp-hardwood forest

L. Podocarp
- [emergent, dominant tall podocarps; typically central North and South Islands]
- kahikatea, rimu, miro, matai, totara, tanekaha, mountain totara, mountain cedar, *Lagarostrobus colensoi*, pink pine, hard beech, tawa, *Weinmannia*, rewarewa, *Nestegis*, rata, broadleaf (*Griselinia littoralis*).

M. Rimu-matai-hardwood
- [frequent podocarps, abundant hardwoods in canopy; throughout]
- rimu, matai, kahikatea, miro, totara, mountain totara, mountain cedar, tawa, rata, *Weinmannia*, hinau, rewarewa, pokaka (*Elaeocarpus hookerianus*), *Nestegis*, broadleaf, mahoe (*Melicytus ramiflorus*), *Fuchsia*.

E. Rimu-taraire-tawa
- [rimu and *Beilschmiedia* forest; northern North Island, <450 m, where kauri absent]
- rimu, miro, mountain totara, taraire (*Beilschmiedia tarairi*) (N), tawa, rata, kohekohe, *Weinmannia*.

D. Rimu-Tawa
- [occasional podocarps in tawa forest; North Island and northern South Island hill country]
- rimu, miro, tawa, rata, pukatea (*Laurelia novae-zelandiae*), hinau, kohekohe, *Weinmannia*, *Ixerba* (N), rewarewa, mahoe, *Toronia toru* (N).

F. Rimu-general hardwood
- [above limits of *Beilschmiedia* spp. throughout]
- rimu, miro, mountain totara, kahikatea, mountain cedar, rata, pukatea, *Weinmannia*, hinau, rewarewa, *Nestegis*, broadleaf.

G. Lowland Steepland and Highland Podocarp-Hardwood
- [low forest in high country, generally above rimu limit, where beeches are absent; throughout]
- rimu, miro, mountain totara, tanekaha, toatoa (N), mountain cedar, *Phyllocladus alpinus*, pink pine, *Lepidothamnus intermedius*, bog pine, *Lagaro-*

stobus, *Weinmannia*, rata, *Ixerba* (N), hinau, broadleaf, pokaka, *Quintinia*, *Olearia lacunosa*, *Pseudopanax crassifolius*.

Podocarp-hardwood-beech forest

H. Rimu-tawa-beeches
- [rimu and tawa with pockets of beech; North Island and northern South Island]
- rimu, miro, mountain totara, tanekaha, black beech, hard beech, red beech, silver beech, tawa, rata, pukatea, hinau, *Weinmannia*.

I. Rimu-general hardwoods-beeches
- [abundant beech, above tawa limit; throughout]
- rimu, miro, totara, mountain totara, matai, mountain cedar, tanekaha, toatoa (N), kahikatea, *Lepidothamnus intermedius*, *Lagarostrobus*, pink pine, black beech, red beech, hard beech, silver beech, mountain beech, *Weinmannia*, rata, pukatea, *Ixerba* (N), *Quintinia*, pokaka.

J. Highland podocarp-hardwood-beeches
- [low forest above rimu limit; main ranges]
- *Phyllocladus alpinus*, miro, mountain cedar, pink pine, mountain totara, toatoa (N), *Lagarostrobus*, hard beech, black beech, mountain beech, red beech, silver beech, *Weinmannia*, broadleaf, *Ixerba* (N), *Quintinia*.

Beech forest

K. Beeches
- [above rimu limit, below on harsh sites; throughout hills and mountains up to timberline of North and South Islands]
- red beech, silver beech, mountain beech, black beech, hard beech, miro, mountain totara, *Phyllocladus alpinus*, pink pine, mountain cedar, *Weinmannia*, *Ixerba* (N), *Quintinia*, broadleaf, *Pseudopanax simplex*, *Fuchsia*, manuka (*Leptospermum*).

Hardwood forest

P. General hardwoods
- [natural or induced secondary, coastal or dryland low forest, below rimu limit, usually without *Beilschmiedia* spp.; throughout]
- tawa, pohutukawa (N), rewarewa, rata, kohekohe, *Vitex lucens* (N), *Litsea* (N), *Weinmannia*, *Hedycarya*, hinau, kowhai, *Nestegis*, *Olearia* spp., *Pseudopanax arboreus*, *Quintinia*, mahoe, broadleaf, *Pittosporum* spp., kanuka, manuka.

Variants

Variants of these basic types are dictated by topographically determined drainage and fertility, by parent materials, age of stand, logging/burning history, and by feral mammal browsing.

Semi-swamp forest is a variant of the Podocarp Type L Forest on fertile, well watered or flooded alluvium (see Norton 1989, Duncan et al. 1990, Norton & Leathwick 1990). The dominants are:

- kahikatea, pukatea, silver beech, pokaka, *Lepidothamnus*, *Plagianthus regius*, *Syzygium*, *Cordyline*, manuka.

On fertile, drier/better-drained **Terrace or Ash Forests**, the following species are typical:

- matai, kahikatea, totara, hinau, *Hedycarya*, *Streblus*, *Pennantia*, *Hoheria angustifolia*, red beech, broadleaf, *Carpodetus*, *Plagianthus regius*.

Wardle (1970, 1984) has further subdivided the **Beech Forests** into 12 general types: Mountain beech of upper slopes, Silver beech of upper slopes, Silver beech-*Ixerba*, Mid-slope silver beech-red beech, Mid-slope beech-*Metrosideros-Weinmannia*, Pure red beech, Pure silver beech of mid and lower slopes, Hard beech-*Agathis*, Beech-*Beilschmiedia*, Hard beech-black beech-podocarp, Pure black beech, Stunted beech-podocarp forest and shrubland.

Dwarf Forests of cold, wet, acid soils usually support:

- stunted cupressoid podocarps, *Phyllocladus alpinus*, mountain or silver beech, epacrids, manuka, asterads and, on less impoverished sites, mountain cedar.

Coastal Forests are almost entirely hardwoods with a number of frost-tender and northern species of tropical affinity:

- *Metrosideros*, *Planchonella*, *Nestegis apetala*, *Litsea*, *Pisonia*, *Vitex*, *Pomaderris*, mangrove in tidal estuaries (all confined to the North Island), *Dysoxylum*, *Macropiper*, *Melicytus*, *Entelea*, *Corynocarpus*, *Dodonaea*, *Muehlenbeckia*, *Hebe*, *Plagianthus divaricatus* (salt marshes), pittosporums, ngaio, coprosmas, asterads, poroporo (*Solanum*), and nikau palm (*Rhopalostylis*).

Dryland Low Forests have a savanna-woodland aspect (Burrell 1965, Wells 1982), often forming a mosaic with short tussock grassland. The woody species are characterised by small-leaved dwarf trees and shrubs including many with divaricating growth forms:

- kanuka, kowhai, NZ brooms, coprosmas, asterads, *Discaria*, *Corokia*, *Pomaderris*, *Cordyline*, *Melicytus alpinus*, pittosporums, and totara and matai at the moister fringes or on deeper soils.

A number of forests have had their conifer component logged out, although subsequent regeneration is taking place. The introduction of browsing mammals has had a devastating impact on both the understorey and canopy hardwood trees of New Zealand forests (Mark et al. 1991, Mark 1989, Allen et al. 1984, Veblen & Stewart 1980, Leutart 1988). Similarly, introduced predators have greatly affected the indigenous invertebrate and bird life (Clout & Hay 1989).

Forest structures

The forest types (above) are based only on the canopy and emergent trees. Structure and diversity within tiers follows the same warm-cold/wet-dry/sheltered-exposed pattern demonstrated by the dominants. The tallest emergents may attain 60 m on fertile, sheltered sites, with 25–40 m being more common. Canopies are typically 10–25 m both in the complex podocarp forests and the simpler beech forests. On exposed, dry, cold or impoverished sites the canopy reduces to 2.5–10 m.

The **conifer-hardwood forest** is the most exuberantly stratified (Fig. 13). Typical components of the layers are:

- **Emergents** – rimu, totara, kahikatea, kauri, miro, matai, mountain totara, *Libocedrus plumosa*, rata, pukatea, rewarewa, and mountain cedar;
- **Canopy** – taraire, tawa, *Weinmannia*, *Vitex*, tanekaha, *Nestegis*, hinau, pokaka, rata trees, cupressoid podocarps;
- **Subcanopy** – kohekohe, nikau, mahoe, *Hedycarya*, *Myrsine*, treeferns (*Cyathea*, *Dicksonia*), broadleaf, araliads, pittosporums, small podocarps, *Ixerba*, *Caldcluvia*, *Toronia*, *Alectryon*, *Quintinia*, *Mida*;
- **Gaps and Gullies** – *Aristotelia*, *Carpodetus*, *Pennantia*, fuchsia, *Hoheria*, *Plagianthus*, pittosporums, *Pseudopanax*, *Cordyline*, *Schefflera*, *Entelea*, *Quintinia*, treeferns;
- **Shrubs** – *Macropiper*, *Pseudowintera*, *Geniostoma*, coprosmas, *Schefflera*, *Pseudopanax*, *Myrtus s.l.*, *Melicope*, tree daisies, *Alseuosmia*, *Rhabdothamnus*, *Phebalium*, *Myrsine*, epacrids;
- **Ground** – *Asplenium*, *Blechnum*, *Polystichum*, *Astelia*, *Microlaena*, *Elatostema*, *Gahnia*, *Uncinia*, *Carex*, *Chionochloa*, *Anemanthele*, *Hymenophyllum*, orchids, herbaceous Apiaceae, Rubiaceae, Urticaceae, Lamiaceae, other lilies and ferns, mosses, liverworts, lichens;
- **Epiphytes, lianes etc.** – *Rubus*, *Parsonsia*, *Metrosideros*, *Ripogonum*, *Passiflora*, *Collospermum*, *Astelia*, *Clematis*, *Freycinetia*, *Muehlenbeckia*, *Senecio sciadophilus*, ferns, bryophytes, lichens.

As many of the dominants are relatively shade-tolerant, the understorey, in unbrowsed condition, is usually occupied by a range of regenerating age-classes, especially when there has been localised opening of the canopy.

Beech Forests are much simpler and may be almost monocultural as far as vascular plants are concerned, at least in the driest or coldest areas.

However, browsing may have exaggerated this apparent simplicity. The major associates are (Wardle 1984):
- mistletoes, coprosmas, epacrids, mountain totara, *Phyllocladus alpinus*, broadleaf, araliads, *Myrsine divaricata*, orchids, ferns, bryophytes, lichens.

In wetter or warmer conditions, unbrowsed beech incorporates increasing contributions from the podocarp-hardwood forest flora (Fig. 14):
- combinations of the above two groups with particular prominance of *Pittosporum*, *Elaeocarpus*, *Griselinia*, *Carpodetus*, araliads, coprosmas and podocarps.

Coastal and Subalpine Low and Dwarf Forests have even canopies of bushy trees with scattered herbs and ferns in coastal, and often dense hummocks of cryptogams in the Subalpine-Subantarctic types (Fig. 15).

Significance of growth forms and forest structure

There are about 150 forest trees and another 100 or so common vascular associates (Wardle 1984). Wardle (1991) gives totals of 358 Warm Temperate and 61 Cool Temperate forest species of which 48 are large trees and >70 are usually less than 10 m tall. The tallest podocarps are 50–60 m and kauri is known to have up to a 7 m girth. *Knightia*, *Laurelia* and *Metrosideros robusta* attain 30 m, with several other hardwoods commonly forming a canopy at 20–25 m.

Growth forms and vegetation structure have been compared to tropical rainforests. True, our podocarp-broadleaved forests appear rich in species and luxuriant, with mounds and curtains of bryophytes, lichens and ferns, and there is the absence of nemoral woodland flowers so characteristic of continental, deciduous woodlands; some trees have plank buttresses, prop roots, pneumatophores, aerial (column) roots, cauliflory, heteroblasty, drip tips, pulvini and domatia. There is limited deciduousness, weak bud protection, and a high incidence of smooth leaf margins; there are many vines (stem twiners, tendrils, hooks, roots), epiphytes (mat, nest, strangler, pendant, shade and sun, epiphylls), vascular parasites and saprophytes (Dawson 1988). Furthermore, there is stratification of the forests into emergents (> 30–40 m), canopy (20–25 m), subcanopy (10–15 m), shrubby-small trees (3–8 m) and ground layer (0–1 m). However, the main divergence from true lowland tropical rainforest is in the fewer species (dozens versus hundreds), the one or few species per tree genus (Wardle 1978), the mixture of conifers and angiosperms versus the pure angiosperm forests of the tropics, and the smaller leaves (mainly microphylls versus mesophylls and some macrophylls). Thus the apparent tropical affinities are largely structural (rather than physiological/dynamic) and reflect the humid, weakly seasonal climate and extensive development of semi-swamp forests with the corresponding adaptations.

The real ecological affinity is with the montane tropics, as Troll (1960) has remarked on before. And this extends to the floristic level in the case of the Malesian tropics, a possible consequence of shared Gondwanic heritage.

An idiosyncrasy of New Zealand woody vegetation is the large number of small-leaved, divaricating shrubs or juvenile states (Halloy 1990) – the filiramulate form of Wardle (1991). One view is that this is a climatic adaptation to establishment in dry, or physiologically droughted soils, subject also to drying winds. Once the tree is established, roots are down to reliable water, there is mutual shelter from wind (or the water table has dropped in swampy soils), a more "normal" adult growth form and foliage may develop (see McGlone & Webb 1981). An alternate or complementary theory, advanced by Atkinson & Greenwood (1989) is that these "inside-out" and often cryptic growth forms are a "fossilised", adaptive defense against browsing by the now extinct ratite (*Dinornis*). We have to be aware that the pristine fauna was vastly more rich than its present emasculated condition (Holdaway 1989); so pollination, dispersal, browsing, nutrient status, and disturbance were all facilitated to a greater extent than we might presently imagine. S. Halloy (pers. comm. 1991) has demonstrated that the "filiramulate" growth form is not unique to New Zealand, indeed is well developed in South America and the Galapagos. In these cases there is an additional defense of spininess, and Halloy points out an historical coincidence with supplementary browsing by marsupials or tortoises, and the late arrival of placental mammals in South America.

Autecology of key forest species

Current thinking is that we treat plant communities as ebbing and flowing mixes of individuals mostly acting independently of one another. The long lists of weakly segregated species in the forest types outlined above would seem to support a non-equilibrium, individualistic paradigm. Alternatively, the environmental contrasts within a relatively small land mass (notwithstanding the much larger area of the continental shelf) may not be sufficient to have established extensive, discrete nodal communities such as dominate the continents (that is, huge uniform areas with relatively narrow zones of transition to other distinctive uniform areas). The homogeneity across New Zealand will have been emphasized by the glacially imposed floristic impoverishment of warm climate specialists, and again by the presently unpredictable climate.

I intend now to examine the environmental envelopes of a range of the dominant or diagnostic forest species and see how their potentials, derived from altitudinal limits, fit with their actual geographical distributions, and whether there is any basis for stable associations of species in tune with a coherent habitat configuration. Finally I shall compare the New Zealand forest envelope with that of overseas evergreen forests.

The scattergrams (Figs. 5–11) demonstrate the potential occupancy of the environment, by dominant or diagnostic forest species. The method was to compile lists of altitudinal ranges (e.g. Wardle 1984, Clayton-Greene & Wilson 1985, Clarkson 1986) for the species considered, for several climatic sectors of the country, and to transform these data into climatic variables (mean January air temperature, precipitation, evapotranspiration, continentality) by using appropriate lapse rates (Meurk 1982, Jane & Green 1984, Norton 1985, N.Z. Meteorological Service 1983, 1985). High-altitude climate records are sparse and of lower quality than for the lowlands, so some estimation was required. At high elevations the temperatures might be treated as relative rather than absolute, but in other cases of unusually low "tree temperatures" we have to remember that tree limit here relates to a 2.5 m definition of a tree, and secondly in the southern islands the low January temperature is compensated by the almost year-round growing season (see Fig. 1 and Meurk & Given 1990). The MI (moisture index) data, although the more difficult to measure, represent a robust parameter and should be fairly represented.

Each diagram attempts to encapsulate 4 principal dimensions of the species niche. A soil fertility index is appended to each species name (*cf.* Wardle 1991), the coordinates of the data points are summer mean temperature and the natural logarithm (Ln) of the Moisture Index, and the symbols express the Conrad (1946) Continentality Index, a measure of seasonality.

First, if we consider the broad climatic conditions of lowland New Zealand, we see January temperatures range from 22°C to 6°C from the northern subtropical islands to the southern subantarctic islands (Fig. 1). Tree limit (January Temperature of 8–10°C) declines from < 1550 m in the north to 120 m on Campbell Island and theoretically below sea level on Macquarie. The main axial ranges divide the mainland into a very wet zone in the west with Moisture Indices (MI) = 2–10 at sea-level, and a dry rainshadow in the east with MI = 0.4–1.4. Continentality ranges from − 5 in the subantarctic islands to 17 in the inland basins of south Canterbury and Otago (Fig. 2). Soils are generally dystrochrepts or shallow, strongly stratified, weakly to strongly acid soils with moderate to low levels of P and cations (Table 1, 2, Fig. 5). Drought conditions in the east may be exacerbated by the shallow, free draining, coarse textured soils of the eastern plains.

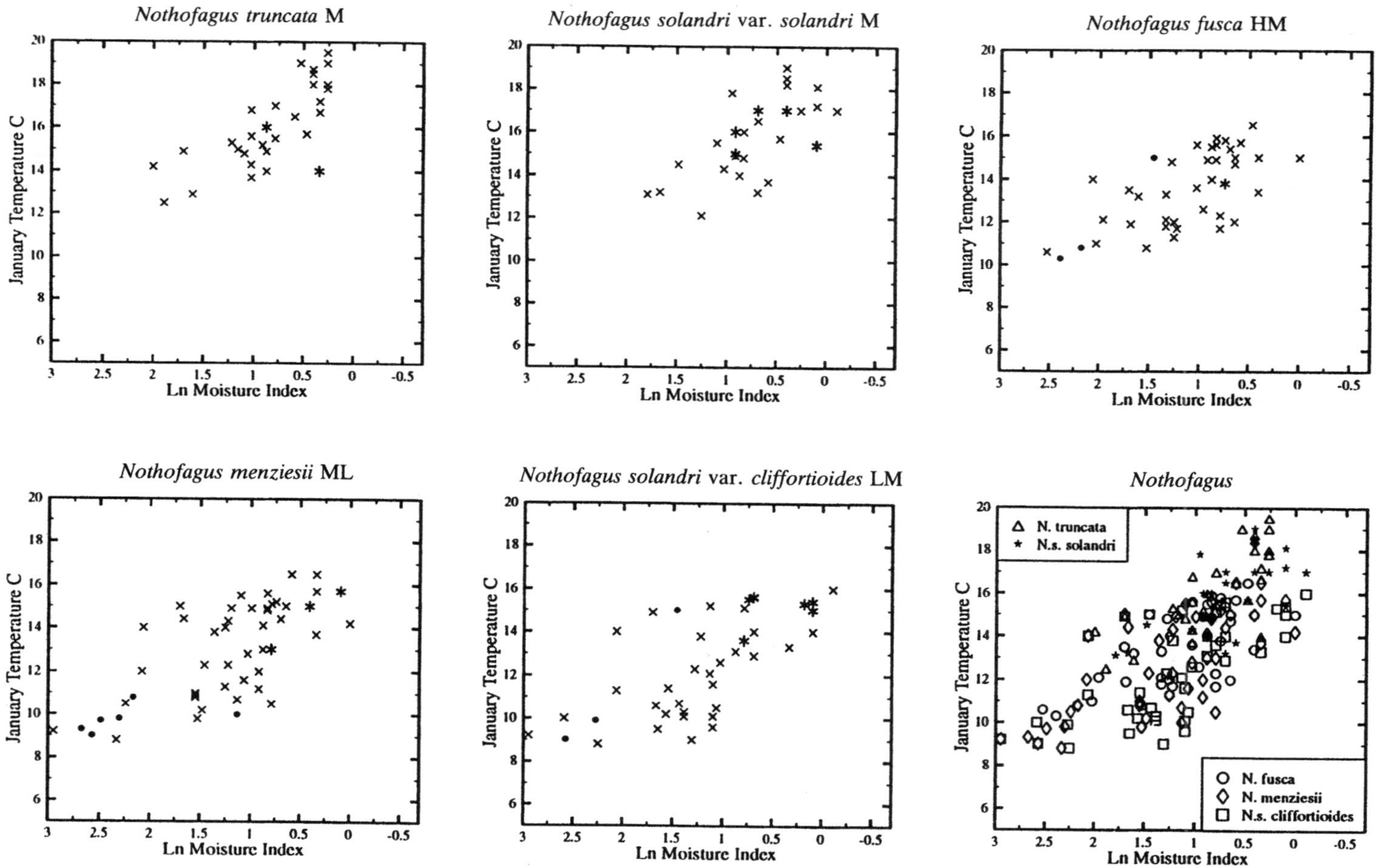

Fig. 7. Scattergrams of southern beeches (as in Fig. 5).

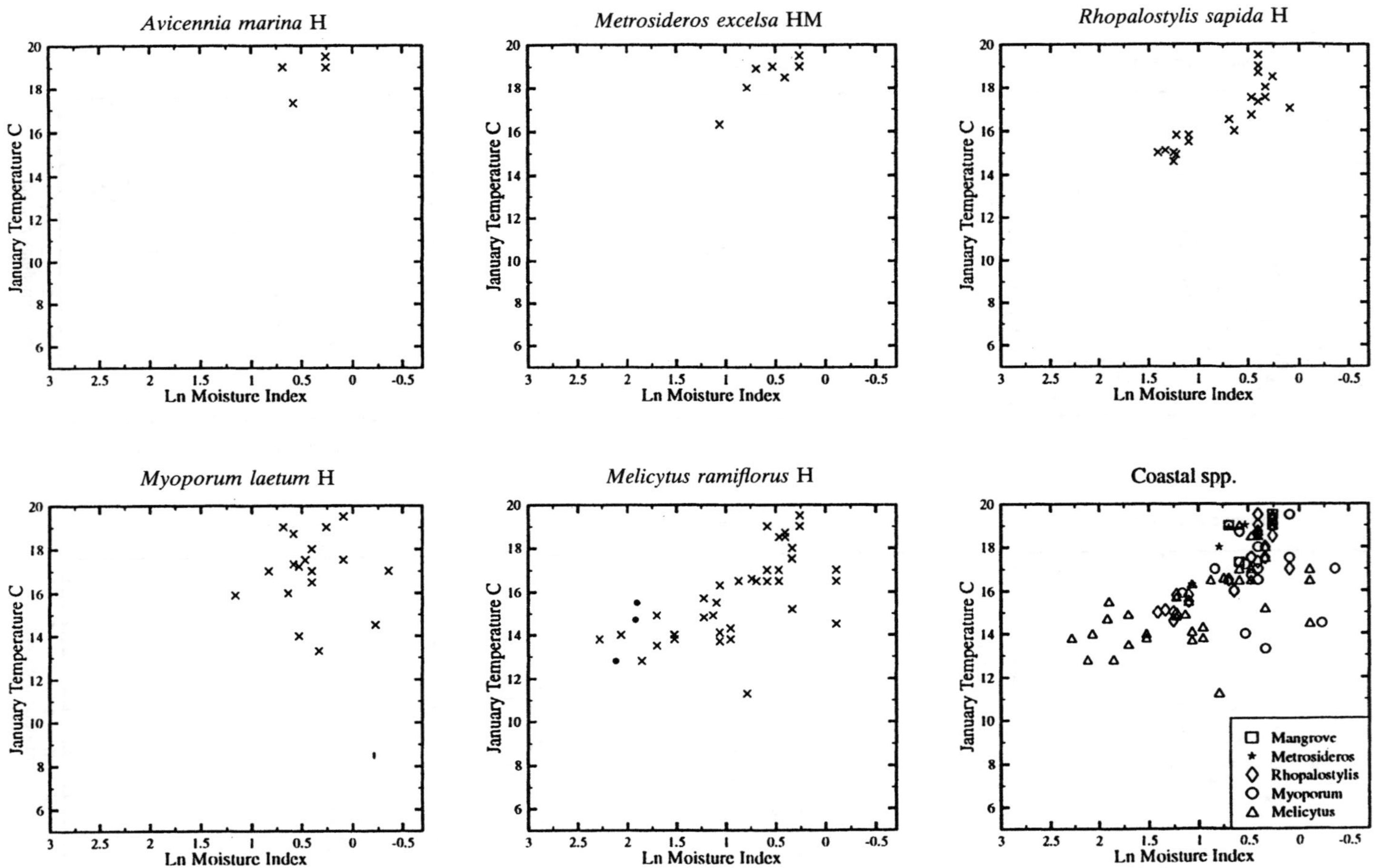

Fig. 8. Scattergrams of coastal bush hardwoods and a palm (as in Fig. 5).

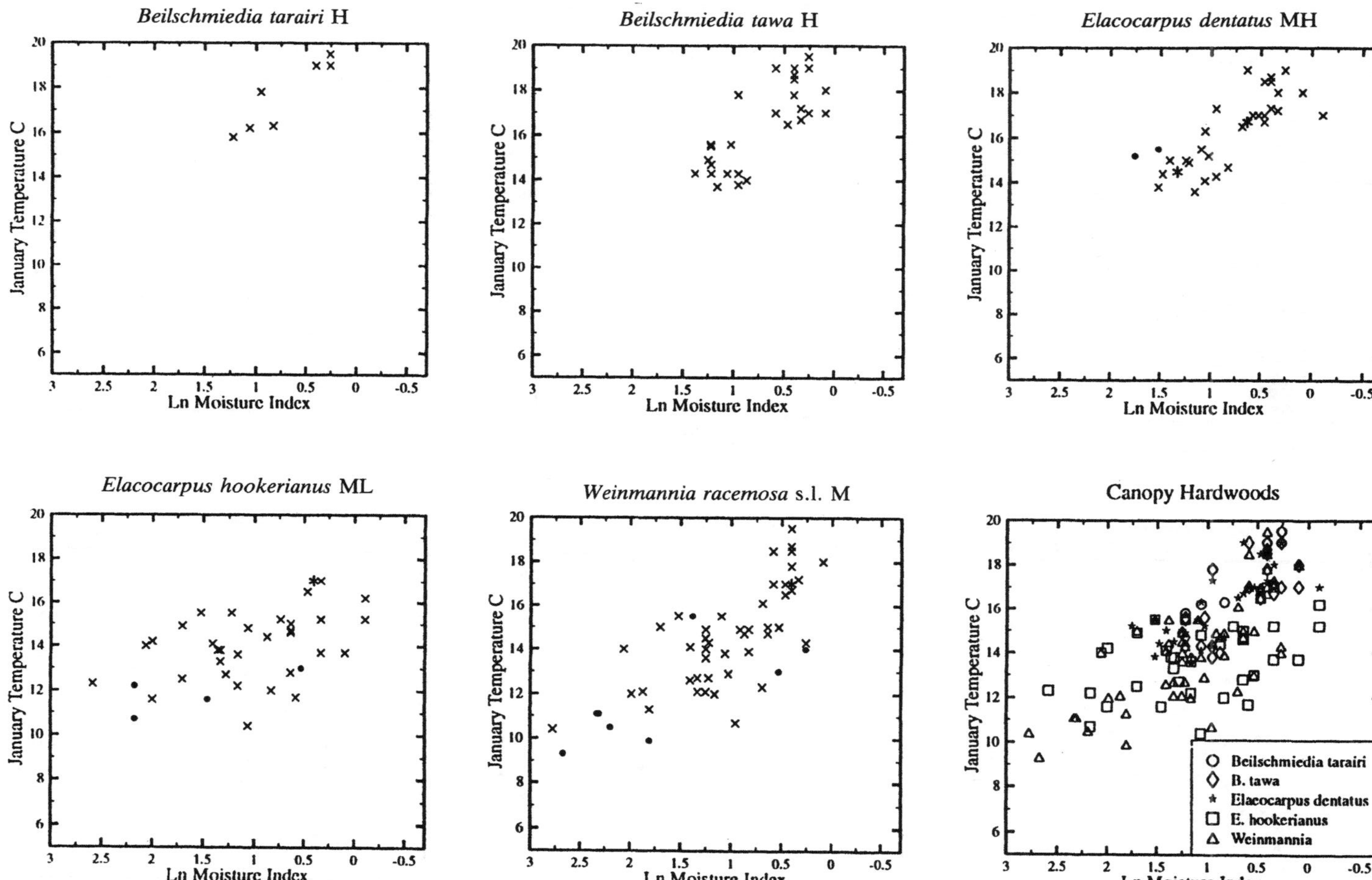

Fig. 9. Scattergrams for canopy hardwoods (as in Fig. 5).

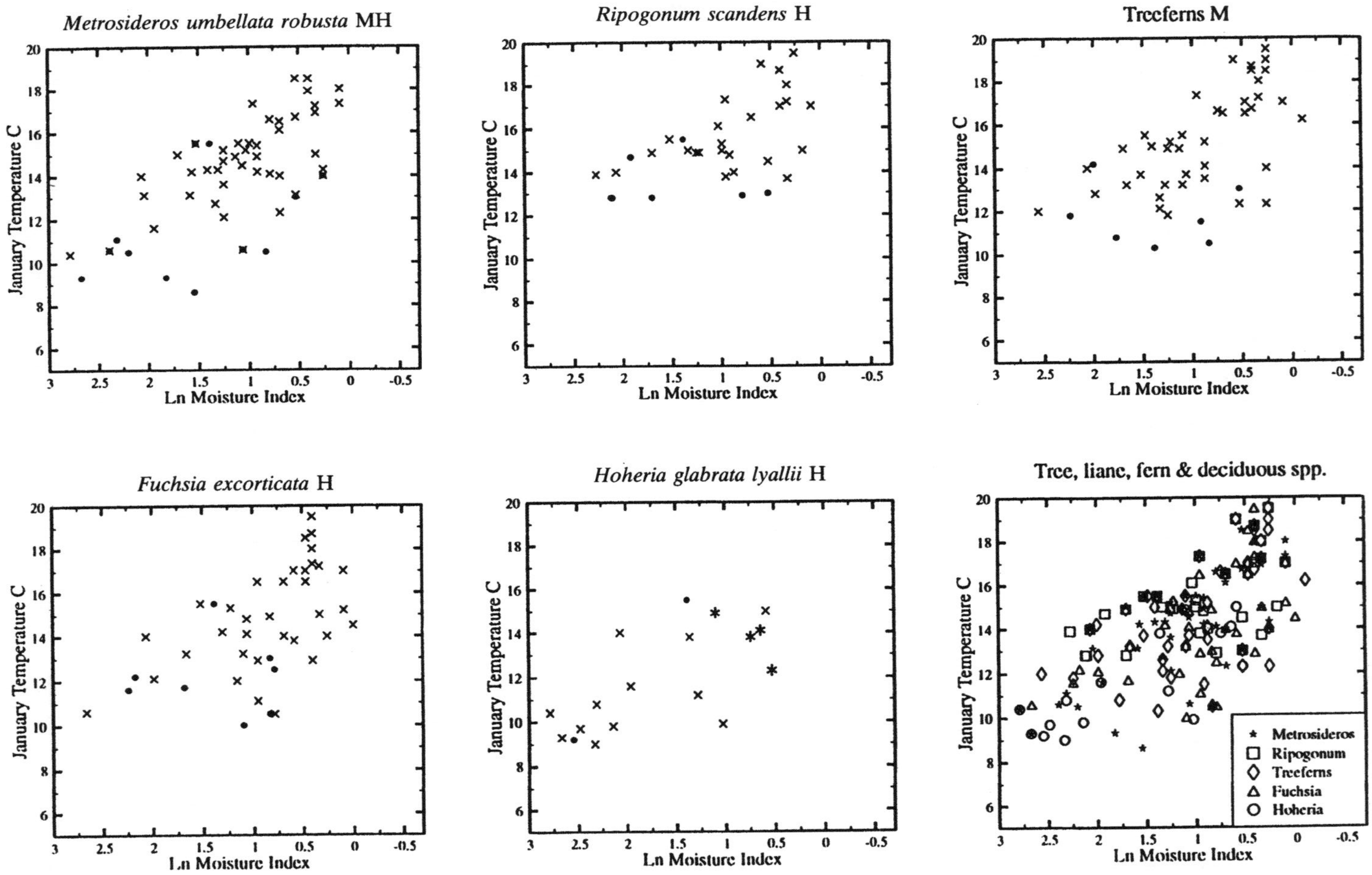

Fig. 10. Scattergrams for tree, liana, tree-fern and deciduous species (as in Fig. 5).

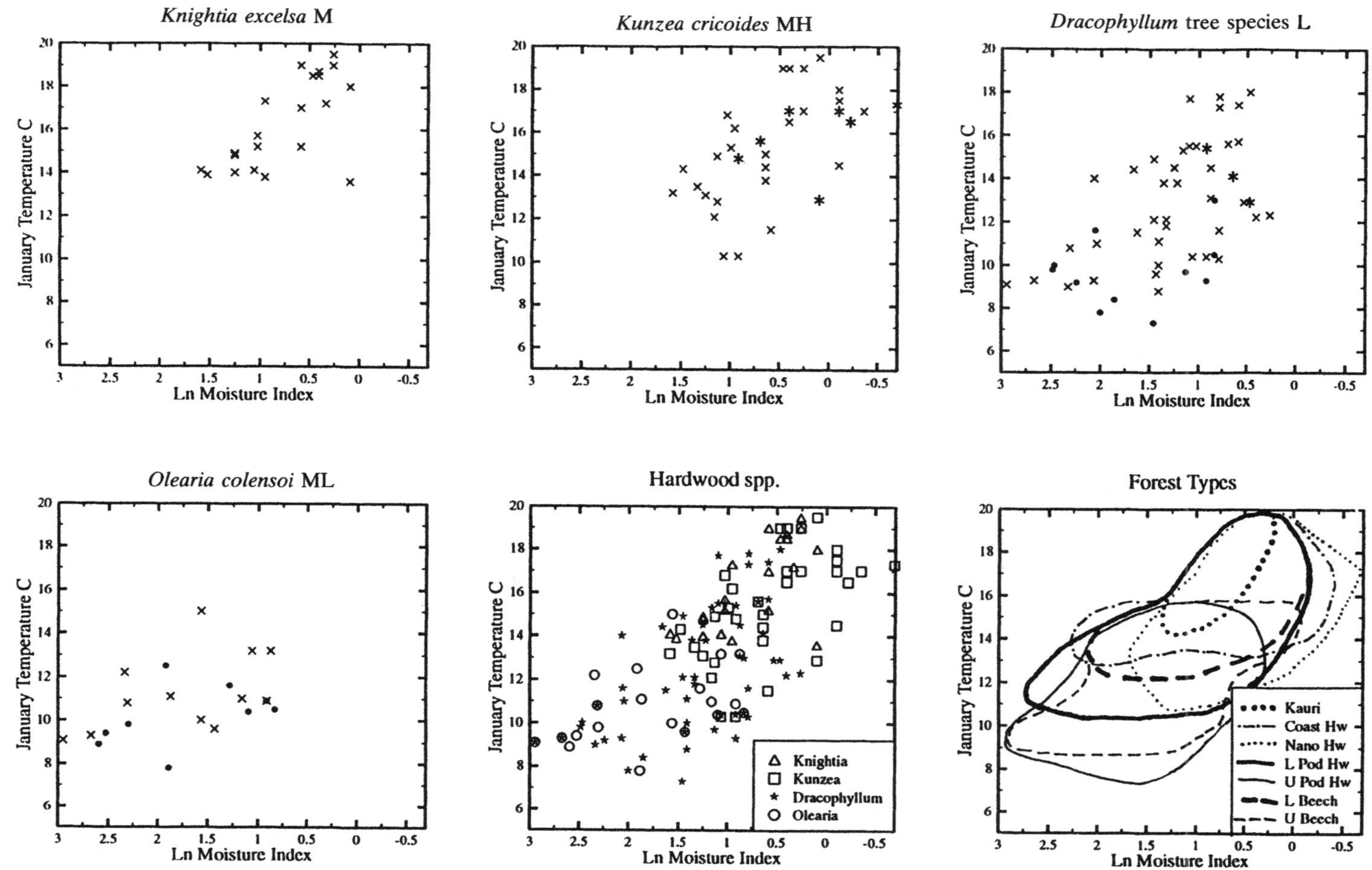

Fig. 11. Scattergrams for various hardwood species, and summary environmental envelopes for the major forest classes in New Zealand (as in Fig. 5). Hw = hardwoods, Pod = podocarps, Nano = nanophyllous small trees, L = lowland, U = upland. The upland podocarp-hardwood forest extends beyond the limits of beech forest, in superoceanic environments, as predominantly hardwood dwarf trees.

The conifers

Agathis australis (kauri) is a giant, distributed in warm, moist areas with moderate seasonality, and tolerates impoverished soils; however its high-elevation populations grow in conditions equivalent to lowland Southland. This conforms to its known cultivable limits.

Dacrydium cupressinum (rimu) is an ubiquitous emergent in moist to superhumid, oceanic climates, on mature to strongly leached soils.

Dacrycarpus dacrydioides (kahikatea), our tallest tree, is a semi-swamp species of warmer fertile sites. The intrazonal wet soils, in which it grows, allows penetration into dry, continental climates.

Libocedrus bidwillii (mountain cedar) is a montane to subalpine emergent in superhumid leaching climates.

Phyllocladus trichomanoides (tanekaha) is a canopy species with a similar niche and unrealised potential to that of kauri.

Prumnopitys taxifolia (matai) is a tall tree, preferring fertile, warmer sites, capable of extending into dry, continental environments. Matai has a divaricating juvenile form.

Prumnopitys ferruginea (miro) is a large, ubiquitous, but less frequent tree than rimu, in moist to superhumid, warm to cool climates.

Podocarpus hallii (mountain totara) is a montane tree of dry to superhumid conditions, on poorer soils than *P. totara* (totara).

Halocarpus biformis (pink pine) is a canopy to emergent montane to subalpine tree or stunted shrub, similar to kaikawaka.

Southern beeches

Nothofagus truncata (hard beech) is a canopy tree of warm, moist climates on poorer soils.

Nothofagus s. solandri (black beech) is similar to hard beech but extends onto drier, seasonal, more fertile sites. Black and hard beeches are probably capable of extending their ranges further to the south.

Nothofagus fusca (red beech) is a tall canopy tree of mesic to oceanic, preferably fertile sites.

Nothofagus menziesii (silver beech), typically of cool, wet, leached soils, occurs also as relicts in favourable microsites on the fringes of "continental" Otago.

Nothofagus s. cliffortioides (mountain beech) is the widest ranging and most cold-tolerant and oligotrophic beech, but, as with silver beech, is excluded from warmer sites.

Coastal trees

Avicennia (mangrove) and *Metrosideros excelsa* (pohutukawa, of cliffs, and closely related to the Hawaiian ohia) are presently confined to the warm north. *Rhopalostylis* (nikau, the southernmost palm) extends into the slightly cooler South Island; while *Myoporum* (ngaio) and *Melicytus* (mahoe) are frost-tender but drought-resistant species that hug the southern coasts or other warm microsites. All these species, and other associates, are excluded from seasonal, inland climates. All occur on relatively rich soils because of the maritime influence and weakly leaching conditions.

Canopy hardwoods

Beilschmiedia tarairi (taraire) is a lauraceous tree of the warmest parts of the country, occurring on relatively rich, moist sites, often associated with lowland kauri forest.

Beilschmiedia tawa (tawa) is a wattle-like tree of relatively warm, fertile sites with reliable rainfall and weak seasonality. Appears to be geographically unfulfilled.

Elaeocarpus dentatus (hinau) is a fairly ubiquitous tree of lowland to lower montane, mesic forests.

Elaeocarpus hookerianus (pokaka) is complementary to the hinau in that it extends into cooler, wetter sites, including semi-swamp forests in dry climates, usually with kahikatea. It has a divaricating juvenile state.

Weinmannia racemosa (kamahi) and the clinally related *W. silvicola* (towhai) represent a ubiquitous aggregate on leached but well-drained hill soils, in moist to superhumid, oceanic climates.

Metrosideros robusta (northern rata) and *M. umbellata* (southern rata) have to some extent a complementary dominance in moist to superhumid environments of northern and southern (or lower and upland) parts of the country respectively. Both are massive trees preferring richer soils, often on well-drained ridges or coastal sites. Southern rata forms a Subantarctic, even-canopied timberline on the Auckland Islands in peat, but with the compensating input of maritime salts. Northern rata is distinguished by its habit of starting life as an epiphyte, commonly on kamahi, and sending "strangler roots" down to make contact with the ground and eventually usurping the place of its host.

Knightia excelsa (rewarewa) is one of the few proteaceous survivors from the Tertiary; it is an emergent to canopy species, often a secondary pioneering species of moist, warm to cool hill country. It shows potential to spread to the bottom of the South Island in oceanic environments, as might be expected from its good growth in Dunedin cultivation.

Other hardwood, deciduous, liana and tree-fern species

Ripogonum (supplejack) is a liana of moist to superhumid, fertile podocarp and coastal forests, up to montane levels.

Tree-ferns (about 20 species of *Cyathea* and *Dicksonia*) are common features of sheltered, moist to superhumid, Subtropical to Subantarctic podocarp or hardwood forests, often in gullies or secondary pioneering situations. The various species separate out according to fertility and saturation, with *Cyathea dealbata* occurring in relatively dry coastal forest and other species appearing in quite dry atmospheres where there is adequate ground water and shelter from frosts. There are no records from the continental end of the spectrum.

Fuchsia excorticata (kotukutuku) is one of the few deciduous small trees of warm to cold, moist to superhumid, oceanic, rich gully sites or in fertile terrace forests with an abundance of ground water.

Hoheria glabrata/lyallii (mountain ribbonwood) are two clinally related, deciduous, high-mountain, timberline species of fresh, rubbly soils. They occur where moisture levels are high, but the more eastern *H. glabrata* is exposed to more continental conditions.

Kunzea ericoides (kanuka), an abundant pioneering species on well-drained sites or subclimax in semi-arid areas such as inland Otago and Canterbury. Subject to the greatest degree of drought and temperature fluctuation. It is a myrtaceous species closely related to Australian kunzeas, but with an apparently long Tertiary record in New Zealand. The similar *Leptospermum scoparium* (manuka) performs a similar seral function on wetter and poorer ground, but both are very hardy.

Dracophyllum spp. growth forms range from cushion plants to small trees; only the tree species are considered here (*D. latifolium* in the northern kauri forests, *D. traversii*, *D. longifolium*, *D. filifolium* in Montane to Subalpine/Subantarctic zones). All require moist to superhumid conditions and are excluded from the warmest and driest areas. Soils are impoverished and highly acid gleys, podzols or peats; and these dwarf trees occur at the coldest extremities of forest.

Olearia colensoi, and other large-leaved asterads of superhumid Subalpine/Subantarctic regions, are similar to and associated with *Dracophyllum* spp., pink pine and kaikawaka, *Phyllocladus* and high altitude stands of manuka.

Species segregation

We can see that there is a vast overlap in the environmental envelopes of most of the species. Of course there are other dimensions to the species niche that cannot be depicted in a 2- or even 4-dimensional space as attempted here. Drainage is a complicating water balance factor that is not easily accommodated. Failing to distinguish between atmospheric and ground moisture can be misleading, such as when attempting to segregate kahikatea from species that genuinely tolerate very dry soils. Semi-swamp forests, with tree/shrub assemblages recognisable throughout the country, transcend relatively dry climates if their swampy habitats occur. On the other hand, the epiphytic vascular, bryophyte and lichen species present, do seem to be related mainly to atmospheric moisture. As one would expect, this element is extremely poorly developed in the drier climates – a case in point being a kahikatea stand within the city of Christchurch.

Another limitation is that the data represent the realised niches of the species. This may depart from the potential to varying degrees depending on dispersal and competition. Only when all species have dispersed to all potential environments will the ideal segregation be observed. The fundamental niche would then be deduced from non-competitive situations.

In the non-ideal world, New Zealand forests present the "kaleidoscopic variety of intergrading communities" alluded to by Wardle (1991). Certainly the composite species diagrams give this impression, but some of the overlap and confusion can be teased apart by considering the parts of ranges (towards the extremes) that do not overlap, as well as the other important niche dimensions of soil fertility and drainage, exposure, seasonality etc. Using Grime's (1979) terminology, species may be competitors, stress tolerators, or ruderals. In the forest context the tall trees that pack the centre and upper centre part of the graphs are the competitors; those at the far right, lower left and at the coast are stress tolerators; and the deciduous and seral species such as treeferns, *Kunzea* and *Leptospermum* might be partial ruderals. When "competitors" are eliminated, other groups may occupy the optimal, central habitats of the diagram in almost any combination, creating the apparent confusion. Since any forest is an aggregate of associations from different times, composed of species with varying life expectancies, responding to disturbance, climate change and successional dynamics, it is easy to see why the equation throws up endless combinations. It will only be at the envelope peripheries, beyond the thresholds of many of the more sensitive species, that segregation will be clear.

Even within the present non-equilibrium situation, there are some apparent threshold values for a number of (potential tree) species clusters. Other associates of the key indicators examined here in detail (Figs. 5–11) are listed above with the forest classes and structural descriptions. Associations and gradients may be further elaborated by reference to Wardle et al. (1983), Nicholls (1983), Wardle (1984), Mark et al. (1988), Leathwick et al. (1988), DeVelice & Burke (1989), Norton & Leathwick (1990), Robertson et al. (1991), Wardle (1991) and other references therein.

Annual MI values of <1 (Ln MI < 0) are tolerated among the big trees only by matai, totara, black beech, and a few species that require subsidies of ground water such as kahikatea (see above), pokaka and hinau. Leptophyllous kanuka has a MI tolerance down to 0.5 The drought-resistant small trees, ngaio and mahoe, surprisingly have mesophylls, although ngaio leaves are coriaceous and mahoe wilts and apparently has physiological tolerance. *Dracophyllum* trees and montane to subalpine *Olearia* and *Hoheria* spp., kamahi, red and silver beeches, tanekaha, mountain cedar and pink pine generally require MI to be greater than 1.5.

Scrubline or dwarf timberline, composed mainly of asterads, epacrids and stunted podocarps, occurs at an estimated minimum January temperature of 7°C in superoceanic environments. A 2.5 m tall dwarf forest of the above groups, sometimes together with mountain cedar, mountain totara, pink pine, mountain beech, silver beech, rata, fuchsia, mountain lacebark (*Hoheria* spp.) or kamahi, is found at the 8–10°C isotherm (again depending on continentality). Montane species indicators such as rimu (see also Norton et al. 1988), miro, pokaka, red beech and treeferns disappear at the 11–12°C mark, whereas Warm Temperate species such as rewarewa, kauri, kahikatea, tanekaha, matai, hard and black beeches, the coastal species, tawa, hinau, and supplejack have only a few outliers below the 14°C isotherm. A few species, including a number of essentially subtropical stragglers not presented here, drop out near 16°C. Taraire, mangrove and pohutukawa fit this category. Complementing these are a number of Subalpine-Montane species, including red, mountain and silver beeches, mountain cedar, pink pine, mountain totara, pokaka, mountain lacebark, and the mountain dracophyllums and olearias, which are excluded from the warmest zone on the mainland.

The species which routinely survive a strongly seasonal climate (continentality > 9) are kanuka, mountain lacebark, mountain, black and silver beeches, mountain totara, matai and kahikatea. On the other hand those which have a preference or tolerance for highly oceanic, saturated conditions (continentality < 4) are rimu, miro, pink pine, silver beech, kamahi, pokaka, rata, supplejack, treeferns, fuchsia, mountain dracophyllums and olearias.

To some extent soil fertility mirrors the moisture and continentality gradients since fertility is negatively correlated with leaching strength and positively with evapotranspiration. Nevertheless, elevated, steep sites with high rainfall are vulnerable to erosion and mass movement, and there are species to take advantage of the fresh, fertile surfaces, e.g. fuchsia, mountain lacebark (both deciduous) and olearias to some extent. The classical lowland species of rich soils are kahikatea, matai, black and red beeches to some degree, all the coastal plants, the beilschmiedias, hinau, rata, supplejack and sometimes kanuka. It is all relative in New Zealand, however, with its preponderance of acid rocks, and being one of the more heavily leached lands on the earth.

After considering these species, and narrowing down their individual likes and dislikes, it seems that there may be some light and pattern in the gloomy recesses of the New Zealand rainforests. This is something that experienced ecologists have subjectively recognised for many decades, but which has often eluded precise quantification and definition. The individualistic approach has much to commend it here, whereas traditional classification, while still valid in dealing with broad forest types (Fig. 11), easily becomes bogged down in a plethora of ever finer categories.

Finally, there is the conclusion, as intimated earlier, that a number of species currently more or less confined to the North Island or even the northern part of it, should be capable of extending their range much further south. This adds grist to the conjecture of disequilibrium.

Concluding remarks

The continentality scores in New Zealand are among the lowest in the world. This translates into low seasonality and correspondingly long growing seasons or high numbers of degree-days. Therefore, the January temperatures do not necessarily give a valid comparison with environments in other parts of the world. With respect to other localities with Temperate, evergreen, broadleaved rainforest, however, the comparison is probably reasonable. If degree-day sums were adequate substitutes, they would be the obvious choice for universal comparisons. The problem with standard screen air temperatures is that they give a biased representation of the thermal conditions and one quite different from that perceived by plant foliage exposed to the elements. The effect of wind (typical of highly oceanic climates) on a leaf is to lower the body temperature; furthermore a directly irradiated leaf will evaporate more moisture from its surface than will a screened dry bulb, with a correspondingly greater depression of experienced temperature, especially in humid, oceanic climates. Finally the unpredictability of such climates will impose a greater stress on plants than the relatively mild average temperatures might suggest. So, until we can transform screen temperatures to reflect the actual plant environment (especially in oceanic climates), much as we do with wind-chill for human environments, there is little advantage in degree-day sums over, say, mean daily temperature for the warmest month (Meurk 1984).

Comparing the climatic envelopes for New Zealand's evergreen broad-leaved rainforests (and of the broadly overlapping though limited extent of deciduous forest) with those of other parts of the world (Box 1981), we can see both conformities and sharp distinctions in terms of continentality, MI

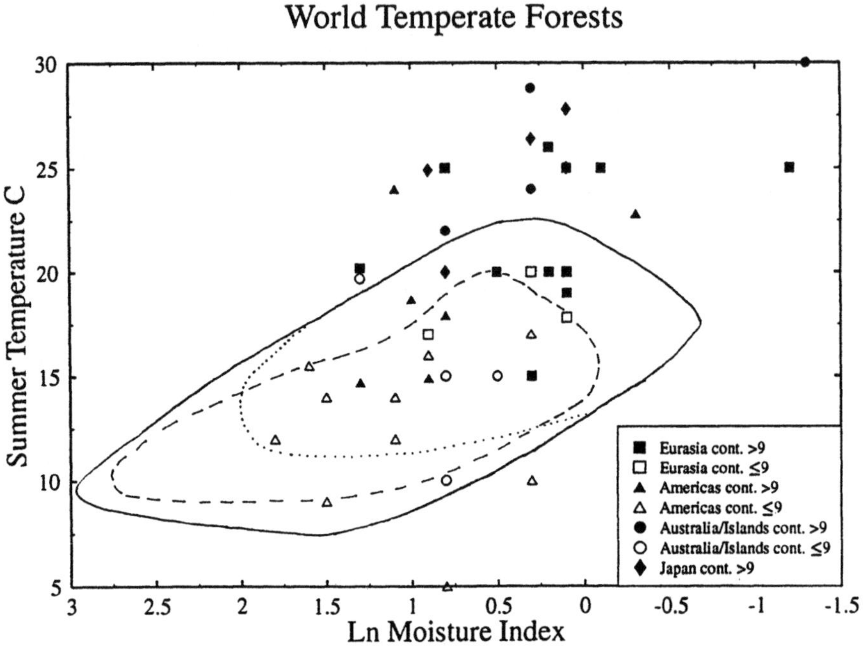

Fig. 12. Scattergram of world temperate evergreen forests, in temperature-moisture space, with overlay of New Zealand's evergreen (continuous line) and mesic, large-leaved deciduous (dashed line) forests. Some small-leaved deciduous species overlap with *Kunzea*. The dotted line separates New Zealand continental sites (scores >9), to the upper right, from oceanic sites to the lower left.

and temperature (Fig. 12). New Zealand summer temperatures (and continentality) are generally lower and moisture levels higher than in other parts of the world. Spurr & Barnes (1980), however, report isohyets of 380–640 mm from north to southern North America as corresponding to the boundary of open woodland. This equates with the tolerances described for kanuka woodland here.

The true Temperate broadleaved rainforests are indeed rare in the world. So the data in the figure are drawn from areas which not only genuinely fit with the New Zealand picture, such as Tierra del Fuego, southeast Australia, Tasmania, Florida, and montane Africa and India, but also from areas with mediterranean climates, and other humid countries such as southern Japan and Britain/Ireland, where at least evergreen trees mingle with the predominant deciduous aspect. For most of these regions, continentality scores are > 15 and frequently > 30. Furthermore, base soil fertility is commonly higher in the continents and, together with seasonal and evolutionary charac-

Fig. 13. Podocarp/mixed hardwood forest interior with trunks of emergent *Prumnopitys taxifolia*, *Dacrycarpus dacrydioides* and *Podocarpus totara* (>1 m DBH) associated with *Elaeocarpus hookerianus*, *Griselinia littoralis* and *Melicytus ramiflorus*. Vines and epiphytic cryptogamic plants abound. The understorey comprises species of *Coprosma*, *Myrsine*, *Neomyrtus*, *Melicope*, *Pseudowintera*, *Pseudopanax*, *Pittosporum*, etc. The ground is dominated by *Blechnum discolor*, with *Astelia* and *Microlaena*. The canopy is >20 m. Site: Peel Forest Park, south Canterbury, 260 m, terrace soil, P = 1050 mm, January mean temperature = 14.5°C. (Photo by G. C. Kelly.)

teristics, results in a more dynamic system. The Himalayas are climatically very similar to New Zealand and further comparison would be worthwhile.

Likewise, comparisons with the tropics, and the montane tropics in particular, have been made (Troll 1960) and commented on above. Suffice it to summarise that New Zealand forests are in a cooler zone than the lowland tropics and therefore lack the ecosystem antiquity (except perhaps in the far north), species diversity, and characteristic tropical taxa. The growth forms and forest structures are comparable insofar as they retain some tropical elements in the flora, that survived the glaciation, and which are perhaps

Fig. 14. Mesic *Nothofagus solandri*/mixed hardwood forest interior. Trunks are 20–40 cm DBH, largely covered by a sooty fungus, with dense (for beech) understorey of *Coprosma*, *Psuedopanax*, *Pittosporum*, *Carpodetus*, *Griselinia*, *Myrsine* and some lianas. The ground has some ferns, beech seedlings, *Uncinia* and *Astelia*. This forest has been fenced from stock for over 15 years. Adjacent browsed stands have almost no understorey apart from the epacrid *Cyathodes juniperina*. The canopy is <20 m. Site: Kowai Bush, mid Canterbury, 440 m, inceptisol on foothills, P = 1050 mm, January mean temperature = 14°C.

adaptive to superhumid or swamp forests. The comparison with the tropics has to some extent been by default of anything better. Lushness has been associated with heat and fertility rather than humidity and lack of severe cold. The isothermal, moist, equatorial alpine zones are the place to seek physiological as well as physical similarity. In the final analysis, it is southern South American vegetation which invites the best fit. These southernmost forests share history, geology, climate, biota and evolutionary responses – and exchange continues via the west wind drift.

An alternative approach to comparing world environments is through "transplant experiments". These have been carried out on a grand scale in New Zealand with huge numbers of exotic trees planted here for production and amenity. A number of these have escaped into the wild, although relatively few penetrate into closed, indigenous forest. Specimens of eucalypt, pine and redwood have been measured at >62 m height (Burstall & Sale 1984). And *Pinus contorta* forms a tree 200 m above natural timberlines of beech (Wardle 1985a). Taxa that have been particularly successful here include: Pinaceae, Cupressaceae, Salicaceae, Rosaceae, Ericaceae, Betulaceae, Theaceae, *Eucalyptus*, *Acacia*, *Populus*, *Taxus*, *Ilex*, *Crataegus*, *Acer*,

Fig. 15. Superoceanic, Subantarctic *Dracophyllum* dwarf-forest interior. Stems are up to 10 cm DBH and densely covered by lichens, bryophytes and some epiphytic ferns (*Hymenophyllum*, *Grammitis*). The understorey comprises *Coprosma* spp. and *Myrsine divaricata*, with *Polystichum vestitum*, *Blechnum*, *Uncinia*, orchids and bryophyte hummocks on the floor. The canopy is up to 5 m. Site: Campbell Island, 15 m, histosol in sheltered ravine, P = 1400 mm, January mean temperature = 9.3°C.

Quercus, *Fraxinus*, *Ulmus*, *Alnus*, *Tradescantia*, *Hedera* and *Clematis*. It is interesting to note that a number of South American trial plants have performed rather indifferently in the South Island, although many of these are from the northern Argentinian interior (S. Halloy pers. comm. 1991). *Feijoa*, *Persea*, *Fuchsia*, *Berberis*, *Tropaeolum*, *Escallonia*, *Erythrina* and *Embothrium* are some of the more successful trees and shrubs, especially in the North Island.

The reverse transplant has been carried out less assiduously, although New Zealand plants are found in cultivation in certain parts of the world which have relatively mild winters. As stated above, there are few Temperate areas in the northern hemisphere which qualify. A few cases can be cited (Beans 1973, Nelson 1989, Meurk 1987) from southern England (E), Ireland (I), Muenchen-Germany (G), Australia (A), California (C), Hawaii (H), Bergen-Norway (N): *Aristotelia fruticosa* (N), *A. serrata* (IE), *Brachyglottis* spp. (IEN), *Carmichaelia* spp. (IE), *Carpodetus* (I), *Cassinia leptophylla* (IEN), *Coprosma* spp. (IEN), *Cordyline australis* (IECA), *Corokia cotoneaster* (IG), *Corynocarpus* (AH), *Discaria toumatou* (IEN), *Dracophyllum* (N), *Dodonaea* (I), *Fuchsia excorticata* (IE), *Griselinia littoralis* (IE), *Halocarpus bidwillii* (IN), *Hebe* spp. (IENG), *Hoheria* spp. (IEN), *Hymenanthera* spp. (NG), *Kunzea ericoides* (IE), *Leptospermum* (IECH), *Metrosideros* spp. (I), *Myoporum laetum* (I), *Myrsine* spp. (I), *Myrtus* spp. (I), *Nothofagus* spp. (IEN), *Olearia* spp. (ING), *Phormium tenax* (IEC), *Pittosporum* spp. (IEAC), *Plagianthus* spp. (I), *Podocarpus hallii* (IEN), *P. nivalis* (IENG), *Pseudopanax* spp. (I), *Solanum aviculare* (IH), *Sophora* spp. (IE), various podocarps (IE), treeferns (I) and other minor species.

This is a relatively small number, and even the hardiest, such as *Cordyline*, *Griselinia*, *Hebe*, *Leptospermum*, *Olearia*, *Phormium*, thrive outside only near the coast of southern England and Ireland, and must be replanted from time to time after periodic outbreaks of especially severe weather. Most of our dominant tree species are lost to winter frosts if they are not given some shelter, even in the most oceanic parts of Europe such as Bergen. Nevertheless, the first four genera above are known, in Ireland, to reproduce spontaneously from seed. This picture seems consistent with the presence of minor evergreen elements in forests of Ireland and southwest England, e.g. *Ilex*, *Hedera*, *Arbutus*, *Rhododendron ponticum*.

To sum up: the term "evergreen broadleaved forest" can be stretched to encompass most of New Zealand's woody vegetation taller than 2.5 m, although the driest woodlands support only leptophyllous or microphyllous species. These forests overlap the humid, cooler extremity of the global forest continuum (Fig. 12). With the greater genetic pool of plants and animals that have been introduced over the past 150 years, and the ongoing attrition of native habitats, the forests of the future are likely to be quite different from the ideal I have described. Pines, eucalypts, redwoods, wattles and some deciduous trees can tolerate, and grow much taller in, both drier and colder environments than the indigenous species (weakened by reduction in their gene pools). Mesic forests are also threatened by shade-tolerant exotics such as sycamore, Douglas fir, and holly, and by browsing mammals.

In attempting to encapsulate the novel circumstances of New Zealand forests, there is always a danger of over-generalising the complexities. There remain many "ifs", "buts" and "maybes" to be tested. This places an onus on New Zealand officials, scientists and the public to preserve the unique ecology and heritage of this country for both recreation and study.

Acknowledgements

This presentation would not have been possible without Rowan Buxton's expert assistance with the computer-graphics. Johnathon Poff helped with Fig. 4. Bruce Clarkson, Warwick Harris, Peter Wardle and Matt McGlone were encouraging and usefully commented on an early draft. I am deeply grateful to Kazue Fujiwara and Elgene Box who assisted me to the 1990 Ecological Congress. Funds for this research were provided by the Hellaby Trust and the Foundation for Research, Science and Technology under contract C-9328 (CDM).

Alphabetized local plant names and Latin equivalents

Araucaria	*Araucaria*
Beech (southern)	*Nothofagus*
Black beech/mountain beech	*Nothofagus solandri*
Bog pine	*Halocarpus bidwillii*
Celery pine	*Phyllocladus alpinus*
Daisies	*Olearia, Brachyglottis*
Heaths	*Dracophyllum, Cyathodes*
Kahikatea	*Dacrycarpus dacrydioides*
Kamahi	*Weinmannia racemosa*
Kanuka	*Kunzea ericoides*
Kauri	*Agathis australis*
Kowhai	*Sophora*
Matai	*Prumnopitys taxifolia*
Miro	*Prumnopitys ferruginea*
Mountain cedar	*Libocedrus bidwillii*
Mountain totara	*Podocarpus hallii*
Pink pine	*Halocarpus biformis*
Podocarp	*Podocarpus*
Rata	*Metrosideros*
Red beech	*Nothofagus fusca*

Red tussock	*Chionochloa rubra*
Rewarewa	*Knightia excelsa*
Rimu	*Dacrydium cupressinum*
Silver beech	*Nothofagus menziesii*
Tanekaha	*Phyllocladus trichomanoides*
Tawa	*Beilschmiedia tawa*
Totara	*Podocarpus totara*

References

Adams, J. A. 1978. Long term aspects of nutrient loss from forest soils and ecosystems. N.Z. Journal of Forestry 23:10–20.

Adams, J. A. & Norton, D. A. 1991. Soil and vegetation characteristics of some tree windthrow features in a south Westland rimu forest. Journal of the Royal Society, N.Z. 24:33–42.

Allen, R. B. 1988. Ecology of *Nothofagus menziesii* in the Catlins Ecological Region south-east Otago, New Zealand (III) Growth and population structure. N.Z. Journal of Botany 26:281–295.

Allen, R. B., Payton, I. J. & Knowlton, J. E. 1984. Effects of ungulates on structure and species composition in the Urewera forests as shown by exclosures. N.Z. Journal of Ecology 7:119–130.

Atkinson, I. A. E. 1981. Vegetation Map of Tongariro National Park, North Island, New Zealand. Scale 1: 50,000. Department of Scientific and Industrial Research, Wellington.

Atkinson, I. A. E. & Greenwood, R.M. 1989. Relationships between moas and plants. N.Z. Journal of Ecology 12, Supplement:67–96.

Balgooy, M. M. J. van. 1971. Plant-geography of the Pacific. Blumea Supplement 6:1–222.

Bannister, P. 1990. Frost resistance of leaves of some plants growing in Dunedin, New Zealand, in winter 1987 and late autumn 1989. N.Z. Journal of Botany 28:359–362.

Batcheler, C. L. 1989. Moa browsing and vegetation formations, with particular reference to deciduous and poisonous plants. N.Z. Journal of Ecology 12, Supplement:57–65.

Bean, W. J. 1973. Trees and Shrubs Hardy in the British Isles. John Murray, London.

Beveridge, A. E. 1973. Regeneration of podocarps in a central North Island forest. N.Z. Journal of Forestry 18:23–35.

Box, E. O. 1981. Macroclimate and Plant Forms: An Introduction to Predictive Modeling in Phytogeography. Dr W. Junk Publishers, The Hague.

Brownsey, P. J. & Smith-Dodsworth, J. C. 1989. New Zealand Ferns and Allied Plants. David Bateman, Auckland.

Burrell, J. 1965. Ecology of *Leptospermum* in Otago. N.Z. Journal of Botany 3:3–16.

Burrows, C. J. 1990. Processes of Vegetation Change. Unwin Hyman, London.

Burns, B. R. & Smale, M. C. 1990. Changes in structure and composition over fifteen years in a secondary kauri (*Agathis australis*)-tanekaha (*Phyllocladus trichomanoides*) forest stand, Coromandel Peninsula, New Zealand. N.Z. Journal of Botany 28:141–158.

Burstall, S. W. & Sale, E. V. 1984. Great Trees of New Zealand. Reed in association with N.Z. Forest Service, Wellington.

Cameron, R. J. 1954. Mosaic or cyclical regeneration in North Island podocarp forests. N.Z. Journal of Forestry 8:337–354.

Clarkson, B. D. 1986. Vegetation of Egmont National Park, New Zealand. National Parks Scientific Series 5. Science Information Publishing Centre, Wellington.

Clarkson, B. D. 1990. A review of vegetation development following recent (< 450 years) volcanic disturbance in North Island, New Zealand. N.Z. Journal of Ecology 14:59–71.

Clarkson, B. R., Patel, R. N. & Clarkson, B. D. 1988. Composition and structure of forest overwhelmed at Pureora, central North Island, New Zealand, during the Taupo eruption (*c.* AD 130). Journal of the Royal Society, N.Z. 18:417–436.

Clayton-Greene, K. A. 1977. Structure and origin of *Libocedrus bidwillii* stands in the Waikato district, New Zealand. N.Z. Journal of Botany 15:19–28.

Clayton-Greene, K. A. & Wilson, J. B. 1985. The vegetation of Mt Karioi, North Island, New Zealand. N.Z. Journal of Botany 23:533–548.

Clout, M. N. & Hay, J. R. 1989. The importance of birds as browsers, pollinators and seed dispersors in New Zealand forests. N.Z. Journal of Ecology 12, Supplement:27–33.

Cockayne, L. 1928. The Vegetation of New Zealand, 2nd ed. Engleman, Leipzig.

Conrad, V. 1946. Usual formulas of continentality and their limits of validity. Transactions, American Geophysical Union 27:663–664.

Dawson, J. 1988. Forest Vines to Snow Tussocks: The Story of New Zealand Plants. Victoria University Press, Wellington.

DeVelice, R. L. & Burke, W. 1989. Gradient analysis of forest vegetation in the Maungataniwha Range, western North Island, New Zealand. N.Z. Journal of Botany 27:27–34.

Duncan, R. P., Norton, D. A. & Woolmore, C. B. 1990. The lowland vegetation pattern, south Westland, New Zealand 2. Ohinemaka Forest. N.Z. Journal of Botany 28:131–140.

Ecroyd, C. E. 1982. Biological flora of New Zealand. 8. *Agathis australis* (D. Don) Lindl. (Araucariaceae) kauri. N.Z. Journal of Botany 20:17–36.

Fleming, C. A. 1975. The geological history of New Zealand and its biota. In: Kuschell, G. (ed.) Biogeography and Ecology in New Zealand, pp. 1–86. W. Junk, The Hague.

Galloway, D. J. 1985. Flora of New Zealand. Lichens. Government Printer, Wellington.

Gibbs, H. S. 1983. Development and plant relationships of lowland forest soils. In: Thompson, K., Hodder, A. P. H. & Edmonds, A. S. (eds.) Lowland Forests in New Zealand, pp. 36–44. University of Waikato, Hamilton.

Grant, P. J. 1989. Effects on New Zealand vegetation of late Holocene erosion and alluvial sedimentation. N.Z. Journal of Ecology 12, Supplement:131–144.

Grime, J. P. 1979. Plant Strategies and Vegetation Processes. Wiley, Chichester.

Haase, P. 1987. Ecological studies on *Hoheria glabrata* (Malvaceae) at Arthurs Pass, South Island, New Zealand. N.Z. Journal of Botany 25:401–409.

Haase, P. 1990. Environmental and floristic gradients in Westland, New Zealand, and the discontinuous distribution of *Nothofagus*. N.Z. Journal of Botany 28:25–40.

Halloy, S. 1990. A morphological classification of plants with special reference to the New Zealand alpine flora. Journal of Vegetation Science 1:291–304.

Holloway, J. T. 1954. Forests and climate in the South Island of New Zealand. Transactions of the Royal Society, N.Z. 82:329–410.

Holdaway, R. N. 1989. New Zealand's pre-human avifauna and its vulnerability. N.Z. Journal of Ecology 12, Supplement:11–25.

Jane, G.T. & Green, T. G. A. 1984. Ecological aspects of climate patterns within the Kaimai Ranges, North Island, New Zealand. N.Z. Journal of Ecology 7:183–196.

Leathwick, J. R., Wallace, S. W. & Williams, D. S. 1988. Vegetation of the Pureora Mountain Ecological Area, west Taupo, New Zealand. N.Z. Journal of Botany 26:259–280.

Leutart, A. 1988. Mortality, foliage loss, and possum browsing in southern rata (*Metrosideros umbellata*) in Westland, New Zealand. N.Z. Journal of Botany 26:7–20.

McGlone, M. S. 1985. Plant Geography and the late Cenozoic history of New Zealand. N.Z. Journal of Botany, 23:723–749.

McGlone, M. S. 1989. The Polynesian settlement of New Zealand in relation to environmental and biotic changes. N.Z. Journal of Ecology 12, Supplement:115–129.

McGlone, M. S. & Webb, C. J. 1981. Selective forces influencing the evolution of divaricating plants. N.Z. Journal of Ecology 4:20–28.

McKelvey, P. J. 1984. Provisional classification of South Island virgin indigenous forests. N.Z. Journal of Forestry Science 14:151–178.

McSweeney, G. D. 1982. Matai/totara flood plain forests in south Westland. N.Z. Journal of Ecology 5:121–128.

Mark, A. F. 1989. Responses of indigenous vegetation to contrasting trends in utilization by red deer in two southwestern New Zealand National Parks. N.Z. Journal of Ecology 12, Supplement:103–114.

Mark, A. F., Baylis, G. T. S. & Dickinson, K. J. M. 1991. Monitoring the impacts of deer on vegetation condition of Secretary Island, Fiordland National Park, New Zealand: a clear case for deer control and ecological restoration. Journal of the Royal Society, N.Z. 21:43–54.

Mark, A. F., Grealish, G., Ward, C. M. & Wilson, J. B. 1988. Ecological studies of a marine terrace sequence in the Waitutu Ecological District of southern New Zealand. Part 1: The vegetation and soil patterns. Journal of the Royal Society, N.Z. 18:29–58.

Meurk, C. D. 1982. Alpine Phytoecology of the Rainshadow Mountains of Otago and Southland, New Zealand. PhD Thesis, University of Otago, New Zealand (unpublshed).

Meurk, C. D. 1984. Bioclimatic zones for the Antipodes, and beyond? N.Z. Journal of Ecology 7:175–181.

Meurk, C. D. 1987. Overseas Travel Report 71. Botany Division, DSIR, Christchurch (unpublished).

Meurk, C. D. & Blaschke, P. M. 1990. How representative can restored islands really be? An analysis of climo-edaphic environments in New Zealand. In: Towns, D. R., Daugherty, C. H. & Atkinson, I. A. E. (eds.) Ecological Restoration of New Zealand Islands, pp. 52–72. Conservation Sciences Publications No. 2. Department of Conservation, Wellington.

Meurk, C. D. & Given, D. R. 1990. Vegetation Map of Campbell Island, 1:25000, Map (1 sheet). Department of Scientific and Industrial Research, Christchurch.

Mildenhall, D. C. 1989. Summary of the age and paleoecology of the Miocene Manuherikia Group, Central Otago, New Zealand. Journal of the Royal Society, N.Z. 19:19–29.

Mitchell, N. D. 1991. The derivation of climate surfaces for New Zealand, and their application to the bioclimatic analysis of the distribution of kauri (*Agathis australis*). Journal of the Royal Society, N.Z. 21:13–24.

Mokma, D. L., Jackson, M. L., Syers, J. K. & Stevens, P. R. 1973. Mineralogy of a chronosequence of soils from greywacke and mica-schist alluvium, Westland, New Zealand. N.Z. Journal of Science 16:769–797.

Molloy, B. P. J. 1964. Soil genesis and plant succession in the subalpine and alpine zones of Torlesse Range, Canterbury, New Zealand. Part 2. Distribution, characteristics, and genesis of soils. N.Z. Journal of Botany 2:143–176.

Molloy, L. 1988. Soils in the New Zealand Landscape The Living Mantle. Mallinson Rendel in association with the New Zealand Society of Soil Science, Wellington.

Moore, L. B. & Edgar, E. 1970. Flora of New Zealand, Volume II. Government Printer, Wellington.

Nelson, E. C. 1989. An archaic duet: New Zealand's contribution to Ireland's garden heritage. Annual Journal, Royal N.Z. Institute of Horticulture 16:4–11.

Newsome, P. F. J. 1987. The vegetative cover of New Zealand. Water and Soil Miscellaneous Publication No. 112, 2 map sheets.

New Zealand Meteorological Service 1983. Summaries of climatological observations to 1980. New Zealand Meteorological Service Miscellaneous Publication 177.

New Zealand Meteorological Service 1985. Climatic Map Series 1:2000000. New Zealand Meteorological Service Miscellaneous Publication 175.

Nicholls, J. L. 1976. A revised classification of the North Island indigenous forests. N.Z. Journal of Forestry 21:105–132.

Nicholls, J. L. 1983. The extent and variability of the native lowland forests. In: Thompson, K., Hodder, A. P. H. & Edmonds, A. S. (eds.) Lowland Forests in New Zealand, pp. 79–92. University of Waikato, Hamilton.

Norton, D. A. 1985. A multivariate technique for estimating New Zealand temperature normals. Weather and Climate 5:64–74.

Norton, D. A., Herbert, J. W. & Beveridge, A. E. 1988. The ecology of *Dacrydium cupressinum*: a review. N.Z. Journal of Botany 26:37–62.

Norton, D. A. & Leathwick, J. R. 1990. The lowland vegetation pattern, south Westland, New Zealand. 1. Saltwater Ecological Area. N.Z. Journal of Botany 28:41–51.

Ogden, J. 1985. An introduction to plant demography with special reference to New Zealand trees. N.Z. Journal of Botany 23:751–772.

Pocknall, D. T. 1989. Late Eocene to Early Miocene vegetation and climate history of New Zealand. Journal of the Royal Society, N.Z. 19:1–18.
Pole, M. 1989. Early Miocene floras from central Otago, New Zealand. Journal of the Royal Society N.Z. 19:121–125.
Robbins, R. G. 1962. The podocarp-broadleaf forests of New Zealand. Transactions of the Royal Society, N.Z. (Botany) 1:33–75.
Robertson, A. W., Mark, A. F. & Wilson, J. B. 1991. Ecology of a coastal lagoon to dune forest sequence, south Westland, New Zealand. N.Z. Journal of Botany 29:17–30.
Rogers, G. M. 1989. The nature of the lower North Island floristic gap. N.Z. Journal of Botany 27:221–241.
Sakai, A. & Wardle, P. 1978. Freezing resistance of New Zealand trees and shrubs. N.Z. Journal of Ecology 1:51–61.
Spurr, S. H. & Barnes, B. V. 1980. Forest Ecology, 3rd Edition. John Wiley and Sons, New York.
Stevens, G., McGlone, M., McCulloch, B. & Ward, V. 1988. Prehistoric New Zealand. Heinemann Reed, Auckland.
Stevens, P. R. & Walker, T. W. 1970. The chronosequence concept and soil formation. The Quarterly Review of Biology 45:333–50.
Tedrow, J. C. F. & Hill, D. E. 1955. Arctic brown soil. Soil Science 80:265–275.
Troll, C. 1960. The relationship between the climate, ecology, and plant geography of the southern cold temperate zone and of the tropical mountains. Proceedings of the Royal Society Series B 152:529–532.
Wace, N. M. 1965. Vascular Plants. In: Van Mieghem, J. & Van Oye, P. (eds.) Biogeography and Ecology in Antarctica, pp. 202–266. Dr. W. Junk, The Hague.
Veblen, T. T. & Stewart, G. H. 1980. Comparison of forest structure and regeneration on Bench and Stewart Islands, New Zealand. N.Z. Journal of Ecology 3:50–68.
Wardle, J. A. 1970. The ecology of *Nothofagus solandri*. N.Z. Journal of Botany 8:494–646.
Wardle, J. A. 1984. The New Zealand Beeches. N.Z. Forest Service, Christchurch.
Wardle, P. 1963. The regeneration gap of New Zealand gymnosperms. N. Z. Journal of Botany 1:301–15.
Wardle, P. 1978. Origin of the New Zealand mountain flora, with special reference to trans-Tasman relationships. N.Z. Journal of Botany 16:535–50.
Wardle, P. 1980. Primary succession in Westland National Park and its vicinity, New Zealand. N.Z. Journal of Botany 18:221–32.
Wardle, P. 1985a. New Zealand timberlines. 3. A synthesis. N.Z. Journal of Botany 23:263–271.
Wardle, P. 1985b. Environmental influences on the vegetation of New Zealand. N.Z. Journal of Botany 23:773–788.
Wardle, P. 1988. Effects of glacial climates on floristic distribution in New Zealand 1. A review of the evidence. N.Z. Journal of Botany 26:541–555.
Wardle, P. 1991. The Vegetation of New Zealand. Cambridge University Press (in press).
Wardle, P., Bulfin, M. J. A. & Dugdale, J. 1983. Temperate broad-leaved evergreen forests of New Zealand. In: Ovington, J. D. (ed.) Temperate Broad-Leaved Evergreen Forests, pp. 33–71. Elsevier Science Publishers B.V., Amsterdam.
Webb, C. J., Johnson, P. N. & Sykes, W. R. 1990. Flowering Plants of New Zealand. DSIR Botany, Christchurch.
Webb, C. J., Sykes, W. R. & Garnock-Jones, P. J. 1988. Flora of New Zealand. Volume IV. Botany Division, DSIR, Christchurch.
Wells, J. A. 1972. Ecology of *Podocarpus hallii* in Central Otago, New Zealand. N.Z. Journal of Botany 10:399–426.
Wilson, H. D. 1987. Vegetation of Stewart Island, New Zealand. N.Z. Journal of Botany 27. Supplement.

10. Land use and human influences in the evergreen broad-leaved forest regions of East Asia, the Mediterranean and Australia

GIUSEPPE PIGNATTI[1] & SANDRO PIGNATTI[2]
[1]*Facoltà di Scienze Forestali, Università – Piazzale Cascine, Firenze, Italy;*
[2]*Dipartimento di Biologia Vegetale, Università di Roma "La Sapienza", Città Universitaria, Roma, Italy*

Abstract

Heavy impacts produce different consequences on evergreen broad-leaved forests. Three examples are described: in the subtropics with humid climate the forest is replaced by grassland; in the Mediterranean Basin after degradation of the forest a vegetation rich in annual species appears; in W. Australia and S. Africa communities of heliophilous evergreen shrubs develop within the forest. The different conditions are discussed on the basis of species diversity and vegetation structure.

Key words: evergreen forests, fire ecology, human impacts on vegetation, mediterranean climate

Introduction

Evergreen broad-leaved forests are often used intensely by humans in subtropical and tropical countries: deforestation, fire, grazing, or forms of more or less temporary cultivation can almost completely destroy the natural vegetation cover. Evergreen broad-leaved forests are an ecosystem with relatively low resilience and have a slow recovery rate; in most cases, after a massive impact, secondary vegetation permanently replaces the original plant cover.

Secondary vegetation types occurring within the range of evergreen broad-leaved forests in different parts of the world will not be described fully here. They are mostly grassland or scrub. The specific composition varies widely on different continents, so a comparison based only on floristic information is of reduced interest. Vegetation changes can best be interpreted as adaptations to local conditions. In consequence, the floristic composition should

E.O. Box et al. (eds.), Vegetation Science in Forestry, 199–210.

be considered a stage of succession. Spatial distribution of plant cover, frequency of species within the communities, and survival strategies appear to be the most relevant variables to describe these aspects of secondary vegetation. In this paper we propose some models that can be used to understand vegetation changes from the ecosystem point of view.

Effects of impacts on evergreen forest vegetation

The first model: decline of diversity

An evergreen broad-leaved forest is degraded (in general by cutting); the tree layer is completely destroyed; the forest is replaced by herbaceous (savanna-like) vegetation. Only under favourable conditions can tree vegetation regenerate and a secondary forest develop. In the tropics and subtropics, savanna usually becames permanent. This succession is often observed in tropical or subtropical humid (monsoon) climates.

The natural forest vegetation of the Ding Hu Shan reserve near Canton (Guangzhou, Guangdong province, South China) was studied with the phytosociological method by Pignatti et al. (1990); this area (23° N latitude) has a tropical monsoon climate with 1800–2200 mm annual rainfall and average yearly temperature of 21 °C.

The primary evergreen broad leaved forest is multistratified, with ca. 15–25 species growing in the tree canopy of mixed stands; there are no dominant species in the canopy layer, and in the phytosociological records none of the tree species has a value over 2 (20–40%). Widespread in the upper tree layer are *Cryptocarya* (*Lauraceae*), *Castanopsis chinensis* (*Fagaceae*), *Erythrophloeum* (*Caesalpiniaceae*) and *Canarium* (*Burseraceae*) together with isolated individuals of *Tsoongiodendron* (*Magnoliaceae*).

The corresponding secondary vegetation develops after reforestation with *Pinus massoniana*, a species not present in the local flora but native in neighbouring areas of South China; *Eucalyptus robusta* is also planted here as a forest pioneer, but with scarce success. The secondary vegetation is seminatural: presently the planted species are slowly declining, and a canopy of native trees, chiefly *Schima* (*Theaceae*) and *Castanopsis fissa* (*Fagaceae*) is developing.

For the Ding Hu Shan mountain area statistical characters of the flora that develops under such conditions were presented by Zhou (1982) and in the phytosociological table of the primary vegetation by Pignatti et al. (1990):

	Flora of Ding Hu Shan (mountain only)	Phytosociological table (original forest)	
Trees	35%	67 species	32.7%
Shrubs	39%	79	38.5%
Herbs	16%	31	15.1%
Vines		28	13.7%
Total	1359 species	205	100.0%

Species adapted to the forest environment are completely dominant, whereas non-forest elements are mostly confined to specialized habitats (pools, creeks, rocks) or near human settlements.

The differences between primary and secondary forest stands in the Ding Hu Shan reserve have been analyzed with statistical methods. Two procedures have been used (Pignatti et al. 1991): investigation of diversity applying the Shannon formula to the number of tree stems, and an Index of Centralisation based on the quotient endemics-subendemics/wide-range species (Fig. 1). The two methods give similar results. The natural forest vegetation has elevated values of evenness (high entropy values) with high centralisation; secondary vegetation, on the contrary, has low entropy and is less centered. In fact, secondary vegetation shows a relatively simple structure with two

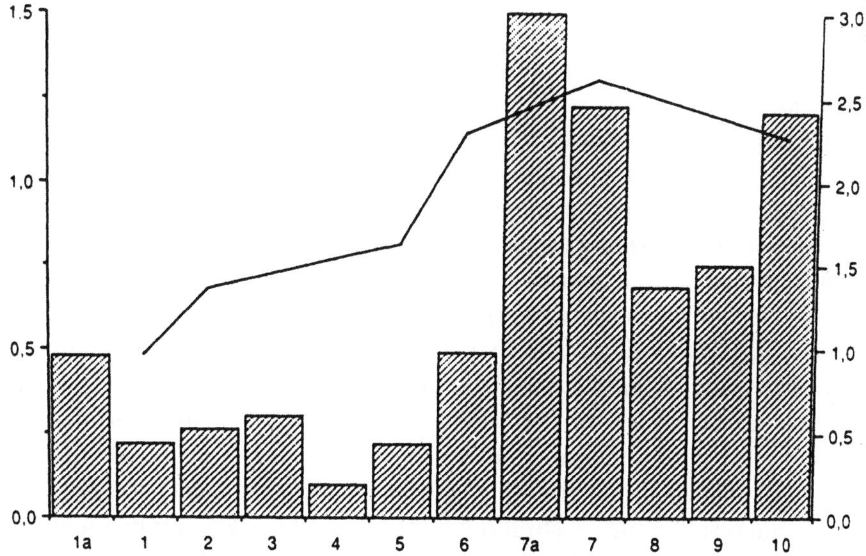

Fig. 1. Values of centralisation (columns) and of diversity (line) in the Ding Hu Shan vegetation by increasing age of the forest vegetation: secondary forest to the left, primary forest on the right side; rel. 1–10 from the year 1987, rel. from previous visits are indicated as 1a and 7a (after Pignatti et al. 1991).

well developed layers, each with dominants: *Schima* in the tree layer, and the fern *Dicranopteris linearis* (a climber with herbaceous consistency) near the soil. Original vegetation has 3 tree layers (high, middle and low), together with shrubs, herbs and climbers of very different shape forming an extremely complex system. The high entropy in original vegetation is not a measure of disorder but possibly signifies that the original forest has a very high information content. The secondary vegetation appears less diverse. Values of centralisation were not calculated until now for any vegetation type: high values mean that natural forests are the product of a long evolutionary process, whereas secondary forests are composed of quite heterogeneous elements. In this example human impact reduces the complexity of the system and its correspondence to the local factors.

This model is apparently widespread in the tropics, and we observed similar patterns in many cases, for example:

- in the Canary Islands a relatively uniform heath of *Myrica faya* is replacing the laurisilva (with a complex tree layer) after cutting and fire;
- in subtropical Argentina (Salta) grasslands are replacing the laurisilva with *Phoebe porphyria*, apparently as an irreversible degradation; and
- grasslands are also replacing Dipterocarp forests in Thailand after cutting.

The second model: diversity increasing by invasion of r-species

An evergreen broad-leaved forest is degraded by human impacts; the tree layer is completely destroyed; the forest is replaced by shrubs (macchia, garriga) and herbaceous vegetation. This succession is observed in the Mediterranean Basin (Fig. 2).

The natural vegetation of the Mediterranean Basin is the evergreen oak forest. The tree layer is composed of a single dominant species: *Quercus ilex*, which in habit and sclerophylly resembles both American (*Q. virginiana*, *Q. agrifolia*) and Japanese (*Q. gilva*, *Q. phillyraeoides*) species of this genus. Usually, no other tree species are present. Shrubs are well represented by several sclerophyllous species (*Arbutus unedo*, *Myrtus communis*, *Phillyrea latifolia*, *Pistacia lentiscus*), and vines (*Asparagus acutifolius*, *Hedera helix*, *Rubia peregrina*, *Smilax aspera*) are also present. Herbs are very few and have low cover values. In the Mediterranean Basin it is assumed that fire can occur for natural causes, but this is surely a very rare event. Instead, impacts of human origin (burning, grazing, cutting) are very frequent and strong, and result in the substitution of the evergreen oak forest by secondary vegetation. Such impacts have occurred since man settled in the area. In a first and very long phase, man lived as gatherer and hunter, causing very extensive but minor changes in this forest. Later, as shepherd and cultivator,

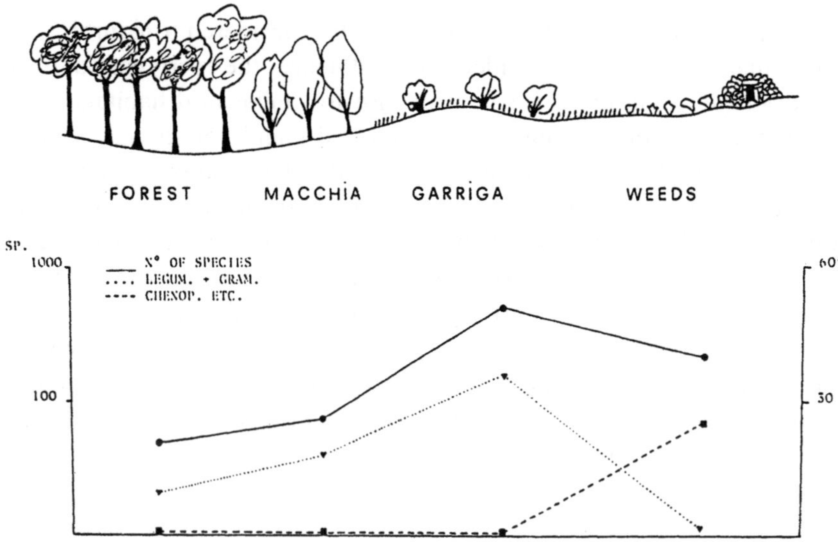

Fig. 2. Floristic diversity in different types of mediterranean vegetation. Number of species present (on the left, logarithmic scale) and percentage (on the right) (after Pignatti 1987).

since about 7000 B.C., the destruction of the original evergreen oak forest became almost complete.

The secondary vegetation in the Mediterranean Basin is well known (*cf.* Braun-Blanquet 1952, Bolos 1962, Rivas and Izco 1977). A large number of associations have been described and can be ordered in several classes of vegetation. The most widespread are listed below (many other classes including weed vegetation, which is probably completely man-made, are not included here).

a) *Ononido-Rosmarinetea* – low evergreen shrubs on limestone substrates
b) *Cisto-Lavanduletea* – low evergreen or semi-evergreen shrubs developing after fire on acid (or acidified) soils
c) *Thero-Brachypodietea* – communities of annual, biennial or perennial herbs on limestone
d) *Tuberarietea guttatae* – small annual herbs developing after fire on acid (or acidified) soils
e) *Chenopodietea* – herbs (mostly annual) developing on nitrate-rich substrates.

Each of these classes includes at least one association occurring in completely natural habitats and many associations known only as a consequence of human impact; e.g. the *Brachypodietum ramosi* (of the class *Thero-Brachypodietea*) occurs on extremely arid limestone cliffs where the succession to forest vegetation is impossible, along the coasts of the Mediterranean Sea from Spain to Greece. Consequently it can be considered a natural association of high-stress habitats. Under man's influence, the evergreen forest has been destroyed. Soil erosion has given rise to similar arid habitats, where secondary anthropogenic vegetation composed mainly of species occurring in the *Brachypodietum ramosi* has developed. The creation of new niches stimulated active speciation, mainly determined by man's activity: this anthropogenic flora is composed largely of annuals with wide ranges, mostly polyploid (results of r-selection). Under such conditions, floristic diversity is increasing (Pignatti 1978, 1979). Species adapted to fire, though, are relatively few in the mediterranean vegetation: most obviously *Q. suber* with a thick layer of insulating bark, and *Cistus* and *Geranium* spp. with seeds germinating after a heat shock. For Southern France the following figures have been calculated (Pignatti 1988):

	No. of species	No. of associations
Natural vegetation		
evergreen oak forest	69	2
macchia	78	2
Secondary vegetation		
garriga	435	28
weeds	120	8

In the natural reserve of Castelporziano near Rome, two forest associations are present (*Viburno-Quercetum ilicis* and *Lathyro-Quercetum cerris*). The flora of these associations is a small portion of the total flora of the site, composed of species from synanthropic environments (data from Anzalone et al. 1990 and Lucchese & Pignatti 1990):

	No. of species	
Viburno-Quercetum ilicis	65	6.9%
Lathyro-Quercetum cerris	114	12.1%
Total flora of the area	945	100.0%

The species occurring in the five classes a–e listed above are predominantly annuals with efficient means of dispersal: high seed production, fruits with wings or adhering to the fur of animals, endozoochory, presence of elaiosomes and dispersal by ants, vegetative propagation, etc. They can be interpreted as the result of r-selection.

As a consequence, the Mediterranean Basin has a particularly rich flora (Malyshev 1975), with almost as many species as tropical areas (the Amazon

Basin, East Africa, SE Asia). This elevated floristic diversity is concentrated in secondary habitats derived from the destruction of the evergreen forest.

The third model: Diversity increasing by dominance of K-species

Repeated natural fires occur in an evergreen broad-leaved forest; the tree layer can survive, but heliophilous shrubs (kwongan, fynbos) become dominant in the understorey. This succession can be observed under mediterranean type climate in Western Australia and in South Africa.

In the portion of Western Australia with a mediterranean-type climate, on well drained soils, forest vegetation is mainly dominated by eucalypts: *Eucalyptus wandoo*, *E. marginata* and *E. diversicolor* in order of increasing rainfall. Natural fires are frequent, and after fire a low shrub vegetation develops. This is the kwongan formation (Pate & Beard 1984), with dominance by extremely sclerophyll shrubs mostly belonging to the *Proteaceae*, *Myrtaceae* and *Epacridaceae*. In areas of low rainfall (400–500 mm) kwongan can become permanent.

The structure of the different *Eucalyptus* communities seems to be strongly adapted to fire and rainfall as follows

- With 400 mm rainfall herbaceous and shrubby vegetation remains discontinuous and fire cannot spread over large areas. The tree layer is composed of *E. salmonophloia* or *E. salubris*, two species not particularly adapted to fire, which occur in patches with several dozen individuals.
- With 500–600 mm rainfall the tree layer is composed of *E. wandoo*, and with 700–800 mm of *E. marginata*. Both species are well adapted to fire, which occurs frequently (in-most cases at least once every 20–30 years). Regeneration after fire is very intense, mostly by resprouting of saplings from subterranean organs (lignotubers), root collars and lateral roots; germination of seeds is scarce. Seedlings and young plants are often killed by fire, but mature trees of *E. marginata* develop a thick fibrous bark which protects the lower parts of the stem. As a consequence, the forest is typically multistratified because of repeated burning alternating with resprouting, scattered germination and growth (Fig. 3).
- With 900–1300 mm of rain the vegetation is composed of *E. diversicolor*. Fires are rare because of the high humidity and occur only once every 2–3 centuries as a catastrophic event, completely killing the trees, which are not fire-resistent. After burning, an intense regeneration takes place, only by germination of seeds which fall from the upper branches of the largest trees. As a consequence, the forest is almost of equal age (Fig. 4).

It is assumed (Singh 1982) that climatic conditions in Western Australia remained more or less constant during the last 30 million years. For this

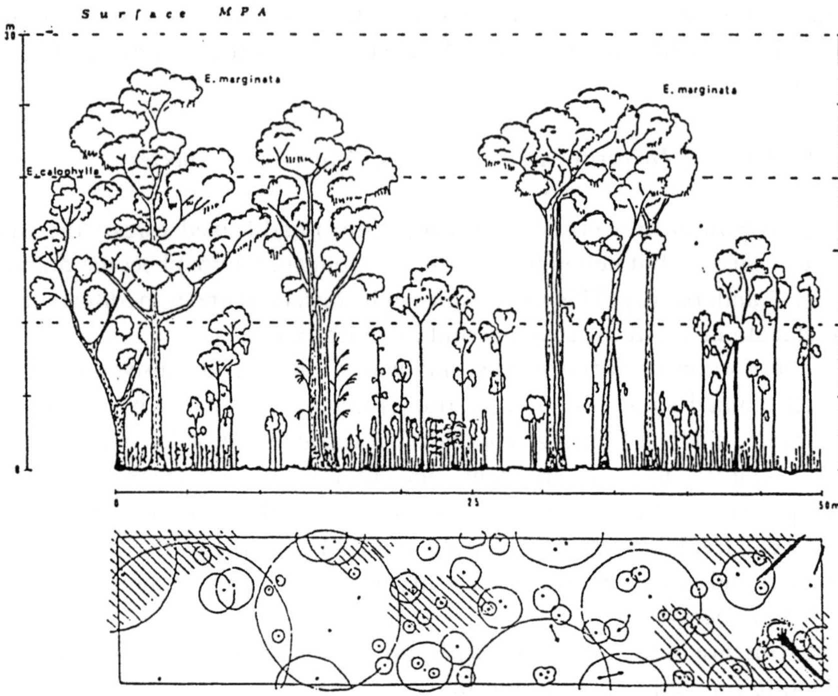

Fig. 3. Virgin stand of *Eucalyptus marginata* in the Darling Range (W. Australia).

reason the vegetation of the area is highly adapted to fire, and this adaptation has stimulated speciation.

Mechanisms of adaptation to fire are still incompletely known, but they have surely had a primary significance in determining the presence of the many *Proteaceae*, *Myrtaceae*, and *Restionaceae*, *Drosera* and *Xanthorrhoea* spp., etc. Species of *Tersonia*, *Gyrostemon*, *Burchardia*, *Anigozanthos*, some *Grevillea* and many orchids appear mostly after fire. Many *Banksia* spp. can germinate only after a thermal shock. Adaptations to fire in kwongan plants are discussed by Pate & Beard (1984).

The SW corner of Western Australia, characterized by a mediterranean type climate, is still incompletely known floristically. The total flora is estimated at ca. 5000–6000 species (Dixon, pers. comm.). Many are still undescribed. Clearing of the natural vegetation for agricultural and pastoral purposes (up to 95% of the area) has surely caused the extinction of many endemics even before they were described. These figures appear very high, especially when compared with the general uniformity of the landscape, consisting of a monotonous tableland with a single mountain system of restricted height, hardly reaching 1000 m elevation.

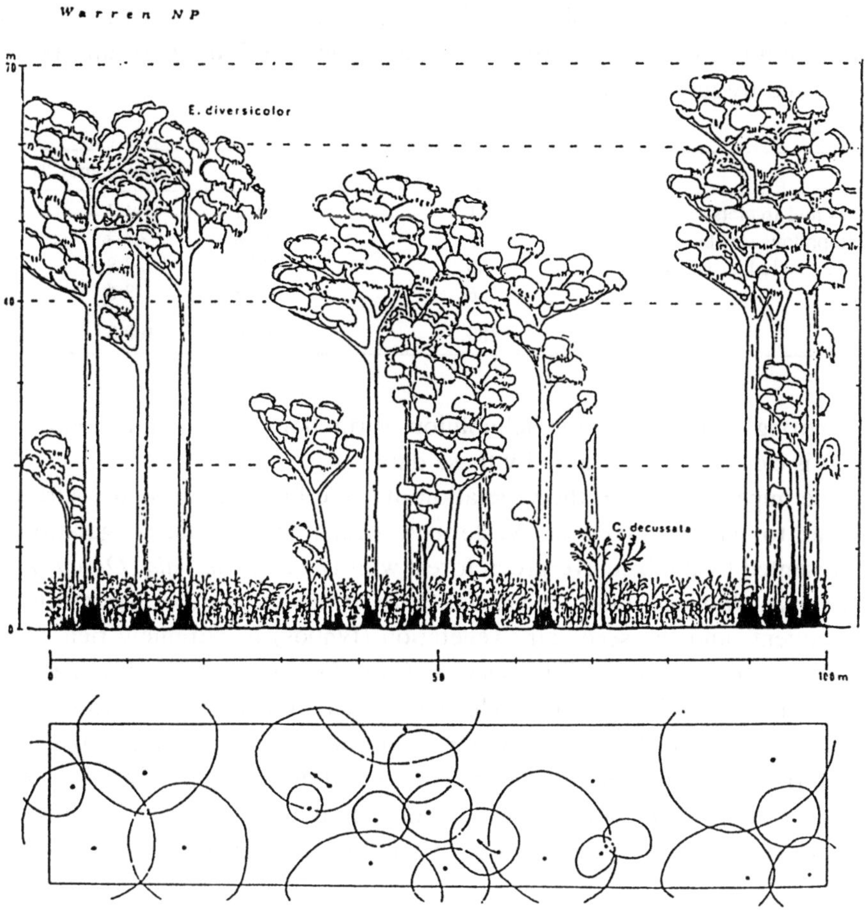

Fig. 4. Virgin stand of *Eucalyptus diversicolor* in Warren National Park (W. Australia).

Some statistical data on the flora of evergreen forests in Western Australia are available. The flora of the Perth district consists of 2057 species (Marchant et al. 1983) which can be distributed over the different environment types as follows (based on random sampling, orig.):

Indigenous	
forests	8.6%
kwongan	46.0
wet areas	12.1
littoral	5.2
other	3.5
introduced	24.7

Forest trees (following Boland et al. 1984) are only 25 species, or 1.2 % of the local flora. In the *E. marginata* forest we observed the following species totals and frequencies in 10 relevés:

	No. of species	%	Frequency %
Trees and treelets	11	6.1	14.7
Heliophilous shrubs	95	52.8	43.9
Herbs (excluding weeds)	68	37.8	37.8
Vines	6	3.3	3.6
Tot.	180	100.0	100.0

The strong connection of the heliophilous shrub flora with the forest environment is clearly demonstrated by these data.

The impact of fire on the vegetation of Cape Province in South Africa is similar to that observed in Western Australia. The climax vegetation is represented by a mixed evergreen forest with *Ocotea*, *Cunonia*, *Olinia*, *Ilex*, *Maytenus*, *Rapanea*, *Olea*, *Cussonia*, *Podocarpus*, etc. Fire of natural origin is frequent, and the secondary vegetation (fynbos) is extremely rich. With an estimated flora of ca. 7000 species, Cape Province is probably the most species-rich place in the world. The age of this flora is estimated to be ca. 7 million years (the beginning of the Pliocene). This is younger than the Western Australian mediterranean ecosystem, but the highly diversified ecological conditions in South Africa, with mountains and plains of different substrates, account for the richness of the floristic elements.

Impacts and conservation

Destruction of the forest cover is usually considered undesirable and much care is given to prevent it. Forests help protect the environment. Forest vegetation reduces erosion, regulates water circulation, increases the reserve of water and nutrients in soil and has a stabilizing effect on meso-climate. Forests are productive and supply important economic benefits. But the secondary vegetation, developing after the destruction of the evegreen forest, can be of economic importance also, mainly as pasture. In addition, secondary vegetation developing in mediterranean climates has a particularly high diversity. In this case, the secondary vegetation can have a high environmental quality. Consequently, conservation strategies for evergreen forests have to be adapted to different situations. In humid climates, where species diversity after impact declines and secondary vegetation is less productive, the main task of conservation is to avoid the degradation of evergreen forests, and eventually reconstruct them (some examples of environmental restoration from Japan, see Miyawaki et al. 1987, others in the laurisilva of the Canary Islands). In mediterranean type climates a mixed strategy has to be chosen: rigid conservation of the forests in the places where they are necessary for landscape protection (watersheds, erosion-threatened slopes) but also conservation of secondary vegetation encouraging traditional uses and allowing recreational activities.

Conclusions

Three general evergreen broad-leaved forest types have been examined and discussed. It seems clear that the vegetation of evergreen forests from different parts of the world reacts in different ways when submitted to heavy human impacts. This is depending in first line on species composition in the tree layer, i.e. a consequence of selective factors operating on the flora through geological time. Actual ecological conditions seem to have only a secondary effect, mainly in dependence of humid vs. dry climates.

Impacts affecting evergreen forest vegetation over a short time (a few thousand year as in the tropical humid climate) cause a reduction both of floristic richness and diversity. Impacts beginning longer ago (hundreds of thousand of years in the Mediterranean basin) allowed the constitution of a secondary vegetation in which species numbers and diversity increase. Here selection has favoured species with efficient reproductive adaptations. Impacts over geological time (7 million years in South Africa and 30 millions in Western Australia) result in a steady-state secondary vegetation, very rich in species and mostly consisting of plants with specific adaptations to fire.

Impacts, completely natural or caused by man, are an important factor of microevolution when they last for a long time. Secondary vegetation that replaces evergreen forests after impact may also warrant conservation.

Acknowledgements

I am grateful to G. Estabrook (Ann Arbor, Michigan) for a critical review of the manuscript. Financial support for field investigations was provided by C.N.R. (Roma) and MURST 40%.

References

Anzalone, B., Lattanzi, E. & Lucchese, F. 1990. La flora della tenuta di Castelporziano (Roma). Accad. Naz. Lincei Quad. 264:133–218.

Boland, J., Brooker, M. I. H., Chippendale, G. M., Hale, N., Hyland, B. P. M., Johnston, K. D., Kleinig, D. A., Turner, & J. D. 1984. Forest Trees of Australia, 4th ed. CSIRO, Melbourne. 687 pp.

Bolos, O. 1962. El paisaje vegetal barcelones. Univ. de Barcelona. 192 pp.

Braun-Blanquet, J. 1952. Les groupements végétaux de la France Mediterranéenne. C.N.R.S. 297 pp.

Lucchese, F. & Pignatti, S. 1990. Sguardo sulla vegetazione del Lazio Marittimo. Accad. Naz. Lincei, Quad. 264:5–48.

Malyshev, L. I. 1975. The quantitative analysis of flora: spatial diversity, the level of specific richness and representativity of sampling areas. Bot. Zhurnal 60: 1537–1550 (in Russian).

Marchant, N. G. 1983. Flora of Perth, 2 vols. Perth.

Miyawaki, A., Fujiwara, K. & Okuda, S. 1987. The status of nature and re-creation of green environments in Japan. In: Miyawaki, A. et al. (eds.), Vegetation Ecology and Creation of New Environments, Tokai Univ. Press, Tokyo.

Pate, J. W. & Beard, J. S. 1984. Kwongan, plant life of the sandplains. Univ. of W. Australia Press, Perth.

Pignatti, G. 1990. Fisionomia e struttura della foresta ad eucalipti dell'Australia Sud-Occidentale. Università della Tuscia, Fac. of Forestry, Thesis, Viterbo (Italy).

Pignatti, S. 1978. Evolutionary trends in mediterranean flora and vegetation. Vegetatio 37: 175–185.

Pignatti, S. 1979. Plant geographical and morphological evidence in the evolution of the Mediterranean flora (with particular reference to the Italian representatives). Webbia 34:243–255.

Pignatti, S. 1988. The relationships between natural vegetation and social system in the Mediterranean basin. In: Miyawaki, A. et al. (eds.), Vegetation Ecology and Creation of New Environments, 35–45. Tokai University Press, Tokyo (1987).

Pignatti, E., Pignatti, S., Huang, C. C., Ding, G. Q. & Huang, Z. L. 1990. Phytosociological investigations in the Ding Hu Shan forest (Guangdong – South China). Rend. Fis. Accad. Lincei s. 9(1):335–356.

Pignatti, E., Pignatti, S., Huang, C. C., Ding, G. Q. & Huang, Z. L. 1991. β-diversity and phytogeographical patterns in the Ding Hu Shan Reserve forest vegetation. Rend. Fis. Accad. Lincei s. 9(2):79–85.

Rivas Martinez, S. & Izco, J. 1977. Sobre la vegetación terofítica subnitrofila mediterranea (Brometalia rubenti-tectori). Anal. Inst. Bot. A. J. Cavanilles Madrid 34:355–381.

Singh, G. 1982. Environmental upheaval. The vegetation of Australasia during the Quaternary, in: Smith, J. M. B. (ed.), A history of Australasian vegetation, pp. 90–108. McGraw-Hill Co., Sydney.

Zhou Yuan-rui 1982. Characters of plants of Ding Hu Shan Region. Trop. and Subtrop. Forest Ecosyst. 1:39–47.

11. West and south-east European deciduous and deciduous-conifer mixed forests of the lowlands

F. KLÖTZLI[1] & K. ZARZYCKI[2]
[1]*Geobotanisches Institut Rübel, Eidgenössische Technische Hochschule, CH-8044 Zürich, Switzerland*; [2]*Krakow, Poland*

Abstract

The deciduous and deciduous-conifer mixed forests occupy vast territories in eastern and western Europe ranging from eastern Poland to the southern Urals (over 2,000 km, eastern types) or further to northern Scotland (western types), respectively. They are situated between the taiga in the north and the steppes and sclerophyllous woodlands in the south. They show characteristics of evergreen broad-leaved forests in the west.

The region has been deeply affected by Pleistocene glaciations and definitely colonized by forest not before the Holocene. The Holocene history of vegetation is shorter in the east in comparison with western Europe, and the Tertiary refuges are in the distances of many hundred km in SW- and SE-Europe (peripheric situation) (Fig. 2). The relief is monotonous in the east (lowland to submontane zone) continuing the same way in the northwest. It is more hilly or mountainous in reaching the Caledonian, Variskian and Alpine mountain ranges. The continentality of climate increases from west to east, summer drought from north to south (east). In the east, the absence of evergreen Atlantic plant species is noted. The number of tree species is not high in the east (<20). The most important are: *Quercus robur*, *Tilia cordata*, *Acer platanoides*, *Fraxinus excelsior* and *Ulmus scabra*; in subcontinental western parts, *Carpinus betulus*, *Quercus petraea*, *Acer pseudoplatanus* and *Picea abies* occur; in southern parts, *Acer tataricum*, *Acer campestre* and thermophilus shrubs appear. In the west, there is more *Fagus silvatica* and about 40 species inclusively; in the south, sclerophyllous *Quercus* spec. prevail (e.g. *Quercus ilex*, *Quercus suber*). It is striking that trees in these vegetation types come into leaf at different times, and internal morphology of herbaceous plants is not homogenous. The tree species are cold resistant, and herbs are sometimes (in the west) or mostly (in the east) protected in winter under snow cover. European plants are dominant in deciduous forests,

E.O. Box et al. (eds.), Vegetation Science in Forestry, 211–254.

boreal elements are numerous in the northern parts, Pontic and south-Siberian elements occur in southern and western parts. Asiatic plants as well as minor endemics are found in the Ural Mountains. There are many taxa corresponding ecologically to North American and eastern Asiatic deciduous forests, but the connections with eastern Asia are even stronger (many closely related woody plants and herbs). The deciduous-conifer mixed forests of the north-western part of the area belong to the class *Vaccinio-Piceetea*, mesophilic deciduous forests to the class *Querco-Fagetea*, and thermophilous woodlands in the south to the order of *Quercetalia pubescentis* or then, passing the border of zonobiome VI to IV (mediterranean) to the class *Quercetea ilicis*. Relative to the character (fertility) of soil and water conditions many forest communities are distinguished (Figs. 5–10). The biomass of oak forests in the east ($400\,t \cdot ha^{-1}$) and net primary production ($9\,t \cdot ha^{-1} \cdot yr^{-1}$) are not much lower than in beech forests in central Europe.

Key words: deciduous forest, deciduous-conifer forest, eastern Europe, western Europe.

Area and history

The deciduous forests occupy a very conspicuous portion of Europe. Practically, they extend from latitude 40° N to 60° N in Scandinavia. They gradually grade through the intermediary of the deciduous-conifer mixed forests into coniferous forests or taiga in the north. In the south they are gradually replaced by sclerophyllous woodlands (Figs. 1, 3, 4).

In this chapter the full range of nemoral forest vegetation is (at least) presented in graphs putting more weight on the eastern European connections (as there is less knowledge on the eastern – most parts of eastern – European deciduous forests; compare Tables 1 and 2).

As all European forests have been treated by Mayer (1986) – and all in the German language – we may follow his lines in this chapter, present his graphs (somewhat modified), and keep text on details short, giving also the ecograms with the main forest plant-communities (Figs. 5–10). Apart from their distribution more information will be given on the other topics, as e.g. history, structure, phenology, community patterns, biomass, limiting factors.

Two main types of deciduous forests are recognizable there: one is essentially developed in the proximity of the Atlantic Ocean being influenced by the oceanic climate (very small thermal variation during the year) and characterized by beech; the other is developed towards the interior of the continent, or rather towards Western Asia respectively, with very great annual temperature variation. It is characterized by oak. These oak forests gradually turn in the south into steppes passing through the steppe-woodland zone.

Mixed deciduous-conifer and deciduous forests extend in eastern Europe

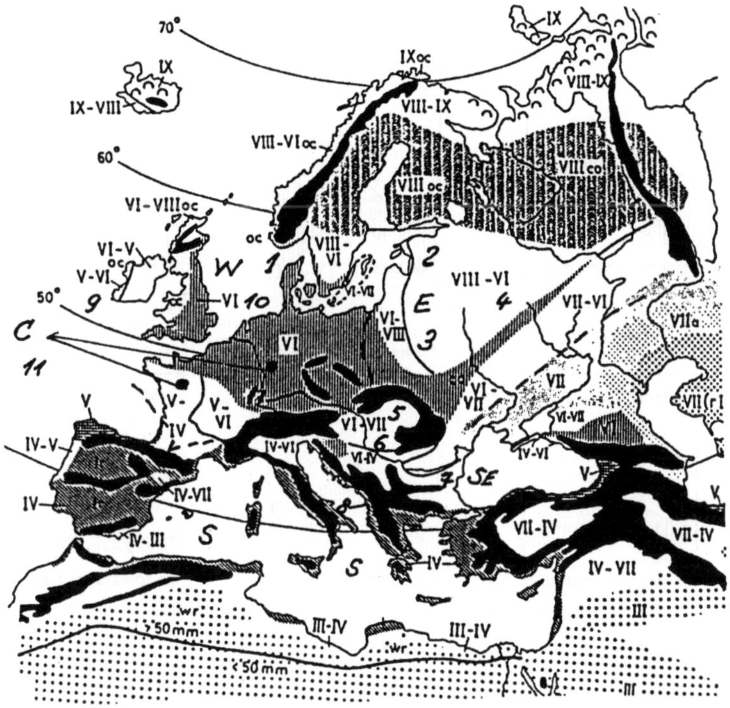

Fig. 1a. Ecological climatical division of Europe. E-Europe: Climatical zonation nearly parallel to the latitudes; in W-Europe: near N-S due to the gulf current; transitional in C-Europe. III, subtropical dry zone (deserts); IV, transitional zone with winter rain and summer drought; V, warm-temperate climate (cold winters); VI, typical temperate climate; VII, semi-arid steppe climate; (VIIa, arid semi-desert climate); VIII, cool-temperate climate and boreal zone, resp.; IX, arctic climate; X, mountain climate (black); VIII–IX, transitional type; oc, oceanic influence; cont, continental influence; wr, predominantly winter rain; a, relatively arid (from Mayer 1986); fr, frequent frosts. Climatical diagram = Cd.

from approximately 56° N in the north to 46° N (in the western part) and 54° N only in the south. The longitudinal extension exceeds 2,000 km (23° E in eastern Poland to 58° E in Ural Mountains).

The region has been deeply affected by Pleistocene glaciations and definitely colonized by forest not before the Holocene (Huntley & Birks 1983). Evidence suggests early-Holocene (9000–7500 B.P.) expansion of deciduous trees and shrubs (*Corylus* pollen type) into the area. *Tilia* expansion is progressive from the south and east. *Quercus* (deciduous) still is in the continuous northward expansion, initially in the west but later in the east also (Fig. 2).

Pairs of isopoll maps at 0 and 6000 B.P. (Huntley & Webb 1988: 364) illustrate the greater range and abundance of *Alnus*, *Corylus* (type) and

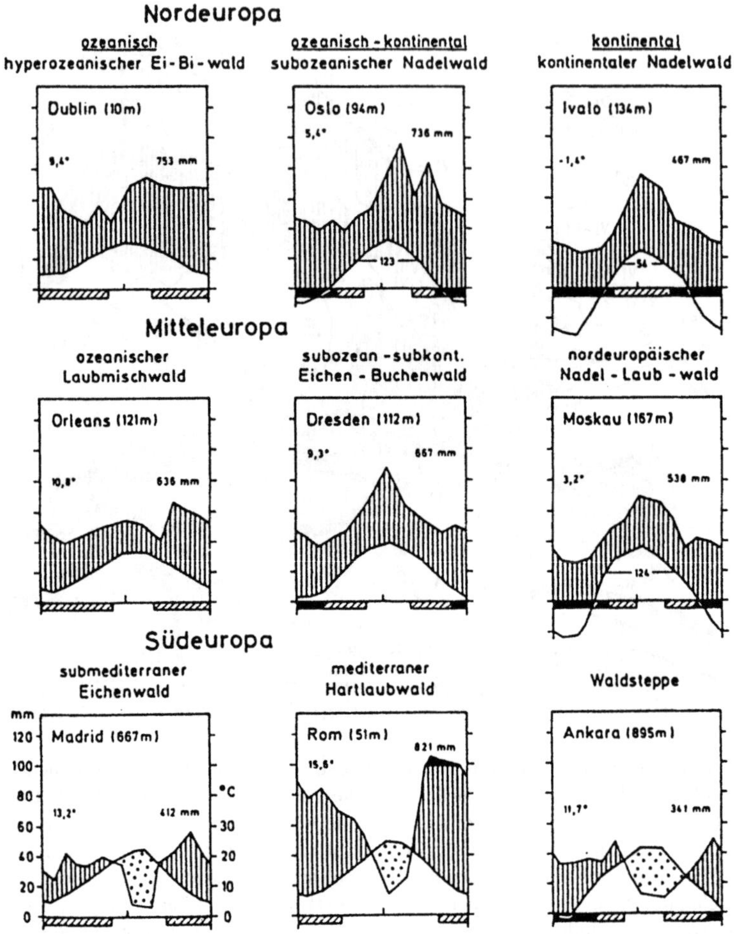

Fig. 1b. Characteristic climate diagrams from northern, central and southern Europe with oceanic to continental character of climate differing by the level and the amplitude of temperature, amount and distribution of precipitation and drought periods (from Mayer 1986).

Quercus (deciduous) in mid-Holocene. *Ulmus* shows a spectacular decline since the mid-Holocene throughout its range. The deciduous forests of the northern hemisphere have been profoundly changed by long-established and repeated clearances for settlement and agriculture, even in the relatively well preserved east European region. The late-Holocene expansion of *Carpinus* coming from the east, its correspondence to the Roman period and especially to the post-Roman land abandonment have been used as a basis for the hypothesis that *Carpinus* extended only when forest disturbance and/or abandonment of cultivated ground allowed it to invade (Huntley & Webb 1988). At present the forests occupy not more than 30% of the former area. Only

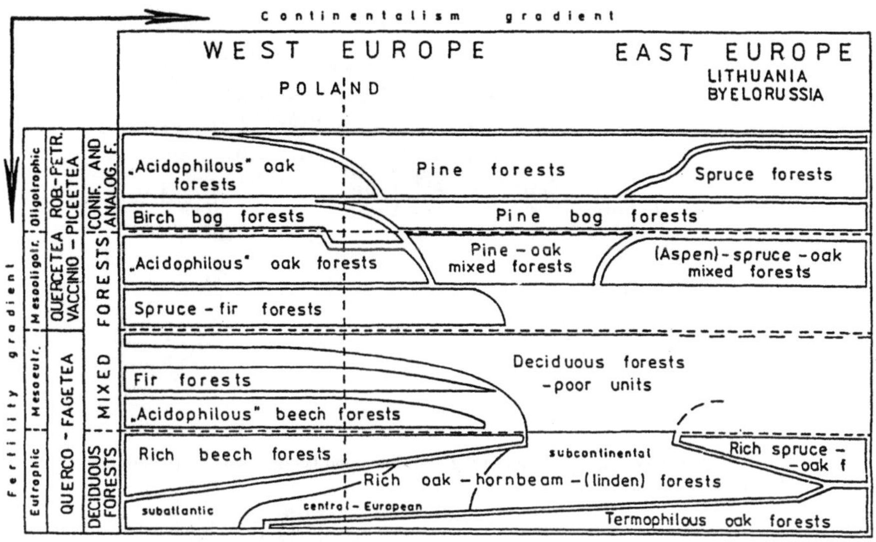

Fig. 1c. Regional differentiation of forest vegetation in subcontinental Europe (Falinski 1988).

remnants remain to piece together their original character. A unique opportunity of getting knowledge of seminatural oak-hornbeam woodlands as well as deciduous-conifer mixed forests is offered by the forest of Bialowieza (Bialowieza National Park), an old forest of more than 100,000 ha, in the Polish plains north-east of Bialystok. The following reserves ("zapowiedniki") are well known: Berezinsky, Les na Vorskle and others. The distribution of deciduous and deciduous-conifer forest in eastern Europe is presented in Fig. 3.

Climate and soil

The climate in the east is fairly dry and warm in summer, but cold in winter. In the west, temperature differences are far more modest. The average annual precipitation in the continental part does not exceed 600 mm nor fall below 400 mm, but exceeds 4,000 mm in some hyper-oceanic areas. The mean July temperature generally is over 18°C in the east and below 17°C in the western parts. The winter in the east is quite severe with more snow than in the more western regions. Continentality of climate is growing from west to east, and the period with snow cover is extending (Jäger 1968, Degorski 1985, Walter & Lieth 1960ff.; Figs. 1, 3, 4, 8, 11).

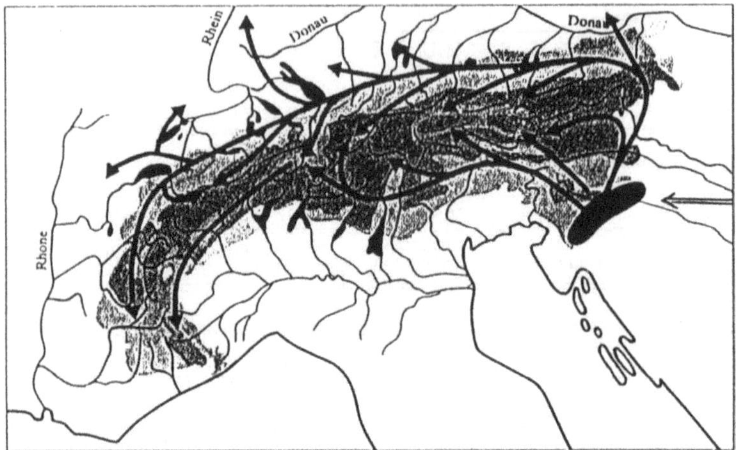

Nacheiszeitliche Einwanderung von Osten und heutige, natürliche Verbreitung der Fichte (*Picea abies*).

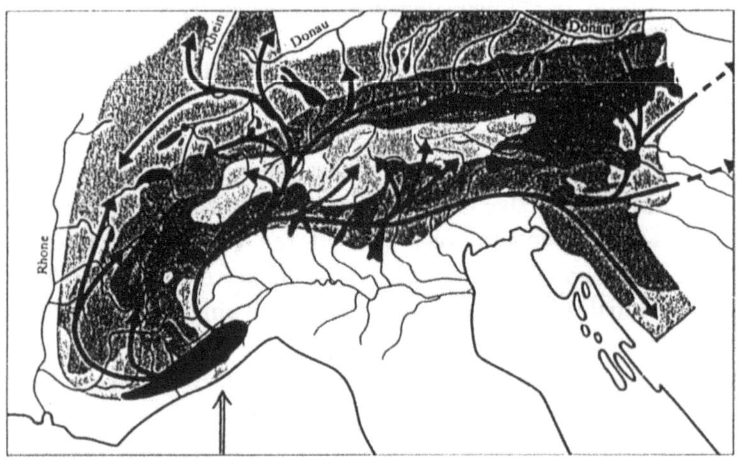

Nacheiszeitliche Einwanderung von Westen und heutige, natürliche Verbreitung der Tanne (*Abies alba*).

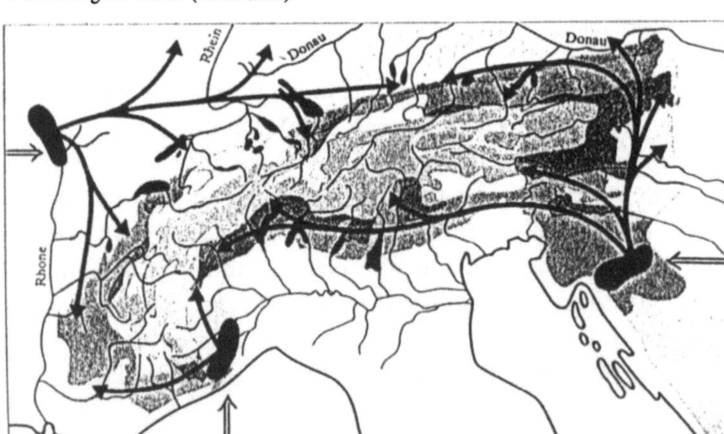

Nacheiszeitliche Wanderwege der Buche (*Fagus silvatica*) von Osten, Westen und Süden.

Fig. 2. Post-glacial migration of spruce, fir and beech from east, south and west, and natural area (from Reisigl & Keller 1989). Last graph for beech (*Fagus silvatica*).

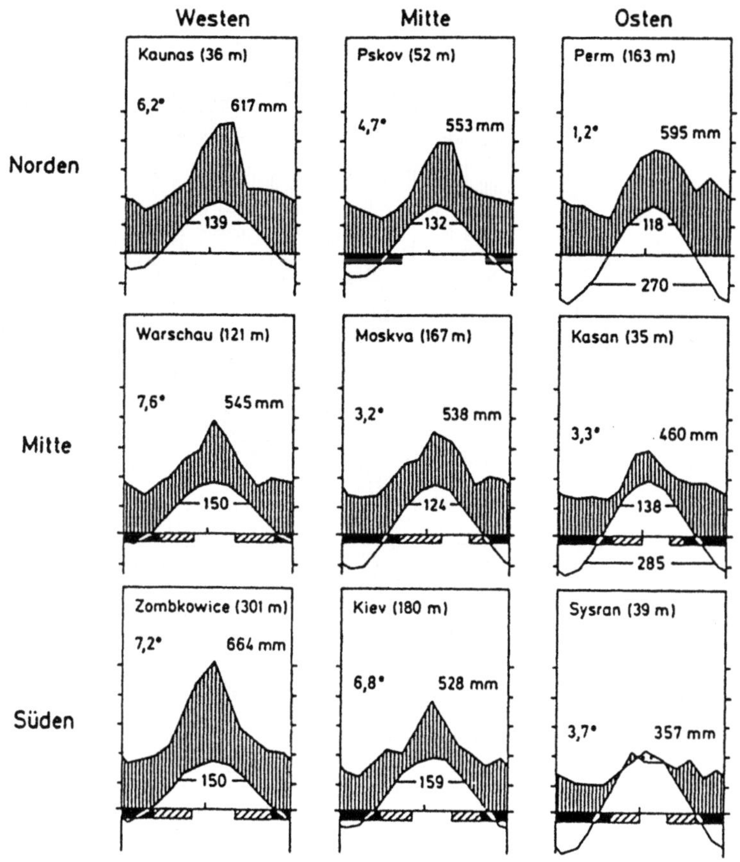

Fig. 3a. Climate diagrams from the eastern European lowland mixed deciduous/coniferous forest area. W → E, more continental (less precipitations, winter colder); N → S, increasing growing season; Kiev, n limit of forest-steppe zone; Sysran (Volga), s limit and transition to steppe (no forest anymore because of summer drought; Astrakhan already in near desert situation). Forest limits near 3–4 mm of precipitation, remnants of woodlands in ravines (from Mayer 1986).

		Hyper-oceanic E-Ireland	Oceanic NW-France	Sub-oceanic E-Germany	Sub-continental E-Poland	Continental C-Russia
Temperature	Year	9–10°C	9–11°C	9°C	6–8°C	3–7°C
	Jan.	5°C	3°C	1°C	−1/−3°C	−2/−5°C
Precipitation	July	16°C	17°C	19°C	16–19°C	18–20°C
(mm)	Year	750	650	650	550–700	(400)500–600

The southern parts show mediterranean type of climate with summer drought (up to 3 months) and winter rains with precipitation of less than 400 mm in south-eastern Spain. Main soil type under deciduous-conifer forest is podzol and brown forest soil, dark grey soil and chernozem under mixed deciduous

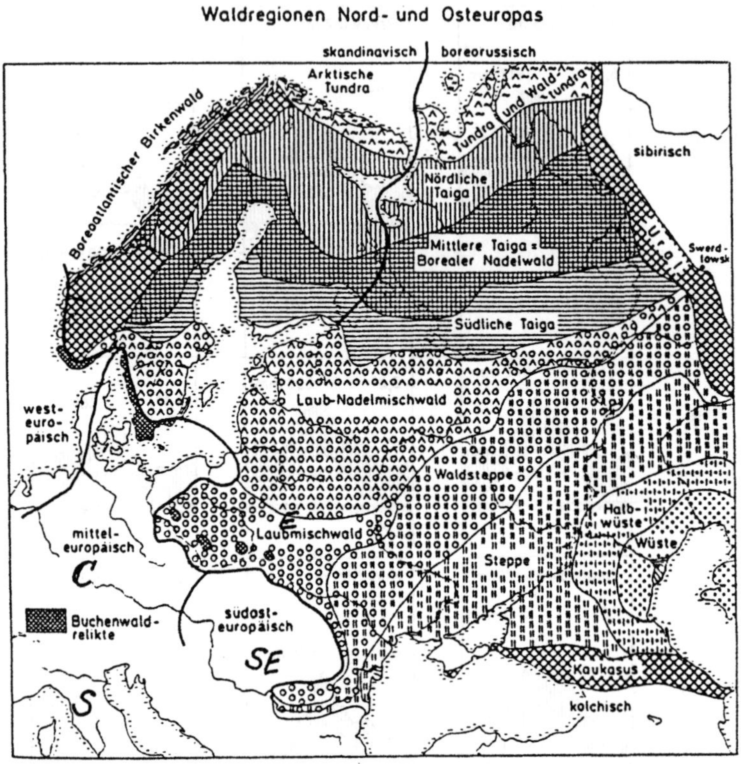

Fig. 3b. Forest regions (zones) in northern, northeastern and eastern Europe: Distinct zonal "belts", W-Siberian species far west of Ural Mts. (from Mayer 1986). For German terms always compare Fig. 9d.

oak forests. In the west, brown forest soils (Cambisols), lessivés (Luvisols) and pseudogleys are most frequent.

Tree and shrub species: Floristic composition

The number of deciduous tree species is not high in the east-European area: c. 20 species in western regions and 12 species in eastern areas mostly, versus more than 40 species in western Europe. The most important are: *Quercus robur*, *Tilia cordata*, *Acer platanoides*, *Fraxinus excelsior* and *Ulmus glabra*. Western, subcontinental areas are characterized by the presence of *Carpinus betulus*, *Acer pseudoplatanus*, *Quercus petraea* and *Fagus sylvatica*; subatlantic *Taxus baccata* is very rare there all reaching the most western areas. The main deciduous trees reach great dimensions in the subcontinental Bialowieza Primeval Forest (more than 40 m of tallness; spruce *Picea abies* may be up

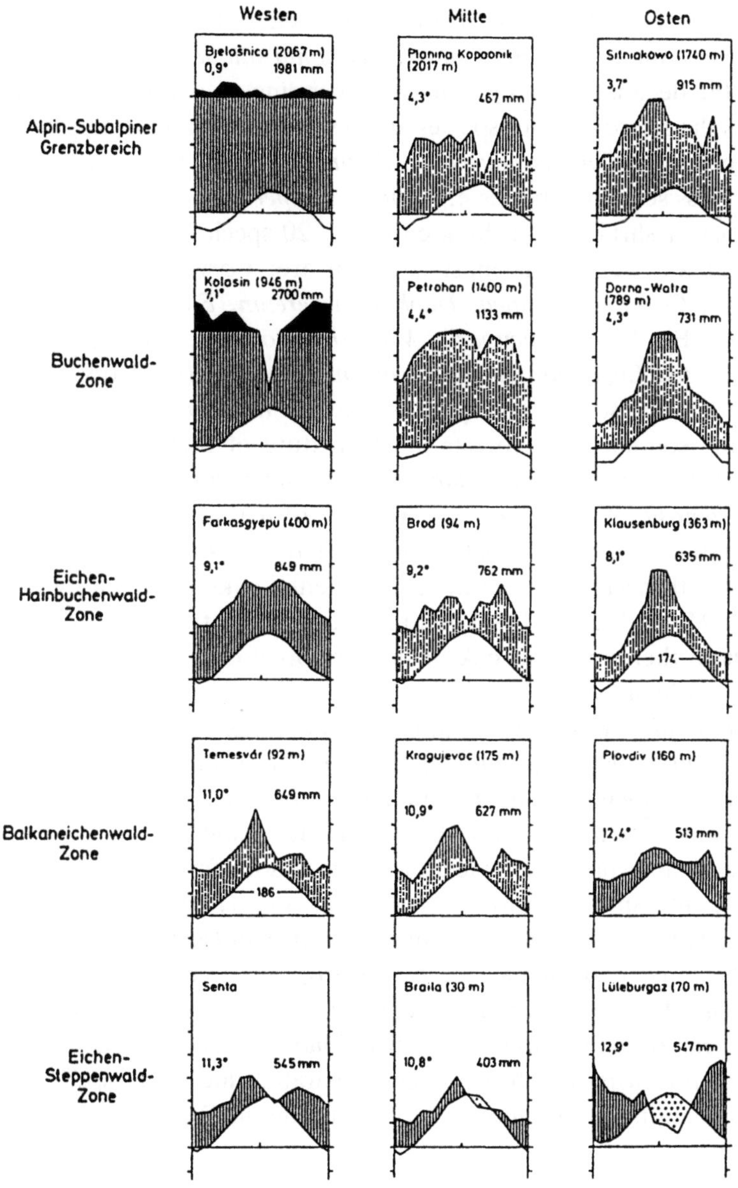

Fig. 3c. SE-European climate diagrams from the western, central and eastern parts of the Balkan areas on different elevations. Summer is less arid in the pannonic (Senta) island-lake occurrence of the *Quercus* forest-steppe zone than in the Danubian transitional area from forest to steppe (Braila) and in Thracian forest steppe (Lüleburgaz). Similar distribution in the slightly moister zone with *Quercus frainetto*. *Quercus*/*Carpinus* forest associations in the forest of Bakony (Farkasgyepü) are relatively moist, in Slavenia (Brod) much dryer, in the Carpathian basin (Klausenburg) not much drier, due to the distinct maximum of precipitation in summer (although the yearly average is lower). Obvious drop in the amount of precipitation from the Illyric region (Kolasin) crossing the Moesic-Balkanic (Petrohan) to the Dacic region (Dorna Vatra, E Carpathians). Also stations in the subalpine belt are distinctively different Bjelačnica, Illyric; Planina Kopdonik, Moesic; Sitniakowa, Balcanic (from Mayer 1986).

to 55 m tall, Falinski 1988). In the south, sclerophyllous species (*Quercus ilex* species complex, *Quercus suber*, *Ceratonia siliqua*, *Phillyrea media* etc.) replace the deciduous ones gradually, depending on parent rock. Limestone sponsors the sclerophyllous species. Also shrubs are more of the sclerophyllous type (e.g. *Rhamnus alaternus*, *Viburnum tinus*, *Lonicera etrusca* etc.; and as a less sensitive species e.g. *Buxus sempervirens*).

The list of shrubs is not long either (c. 20 species in the east, about 40 species in the west). The most common are: *Corylus avellana*, *Lonicera xylosteum*, *Cornus sanguinea*, *Daphne mezereum*, *Euonymus europaea*, *E. verrucosa* (E), *Sorbus aucuparia*, *Virburnum opulus*, *Crataegus* div. spec., *Rosa arvensis*, *Ligustrum vulgare*, *Viburnum lantana* (the latter are more in the west and the south), and typically oceanic *Ilex aquifolium*.

For the southern oak forests the admixture of thermophilous shrubs is typical (*Viburnum lantana*, *Chamaecytisus ruthenicus*, *Caragana frutescens*, *Prunus spinosa*, *P. fruticosa*). The number of herbaceous plants is between 10–20 species per acre in oak-bush forest in the southern Ural (Igoshina 1964), 40–50 species per acre in subcontinental oak-hornbeam forest (Falinski 1988, Matuszkiewicz 1980) and may reach more than 50 species per acre in southern thermophilous oak or in thermophilous western beech and oak forests (Igoshina 1964).

Stellaria holostea, *Carex pilosa*, *Galium odoratum*, *Circaea lutetiana*, *Lamium montanum*, *Brachypodium silvaticum*, *Carex silvatica* and many other plants are characteristic of mesophilous woodlands (class *Querco-Fagetea*) and are widespread (Laivins 1991) (see Tables 1 and 2).

Some plants which are typical of the deciduous forest in central Europe, penetrate into spruce forests in the boreal zone (*Festuca altissima*, *Milium effusum*, *Pulmonaria obscura*, *Asarum europaeum* (Karpov 1973)).

In the most eastern part of east European deciduous forest endemic species have been reported. These endemics are represented by a few species of European origin (*Lathyrus litvinovii*, *Knautia tatarica*, *Cicerbita uralensis*, *Anemone uralensis*) which have been separated due to the isolation of the Ural population of their ancestral forms during the Pleistocene (Gorchakovsky 1969).

Structure

The structure of deciduous forest is remarkably similar in spite of their varied habitats and species composition. It is a typical multi-layered plant community often consisting of one or two tree strata, a shrub stratum and a herbaceous stratum. There is no continuous moss layer owing to the masses of fallen leaves and the relatively thick herb layer. Near its northern and

Table 1. The distribution of the typical plants of mixed oak woods from E. Poland to S. Ural (west to east gradient)

	A	B	C		D		E		F		
Trees											
Picea abies	5	3	5								
Pinus sylvestris	(+)	(3)	(+)								
Carpinus betulus	5	5	5	5	1						
Quercus petraea	(+)				1						
Fagus sylvatica				1							
Acer pseudoplatanus				(+)	(+)						
Acer tataricum				(+)	1						
Acer campestre				(+)	1						
Quercus robur	4	5	2	5	5	4	3	4	5	1	2
Tilia cordata	4	5	5	(+)		(+)	3		5	5	5
Acer platanoides	5	5	4		(1)		(+)	+	4	5	5
Fraxinus excelsior	3	2	1	(+)				1	4	5	5
Ulmus glabra	2	2			(+)			+	4	5	5
Prunus avium	(1)			5	1						
Pirus communis	3			1	(+)			+			
Malus sylvestris	2	2	3	1				+	3		
Betula pendula			2	3						1	2
Populus tremula				1						1	2
Ulmus laevis	(+)				1						
Ulmus campestris	(+)				2			+			
Shrubs											
Corylus avellana	5	4	1	2	3	4	3	+	(+)		
Cornus sanguinea		(+)		5	5			+			
Crataegus monogyna				4	5			+			
Daphne mezereum	4	1		4	1				(+)		
Euonymus europaea		1			3			+			
Euonymus verrucosa	2	4		1	5		+	+		1	
Lonicera xylosteum		4		2	4	1	+	+			2
Prunus padus									5	5	3
Prunus spinosa				4	3			+			
Rhamnus cathartica				1				+			
Ribes rubrum				1							
Rosa canina				2	4						
Rosa tomentosa				1	1						
Rosa cinnamomeum											
Rubus caesius				4	1						
Rubus idaeus									4	5	3
Sorbus aucuparia	3	2	3			+	+		5	5	4
Viburnum opulus		1	2		1	+					3
Viburnum lantana				4	5						
Characteristic species of oak-hornbeam forests (Carpinion and Tilio-Carpinetum)											
Stellaria holostea	5	5	5	5	5	1	+	+	4	5	5
Carex pilosa	3			1	5		4	+	+		
Melampyrum nemorosum		3	1		1						
Hepatica nobilis	5	5	4								
Isopyrum thalictroides	2				2						
Galium schultesii											
Lathrea squamaria	3										
Ranunculus cassubicus			4								

Table 1. Continued.

	A	B	C		D		E		F		
Species occurring in deciduous forest in S. Ural											
Anemone altaica									5	5	4
Aconitum septentrionale									5	5	
Stellaria bungeana									5		1
Crepis sibirica									5	5	3
Cacalia hastata									4	5	4
Primus veris ssp. *macrocalyx*									5		1
Bupleurum longifolium ssp. *aureum*										3	2
Cicerbita macrophylla ssp. *uralensis*										2	2
Pleurospermum uralense										2	
Knautia tatarica											2
Widespread herbaceous plants (Querco-Fagetea)											
Aegopodium podagraria	5	5	4	3	5				5	5	5
Galium odoratum	5	5	3	3	5		1		5	5	5
Milium effusum	5	2	5		3				5	5	4
Athyrium filix-femina	4		3	3		1			5		2
Pulmonaria obscura	4	5	4	2	3	2			5	5	
Paris quadrifolia	5	1		1		1			5	3	3
Asarum europaeum	2	5	3		5	3		+	5	5	5
Mercurialis perennis		3			3		1				
Stachys sylvatica	2	1	4	1	1				5	5	2
Bromus benekenii				2	1	+	+	+	5	2	5
Viola mirabilis	1	4	3	3	4	+	+	+	5	5	4
Melica nutans	1	2	2		5	+		+		3	2
Lathyrus vernus	4	5	4	5	5		+	+	5	5	2
Dryopteris filix-mas	4	2	3	4		+	+		5	5	4
Poa nemoralis		4	4	3	5		1	1	1		

5–1: Constancy in phytosociological tables, (+): Occurring sporadically, A: E. Poland (Bialowieza), B: Lithuania, C: Latvia (Laivins 1991), D: W. Ukraine, E: Central Russia (Walter 1940), F: S. Ural (Gorchakovsky 1969).

eastern limits the oak gradually loses its typical form and associations, so that the structure of the forest is modified. For example, the oak forest of southern Ural at 600–800 m a.s.l. may reach only 4–10 m, in some cases it is only 2–3 m tall (Igoshina 1964, Atlas SSSR 1969).

In subcontinental oak-hornbeam forest (Matuszkiewicz 1980), hemicryptophytes (45–60%), phanerophytes (15–20%) and geophytes (15–20%) are numerous, and a very small role is played by chamaephytes (2–3%,) and therophytes (1–2%).

Geoelements of vascular plants

In the zone of deciduous and deciduous-conifer forests different geoelements (Walter & Straka 1970, Kleopov 1990) are represented:

Table 2. Differential species in oak plant communities of Central Russia (the analysis by means of counting similarity coefficients according to Czekanowski) (Shatkauskas 1975)

I. Alliance of associations	
	Quercus robur – Fraxinus excelsior – Acer campestre*
1.A.	*Pyrus communis – Prunus spinosa – Euonymus europaea*
3.W.	*Pyrus communis – Prunus spinosa – Ulmus carpinifolia – Melica picta*
3.B.W.	*Prunus spinosa – Ulmus carpinifolia – Melica picta*
3.B.	*Euonymus europaea* – Melica picta
II. Alliance of associations	
	Quercus robur – Tilia cordata – Lonicera xylosteum
12.A.	*Festuca altissima – Cicerbita uralensis – Knautia tatarica*
2.A.	*Fraxinus excelsior – Acer campestre – Mercurialis perennis – Allium ursinum – Galeobdolon luteum*
5.A.	*Fraxinus excelsior – Festuca altissima – Mercurialis perennis*
8.A.	*Festuca altissima – Mercurialis perennis*
4.A.	*Fraxinus excelsior – Festuca altissima – Mercurialis perennis – Neottia nidus-avis – Dentaria bulbifera*
9.A.	*Pleurospermum uralense*
III. Alliance of associations	
	Quercus robur – Tilia cordata – Betonica perancta
11.A.	*Festuca altissima – Elymus caninus – Pleurospermum uralense*
6.A.	*Carex montana*
IV. Alliance of associations	
	Quercus robur – Pinus sylvestris – Chamaecytisus rhutenicus – Carex supina
6.B.	*Tilia cordata – Carex montana*
7.A.	*Carex montana*
10.A.	*Vincetoxicum hirundinaria*

* Nomenclature acc. to "Flora SSSR" (I–XXX).

1. Central-European vascular plants reaching the Ural (e.g. *Festuca gigantea*, *Carex pilosa*, *Mercurialis perennis*)
2. Central-European – south Siberian plants (e.g. *Dryopteris filix-mas*, *Carex silvatica*, *Actaea spicata*)
3. Sub-pontic element (e.g. *Potentilla alba*, *Anemone sylvestris*)
4. Pontic species (e.g. *Cytisus ratisbonensis*, *Adonis vernalis*)
5. Pontic-mediterranean (e.g. *Stipa pennata*, *S. capillata*, *Dictamnus albus*).

Sub-zones

In both parts of Europe, in the eastern and the western parts, the zone of deciduous and deciduous-conifer mixed forests exhibits marked diversity both in the N-S and W-S directions (Fig. 4). These are:

1. northern region – mixed evergreen-deciduous forests (boreal – nemoral; Fig. 5).

2a. east-central part – deciduous forest (*Carpinus – Quercus* between the Bug River and Dniepr River; *Fraxinus excelsior – Quercus robur* between the Dnieper River and Volga River, and *Ulmus – Quercus robur* to the Ural Mountains; Figs. 5, 6).
2b. west-central part – deciduous forest. *Fagus/Quercus* and *Fraxinus/Quercus* (compare Hofmann 1985, for eastern Germany: 50% beech, 13% oak/hornbeam, 14% oak-forest; Figs. 6, 7, mountains Fig. 10).
3. southern part: thermophilous oak and mixed sclerophyllous and deciduous woodlands (Fig. 8, mountains Fig. 10; for comparison see Table 3).

Phenology

Very impressive is the spring aspect of flowering and pronounced summer depression of activity of the ground flora (Gorchakovsky 1969). The group of spring geophytes consists of *Anemone nemorosa*, *A. ranunculoides*, *Isopyrum thalictrioides*, *Ranunculus ficaria*, *Gagea* ssp., *Corydalis* ssp., *Allium ursinum*. It is clear from the varying behaviour of trees and herbs that the Quaternary glaciations have had a profound effect on the species that compose them. It is striking that trees in these vegetation types come into leaf at different times: *Populus*, *Betula*, *Alnus*, *Acer*, *Carpinus*, *Prunus* are early-leafing trees; *Fraxinus*, *Quercus*, *Ulmus* are later-leafing trees.

There exists now, as a result, a general evolutionary disequilibrium between the climatic conditions in which they currently grow and their phenological characteristic, i.e. their flowering, fruiting, leafing and so on. Their patterns of foliation and their internal morphology are only loosely attuned to the climatic conditions. *Asperula* (*Galium*) *odorata* and *Mercurialis perennis* are aperiodic species (occurring all the year in greenhouse being protected from the cold of winter), while periodic species, well harmonized in their phenology with the cold season (*Polygonatum* type) are represented by *Corydalis solida*, *Convallaria majalis* and *Dentaria bulbifera*. The plants of *Leucoium* type of mediterranean origin, distributed in deciduous forests in Central Europe, are practically absent in Eastern Europe.

Community patterns

In Central and Eastern Europe, and especially in its northern part, two principal types of forests occur side by side: acidophilous coniferous forests and subneutrophilous broad-leaved deciduous forests. The coniferous type, poor in species and limited to more or less oligotrophic soils, has much affinity to the boreal taiga. The deciduous type is floristically much richer

Table 3. The temperate deciduous and deciduous-evergreen mixed forest in Europe

Factors, parameters, processes	Western- and south(-western) Europe	Eastern Europe
Quaternary history	Mostly Holocene (except southern France and further south)	Holocene only
Geographic pattern	Mostly peripheric in relation to Tertiary refugia (Tertiary refugia in S France and further south)	Peripheric in relation to Tertiary refugia
Relief	Lowland and submontane, often montane to subalpine in mountains	Lowland and submontane
Climate	(Sub-)atlantic to (sub-)mediterranean	Subcontinental to continental, long snow cover
Soils and subsoils	Often pleistocene deposition, mesic to eutrophic brown forest soil and lessívé; podzolic in the caledonian/variskian areas and higher mountains; often gley, pseudogley in the west; terra rossa in the south, etc.	Mainly pleistocene deposition, mesic to eutrophic, brown forest soil, podzolic, chernozem
(main) Trees	c. 40 tree species, *Fagus silvatica*, *Quercus robur*, *Quercus petraea*, *Carpinus betulus*, *Fraxinus excelsior*, *Acer pseudoplatanus*, (*Taxus baccata*) (*Alnus glutinosa*)	c. 20 tree species only; cold resistant, *Quercus robur*, *Tilia cordata*, *Acer platanoides*, *Picea abies*, *Pinus silvestris*, *Betula pubescens*, *Populus tremula*
Forest types	Sclerophyllous (in the extreme south-west): *Quercetea ilicis*; Deciduous: *Quercetalia pubescentis* (in the south-west): *Quercetea roboris-petraeae*, *Querco-Fagetea* (whole area); Coniferous-deciduous (in north-west and north, and in mountains): *Vaccinio-Piceetalia*	Coniferous-deciduous (in north-east): *Vaccinio-Piceetalia*; Deciduous: *Querco-Fagetea* and *Quercetalia pubescentis* (in south)
Floristic complexity	Higher	Lower
Human impact	Very long and intensive	Shorter (many hundreds years), intensive in last centuries
Role of fire	Very important in the south	Without importance
Animals, insects	*Cervidae* (only in the west)	Bison
Biomass ($t \cdot ha^{-1}$)	300	>400
Net primary production ($t \cdot ha^{-1} \cdot yr^{-1}$)	10	9

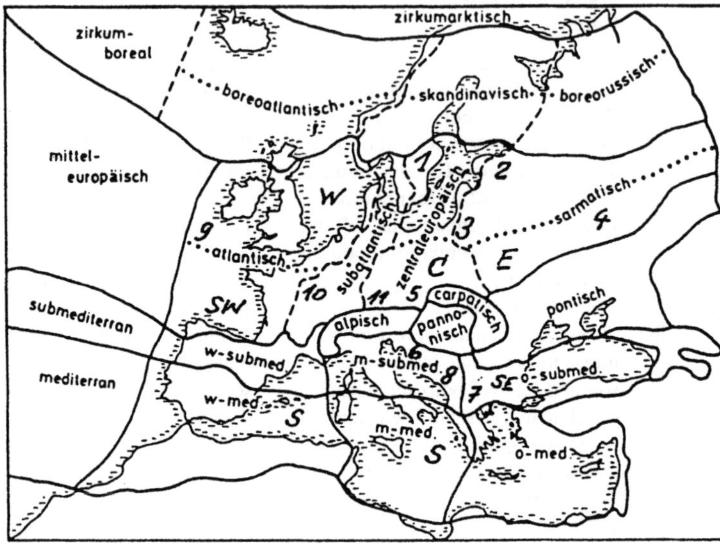

Fig. 4a. Floristic regions and provinces in Europe (from Mayer 1986).

and occupies eutrophic or mesotrophic sites and is optimally developed in the nemoral zone reaching the western most parts of Europe and the ecotone towards the boreal forests with similar structures. Pure deciduous forests are limited by climatic and edaphic factors, namely according to latitude and continentality (Klötzli 1976; Fig. 11):

- temperature sum in growing season, and especially in autumn;
- late and extreme frost periods;
- regular or prolonged drought or wet periods (modified by winds, mist, hoar frosts, wet snow and pathogenic influences), but also
- water and nutrient availability (influenced by building-up of raw humus and/or leaching) (Fig. 12).

In southern areas, they are gradually replaced by sclerophyllous woodlands. The availability of water (incl. soil texture) and the length of the drought period are decisive for their existence (northern parts up to 44° N about 13–15°C annual mean temperature and up to 900 mm annual precipitation, 3–4 months of drought, according to the definition of drought in climate diagrams *sensu* Walter & Lieth (1960ff.); in more southern areas, about 15 up to 20°C, down to 400 mm and 4–5 months of drought in the plains). Just for comparison a few words on typical Mediterranean trees will follow.

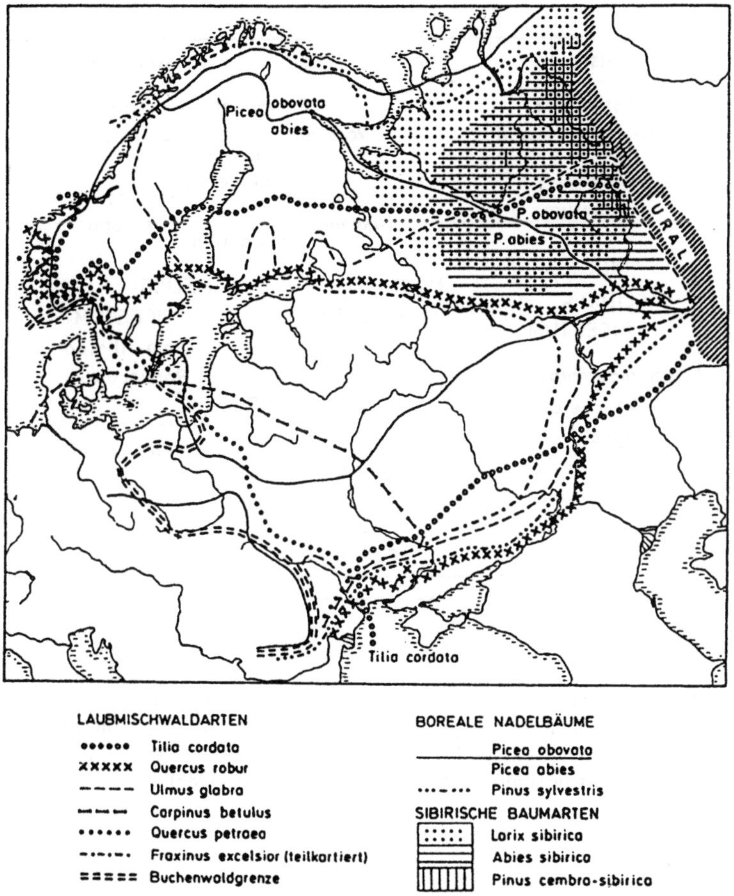

Fig. 4b. Distribution of some deciduous trees ("Laubmischwaldarten") and of some boreal and Siberian coniferous trees ("boreale Nadelbäume") showing increasing continentality towards the east (from Mayer 1986).

The sclerophyllous leaf and tree is well adapted to coping with effects of:
- relatively warm, humid winters and hot dry summers;
- drying soils and subterranean water provisions;
- relatively suitable temperature for assimilation the year over ("be prepared" strategy) and regular fires.

The typical sclerophyllous tree species are deep-rooted, small-leaved and thick-barked, as e.g. a number of *Quercus* species (*Quercus ilex*), *Phillyrea media*, *Olea oleaster* and *Ceratonia siliqua*. Most of these woodlands have been transformed to maquis (chaparral), fire-prone scrub (garrigue) or open

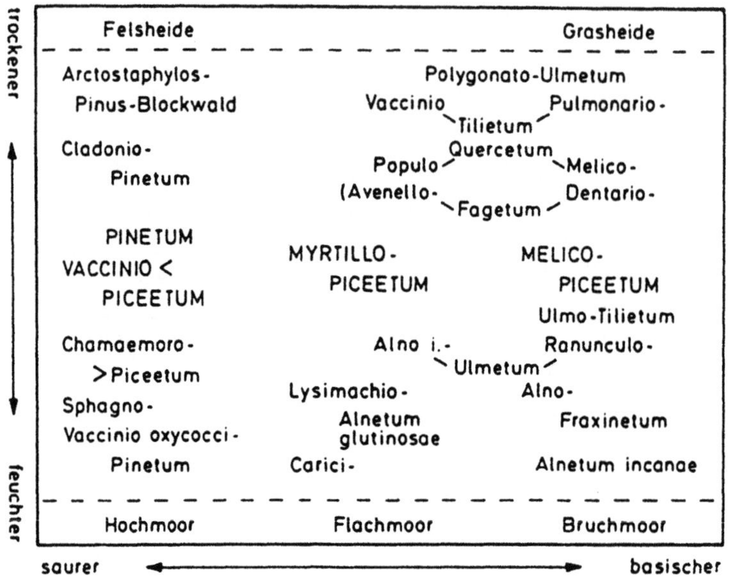

Fig. 5a. FAC of the Scandinavian subboreal lowland mixed deciduous/coniferous forest area. Pine forests only on poor, dry or very wet sites. *Picea* forests are dominant on all mesic sites, regardless of base regime, *Fraxinus/Ulmus/*(etc.). *Picea* forests mostly on more nutrient richt and warm sites. – *Fagus* in southern parts only (from Mayer 1986). FAC = forest association complex.

pastures, thereby taking up malacophyllous species which are well-adapted to fire and drought.

Between the deciduous forest and the sclerophyllous woodlands there is a large ecotone where trees of sclerophyllous ("mediterranean") character are gradually replaced by deciduous ones, herbs and shrubs of more southern descent staying in the understorey. However, in the northern parts deciduous trees are promoted by siliceous rocks (details see e.g. in Mayer 1986, Walter & Breckle 1986). A similar case can be traced in the ecotone between laurophyllous (or "Laurineous" or "Evergreen Broadleaved") and deciduous forest, e.g. in middle Chile (in the montane belt of the Chilean Andes south of 40° latitude), Japan or mountains of western China (Klötzli 1987). And presently, more southern deciduous forests are slightly colonized subspontaneously by laurophyllous species escaping from gardens which is a specific reaction of a series of mild winters since 1980. Such an ecotone with sclerophyllous species in deciduous forest is, globally seen, a rather rare case. In California and the eastern Mediterranean, conifers take over the role of

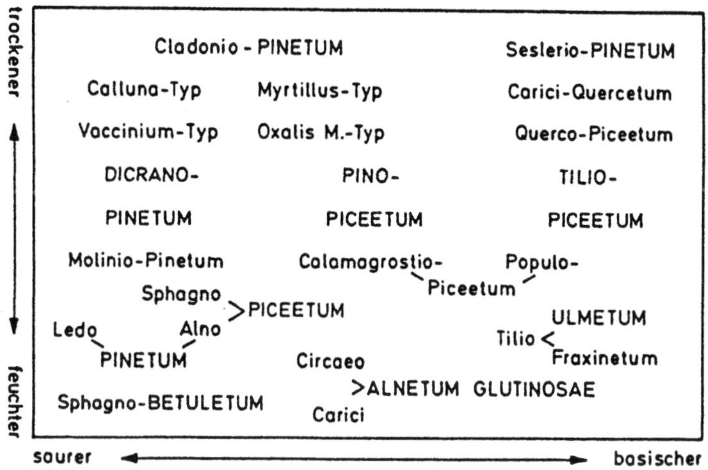

Fig. 5b. FAC of the northeast-european (Estonian) mixed deciduous/coniferous forest area. On mesic sites coniferous trees dominating. *Fraxinus/Ulmus/Tilia* etc.) on base-rich dry and moist sites (from Mayer 1986).

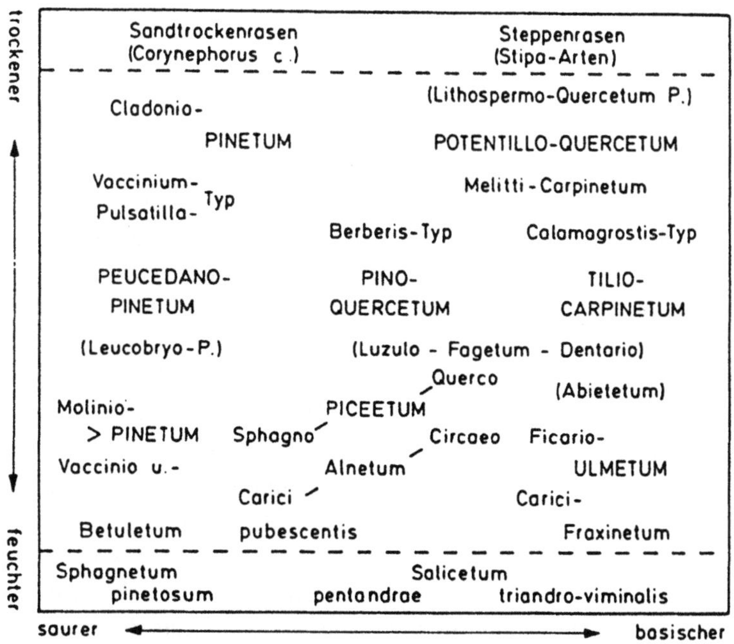

Fig. 5c. FAC of the subcontinental (Polish) mixed deciduous/coniferous forest area. Predominant *Tilia*-rich *Quercus robur/Carpinus-*, *Pinus/Quercus-* and subcont. *Pinus* forest from base rich to base poor sites (from Mayer 1986).

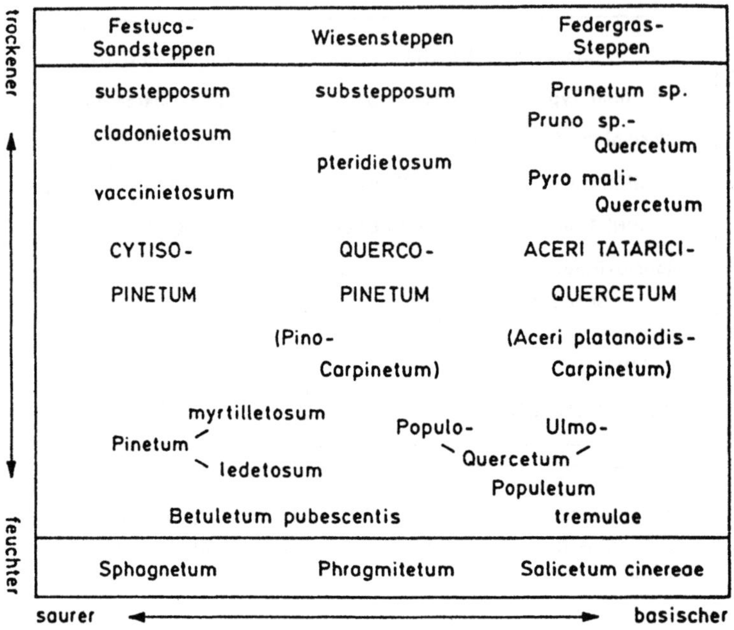

Fig. 5.d. FAC of the continental (Russian) mixed forest zone. From *Acer tataricum/Quercus*-steppe forest towards *Stipa*-steppe continuous "dissolution" of forest and bush-rich stages (from Mayer 1986).

dominant tree and shrub species. Deciduous oaks are (or would be under natural conditions) not very conspicuous. Obviously, there are certain similarities in the north-western United States with a number of deciduous oaks in the ecotone towards the coniferous rain forest near Vancouver and Seattle (Vankat 1979) with similar structures as e.g. laurophyllous woodland, or then in middle Chile with some *Nothofagus* species in the mountain belt (Klötzli 1983).

A true submediterranean ecotone with dominant deciduous trees (e.g. *Quercus pubescens* species group) is unique and typical for the western mediterranean region (less so in the eastern), probably a sign of pronounced (winterly) oceanity. It is nevertheless not comparable to the climatic gradient along the north-western coast of the United States with mediterranean type of climate gradually shifting to a cool laurophyllous type with much higher precipitation. In southern and south-eastern Australia, the shift clearly is from more or less sclerophyllous to laurophyllous forests due to mild winters and even distribution of precipitation.

In northern to eastern areas with less regular climatic conditions coniferous evergreen trees showing an adaptation in the way of a "be prepared" strategy

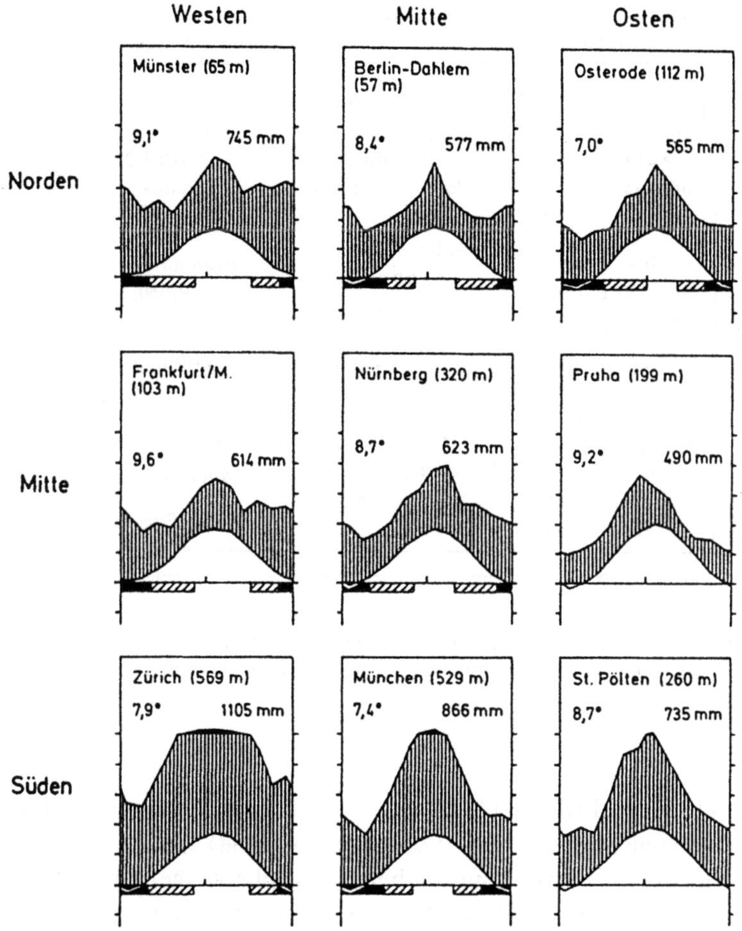

Fig. 6a. Climate diagrams from the middle-European lowland deciduous forest area. N → S, more precipitations towards the Alps; W → E, more continental, max. of precipitation in summer, greater differences in temperatures, colder winters (from Mayer 1986).

(needles "ready" in any warmer period), are gradually or abruptly mixed with the purely deciduous trees. The length of winter frost, late spring or early autumnal frosts, autumn warmth (for bud maturation) and base content of the rooted soil are decisive for the existence of pure or mixed forests; oasis-like isles, e.g. on boulder heaps, go much further to the north or then to the east, surrounded by partly mixed spruce or pine woodlands. Such ecotones from deciduous to coniferous forests have been observed and described from more oceanic (but not hyperoceanic) parts of the northern hemisphere. Examples for North America (e.g. Great Lakes area, S Manitoba) are given by Wilde (1962) and Mueller-Dombois (1964), for Europe

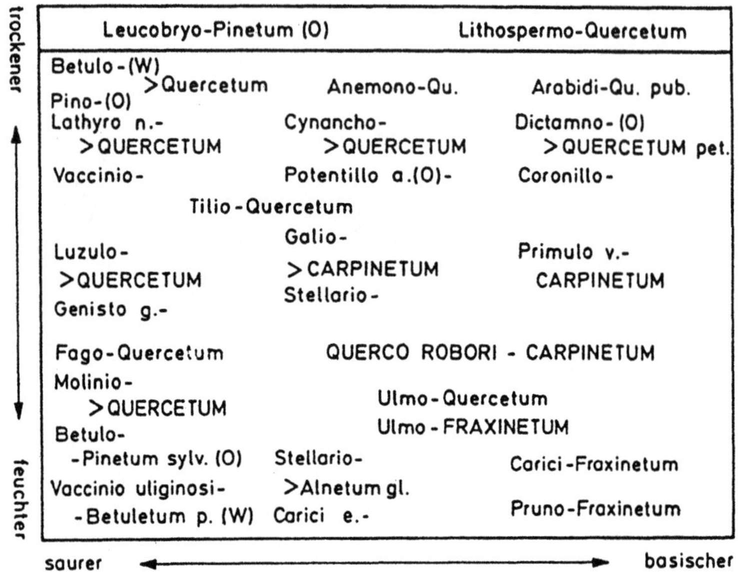

Fig. 6b. FAC of the central-European lowland mixed oak-forest zone. Richer soils with oak-hornbeam, oak being more competitive on acid soils. Pine-rich sites only in the east (from Mayer 1986).

by Sjörs (1950) and Klötzli (1975; especially for the northwestern part of the ecotone) (see also Figs. 1 and 4; see also Box 1988).

In hyperoceanic parts of the southern hemisphere, there is no true equivalent with conifers. But *Podocarpaceae* are often dominant or codominant in the milder rainforest areas (compare e.g. the rougher areas which are completely dominated by laurophyllous species (New Zealand) or by deciduous elements and/or mixed with *Nothofagus* species (South America)). Only special sites carry coniferous stands, namely bogs or rocks or then, under steppe conditions (Klötzli 1984a, Walter & Breckle 1986).

Under limiting conditions seedling establishment may be prevented or restricted by the influence of edaphic factors, non-compatible seedbed conditions or fungal attack in raw humus before the main root has reached mineral soil. Soil structure and texture, soil development, soil chemistry, soil biology (mycorrhiza) may be essential under specific limiting living conditions. In extremely western areas (e.g. Ireland) and in some southern valleys of the Alps ("Insubria"), the area of laurineous forest (evergreen broad-leaved forest; Box 1988, Hübl 1988, Klötzli 1987) is touched by the prevalent admixture of evergreen non-coniferous trees and shrubs (e.g. *Ilex*, *Hypericum androsaemum*) subspontaneously occurring *Rhododendron ponticum* and some other shrubs of that same leaf consistency from eastern Asia

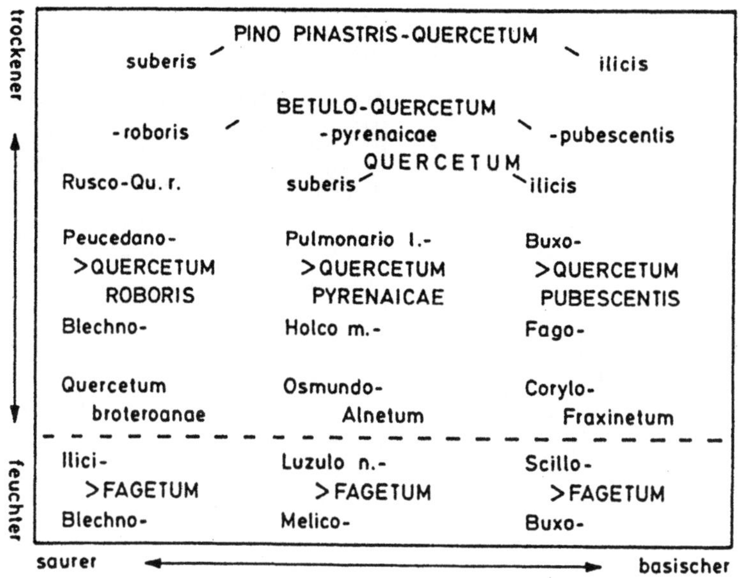

Fig. 7a. FAC of the south-atlantic lowland mixed oak-forest zone. Three different oaks are concentrated on three different sites along the acidity gradient (from Mayer 1986).

(*Cotoneaster*, *Eleagnus*, *Prunus*, *Quercus*), and furthermore many evergreen herbs in the understorey. Thereby, we find similar shifts as in southeastern North America or eastern Asia. However, no species of the laurophyllous tree layer are present. The last glaciation is responsible for their extermination from Europe and their partly survival in Macaronesia.

In the whole area of western Europe, beech often is one of the chief dominant trees, limited by hyperoceanic or submediterranean climate and sponsored by mountain chains (Pyrenees, Alps, Dinarides and other mountains of the Balkan), from southern Sweden (Scandes) to the higher hills of Burgundy and the Plateau-like western Jura Mountains, and from the Polish western border to Brittany and the southern U.K. Quite often beech has been planted beyond those limits and regenerates subspontaneously (Stockholm area, Scotland, Ireland).

Beech forest is normally dominant on mesic sites all over the area. Beech is one of the best adapted shade-tolerant species having regular mast years. However, in the more oceanic areas, beech forests are confined to more acid (and montane) sites. The richer soils are dominated by deciduous broad-leaved trees of non-*Fagaceae* species (*Acer*, *Fraxinus*, *Ulmus* etc.) and partly by oak species and birch (and pine in the east), the wetter ones by ash, alder and/or birch. In the more continental areas oak and/or pine are the prevailing

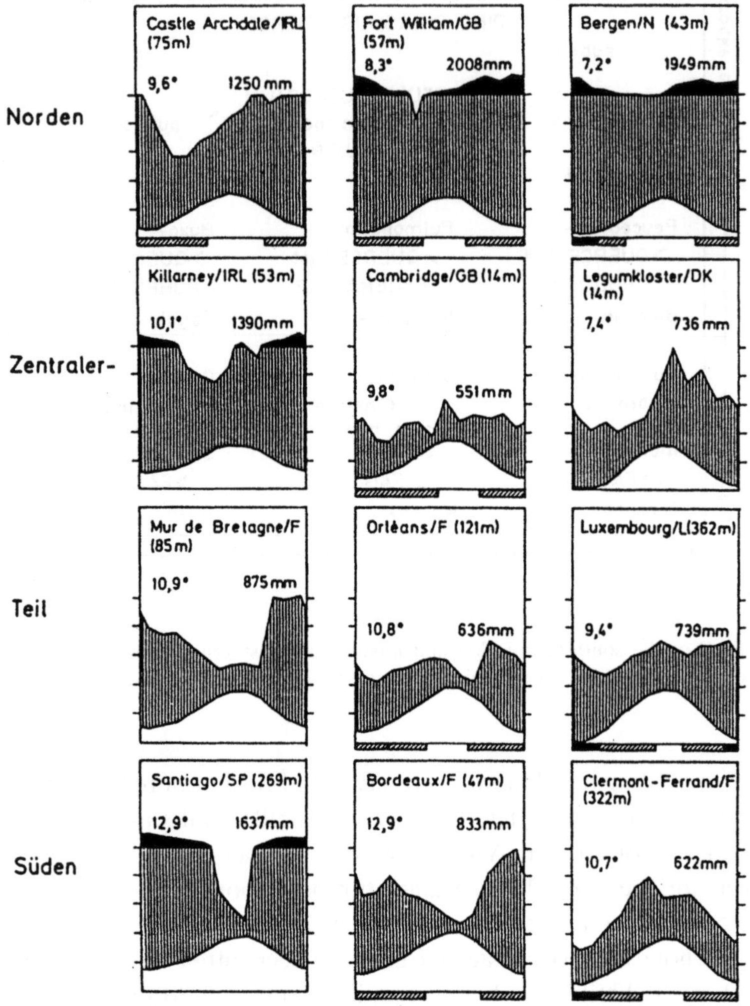

Fig. 7b. Climate diagrams of stations in Atlantic lowland sites. W–E: from hyperatlantic to subatlantic, only little decrease in precipitation, distribution of precipitation from double maximum curves in autumn and spring to summer maxima in the east; simultaneously average annual temperature, length of growing season and cold winter periods are increasing (especially under submediterranean influence) (from Mayer 1986).

species on more acid and dry sites. Fire sponsors the development of more conifer-rich stands (Fig. 7; for more details on plant communities see Mayer 1986).

Altitudinal zonation in the more humid parts of the higher mountain chains (Alps, Carpathians, northern Apennines) is characterized by the grad-

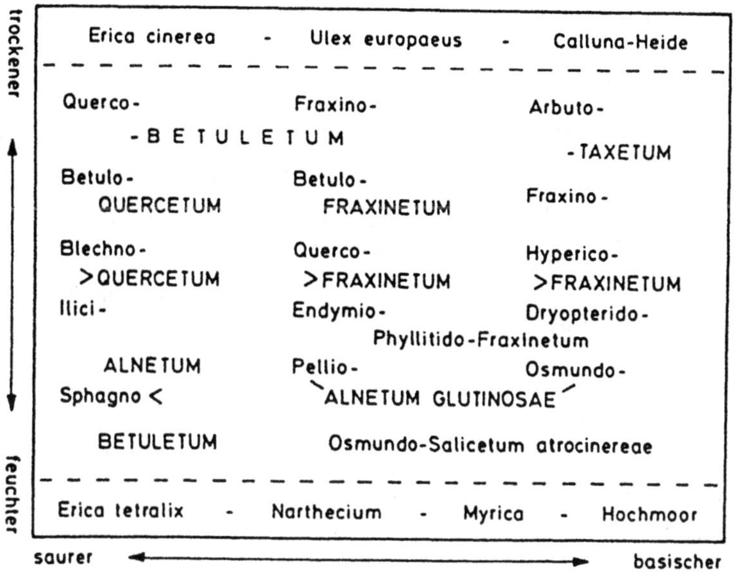

Fig. 7c. FAC of the hyperatlantic lowland mixed *Quercus-Fraxinus* forest zone. Mixed *Fraxinus* forests on mesic sites, *Quercus* forests confining to acid sites. Competitive *Alnus-forest* on humid sites as well (from Mayer 1986).

ual admixture of fir (without continental valleys where pine is dominant), and in higher elevations and in more continental areas by spruce, pine species (cembra pine at the timberline) and larch (Fig. 9). Under more arid mediterranean (eastern) climate types pine, partly fir species, partly *Juniperus* reach the timberline (not treated in this paper; Fig. 19). In the Pyrenees these different belts and zones meet, depending on aspect and position on the mountain chain from west to east or north to south.

By centuries-old anthropogenic influence there is a general shift towards stands with dominating species ready to sprout after coppicing (formation of oak-hornbeam forests on such stands in central Europe), and in beech areas spruce stands have been installed (in eastern areas with oak pine was sponsored). For about 50 years fir is under manyfold influence and now heavily influenced by dieback of the European type. Most European woodlands are far from getting a chance to get to a mature state. Change in competition between the species by management, influence of certain browsing animals (e.g. roe deer on fir), fire under human control (mediterranean areas, pine woodlands) or other cultural influences are responsible for the development of anthropogenic submature stages.

All these forest and woodland types have been described by phytosociolog-

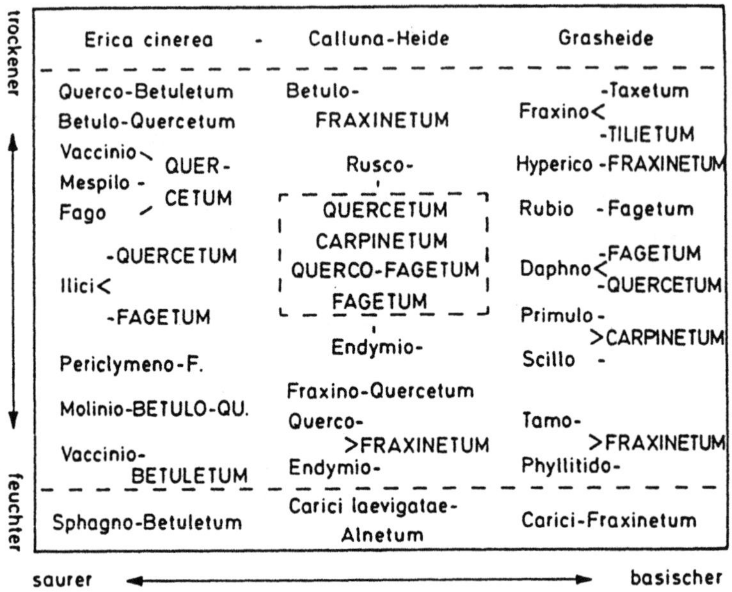

Fig. 7d. FAC of the Atlantic lowland mixed *Quercus* forest zone, *Quercus*, *Carpinus* and *Fagus* associations on poor, dry and mesic sites, *Fraxinus* on mesic, humid but also on dry sites (from Mayer, 1986).

ical methods and are listed in different synthetic publications country-wide or subcontinent-wide (references in Mayer 1986; also for the new forest conditions; summary in Klötzli 1992).

In principle, in the deciduous area two to three general groups of forested climax plant communities are recognized and placed in two syntaxonomic units, one bearing coniferous trees, the other being more or less deciduous: viz. the pure deciduous and the mixed forests are recognized as two syntaxonomic units: class *Vaccinio-Piceetea* and class *Querco-Fagetea*. The acidophilous deciduous-conifer mixed forests (*Pino-Quercetum*, *Querco-Pinetum* etc.) belong to the first class; mesophilous oak-lime-hornbeam, thermophilous oak forest, and floodplain ash-elm forest (*Tilio-Carpinetum*, *Potentillo albae-Quercetum*, *Ficario-Ulmetum* etc.) belong to the second class (Figs. 5, 8). All other western types, except those on very acid soils, are put apart (*Quercetea roboris-petraeae*; Fig. 7). The deciduous oak forests of southern and south-eastern Europe partly belong to the xerothermic group (*Quercetalia pubescentis*) (Horvat et al. 1974; Fig. 8).

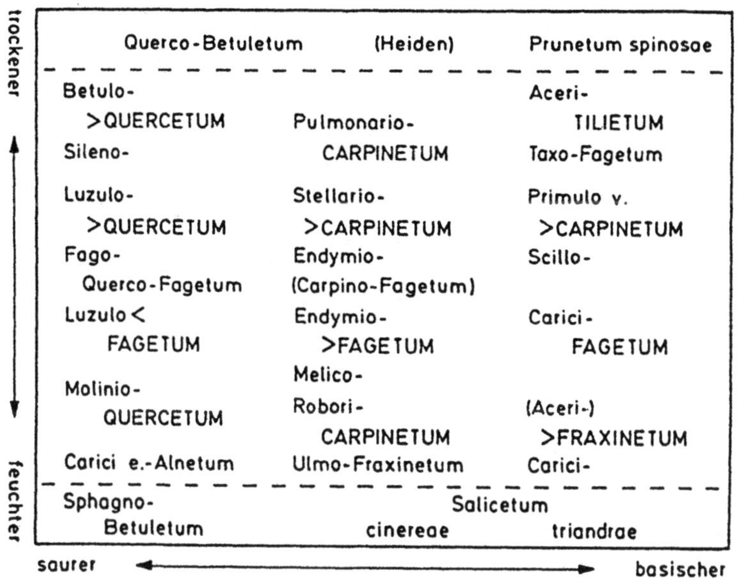

Fig. 7e. FAC of the subatlantic lowland mixed *Quercus-Fagus* forest zone. *Fagus* and *Carpinus*, and/or *Quercus* stands may alternate on fresh and dry sites (from Mayer 1986).

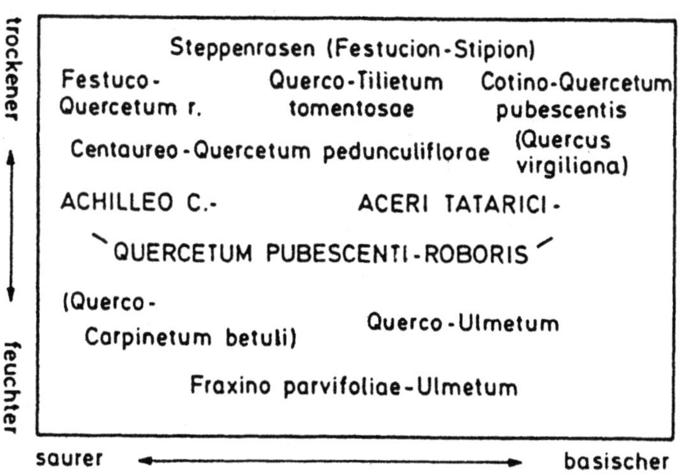

Fig. 8a. FAC of the continental *Quercus*/steppe forest zone (*Acer tataricum*, *Qu. pubescens*). Variability of associations is limited by the extreme climatic conditions very near to the forest limit (from Mayer 1986).

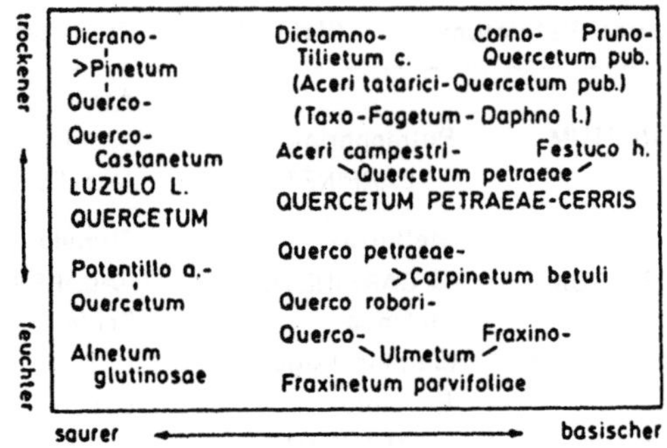

Fig. 8b. FAC of the Pannonic mixed *Quercus* (*petraea*, *cerris*) zone (from Mayer 1986).

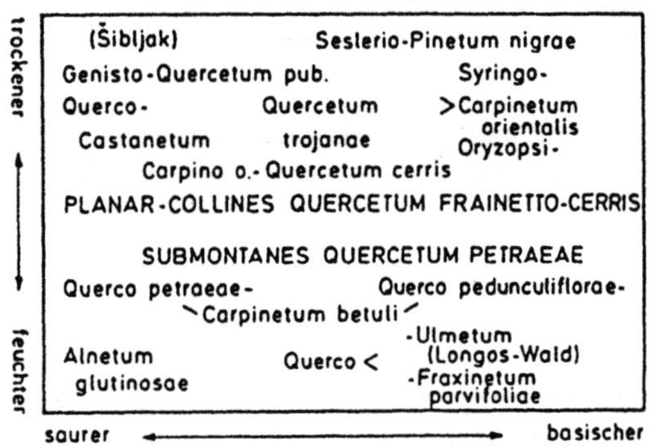

Fig. 8c. FAC of the subcontinental mixed *Quercus* (*frainetto*, *cerris*) zone (from Mayer 1986).

Biomass, production and mineral cycles

An overview on biomass and production has been given by Lieth & Whittaker (1975) putting the range of nemoral forests in areas with 400–*1000*–2500 g/m^2 · yr dry matter (± from E to W; Miami or Thornthwaite Memorial Model on the basis of evapotranspiration; compare Hofmann 1985, phytomass [300–] 810 [–1350] g/m^2 · yr.; total phytomass 8.1 (max. 13, min. 3) t, wood 4.3 t dry matter/ha · yr.). LAI amounts to 3–12 m^2/m^2, chlorophyll content to 2–6 g/m^2, mature biomass 42–46 kg/m^2, annual energy fixation, mean, to 4.6 × 10^6 cal/m^2 (growing season 4.4–*6.4*–8.4 months/yr. from E to W).

The biomass of *Quercus robur* stands (Western Ukraine) are growing

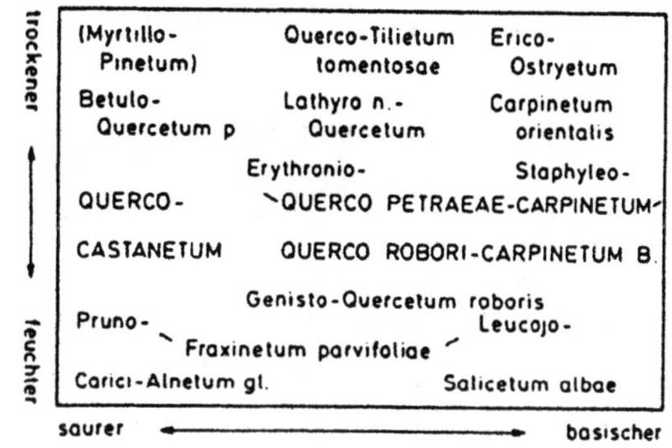

Fig. 8d. FAC of the Illyric lowland *Quercus/Carpinus* zone (from Mayer1986).

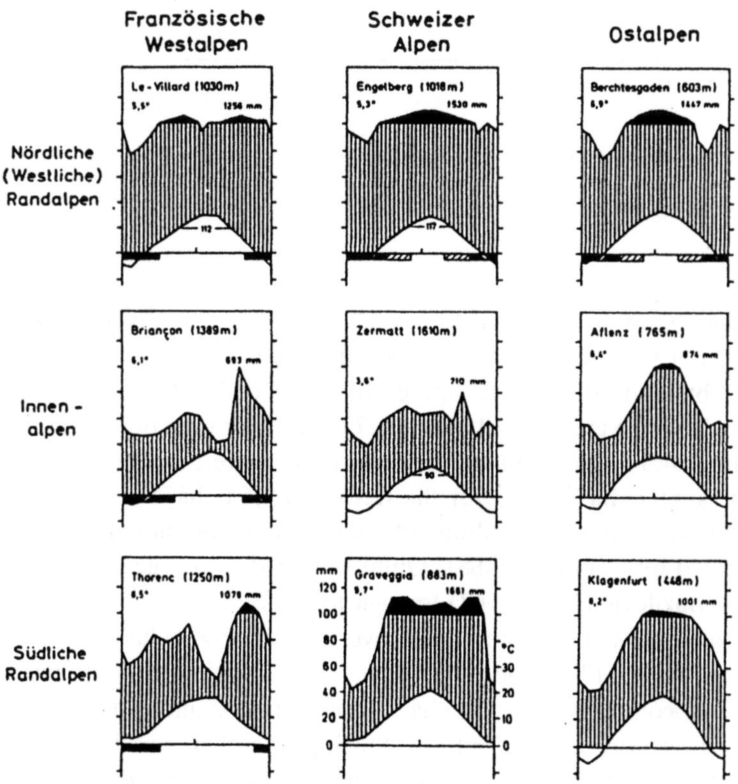

Fig. 9a. Climate diagrams of stations in average elevations in the French western, Swiss and eastern Alps on three transects from the Northern Prealps to the Central and Southern Prealps. A depression in the precipitation curve documents a mediterranean-atlantic influence, the maximum in precipitation in the east is typically subcontinental. Stations in the interior are drier than the peri-Alpine stations, and towards the east this dry climate of the interior is less obvious (from Mayer 1986).

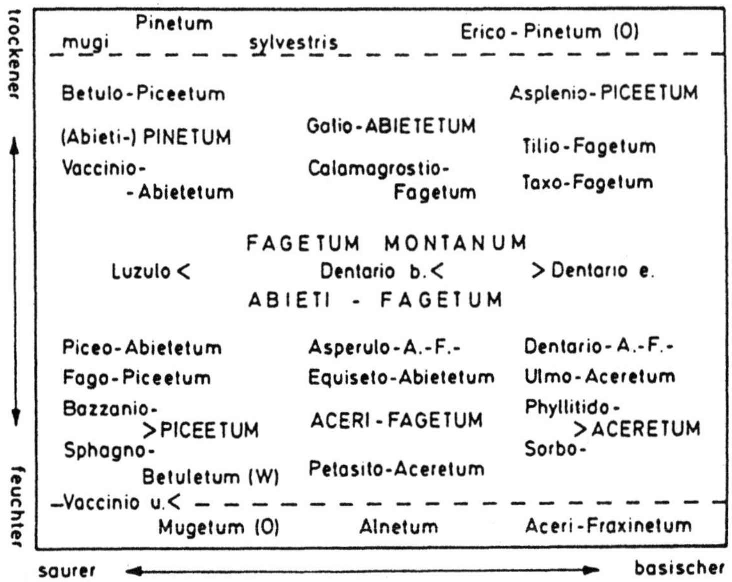

Fig. 9b. FAC of the montane *Fagus* and *Fagus-Abies* belt (similar to the division in the submontane belt) with *Picea and Pinus*, but also *Fraxinus*, *Ulmus*, *Acer*, *Tilia* etc. in the northern Alps (from Mayer 1986).

from approximately 120 t · ha^{-1} (33 ys) to 240 t · ha^{-1} (10^6 ys). However, the IBP project carried out on primeval mixed oak forest in the middle Dnieper region, obtained comparable figures to those of North America of about 431 t · ha^{-1} which are divided into 306.7 t · ha^{-1} for the above-ground phytomass and 124.9 t · ha^{-1} for the below-ground material. Primary production is given as 8.9 t · ha^{-1} · yr^{-1} for the trees and 9.6 t · ha^{-1} · yr^{-1} for the total plant material representing about 1% of the incoming radiation (Goryschina 1974 quoted by Collinson 1988). Quite similar data are generally quoted by Rodin & Bazilevic (1966, 1967) for oak forests.

Biomass: 400 t · ha^{-1}; perennial overground: 300 t · ha^{-1} (75%); green parts: 4 t · ha^{-1} (1%); root: 96 t · ha^{-1} (24%). Net primary production: 9 t · ha^{-1} in comparison to 13 t · ha^{-1} in beechwood and 32 t · ha^{-1} for tropical forest. According to Zimka & Stachurski (1976) oak-hornbeam forest represents an intermediate type of nutrient transfer (by the withdrawal of the degree of C and N accumulation in system) between mixed forest and wet alder woods (Fig. 13).

Very detailed results were drawn from the "Solling Project" near Göttingen (Germany) (Ellenberg et al. 1986). These investigations were in connection with the IBP between 1966–1986 on the site of the most frequent beech-dominated plant community in western Europe (i.e. *Luzulo-Fagetum*).

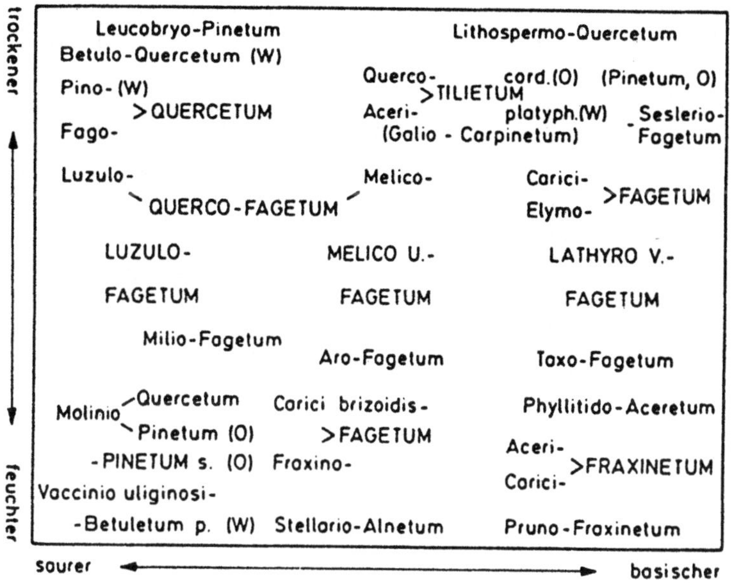

Fig. 9c. FAC of the submontane *Fagus* belt. Similarly, the main *Fagus* associations (e.g. *Luzulo-Fagetum*) are replaced by *Pinus*, *Quercus* and *Fraxinus* associations on more extreme site conditions, *Pinus* occurring under more subcontinental conditions (from Mayer 1986). (Aro-Fagetum should be placed under Taxo-Fagetum.)

Apparently, net primary production is very similar for nearly natural beech forest, spruce forest (planted), *Trisetum flavescens*-rich hay meadow and *Lolium perenne* field on the same site, namely about $32 \cdot 10^{10}$ J · ha^{-1} · yr^{-1}.

Biomass for beech forest (vs. spruce forest in brackets) is about 310 t · ha^{-1} at an age of 130 years (269 t · ha^{-1}/95 ys) with yearly production of about 10 t · ha^{-1} · yr^{-1} (6.5–10.5 t · ha^{-1} · yr^{-1}). Total herb layer biomass is only about 20–30 kg · ha^{-1} · yr^{-1}, compared with that for alder swamp forest (950 kg · ha^{-1} · yr^{-1}, Poland), pine–oak forest (800 kg · ha^{-1} · yr^{-1}, Poland and Sweden), lime-hornbeam (150 kg · ha^{-1} · yr^{-1}) or dry pine woodland (125 kg · ha^{-1} · yr^{-1}). The efficiency in transforming photosynthetically active radiation (PhAR) into biomass during the growing season (1–5 till 30–9) is about 1.4% only (Fig. 14).

Limiting factors – hypothesis and suggestions

The oak in Russia is the representative of the central European deciduous forest. It grows best, according to Morozov, in the south-west and Poles'ye (Ukraine). In the east it suffers much from spring frost. And probably, it

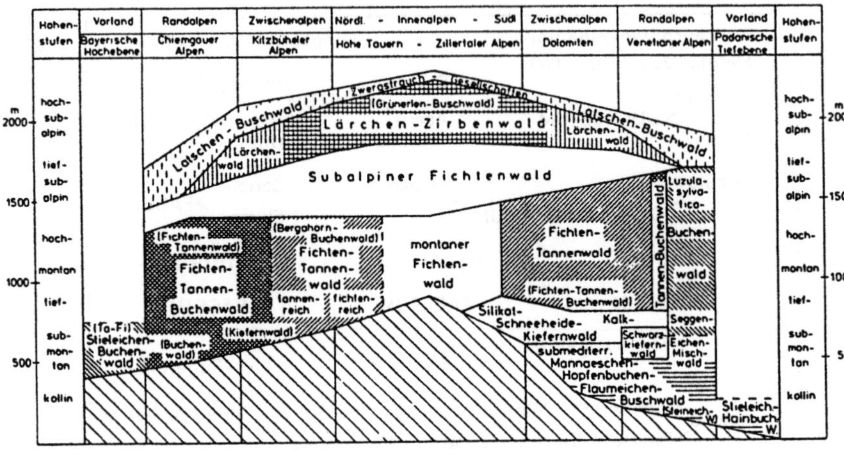

Fig. 9d. Profile of forest vegetation in the middle Eastern Alps. Distinct vertical and horizontal divisions (from Mayer 1986). **Translations of tree species:** Aleppokiefer – *Pinus halepensis*; Bergahorn – *Acer pseudoplatanus*; Bergkiefer – *Pinus montana* (tree); Buche – *Fagus silvatica*; Edelkastanie – *Castanea sativa*; Esche – *Fraxinus excelsior*; Fichte (Fi) – *Picea abies*; Flaumeiche – *Quercus pubescens*; Grünerle – *Alnus vividis*; Hainbuche – *Carpinus betulus*; Hopfenbuche – *Ostrya carpinifolia*; (Wald-) Kiefer – *Pinus silvestris*; Lärche (Lä) – *Larix decidua*; Latsche – *Pinus montana* (shrub); Linde – *Tilia*; Mannaesche – *Fraxinus ornus*; Steineiche – *Quercus ilex*; Stieleiche – *Quercus robur*; Tanne (Ta) – *Abies alba*; Traubeneiche – *Quercus petraea*; Weihrauch-Wachholder – *Juniperus thurifera*; Zeder – *Cedrus*; Zirbe – *Pinus cembra*. **Other explanations:** Mischwald – mixed forest; Buschwald – shrub; Selve-fruit stand; Zwergstrauch – dwarf shrubs (*Vaccinium*); Hochstauden – tall forbs; Segge – *Carex*; Schneeheide – *Erica carnea*; – reich – rich in; – hoch, – tief – high(er), low(er); Vorland – lowlands near mountains; Randalpen – outer chains of Alps; Innenalpen – inner chains of Alps; Zwischenalpen – inner valleys; Silikat – siliceous; Kalk – calcareous.

has developed here a race which flowers and unfolds its leaves two to three weeks later in spring. Perhaps this race can penetrate further east than the ordinary form (compare Fig. 11).

According to the hypothesis accepted by Walter (1986) and Ellenberg (1988), *Fagus* does not extend as far into the continental interior of the European (former) USSR as the less drought-tolerant trees *Tilia cordata* and *Quercus robur*, suggesting intolerance of extreme winter cold or of some other aspect of continental climate (*Fagus* trees were unable to compete with these species in the conditions that then prevailed in the east-European lowland; compare also Klötzli 1984b).

According to the opinion of Morozov the oak in Russia is a species which is characteristic of the neighbourhood of the steppes. The oak region is a field of battle between forest and steppe, where under natural conditions the forest will get the upper hand. In accordance with this view the whole oak

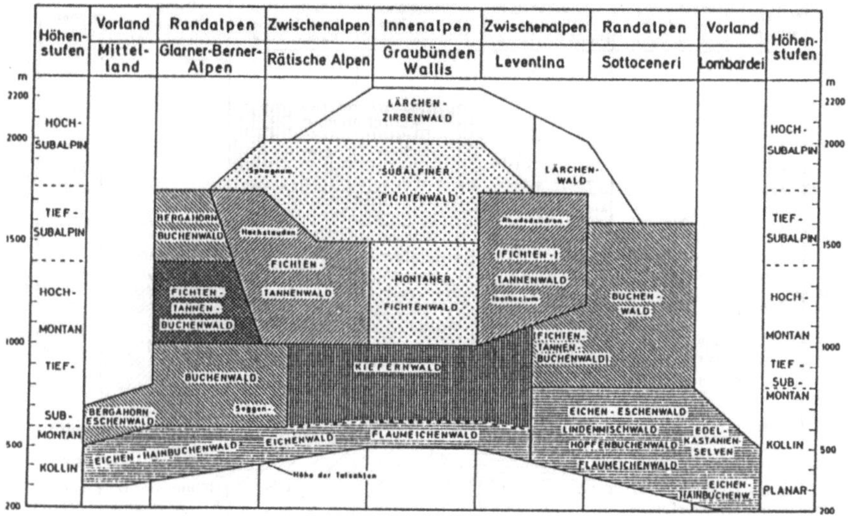

Fig. 9e. Profile of forest vegetation in the Swiss Alps. Common to the eastern conditions are the positions of outer (N & S) *Fagus*-, intermediate *Abies*- and inner *Picea*-rich associations typically developed central-alpine *Pinus* forests (from Mayer 1986).

region of European Russia incl. Belarussia and Ukraine may be divided into three zones: the belt of continuous oak forest, a belt with island-like clumps, and a steppe belt (fragmentary forests in ravines only; cf. different conditions in similar regions in North America, e.g. Wisconsin showing the "oak-savanna" type in contrast to European conditions). The oak forest, like spruce forest, occurs on clay soils, but the oak does not tolerate complete podzolisation and requires greater warmth than the spruce. The oak penetrates far to the north in the coniferous forest region. But there it reaches the river valleys, where the soil is warmer and less podzolised, giving way to the spruce on the elevated clay soils.

Then the zonation of forest vegetation in eastern Europe may be the result of the response of plants to both temperature and moisture gradient. Subatlantic trees and shrubs (*Taxus baccata*, *Ilex aquifolium*, *Hedera helix*, *Ulex*, *Sarothamnus*) are avoiding the eastern continental regions with their cold winters. Leaf-shedding east European species (*Quercus robur*, *Tilia cordata*, *Acer platanoides*) are resistant to cold. In extreme continental Siberia the only broad-leaved deciduous trees are small-leaved birch (*Betula*), the aspen (*Populus tremula*) and the ash (*Fraxinus excelsior*). An increase in resistance to cold by hardening can also be demonstrated on herbaceous plants, even though they are not exposed to extremely low temperature

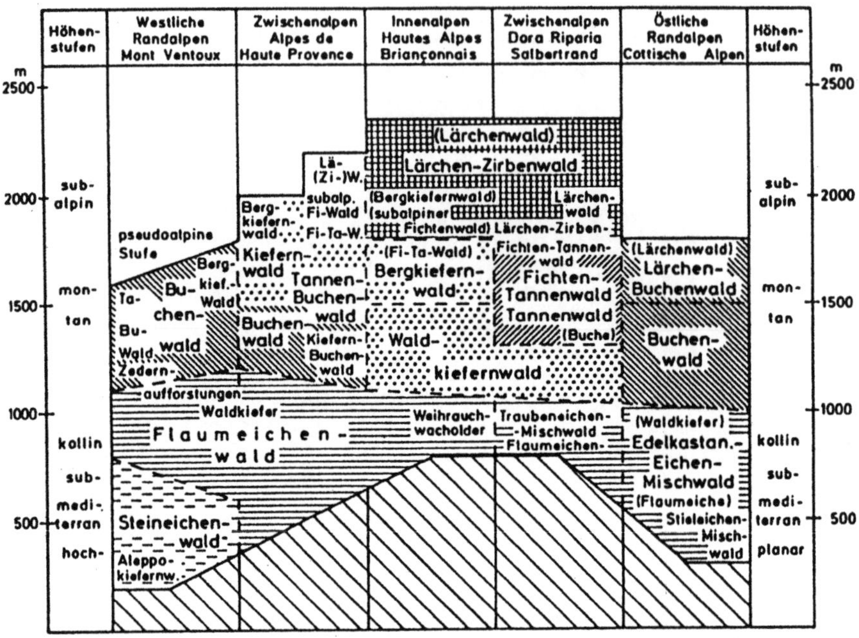

Fig. 9f. Profile of forest vegetation in the middle French Alps with the drought valley called Briançonnais. Great differences usually on sunny and shady slopes. Some relictual associations (from Mayer 1986).

because of their covering of litter and snow (Walter & Breckle 1986). Climate and soil (oligotrophic, mesotrophic and eutrophic status) determine the composition and structure of forest communities. There is a tendency in both the western and eastern part of a region of a slow succession of broad-leaved forests to the *Tilia cordata*-dominated stands (*cf.* "community patterns").

Forest communities versus large herbivorous mammals

Meso-eutrophic *Tilio-Carpinetum* is the main biotop for *Bison* and *Cervus*, meso-oligotrophic *Pino-Quercetum* for *Capreolus*. *Sus* penetrates both biotopes in subcontinental Bialowieza Forest (Fig. 5c). In the western and central parts often too high numbers of roe deer and/or red deer are actually co-steering tree regeneration due to their heavy browsing pressure (especially towards *Abies*, *Fraxinus*, *Acer*, *Taxus*; details in e.g. Ellenberg 1987).

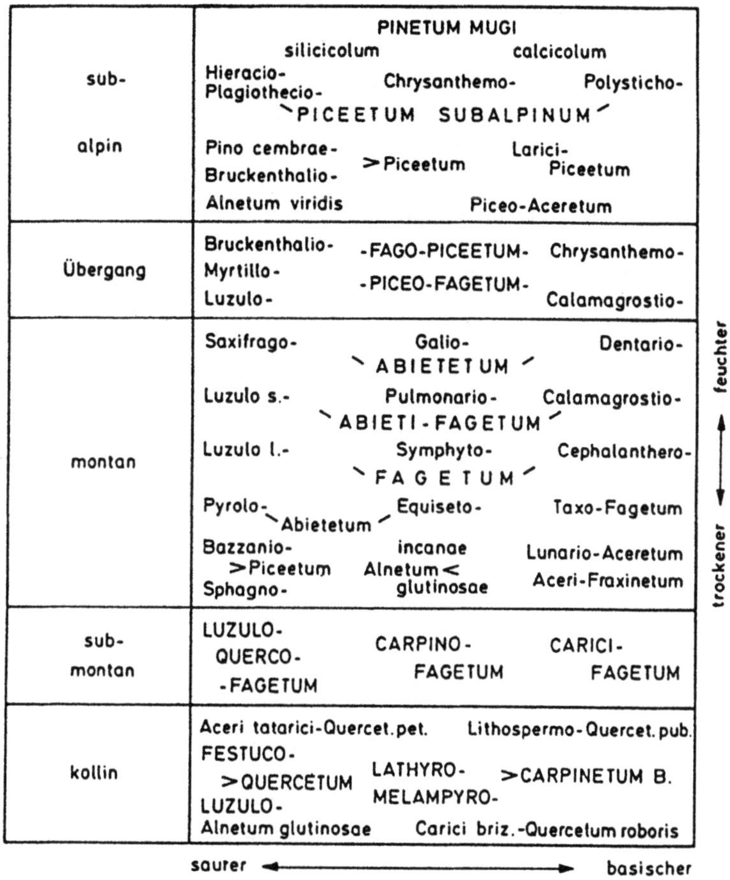

Fig. 10a. FAC of the lowland, montane and subalpine belt of the Carpathian (Dacic) *Fagus* regions (from Mayer 1986).

Remarks: Comparison

The broad-leaved deciduous forests in Europe are chiefly composed of "boreomeridional" species ("temperate Arten" *sensu* Meusel et al. 1965) with rather narrow, mostly European ranges. Their affinities to eastern Asia and North America are very clear. They mostly concern taxa of supraspecific rank (sections, genera) and only very few species. In lowland deciduous forest of eastern Europe, local (unicontinental) species prevail; but they occur together with some western Asiatic species.

The nemoral forest of south-eastern Canada and the subcontinental one

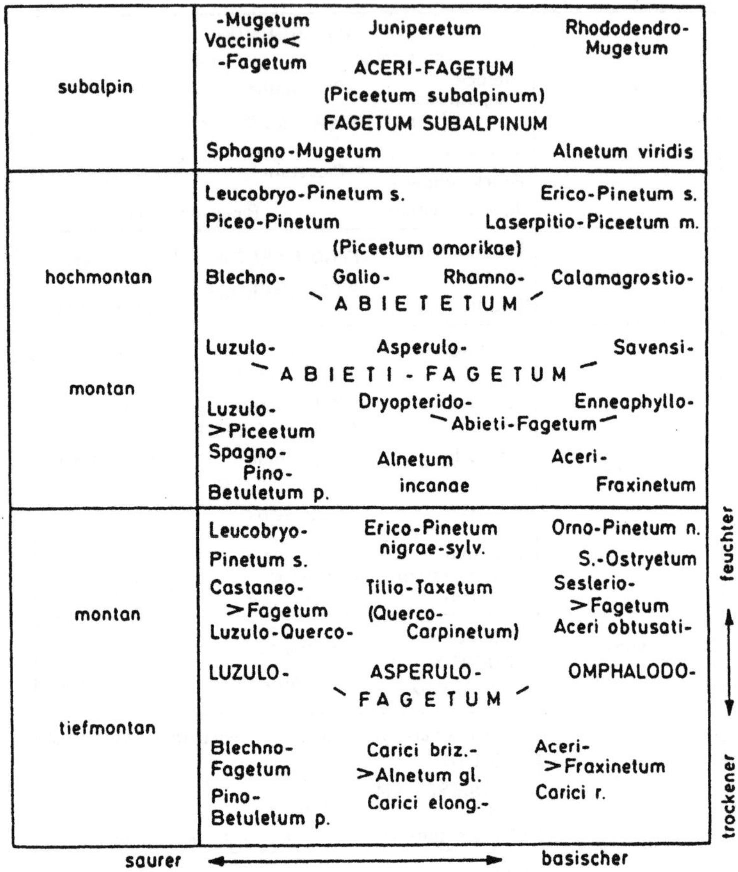

Fig. 10b. FAC of the montane and subalpine belt of the Illyric *Fagus* regions (from Mayer 1986).

situated in Poland have very few species in common. However, the generic affinities between the deciduous forest of the two regions compared, are very clear. Coefficients of generic affinity between four nemoral analogous forest communities from Poland and south-eastern Canada are 35–39% (Kornas 1972). This is due, above all, to the presence of numerous pairs of corresponding taxa of trees (the genera: *Carpinus*, *Fraxinus*, *Padus*, *Quercus*, *Tilia*, *Ulmus*), shrubs (e.g. *Sambucus*, *Viburnum*), and herbs (e.g. *Actaea*, *Asarum*, *Circaea*, *Hepatica*, *Impatiens*, *Sanicula*).

Besides similarities there are also some interesting phytogeographical differences between European and North American nemoral communities. In Europe there are many biocontinental species – a group which is scarcely represented in North America. Nearly all of them occur in Asia; they are

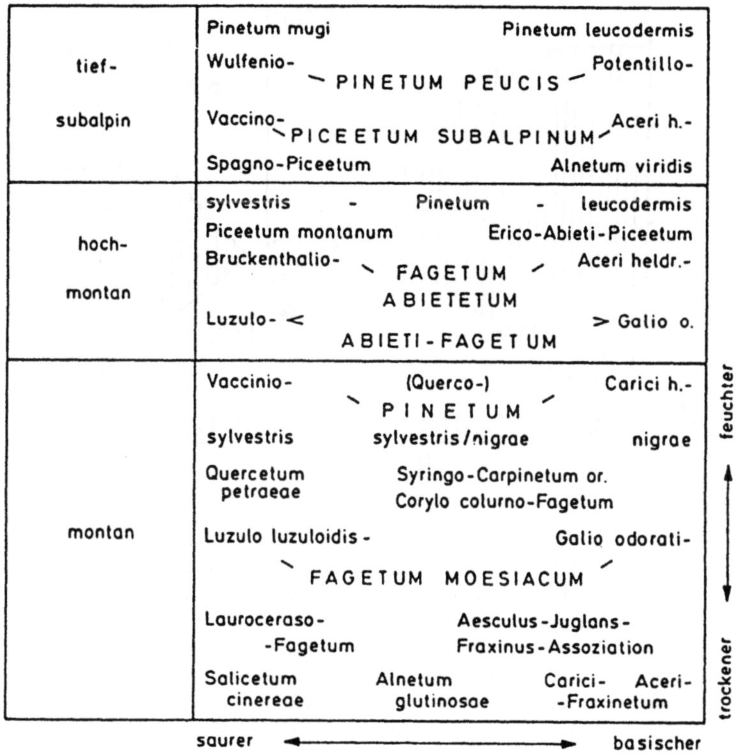

Fig. 10c. FAC of the montane and subalpine belt of the Moesic *Fagus* regions. A small *Picea* belt is in the lower subalpine belt, higher up to the timber line *Pinus peuce* is dominating. In the *Fagus* belt there are some relicts, some endemics and thermophilous forest associations (from Mayer 1986).

either in a few places within isolated, relic deciduous forests in the mountains, especially in the Ural and Altai Mountains (e.g. *Actaea spicata*, *Asarum europaeum*, *Lathyrus vernus*, *Sanicula europaea*, *Stellaria holostea*, etc.), or scattered through Siberia, thus connecting the European and eastern Asiatic centers of nemoral flora (e.g. *Galium odoratum*, *Melica nutans*, *Paris quadrifolia*, etc.). In North America in the never-glaciated territories, a most remarkable feature is the presence of residual Arctotertiary genera which occur also in the eastern Asiatic refugium. They are completely absent from Europe and western Asia (e.g. *Liriodendron*, *Magnolia*, *Chionanthus*, etc.). The representatives of genera and families of tropical origin are also missing in Europe.

Numerous species of the ancient Mediterranean genetic elements (e.g.

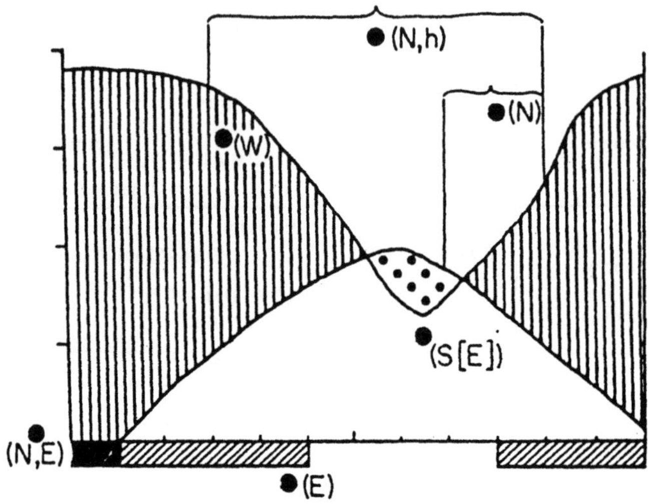

Fig. 11. Model-like climate diagram to visualise *Fagus* limits and those of other deciduous trees. The thick dots point out the decisive climate limiting. At these limits there is a direct influence on tree and shrub layer. Northern and elevational limit (N, h); duration of growing season, or then duration of a warm autumn period (= available warmth); western limit (W); wet period in early summer; norther and eastern limit (N, E); extreme frost (periods); eastern limit (E); late frost; southern (and eastern) limit (S[E]); summer drought. Modification of climate characteristics by mist, hoar frost, wind, wet snow (duration and pressure). Direct influence on edaphic factors is possible through leaching of nutrients (dependent on precipitation regime) or the accumulation of humus (absorption of nutrients, resistance against wetting, temperature regime) (from Klötzli 1976).

the genera *Arum*, *Galanthus*, *Helleborus*, *Leucoium*, *Symphytum*, etc.) are typical of the deciduous forest in central and western Europe. They are remnants of a former autochthonous flora which developed in Europe before the Miocene invasion of the Arctotertiary nemoral flora. This group is practically absent from east European deciduous forest. The very interesting fact, discovered by comparing temperate forest floras of North America and Europe, is that the corresponding taxa within them are always "ecologically corresponding". This means that "both counterparts have similar ecological values, i.e. they occur in analogical habitats and in analogical plant communities on both continents" (Medwecka-Kornas 1961).

General conclusion

The nemoral deciduous forests in the Holarctic are remnants of a formerly continuous Arcto-tertiary temperate zone which has been pushed southwards and divided up into isolated refugia by the deterioration of the northern

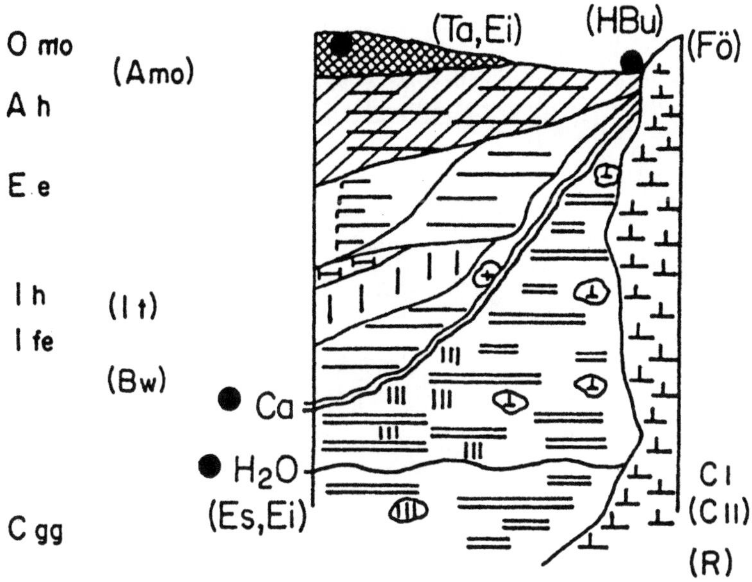

Fig. 12. Model-like soil profile to visualise limits of deciduous forests. Soil development from Ranker to Podsol from right to left (not signifying that soil development is following all these steps). Signatures acc. to customs of ETH/Richard, mscr. and Frei 1975, in Klötzli 1976). The thick dots are giving the decisive soil characteristics for the limits of deciduous species. On these limits there is a direct influence of shrub and tree layer. Soildepth: *Ex.*: *Carpinus betulus* and *Quercus petraea* eliminate *Fagus sylvatica* on shallow or excessively drained soils. Extremely shallow soils or lithosols with *Pinus silvestris* (incl. unstable marne slope). Raw humus layer: *Abies alba* and *Quercus robus* dominate *Fagus* on thicker raw humus layers due to inappropriate seed conditions, nutrient and water availability, competition, pathogenic factors. Ca limit (in the profile): availability of bases determines the appearance of e.g. *Acer campestre*, *A. platanoides*, *Prunus avium*, *Ulmus scabra*, *Fraxinus excelsior* and finally *Fagus*. (Ground-) Water level (or then characteristics of water movement, vertical and lateral, seepage, paludification): *Fraxinus* and *Quercus robur* push *Alnus glutinosa* to wetter, *Fagus* to dryer areas. "wichtigste mögliche Horizonte" = most important possible horizons. Ta, *Abies*; Ei, *Quercus*; Hbu, *Carpinus*; Fö, *Pinus*; Es, *Fraxinus* (from Klötzli 1976).

latitudes in the Pliocene and early Pleistocene. This long-lasting and complete isolation has resulted in advanced evolutionary divergence of nemoral floras.

"Nearly all series of closely related taxa in the forests of temperate Eurasia and North America consist of ecologically corresponding components which have similar ranges of tolerance, occupy similar habitats and grow in strictly analogous plant communities. The ecological constitution of such groups must be very ancient and very rigid. Even those nemoral taxa which are separated into morphologically distinct species have undergone no noticeable

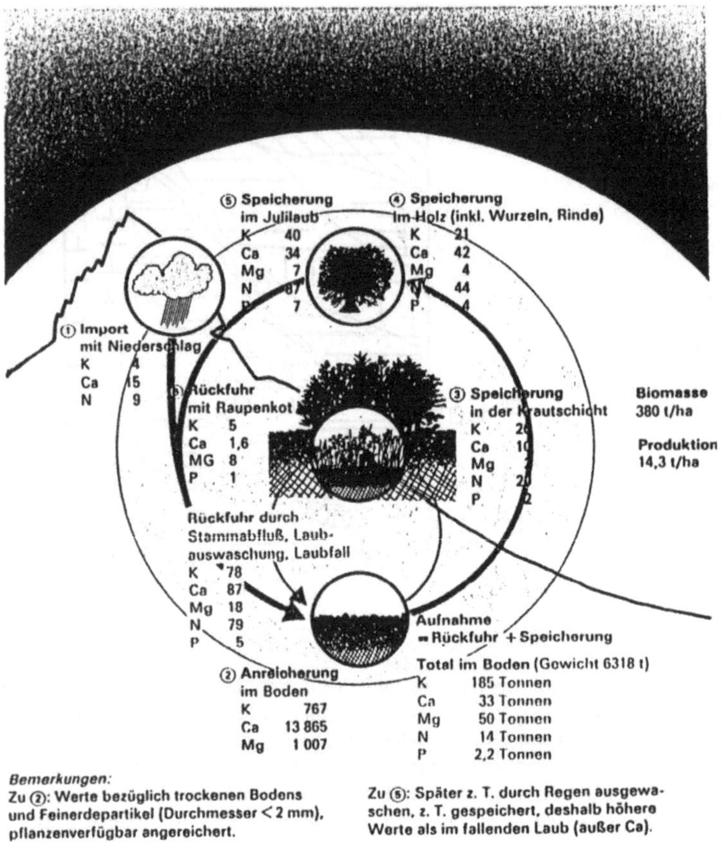

Fig. 13. Nutrient cycle in an oak-ash forest with an understorey of hornbeam/hazel in Wavreille-Wève, Belgium. 60–80% of all nutrients are continuously on the move. (From Duvigneaud & Denaeyer-de Smet 1969, in Klötzli 1992, p. 171, Fig. 242.) 1. Import with precipitation; 2. Accumulation in soil; 3. Accumulation in herb layer; 4. Accumulation in wood (incl. roots, bark); 5. Accumulation in leaves (July); 6. Sedimentation in feces of caterpillars and sedimentation by stemflow, outwash of leaves, leaf fall. Input in soil = sedimentation and accumulation (total weight of soil: 6318 t). ad 2, values related to dry soil and fine earth ($\phi < 2$ mm), available to plants; ad 5, later, partially washed out by rain, partially accumulated, therefore higher values than in falling leaves (except for Ca).

ecological changes since their isolation some 10–15 million years ago" (Kornas 1972, Jäger 1972, Zarzycki 1981).

The zone of deciduous broad-leaved forests of the western side of the Old World (and especially the deciduous forests of eastern Europe) is poor in woody plants; this zone is young. Marked variability and formation of new taxa (microspecies) and ecotypes like *Betula*, *Sorbus* and *Rubus* may suggest

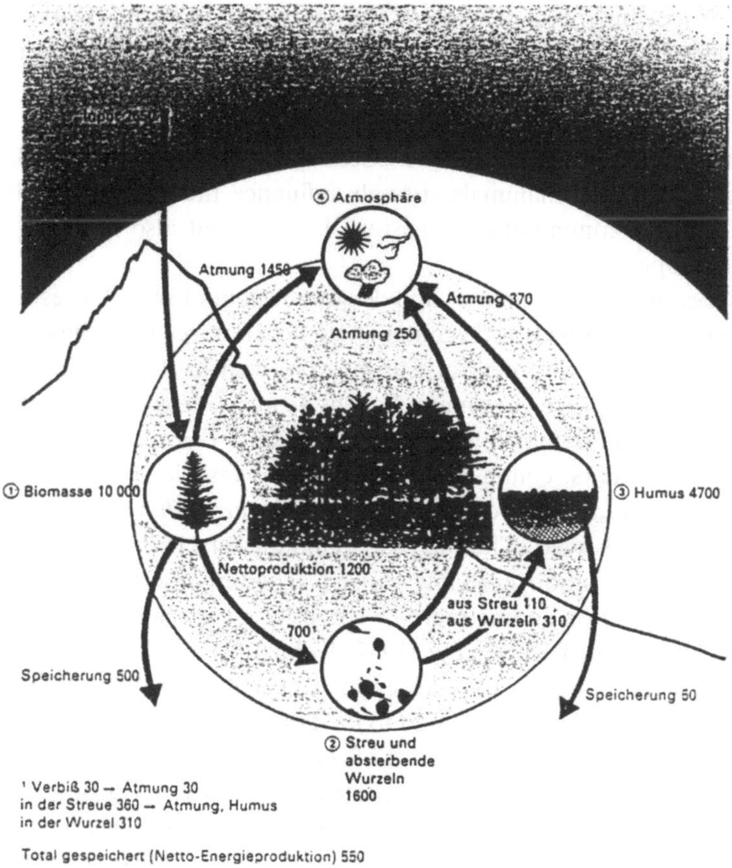

Fig. 14. Energy balance of a forest. Production and use of biomass in a near-climax deciduous forest in North America. Numbers for trees, litter, humus etc. in g dry substance per m^2 soil surface. (From Duvigneaud 1974, in Klötzli, 1992, p. 177, Fig. 2.47.) 1. Biomass; 2. Litter and dieing roots; 3. Humus, from litter 110; from roots 310; Speicherung = accumulation; Atmung ≡ respiration; 1. Verbiss = browsed original material.

(Hübl 1988) that the deciduous forests of eastern and western Europe have not yet reached the full "saturation" in species after the impoverishment evoked by the ice age. Biomass and net primary production are not much lower than in west European and North American deciduous forests (compare also Figs. 1 and 4).

The Pleistocene and Holocene history of eastern Europe as well as the character of the monotonous relief and soils, and the continentality of the climate determine the contemporary composition and structure of the deciduous and deciduous-conifer mixed forest of the region. The same is true for the (hyper-)oceanic parts of western Europe where more marked and often sharp ridges (Alps, Carpathians, Pyrenees) often prevail, but also hilly regions such as the Variskian/Hercynic mountains (Black Forest, Vosges Mountains, Harz, Teutoburger Wald, etc.) in Germany and France, and the Caledonian mountains in Scandinavia, Scotland, Wales, Ireland and Brittany.

Large herbivorous mammals strongly influence the structure and regeneration of forest communities of eastern Europe, but also in most parts of western Europe.

The subcontinental deciduous and deciduous-conifer mixed forests in eastern Poland were relatively little contaminated (pH of tree bark, heavy metals, etc.) in comparison with central Europe (Grodzinska 1982). Nevertheless, the dying-out of spruce and many lichens (esp. *Usneaceae*) as well as the growing role of aerophytic algae (e.g. the genera *Trebouxia*, *Apatococcus*, etc.; Mrozinska, pers. comm.) and of expansive vascular plants like Asiatic *Impatiens parviflora*, American *Lupinus polyphyllus*, *Tilia cordata* and *Carex brizoides* were observed. The increased amounts of nitrogen and acidifying substances in deposition probably caused an altered soil chemistry and a change in competition among more or less tolerant species. To this compare the critical remarks of Kuhn et al. 1987, Kuhn 1993.

As the result of the reactor accident at Chernobyl on April 26th, 1986, a variety of radionuclides was released to the atmosphere. At present it appears (Goldman 1987) that about one-third of the radiocesium released is in European Russia/Ukraine, one-third in central and western Europe, and one-third in Asia and the rest of the Northern Hemisphere. This new artificial factor will probably have a strong, direct and indirect impact on the ecosystems in future in eastern Europe as well as in central Europe, (Ronneau et al. 1987, Wheeler 1988, and many researchers in progress).

References

Atlas SSSR 1969. Glavn. upravl. geod. i kartogr. pri Sovete Ministrov USSR, 2 izd., Moscow.

Box, E. O. 1988. Some similarities in the climates and vegetation of Central Honshu and Central Eastern North America. Veröff. Geobot. Inst. ETH, Stiftung Rübel, Zürich 98:141–168.

Collinson, A. S. 1988. Introduction to World Vegetation, 2nd ed. Hyman, London. 325 pp.

Degorksi, M.L. 1985. An investigation into the spatial variability of continentality in West and Central Europe by the Ellenberg method. Documents phytosoc. N.S. 9:337–349.

Ellenberg, H. 1988. Vegetation Ecology of Central Europe, 4th ed. (1st Engl. ed.) Cambridge Univ. Press, Cambridge, New York. 731 pp., in German: 1986, Ulmer, Stuttgart, 989 pp.

Ellenberg, H., Mayer, R. & Schauermann, J. 1986. Ökosystemforschung. Erg. Solling-Projekt. Ulmer, Stuttgart. 507 pp.

Ellenberg, H. 1987. Eutrophierung – Veränderungen der Waldvegetation – Folgen für den Rehverbiss und dessen Rückwirkungen auf die Vegetation. Schweiz. Z. Forstwes. 139:261–282.

Falinski, J. B. 1988. Vegetation dynamics in temperate lowland primeval forests. Geobotany 8. Dr. W. Junk Publishers, The Hague.

Goldman, M. 1987. Chernobyl: A radiobiological perspective. Science 238:622–623.

Gorchakovsky, P.L. 1969. The principal problems of historical phytogeography of the Ural. UFAN SSSR, Swerdlowsk (in Russian).

Grodzinska, K. 1982. Monitoring of air pollutants by mosses and tree bark. In: Steubing, L. & Jager, H.-J. (eds.) Monitoring of Air Pollutants by Plants. Dr. W. Junk Publishers, The Hague.

Hofmann, G. 1985: Die potentielle natürliche Nerttoprimärproduktion an oberirdischer Pflanzentrockenmasse – ihre Quantifizierung und Kartierung für das Gebiet der DDR. Beitr. Forstwirtsch. 19(3):110–116.

Horvat, I., Glavac, V. & Ellenberg, H. 1974. Vegetation Südosteuropas. Geobotanica selecta 4. G. Fischer, Stuttgart. 768 pp.

Hübl, E. 1988. Die sommergrünen Wälder Japans und Westeurasiens, ein floristisch-klimatographischer Vergleich. Veröff. Geobot. Inst. ETH, Stiftung Rübel, Zürich 98:225–298.

Huntley, B. & Birks, H. J. B. 1983. An Atlas of Past and Present Pollen Maps for Europe: 0–13,000 Years Ago. Cambridge University Press.

Huntley, B. & Webb T. 1988 (eds.) Vegetation History. Hb. Veg. Sci. 7, Kluwer Academic Publishers, Dordrecht. 803 pp.

Igoshina, K.N. 1964. Vegetation of the Urals. Geobotanika 16:83–230 (in Russian).

Jäger, E. 1968. Die klimatischen Bedingungen des Areals der dunklen Taiga und der sommergrünen Breitlaubwälder. Ber. Deutsch. Bot. Ges. 81:397–408.

Jäger, E. 1972. Comments on the history and ecology of continental European plants. In: Valentine, D. H. (ed.) Taxonomy, Phytogeography and Evolution, pp. 349–362. Academic Press, London and New York.

Karpov, V.G. 1973. Structure and productivity of spruce forest of south-taiga. AN SSSR. Bot. Inst., Izd. "Nauka", Leningrad (in Russian).

Kleopov, Y. D. 1990. Analysis of the flora of deciduous forest in the European part of the Soviet Union. Naukova Dumka, Kiev (in Russian).

Klötzli, F. 1975. Zum Standort von Edellaubwäldern im Bereich des südlichen borealen Nadelwaldes. Mitt. Eidg. Anst. Forstl. Versuchswes. 51:49–64.

Klötzli, F. 1976. Grenzen von Laubwäldern in Europa. Ber. Dtsch. Bot. Ges. 89:371–380.

Klötzli, F. 1983. Standörtliche Grenzen von Fagaceen – ein Vergleich in beiden Hemisphären. Tuexenia 3: 47–65.

Klötzli, F. 1984a. Die Nische der Pinus-Vikarianten in der südlichen Hemisphäre – ein Kapitel globaler Konvergenz. Bh. Schweiz. Z. Forstwes. 72:89–101.

Klötzli, F. 1984b. Neuere Erkenntnisse zur Buchengrenze in Mitteleuropa. In: Festschr. P. Fukarek, Akad. Nauka umjetn. Bosne Herc., rad 72, Odj. prir. mat. nauka 21, Sarajewo:381–395.

Klötzli, F. 1988. On the global position of the evergreen broad-leaved (non-ombrophilous) forest in the subtropical and temperate zones. Veröff. Geobot. Inst. ETH, Stiftung Rübel, Zürich 98:169–196.

Klötzli, F. 1992. Ökosysteme. 3. Aufl. UTB 1479. G. Fischer, Stuttgart, Jena, New York. 447 pp.

Kornas, J. 1972. Corresponding taxa and their ecological background in the forests of temperate Eurasia and North America. In: Valentine D.H. (ed.) Taxonomy, Phytogeography and Evolution, pp. 37–59. Academic Press, London, New York.

Kuhn, N. 1993. Ursachen floristischer und ökologischer Vorgänge in Waldbeständen. Schweiz. Z. Forstwes. 144:347–367.

Kuhn N., Amiet, R. & Hufschmid, N. 1987. Veränderungen in der Waldvegetation der Schweiz infolge Nährstoffanreicherungen aus der Atmosphäre. Allg. Forst-u. J. Ztg. 158:77–84.

Laivins, M. 1991. Klassifikation der Linden-Hainbuchengesellschaften (Tilio-Carpinetum Traczyk 1962) in Litauen und Lettland. Veröff. Geobot. Inst. ETH, Stiftung Rübel, Zürich 106: 35–52.

Lieth, H. & Whittaker, R. H. (eds.) 1975. Primary Production of the Biosphere. Ecol. Stud. 14. Springer, New York, Hamburg, Berlin. 339 pp.

Matuszkiewicz, W. 1980. Synopsis und geographische Analyse der Pflanzengesellschaften von Polen. Mitt. for.-soz. Arbeitsgem. N.F. 22:19–50.

Mayer, H. 1986. Europäische Wälder. Ein Überblick und Führer durch die gefährdeten Naturwälder. G. Fischer, Stuttgart, New York. 385 pp.

Medwecka-Kornas, A. 1961. Some floristically and sociologically corresponding forest associations in the Montreal Region of Canada and in Central Europe. Bull. Acad. Polon. Sci., Ser. Sci. Biol. 9:255–260.

Meusel H., Jäger, E. & Weiner, E. 1965. Vergleichende Chorologie der zentral-europäischen Flora. Jena.

Mueller-Dombois, D. 1964. The forest habitat types in southeastern Manitoba and their application to forest management. Can. J. Bot. 42:1717–1744.

Reisigl, H. & Keller, R. 1989. Lebensraum Bergwald. Alpenpflanzen in Bergwald, Baumgrenze und Zwergstrauchheide. G. Fischer, Stuttgart/New York. 144 pp.

Rodin, L. E. & Bazilevich, N. I. 1966. The biological productivity of the main vegetation types in the Northern Hemisphere of the Old World. For. Abstr. (London) 27: 357–372.

Rodin, L. E. & Bazilevich, N. I. 1967. Production and Mineral Cycling in Terrestrial Vegetation. Oliver and Boyd, Edinburgh and London. 288 pp.

Ronneau, C., Cara, J. & Apers, D. 1987. The deposition of radionuclides from Chernobyl to a forest in Belgium. Atmosph. Environ. 21:1467–1468.

Shatkauskas, A. V. 1975. Some geographical pecularities of East-European broad-leaved forests. Bot. Z. 60(5):646–662 (in Russian).

Sjörs H., 1956. Nordisk växtgeografi. Scand. Univ. Books. Sv, Bokförl. Bonniers, Stockholm. 229 pp.

Vankat, J. L. 1979. The Vegetation of North America. An Introduction. Wiley & Sons, New York, Chister, Brisbane, Toronto. 261 pp.

Walter, H. 1968. Die Vegetation der Erde. II.: Die gemässigten und arktischen Zonen. VEB Gustav Fischer, Jena.

Walter, H. & Breckle, W. 1986. Ökologie der Erde. Bd. 3. G. Fischer, Stuttgart. 587 pp.

Walter, H. & Lieth, W. 1960ff. Klimadiagramm-Weltatlas. VEB, G. Fischer, Jena.

Wheeler, D.A. 1988. Atmospheric dispersal and deposition of radioactive material from Chernobyl. Atmosph. Environ. 22:853–863.

Wilde, S. A. 1962. Forstliche Bodenkunde. P. Parey, Hamburg, Berlin. 239 pp.

Zarzycki K. 1981. Notes on the contemporary vegetation of the southeastern United States and Pliocene vegetation of the West Carpathian Mountains in Europe. An attempt at comparison. Veröff. Geobot. Inst. ETH, Stiftung Rübel, Zürich 77:73–111.

Zimka, J.R. & Stachurski, A. 1976. Vegetation as a modifier of carbon and nitrogen transfer to soil in various types of forest ecosystems. Ekol. pol. 24, 4:493–514.

12. Deciduous broad-leaved forests in North-Central China

CHEN LING-ZHI
Institute of Botany, Academia Sinica, Beijing 100044, China

Abstract

Deciduous broad-leaved forest is the zonal vegetation type in the warm-temperate region of China. Sample data from these forests were analysed by two-way indicator-species analysis (TWINSPAN) and by detrended correspondence analysis (DCA). Nineteen forest types were recognized, including six types of oak forest and four types of birch forest. Indicator species are shown in the branches of the TWINSPAN diagrams herein. DCA results suggest relationships between the forests and their site conditions. Data from 120 climatic stations in north-central China are used in a multivariate regression for estimating the climatic affinities of various deciduous forests, according to the latitude, longitude and altitude of each sample plot. The climatic information interprets the DCA result reasonably. Similarity in the floristic composition of different forests is related to the homogeneous climate. The characteristics of major forests are described briefly.

Key words: DCA, deciduous broad-leaved forest, north-central China, TWINSPAN

Introduction

Deciduous broad-leaved forest is the zonal vegetation type in the warm-temperate climatic zone (see Hou 1983, *cf.* Song 1988) of north-central China. It is located in the region from 32°30′ to 42°30′ N and between 103°30′ and 124°10′ E (Ed. Comm. Veg. China 1980). This forest type is found as a secondary forest type in the temperate zone and as a vertical, oro-zonal type in the mountains of the subtropical zone.

In north-central China there is a series of mountains from north to south: the Qian-shan, Yan-shan, Wu Tai, Tai Hang, Lu Liang, Zhong Tiao, and

E.O. Box et al. (eds.), Vegetation Science in Forestry, 255–271.

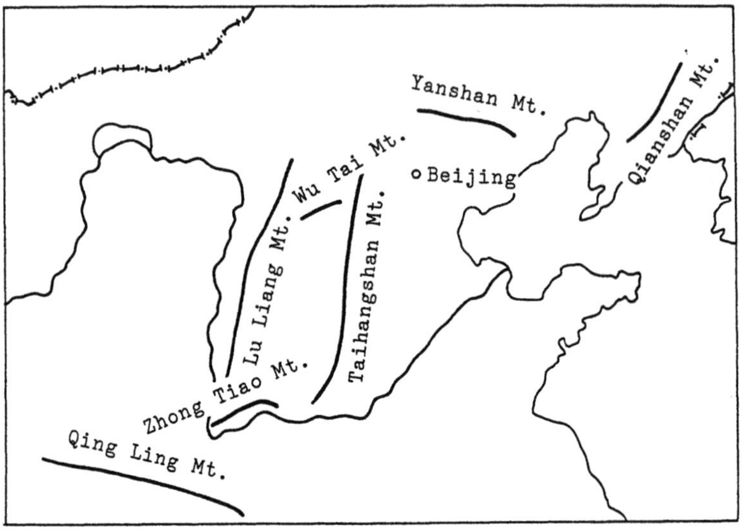

Fig. 1. The geographic location of the main mountain ranges in North Central China.

Qing Ling ranges. The average elevation of these mountains is above 1500 m, and the highest elevations reach 3000 m above sea level. In the east, the Liaodong and Jiaodong peninsulas jut into the Yellow and Buo Seas. The topography of north-central China is thus composed of plains, hills, mountains, and the peninsulas formed by mountain ranges (Fig. 1).

The climate in this region is warm and subhumid. The annual mean temperature is 8–14°C and gradually increases from north to south (Fig. 2). The mean temperature of the coldest month ranges from −13°C to −2°C, with an absolute minimum temperature of −30°C. The mean temperature of the warmest month reaches 25–28°C, with an absolute maximum temperature of 40°C. The frost-free period is about 180–240 days and increases from north to south and from west to east. The annual precipitation is about 600–900 mm, decreasing from southeast to northwest. The seasonal distribution of the precipitation is uneven, that of winter and spring being only 3–7% and 10–14% of the annual total respectively. Rainfall is most abundant during the summer, which accounts for nearly 60–70% of the total. Usually it is dry from October to April and wet from May to September. The climate of the Liaodong and Jiaodong peninsulas is warmer, with a longer frost-free period and more moisture than in other areas at the same latitude.

According to fossil-pollen analyses in north-central China, for example from Beijing, the early Cretaceous flora (113 million years ago) was composed mainly of Lygodiaceae, Pinaceae and Ginkgoaceae. In the late Cretaceous period the Angiosperms appeared. The climate was wet and hot,

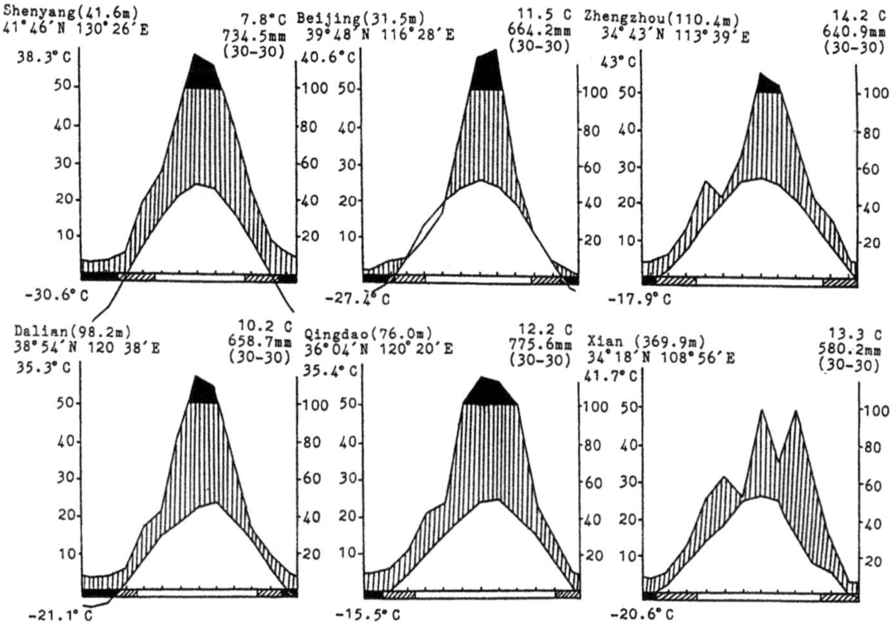

Fig. 2. Climate diagrams of the warm-temperate deciduous broad-leaved forest region.

and tropical and subtropical vegetation appeared in that era. The annual mean temperature was about 10°C higher than that of today. The Palaeogene flora was represented by subtropical coniferous and deciduous forests, composed of such genera as *Quercus*, *Castanea*, *Ginkgo*, *Castanopsis*, *Fagus*, *Juglans*, *Carya*, *Pterocarya*, and *Liquidambar*, as well as a few evergreen elements such as *Magnolia* and *Myrica*. The climate at this time was also wet and hot, with a mean annual temperature about 5°C higher than at present.

The Neogene flora was composed mainly of *Quercus*, *Lonicera*, *Pinus*, *Ulmus*, *Betula*, *Corylus*, *Carpinus*, and Compositae, Cyperaceae and Gramineae. Subtropical families and genera decreased. A few subtropical species of *Liquidambar*, *Pterocarya*, *Osmunda*, etc. were still in existence. The climate was warm and dry, with annual temperature similar to or only slightly higher than at present. As for the late Pliocene, herbaceous plants such as Chenopodiaceae, Gramineae and Compositae, and especially *Artemisia*, were predominant, and deciduous broad-leaved forests and grassland were the major vegetation of this period.

During the Quaternary, the flora of north-central China varied greatly between the glacial and inter-glacial periods. During the glacial periods, *Picea* and *Abies* forests were dominants in the plains. During the last glacial period, the annual temperature was about 7°C lower than at present. After glaciation, as the temperature rose, *Pinus* and broad-leaved deciduous forest and grassland increased, and some subtropical trees such as *Diospyros*, *Cotinus*, *Pterocarya*, *Morus*, etc. returned (Kong et al. 1976, Zhou 1984).

The paleobotanical data have shown that deciduous forest is a vegetation adapted to a warm temperate climate. The natural deciduous forests in the plains were destroyed by human settlement and only occur in limited areas in the hills and mountains.

The floristic component of deciduous broad-leaved forest is dominated by northern temperate elements, for example Gramineae, Rosaceae, Compositae, Liliaceae, Ranunculaceae, Umbelliferae, Leguminosae, Scrophularoaceae, Labiatae, Betulaceae, Fagaceae, Pinaceae, Aceraceae, Ulmaceae, Salicaceae, etc. (Ed. Comm. Phys. Geogr. China 1983).

Natural deciduous broad-leaved forests in north-central China can be separated into 8 types dominated by *Quercus*, 4 types dominated by *Betula*, 2 types by *Populus*, 2 types by *Salix*, 5–6 types of mixed deciduous broad-leaved forest, and 1 type of *Alnus* forest. In this paper, only some of these forest types will be discussed.

Classification of deciduous broad-leaved forests

There were 166 sample plots within the study area, which included the Yanshan, Tai Hang, Lu Liang, Zhong Tiao and Qing Ling Mountains and the Jiaodong Peninsula. In each sample plot, all species were characterized by six coverage grades, 1 for <1% to 6 for >75%. The sample data were subsequently analysed by two-way indicator-species analysis (TWINSPAN, Hill 1979) and by de-trended correspondence analysis (DCA, Hill & Gauch 1980).

TWINSPAN is a classification procedure which places each sample plot into a sequence based on similarity to other plots, such that those close together in the resulting dendrogram are very similar floristically and those remote from each other are very different from each other. Nineteen forest types were identified, with indicator species shown on the branches of the TWINSPAN dendrogram (Fig. 3). *Robinia pseudoacacia* plantations are different from the other forests. *Quercus aliena* var. *acuteserrata* and *Pinus armandii* forests are closely related, and their floristic compositions are similar except for the dominant species. There are significant relationships between *Quercus variabilis* and *Qu. dentata* forests, and between *Qu. aliena* and mixed *Carpinus turzaninowii* and *Ulmus laciniata* forests. The composition of *Qu. liaotungensis* forest is close to those of *Populus davidiana*, *Betula platyphylla* and *Betula dahurica* forests. *Qu. liaotungensis* forms nearly pure

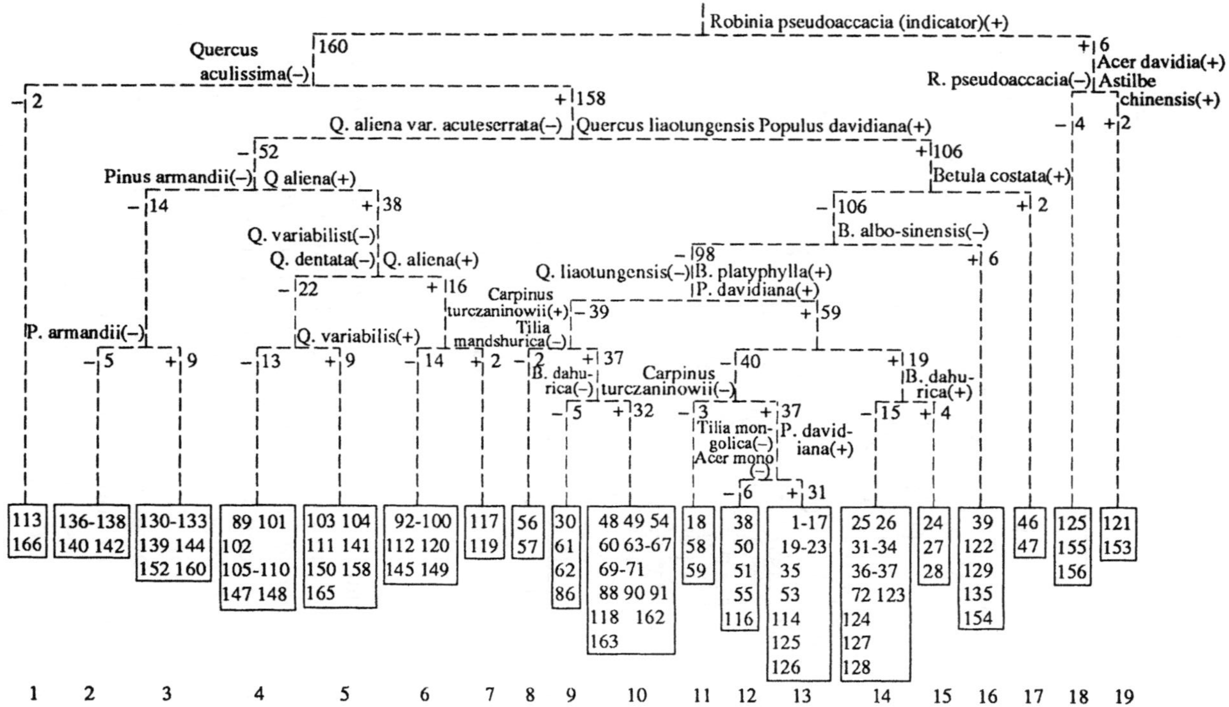

1. Quercus acutissima forest 2. Pinus armandii forest 3. Q. aliena var. acuteserrata forest 4. Q. dentata forest 5. Q. variabilis forest 6. Q. aliena forest 7. Mixed Carpinus turczaninowii and Ulmus laciniata forest 8. Tilia mandshurica forest 9. Mixed Quercus liaotungensis, Betula dahurica. Tilia mongolica forest 10. Q. liaotungensis forest 11. Mixed Carpinus turczaninowii, P. davidiana, Q. liaotungensis forest 12. Mixed Tilia mongolica, Acer mono, P. davidiana, Fraxinus rhynchophylla forest 13. Populus davidiana forest 14. Betula platyphylla forest 15. Betula dahurica forest 16. Betula alho-sinensis forest 17. Betula costata forest 18 Robinia pseudoacacia plantation 19. Mixed Acer davidii. Fraxinus sp. forest

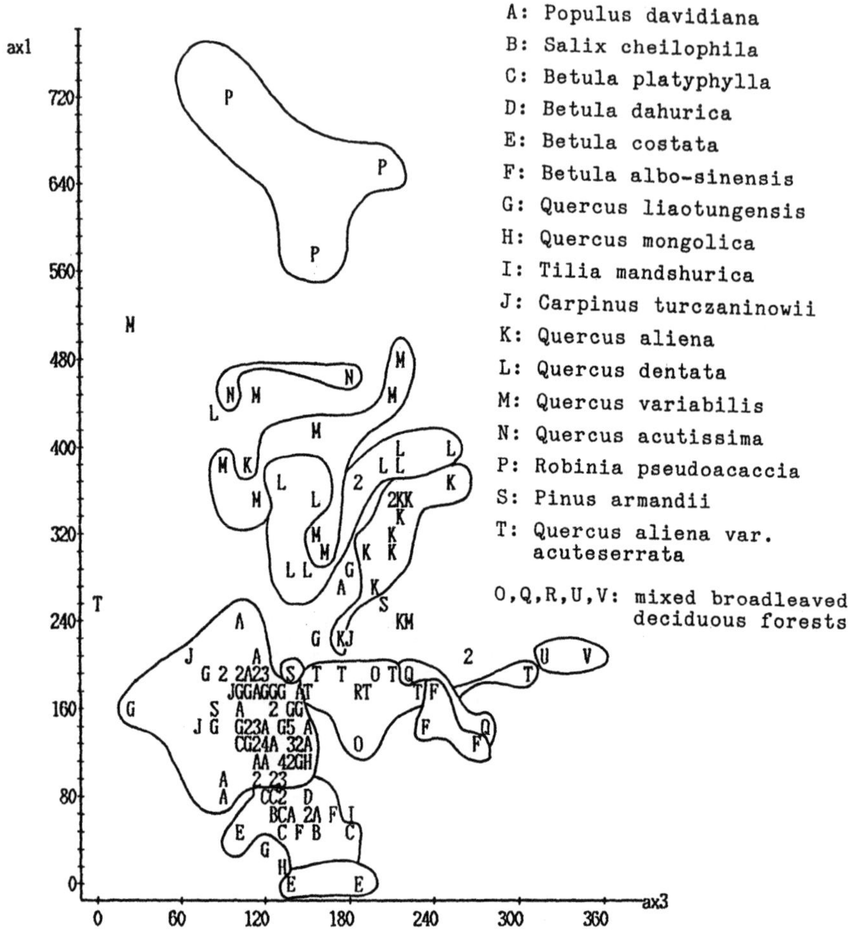

Fig. 4. The orgination of deciduous broad-leaved forests using DCA.

stands on drier sites and is a major species mixed with *Betula dahurica*, *Tilia mongolica* and *Acer mono* to form a mixed deciduous broad-leaved forest on mesic sites. *Populus davidiana* forest and mixed *Populus-Quercus-Tilia-Acer-Fraxinus* forest occur frequently on the same sites.

Four types of *Betula* forest are clearly identified. The similarity between *B. dahurica* and *B. platyphylla* forests is very close. Usually *B. costata* forest occurs near the top of mountains and *B. albo-sinensis* forest occurs in the southwestern mountains of north-central China.

The same data were used for the de-trended correspondence analysis (DCA), which is an ordination procedure. The ordination result for the first and third axes of various forests is shown in Fig. 4. Axis 1 appears to represent temperature, showing (from warm to cool sites): *Robinia psuedoacacia* plantation, *Quercus acutissima*, *Qu. variabilis*, *Qu. dentata*, *Qu. aliena*, *Qu. aliena* var. *acuteserrata*, *Qu. liaotungensis*, *Populus davidiana*, mixed *Carpinus turczaninowii-Ulmus lacianiata*, *Betula dahurica*, *B. albo-sinensis*, *B. platyphylla*, and *B. costata* forests, in that order. Actually, the sites of some oak forests and poplar forests overlap. Axis 3 appears to represent moisture availability, from dry to moist, showing that the sites of *Quercus liaotungensis* and *Populus davidiana* forests are drier than the others. *Qu. aliena* var. *acuteserrata*, *Betula albo-sinensis*, and mixed deciduous broad-leaved forests occur on the moist sites. Some oak forests have relatively broad amplitudes of tolerance to different climatic conditions.

The data from 120 climatic stations in north-central China were used in a multivariate regression to estimate the climatic ranges of various deciduous broad-leaved forests, according also to the latitude, longitude and altitude of each sample plot (see Table 1). The mean annual temperature of *Quercus acutissima* forest and *Betula costata* forest are the highest and lowest respectively. The mean annual temperature of the *Qu. aliena* var. *acuteserrata* sites is higher than that of *Qu. aliena* forest, but the mean July temperature of these two oak forest types is reversed. The DCA ordination (Fig. 3) shows the sites of *Quercus aliena* forest to be warmer than those of *Qu. aliena* var. *acuteserrata*, but *Qu. aliena* can apparently also tolerate a relatively cold climate. The annual range of mean temperature for *Populus davidiana* and *Betula albo-sinensis* forest sites is larger than for the other birch sites. The mean January temperature of *Betula albo-sinensis* forest sites is higher than for other birch and poplar sites, and precipitation and relative moisture are also higher. The precipitation of *Populus davidiana*, *Betula dahurica* and *Quercus liaotungensis* sites is lower than for the other sites. The climatic data interpret the DCA results reasonably. The similarity between the floristic composition of different forests is also related to the relatively homogeneous climate.

Forest types

1. *Quercus* (oak) forests

The natural *Quercus* forests in the plains were obliterated by human settlement. Those on the hills and mountains have had a long and continuing history of cutting, burning and grazing, and are highly modified. Many such forests have regenerated naturally, some have been planted. Most of the oak forests have a coppice physiognomy, with trees and shrubs of uniform or varying age. Thus, oak forests have a limited occurrence on the hills and mountains of north-central China.

Oak forests consist mainly of medium-sized trees and are characterized

Table 1. Climatic characteristics of various deciduous broadleaved forests

Mean temperature	Mean temperature in Jan.	Mean temperature in July	Precipitation	Relative moisture
Quercus acutissima forests				
9.6°C–10.5°C	−5.2°C–−9.3°C	22.9°C–24.5°C	491–811 mm	60–67%
Quercus variabilis forests				
7.9°C–12.2°C	−1.9°C–−10.3°C	20.6°C–24.5°C	480–844 mm	60–67%
Quercus dentata forests				
7.0°C–11.9°C	−2.2°C–−12.1°C	18.6°C–24.8°C	474–934 mm	59–67%
Quercus aliena var. *acuteserrata* forests				
6.7°C–10.0°C	−4.4°C–−5.0°C	17.5°C–21.6°C	772–964 mm	64–67%
Quercus aliena forests				
5.2°C–9.2°C	−4.9°C–−13.7°C	19.6°C–22.1°C	511–911 mm	58–67%
Quercus liaotungensis forests				
3.4°C–8.7°C	−6.2°C–−14.6°C	17.5°C–22.2°C	518–753 mm	58–62%
Mixed deciduous broad-leaved forests				
7.7°C–9.3°C	−4.3°C–−5.8°C	18.5°C–20.3°C	896–943 mm	66%
4.6°C–6.4°C	−8.4°C–−10.1°C	16.1°C–18.2°C	846–904 mm	63%
Populus davidiana forests				
2.8°C–8.6°C	−10.3°C–−14.2°C	15.7°C–23.5°C	502–736 mm	58–60%
Betula dahurica forests				
4.0°C–5.0°C	−11.9°C–−14.0°C	18.2°C–20.6°C	595–663 mm	58–60%
Betula albo-sinensis forests				
3.9°C–8.8°C	−4.8°C–−10.7°C	15.4°C–19.6°C	923–971 mm	63–67%
Betula platyphylla forests				
2.1°C–5.3°C	−10.3°C–−16.7°C	15.8°C–19.6°C	598–912 mm	58–63%
Betula costata forests				
1.9°C–3.9°C	−14.7°C–−16.1°C	15.8°C–18.3°C	668–727 mm	58–59%

by the dominance of deciduous broad-leaved species whose leaves are shed in winter. The dominant trees are various species of *Quercus*, such as *Quercus mongolica*, *Qu. liaotungensis*, *Qu. aliena*, *Qu. variabilis*, *Qu. dentata*, *Qu. acutissima*, *Qu. aliena* var. *acuteserrata*, and *Qu. baronii*, the last species of which is semi-evergreen.

The distribution of the various oak forests depends on climatic and soil factors. *Quercus mongolica* forest, as a dominant vegetation type, is present in the northern part of the warm-temperate broad-leaved deciduous forest zone, including the hills and lower mountains within an altitude range of about 200 to 700 m in eastern and southwestern Liaoning (Dong 1985) and in the mountains at about 800 m in northeastern Hebei Province. *Qu. mongolica* forest occurs generally in the area 40–53° N, 114–133° E. It also occurs in the hills of eastern Shandong Province as small, scattered stands. Actually,

Quercus mongolica forest is also widespread throughout the northeastern temperate mixed coniferous-deciduous forest region of China.

Quercus mongolica is a major dominant species in this type of oak forest. *Qu. liaotungensis*, *Fraxinus rhynchophylla*, *Celtis koraiensis*, *C. bungeana*, and *Acer truncatum* are found as prominent subordinate trees and are mixed in the overstorey. *Qu. dentata*, *Qu. variabilis*, *Qu. aliena*, *Betula dahurica*, *Ulmus macrocarpa*, *Populus davidiana*, *Tilia amurensis* and *T. mongolica* sometimes also occur. The evergreen conifer *Pinus tabulaeformis* occurs admixed in the overstorey; *P. densiflora* appears in the *Qu. mongolica* forests on the peninsulas.

Deciduous shrubs, for instance *Lespedeza bicolor*, *Spiraea trilobata*, *S. pubescens*, *Corylus heterophylla* and *Rhododendron mucronulatum*, are dominants in the understorey in different microhabitats. Some shrub species, however, are found only in *Quercus mongolica* forests on the hills of the Liaodong Peninsula, such as *Lindera obtusiloba*, *Symplocos paniculata*, *Zanthoxylum schinifolium*, and *Rhus chinensis*. In addition to these species, *Prunus sibirica*, *Andrachne chinensis*, *Myripnois dioica* and *Lespedeza floribunda* occur frequently.

The ground cover consists of sedges and other herbs. Dominants include *Carex lanceolata* and *C. siderosticta*, and the associated forbs *Sanguisorba officinalis*, *Convallaria majalis*, *Polygonatum odoratum*, *Cacalia aconitifolia* and grasses *Spodiopogon sibiricum*, etc. constitute a well developed ground layer.

Quercus liaotungensis forest extends over large areas of north-central China but is mainly present in the middle of the region (Chen 1985, Zhu 1982). Passing from north to south, the elevational occurrence of the oak forests gradually increases. *Qu. liaotungensis* forest occurs on the hills at 80–600 m in the north and in the mountains at 1000–1700 m in the middle part of the region. It attains maximum altitudes of 1800–2300 m on north slopes of the western Qing Ling Mountains, where it is found in the transition between the warm-temperate deciduous broad-leaved forest and the subtropical evergreen broad-leaved forest belts. *Qu. liaotungensis* forest displays a diversity of species in the tree, shrub and ground layers. *Qu. liaotungensis* is the dominant species, with canopy density of 0.6 to 0.7 and tree height of 10–18 m. The composition and structure of this oak forest is similar to that of *Qu. mongolica* forest, but there are more species, such as *Ulmus davidiana*, *U. propinqua*, *Maackia amurensis*, and *Juglans mandschurica*, which grow in the overstorey of *Qu. liaotungensis* forest in the middle and northern parts of the region. *Qu. aliena* var. *acuteserrata*, *Toxicodendron verniciflua* and *Acer ginnala* are restricted to the southern part of the region, due to their requirement of a warm, moist climate with a long frost-free period.

Pinus tabulaeformis and *Platycladus orientalis* occasionally are mixed in the overstorey. *Pinus armandii* is only present in the oak forests in the southwestern part of the region.

Qu. liaotungensis forest shares common shrubs with *Quercus mongolica* forest. Besides these, the understorey may also be dominated by *Spiraea frischiana*, *Ostryopsis davidiana*, *Elaeagnus umbellata*, *Rosa hugonis* and/or *Sinarundinaria nitida* (shrub bamboo), with a large number of accompanying species, such as *Abelia biflora*, *Deutzia grandiflora*, *Campylotropis macrocarpa*, *Rhododendron micranthum*, *Rhamnus globosa*, *Rh. parvifolia*, *Cotoneaster multiflorus*, or *C. zebelli*, on different sites.

Carex lanceolata, *C. siderosticta* and/or *C. humilis* var. *nana* are dominants in the ground layer. There is a variety of forbs and grasses associated with these sedges, for example *Bupleurum pekinensis*, *B. falcatum*, *Atractylodes chinensis*, *Convallaria majalis*, *Rhaponticum uniflorum*, *Scorzonera glabra*, *Vicia unijuga*, *Potentilla chinensis*, *Sedum aizoon*, *Polygonatum sibiricum*, *Sanguisorba officinalis*, *Cleistogenes squarrosa*, *C. caespitosa*, *Spodiopogon sibiricus* and *Arundinella hirta*. *Selaginella sinensis* often covers the ground.

Toward the west, this type of oak forest may penetrate to the mountains of the temperate forest-steppe region of China.

Quercus aliena forest is a major vegetation type at an elevation of 1000–1500 m on mountains in the southern part of the region. In the central part it appears at 600–1350 m, and to the north it occurs at 150–600 m elevation but generally is found only as scattered stands. *Qu. aliena* is the dominant in this oak forest, but the floristic composition is close to that of the two above-mentioned types and to the *Qu. variabilis* forest. *Pistacia chinensis* and *Qu. glandulifera* var. *brevipetiolata* are restricted to the oak forests in the southern part of the region. *Vitex negundo* var. *heterophylla*, *Prunus sibirica*, *Spiraea trilobata* and *Carex humilis* var. *nana* are dominants in the shrub and herbaceous layers, respectively. The accompanying shrubs, herbs and grasses are similar to those of the *Qu. liaotungensis* and *Qu. variabilis* forests, but in addition to these, *Cotinus coggygria* var. *cinerea* and *Euonymus* spp. grow frequently in the shrub layer.

Quercus variabilis forest is located in the northern part of the subtropical region, but the species also appears widely in the warm-temperate and subtropical regions. The northern boundary is at 39–40° N, but it mainly occurs at 33–34° N, 105°53′ to 124° E. The vertical distribution of *Qu. variabilis* forest in the various regions is different: 200–400 m in the north, below 800 m in the central part, and at 550–1300 m in the southern part of north-central China. *Qu. variabilis* forest often occurs lower than the *Qu. aliena* forest in the north and below *Qu. aliena* var. *acuteserrata* forest in the southern part of north-central China. Those oak forests at lower altitudes, of course, come under more severe human disturbance.

The age of *Quercus variabilis* forest is about 20–40 years. The tree reaches

9–18 m and crown density of 0.6 to 0.8. Besides the major species, *Quercus variabilis*, the most frequent accompanying species include *Quercus aliena*, *Qu. dentata*, *Qu. acutissima*, *Ailanthus altissima*, *Broussonetia papyrifera*, *Albizzia kalkora*, and *Acer truncatum*. *Pinus tabulaeformis* and *P. bungeana* are sometimes mixed with the oak trees. *Dalbergia hupehana*, *Rhus chinensis*, *Melia azedarach* and several shrubs, *Forsythia suspensa*, *Lindera obtusiloba*, *Symplocos paniculata*, and *Zanthoxylum schinifolium*, are restricted to *Quercus variabilis* forests of the peninsulas or southern part of the region. The shrub layer is composed of *Lespedeza bicolor*, *Forsythia suspensa*, *Grewia biloba* var. *parviflora*, *Vitex negundo* var. *heterophylla*, *Cotinus coggygria* var. *pubescens*, etc. The cover of the shrub layer is 5–40%, depending on the canopy density.

The cover of the herb layer is about 15%. *Carex lanceolata*, *C. humilis* var. *nana*, *Arthraxon hispidus* and *Dendranthema zawadskii* are the dominants in the herb layer. Vines, such as *Akebia trifoliasta*, *Schizandra chinensis*, *Smilax scobinicaulis*, *Pueraria lobata*, and *Vitis piasezkii* appear in the more southern *Quercus variabilis* forests of north-central China (Table 2).

Quercus dentata forest is of limited extent at the lower altitudes of the warm-temperate deciduous broad-leaved forest region. The floristic componments are close to those of the *Quercus aliena* forests in the northern areas and to those of the *Qu. variabilis* forests of the middle and southern parts of the region.

Quercus acutissima forest appears mainly on the hills and low mountains in Shandong Province and the Liaodong Peninsula. The elevational range of this type of oak forest is below 1100 m in Shandong and at 80–400 m in the southeastern Liaodong peninsula. This type appears widely, though, in the subtropical region of China. There are similarities in floristic composition and structure between *Qu. acutissima* and *Qu. variabilis* forests. *Pinus densiflora* also appears in the *Qu. acutissima* forest.

Quercus aliena var. *acuteserrata* forest occurs in the Qing Ling Mountains, on the boundary between the warm-temperate and subtropical zones (at 33°31′ to 35°11′ N and 103°53′ to 111°57′ E), at an elevation of 1100–1800 m. *Qu. aliena* var. *acuteserrata* forests only appear in the southern part of north-central China and in the subtropical region. In montane areas, *Qu. aliena* var. *acuteserrata* forest is found above the *Qu. variabilis* forest but below the *Qu. liaotungensis* forest. *Pinus armandii* forest often appears at similar elevations as the *Qu. aliena* var. *acuteserrata* forest.

The age of the *Qu. aliena* var. *acuteserrata* forest is about 20–60 years. About 150 higher plants were found in these forests. The average tree height was 10–18 m and average diameter at breast height 8–20 cm, with canopy density 0.6 to 0.7. The tree layer consists of *Qu. aliena* var. *acuteserrata*, *Pinus armandii*, *Carpinus cordata*, *C. turczaninowii*, *Betula platyphylla*, *Pop-*

Table 2. Composition of *Quercus variabilis* forest (10 plots each 400 m^2)

Species	Coverage	Constancy
Quercus variabilis	6	5
Q. dentata	2–3	3
Q. acutissima	2	3
Q. aliena	3–4	2
Fraxinus mandshurica	1	2
Albizzia kalkora	1	3
Prunus padus	1	3
Vitex negundo var. *heterophylla*	3–4	3
Lespedeza bicolor	3–4	3
Campylotropis macrocarpa	2	3
Grewia biloba var. *parviflora*	2–3	2
Lespedeza floribunda	2	3
L. inschanica	2	3
L. dahurica	1	3
Sophora flavescens	2	3
Spiraea pubescens	2	3
Rubus idaeus	2	2
Philadelphus pekinensis	2	2
Celastrus orbiculatus	2	2
Artemisia gmelinii	2	5
Carex lanceolata	2–4	3
Bothriochloa ischaemum	2–4	3
Carex humilis var. *nana*	2–3	3
Anthraxon hispidus	2	2
Biden parviflora	1	2
Spodiopogen sibiricus	2	3
Clestogenes hancei	1	2
Potentilla chinensis	1	3

Dalbergia hupeana, *Rhus chinensis*, *Melia azedarach*. *Ailanthus altissima*, *Broussonetia papyrifera*. *Forsythia suspensa*, *Lindera obtusiloba*, *Symplocus paniculata*, *Zanthoxylum schinifolium*. *Akebia trifoliata*, *Smilax scobinicaulis*, *Pueraria lobata*. The forest samples included a total of 102 species.

ulus davidiana, *Toxicodendron verniciflua* and *Pinus tabulaeformis*. Small trees, to 5–8 m, include *Lindera obtusiloba* and *Acer hersii*. *Acer geriseum*, *A. maximowiczii*, *Rhus potaninii*, *Cornus walteri* and *Tilia heryana* var. *subglabra* appear as an intrusive layer.

The natural regeneration of *Qu. aliena* var. *acuteserrata* is quite successful. The cover of the shrub layer is 40–50%, consisting of *Lespedeza* spp., *Spiraea* spp., *Corylus* spp., *Nellia* spp., *Campylotropis* sp., *Euonymus phellomona*, etc. Vines are abundant. The cover of the herb layer is below 25%. Major species in the herb layer are from the genera *Carex*, *Deyeuxia*, *Desmodium*, *Thalictrum*, *Epimedium*, *Sausurrea*, etc., as shown in Table 3.

According to data on the paleovegetation of the north slope of the Qing Ling Mountains, *Quercus* species appeared in the mid-Tertiary or earlier but

Table 3. Composition of *Quercus aliena* var. *acuteserrata* forests (10 plots each 400 m^2)

Species	Coverage	Constancy
Quercus aliena var. *acuteserrata*	5–6	5
Pinus armandii	2–4	3
Populus davidiana	2–4	3
Betula platyphylla	2	3
Toxicodendron verniciflum	2	3
Carpinus cordata	2–4	1
C. turczaninowii	2	1
Quercus aliena	2–3	1
Q. variabilis	2	1
Q. liaotungensis	2	1
Q. glandulifera var. *brevipetiolata*	3	1
Q. baronii	2	1
Castanea sequinii	2	1
Celtis koraiensis	2	1
Tilia henryana var. *subglabra*	2	1
Lespedeza bicolor	2–3	5
L. buergeri	2–3	3
L. formosa	2–3	3
Forsythia suspensa	2–4	3
Spiraea fritschiana	2	3
S. japonica var. *fortunei*	2	3
Nellia sinensis	1– 2	3
Rhus potaninii	2–3	2
Corylus mandshuricus	2	2
C. heterophylla	2	2
Styrax japonica	2	2
Acer maximowiczii	2	2
A. geriseum	1–2	2
A. hersii	2	1
Lindera glauca	2–3	1
L. obtusiloba	2	1
Carex siderosticta	2–4	4
Smilax stans	2	4
Carex lanceolata	2–4	3
Deyeuxia sylvatica	2	5
Saussurea mongolica	2	3
Phlomis umbrosa	2	3
Thalictrum baicalense	2	3
Dendranthema zawadskii	1	2
Convallaria majalia	1	2

The forest samples included a total of 150 species

Celtis and *Ulmus* played the key roles in the vegetation during this period. After the mid-Pleistocene, oak forest was dominant in the Qing Ling Mountains. *Qu. aliena* var. *acuteserrata* evolved from *Qu. aliena* and appeared in the late Tertiary, in the subtropical and southern warm-temperate zones.

Due to the long-term effects of the moist, cool climate, the ecological range of *Qu. aliena* var. *acuteserrata* was narrow.

2. *Betula* and *Populus* forests

Betula forests and *Populus* forests occur widely in north-central China but are secondary forests. *Betula platyphylla*, *B. dahurica*, *B. costata*, and *B. albo-sinensis* dominated forests can be found in different mountain areas.

Populus davidiana may be found in pure stands or mixed with *Betula platyphylla* or *Quercus liaotungensis* at different elevations in the mountains, generally in the range 1200–2200 m. The canopy density of this forest is 0.6–0.7. The most frequent species is *B. platyphylla*; others include species of *Quercus*, *Acer*, *Tilia*, and *Carpinus*. *Picea* and *Larix* appear in the *Populus* forests at higher elevations. The cover by the shrub layer is 30–50%, composed of *Corylus heterophylla*, *C. mandschurica*, *Lespedeza bicolor*, *Spiraea trilobata*, *S. pubescens*, *Rosa xanthina*, *R. bella*, *Abelia biflora*, and *Cotoneaster* spp. and *Lonicera* spp., etc. The cover by the herb layer is 20–40%, consisting of *Carex* spp., *Phlomis umbrosa*, and species of *Polygonatum*, *Thalictrum*, *Potentilla*, *Pyrola*, etc. (see Table 4).

Betula platyphylla forests are widespread at 1200–2500 m in north-central China. Other types of *Betula* forest have limited ranges. The floristic composition of *Betula* forests is similar to that of *Populus* forests (see Table 5). Both types are secondary forests which follow the destruction of *Quercus* forests at middle elevations in the mountains and which follow *Picea* forests at higher elevations (above 1800–1900 m).

Conclusion

There are many types of deciduous broad-leaved forest in north-central China, but some types are difficult to find in hills and low mountains today (and cannot be found at all in the plains). Oak forests are being replaced by shrubland, grassland and farmland; other oak forests become coppice stands. The deciduous broad-leaved forests are usually clear-cut for cultivating pure pine or larch plantations. The species diversity in these forests is decreased, and soil erosion in the mountains is increased. People ignore the cultivation of valuable hardwood (broad-leaved) trees. The conservation and restoration of deciduous broad-leaved forests are very important in north-central China.

Table 4. Composition of *Populus davidiana* forests (20 plot each 400 m^2)

	500–1500 m altitude		1960–2160 m altitude	
Species	Coverage	Constancy	Coverage	Constancy
Populus davidiana	5–6	5	4–5	5
Betula platyphylla	2–3	2	4–5	5
Acer mono	2	4		
Quercus liaotungensis	2–4	3		
Tilia mandshurica	2	2		
Tilia mongolica	2	2		
Salix caprea	2	2		
Betula dahurica	2–3	2		
Fraxinus rhynchophylla	2–3	2		
Betula albo-sinensis	2	1	2–3	5
Larix principis-rupprechtii			2	2
Sorbus pohuashanensis	2	1	2	3
Salix heterochroma			1	2
Lespedeza bicolor	2	4		
Corylus heterophylla	2–5	4		
Corylus mandshuricus	2	1	2–4	4
Spiraea trilobata	2	4		
Spiraea pubescens	2–4	4		
Cornus alba			2–4	3
Lonicera chrysantha	1	1	2	5
Viburnum mongolicum			2	5
Cotoneaster acutifolius			2–3	4
Rosa bella			2–3	5
Abelia biflora	2	2		
Rosa xanthina	1	1		
Carex lanceolata	2–4	5	2–4	5
Saussurea nivea	2	4		
Dendranthema zawadskii	2	2		
D. chanetii			2	4
Artractylis chinensis	2	3		
Thalictrum thunbergii	1	3		
Vicia pseudo-orobus	1	1	2–3	3
Vicia unijuga			2	4
Aster ageratoides			2	4
Maianthemum bifolium	2	1	2	4
Convallaria majalis	1	1	2–3	3
Clematis macropetala			1	5
Spodiopogon sibiricus	2	3		
Galium bungei			2	5

The forest samples included a total of 124 species.

Acknowledgements

I am grateful to Miss You Jia, Dr. Gao Qiong and Dr. Dai Xiao-bing for assistance with computing, and to Miss Li Lu-ping for kind help in preparation of the figures. Thanks also to Prof. P. F. Maycock for helpful sugges-

Table 5. Composition of *Betula platyphylla* forest (10 plots each 400 m^2)

Species	Coverage	Constancy
Betula platyphylla	4–6	5
Acer mono	2	3
Populus davidiana	2–4	4
Quercus liaotungensis	2	3
Salix caprea	2	3
Betula albo-sinensis	2–3	2
Sorbus pohuashanensis	2–3	2
Betula dahurica	2–3	3
Corylus mandshuricus	1–6	5
Spiraea pubescens	1–4	5
Abelia biflora	2–4	5
Lespedeza bicolor	2	3
Syringa pekinensis	2	2
Viburnum mongolicum	2	3
Cornus alba	2	3
Cornus bretschneidera	2	2
Lonicera chrysantha	2–3	2
Carex lanceolata	2–4	4
Phlomus umbrosa	2	4
Vicia unijuga	2	4
Veratrum nigrum	2	3
Gymnocarpium continentalis	2–3	3
Sanguisorba officinalis	2	2
Maianthemum bifolium	2–3	2
Aquilegia yabeana	2	2
Aster ageratoides	2	2
Carex siderosticta	2	2
Clematis macropetala	2	3
Convallaria majalis	2–3	3
Iris ruthenica var. *nana*	2	2
Agrimonia pilosa	1	2

The forest samples included a total of 114 species.

tions. This study was supported in part by the Laboratory for Quantitative Vegetation Ecology, Institute of Botany, Chinese Academy of Sciences.

References

Chen Ling-zhi et al. 1985. (The oak forest in the Beijing mountain area.) Acta Phytoecol. Geobotan. Sinica 9(2) (in Chinese).

Dong Hou-de et al. 1985. (The compilation and use of the vegetation map of Liaoning Province.) Liaoning University Press (in Chinese).

Editorial Committee for "Physical Geography of China" (eds.) 1983. (Physical Geography of China: Plant Geography.) Science Press, Beijing (in Chinese).

Editorial Committee for the "Vegetation of China" (eds.) 1980. (Vegetation of China). Science Press, Beijing. 1375 pp. (in Chinese).

Hill, M. O. 1979. TWINSPAN – a FORTRAN program for arranging multivariate data in an ordered two-way table by classification of the individuals and attributes. Cornell University, Ithaca (New York).

Hill, M. O. & Gauch, H. G. 1980. De-trended correspondence analysis, an improved ordination technique. Vegetatio 42:47–58.

Kong Zhao-chen et al. 1976. (Vegetational and climatic changes in Peking during the past one hundred million years.) Acta Phytotax. Sinica 14(2) (in Chinese).

Zhou Kun-shu 1984. (Quaternary succession of vegetation and climatic evolution in north China.) Geological Science (China) 2 (in Chinese).

Zhu Zhi-cheng 1982. (A preliminary study on the *Quercus liaotungensis* forests in the Qing Ling Mountains and the Loess Plateau of northern Shanxi Province.) Acta Phytoecol. Geobotan. Sinica 6(2) (in Chinese).

13. Composition and distribution of deciduous broad-leaved forests in Korea[1]

YANG-JAI YIM
Department of Biology, Chung-Ang University, Seoul 156, Republic of Korea

Abstract

The distributional range and floristic composition of deciduous broad-leaved forest in Korea were examined based on plant communities classified by the Zürich-Montpellier (Z-M) method and on vegetation types recognized by ordination techniques. Plant communities by the Z-M method and vegetation types by ordination techniques coincided clearly in two-dimensional plots of elevation versus topographic moisture, with minute differences between different regions. Plant communities in Korea were grouped into five groups in relation to five vegetation zones (plant formation level), which also represent thermal zones. In the Korean peninsula, with its dry, cool climate, the character species of plant communities in deciduous broad-leaved forest are separated clearly from each other, but on warmer, wetter Cheju-do island they often occur together in the same stand. In the peninsula, the *Carpinus laxiflora* community inhabits the cooler areas and the *C. tschonoskii* community the warmer areas; in Cheju-do the two are often mixed to form one community, a *C. laxiflora-C. tschonoskii* community. *Quercus mongolica* is the character species of plant communities with the largest area in the deciduous broad-leaved forest of the peninsula. In Cheju-do, however, this species is replaced by *Q. grosseserrata* to form a *Q. grosseserrata* community similar in physiognomy to the *Q. mongolica* community.

Key words: broad-leaved forest, character species, classification, ecological niche, ecological species group, hydro-thermic chart, ordination, plant community, thermal zone, vegetation map, vegetation zone

[1] Note on spelling: Mt. Sorak (in main text) is an alternative spelling of Mt. Seolag (in most bibliographic entries).

E.O. Box et al. (eds.), Vegetation Science in Forestry, 273–298.

Introduction

The deciduous broad-leaved forests of East Asia occur at relatively lower latitude than those of Europe and most of eastern North America, a reflection of the cold climatic conditions caused by the large Asian land mass and its topographic features. Most mountain ranges in East Asia run east to west, while those of the other two continents run largely north to south. Such differences have affected the migrations of plant species on each continent. The migration histories of modern vascular plant species over geological time are reflected in the floristic composition of the deciduous broad-leaved forests. The floristic similarity between East Asia and eastern North America has been interpreted through geological events such as the temporary occurrence of the land bridge between Siberia and Alaska, and by convergent evolution under similar climatic conditions after the separation of the two continents (Tanai 1974). The modern deciduous broad-leaved forests of East Asia were formed 10,000 or more years ago, while those of eastern North America or Europe only 6000 years ago or later (Davis 1976, Tsukada 1974). The environmental characteristics of their continents are reflected even in the distribution of beech (*Fagus*) species or beech-dominated forests, which are one of the representative forest of the deciduous broad-leaved forest zone (Yim 1983).

Much of the Korean peninsula is occupied by *Quercus mongolica* forest. This species does not occur in Japan (Satake et al. 1989). In Japan and China, warm-temperate evergreen broad-leaved forests (Camellietea class) are usually contiguous with the beech forests (Fagetea class) as phytosociological units (Miyawaki et al. 1977), while in the Korean peninsula the evergreen broad-leaved forests are next to *Quercus mongolica* forests, with a transition zone of *Carpinus tschonoskii* and *C. laxiflora* forest (Yim & Baik 1985, Kim & Yim 1986a, 1988a, Yim et al. 1990).

The Korean peninsula was linked to the Japanese Archipelago during the glacial age (Tsukada 1974) by a stepping-stone series of many small islands, while the peninsula itself, of course, belonged to the East Asian continent. Except for Ulreung-do island, however, beech cannot be found in Korea, even on Cheju-do (1950 m high) with its various belts from warm-temperate to subalpine. Since beech fossils occur in Miocene strata in Korea, the current absence of beech in Korea suggests a major climatic change (Yim 1983).

Similarities or dissimilarities in the floristic composition of vegetation in different regions largely depends on the climatic conditions and the length of evolutionary time. Although the floristic composition of deciduous broad-leaved forest in Korea is quite different from that in other regions, the floristic differences at the level of genera or physiognomy are not so great.

In addition, plant communities classified by floristic composition (Z-M method) can be arranged well along climatic gradients (Kim & Yim 1988b).

The classification of plant communities by ecological species group or species combinations and the recognition of vegetation types by dominant species are not the same thing, and the community-unit and vegetation-continuum concepts should be evaluated according to the environmental conditions of the given area or to the aims of the vegetation recognition. Each concept represents an oblique view of concrete entities. Considering a particular community as a structural-functional system (Whittaker 1975) and as a product of long evolutionary time, it is reasonable that both classification and ordination techniques will be useful and complementary for the recognition of vegetation. The vegetation of Korea was thus studied by various methods, attempting to integrate results on the relationship between vegetation types and climatic conditions for recognition of the deciduous broad-leaved forest in East Asia.

Climatic limits on the distribution of broad-leaved forest

The climate of East Asia is characterized by the Siberian monsoon and the tropical monsoon, bringing a hot, wet summer and cold, dry winter over the whole region from tropical to subarctic. This is basically different from the situation in other regions on Earth (Kira 1976).

According to Yim (1983), the modern species of *Fagus* occur over climatic zones with mean annual temperature ranging from 4.5 to 20°C and average annual precipitation above 600 to 1000 mm as a lower limit, with no pronounced dry season. This thermal range spans the latitudes 23–45° N in East Asia, 30–50° N in eastern North America, and 35–60° N in Europe. These latitudes are a useful indicator for the comparison of climatic conditions on the three continents of the Northern Hemisphere.

In the Korean peninsula all vegetation maps presented agree in recognizing the subalpine conifer forest zone in the northern mountainous area, the warm-temperate evergreen forest zone along the southern coast, and the temperate deciduous forest zone in-between (Yim 1977b).

The correlation between the distribution of vegetation and climatic patterns in the Korean peninsula, especially thermal climate in terms of Kira's (1945) Warmth Index (WI) and Coldness Index (CI)(units:°C-month), was reported by Yim & Kira (1975, 1976; Fig. 1) and by Yim (1977a, 1977b). Parallel classifications of vegetation and climate were presented for Japan (Kira 1977) and China (Fang 1988).

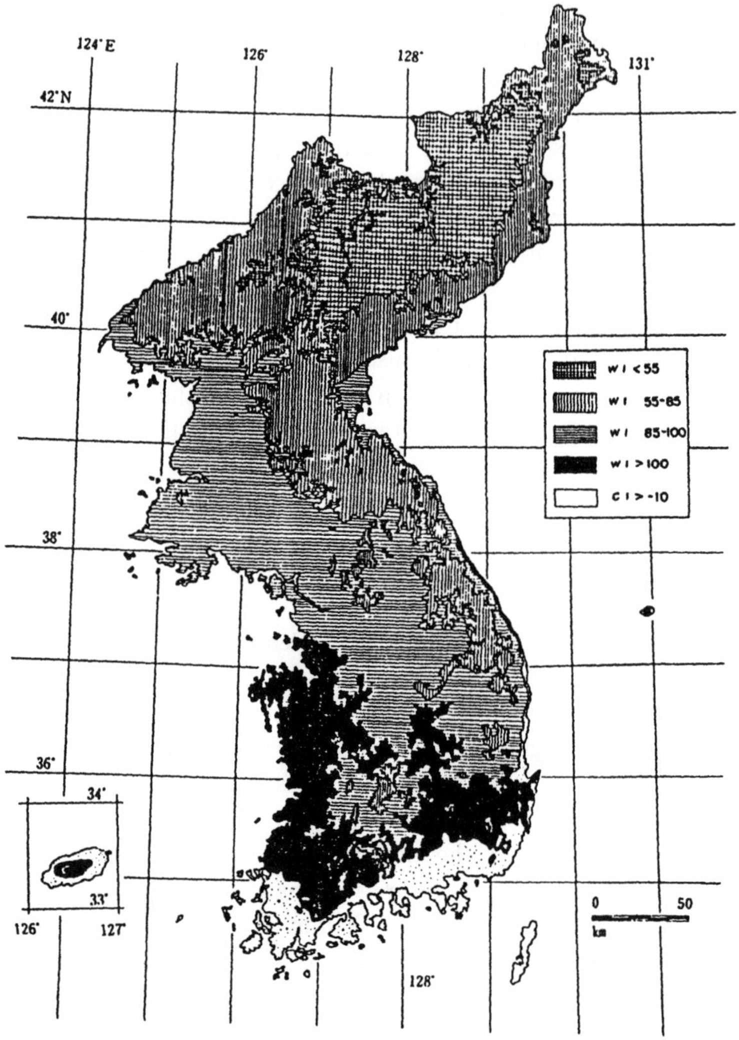

Fig. 1. The thermal zones of Korean peninsula (from Yim & Kira 1975).

The distribution of forest types in the Korean peninsula coincides with thermal patterns as follows (Yim 1977b):

a) Warm-temperate evergreen lucidophyll forest: northern limit at CI = −10
b) Warm-temperate deciduous broad-leaved forest: B1. Southern subzone: northern limit at WI = 100–105, B2. Northern subzone: northern limit at WI = 90

c) Cool-temperate deciduous broad-leaved forest: northern limit at WI = 45–55

d) Subarctic coniferous forest: southern limit at WI = 45–55.

These vegetation zones were interpreted reasonably by thermal gradient analysis using the distribution of tree species given by data of a 1923–26 intensive survey in 52 districts covering all of Korea (Fig. 2), as well as in many other papers (Yim 1977a). Data on the three-dimensional distribution of 50 tree species with high frequency in 93 localities were matched with the corresponding value of Warmth Index, and the optimal thermal range of distribution was determined by WI frequency curves (Yim 1977a).

The deciduous broad-leaved forest zone of Korea belongs to the Korean Province of the Sino-Japanese Region, as recognized by Good (1953). This province is divided into subprovinces based on the distribution of some key genera. The division largely corresponds to the above described vegetation zones (Lee & Yim 1978).

Distribution of plant communities

The distributional ranges of plant communities in Korea can be clarified when plotted against altitude and latitude (Figs. 3–4). By this method, 14 communities were separated into five groups and two subgroups, as follow:

Group I	*Camellia japonica* community	
Group II	Subgroup IIa:	*Carpinus tschonoskii* and *Pinus thunbergii* communities
	Subgroup IIb:	*Carpinus laxiflora*, *Quercus serrata*, *Q. variabilis*, *Q. aliena*, and *Pinus densiflora* communities
Group III	*Quercus mongolica* community	
Group IV	*Abies koreana*, *Taxus cuspidata*, and *Pinus koraiensis* communities	
Group V	*Betula ermani* and *Rhododendron schlippenbachii* communities.	

Group I occurs only above CI = − 10. Group IIa occurs above WI = 100, while group IIb occurs only within the range WI = 80–100. Group III occurs within the range WI = 55–80, group IV within the range WI = 45–55, and group V only below WI = 45 (Fig. 5, Table 1).

In the triangle diagram (Fig. 6) with three axes representing thermal climate (Warmth Index and/or Coldness Index), altitude and latitude, the five community groups I-V were translated into thermal zones A-E, respectively. This means that the distributional ranges of the plant communities are interpreted in terms of thermal climate. The distributional range of deciduous broad-leaved forests in Korea is limited within the range WI = 55–100, i.e. thermal zones B (warm-temperate deciduous broad-leaved for-

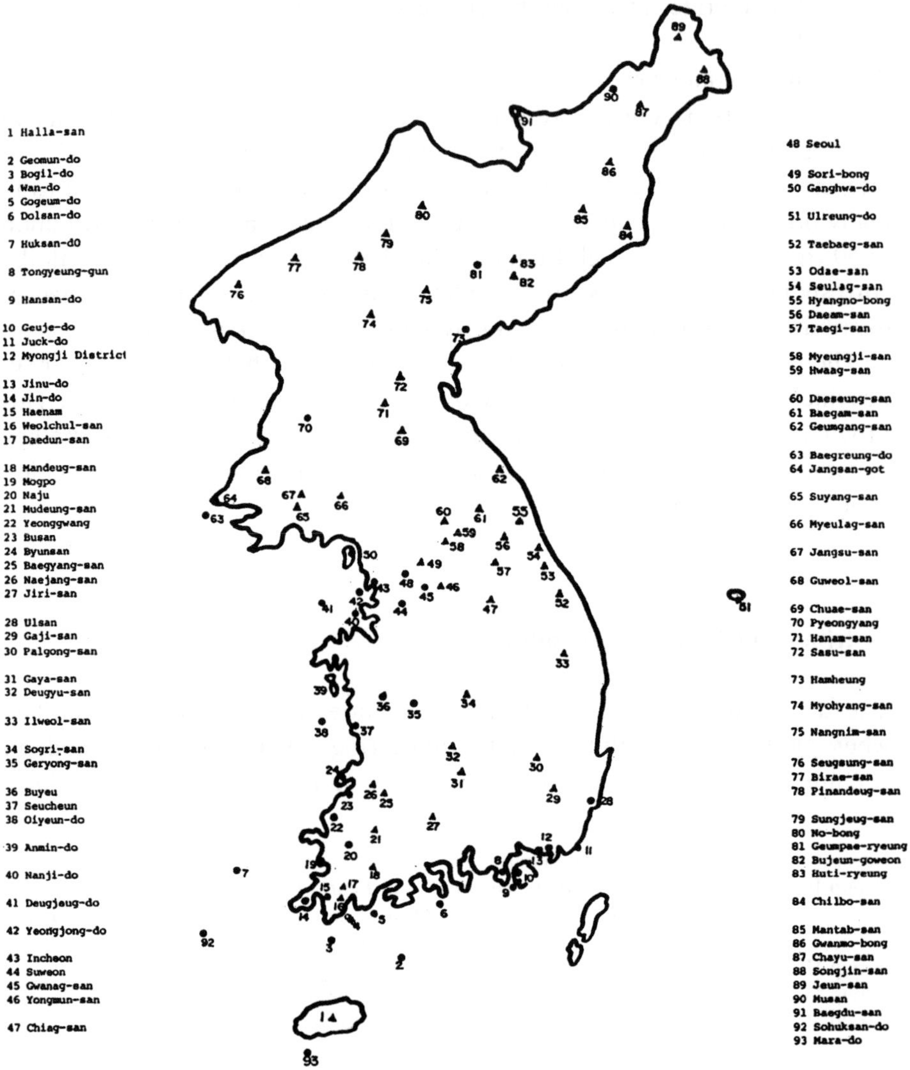

Fig. 2. Map of the Korean Peninsula showing approximate locations of the places where the distribution data of 50 tree species were available.

est) and C (cool-temperate deciduous broad-leaved forest, Yim 1977b), which is the largest area of the three main plant formations of Korea.

On the other hand, according to the results of Kim & Yim (1988b) at Mt. Naejang, different species populations along a soil-moisture gradient showed bell-shaped frequency curves with single peaks in different classes. The peaks of the different species, from mesic to xeric sites, showed the following sequence: *Zelkova serrata*, *Celtis sinensis*, *Lindera erythrocarpa*, *Cornus con-*

Table 1. Distribution ranges of 50 tree species selected in terms of warmth index

Species	Whole range (°C · month)	Optimal range (°C · month)
Pinus pumila Regel	11–56	28–47
Abies nephrolepis Maxim.	14–75	34–65
Betula ermanii Cham.	8–121	30–63
Picea jezoensis Carr.	18–79	34–59
Taxus cuspidata Sieb. & Zucc.	13–90	37–62
Pinus koraiensis Sieb. & Zucc.	21–121	45–81
Salix hultenii Floderus	19–100	45–86
Betula davurica Pallas	25–102	45–88
Quercus mongolica Fisch.	18–111	46–90
Betula chinensis Maxim.	30–97	46–82
Acer mono Maxim.	28–121	48–88
Tripterygium regelii Sprague & Takeda	11–112	49–84
Prunus padus Linn. var. *seoulense* Nakai	24–100	49–86
Betula schmidtii Regel	23–98	50–84
Maackia amurensis Rupr. & Maxim.	30–112	51–89
Acanthopanax sessiliflorus Seem.	30–111	51–87
Juglans mandshurica Maxim.	41–98	52–84
Fraxinus rhynchophylla Hance	20–121	54–91
Ulmus laciniata Mayr	30–98	54–78
Pyrus ussuriensis Maxim.	30–99	56–88
Carpinus cordata Blume	40–121	57–87
Ulmus davidiana Planch. var. *japonica* Nakai	37–112	58–91
Magnolia sieboldii K. Koch	45–110	58–85
Cornus controversa Hemsl.	45–114	59–92
Fraxinus sieboldiana Blume	47–122	59–92
Pinus densiflora Sieb. & Zucc.	30–122	60–95
Lindera obtusiloba Blume	47–127	63–95
Quercus serrata Thunb.	40–122	67–92
Styrax obassia Sieb. & Zucc.	40–121	67–92
Rhus javanica Linn.	56–122	67–94
Carpinus laxiflora Blume	41–121	67–94
Salix koreensis Anders.	45–111	68–92
Salix pseudo-lasiogyne Leveille	43–105	68–87
Cornus walteri Wangerin	46–112	69–93
Platycarya strobilacea Sieb. & Zucc.	62–123	71–94
Lindera erythrocarpa Makino	46–121	71–102
Quercus aliena Blume	50–121	75–91
Celtis sinensis Pers.	46–127	73–102
Castanea crenata Sieb. & Zucc.	62–122	75–98
Picrasma quassioides Bennett	47–112	74–89
Quercus dentata Thunb.	50–123	75–97
Zelkova serrata Makino	63–123	76–99
Juniperus rigida Sieb. & Zucc.	41–112	77–93
Callicarpa japonica Thunb.	47–122	77–98
Carpinus tschonoskii Maxim.	47–122	77–103
Quercus variabilis Blume	58–122	77–95
Phyllostachys bambusoides Sieb. & Zucc.	72–121	–
Quercus acutissima Carr.	58–112	85–99
Pinus thunbergii Parl.	67–123	93–104
Camellia japonica Linn.	68–125	100–125

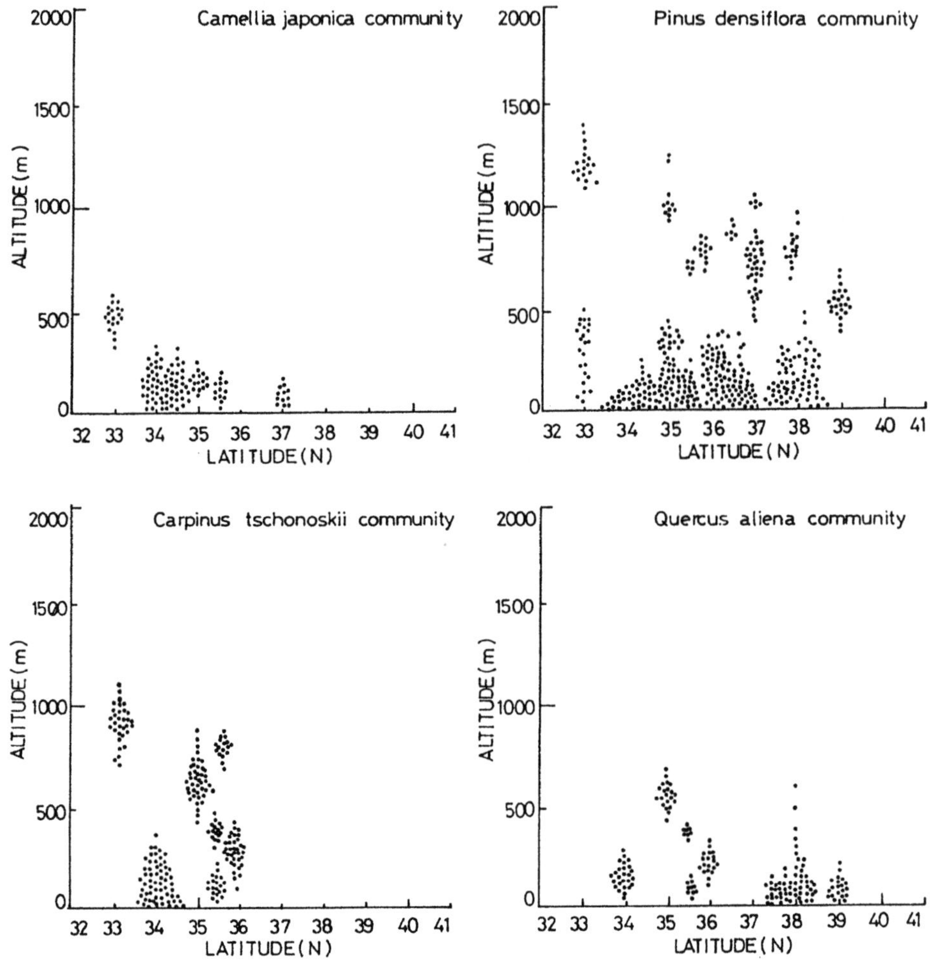

Fig. 3. The distribution in two dimensional charts of altitude and latitude of 14 plant communities classified by Z-M method in Korea.

troversa, *Acer mono*, *Carpinus tschonoskii*, *Quercus aliena*, *Daphniphyllum macropodum*, *Torreya nucifera*, *Carpinus laxiflora*, *Quercus serrata*, *Q. variabilis*, *Q. mongolica*, and *Pinus densiflora* for tree species, and *Acer pseudo-sieboldianum* var. *koreanum*, *Lindera obtusiloba*, *Styrax obassia*, *St. japonica*, *Acer pseudo-sieboldianum* and *Rhododendron schlippenbachii* for shrub species (Fig. 7).

Based on relative locations of the peaks for these tree and shrub species, 10 ecological groups (*sensu* Mueller-Dombois & Ellenberg 1974) were identified:

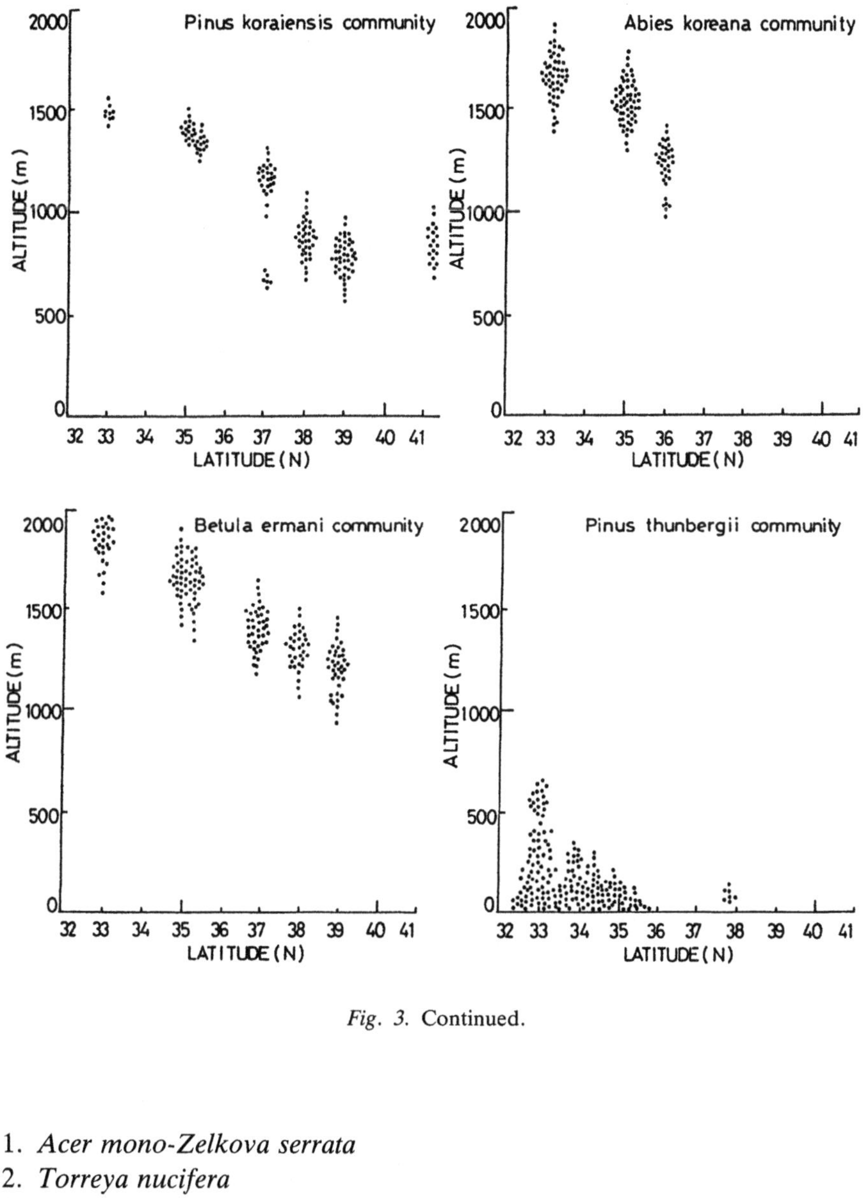

Fig. 3. Continued.

1. *Acer mono-Zelkova serrata*
2. *Torreya nucifera*
3. *Cornus controversa-Lindera erythrocarpa*
4. *Quercus aliena-Carpinus tschonoskii*
5. *Carpinus tschonoskii*
6. *Daphniphyllum macropodum*
7. *Carpinus laxiflora*
8. *Quercus variabilis*
9. *Rhododendron schlippenbachii-Quercus mongolica*
10. *Rhododendron mucronulatum-Pinus densiflora.*

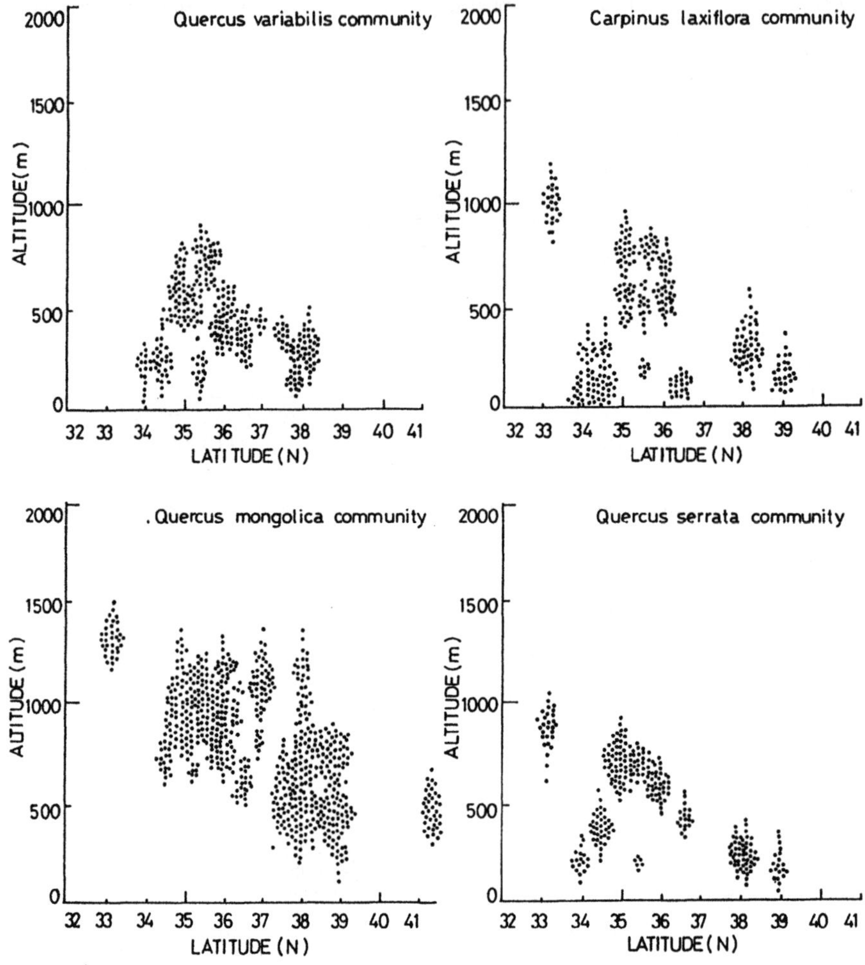

Fig. 3. Continued.

These groups coincide with communities in 10 phytosociological units and with the species associations of 10 correlation groups at Mt. Seonun (Kim & Yim 1986a, 1988b).

Floristic composition of plant communities

Among the many communities recognized by classification (Z-M method) and ordination techniques in deciduous broad-leaved forest (Yim & Baik 1985, Kim & Yim 1986a, 1986b, 1988a, 1988b, Yim et al. 1990, Koh 1990), plant communities dominated by the following seven species are especially

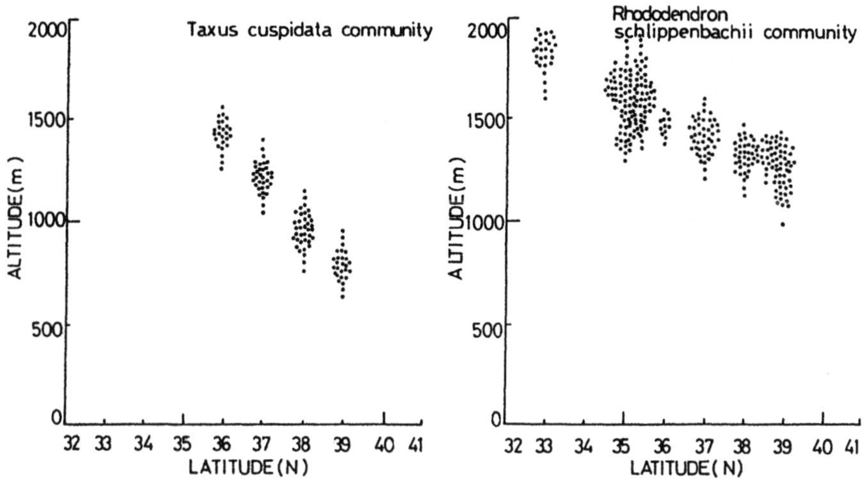

Fig. 3. Continued.

noticeable: *Carpinus tschonoskii*, *C. laxiflora*, *Quercus serrata*, *Q. variabilis*, *Q. aliena*, *Q. mongolica*, and *Zelkova serrata*. The distributional ranges of their character species or dominant species populations are shown diagramatically in the mosaic charts (Figs. 8–9) in relation to elevation (thermal gradient) and topographic moisture, based on data from Mt. Seonun (Kim & Yim 1986b), Mt. Naejang (Kim & Yim 1988b), Mt. Sorak (Koh 1990), Mt. Moak, Mt. Chiri, etc. The *Quercus mongolica* community can be found on relatively xeric mountain slopes, *Q. variabilis* on steep xeric slopes, and *Q. serrata* and *Q. aliena* on more mesic slopes. The *Zelkova serrata* community is limited to mesic slopes or to mesic sites along streams. Hornbeam forest on mesic sites is separated into *Carpinus laxiflora* and *C. tschonoskii* types based on thermal conditions, the former in cooler and the latter in warmer situations.

On Mt. Halla (Cheju island), however, *C. laxiflora* and *C. tschonoskii* usually occur together to form a single mixed community. These two hornbeam species and *Q. serrata* often occur within the same stand as co-dominants. Such a phenomenon suggests that the unification or separation of plant communities and the flexibility of floristic composition depend on changing environmental conditions. When some factor acts as a limiting factor for different distributions of two species, these species may be separated; when there is no such limiting factor, the two species may co-exist in the same stand (Yim 1989).

There are many differences in environmental conditions between Cheju-do island and the Korean peninsula. It seems that climate, however, is the

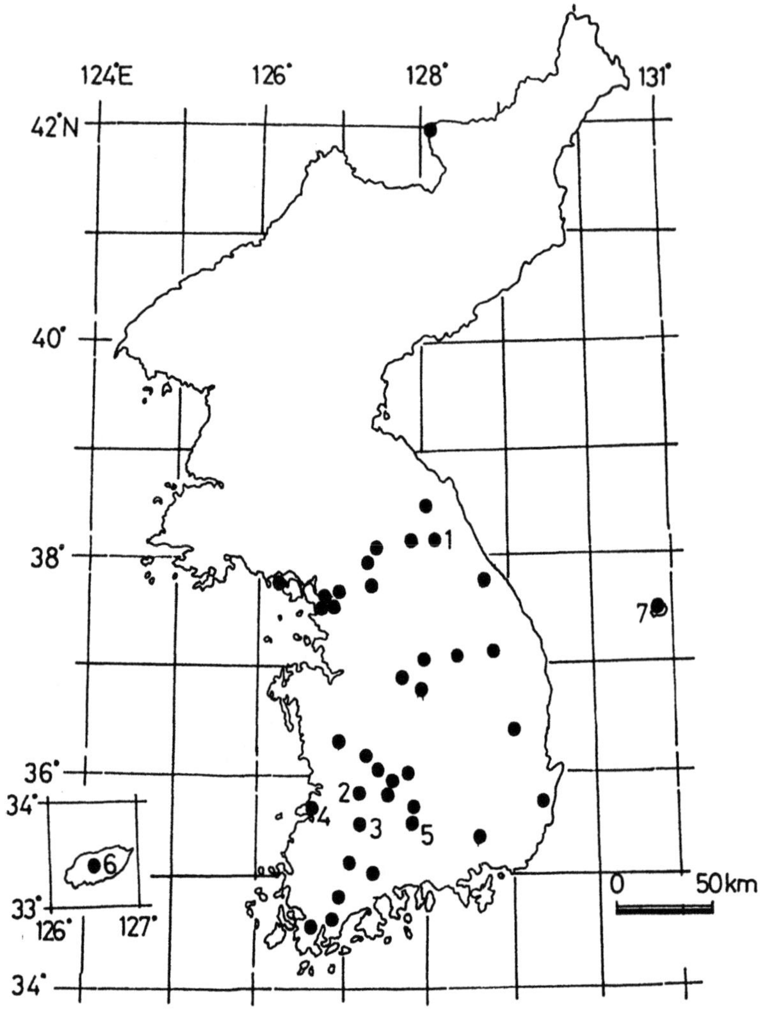

Fig. 4. Locations of intensive study areas in Korea. Mountains mentioned in text among them: 1, Mt. Sorak (1308 m); 2, Mt. Daedun (818 m); 3, Mt. Naejang (763 m); 4, Mt. Seonun (444 m); 5, Mt. Chiri (1915 m); 6, Mt. Halla (1950 m); 7, Ulreung-do (island, 984 m).

key factor affecting the floristic composition and distribution of their plant communities. Cheju-do belongs to the humid (B4) zone (northern part) and perhumid zone (southern part) defined by Thornthwaite's (1948) moisture index, whereas the central part of the peninsula belongs to the humid zone (B3). Cheju-do corresponds to a continentality value of 130, the peninsula to only 110 (Yim & Kira 1975).

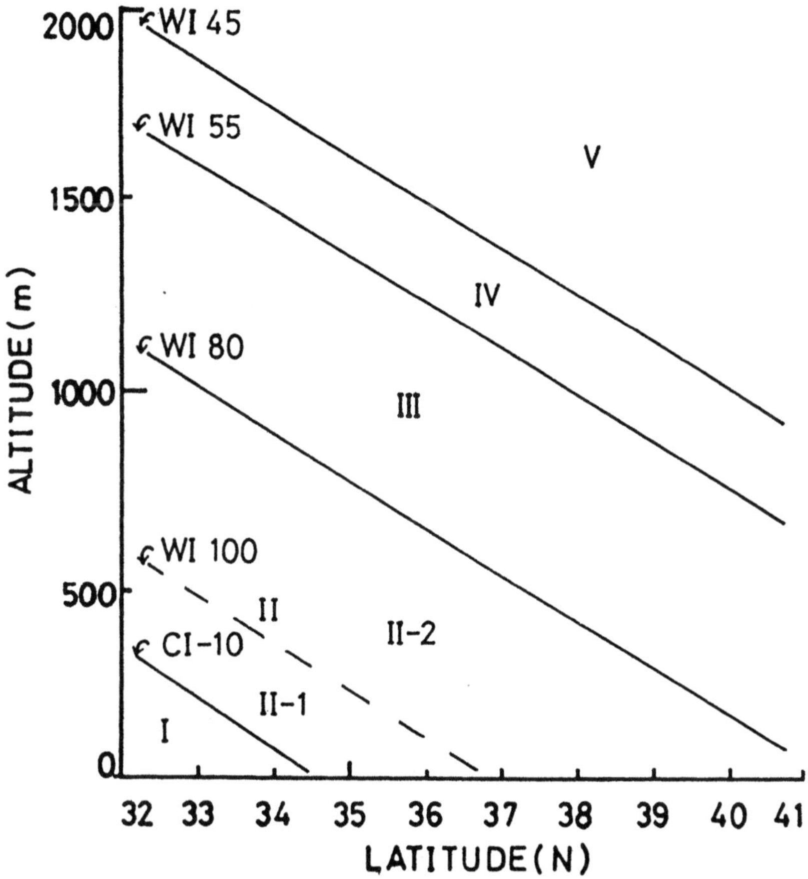

Fig. 5. The distributional ranges in warmth and coldness index of five groups and two subgroups separated in 14 communities. See Fig. 3.

Species association

Kim & Yim (1988a) found four major groups of positively correlated species at Mt. Naejang (Fig. 10) which can be regarded as alliances. There are three species intermediate between the four groups: *Fraxinus rhynchophylla*, *Smilax china* and *Quercus serrata*. These groups might be regarded as one order, because the four groups are not completely independent from each other. Some groups were divided into several subgroups, which might be recognized as associations:

Group 1	*Pinus densiflora*, *Juniperus rigida*, *Rhododendron yedoense* var. *poukhanense*, *Festuca ovina*, *Millium effusum* and *Rhododendron mucronulatum*	
Group 2	Subgroup 1:	*Meliosma myriantha*, *Stephanandra incisa* and *Carpinus tschonoskii*
	Subgroup 2:	*Sapium japonicum*, *Rhus trichocarpa*, *Viburnum dilatatum* and

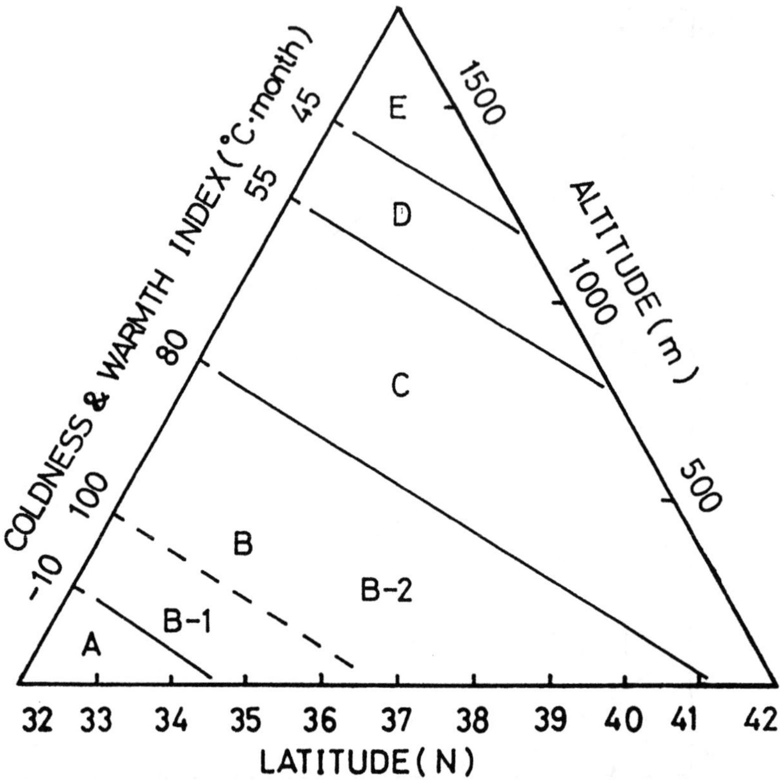

Fig. 6. The distribution of different plant community groups (= forest zones) in the triangle diagram of altitude-latitude-thermal climate. A, warm-temperate evergreen lucidophyll forest; B, cool-temperate deciduous broad-leaved forest; B-1, southern part; B-2, middle part; C, cool-temperate deciduous broad-leaved forest northern part; D, subartic coniferous forest; E, shrubby forest.

		Carpinus laxiflora
	Subgroup 3	*Lespedeza bicolor*, *Indigofera kirilowi* and *Quercus variabilis*
Group 3	*Zelkova serrata*, *Acer pseudo-sieboldianum* var. *koreanum* and *Zanthoxylum piperitum*	
	Subgroup 1:	*Acer mono*, *Celtis sinensis*, *Polystichum tripteron* and *Hydrangea serrata* for. *acuminata*
	Subgroup 2	*Torreya nucifera*, *Adenocaulon himalaicum* and *Disporum sessile*
	Subgroup 3	*Lindera erythrocarpa*, *Cornus controversa*, *Arisaema amurense* and *Alangium platanifolium* var. *macrophyllum*
Group 4	*Quercus mongolica*, *Acer pseudo-sieboldianum*, *Fraxinus sieboldiana*, *Rhododendron schlippenbachii*, *Melampyrum roseum* and *Ainsliaea acerifolia*.	

Group 1 was found on areas destroyed by human activity and on hills and exposed ridgetops, i.e. dry, poor montane habitats. The distributional range

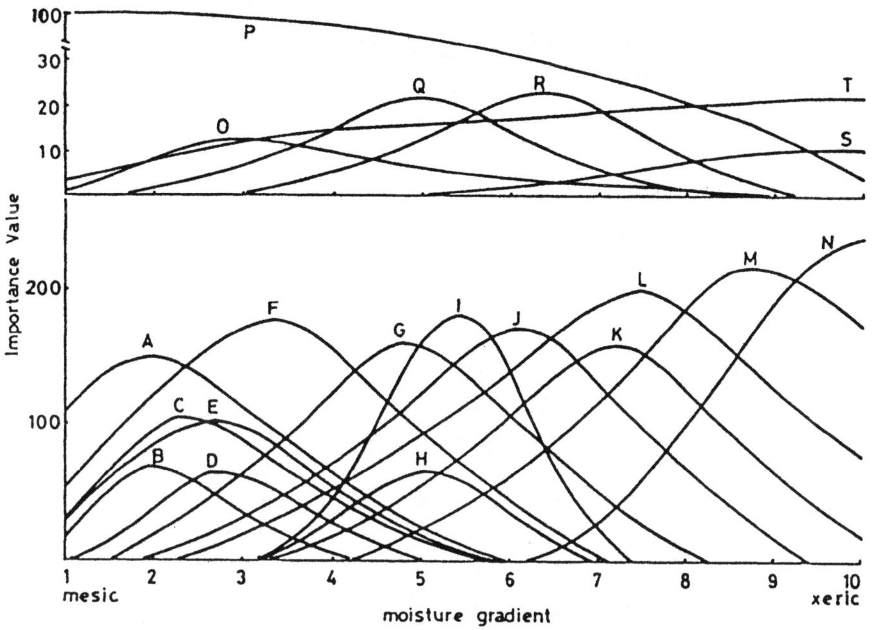

Fig. 7. Importance value curves of species population along moisture gradient in Mt. Naejang (from Kim & Yim 1988a). Upper side: shrub species, underside; tree species. A, *Zelkova serrata*; B, *Celtis sinensis*; C, *Lindera erythrocarpa*; D, *Cornus controversa*; E, *Acer mono*; F, *Carpinus tschonoskii*; G, *Quercus aliena*; H, *Daphniphyllum macropodum*; I, *Torreya nucifera*; J, *Carpinus laxiflora*; K, *Quercus serrata*; L, *Quercus variabilis*; M, *Quercus mongolica*; N, *Pinus densiflora*; O, *Lindera oblusiloba*; P, *Acer pseudo-sieboldianum* var. *koreanum*; Q, *Styrax obassia*; R, *Stylax japonica*, S; *Rhododendron schlippenbachii*; T, *Acer pseudo-sieboldianum*.

of this group coincides with that of the Rhododendro-Pinion densiflorae in the classification by the Z-M method.

Group 2 was found on mesic, fertile sites at elevations below 600 m and coincides with the Carpinion laxiflorae. Subgroup 1 occured on lower slopes (*cf.* Carpinetum tschonoskii), subgroup 2 occured on middle slopes (*cf.* Carpinetum laxiflorae), and subgroup 3 on sunny, xeric hillsides (*cf.* Quercetum variabilis).

Group 3 was found on well drained, rocky slopes near streams and coincides with the Corno-Zelkovion serratae. Subgroup 1 occured on lower slopes and mesic rocky sites corresponding to the Acero-Zelkovetum serratae; subgroup 2 on lower slopes (*cf.* Torreyo-Zelkovetum serratae); and subgroup 3 on lower slopes and in ravines, i.e. shady, humid sites, corresponding to the Corno-Linderetum erythrocarpae.

Group 4 developed on xeric, sterile sites at elevations above 600 m, coinciding with the Acero-Quercion mongolicae. The results of correlation analy-

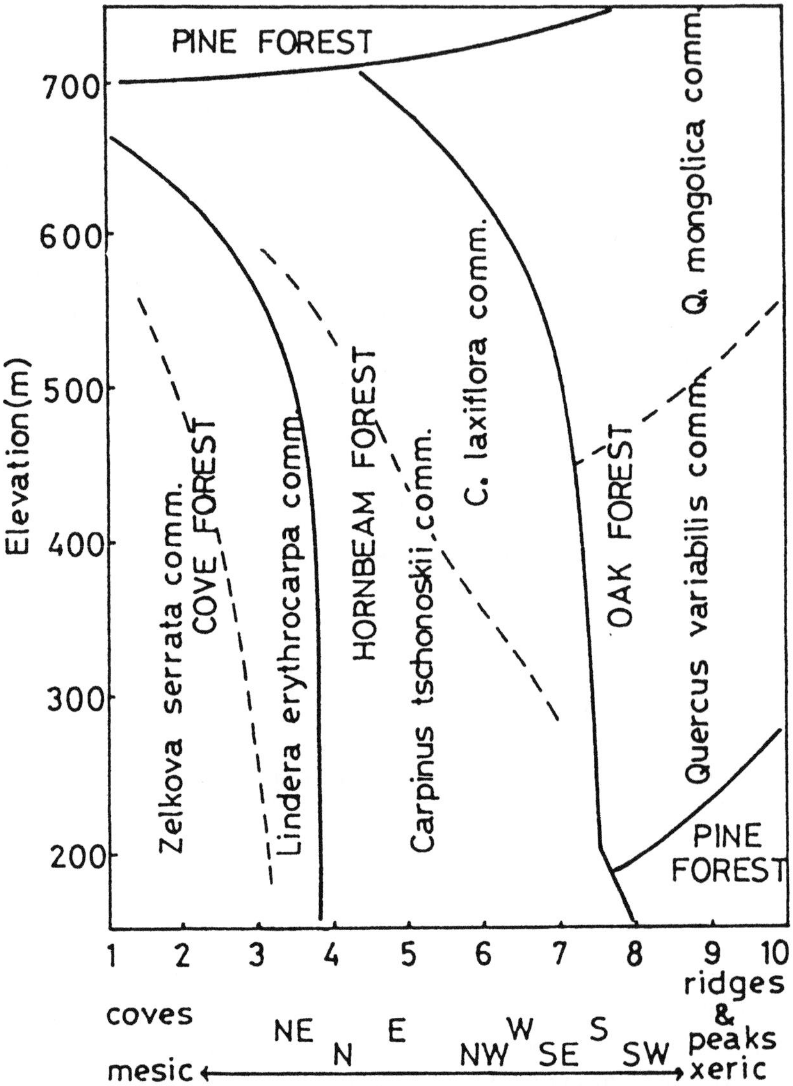

Fig. 8. Mosaic chart for species population of Mt. Naejang in Korea (from Kim & Yim 1988b). The vertical axis is the complex gradient of temperature and other factors related to elevation; the horizontal axis is the complex gradient of moisture and other factors from moist or mesic situations on the left to dry or xeric on the right, as affected by topographic position.

sis between species reflected the phytosociological classification units based on floristic composition.

Species sequence

The species sequence in the dominance-diversity curve for all of Mt. Sorak showed the following, in order, as the most important of 112 deciduous

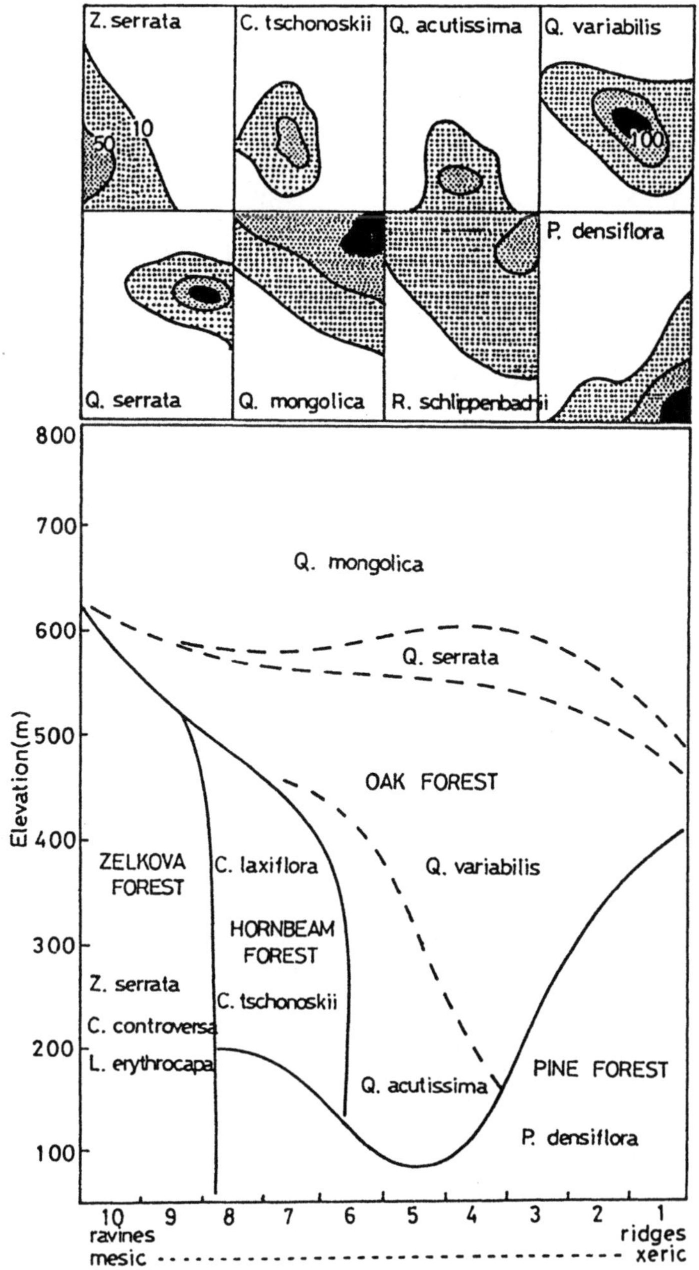

Fig. 9. Mosaic chart for species population of Mt. Moak (from Kim & Yim 1992).

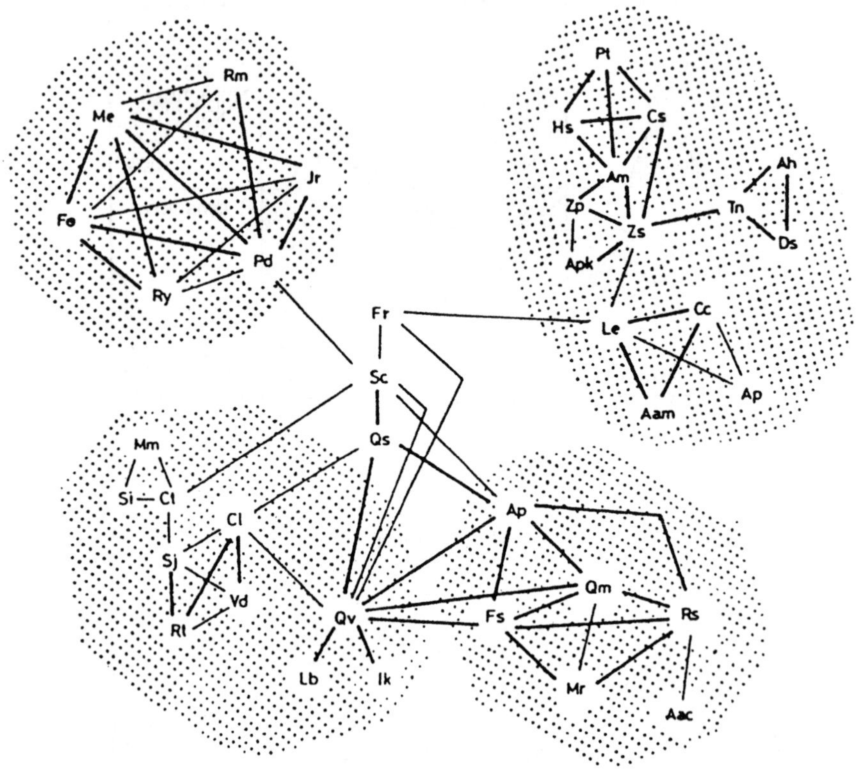

Fig. 10. Species constellations showing positive correlations between 39 species in Mt. Naejang in Korea (from Kim & Yim 1988b). Single lines represent 5% ⩾P ⩾ 1%, bold lines P < 1% for chi-square values. Fr, *Fraxinus rhynchophylla*; Pd, *Pinus densiflora*; Ap, *Acer pseudo-sieboldianum*; Sj, *Sapium japonicum*; Zp, *Zanthoxylum piperitum*; Rm, *Rhododendron mucronulatum*; Mr, *Melampylum roseum*; Ik, *Indigotera kirilowi*; Vd, *Viburnum dilatatum*; Ct, *Carpinus tschonoskii*; Rt, *Rhus trichocarpa*; Le, *Lindera erythrocarpa*; Hs, *Hydrangea serrata* for. *acuminata*, Ds, *Disporum sessile*, Cc, *Cornus controversa*; Sc, *Smilax china*; Cs, *Celtis sinensis*; Qm, *Quercus mongolica*; Jr, *Juniperus rigida*; Fs, *Fraxinus sieboldiana*; Am, *Acer mono*; Cl, *Carpinus laxiflora*; Fo, *Festuca ovina*; Rs, *Rhododendron schlippenbachii*; Qv, *Quercus variabilis*; Apk, *Acer pseudo-sieboldianum* var. *koreanum*; Mm, *Meliosma myriantha*; Ry, *Rhododendron yedoense* var. *poukhanense*; Lb, *Lespedeza bicola*; Ah, *Adenocaulon himalaicum*; Ap, *Alangium platanifolium* var. *macrophyllum*; Si, *Stephanandra incisa*; Pt, *Polysticum tripteron*; Aac, *Ainsliaea acerifolia*; Me, *Milium effusum*; Zs, *Zelkova serrata*, Aam, *Arisaema amurense*; *Th*, *Torreya nucifera*; Qs, *Quercus serrata*.

broad-leaved tree species: *Quercus mongolica*, *Q. serrata* and *Carpinus laxiflora* (Choi & Yim 1984). The curves for different valleys reflected the nature of the physical environment and the plant communities at those sites. The curves were grouped into two types, a log-normal distribution in many deciduous broad-leaved forests (Fig. 11) and a geometric series in pine forests on disturbed or rocky sites, based on some hypotheses such as the random-

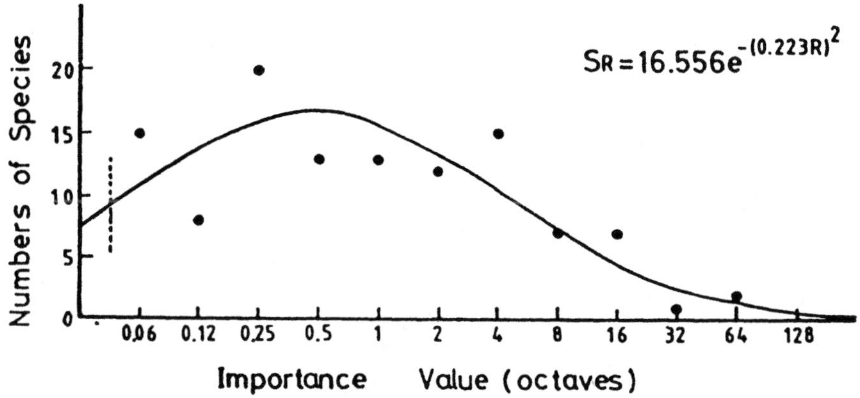

Fig. 11. The lognormal distribution of species importance value by Preston (1948) in Mt. Sorak. Points are numbers of species plotted against their importance value within the octaves of importance value. The dashed vertical line is the "veil line", to the left of which no data on species numbers and importance value are available.

niche hypothesis (MacArthur & MacArthur 1961, Whittaker 1965, 1972, 1975), the niche pre-emption hypothesis (Motomura 1932, Whittaker 1965, 1972, 1975), the log-normal distribution (Preston 1948, 1962), and logarithmic series (Fisher et al. 1943, Pielou 1969).

At Mt. Halla various types were also recognized (Fig. 12): one community of *Carpinus laxiflora*, *C. tschonoskii*, *Quercus serrata*, *Q. grosseserrata* and *Pinus densiflora*, and another of *Abies koreana*, as classified by the Z-M method based on data from 81 sample plots. The former curves coincided with the ideal curve calculated by Preston's log-normal distribution theory, with small variations among the five communities; the latter coincided with Motomura's niche pre-emption hypothesis.

It seems that plant communities classified by the phytosociological method based on the unit concept coincide with vegetation types recognized by species population analysis based on the continuum concept, and that the dominance-diversity relationships reflect the differences in coenoclines among their plant communities.

Vegetation map of Korea

Vegetation maps of Korea have been published by Honda (1922), Uyeki (1933), Lautensach (1935), Chung & Lee (1965), Yim (1968) and others, though these maps were not always accompanied by detailed discussion and description.

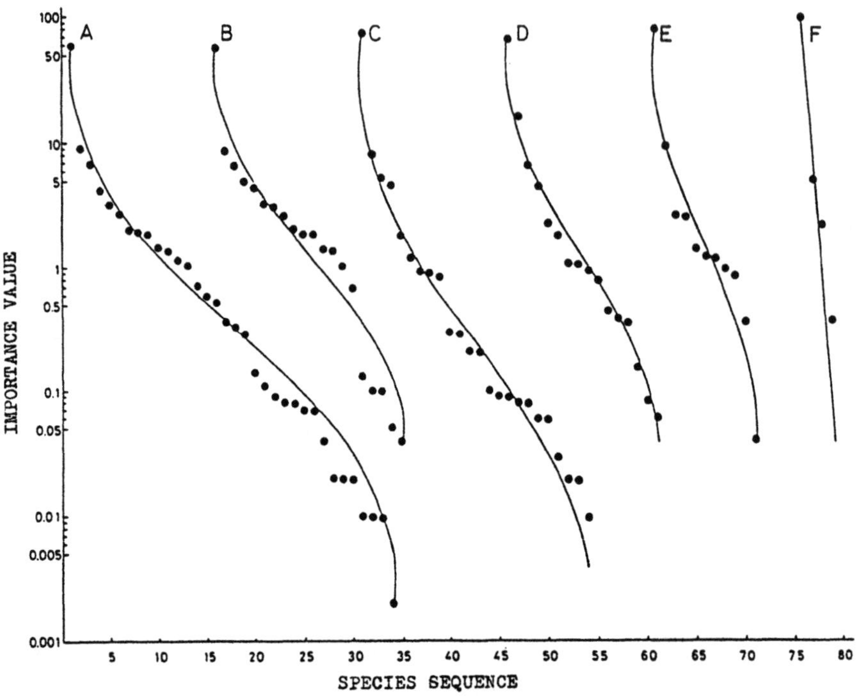

Fig. 12. Dominance-diversity curves of different communities in Mt. Halla national park. A, *Carpinus laxiflora* community; B, *Carpinus tschonoskii* community; C, *Quercus serrata* community; D, *Quercus grosseserrata* community; E, *Pinus densiflora* community; F, *Abies koreana* community.

Since the 1970's, numerous studies on vegetation in Korea have been performed on the 20 national parks and many provincial parks. The area of Mt. Sorak National Park is covered with 18 typical plant communities recognized from the field data of 450 quadrats (Yim & Baik 1985). According to these, most of the mountain is covered with a *Quercus mongolica* community, while a *Carpinus laxiflora* community is found at the foot of the mountain, just reaching its northern limit as a community (Fig. 13). In Mt. Naejang National Park (Fig. 14), a *Quercus mongolica* community occurs above 600 m, *Q. variabilis* and *Q. serrata* communities at 400–600 m, *Carpinus laxiflora* communities at 200–400 m, *C. tschonoskii* and *Q. aliena* communities below 200 m, and *Zelkova serrata* communities along streams and on mesic rocky sites (Kim and Yim 1988c). Similar results were obtained from the three provincial parks, Mt. Seonun (Kim & Yim 1987, 1988c), Mt. Moak (Kim & Yim 1989), and Mt. Daedun (Kim et al. 1988).

In Mt. Chiri National Park (Yim & Kim 1992), a *Quercus mongolica* community occurs at 850–1450 m, *Q. serrata*, *Q. variabilis*, and *Carpinus*

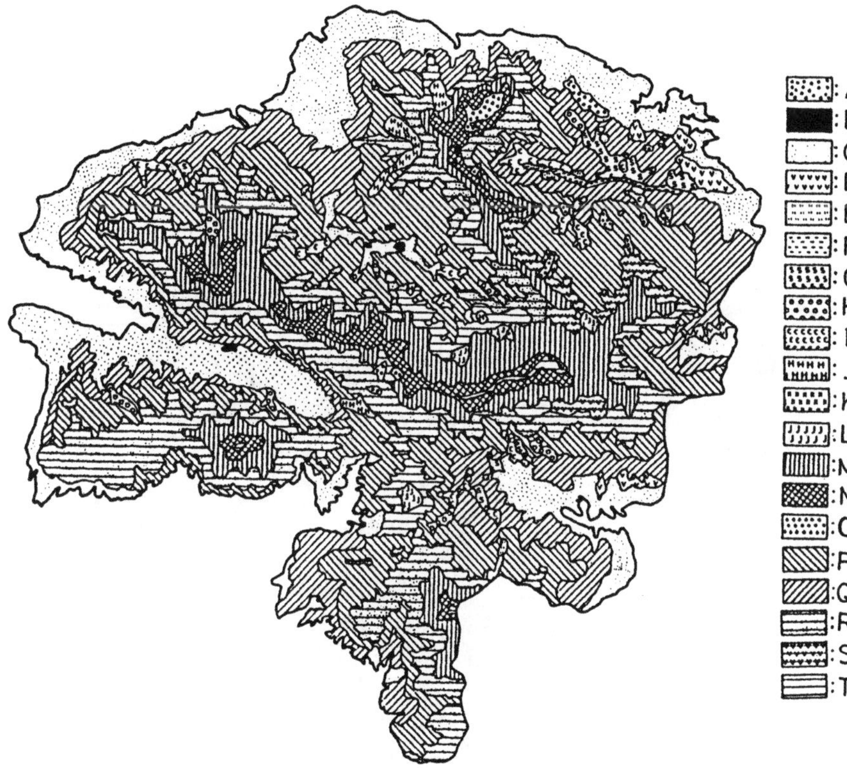

Fig. 13. Actual vegetation map of Mt. Sorak. A, Rocks; B, Old field; C, *Pinus densiflora* community; D, *Quercus variabilis* community; E, *Quercus aliena* community; F, *Populus maximowiczii* community; G, *Quercus serrata* community; H, *Cornus controversa* community; I, *Carpinus laxiflora* community; J, *Abies holophylla* community; K, *Pinus koraiensis* community; L, *Thuja koraiensis* community; M, *Betula ermani* community; N, *Abies nephrolepis* community; O, *Pinus pumila* community; P, *Quercus mongolica-Lespedeza maximowiczii* community; Q, *Quercus mongolica-Sasa borealis* community; R, *Quercus mongolica* community; S, *Quercus mongolica-Lingularia ficheri* community; T, *Quercus mongolica-Vaccinium koreanum* community.

laxiflora communities at 500–850 m, a *C. tschonoskii* community at 400–500 m, and *Q. aliena* and *Zelkova serrata* communities along mesic streamsides.

In Cheju-do, on Mt. Halla (Fig. 15), *Quercus grosseserrata*, which is a hybrid of *Q. mongolica* and *Q. serrata* (Lee 1979), occurs on slopes at 1200–1400 m. This species occupies an ecological niches similar to that of *Q. mongolica* in the Korean peninsula (Yim et al. 1990). *Carpinus laxiflora* and *C. tschonoskii* communities occur on slopes at 800–1200 m and 700–1100 m, respectively.

Based on the results described above, maps of the actual vegetation were

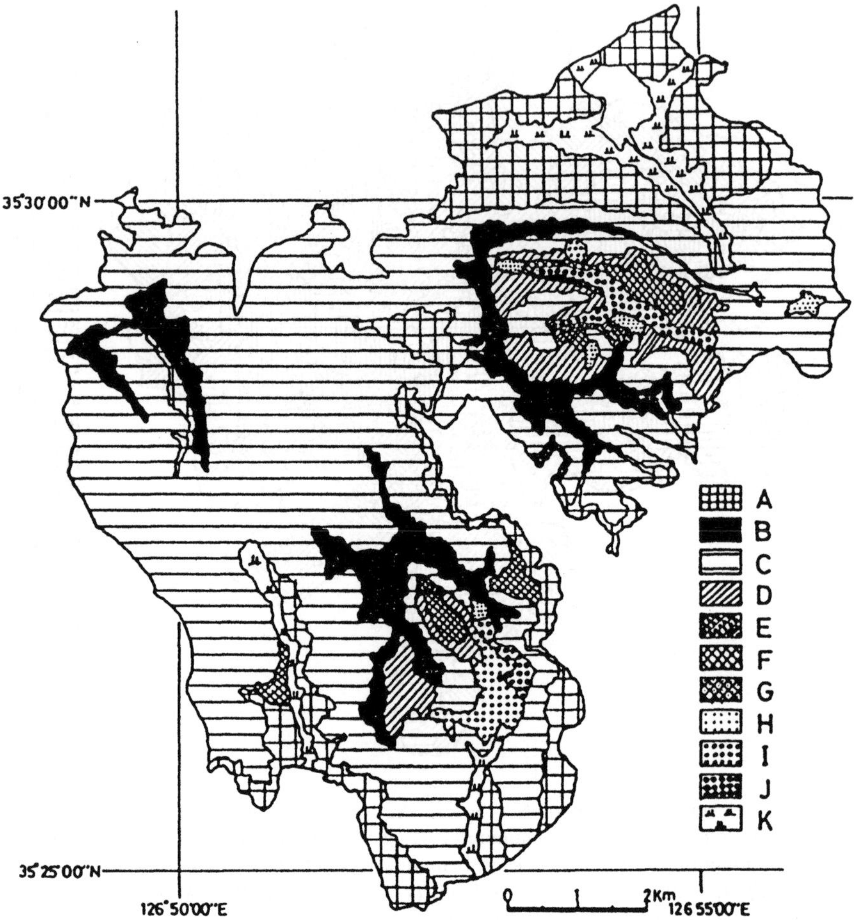

Fig. 14. Actual vegetation map of Mt. Naejang. A, *Rhododendro mucronulati – Pinetum densiflorae*; B, *Rhododendro – Quercetum mongolicae*; C, *Quercetum variabilis*; D, *Carpinetum laxiflorae*; E, *Daphniphyllum macropodum* community; F, *Carpinetum tschonoskii*; G, *Quercus aliena – Carpinus tschonoskii* community; H, *Corno – Linderetum erythrocarpae*; I, *Torreyo – Zelkovetum serratae*; J, *Acero – Zelkovetum serratae*; K, Crop land.

made at scales of 1:25,000 or 1:50,000 (Küchler & Zonneveld 1988). A map of the potential natural vegetation of Korea should be available in the near future.

Summary

1. The forest vegetation zones of Korea coincide with thermal zones.

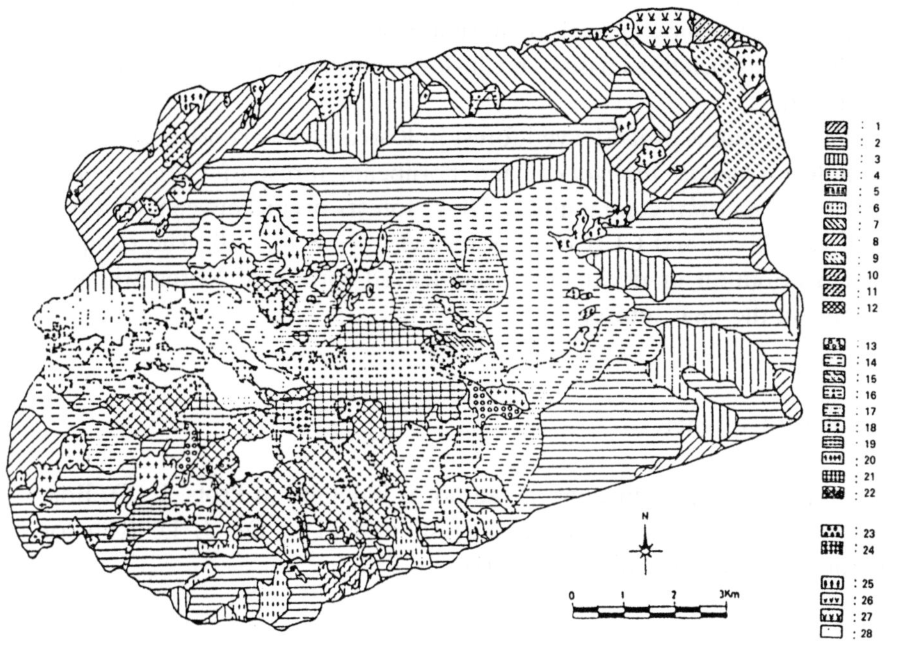

Fig. 15. Actual vegetation map of Mt. Halla national park. 1, *Quercus serrata-Carpinus tschonoskii* community; 2, *Carpinus laxiflora-Carpinus tschonoskii* community; 3, *Carpinus laxiflora-Quercus serrata* community; 4, *Carpinus laxiflora-Quercus grosseserrata* community; 5, *Carpinus laxiflora-Pinus thunbergii* community; 6, *Carpinus laxiflora-Pinus densiflora* community; 7, *Carpinus tschonoskii-Carpinus laxiflora* community; 8, *Quercus serrata-Carpinus laxiflora* community; 9, *Carpinus tschonoskii-Quercus serrata* community; 10, *Quercus serrata-Pinus densiflora* community; 11, *Quercus serrata-Pinus thunbergii* community; 12, *Rhododendron yedoense* var. *poukhanense-Rhododendron mucronulatum* var. *ciliatum* community; 13, *Quercus grosseserrata* community; 14, *Quercus grosseserrata-Carpinus laxiflora* community; 15, *Quercus grosseserrata-Abies koreana* community, 16, *Quercus grosseserrata-Pinus densiflora* community; 17, *Quercus acutissima-Quercus serrata* community; 18, *Pinus densiflora* community; 19, *Pinus densiflora-Carpinus laxiflora* community; 20, *Pinus thunbergii* community; 21, *Abies koreana* community; 22, *Abies koreana-Rhododendron yedoense* var. *poukhanense* community; 23, *Juniperus chinensis* var. *sargentii* community; 24, *Juniperus chinensis* var. *sargentii-Empetrum nigrum* var. *japonicum* community; 25, *Pinus thunbergii* plantation; 26, *Miscanthus sinensis* community; 27, Grass land; 28, *Sasa quelpaertensis* community.

2. Plant communities classified by the Z-M method coincide with vegetation types recognized by species population ordination.
3. Mosaic charts of elevation (thermal gradient) versus topographic moisture showed *Quercus mongolica* communities on relatively xeric slopes, *Q. serrata* and *Q. aliena* communities on mesic slopes, and *Zelkova serrata* communities along rocky streamsides.
4. Character species of different habitats on the peninsula (dry, cool climate) often co-exist in the same habitat on Cheju-do (warm, wet climate).

5. The dominance-diversity curves of plant communities in broad-leaved forest zones show typical log-normal curves.
6. Species associations based on Chi-square tests corroborate phytosociological units.

The deciduous broad-leaved forest in Korea mainly consists of oak forest, including *Quercus mongolica*, and hornbeam forest of *Carpinus laxiflora* and *C. tschonoskii* communities. These forests occur between WI (Warmth Index) = 45–55 and CI (Coldness Index) = − 8 to − 10. The composition and distribution of deciduous broad-leaved forests of Korea will become clearer in the near future.

Acknowledgements

In preparing this manuscript the author is indebted to Profs. E. O. Box and K. Fujiwara for their valuable suggestions and to Dr. J. U. Kim for his cordial help with the literature survey.

References

Choi, K. R. & Yim, Y. J. 1984. On the dominance-diversity in the forest vegetation of Mt. Seolag. Korean J. Bot. 27:25–32.

Chung, T. H. & Lee, W. T. 1965. A study of the Korean woody zones and favorable regions for growth and proper species. Sung Kyun Kwan Univ. Journal 10:329–434 (in Korean).

Davis, M. B. 1976. Pleistocene biogeography of temperate deciduous forests. Geosci. and Man 13:13–26.

Fang, J.-Y. 1988. Climate and Distribution of Vegetation in China. Ph.D. thesis, Osaka City University. 172 pp.

Fisher, B. A., Corbert, S. S. & Williams, C. B. 1943. The relation between the number of species and number of individuals in a random sample of an animal population. J. Animal Ecol. 12:42–58.

Good, R. E. 1953. The Geography of the Flowering Plants, 2nd ed. Longmans, Green and Co, New York.

Honda, S. 1922. Forest Zones of Japan, revised 4th ed. (first published in 1912). Tokyo (in Japanese).

Kim, J. U. & Yim, Y.-J. 1986a. Classification of forest vegetation of Seonun-san area, southwestern Korea. Korean J. Ecol. 9:209–223.

Kim, J. U. & Yim, Y.-J. 1986b. A gradient analysis of the mixed forest of Seonun-san area in southwestern Korea. Korean J. Ecol. 9:225–230.

Kim, J. U. & Yim, Y.-J. 1987. Actual vegetation and potential natural vegetation of Seonun-san area, southwestern Korea. Korean J. Ecol. 10:159–164.

Kim, J. U. & Yim, Y.-J. 1988a. Phytosociological classification of plant communities in Mt. Naejang, southwestern Korea. Korean J. Bot. 31:1–31.

Kim, J. U. & Yim, Y.-J. 1988b. Environmental gradient analyses of forest vegetation of Mt. Naejang, southwestern Korea. Korean J. Bot. 31:33–39.

Kim, J. U. & Yim, Y.-J. 1988c. Actual vegetation and potential natural vegetation of Naejang-san National Park, southwestern Korea. Korean J. Ecol. 11:145–152.

Kim, J. Y. & Yim, Y.-J. 1989. Flora and vegetation maps of Mt. Moak Provincial Park, Korea. Chung-Ang Univ. J. Natural Science 3:83–100.
Kim, J. Y., Yim, Y.-J. & Kil, B. S. 1988. Classification and pattern analysis of the forest vegetation in Daedun-san Provincial Park, Korea. Korean J. Ecol. 11:109–122.
Kira, T. 1945. A new classification of climate in eastern Asia as the basis for agricultural geography. Hort. Inst. Kyoto University, Kyoto (in Japanese).
Kira, T. 1976. (Terrestrial ecosystems – a general review.) Ecology Course 2. Tokyo: Kyoritsu Shuppan (in Japanese).
Kira, T. 1977. A Climatological Interpretation of Japanese Vegetation Zones. In: Miyawaki, A. & Tüxen, R. (eds.) Vegetation Science and Environmental Protection, pp. 21–30. Maruzen Co, Tokyo.
Koh, J. K. 1990. Studies on the Interrelationships among the Plant Communities by the Analysis of Species Distribution at Mt. Seolag. Ph.D. thesis, Chung-Ang University (Seoul). 84 pp.
Küchler, A. W. & Zonneveld, I. S. 1988. Vegetation Mapping. Kluwer Academic Publishers, Dordrecht. 635 pp.
Lautensach, H. 1935. Cited in: Weltforstatlas: China, Korea and Taiwan (Bundesforschungsanstalt fuer Forst- und Holzwirtschaft, ed.). Fritz Halled Verlag, Berlin (1957).
Lee, W. T. & Yim, Y.-J. 1978. Studies on the distribution of vascular plants in the Korean Peninsula. J. Korean Plant Tax. 8:1–33 (in Korean).
Lee, T. B. 1979. Illustrated Flora of Korea. Hyangmoonsa, Seoul. 791 pp. (in Korean).
MacArthur, R. H. & MacArthur, J. W. 1961. On bird species diversity. Ecol. 42:594–598.
Miyawaki, A., Suzuki, K. & Fujiwara, K. 1977. Human impact upon forest vegetation in Japan. Naturaliste Canad. 104:97–107.
Motomura, I. 1932. A statistical treatment of associations. Jap. J. Zool. 44:379–383.
Mueller-Dombois, D. & Ellenberg, H. 1974. Aims and Methods of Vegetation Ecology. John Wiley & Sons, New York. 547 pp.
Pielou, E. C. 1969. An Introduction to Mathematical Ecology. Wiley Interscience, New York. 286 pp.
Preston, F. W. 1948. The commonness and rarity of species. Ecol. 29:254–283.
Preston, F. W. 1962. The canonical distribution of commonness and rarity. Ecol. 43:185–215, 410–432.
Satake, Y., Hara, H. Watari, S. & Tominari, T. 1989. Wild Flowers of Japan: Woody Plants. I and II. Heibonsha Ltd., Tokyo. 321 and 304 pp. (in Japanese).
Tanai, T. 1974. Evolutionary trend of the genus *Fagus* around the northern Pacific Basin. In: Symposium on Origin and Phytogeography of Angiosperms. Birbal Sahni Institute of Paleobotany, Spec. Publ. 1:62–83.
Thornthwaite, C. W. 1948. An approach toward a rational classification of climate. Geogr. Rev. 38:55–94.
Tsukada, M. 1974. Paleoecology I. Ecology Course 27. Kyoritsu Shuppan, Tokyo. 166 pp. (in Japanese).
Uyeki, H. 1933. On the forest zones of Korea. Acta Phytotax. Geobotan. 2:73–85 (in Japanese).
Whittaker, R. H. 1965. Dominance and diversity in land plant communities. Science 147:250–260.
Whittaker, R. H. 1972. Evolution and measurement of species diversity. Taxon. 21:213–251.
Whittaker, R. H. 1975. *Communities and Ecosystems*. Macmillan Co, New York. 385 pp.
Yim, K. B. 1968. Jorimhak Wonron (Principles of Silviculture). Hyangmoonsa, Seoul (in Korean).
Yim, Y.-J. 1977a. Distribution of forest vegetation and climate in the Korean peninsula. III. Distribution of tree species along the thermal gradient. Jap. J. Ecol. 27:177–189.
Yim, Y.-J. 1977b. Distribution of forest vegetation and climate in the Korean peninsula. IV. Zonal distribution of forest vegetation in relation to thermal climate. Jap. J. Ecol. 27:269–278.
Yim, Y.-J. 1983. On the distribution of beech (*Fagus*, Fagaceae) and beech-dominated forests in the Northern Hemisphere. Korean J. Ecol. 6:153–166.

Yim, Y.-J. 1989. Vegetation Mapping in Korea. In: Proceedings UNESCO MAB Regional Seminar on BICEM, Tokyo, pp. 162–188.
Yim, Y.-J. & Kira, T. 1975. Distribution of forest vegetation and climate in the Korean peninsula. I. Distribution of some indices of thermal climate. Jap. J. Ecol. 25:77–88.
Yim, Y.-J. & Kira, T. 1976. Distribution of forest vegetation and climate in the Korean peninsula. II. Distribution of climate humidity/aridity. Jap. J. Ecol. 26:157–166.
Yim, Y.-J. & Baik, S. D. 1985. The Vegetation of Mt. Seolag. Chung-Ang Univ. Publications, Seoul. 200 pp. (in Korean).
Yim, Y.-J. & Kim, J. U. 1992. The Vegetation of Mt. Chiri. Chung-Ang Univ. Publications, Seoul. 471 pp. (in Korean).
Yim, Y.-J., Kim, J. U., Lee, N. J., Kim, Y. B. & Paek, K. S. 1990. Phytosociological classification of plant communities in Mt. Halla National Park, Korea. Korean J. Ecol. 13:101–130 (in Korean).

14. Management and ecological perspectives on the oak forests of mountainous northeastern China

YANG SI-HE & WEN SHIYUN
Institute of Applied Ecology, Academia Sinica P.O. Box 417, Shenyang, China

Abstract

The largest remaining forested area of northeastern China is dominated by secondary forests in which oaks are dominant. The forests of this mountainous terrain cover approximately 28 million hectares. Due to pressure for economic exploitation, these oak forests have steadily decreased in stand area and stand quality. Present management practices include clearcutting, stand thinning, under-canopy planting, and coppicing for silk worm production. Recent studies suggest that although wood production is important, the ecological benefits of maintaining these forests outweigh the benefits of harvesting all the trees for timber production. We suggest that restrictions be placed on clearcutting in these forests, and that future plantings be of mixed coniferous and broad-leaved species.

Key words: clearcutting, coppice, mixed conifer-broadleaved forest, mongolian oak, *Quercus*, secondary forest

Introduction

Northeastern China contains the largest total area of forest in the country, much of it being oak forest. The area includes the Great Xingan (Da Xing An Ling), Lesser Xingan (Xiao Xing An Ling) and Changbai Mountain forest regions. The forest covers in excess of 31 million hectares (Planning Office of Forestry Ministry 1981), 28 million hectares of which is secondary forest (Han Linfeng 1982). A portion of these forests remains as virgin forest, but the amount of virgin forest has declined sharply over the past several decades. The montane forests of northeastern China play an important role in protecting the environment of the region, and by providing suitable habitat for animals and plants. Rapid growth of the human population plus rapid

E.O. Box et al. (eds.), Vegetation Science in Forestry, 299–305.

Table 1. Altitudinal distribution of two oak-dominated vegetation types in three mountain ranges at different latitudes in northeast China

	Mountain range and latitude		
	Changbai Mt. 42° N	Zhang Gongcae Ridge 45° N	Lesser Xingan Mts. 48° N
Conifer-broadleaf mixed forest	500–1200 m	600–900 m	300–700 m
Broadleaf forest	<500 m	<600 m	<300 m

industrial growth has progressively increased timber production at the cost of decreased reserves of both virgin and secondary forest. The purpose of this paper is to clarify and suggest alternative options for future oak forest management.

Oak forest distribution and composition

Oak forests are widely distributed in northeastern China, making up about one fifth of the forested area (approximately 6 million hectares). Several species of oak form either pure and mixed stands from the Heilongjiang River (53° N) to the Liaodong Peninsula (39° N). These oak stands are found in three vegetation zones: 1) cool temperate coniferous forest, 2) temperate mixed (coniferous and broadleaved) forest, and 3) warm temperate deciduous (broadleaved) forest. The elevational distribution of the communities varies with latitude (Planning Office of Forestry Ministry 1981; see Table 1).

Stand composition in these forests varies in the three vegetation zones. The cool temperate coniferous forest zone extends throughout the Great Xingan Mountains (49°20′–53°34′ N latitude and 119°30′–127°29′ E longitude). Topographically, the area consists of gentle mountains with elevations of 700–1100 m above sea level. In stands dominated by Dahurian larch (*Larix gmelinii* Rupr.), the Mongolian oak (*Quercus mongolica* Fisch. ex Turcz.) occurs as an understory tree. Mongolian oak may occur alone in the understory or with other deciduous species such as *Betula dahurica* Pall. and *Tilia amurensis* Rupr. Pure stands of oak are often found on poor sites after a stand of larch has been clear cut. The shrubby undergrowth in cool temperate coniferous forests consists mainly of *Rhododendron dauricum* L., *Corylus heterophylla* Fisch ex Bess. and *Lespedeza bicolor* Turcz.

The temperate mixed (coniferous and broadleaved) forests generally occur below 1300 m elevation. This area includes Lesser Xingan Mountains, Wanda Mountain, Zhang Guongcai Ridge and Changbai Mountain (40°15′–50°15′ N, 126°–135°30′ E). Korean pine (*Pinus koraiensis* Sieb. et Zucc.) dominates

Table 2. Distribution of oak (*Quercus*) species in northeast China

Species	Northern limit	Distribution	Stand structure
Q. mongolica	55.3° N	Liaoning, Jilin, Heilongjing and Inner Mongolia	Pure and mixed
Q. dentata	46.3° N	Liaoning	Mixed
Q. liaotungensis	44.5° N	Liaoning, Jilin, Heilongjing and Inner Mongolia	Pure and mixed
Q. variabilis	41.5° N	Liaoning	Mixed
Q. acutissima	41.0° N	Liaoning	Pure and mixed
Q. aliena	40.1° N	Liaoning	Mixed

the virgin stands. Other conifer and broadleaved species that also contribute to the virgin communities include *Picea jezoensis* Carr., *Abies holophylla* Maxim, *A. nephrolepsis* Maxim, *Larix olgensis* A. Henry, *Betula platyphylla* Suk., *Tilia amurensis* Rupr., *Fraxinus mandshurica* Rupr., *Phellodendron amurense* Rupr., and *Quercus mongolica* (Mongolian oak) Fisch. Once the virgin forests are cut, a secondary forest develops with Mongolian oak as the dominant species. Pure oak stands are not uncommon. The undergrowth in mixed and pure oak forests consists mainly of *Corylus mandshurica* Maxim. ex Rupr., *Syringa reticulata* var. *mandshurica* Hara and *Acer ukrunduense* Trautv. et Mey.

The warm-temperate deciduous broadleaved forest zone covers the eastern part of Liaoning Province and Liaodong Peninsula (39°–42°30′ N, 121°30′–125°30′ E). Virgin sections of this forest are very rare. Secondary forests are dominated by oak species, both pure stands and mixed. Other species in the mixed stands are *Robinia pseudoacacia* L., *Acer mono* Maxim., *Tilia mandshurica* Rupr. et Maxim., and *Juglans mandshurica* Maxim.

Along with *Quercus mongolica*, other oak species found in the forests are *Q. liaotungensis* Koidz., *Q. acutissima* Carr., *Q. dentata* Thunb. and *Q. aliena* Blume. Although these species are considered widespread, many of them have reached their northernmost limit within the forests of northeastern China (The Research Group on Deciduous Oaks 1988). Table 2 shows the latitudinal limits of various oak species. The undergrowth of the oak forests includes *Lespedeza bicolor* Turcr., *Corylus heterophylla* Fisch. ex Bess. and *Maackia amurensis* Rupr. et Maxim.

Management and utilization

Secondary oak forests in northeastern China are the result of the destruction of virgin pine and mixed broadleaved forests (Zhao Dachang 1980). A variety of oak forest communities now exist.

Table 3. Increase in growing stock volume in mixed oak stands five years after thinning (m^3/ha)

Plot	Control		Thinned	
Year	1959	1964	1959	1964
Total volume	26.64	68.69	26.17	76.12
Volume increment		42.05		49.95
Percent increase		157.85		190.87

Stand thinning

A large portion of the forests have dense canopies. The shade caused by the dense canopies restricts the growth of saplings into the canopy. Selective thinning of stands releases the suppressed trees from competition and accelerates their growth rates. Often during manual thinning, unhealthy or undesirable trees are selectively removed leaving more valuable trees. Table 3 shows the results of a thinning experiment in the Lesser Xingan Mountain area where 30% of the stems were selectively removed. After 5 years, an increased growth rate is evident in the thinned plot as compared to the control plot (Han Linfeng 1982). Extensive areas of forest in this region are selectively thinned. We suggest, however, that the amount of thinning in a stand be designed to both maintain the productivity of the forest community and improve the stand quality.

Natural rejuvenation

Heavy-use areas of the forests tend to be near areas heavily populated by people, typically upper slopes and ridges. These areas have lower productivity and recover very slowly. For stand rejuvenation to occur, it may be necessary that logging, fuel wood cutting, and grazing be prohibited for set lengths of time to allow reestablishment of the natural plant communities. Toward this end, we suggest that a rotation plan be adopted where the forest is divided into sections. Each section would get a turn at being off limits for a rejuvenation period of 7–10 years.

Clear-cutting and selective planting

The oak stands on the lower slopes of the mountains are considered of lesser economic value because they contain less desirable species. These stands are normally designated for clear-cutting. The maximum clear-cut plot size is 5 hectares. Once clear-cut, these stands are planted with fast growing species such as *Larix olgensis* Henry and *Pinus sylvestris* L. var. *mongolica* Litv.

This planting strategy has been shown to have negative effects on forest productivity. For example, Han Linfeng (1982) found that pure larch stands planted on clear-cut oak sites acidified the soil, depressing tree growth.

Planting conifers in oak understories

The stand structure of forests with open subcanopies can be improved by introducing conifer seedlings into the understory. Researchers at forest farms in Jilin and Heilongjing Provinces have experimented with planting Korean pine seedlings under oak canopies since the 1950's. These researchers have been successful at improving stand structure by establishing mixed conifer-broadleaved stands.

Silk worm culture in oak coppice stands

The tussah silk worm was brought to northeastern China approximately 300 years ago. The practice of culturing silk worms on oak trees is still a profitable and important industry in the four provinces in northeastern China. At present, approximately 400,000 hectares of coppice oak forest are used for silk worm production. Management of these stands consists of rotational coppicing, pruning, cleaning out undergrowth, and fertilizing to promote new leaf growth for the worms to feed on. Recently, there has been a decline in the quality of the forests due to coppicing too frequently and over-grazing by silk worms. Efforts have been made to improve the ecological state of the forests and increase production (Sun Wu 1984).

Perspectives on oak forests

The oak forests of northeastern China harbor many species of plants and animals. The forests have gained importance for wildlife protection, rare plant cultivation, and the collection of medicinal herbs. It is in these forests that we find the "three treasures of the northeast": ginseng, marten fur, and pilose antlers. These traditional local products are more important than wood production as a main-stay of the regional economy.

The oak forests have deteriorated due to heavy use. Many plant and animal species have diminished in abundance and soil erosion is a cause for great concern. Oak forests are commonly situated near the headwater streams of large rivers. These rivers supply reservoirs and hydrologic power in the heavily populated industrialized cities which they flow through. The presence of the oak forests is important for maintaining water quality for agriculture, industrial production, and personal use in the populated cities.

Proper management of oak forests depends not only on good policy and planning, but also on understanding the ecology of the oak forest community. The present plan for the preservation of oak forests emphasizes land protection and stand improvement. Restoration of mixed conifer-broadleaved stands from the oak forests is encouraged as an integral part of the plan. In the future, mountain oak forests will likely play a more important role in the region than they do now.

Conclusions

1. The oak forests in the mountains of northeastern China are important for the preservation of biologic resources and for the protection of the environment.
2. Clear-cutting oak forests and replacing them with quick-growing larch stands has been shown to have negative effects on the environment. From an ecologic point of view, clearcutting should be restricted.
3. Silk worm culture is a good example of how the oak forests can be economically productive and still preserve the ecology of the natural system. We suggest this as an alternative to wood production.
4. A plan for rational use and proper management of oak forests has been presented. We suggest that the preservation of oak resources still needs to be emphasized in forest management.

Acknowledgements

The authors wish to express sincere thanks to Mr. Lin Jihui and Mrs. Yin Zhongfu for their help in field investigation and manuscript preparation. Kathleen Baker-Brosh and Robert Peet provided editorial and linguistic assistance. The authors are greatly indebted to them.

References

Dachang, Z. 1980. The Altitudinal Distribution of Vegetation on Chagbai Mountain, *Research of Forest Ecosystem*, Volume 1. Changbai Mt. Research Station, Academia Sinica. 69 pp. (in Chinese).

Linfeng, H. 1982. Forestry of Northeast China. China Forestry Press, Beijing, p. 394 (in Chinese).

Planning Office of Forestry Ministry 1981. The Mountainous Forests in China. China Forestry Press, pp. 14–79 (in Chinese).

The Research Group of Deciduous Oaks 1988. A synoptic summary of the research on Chinese deciduous oaks. *Journal of Beijing Forestry University* 10(3): 75–83 (in Chinese).

Wu, S. 1984. Causes of ecological imbalance in Tussah-oak forests. Natural Resources 84(3): 9. Academic Press, Beijing (in Chinese).

15. Deciduous and deciduous-evergreen forests in Northeastern China[1]

CHOU YI-LIANG & LIU JIAN-WEI
Botanical Institute, Northeast Forestry University, Haerbin 150040,
The People's Republic of China

Key words: deciduous coniferous forest, evergreen-deciduous mixed forest, northeastern China

Introduction

Northeast China is located on the eastern edge of Eurasia, near the Japan sea. Many forest types occur there, influenced by the southeastern summer monsoons, which gradually diminish from east to west, and by the cold, dry winter monsoon winds from Mongolia-Siberia. Two types of forest show a north-south zonation, the deciduous conifer forests in the north and the deciduous-evergreen mixed forests in the southeast (Fig. 1).

1. Deciduous coniferous forests

The deciduous conifer (*Larix*) forests occur mainly in the northern Daxinganling Mountains, north of 40°29′ N, under the influence of the cold-temperate (boreal continental) climate. This region is located far from the sea and is separated from it by the extensive, more eastern mountains, the Wanda, Zhangguangchailing, Laoyeling, and Changbaishan Mts. Thus it shows typical continental characteristics, being severely cold in the long winter, but dry and with little snow.

The deciduous conifer forests in these areas, which are composed of the single dominant species *Larix gmelini*, are considered to be the zonal forests of the cold-temperate zone. There forests belong to the southern part of the

[1] Zonal terminology represents Chinese usage. Summergreen deciduous forest regions are called 'temperate' or 'cool-temperate' (not warm-temperate) in Japan, Korea, and in most N. American, European, and global terminology. *Larix* forests would be called 'boreal' (rather than cold-temperate).

E.O. Box et al. (eds.), Vegetation Science in Forestry, 307–315.

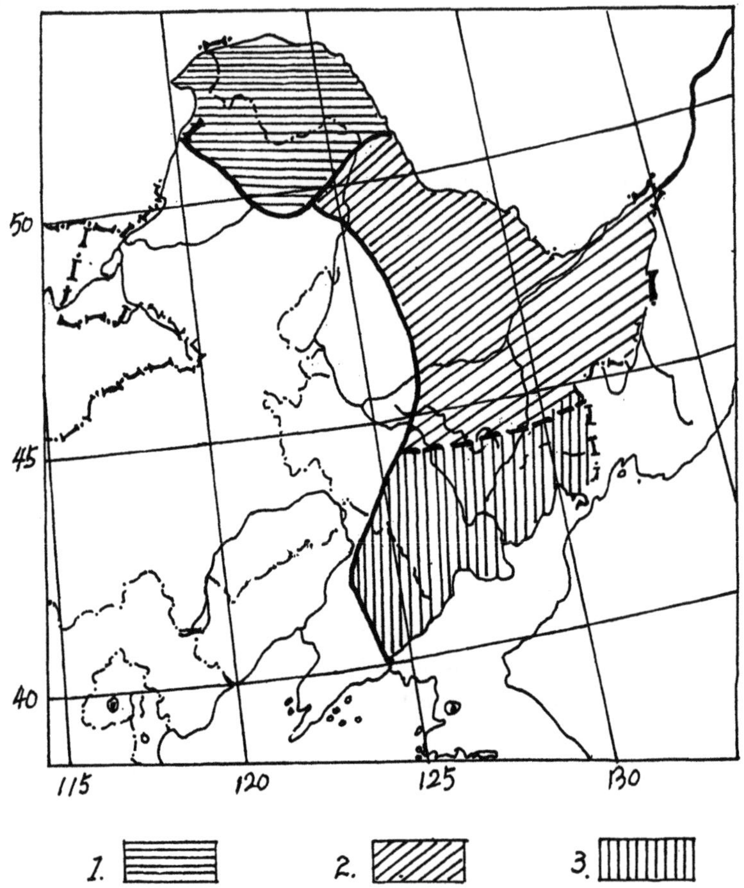

Fig. 1. Forest regions in Northeastern China. 1) Deciduous coniferous forests. 2) Temperate evergreen conifer-deciduous broad-leaved mixed forests. 3) Nearly warm-temperate evergreen conifer-deciduous broad-leaved mixed forests.

Siberian "light taiga," in the Eurasian pine-forest area, and occur on the south slopes of the Daxinganling Mountains, appearing there as montane forests at the edge of the temperate-zone grassland areas. Because of the climate, the deciduous forests are mono-dominant. Floristic analysis shows that East-Siberian elements constitute up to 50% of the total, along with elements from the adjacent temperate-zone coniferous/broad-leaved mixed forests (38%), with almost no endemics.

During the Tertiary Period, subtropical forests occurred in these deciduous conifer-forest areas as well as in the adjacent zone of temperate conifer-broadleaved mixed forests. Although these zones are adjacent and both experienced glaciation, the mixed forests contain many more Tertiary relict species. This is perhaps because of the more favorable climate, with greater maritime influence and with warm, moist monsoonal influence in summer. The deciduous conifer forests, in the drier, colder continental climate to the north, have a poorer flora and lack the relict species. These cold-temperate *Larix* forests conform to the longitudinal zonation pattern, having formed zonal cold-temperate deciduous coniferous forest with the rise of sea level after glaciation.

1.1. *Southern cold-temperate deciduous conifer forests*

This type of forest occurs in the lowlands at 450–600 m elevation, increasing from north to south. Because the elevation is mostly higher, however, this forest type is uncommon. The main soil is the dark-brown coniferous type.

Despite the strong influence of the mixed forests (evergreen conifers plus deciduous broad-leaved trees), the southern cold-temperate forests are still dominated by a single species, *Larix gmelini*, though often mixed with deciduous broad-leaved trees, mainly *Quercus mongolica* and *Betula davurica*, which are larger and drought-resistant. Other deciduous species, like *Populus davidiana*, *Tilia amurensis*, and *Fraxinus mandschurica*, are less abundant and produce mainly a secondary forest canopy. In the understory of the forest there are temperate-zone shrubs and grasses, such as *Lespedeza bicolor*, *Corylus heterophylla*, *Corylus mandschurica*, etc. Herbs include *Atractylodes japonica*, *Vicia pseudo-orobus*, *Pteranum agulium*, and even the typical moss *Clomacium dendrodes* of the temperate-zone mixed forests. In addition, there are a few vines, such as *Clematis serratifolia*, *Cl. sibirica* and less developed *Schisandra chinensis*. These plants enhance the temperate-zone characteristics of this cold-temperate deciduous coniferous forest.

1.2. *Typical (Siberian) cold-temperate deciduous conifer forests*

This type of forest occurs widely at elevations 450–820 m, with the width of the range decreasing from 600 to 1050 m from the north to the south (Fig. 2). The soils are of a typical brown coniferous-forest type, shallow but with good aeration.

In physiognomy and composition these forests are similar to the southern light taiga of Siberia. The main feature is the single-layered structure, with the grass and moss in the understory not well developed. The drought-resistent understory includes mainly *Rhododendron dauricum* but also *Ledum palustre*, *Vaccinium vitis-idaea*, *V. uliginosum*, etc. Secondary *Betula platyphylla* forest or *B. platyphylla-Larix gmelini* mixed forest follows disturbance. Over succession these forests develop into *Larix gmelini* secondary forests, which are the major type in the northern Daxinganling Mts. and are

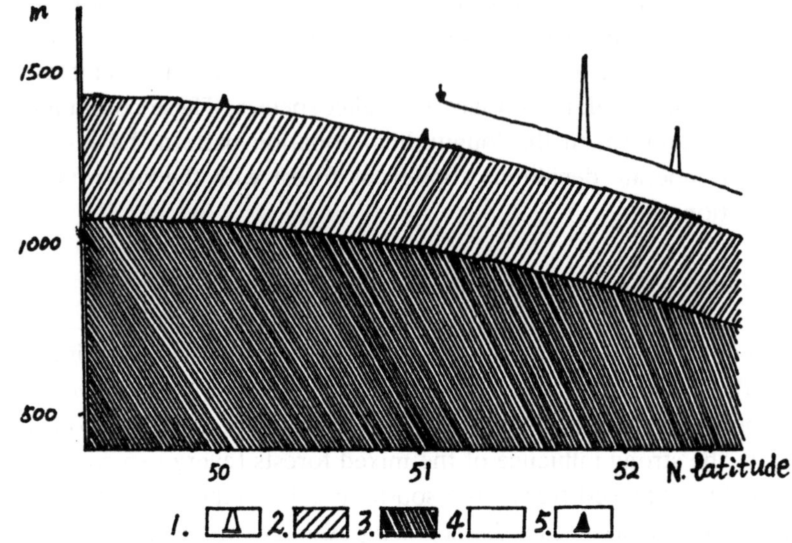

Fig. 2. Vertical distribution of *Larix gmelini* forests in the northern Daxinganling Mountains. 1) Profile of forests of Auokelitueishan Mt. and Baihalai Mt. in the subalpine belt. 2) Montane cold-temperate open coniferous forests. 3) Deciduous coniferous forests above mid-mountain. 4) Deciduous coniferous forests at mid-mountain. 5) Deciduous coniferous forests below mid-mountain.

very important for wood production. In the typical Siberian deciduous conifer forests, good natural recovery depends on the manner in which cutting and logging are practiced.

1.3. *Humid cold-temperate deciduous-conifer forests*

This type of forest occurs over narrower ranges at higher elevation, from 820–1100 m in the north to 1050–1380 m in the south (Fig. 2). It grows in the coldest, wettest areas of the Daxinganling Mountains. The soil is the podzolic brown coniferous type. The composition and physiognomy show the usual characteristics of the sciophyllic dark evergreen coniferous forests with *Picea asperata* and species of *Abies*. This forest type is similar to the central light taiga of higher latitudes in Siberia. Its main feature is that there is a single dominant species, *Larix gmelini*, in the tree layer, sometimes with a few scattered deciduous broad-leaved trees such as *Sorbus pohuashanensis* and *Betula ermanii*. *Trientalis europaea* is a typical herb of the understory. At the same time the understory is rich in mosses, with cover reaching 90% or more, mainly by *Hylocomnium splendens*, *Ptilium crista-castrensis* and *Girgensohnia ruthenica*. Common periphytes like *Alectoria jubata* and even

Usnea spp. appear on the stems and branches, as usually appear in the sciophilous coniferous forests.

This *Larix gmelini* forest usually grows poorly and natural regeneration is difficult, as is common at high elevation and on thin soil. These forests, however, are important in landscape management, as protection forests for water conservation and for holding the soil.

1.4. *Cold-temperate deciduous open conifer forests*

These low, open forests appear infrequently on the summits of a few peaks, at 1100–1350 m in the Baihalai Mts. and at 1300–1450 m in the Yingilishan Mts. in the south (Fig. 2). The soils are thin lithosols or even bare rock. The climate is very cold, with strong winds. Other trees cannot survive, but *Larix gmelini* grows slowly to form open forests which are similar to the northern, open light taiga of eastern Siberia at the same latitudes, which have similar physiognomy and composition. The cover is not more than 30–40% at the upper treeline of the northern Daxinganling Mountains. Due to the harsh environment there, the growth of *Larix gmelini* is limited to short trunks, called "little-old-trees" in the local areas. The tops are easily broken off, and flagged krummholz occurs in wind-exposed areas. Wind-resistant *Betula ermanii* invades, in shrub form, into the open forests. The more moisture, the more *Betula ermanii* occurs.

The other feature of the open forest is that shrub-form *Pinus pumila* can dominate the understory, usually in clumps growing up to 2–8 m. Sometimes it is mixed with *Betula middendorfii*, *Vaccinium uliginosum*, and/or *V. vitis-idaea*. The forest regenerates from the seeds of *Larix gmelini* and well developed *Pinus pumila*, but there are relatively few seedlings and young trees, so the regeneration of *Larix gmelini* is difficult. This open forest has no value for wood and is maintained as protection forest.

2. Deciduous-evergreen mixed forests

Deciduous-evergreen mixed forest, containing evergreen coniferous and deciduous broad-leaved trees, covers extensive mountain areas in northeastern China from 40°15′ to 50°20′ N latitude and from 120° to 135°30′ E longitude (Fig. 1). This includes the Xiaoxinganling, the Wanda-shan (Nadanhaling), the Zhangguangchailing, the Laoyeling, and the Changbaishan mountain ranges, with the highest peaks below 1900 m except for Changbaishan (2691 m), which is the highest peak in northeastern China.

The climate is a moist-temperate monsoon type, influenced by the Japan

Sea. The abundant precipitation and suitable temperature provide a better growth environment, so the forests and flora are rich and diverse.

The evergreen conifer-deciduous broadleaved mixed forests represent the distributional center of the Manchurian Forest Region, which includes the Amur valley, eastern coastal areas of Siberian Russia, and the northern part of the Korean peninsula. This type of forest is one of the most important for timber production in China. *Pinus koraiensis* is the dominant species in the mixed forests. Other evergreen coniferous trees include *Abies holophylla*, *Taxus cuspidata*, and *Thuja koraiensis*. The main deciduous broad-leaved trees are *Acer triflorum*, *Acer pseudo-sieboldianum*, *Acer tegmentosum*, *Acer mandschurica*, *Phellodendron amurense*, *Fraxinus mandschurica*, *Maackia amurensis*, *Juglans mandschurica*, *Populus ussuriensis* and *Populus koreana*, etc. In addition, vines like *Vitis amurensis* and *Schisandra chinensis*, and the endemic *Panax ginseng*, together with relict species, emphasize the Tertiary origin. This is related not only to the lesser influence of the glacial period and the moist subtropical climate over longer periods of history, but also to the modern influence by the warm, humid summer monsoon. These factors together permit the survival and development of the relict plant species.

With respect to altitudinal distribution, these forests do not occur above 700 m in the northern part of their range or above 1160 m in the southern part. They are replaced upward by cold-temperate evergreen conifer forests.

The deciduous-evergreen mixed forests extend across 10° of latitude and are thus extensive. They can be divided into two types, related to species dominance and environmental factors. South of 43°40′ N the forest is considered to be an evergreen conifer-deciduous broadleaved mixed forest (Fig. 1), occurring below 700 m. North of 43°40′ N the evergreen conifer-deciduous broadleaved mixed forest is considered to be temperate.

2.1. *Temperate evergreen-deciduous mixed forests*

In the mixed forests, *Pinus koraiensis* is the most important evergreen conifer and is found growing with about a dozen broad-leaved trees, such as *Quercus mongolica*, *Fraxinus mandschurica*, *Tilia amurensis*, *Tilia mandschurica*, *Ulmus propinqua*, *Ulmus laciniata*, *Phellodendron amurense*, *Betula costata*, *Acer mono*, *Populus koreana*, *Populus ussuriensis*, *Maackia amurensis*, *Juglans mandschurica*, etc. The understory shrubs are scattered, except in forest gaps where they are more common. There are both shade-tolerant shrubs, such as *Corylus mandschurica*, *Philadelphus schrenkii*, *Lonicera chrysantha*, *Eleutherococcus senticoccus*, *Deutzia amurensis*, and *Deutzia glabra*, and light-demanding shrubs such as *Euonymus pauciflora*, *Berberis amurensis*, *Viburnum sargentii*, and *Viburnum burjaetium*. *Syringa amurensis*, *Lonicera ruprechtiana*, and *Padus asiatica* occur on deep, suitably moist soil. The

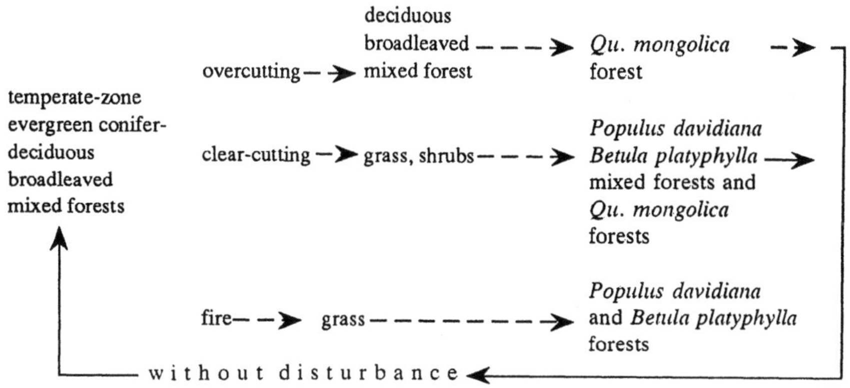

Fig. 3. Succession of temperate-zone mixed evergreen-conifer and broadleaved deciduous forests following disturbance.

main herbs are *Carex quadriflora*, *Carex ussuriensis*, *Carex callitrichos*, and *Athyrium spinulosus*; others include *Phryma leptostachya*, *Melica nutans*, *Diarrhena mandschuricum*, etc. *Plectranthus excisus*, *Lamium album*, and *Bupleurum longiradiatum* occur in forest gaps, and *Dryopteris crassirhizoma*, *Athyrium acrostichoides*, *Athyrium sinense*, and *Carex camylorhina*, etc. occur below mid-mountain. Some are relict Tertiary species, such as *Fraxinus mandschurica*, *Phellodendron amurense*, and *Juglans mandschurica*. Vines are also well-developed, such as *Vitis amurensis*, *Schisandra chinensis* and *Actinidia kolomicta*. All these are characteristic of subtropical landscapes (in the Chinese sense). On the other hand, some cold-temperate conifers of the Eurasian coniferous forests also appear, including *Picea jezoensis* var. *microsperma*, *Picea jezoensis* var. *komarovii*, *Picea koraiensis*, and *Abies nephrolepis*. Thus, the evergreen-deciduous mixed forests show characteristics transitional between more southern subtropical forests and northern cold-temperate forests.

When such forests are damaged, they are succeeded by various secondary deciduous broad-leaved forests and usually remain as deciduous broad-leaved forests, with pioneer trees like *Quercus mongolica*, *Betula platyphylla*, and *Populus davidiana* forming secondary forests, which are very widespread. The pattern of succession is shown in Fig. 3.

The normal process of succession suggests that deciduous broad-leaved forests dominated by *Populus davidiana* and *Betula platyphylla* generally recover (naturally) if managed or protected. Following cutting or fire, these forests are destroyed and their sites become subject to heavy soil erosion and drying. It is difficult for *Pinus koraiensis* forest to become reestablished, and such sites usually develop into deciduous *Quercus mongolica* forest.

2.2. *Nearly-warm-temperate evergreen-deciduous mixed forests*

Although *Pinus koraiensis* is the dominant species in this type of forest, temperate evergreen conifers are also abundant, such as *Abies holophylla* as well as *Taxus cuspidata* and *Thuja koraiensis* from the warm-temperate zone. These trees of the warmer zone are more abundant than in the northern temperate deciduous-evergreen forests. For example there are 12 species of *Acer*, many more than the 5 species found in the temperate evergreen-deciduous forests.

Moreover, there are many deciduous broad-leaved trees from typical warm-temperate deciduous broad-leaved forests, such as *Kalopanax septemlobus*, *Styrax obassia*, *Magnolia sieboldii*, *Bothrycaryum controversum*, as well as a dozen well-represented vine species, especially *Actinidia kolomikta*, *Trytospermum regelii* and *Aristolochia mandschuriensis*. All of the above emphasizes the southern subtropical character of these forests, but evergreen conifers from the cold-temperate zone are still also found in some areas.

Many endemic herbs, such as *Panax ginseng*, *Asarum heterotropoides*, *Astilboides tafularis* and *Gastrodia alata*, also occur. These have their distributional center in the evergreen-deciduous mixed forests of the nearby nearly-warm-temperate zone.

Most of these forest types have been cleared for farmland or have become secondary deciduous broad-leaved forests because of human activities. The secondary forests of *Quercus mongolica* and *Populus davidiana* are much the same as the secondary types of the temperate-zone evergreen-deciduous mixed forests, but the species composition is more diverse. In *Quercus* forests, for example, there are also *Quercus dentata* and *Qu. liaotungensis* as well as *Qu. mongolica*.

In this kind of forest, other typical endemic plants are *Carpinus erosa*, *Fraxinus rhynchophylla*, *Acer pseudo-sieboldianum*, *Acer micro-sieboldianum*, *Acer barbinerve*, *Acer tschonoskii* var. *rubripes*, *Acer mandschuricum* and *Acer triflorum*, as well as *Kalopanax pictum*, which forms the subcanopy.

References

Chou Yiliang & Li Jingwen 1964. (Main vegetation types and their distribution in the Eastern Mountains of Northeastern China.) Acta Phytoecologica et Geobotanica Sinica, vol. 2(2):190–204 (in Chinese).

Chou Yiliang & Zhao Guangyi 1964. (The natural secondary forest, its distribution and succession in Xiaoxinganling and Changbaishan forest areas.) Journal of Northeast Forestry University 1964(3) (in Chinese).

Chou Yiliang (chief editor) 1986. (Ligneous Flora of Heilongjiang) Heilongjiang. Science Press, Haerbin (in Chinese).
Chou Yiliang 1990. (Forests of China.) Science Press, Beijing (in Chinese).
Chou Yiliang (chief editor) 1990. (Vegetation of Dahinggan Ling in China.) Science Press, Beijing (in Chinese).
Wu Zhengyi (Chief editor) et al. 1980. Zhongguo Zhibei. (Vegetation of China). Science Press, Beijing. 1375 pp. (in Chinese).

16. Subtropical montane deciduous forests in Southern China

CHEN WEI-LIE
Institute of Botany, Academia Sinica, Beijing 100044, China

Abstract

Evergreen broad-leaved forests are the zonal vegetation in southern China, but deciduous broad-leaved forests also occur in this region, growing naturally at elevations of 1300–2200 m in the mountains. The main types are dominated by *Fagus lucida*, *F. engleriana* and *F. longipetiolata*. These forests are not ecological counterparts of the *Fagus sylvatica* forests of Europe:

1. They grow in wet, temperate conditions on montane yellow soils or brown earths.
2. The main components of the communities are subtropical floral elements of China, with many evergreen species, especially of the genera *Sinarundinaria* and *Rhododendron*.
3. The structure of the communities is complex; the trees form a dense cover, the shrub layer is important, and there are epiphytes and vines, for example *Hedera*, *Actinidia*, *Schizandra*, *Smilax*, etc.
4. Natural regeneration in this community is rather difficult; there are few tree seedlings under the forest canopy.

In addition, there are many other types of deciduous forest on low mountains or hills in the subtropical region. These involve many tree species, such as *Davidia involucrata*, *Betula luminifera*, *Liquidambar formosana*, *Pteroceltis tatarinowii*, *Quercus variabilis*, and *Alnus nepalensis*, etc.

Key words: *Fagus*, southern China, subtropical deciduous forest

Introduction

Evergreen broad-leaved forests dominated mainly by species of Fagaceae are the zonal vegetation in subtropical southern China. At altitudes above

E.O. Box et al. (eds.), Vegetation Science in Forestry, 317–323.

1300 m, however, there are deciduous broad-leaved forests, often containing evergreen broad-leaved species. The typical type is beech (*Fagus*) forest.

Normally, subtropical deciduous forests represent secondary vegetation developed in situations where evergreen broad-leaved forest formerly existed. After destruction of the evergreen forest, though, the environment is changed to some extent. Thus, deciduous forests can be regarded as successional, whereas the (climatic) climax vegetation on these sites is evergreen broad-leaved forest.

Chinese *Fagus* forests have been studied by Wang (1965), Zhou (1965), and Jian (1975) in Guizhou and Anhui provinces, and later, also in Guizhou, by Huang (1982, 1988) and Zhou (1983). *Fagus* forests were also reported on by Zhou (1980), in Hubei province, and were analyzed in more detail in "The Vegetation of China" (Wu & Comm. 1980).

General aspects

There are about 25 types of deciduous broad-leaved forest in the subtropical area of China (see Table 1). They appear in almost every part of southern China but are concentrated in the central subtropical and the western part of the northern subtropical regions (see map by Hou 1983).

From analysis of the distribution and composition of deciduous forests in these areas it can be concluded that:

1. The distributional ranges are wide. The dominant species are similar to those in temperate deciduous forests, but the forest compositions have a transitional, secondary character.
2. The forests are limited to elevations between 1300 and 1500 m in the mountains. In the phytogeographical sense, the dominant species are subtropical montane elements. The subtropical deciduous forests are a natural vertical vegetation belt in the mountains.
3. Certain special types of subtropical deciduous forest occur in lower mountains and hills of the region, generally in small, scattered areas. The dominant species are often Tertiary relics.

As for community structure, the canopy layer is composed almost completely of deciduous species. The first and second sub-layers are mainly composed of evergreen broad-leaved species.

Considering all types of subtropical deciduous forest, *Fagus* forest is of special interest. Many authors have included it in the category of semi-evergreen broad-leaved forest (mixed deciduous and evergreen broad-leaved trees). This conclusion was made on the basis of analysis of IVI values (Importance Value Index, see Wang 1963). There are, however, also pure *Fagus* forests, in which the tree layer consists mainly of deciduous *Fagus* and

Table 1. Subtropical montane deciduous forest types in southern China

Dominant Species	Range within south China	Altitudinal range (m)
Quercus aliena	North part	1000–1100
Qu. aliena var. *acuteserrata*	Qinlin	1000–1800
Qu. acutissima	Whole region	400–1800
Qu. baronii	Northwest	
Qu. dentata	Qinlin	800–1300
Qu. serrata var. *brevipetiolata*	Northwest	1000–1600
Qu. stewardii	Huangshan	1500
Betula albo-sinensis	Northwest	2000–2500
B. albo-sin. v. *septentrionalis*	Qinlin	2000–2800
B. luminifera	North-central	800–1000
B. platyphylla	Northwest	3200–3500
Populus davidiana	Whole region	300–1900
Alnus nepalensis	Whole region	500–3000
A. trabesculosa	East part	>500
Canthium dicoceum	South part	800
Pteroceltis tatarinowii	East part	100–200
Liquidambar formosana	Whole region	600
Juglans cathayensis	North-central	800
Castanea segunii	Central part	>1000
Robinia pseudoacacia (exotic)	Whole region	500
Fagus engleriana		1300
F. longipetiolata		1300–1500
F. lucida		1500–2200
Davidia involucrata		
Liriodendron chinense		
Teracantron sinense		
Cercidiphyllum japonicum		

the undergrowth is sparse, with only a more dense growth of *Sinarundinella*. The Importance Value Index certainly does not provide answers to all compositional questions. For example, a species within a subcanopy layer may be very abundant and uniformly distributed. Thus it may have a large IVI value but also relatively short stature and small diameter. Such trees actually have less significant effects in the community than do the larger trees. In other words, even if there are more evergreen trees with greater cover values in the subcanopy layers, the canopy layer will have the greatest influence on the character of the community if it is continuous and has a high cover value. Study of pollen history, age structure and successional relationships may provide detailed evidence for this problem.

Environmental characteristics of Fagus forest

Fagus requires temperate, moist environmental conditions. Within its distributional areas, the mean annual temperature is 8–12°C, average annual

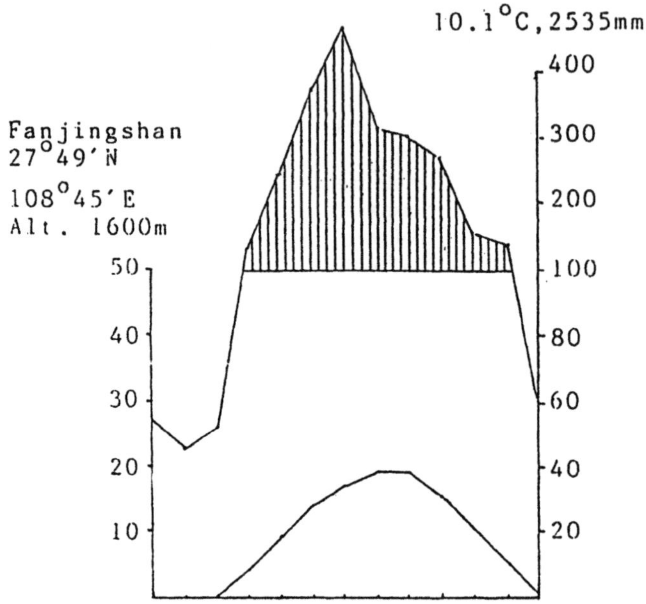

Fig. 1. The climatic diagram of the Fanjingshan.

Table 2. Montane yellow-brown earth soils of the *Fagus lucida* forests of the Kuankuoshui area (altitude 1600 m)

Soil layer (cm)	pH		Organic	C/N ratio	Total PO (%)	
	(water)	(KCl)	%		N	25
0–2						
2–4	4.7	3.9	44.4	18	1.45	0.22
4–10	5.4	4.5	8.9	16	0.32	0.1
10–30	5.4	4.6	6.3	18	0.21	0.08
30–60	5.7	4.8	3.4	9	0.21	0.06
60–100	5.8	5.0	2.4	16	0.09	0.08

precipitation 1200–2500 mm, and average relative humidity around 80%. This region is also subject to monsoonal influence from the south. Because of the altitude, it is frequently overcast and the forests receive less sunshine. A representative climate diagram, for Fanjing-shan (Guizhou province), is shown in Fig. 1.

The soils are montane yellow or yellow-brown earths, with pH in the range 4–5.8 (see Table 2). Soil depth is typically 70 cm, and the surface organic content is up to 30%. These soils are a very suitable substrate for the growth of *Fagus* forests.

Table 3. Structure of the *Fagus lucida* forest at Fanjing-shan (altitude 1520–1570 m)

Species	Height (m)	DBH (cm)	Cover (%)	Seasonal habit
Trees				
Fagus lucida	24	50	70	D
Cyclobalaopsis stewardiana	12	20	35	E
C. glauca	8	15	5	E
Neolitsea confertifolia	2.5	–	4	E
Quercus engleriana	10	12	5	E
Acer sinensis	8	–	5	D
Shrubs				
Sinarundinaria chungi	0.7	–	70	E
Clethra monostachya	1.8	–	5	D
Phoebe neuranthoides	3	–	7	E
Camellia cuspidata	0.5	–	7	E
Eurya brevistila	0.8	–	2	E
Symplocos lancifolia	2	–	4	E
S. anomala	–	–	2	E
Lindera fragrans	3	–	3	E
Rubus swinhoei	0.5	–	0.1	D
Herbs				
Plagiogyria sp.	0.3	–	0.2	
Disporum sessile	0.2	–	–	
Lycopodium japonicum	–	–	–	

Total coverage 60%.
D = deciduous, E = evergreen.

Compositional characteristics of Fagus forest

Fagus is an ancient genus, well represented in the Tertiary period. It has an extensive geological history. There are mainly three species of *Fagus* in the forests of southern China: *Fagus lucida*, *F. longipetiolata*, and *F. engleriana*. These occur in the mountains of Guangxi, Guizhou, Sichuan, Jiangxi and Zhejiang. There is, however, no extensive area of continuous *Fagus* forest. *Fagus lucida* has the widest distribution.

The structure and composition of *F. lucida* forest in the region of Fanjing-shan is shown in Table 3. *Fagus* is the dominant species, whether analyzed on the basis of height, breast-height diameter, or coverage. In addition, results from Kuankuoshui, in Guizhou province, also show *Fagus lucida* as the dominant species in the tree layer of deciduous forests there.

Subtropical elements comprise up to 85% of the *Fagus lucida* forests. *Sinarundinaria* is an obvious feature in the shrub layer. The composition is evidence that this forest is a natural forest type in montane belts of the subtropical zone. By comparison, *Fagus sylvatica* forests in Europe have been analyzed thoroughly by Ellenberg (1986). These forests are simple in vertical structure, with no evergreen tree species, and occur at very low altitudes. These forests are, thus, not ecological vicariants of the *Fagus* forests of the subtropical mountains in southern China.

Fagus lucida forms various communities, with different companion species but similar basic structure. *F. engleriana* forests in Shen Nongjia (Hubei province) have similar composition and structure and can thus also be included as deciduous broad-leaved forests. The *F. longipetiolata* communities also have similar structure but occur at lower elevations than does *F. lucida*. *Fagus longipetiolata* forests are the most widely distributed *Fagus* forests in the subtropical mountains of southern China, occurring from eastern China to Sichuan in the west. The second most widespread type is the *Fagus engleriana* forest, which occurs mainly in the west. *Fagus lucida* forest, in the middle of the region, has the most complicated composition.

Conclusion

1. The *Fagus lucida* communities should be considered deciduous broad-leaved forests of the subtropical mountains. *Fagus* forests are the typical deciduous broad-leaved forests in the subtropical mountains in southern China.
2. *Fagus* forests in China represent a special vegetation type of subtropical mountains (see also Peters, this volume). This type of forest is different ecologically from the *Fagus sylvatica* forests found in Europe.
3. *Fagus* forest is protected in China, but research should be continued in greater detail.

References

Ellenberg, H. 1986. Vegetation Mitteleuropas mit den Alpen. Verlag Eugen Ulmer, Stuttgart.

Han, Y.-L. et al. 1988. Anhui Zhibei (Vegetation of Anhui). Anhui Science Press. 321 pp. (in Chinese).

Huang, W.-L. 1982. Scientific Survey of Mt. Fanjing-shan. China Environmental Science Press. 277 pp. (in Chinese).

Huang, W.-L. 1988. Guizhou Zhibei (Vegetation of Guizhou). Guizhou People's Press. 502 pp. (in Chinese).

Jian Z.-P., Ying, J.-S. & Ma, C.-G. 1975. The significance of the distribution of *Fagus* forest on Mt. Fanjing-shan, Guizhou. Acta Phytotax. Sinica 1(13):5–18 (in Chinese).

Wang, X.-P., Weng, J.-J., Chen, W.-L. & Liu, Y.-A. 1965. Generalization of vegetation and its rational utilization at Kuankuoshui forest areas in Jingyang County of Guizhou. Acta Phytoecol. Geobotan. Sinica 2(3):264–268 (in Chinese).

Wu, Zh.-Yi & Commitee (eds.) 1980. Zhongguo Zhibei (Vegetation of China). Science Press, Beijing. 1375 pp. (in Chinese).
Zhou, J.-L. 1965. (Study of Mt. Wang-shan.) Science and Technology Press, Shanghai (in Chinese).
Zhou Z. 1980. (The Plants of Shen Nongjia.) Hubei People's Press. 467 pp. (in Chinese).
Zhou Z.-X. 1983. (Scientific Survey of Kuankuoshui Forest Areas.) Guizhou People's Press. 236 pp. (in Chinese).

17. Architecture and development of Mexican beech forest

ROB PETERS
Department of Forestry, Wageningen Agricultural University, P.O. Box 342, 6700 AH Wageningen, The Netherlands

Abstract

Found in the foggy mountain zone of the Mexican tropics, *Fagus mexicana* dominates some of the southernmost *Fagus* forests in the world. These forests are located on north slopes facing the Gulf of Mexico. Canopy beeches are deciduous during two weeks in January-February, but suppressed beeches retain green leaves till they flush again. Hence, suppressed beeches may benefit from a long growth season. The architecture and development of a stand near Zacualtipán (Hidalgo) was studied using three-dimensional recordings. The forest is a mosaic of patches in different phases of development. On the slope the forest is dominated by beech only, near the ridge evergreen *Quercus ocoteaefolia* co-dominates. Shrub and herb layers are little developed. There are high densities of beech in all size classes. Regeneration of beech and oak is successful, dominating all phases of forest development, i.e. the present dominance probably will be maintained. Growth rates of individual beeches under different light conditions were studied using increment cores. Individual beeches show alternating low and high growth rates, related to periods of suppression and release. Hence, they reach the forest canopy step-wise, frequently being in or near brief canopy gaps.

Key words: Cloud forest, Dynamics, Eco-unit, *Fagus mexicana*, *Quercus ocoteaefolia*, Regeneration, Stand architecture

Nomenclature: Standley (1920–6) for Mexico; Fernald (1950) for U.S.A.; Zheng (1983, 1985) for China.

E.O. Box et al. (eds.), Vegetation Science in Forestry, **325–343.**

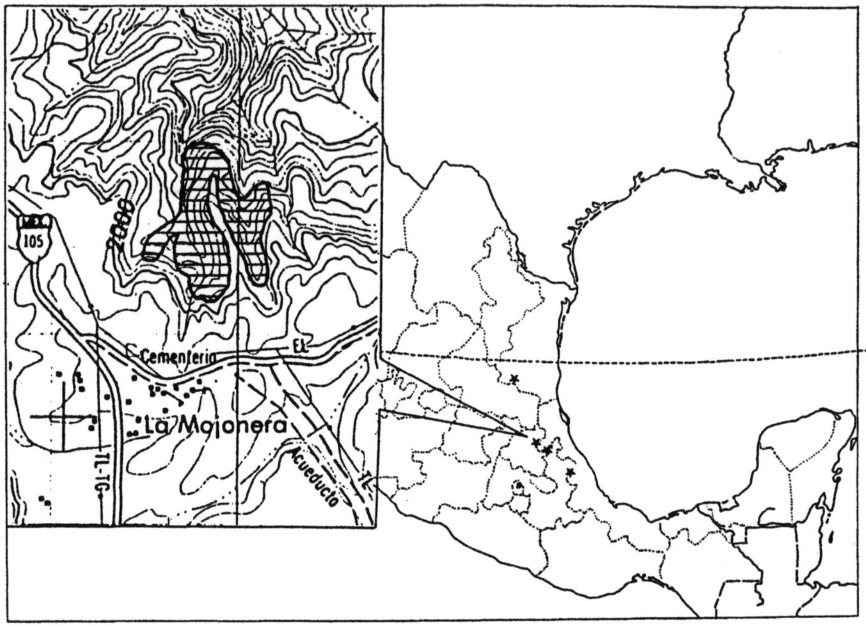

Fig. 1. Distribution of *Fagus mexicana* in Mexico, with the location of the beech forest near Zacualtipán, Hidalgo.

Introduction

In the montane belt of tropical Mexico a few isolated forest stands are dominated by beech trees. Martínez (1940) described the species as *Fagus mexicana*, while others treat it as *Fagus grandifolia* var. *mexicana* (Little 1965, Graham 1973). Mexican beech forests are known in four locations only (Fox and Sharp 1954) (Fig. 1), where they occupy small, unprotected areas. The beech forests are found on the Gulf of Mexico side of the Sierra Madre Oriental, on steep northeast-facing slopes, at an altitude between 1400 m and 2000 m. With their latitude of about 20° N, they represent the southernmost *Fagus* forests in the world. Only the *Fagus longipetiolata* forests in northern Vietnam are found at similar latitude (21°30′ N; Thái Van Trùng 1970).

Floristically the mexican beech forests are closely related to the beech forests in the eastern U.S.A. (Miranda & Sharp 1950, Graham 1973, Puig 1976, Rzedowski 1981). This similarity concerns mainly trees, e.g. *Liquidambar styraciflua*, but also herbs like *Epifagus virginiana*. The mixture of deciduous beech with evergreen broad-leaved tree species like *Quercus ocoteaefolia* and *Magnolia schiedeana* is similar to that of other low-latitude beech forests, e.g. in northern Florida (Blaisdell et al. 1973, Platt & Schwartz 1990) and

Table 1. Main woody species in three mixed forest of beech and evergreen broad-leaved trees. A: forest canopy layer; B: sub-canopy layer (Peters, unpublished data). Note that the Fagaceae are dominating these forests

Zacualtipán (Mexico) 20° N, 1900 m	Tallahassee (Florida) 29° N, 50 m	Miaoershan (south China) 26° N, 1600 m
A		
Fagus mexicana	*F. grandifolia*	*F. lucida*
Quercus ocoteaefolia	*Q. michauxii*	*Castanopsis lamontii*
	Q. nigra	*Lithocarpus hancei*
Magnolia schiedeana	*M. grandiflora*	*Maglietia chingii*
Liquidambar styraciflua	*L. styraciflua*	
B		
Cleyera theaeoides	*Ostrya virginiana*	*Cinnamomum burmanii*
Symplocos limonsillo	*Carpinus caroliniana*	*Rehderodendron kwangtungense*
Miconia anisotricha	*Cornus florida*	*Daphniphyllium glaucescens*
Ilex discolor	*Ilex opaca*	*Ilex ficoidea*
13 woody species	26 woody species	25 woody species

in southern China (Wang & Li 1986; Table 1). In all these beech forests the Fagaceae (deciduous *Fagus*, evergreen *Quercus*, *Castanopsis* and *Lithocarpus*) are dominating the forest canopy.

During the Tertiary, temperate elements like *Fagus* and *Liquidambar* migrated south from the eastern U.S.A. to Mexico. They first occurred in the vegetation of Veracruz during the Middle Miocene (Graham 1973). Although it is unclear when, the dry environments that were present in central North America during the Middle Eocene may have extended south (Tiffney 1985), and an arid zone developed in the coastal plain of Texas and Tamaulipas which prevented further exchange of temperate elements.

Since only a few small isolated stands of Mexican beech forest are left at present, it becomes urgent to analyze the dynamics of these stands. Are these stands remnants of a Tertiary forest zone that is slowly disappearing from the Sierra Madre Oriental? How does beech respond to local climate, e.g. by growth or leaffall? Is beech regenerating unsuccessfully, and are 'beech' stands slowly changing into an evergreen broad-leaved forest? These beech stands may be threatened by grazing and cutting, but if they need to be preserved, it is important to analyze whether beech can survive under natural conditions.

Since the work of Watt (1947) in beech forest it has been increasingly appreciated that beech forest communities are mosaics of patches in different phases of development (e.g. European beech forest: Lemée 1978, Koop 1989; Japanese beech forest: Nakashizuka & Numata 1982a, Hara 1983, Tanaka 1986, Peters & Ohkubo 1990). Natural disturbances create canopy

Photo 1. A Mexican beech forest eco-unit in the aggradation phase of development, near Zacualtipán, Hidalgo.

gaps and initiate the development of new forest patches. Oldeman (1983, 1990) defined the term forest eco-unit as the living system which is a forest patch that has started to grow on one surface at one moment. Four phases are recognized in the development of an eco-unit. During an innovation phase (reorganization phase, Bormann & Likens 1979) tree juveniles establish themselves until canopy closure is achieved. The eco-units are then aggrading and are dominated by potential trees (Photo 1). A potential tree has a potential for future development. Biostatic eco-units are dominated by trees of the present (Photo 2). A 'tree of the present' has reached its maximum size and co-determines forest architecture (Hallé et al. 1978). Finally, degrading eco-units are not dominated by vital trees and contain senescent ones (Oldeman 1990). Tall trees of the present form the forest canopy, with 'canopy gaps' between them. Within one canopy gap one or more aggradation and/or innovation phases are found. The forest eco-unit may be seen as the basic scale on which forest reproduction, growth and development take place, and as such it provides a correct scale for analyzing the processes of regeneration and development in Mexican beech forest.

In one of the four locations of beech, near Xalapa in Veracruz, the beech forest showed no regeneration due to (external) grazing pressure and was clearly giving way to pasture. On the other hand, a beech forest near Zacual-

Photo 2. A mid-slope Mexican beech forest eco-unit in the biostatic phase of development, near Zacualtipán, Hidalgo.

tipán, Hidalgo, seemed to regenerate well. Here a case study was carried out. Regeneration of different species and their presence in different phases of eco-unit development were studied. Would the dominance of beech in these forests be maintained under the present conditions? The effect of latitude on forest composition, regeneration and growth was studied, and this forest was compared with other beech forests at different latitudes.

Study site and method

The study site is located near Zacualtipán in Hidalgo (20°38′ N, 98°43′ W), at an altitude between 1800 m and 2000 m, on two sides of a ridge facing north, its slopes ranging from 10° to 40° (Fig. 1, Photo 2). Beech is found only in a limited area of 45 ha, where it dominates the slopes and co-dominates the ridges. Though there is a rather clear border to this area, no beech being found outside, this border cannot be explained by obvious environmental factors, i.e. aspect or moisture. Human use of surrounding, sometimes more accessible forests, may have been an important cause in limiting the beech.

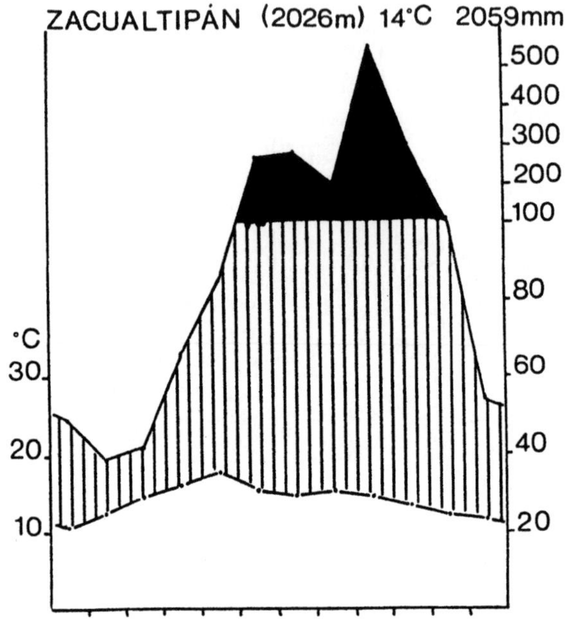

Fig. 2. Climatic diagram for Zacualtipán (after Puig 1976).

The climatic diagram for Zacualtipán (Fig. 2) explains why beech and other temperate elements can live here. The climate is very different from that of the nemoral (typical humid-temperate) zone (Walter 1984), where leaf shedding is an adaptation to the cold winter. During the summer months the average temperatures are similar to those in cool-temperate zones, but the daily variation is large (Walter 1984). Whereas in the temperate zone the winter is a cold season with physiological drought, here the winter is a dry season with relatively high temperatures, frequent fog and only occasional frost or snow. During the dry season, the abundant fog keeps the humidity level above a critical minimum for many species (Miranda & Sharp 1950).

The study site was visited in October 1989 and in January 1990. Within the area, the presence of woody species was studied using three transects, each running from ridge to valley bottom (total length 370 m, width 10 m). In ten gaps, nine caused by single treefall and one multiple treefall, the establishment of juvenile beeches, as compared to other species, was analyzed.

Individual trees and eco-units were studied in three plots. The plots were three-dimensional representations of a part of the forest. One was selected in a mixed stand 10 m from a ridge and two were selected in pure stands on slopes, one with and the other without a large canopy gap (diameter 25 m). The plots, ca. 50 m long and ca. 30 m wide, were established in the direction of the slope. A crown projection map and a profile diagram of trees and shrubs taller than 0.5 m were drawn to scale (*cf.* Oosterhuis et al. 1982). Heights, diameters at 0.5 m height, crown sizes and internal densities of the crowns of individual trees were recorded. Increment cores (at 0.5 m height) were taken from six trees of the present and 17 potential trees to assess radial growth. The potential trees were growing under various light conditions. Five light classes were distinguished (CC = percentage of a crown that is covered from above in a vertical projection): 1. CC = 0%, crown emergent; 2. CC = 0%, crown non-emergent; 3A. CC < 50%; 3B. CC > 50%; 4. CC = 100%, receiving side light; 5. CC = 100%, no side light (Synnott 1979). A potential tree grows in light classes 2 or 3A, and is suppressed in light classes 4 or 5. The relation between light class and radial growth rate was analyzed.

In potential trees the vertical expansion (height growth) is important and shows a more or less constant ratio to the horizontal expansion, i.e. diameter and crown growth. In trees of the present the vertical expansion has become much less important and emphasis is on the horizontal expansion. This is visualized in height-diameter and height-crown projection diagrams. Biostatic eco-units may fuse and become one architectural aggregate (Oldeman 1990). Hence, they are analyzed as if they were one eco-unit. In the upper height of aggrading eco-units three classes are distinguished and named 'early', 'medium' and 'late', with height ranges of less than 10 m, 10 m to 20 m, and over 20 m respectively. Areas dominated by shrubs belong to a different class of eco-unit. Areas without trees or shrubs taller than 0.5 m are in an innovation phase. In the plots each point belongs to only one phase of eco-unit development. The area covered by each of the different phases of eco-unit development was measured on the crown projection maps.

Results

On the slope the forest is dominated by beech only, while near the ridge evergreen *Quercus ocoteaefolia* co-dominates (Table 2). The vertical projection of single-tree canopy gaps ranged in size from 12 m^2 to 206 m^2 (average 91 m^2). In two gaps within 50 m from the ridge, the gapmakers were beech and three oaks were the dominating potential trees. In the seven mid-slope gaps the gapmakers were six beeches and one oak, and beeches were the dominant potential trees in the gaps. The multiple-tree gap (316 m^2) was caused by the fall of four beech trees, and the dominant potential trees were eight beeches and one oak. Other tall tree species were unimportant in the gaps. In and near all gaps except one, beech juveniles were present.

Table 2. Woody species composition on a ridge (135 m × 10 m) and on a slope (235 m × 10 m), based on 10 m-wide transects. Indicated are the numbers of individuals in the canopy and in the undergrowth (in parentheses)

Tree height	Ridge 15–25 m	Slope 25–40 m
TALL TREES:		
Fagus mexicana Martínez	12 (4)	41 (63)
Quercus ocoteaefolia	25 (36)	8 (23)
Clethra macrophylla Martens & Galeotti	4 (3)	(2)
Pinus patula	6	
Magnolia schiedeana Schlechter	(2)	2 (18)
Turpinia pinnata Schlechter		(5)
Liquidambar styraciflua	1	
Castanea sp.		(1)
SMALL TREES & SHRUBS:		
Cleyera theaeoides (Swartz) Choisy	(10)	(22)
Symplocos limnocillo Humb. & Boupl.	(1)	
Litsea glaucescens H.B. & K.	(1)	
Ilex discolor Hemsl.	(4)	(2)
Miconia anisotricha (Schlechter) Triana		(16)
Carpinus carolinensis		(1)
Sambucus sp.		(1)

Individual beech trees

While height growth slowed down, diameter and horizontal crown increments were maintained when the trees had reached a height of 25 m (near the ridge) or 30 m (on the slope) (Figs. 3 and 4). Other tall tree species near the ridge showed the same pattern. Hence, trees of the present were decided to be taller than 25 m near the ridge and taller than 30 m on the slope. Usually the trees of the present have a diameter larger than 40 cm (Fig. 3). The diameter and crown projection of beeches and other tall trees near the ridge remain smaller than those of beeches on the slopes (Figs. 3, 4 and 5).

The average radial growth rates for trees growing in different light classes showed a clear decrease from trees of the present, to free-growing potential trees, to suppressed potential trees (Table 3). On this basis a growth rate of 0.8 mm/yr was chosen as a threshold indicating release. Periods of suppression were defined as intervals in which there were four or more years of growth below 0.8 m m/yr, during which there were no periods of three or more years of consecutive growth greater than 0.8 mm/yr (*cf.* Canham 1990). For the 17 potential trees, the length of the periods of suppression or free growth was generally (80%) less than 20 years (Figs. 6 and 7). This indicates that improved light conditions, caused by canopy gaps, although lasting a short time, recur regularly. In order to find an indication of the time span that beeches could spend in the biostatic phase, the time that passed since

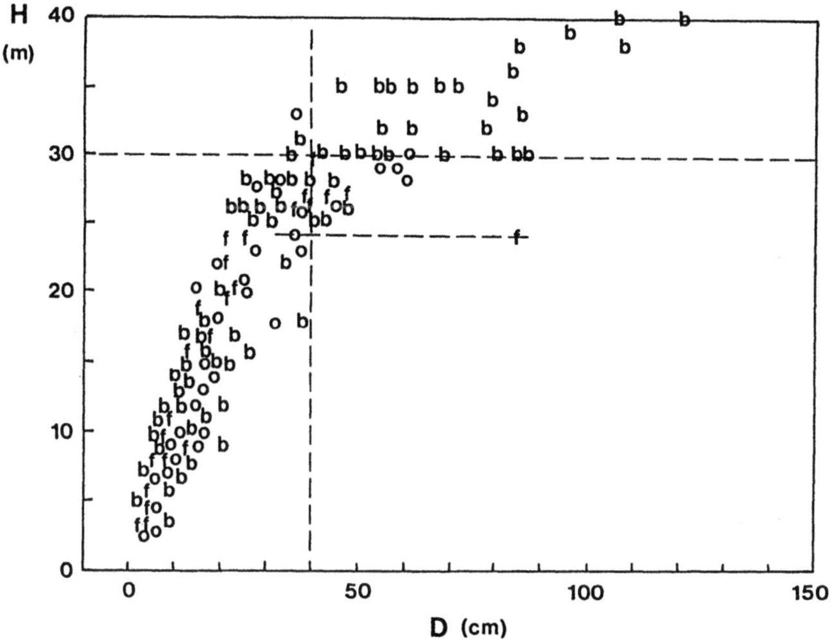

Fig. 3. Height-diameter (H/D) relationship of *Fagus mexicana* (b = on slope; f = near ridge), and other tall tree species (o).

the diameter reached 40 cm (*cf.* Fig. 3) was calculated for five beeches. This was maximally 130 years.

As local people suggested Mexican beech to be an evergreen tree, the forest was revisited during the last week of January. At that time, the forest floor was covered by fresh leaves, the canopy beech trees displayed colored leaves and they had dropped approximately 2/3 of their leaves, as judged by their estimated crown density (Table 4). The majority of the smaller beeches were still green. Bud scales were found on the forest floor, and swelling buds were seen in the tops of the canopy trees. Several saplings had already flushed while still carrying green leaves from the previous year. It is suggested that the deciduous period for the canopy beeches lasts 2 to 4 weeks, whereas the beeches in the undergrowth may be essentially evergreen.

Forest stand

Both forest types, the mixed forest on the ridge and the pure beech forest on the slope, have a similar distribution of eco-units in different phases of development (Fig. 8, Table 5). With increasing height, i.e. from early to late, aggradation phases cover larger areas. The aggradation phases are

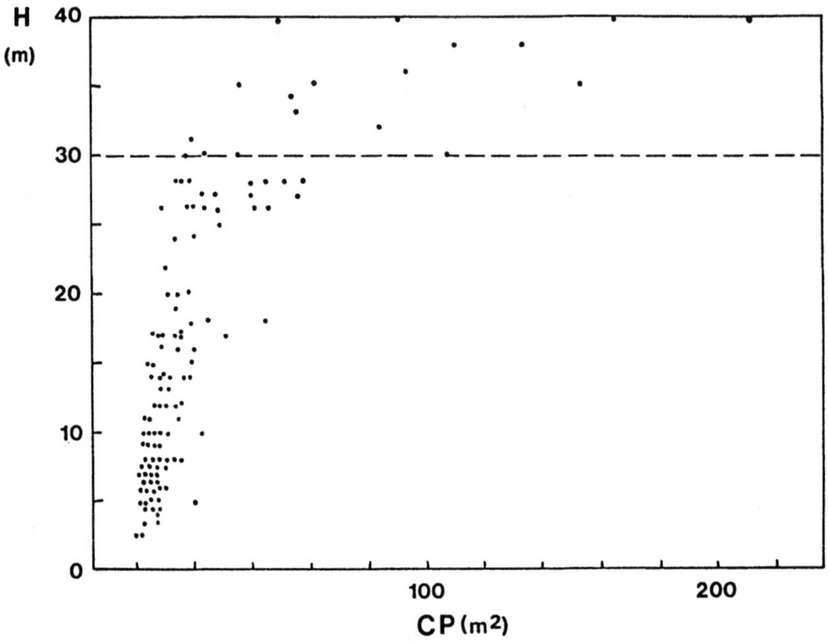

Fig. 4. Relationship of tree height to crown projection (H/CP) in *Fagus mexicana*.

dominated by beech alone on the slope and co-dominated by oak near the ridges. This indicates that the present dominance may be maintained into the future. It should be noted that there were abundant potential beeches not only in the aggradation phase but also in the biostatic phase. On the ridge, potential beeches were evenly distributed over all height classes, but on the slopes there was a higher density in the lower height classes (Table 6). On the ridges, oaks also were evenly distributed in all height classes, but other tree species were relatively unimportant (Table 6). Beech juveniles smaller than 1 m were found in age classes from 1 to 14 years, with a clear peak at 6 years (43%, n = 42). This indicates a yearly establishment of beech juveniles, with changing rates of success. Many juveniles were damaged by deer browsing. Both the ridge and slope forest showed deficient shrub and herb layers (Table 7). Some shrub-type eco-units, small in size and height (ca. 10 m), were distinguished, but cover only 2% of the area (Table 5).

Discussion

The architectural analysis of the plots shows that the forest is a mosaic of eco-units in different phases of development. The biostatic phases increase

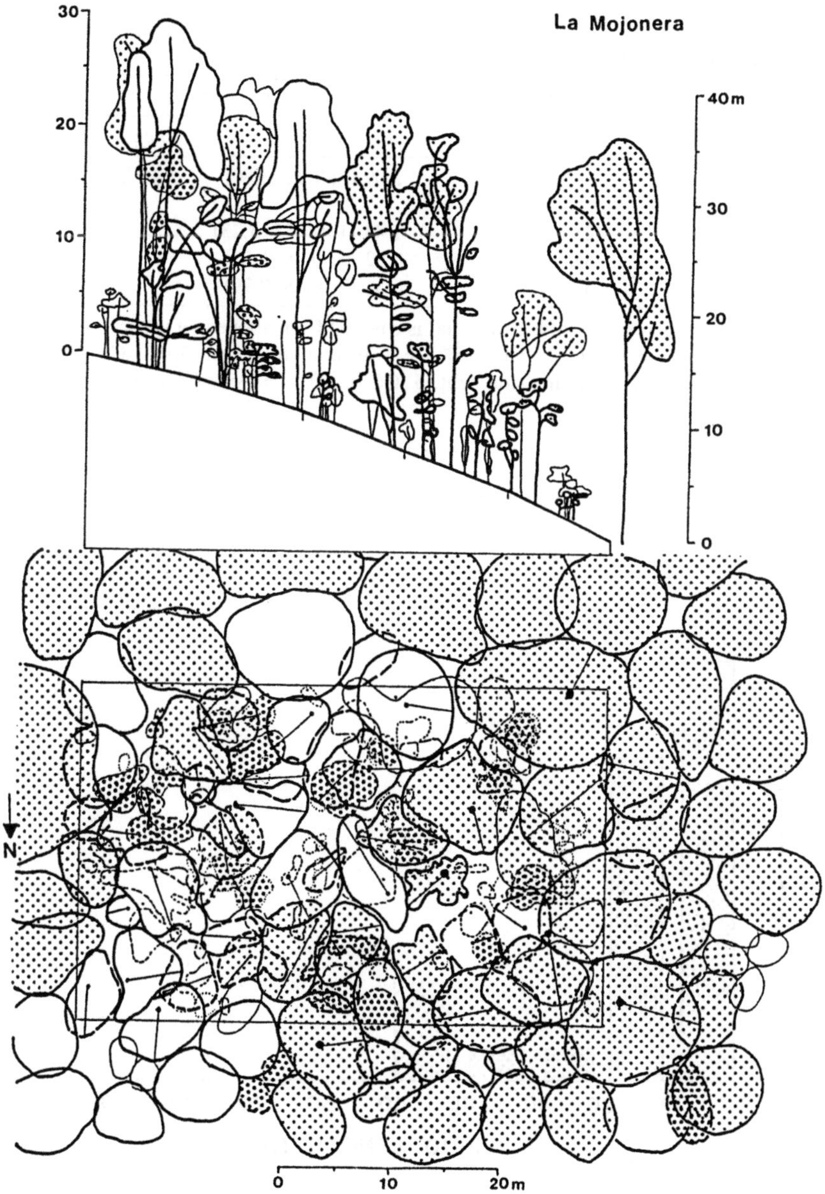

Fig. 5. Plot 2, near a ridge. Above: Profile diagram in 10 m wide belt for tall trees (≥5 m) and in 2 m wide belt for small trees and shrubs (5 m). Below: Crown projection map, showing *Fagus mexicana* trees (shaded).

Table 3. Radial growth rates of *Fagus mexicana* trees. Indicated in parenthesis are number of individuals and range of values

Tree status	Light class*	Average growth (mm/yr)	
		last 5 yrs	since d = 40 cm
Tree of the present	2, 3A	2.09 (6; 1.12–3.80)	2.46 (5; 1.76–4.03)
Potential tree			
free	2, 3A	1.44 (10; 0.55–3.23)	
intermediate	3B	0.92 (5; 0.48–1.57)	
suppressed	5	0.54 (2; 0.28–0.79)	

* Light classes (CC = percentage of a crown that is covered from above in a vertical projection): 1. CC = 0%, crown emergent; 2. CC = 0%, crown non-emergent; 3A. CC < 50%; 3B. CC > 50%; 4. CC = 100%, receiving side light; 5. CC = 100%, no side light (Synnott 1979).

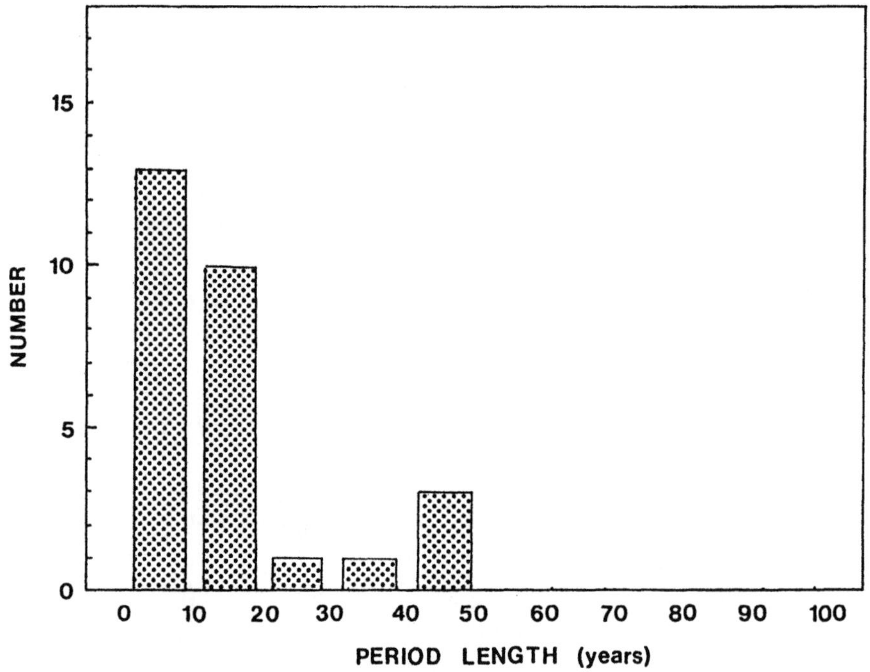

Fig. 6. Duration of periods of free growth of beech individuals.

their area by crown expansion and ingrowth of potential trees in the canopy. Innovation phases occupy just a small area if defined by vertical crown projection of neighboring trees. However, the establishment of juvenile trees also takes place outside the vertical gaps. Potential beeches (H > 5 m) were found on all sides of vertical gaps. Beech juveniles (H < 1 m) were found to survive about 14 years under biostatic phases (*cf.* about 10 years for *Fagus*

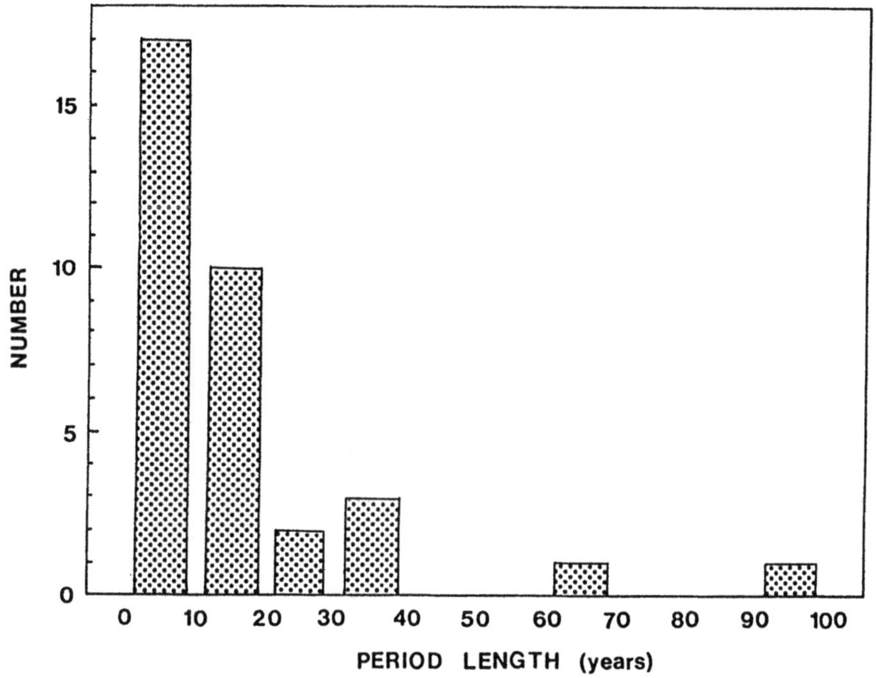

Fig. 7. Duration of periods of suppressed growth of beech individuals.

Table 4. Leaffall among beech trees as expressed by percent of average crown density

	Oct '89	Jan '90
Trees of present	73	27
Potential trees – free	74	30
suppressed	42	37

crenata seedlings in Japanese beech forest, Nakashizuka 1982b). This may be sufficient to survive until a period of increased light (caused by a nearby gap) occurs and to maintain a seedling bank. A periodic very successful establishment of beech seedlings, which occurred here six years ago, may have been initiated either by some local disturbance or by a moist year, as was observed in beech-sugar maple (*Fagus grandifolia-Acer saccharum*) forest in southern Michigan (Poulson, pers. comm.). In the study site, juvenile beeches are not hindered by strong competition for light by herbs or shrubs, because the latter have a low cover density (Table 7). In conclusion, beech regenerates well under the given environmental conditions of Zacualtipán.

The crown projection map shows that many small gaps, i.e. all the area

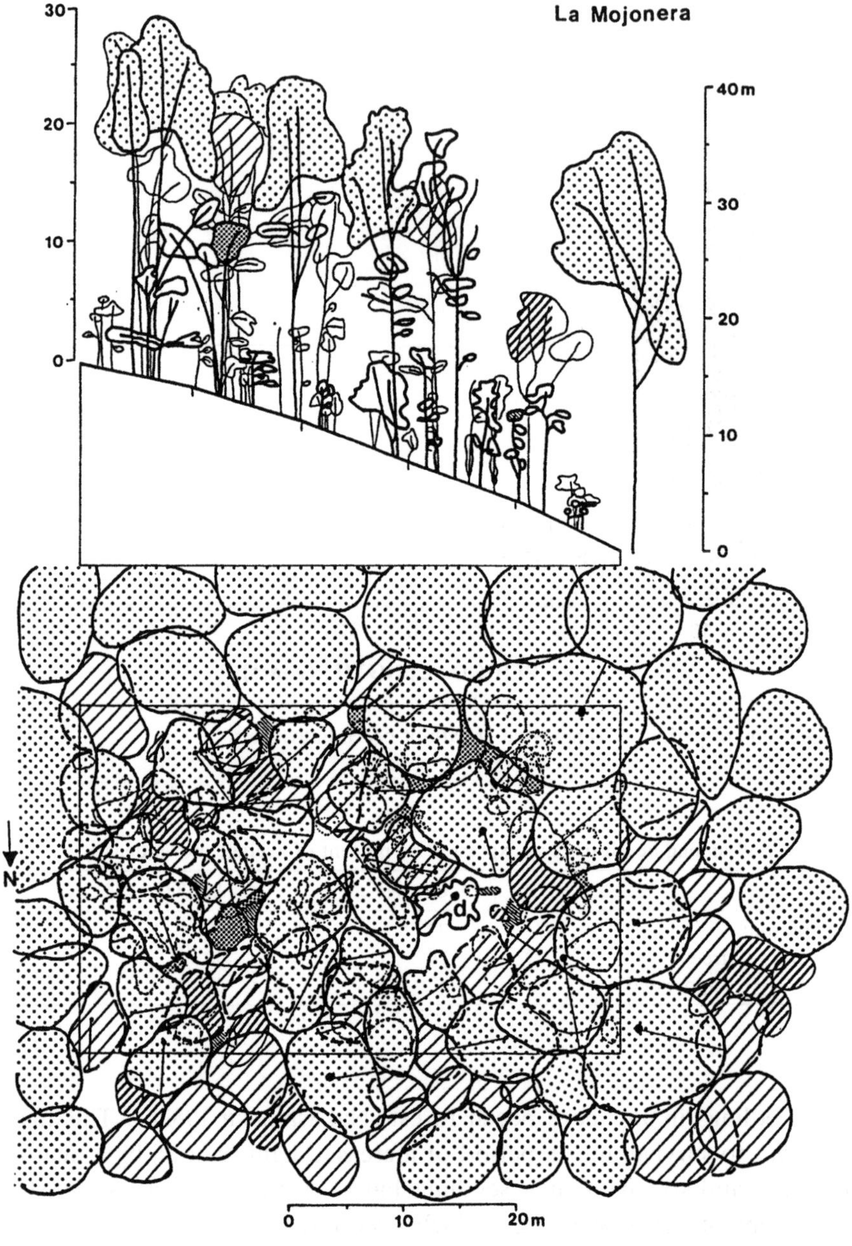

Fig. 8. Plot 2, the different phases of eco-unit development: innovation (not colored), early aggradation (dark diagonal shading), medium aggr. (medium shading), late aggr. (light shading), biostatic (sparsely dotted) and degradation (d), and a shrub phase (densely dotted). Areas are measured within the rectangular plot (*cf.* Table 5).

Table 5. Area coverage (%) of different phases of eco-unit development in a mixed stand and a pure beech stand (*cf.* Fig. 8). FM = *Fagus mexicana*, QO = *Quercus ocoteaqefolia*

Pure (no. 1) (0.15 ha)				Mixed (no. 2) average (0.15 ha)				Average
Phases	FM	Others	Total	FM	QO	Others	Total	
IN	–	–	4	–	–	–	5	4
SS	–	–	1	–	–	–	3	2
EA	2	0	2	1	0	0	1	1
MA	5	1	5	4	1	2	6	6
LA	16	0	16	8	5	0	14	15
BI	71	1	72	47	16	5	69	71
DG	0	0	0	2	0	0	2	1

IN = innovation; SS = shrub & small tree; EA = early aggradation; MA = medium aggr.; LA = late aggr.; BI = biostasis; Dg = degradation.

Table 6. Presence of beech trees (B) and other tall tree species (O) in different height layers (in number of stems/ha). Asterisk: plot area (ha) on which these data are based

Plot height (m)	1 (pure) (0.27)*		2 (mixed) (0.15)*			Japan[1] (0.21)*		China[2] (0.20)*		Florida[3] (0.42)*		Belgium[4] (0.50)*	
	B	O	B	Q[6]	O	B	O	B	O	B	O	B	O
0.5–5[5]	545	337	67	20	0	71	214	5	20	57	1752	4780	0
5–10	278	44	47	27	7	38	14	5	30	50	31	6	0
10–15	115	15	40	20	0	71	0	5	60	12	17	8	0
15–20	52	0	40	13	7	81	0	5	25	14	24	6	0
20–25	19	0	47	40	20	219	0	30	0	17	55	4	4
25–30	85	0	53	27	20	14	0	40	0	19	81	24	0
30–35	30	7	40	20	0					14	21	68	8
35–40	37	0	13	0	0							24	0
Total[7]	616	66	280	167	54	423	14	90	135	126	219	140	12

[1] *Fagus crenata*: Peters & Ohkubo (unpubl.), Kayanodaira, Japan. (*cf.* Nakashizuka & Numata 1982a).
[2] *Fagus lucida*: Peters & Cao (unpubl.), Fanjingshan, Guizhou, P.R. China. Relatively many small tree species dominated the undergrowth (here excluded).
[3] *Fagus grandifolia*: Peters (unpubl.), Tall Timbers, Tallahassee, Florida, U.S.A.
[4] *Fagus sylvatica*: Peters (unpubl.), Masblette, Ardennes, Belgium. Typical for European beech forest with few potential beeches under biostatic phases, but many in aggradation phases (*cf.* Ellenberg 1986).
[5] Below 5 m counted for a smaller area (5% to 10%) and extrapolated.
[6] *Quercus ocoteaefolia.*
[7] Excluding 0.5–5 m (to eliminate the extrapolation error).

not covered by biostatic eco-units, are found in the forest canopy (Fig. 8). From the radial growth analysis it became clear that improved light conditions are briefly but regularly recurring events. This is a factor favoring the survival of potential trees.

At low latitude the deciduous period seems to be shortened. While this period lasts two weeks at 20° N, it becomes two months in beech-magnolia

Table 7. Ground cover per phase of development in the three belt transects (in%)

		Shrubs (<5 m)	Herbs
Pure (no. 1):	Biost.	8	
	Aggr.	6.4	
	Innov.	0	
Mixed (no. 2):	Biost.	6.3	4
	Aggr.	13	5.3
	Innov.	8	11
Pure (no. 3):	Biost.	13	1
	Aggr.	24	3
	Innov.	1	23

forests at 29° N in Florida (Platt, pers.comm.) and six months in beech-sugar maple forest at 41° N in Michigan (Poulson, pers. comm.). For Mexican beech the longer growing season permits a higher total net production, sufficient to sustain growth and maturation of the woody axes and formation of reserves for leaf formation the next spring (Walter 1984). New leaves can even be formed while being supplied by active leaves of the previous year. This long growing season may also be a factor favoring a high survival probability for suppressed potential trees.

In the temperate zone, temperature and day-length are the main factors determining the deciduous period (Kramer & Kozlowski 1979), but in the tropical zone the period is related to drought (Frankie et al. 1974, Carabias-Lillo & Guevara-Sada 1985). In Mexican beech forest, winter temperatures are high as compared to the temperate zone (Fig. 2). The winter is a dry season, but fog maintains the humidity at a level permitting photosynthesis. The dry season affects canopy trees only. Beeches in the undergrowth drop their leaves later, probably after the unfolding of new leaves. This time lag in leaffall between canopy and undergrowth trees is also observed elsewhere, e.g. in European beech (Peters, pers. observ.) and in wet tropical forest (Frankie et al. 1974). It is suggested that potential beech trees, still possessing green leaves, profit from extra photosynthesis during warm January days (Fig. 2). In Europe, measurements on *Fagus sylvatica* indicated that high temperatures in autumn coincide with increased assimilation rates (Mitscherlich 1975), and shade leaves in the undergrowth were photosynthetically active two weeks longer than sun leaves in the forest canopy (Schulze 1970).

The dynamic change of light conditions under the forest canopy, and the latitudinal effects of the long growing season and evergreen behavior of potential beech trees all seem to favor the survival of many beech trees of the present and potential beech trees next to and under each other. Especially in the pure beech forest of the mid-slope plots, the number of beeches taller than 5 m per ha is much higher than in typical beech forests elsewhere (Table

6). Whereas in the mid-slope plots there are many potential beech trees between 5 m and 15 m tall, in other beech forests potential beech trees are relatively few and there are more competing potential trees of other tall tree species (Table 6; Wang et al. 1965, Nakashizuka & Numata 1982a).

The successful regeneration of beech seedlings provides the forest with a constant input of young beech trees. The large number of potential beech trees makes for a constant ingrowth of beech in the forest canopy. On the slope it may therefore be expected that beech will maintain its dominance.

On the drier and nutrient-poorer ridges, beech seems to be competitively weaker and oak co-dominates. This relation between beech and oak is very similar to that between *Fagus sylvatica* and *Quercus petraea/robur* in Europe, where beech dominates the mesic, low-nutrient Fago-Quercetum forest. On the more nutrient-poor sites oak becomes co-dominant, and when the site becomes very dry, beech disappears altogether (Noirfalise 1984, Ellenberg 1986).

Although this Mexican beech forest is floristically similar to some mixed evergreen-deciduous broad-leaved forests of southern China (Table 1), the forest structure and dynamics are quite different. Evergreen broad-leaved trees and shrubs, and dwarf bamboo strongly dominate sub-canopy layers in these Chinese beech forests. Whereas Mexican beech regenerates well, Chinese beech regenerates poorly compared to the codominant evergreen broad-leaved tree species (Chen Wei-Lie, Chapter 16; personal observation). Both the Mexican beech forest and the beech-magnolia forest in Florida, may serve as examples of codominance of beech and evergreen broad-leaved trees, and these forests may show the requirements for regeneration of the Chinese beeches, i.e. absence of dwarf bamboo and lower densities of evergreen broadleaved shrubs in the undergrowth.

Conclusion

This Mexican beech forest is a mosaic of eco-units which form a dynamic pattern of long-lasting biostatic eco-units that increase (or decrease) in size, next to aggrading eco-units that usually appear and disappear briefly but regularly. Beech regenerates well and will remain dominant on the slopes, whereas oak will remain co-dominant on the ridges. The effect of latitude is apparent from the high density of beech trees taller than 5 m on a small area. Even though this Mexican beech forest is limited to such a small isolated area, it can maintain itself and, without outside interference, may well survive far into the future.

Acknowledgements

The author would like to thank Dr. S. Guevara Sada and P. Zamora C. for their help and advice. The Instituto de Ecologia in Xalapa, Veracruz, provided a base of operations during my stay in Mexico. I also thank Prof. R. A. A. Oldeman, Dr. F. J. J. M. Bongers and P. van der Meer for their valuable advice and comments, and E. A. P. De Bruijn for his help in preparing the figures. The research was made possible through a grant from the Netherlands Organization for Scientific Research (NWO) (L 84–316.89).

References

Blaisdell, R. S., Wooten, J. & Godfrey, R. K. 1973. The role of magnolia and beech in forest processes in the Tallahassee, Florida, Thomasville, Georgia area. Proceedings, Annual Tall Timbers Fire Ecology Conference, pp. 363–397.

Bormann, F. H. & Likens, G. E. 1979. Pattern and process in a forested ecosystem. Springer, Heidelberg. 253 pp.

Canham, C. D. 1990. Suppression and release during canopy recruitment in *Fagus grandifolia*. Bull. Torrey Bot. Club 117(1):1–7.

Carabias-Lillo, J. & Guevara-Sada, S. 1985. Fenología de una selva tropical húmeda. In: Gomez-Pompa, A. & Del Amo-R., S. (eds.) Investigaciones sobre la regeneración de selvas altas en Veracruz, México, pp. 27–66. Editorial Alhambra Mexicana, S.A. de C.V. p. 421.

Ellenberg, H. 1986. Vegetation Mitteleuropas mit den Alpen, in oekologischer Sicht, 4 ed. Stuttgart-Ulmer Verlag. 989 pp.

Fernald, M. L. 1950. Gray's Manual of Botany, Eight Edition. American Book Company. 1632 pp.

Fox, W. B. & Sharp, A. J. 1954. La distribución de *Fagus* en México. Boletín Sociedad Botánica de México 17:31–33.

Frankie, G. W., Baker, H. B. & Opler, P. A. 1974. Comparative phenological studies in tropical wet and dry forests in the lowlands of Costa Rica. J. Ecol. 62:881–919.

Graham, A. 1973. History of the arborescent temperate element in the Northern Latin American biota. In: Graham, A. (ed.) Vegetation and Vegetational History of Northern Latin America, pp. 301–314. Elsevier, Amsterdam.

Hallé, F., Oldeman, R. A. A. & Tomlinson, P. B. 1978. Tropical Trees and Forests. An Architectural Analysis. Springer Verlag, Berlin. 441 pp.

Hara, M. 1983. A study of the regeneration process of a Japanese beech forest. Ecological Review 20(2):115–129.

Koop, H. 1989. Forest Dynamics. Silvi-star: A Comprehensive Monitoring System. Springer Verlag, Berlin. 229 pp.

Kramer, P. J. & Kozlowski, T. T. 1979. Physiology of Woody Plants. Academic Press, New York. 811 pp.

Lemée, G. 1978. La hêtraie naturelle de Fontainebleau. In: Structure et fonctionement des écosystèmes terrestres, pp. 75–128. Problèmes d'écologie. Masson, Paris.

Little, E. L. 1965. Mexican beech, a variety of *Fagus grandifolia*. Castanea 30(3):167–170.

Martínez, M. 1940. Una nueva especie forestal (*Fagus mexicana* Sp. nova). Anales del Instituto de Biologia Univ. México 11:85–89.

Miranda, F. & Sharp, A. J. 1950. Characteristics of the vegetation in certain temperate regions of eastern Mexico. Ecology 31(3):313–333.

Mitscherlich, G. 1975. Wald, Wachstum und Umwelt. III. Boden, Luft und Produktion. J.D. Sauerländer's Verlag, Frankfurt am Main.
Nakashizuka, T. & Numata, M. 1982a. Regeneration process of climax beech forest. I. Structure of a beech forest with undergrowth of *Sasa*. Jap. J. Ecol. 32:57–67.
Nakashizuka, T. & Numata, M. 1982b. Regeneration process of climax beech forests. II. Structure of a forest under the influence of grazing. Jap. J. Ecol. 32:473–482.
Noirfalise, A. 1984. Forêts et stations forestières en Belgigue. Les Presses Agronomique de Gembloux. 234 pp.
Oldeman, R. A. A. 1983. Tropical rain forest, architecture, silvigenesis and diversity. In: Sutton, S. L. et al. (eds.) Tropical Rain Forest: Ecology and Management, pp. 139–150. Blackwell, Oxford.
Oldeman, R. A. A. 1990. Forests: Elements of Silvology. Springer, Berlin. 624 pp.
Oosterhuis, L., Oldeman, R. A. A. & Sharik, T. L. 1982. Architectural approach to analysis of North American temperate deciduous forests. Can. J. For. Res. 12:835–847.
Peters, R. & Ohkubo, T. 1990. Architecture and development in *Fagus japonica-Fagus crenata* forest near Mount Takahara, Japan. J. Veg. Sci. 1:499–506.
Platt, W. J. & Schwartz, M. 1990. Temperate hardwood forests. In: Myers, R. & Ewel, J. (eds.) Ecosystems of Florida, pp. 194–229. Academic Press, Orlando, Florida.
Puig, H. 1976. Végétation de la Huasteca, Mexique. Etudes Mesoaméricaines, Vol.V. Mission Archéologique et Ethnologique Française aux Mexique. 531 pp.
Rzedowski, J. 1981. Vegetación de México. Limusa, México. 432 pp.
Schulze, E.-D. 1970. Der CO_2-Gaswechsel der Buche (*Fagus silvatica* L.) in Abhängigkeit von den Klimafaktoren im Freiland. Flora 159:177–232.
Standley, P. C. 1920–6. Trees and Shrubs of Mexico, 5 vols. Government Printing Office, Washington. 1721 pp.
Synnott, T. J. 1979. A manual of permanent plot procedures for tropical rain forest. Tropical Forestry Papers n° 14, Oxford. p. 67.
Tanaka, N. 1986. Structure and regeneration of the *Fagus crenata-Abies mariesii forest*. Bull. Tokyo Univ. Forest, 75:141–197.
Thái Van Trùng 1978. Oecogenesis and Classification of Forest Vegetation of Vietnam, 2nd ed. Science and Technic, Hanoi, 276 pp. (English summary).
Tiffney, B. H. 1985. The Eocene North Atlantic land bridge: Its importance in Tertiary and modern phytogeography of the Northern Hemisphere. J. Arnold Arb. 66:243–273.
Walter, H. 1984. Vegetation of the Earth, 3rd ed. Springer, Berlin. 318 pp.
Wang, X. P. et al. 1965. General description of the vegetation in Kuen-kousui forest area, Suiyang county, Guizhou. Acta Phytoecologia et Geobotanica 3(2):264–286.
Wang, X. P. & Li, X. X. 1986. General situation of vegetation in Miaoershan reserve of Xingan county, Guangxi. Guihaia 6:(1–2):79–91 (English summary).
Watt, A. S. 1947. Pattern and process in the plant community. J. Ecol. 35:1–22.
Zheng, W. J. (ed.) 1983, 1985. Silva Sinica, Vols. 1 and 2. China Forestry Publishing Company, Beijing. 2398 pp.

18. Insights on deciduous forests from a European perspective

ROBERT NEUHÄUSL
Botanical Institute, Academy of Sciences of the Czech Republic, CZ-252 43 Průhonice, Czech Republic

Abstract

European summergreen broad-leaved forests represent the zonobiome of the temperate nemoral climate. These occupy the nemoral zone lying between the boreal coniferous forests and the winter-rain Mediterranean evergreen forests (in the West) or arid-temperate steppes (in the East). Pure broad-leaved forests occur on mineral-rich soils from the lowland to the submontane belt; these are replaced by mixed coniferous-deciduous forests in the montane belt, on poor or dry soils, and in areas with subcontinental climate. European deciduous forests require a warm growing season (at least 120 days with mean daily temperature above 10°C) with sufficient rainfall and a mild winter. The main vegetation types are acidophilous *Quercus* (oak) forests, *Quercus-Fraxinus* and *Quercus-Carpinus* forests, *Tilia-Quercus* forests and *Fagus* forests (with *Fagus sylvatica* or *F. orientalis*). Submediterranean xerothermic deciduous forests consist of different oak species (*Quercus pubescens*, *Q. cerris*, *Q. frainetto*, etc.) with admixtures of *Carpinus orientalis*, *Fraxinus ornus*, *Acer tataricum*, etc., partly with evergreen shrubs. The distribution of the vegetation types can be correlated with climatic and edaphic conditions. The position of summergreen broadleaved forests in Europe and Asia is compared.

Key words: climatic correlation, geographic distribution, major vegetation types, nemorale zone, soil conditions

Introduction

Summergreen broad-leaved forests are characteristic of the climatic zone with a warm growing season of 4–6 months (at least 120 days with mean

E.O. Box et al. (eds.), Vegetation Science in Forestry, 345–352.

daily temperature above 10°C), adequate rainfall (with maximum occurring in summer) and a mild winter lasting 3–4 months (Walter 1979). The deciduous tree species avoid the extreme maritime as well as continental regions. In Europe, in comparison with other continents, summergreen broad-leaved forests represented very important and diversified formations in the primary vegetation. Potential natural deciduous broad-leaved forests, including mixed forests, occupy the nemoral zone lying between the boreal coniferous forests and the winter-rain Mediterranean evergreen forests (in the West) or arid-temperate steppes (in the East). In Southeast Europe, in contact with the mountain ranges of the Crimea and Caucasus, an isolated belt of deciduous broad-leaved forest contacts the South Pontic (Euxinian) deciduous forests of Asia Minor.

Major vegetation types

According to some structural features, bioclimatic conditions and substantial differences in floristic composition, the European deciduous broad-leaved forests, including mixed needle leaved-broad leaved forests, may be divided into two subformations:

1. Mesophytic deciduous broad-leaved forests, and
2. Xerothermic deciduous broad-leaved forests (Neuhäusl et al. 1989).

Mesophytic deciduous forests are mostly closed tall forests with a gradual tree layer, open understorey and thin to dense herb cover. Communities on euthropic wet or damp soils are rich in herbs or forbs, those on oligotrophic or dry soils rich in low graminoids. Tall closed stands are mostly without herb layer because of low light and soil dryness; their soil surface is covered by a continuous litter layer. Mesophytic deciduous forests present a great range of primary productivity, from very low ($<10\,t.ha^{-1}/yr$) on nutrient-poor and dry soils to very high ($>20\,t.ha^{-1}/yr$) on rich, damp soils (*cf.* Duvigneaud 1980).

In the subzone with Central and East European mild climate mesophytic deciduous forests occupy lowlands and hilly terrain, including the submontane belt. These reach the montane belt in regions with Atlantic climate and withdraw into mountains with cooler and moister climate in the southern subzone with sub-mediterranean climate. Mixed coniferous-broad leaved forests cover the montane belt of Central European mountains. The northern limit of mesophytic deciduous forests is brought about by the shortness of the growing period, the southern limit by a long period of summer dryness.

Mesophytic deciduous and mixed needle leaved-broad leaved forests in Europe may be divided into seven major units, according to the dominance of particular tree species or species combinations.

1. Acidophilous oak (*Quercus*) and mixed oak forests, poor in species, occur only on oligotrophic acid soils in lowlands and hilly terrain, in the south (sub-mediterranean climate) at higher altitudes (submontane to montane belt). In areas with oceanic or suboceanic climate, acidophilous oak forests occur with an admixture of birch (*Betula*), in the subcontinental area of distribution they are replaced by mixed pine-oak forests.
2. Mixed oak-ash (*Quercus-Fraxinus*) forests are characteristic of rich and wet soils, and occur only in areas with oceanic climate. These represent the western wing of the European mesophytic woodland.
3. Mixed oak-hornbeam (*Quercus-Carpinus*) forests occupy mesoeutrophic soils in the middle part of the European nemoral zone, with suboceanic to subcontinental climate. These represent the climax of the lowland and hilly country, in the south they withdraw into the submontane belt.
4. Lime-pedunculate oak (*Tilia-Quercus robur*) forests represent an analogue of oak-hornbeam forests but occur in Eastern Europe with continental climate. The simplified structure of the tree layer is conditioned by florogenetic reasons (lime-oak forests occur behind the area of distribution of *Carpinus betulus* and *Quercus petraea*).
5. Beech (*Fagus*) and mixed beech forests occupy oligotrophic as well as eutrophic soils but avoid more extreme dry or wet soils. Beech forests need an oceanic or suboceanic climate or other climate with higher precipitation, as in higher altitudes in the middle and southern part of the nemoral zone, penetrating into the upper montane belt in the Mediterranean zone. Therefore, the acidophilous beech forests may be found in all altitudinal belts from lowland to the upper montane, the eutrophic beech forests to the subalpine belt. European eutrophic beech forests show a clear altitudinal differentiation. Both on the east-west precipitation gradient and a north-south temperature gradient, the beech forests withdraw to the higher altitudes with cooler and moister climate, giving way to more drought-resistant communities dominated by oaks, hornbeam, lime, etc.
6. Montane Oriental beech (*Fagus orientalis*) forests represent the Euxinian and Caucasian analogue to western, central and southern European beech forests. These are linked to the montane altitudinal belt and have no parallel in the lowland.
7. Caucasian mixed hornbeam-oak (*Carpinus-Quercus*) forests occurring in the colline-submontane belt of southeastern European mountain ranges are analogous to the Central European oak-hornbeam forests (item 3). The occurrence of these units is brought about by the combination of climatic and soil conditions, partly by florogenetic reasons.

Xerothermic deciduous broad-leaved forests, including mixed needle leaved-broad leaved forests, represent low to medium-tall, more or less open

Table 1. Xerothermic deciduous broad-leaved forests and mixed needle leaved-broad leaved forests of Europe

First division	Interpretation
1. Subcontinental, partly periodically moist, weakly acidophilous oak and pine-oak forests (*Quercus robur*, *Q. petraea*, *Q. dalechampii*, *Q. pubescens*, *Pinus sylvestris*)	Extrazonal, subcontinental climate, special soil conditions
2. Subcontinental mixed maple-oak forests (*Quercus robur*, *Q. petraea*, *Q. pedunculiflora*, *Q. pubescens*, *Q. virgiliana*, *Q. cerris*, *Acer tataricum*, *A. campestre*)	Subzonal, Southeast European transition to steppe
3. Subcontinental-submediterranean and uppermediterranean mixed sessile oak (*Quercus petraea*, *Q. dalechampii*, *Q. polycarpa*), bitter oak (*Quercus cerris*) and Balkan oak (*Quercus frainetto*) forests	East and Central submediterranean, montane sites in Mediterranean Zone, subcontinental climate
4. Submediterranean and uppermediterranean mixed oak forests and shrub-rich forests (*Quercus pubescens*, *Q. faginea*, *Q. pyrenaica*, *Q. brachyphylla*, *Fraxinus ornus*, *Ostrya carpinifolia*, *Carpinus orientalis*, *Castanea sativa*)	Eu-submediterranean, montane sites in Mediterranean Zone, submediterranean climate

stands of thermophilous, drought-resistent deciduous trees, in continental areas, mixed with xerophilous ecotypes of pine. All xerothermic forests have been used as coppice, partly burned in order to get more space for grazing. Open stands are rich in shrubs, and the herb layer is dominated by grasses and light-loving forbs. The primary productivity is lower than of mesophytic deciduous forests. In the east, steppe woodlands occupy the transition between deciduous forests and meadow steppes. The transition belt to Mediterranean sclerophyllous forests consists of low open woodlands or shrubs. Xerothermic deciduous broad-leaved forests occur in the southern subzone with submediterranean climate, noted for milder winters and warmer summers, with a dry period which frequently lasts for eight or more weeks. These penetrate into the subzone of mesothytic deciduous forests on special sites with low precipitation and/or specific soil conditions (periodically moist-dry soils, rendsina soils, etc.).

Xerothermic deciduous broad-leaved forests of Europe may be divided into 4 ecologically distinguished groups with specific tree species combinations (Table 1).

1. Subcontinental, partly periodically moist, weakly acidophilous oak and pine-oak (*Pinus-Quercus*) forests occur mostly extrazonally (within mesophytic forests), in areas with subcontinental climate, on special soils (periodically wet-dry, dry habitats on limestone, marl, loess, etc.).
2. Subcontinental mixed maple-oak (*Acer-Quercus*) forests which occur on

the northern border of the submediterranean subzone and form a link with the temperate steppe in Eastern Europe.

3. Subcontinental-submediterranean mixed sessile oak (*Quercus petraea*, *Q. dalechampii*, *Q. polycarpa*) forests, bitter oak (*Q. cerris*) forests and Balkan oak (*Q. frainetto*) forests including upper mediterranean (extrazonal) analogues in the montane belt of the Mediterranean zone. This group is characteristic of the eastern and middle part of the submediterranean subzone with subcontinental climate. The dominance of particular tree species depends on summer dryness. With increasing continentality the species composition changes from sessile oak to Balkan oak.
4. Submediterranean mixed oak forests and shrub-rich forests with downy oak (*Quercus pubescens*), and typical submediterranean tree species. Analogous communities occur in the Mediterranean zone but at higher altitudes. Forest types of this group correspond to the typical submediterranean climate and occur on the southern border of the submediterranean subzone. Close to the Mediterranean Sea the deciduous woodlands with shrubs are replaced by sclerophylous evergreen woodlands.

Comparison with East Asian deciduous broad-leaved forests

European summergreen broad-leaved forests and their community patterns are quite different from those in East Asia. The florogenetically conditioned differentiation with a large distribution discontinuity of common genera is intensified by the occurrence of the summer monsoon in East Asia. Macroclimatic differences of oceanic areas in Europe and Asia are brought about by warm and cold ocean currents (Gulf Stream in Europe, Oya-shio current in East Asia).

The broad-leaved deciduous forest zones in Europe and Asia have a similar shape. They occupy areas with oceanic climate (Western Europe, Eastern Asia), i.e. the belt 57°–44° N in Europe and 45°–35° N in Asia (Table 2). In Europe, *Fagetalia sylvaticae* and *Quercetalia pubescentis* represent the major community types; in Asia *Fagetea crenatae* (especially *Saso-Fagion crenatae*) is characteristic of the analogous segment (Miyawaki 1979). The middle part of this zone, areas with subcontinental climate, lies between 55° and 44° N in Europe and 44° and 33° N in Asia. *Carpinion* and *Quercetalia pubescentis* occur in Europe, *Quercus mongolica* forests in Asia. On both continents the admixture of pine and other conifers is characteristic of this section. Under continental climate, deciduous forests occupy only a very narrow belt, in Europe between 56° and 53° N, in Asia between 39° and 35° N. Only a limited number of deciduous trees can withstand conditions of this section (in Europe *Quercus robur*, *Tilia cordata*, *Acer platanoides*).

Table 2. Major differences between European and Asian broad-leaved deciduous forests

	Oceanic	Subcontinental	Continental
Europe			
N limit	57° N	55° N	56° N
S limit	44° N	44° N	53° N
Characteristic vegetation	*Fagetalia* *Quercetalia pub.*	*Carpinion* *Quercetalia pub.*	*Quercus robur* *Tilia cordata* woods
Asia			
N limit	45° N	44° N	39° N
S limit	35° N	33° N	35° N
Characteristic vegetation	*Fagetea crenatae* (*Saso-Fagion*)	*Quercus mongolica* (*Pinus* ssp.)	

East Asian deciduous forests, as compared with their European counterparts, are:

1. rich in tree and shrub species,
2. rich in evergreen species (mostly graminoids) in the herb layer,
3. less differentiated along a north-south gradient (the submediterranean subzone is missing),
4. less differentiated in altitudinal belts (some tree species show great altitudinal spans),
5. more differentiated concerning dominant or frequent tree species (e.g. *Quercus*, *Carpinus*, *Tilia*, *Acer*, etc.),
6. more tolerant to soil nutrient status and acidity, linked to soil moisture conditions and mesoclimatic humidity,
7. less recognized, from a typological and synecological point of view.

Conclusions and summary

European summergreen broad-leaved forests occupy the nemoral zone with temperate nemoral climate, between the boreal coniferous forests and the winter-rain Mediterranean evergreen forests, in East Europe the temperate steppes. With exception of extreme habitats (rocks, wet or extremely poor soils, serpentine substrate, etc.), summergreen forests cover all altitudinal belts from lowland to the montane belt. Mixed coniferous-deciduous forests replace the purely broad-leaved forests in the montane belt, on poor or dry soils, and in areas with subcontinental climate (East Europe). Relatively long, warm growing seasons (at least 120 days with mean daily temperature above 10°C), with sufficient rainfall and mild winter, are characteristic of European deciduous broad-leaved forests.

Two subformations of deciduous forest may be distinguished in Europe, based on structural features, floristic composition and relevant climatic conditions:

1. mesophytic types represented by mostly closed tall forest and typical nemoral or boreo-nemoral flora, and
2. xerothermic deciduous forests constituted by low to medium-tall, open stands with a rich submediterranean flora.

Mesophytic deciduous forests may be divided into seven major vegetation types brought about by combinations of climatic and soil conditions, partly by floristic history.

1. Acidophilous oak and mixed oak forests, poor in species, occur on oligotrophic acid soils from Western to Eastern Europe.
2. Mixed oak-ash forests are characteristic of rich and wet soils, and occur only in areas with oceanic climate.
3. Mixed oak-hornbeam forests occupy meso-eutrophic soils in the middle part of the European nemoral zone, with suboceanic to subcontinental climate.
4. Lime-pedunculate oak forests represent an analogue of oak-hornbeam forests in Eastern Europe.
5. Beech and mixed beech forests occur on both eutrophic and oligotrophic soils in areas with oceanic and suboceanic climate or in mountains with higher precipitation. These avoid more extreme dry and wet soils.
6. Montane Oriental beech forests represent the Euxinian and Caucasian analogues of beech forests in Western Europe.
7. Caucasian mixed hornbeam forests are analogous to the Central European oak-hornbeam forests.

The northern limit of mesophytic deciduous forests is brought about by the shortness of the growing period, the southern limit by a long period of summer dryness. These forests withdraw into mountains (with cooler and moister climate) in the submediterranean subzone.

Xerothermic deciduous broad-leaved forests of Europe may be divided into 4 ecologically distinguished groups characterized by particular tree-species dominance or combinations.

1. Subcontinental, partly periodically moist, weakly acidophilous oak and pine-oak forests occur mostly extrazonally in areas with subcontinental climate on special soils.
2. Subcontinental mixed maple-oak forests occur on the northern border of the submediterranean subzone.
3. Subcontinental-submediterranean mixed sessile-oak forests, bitter-oak forests and Balkan-oak forests occur in South and Southeast Europe with subcontinental climate.
4. Submediterranean mixed oak forests and shrub forests, dominated by

downy oak and other submediterranean tree species, are characteristic of the southern border of the submediterranean subzone.

Xerothermic deciduous broad-leaved forests are linked with a submediterranean climate noted for milder winters and warmer summers with a long dry period.

European summergreen broad-leaved forests are quite different from those in East Asia. A large discontinuity in the distribution of common genera (*Quercus*, *Acer*, *Tilia*, *Fagus*, etc.), macroclimatic differences (summer monsoon in East Asia, influence of warm Gulf Stream in Europe, etc.) and the specific ecosystem evolution cause different vegetation patterns of deciduous forest in Europe and East Asia. The European deciduous forests are poorer in tree and shrub species as well as in evergreen species, more uniform concerning dominant and frequent tree species, and more differentiated along north-south and altitudinal gradients. East Asian deciduous forests seem to be more tolerant to soil nutrient conditions and acidity, and the community distribution is more correlated with soil moisture and mesoclimatic humidity.

References

Duvigneaud, P. 1980. La synthèse écologique, 2nd. ed. Doin editeurs, Paris.

Miyawaki, A. 1979. Vegetation und Vegetationskarten auf den Japanischen Inseln. In: Miyawaki, A. & Okuda, S. (eds.) Vegetation und Landschaft Japans, pp. 49–70. Yokohama.

Neuhäusl, R., Gribova, S. A. & Bohn, U. 1989. Natürliche Vegetation Europas. Gesamtlegende zur Vegetationskarte 1:2,5 Mio. Bundesanstalt für Naturschutz und Landschaftsökologie, Bonn.

Walter, H. 1979. Vegetation of the Earth and Ecological System of the Geo-biosphere, 2nd. ed. New York, Heidelberg, Berlin.

19. The spruce forests of northern China

CHEN LING-ZHI
Institute of Botany, Academy Sinica, Beijing 100 044, China

Abstract

The mountains of China harbor many types of spruce (*Picea*) forest, and the geographical regions of China can be distinguished according to the species of spruce which dominate their forests. In the southwestern region of China, there are four different species of spruce that dominate the forests: *Picea asperata*, *P. purpurea*, *P. likiangensis*, and *P. brachytyla*. From northeast to north-central China, various spruce species appear: *Picea koraiensis*, *P. jezoensis*, *P. wilsonii*, and *P. meyeri*. On the Altai and Tianshan mountains of northwest China, *Picea schrenkiana* and *P. obovata* forests occur. Forests containing *Picea morrisonicola* are only present in Taiwan. In addition to this geographical differentiation of spruce forests in China, there is also elevational differentiation of spruce forest types. The highest elevation at which spruce forests are found in the mountains of China follows a north-south gradient; spruce forests occur at 360–700 m above sea level in northeast China, 1500–2000 m in north-central China, 1500–2800 m in northwest China and at 2400–3800 m in southwest China.

The spruce forests of China usually consist of canopy and subcanopy layers, scattered shrub and herb layers, and a dense moss layer. There are several tree and shrub genera that are common in the spruce forests of China. There include *Abies*, *Acer*, *Lonicera*, *Rosa*, *Sorbus*, *Salix*, *Euonymus*, and *Sorbaria*, among others. Herb genera common to all spruce forest communities in China include *Oxalis*, *Equisetum*, *Cacalia*, *Circaea*, *Pyrola*, *Lycopodium*, *Paris*, and *Convallaria*.

Key words: biomass, China, composition, life form, *Picea*, spruce

E.O. Box et al. (eds.), Vegetation Science in Forestry, 353–365.

Introduction

The spruce forests of China are widely distributed in the subalpine mountains extending from northern to southwestern China. We recognize 15 distinct types of spruce forests, each including one or several unique combinations of spruce species. Seven of the forest types occur in northern China. One spruce forest type, dominated by *Picea asperata*, is distributed from westcentral to southwestern China. The other forest types appear only in southwestern China and in Taiwan. These include forests of *Picea likiangensis*, *P. likiangensis* var. *lintziensis*, *P. likiangensis* var. *balfouriana*, *P. purpurea*, *P. brachytyla*, *P. brachytyla* var. *complanatax*, *Abies kawakamii*, and *Picea morrisonicola*.

Before the most recent glacial period, *Picea* and *Abies* were the major dominants of forests from southwestern to northeastern China, and from the northcentral plains to the Northwest Plateau. Analysis of pollen and macrofossils indicates that a large amount of *Picea* and *Abies* appeared in the southwest, northwest, north-central, and southeast during the Pleistocene. The climate change brought on by the ice ages strongly affected the migration of vegetation so that now, *Picea* forests are mainly located in the subalpine mountains in the southwestern and northern parts of China.

The distribution of spruce forests

Table 1 and Fig. 1 show the distribution of *Picea* forests in northern China. *Picea koraiensis* and mixed *P. jezoensis-Abies nephrolepsis* forests occur in the Great Xing An Mtns. (Da Xing An Ling), Lesser Xing An Mtns. (Xiao Xing An Ling) and Changbaishan Mountains of northeastern China. *Picea koraiensis* forest occurs at lower elevations in the valleys and lowlands between 200–700 m. A mixed *Picea jezoensis-Abies nephrolepsis* forest type appears as an altitudinal band on the lee slope of the Great Xing An, Little Xing An, and Changbaishan Mountains, between 800 and 1800 m elevation, and also in the northern region of the Qingshan Mountains between 1100–1200 m. A different mixed-forest type of *Picea meyeri-P. wilsonii* is primarily distributed in north-central China in the Wu Ling, Xiao Wu Tai, Wu Tai, and Lu Liang Mountains, including the regions of Guan Di and Guan Qin at elevations between 1500–2400 m. This forest type dominated by *Picea meyeri* also occurs in the southern parts of the Great Xing An Mountains as well in Inner Mongolia. The *Picea wilsonii* forest type extends westward and includes the northern region of Qing Ling Mountain and the Yellow River Valley of the Qinghai Province. In the eastern region of the Qi Li Mountains, a small forest of *Picea crassifolia* is located on the lee slope between 2500–3950 m. On the northern slope of the Tianshan Mountains, a forest dominated by *Picea schrenkiana* is found. This forest type is also found in the western portion of the Kun Lun Mountains and the western mountains of the Zhun Ge Er Lowland. *Picea schrenkiana* is also widely distributed in the mountains of northwestern China. The valleys and lower elevations of the southwestern slopes the Altai Mountains contain the only domain of the *Picea obovata* forest. In general, spruce forests increase in elevation from the northeast to the northwest.

The climate of spruce forests

According to the Editorial Committee on The Vegetation of China (1980), there are seven types of *Picea* forest in northern China located in five different vegetation regions. All spruce forests are characterized by long, cold winters. During the winter, the air is normally dry. Due to the cold temperature and lack of water, spruce growth stops during the winter months. Although winter is harsh, summer can be quite warm (Table 2) and water, light, and temperature levels are sufficient for active photosynthesis from May through September.

The *Picea koraiensis* forest type is located mainly in the cold temperate coniferous forest region, where the mean annual temperature is −2.2°C to −5.5°C, a lower temperature than is found in any of the other spruce forest types. The soil is peaty and gleyed. The *P. koraiensis* forest occurs in the same vegetation region in northern China as the mixed *Picea jezoensis-Abies nephrolepsis* forest-type. However, the *Picea jezoensis* forest-type is situated in the temperate-mixed-coniferous-and-broadleaved-forest region where slightly higher temperatures prevail than are found in the *P. koraiensis* forests. This area has the most abundant precipitation in northern China. Soils are in the brown forest soil type. *Picea meyeri* and *P. wilsonii* forests are located in the warm-temperate, deciduous broad-leaved forest region. The climate of this region is warmer and drier than that of the former region. Soils are of the brown forest soil type and the gray brown type. The other three spruce forest types in northern China are situated in the moist valleys and the middle and high elevation areas of the temperate steppe and desert regions respectively. Due to low precipitation, these forests do not occur on low elevation sites. Soils of the gray drab forest type predominate.

The floristic composition of the spruce forest

Although spruce forests differ in floristic composition, there are many similarities at the generic level (Table 3). *Abies nephrolepis* grows in *Picea*

Table 1. The distribution of spruce forests in the northern part of China

P. koraiensis forest	*P. jezoensis* forest	*P. meyeri* forest		*P. wilsonii* forest	*P. crassifolia* forest	*P. schrenkiana* forest	*P. obovata* forest
Location (northeastern part)		(north central to western parts)			(western part)	(northwestern part)	
Great Xing An		Yanshan			East section	Northern slope	South western
Mt. 360–700 m	800–1200 m	Mt. 1500–1800 m		1500–1800 m	of Qi Li Mt.	and southern	slope of Alta
Little Xing An		Xiao Wu Tai			on lee slope	of Tian Shan Mt.	Mt. 1300–
Mt. 200–500 m	800–1100 m	Mt. 1500–1800 m		1500–1800 m	(limited area)	west part of	1800 m
Changbaishan		Wu Tai Mt.	1600–2400 m			Kunlun Mt.	located in
Mt. 300–500 m	1400–1800 m	Guan Di Mt.	on lee slope	1600–2400 m	2500–3900 m	west Mt.	the valley
located in the	North section	Guan Qin Mt.				of Zhun GeEr	of low part
valley and	of Qianshan Mt.	Lu Liang Mt.		North section		1500–3000 m	of the slope
lowland.	1100–1200 m	South part of		of Qing Ling Mt.			
	located on the	Great Xing An		2200–3200 m			
	lee slope of	Mt.		(larger area)			
	the mountain	1500–2100 m		The valley of			
		(Fragment)		Yellow River			
		South part of		in Qing Hai			
		Inner Mongolia		Province			
		1300–1500 m		2200–2500 m			
		on the sands		(small area)			

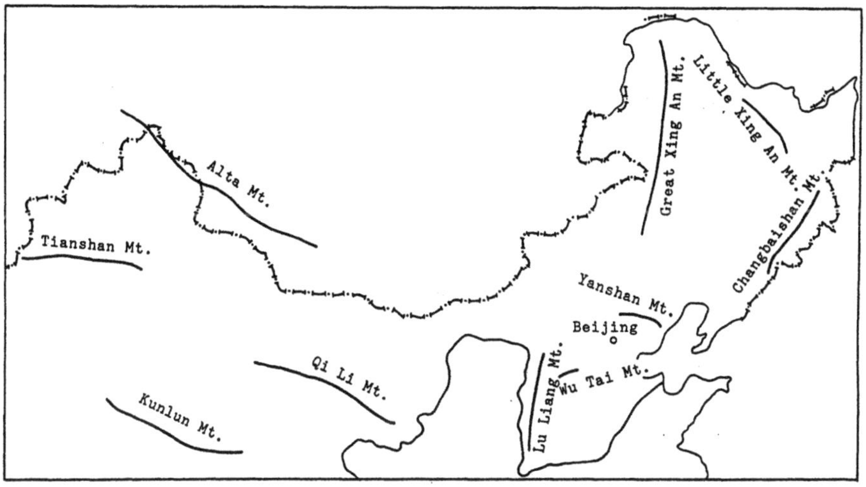

Fig. 1. The geographic location of the main mountains in the north part of China.

jezoensis forests in which it is often a codominant, and in *Picea koraiensis* forests. *Alnus* grows only in *Picea koraiensis* forests which occur on moist sites. The canopy layer of *Picea* forests includes species of *Larix*, *Betula*, and *Populus*. *Larix gmelini*, *L. olgensis* var. *changbaiensis*, *L. pricipis-rupprechti* and *Larix sibirica* occur in distinct spruce forests. There are eight species of birch in the spruce forest, but *Betula pendula*, *B. tianshanica*, *B. turkestanica* and *B. jamolenkana* grow only in the spruce forests of northwest China. *Populus davidiana* is widely distributed in various spruce forests, but it is replaced by *Populus tremula*, *Populus laurifolia*, and *Populus nigra* in *Picea schrenkiana* and *Picea obovata* forests, the latter two spruce forest types being dissimilar from all other spruce forest types. *Acer ukurunduense*, *Acer tegmentosum*, *Acer barbinerve*, *Eleuthrococcus senticosus*, and *Echinopanax elatus* are present only in the *Picea jezoensis* forest, and are absent in all other spruce forests. The floristic composition of the *Picea jezoensis* forest type, including the moss species, is most abundant in these spruce forest types.

The understory of the spruce forest in general consists mainly of species of *Lonicera*, *Rosa*, *Ribes*, *Rubus*, *Contoneaster*, *Spiraea*, *Sorbus*, *Cornus*, *Salix*, *Padus*, and *Euonymus*. Seven species of *Lonicera* and *Ribes* and five species of *Rosa* grow in the different spruce forests. Among them, *Lonicera chrysantha*, *Rosa bella*, *Rosa xanthiana*, *Spiraea pubescens* and *Ostryopsis davidiana* occur in the *Picea meyeri* and *P. wilsonii* forest types. Three species of *Caragana* occur only in the *Picea crassifolia* forest.

The components of the herb layer are many and varied; the great majority

Table 2. The climatic conditions of various *Picea* forests

P. koraiensis forest	*P. jezoensis* forest *P. koraiensis* forest	*P. meyeri* forest *P. willsonii* forest	*P. crassifolia* forest *P. willsonii* forest	*P. schrenkiana* forest *P. obovata* forest
Locates in the cold temperate coniferous forest region E127°20′ to west N49°20′ to north	Locates in the temperate mixed coniferous and deciduous forest region N40°15′–50°20′ E126°–135°30′	Locates in the warm temperate deciduous broadleaved forest region N32°30′–42°30′ E103°30′–124°10′	Locates in temperature steppe region, south part N34°–41° E102°–114°	Locates in temperate desert region N36° to north, E105° to west
Annual mean temperature −2.2 to −5.5	2 to 8°C	9 to 14°C	6 to 8°C	4 to 12°C
Mean temperature of cold months −28–38°C	−10 to −2.5°C	−13 to −2°C	−7 to −8°C	−6 to −20°C
Mean temperature of hot months 16–20°C	21 to 24°C	24 to 28°C	18 to 24°C	20 to 30°C
Annual precipitation 350–550 mm (7–8)	600 to 1000 mm (6–8)	500 to 900 mm	330 to 500 mm	210 to 250 mm
Climatic station Ai Hui (165.8 m)	Climatic station Yi Chun 231.3 m	Climatic station Xi An 396.9 m	Climatic station Qi Li 2784.4 m	Climatic station Tian Shan Mt. 2300 m
N50°15′ E127°27′	N47°43′ E128°54′	N34°18′ E108°56′	N38°11′ E100°18′	N40°45′ E77°55′
Mean temperature −2.1°C	0.3°C	13.3°C	0.6°C	1.9°c
Maximum temperature 37.7°C	34.4°C	41.7°C	12.8°C	14.3°C
Minimum temperature −40.7°C	−43.1°C	−20.6°C	−13°C	−11°C
Precipitation 515.5 mm	636.4 mm	604.2 mm	393.3 mm	624.5 mm (200–300 mm in Basin)
Soil: Peaty gleyed soil	Soil: brown forest soil	Soil: brown forest soil or gray brown soil	Soil: brown forest soil	Soil: gray drab forest soil

Table 3. The composition of spruce forests

P. koraiensis forest	*P. jezoensis* forest	*P. meyeri* forest	*P. willsonii* forest	*P. crassifolia* forest	*P. schrenkiana* forest	*P. obovata* forest
	P. jezoensis	*P. meyeri*	*P. mayeri*		*P. schrenkiana*	*P. obovata*
P. koraiensis	*P. koraiensis*	*P. willsonii*				
Larix dahurica	*Larix olgensis* var. *changbaiensis*	*L. principis-rupprechti*	*P. crassifolia*	*P. crassifolia*	*Larix sibiricus*	*L. sibiricus*
				P. purpurea		
	Pinus koraiensis				*Sabinia* sp.	
Abies nephrolepis	*Abies nephrolepis*	*Betula dahurica*	*Betula dahurica*	*Sabinia przewalskii*	*B. pendula*	*B. pendula*
Betula platyphylla	*B. ermanii*	*B. platyphylla*	*B. platyphylla*		*B. tianshanica*	
					B. turkestanica	
	B. costata	*B. albo-sinensis*		*Betula albo-sinensis*	*B. jamolenkoana*	
Populus davidiana	*Populus koreana*	*Populus davidiana*		*Populus davidiana*	*P. tremula*	*P. laurifolia*
	P. davidiana					*P. nigra*
	Acer ukurunduense				*Malus sieversii*	
	Ulmus propinqui					
	Tilia amurensis					
	Acer tegmentosum					
Cornus alba	*Lonicera maximoiwizii*	*Lonicera chrysantha*	*Lonicera chrysantha*	*Lonicera syringantha*	*Lonicera altmanii*	*Lonicera angustifulia*
Alnus sibirica	*L. caerulea* var. *edulis*	*L. hispida*		*L. hispida*	*L. hispida*	*Salix* sp.
Betula fruticosa		*L. caerulea* var. *edulis*			*L. kareline*	
Rosa dahurica	*Rosa dahurica*	*Rosa xanthiana*	*Rosa bella*	*Potentilla fruticosa*	*Rosa albertii*	*Rosa albertii*
Potentilla fruticosa	*Padus asiatica*	*R. bella*	*Ribes procumbens*		*R. suavis*	*Rosa suavis*
Ribes procumbens	*Ribes maximoviczianum*	*Ribes burejense*	*Ribes alpestre*	*P. glabra*	*Ribes meyeri*	*Padus racemosa*
	Ribes triste	*Ribes mandshuricum*	var. *giganteum*	*Caragana jujuba*		

Table 3. Continued

P. koraiensis forest	*P. jezoensis* forest	*P. meyeri* forest	*P. willsonii* forest	*P. crassifolia* forest	*P. schrenkiana* forest	*P. obovata* forest
Sorbaria sorbifolia	Grossularia burejensis	Rubes idaeus	Rubus idaeus	C. tangutica	Euonymus seminoruii	Rubus idaeus
		Syringa villosa	Lespedeza bicolor	C. rosea		
	Crataegus maximowiczii	Cotoneaster acutifolia		Cotoneaster acutifolia		
	Eleutherococcus senticosus	Lespedeza bicolor	Cotoneaster acutifolia	Sorbus hupenensis	Sorbus tianschanica	Contoneaster sp.
		Spiraea pubescens		Spiraea alpina		
	Sorbus pohuashanensis	*Ostryopsis davidiana*	*Sorbus discolor*	*Cotoneaster multiflorus*		*Vaccinum vites-idea*
			Corylus mandschrica		*Salix xerophylla*	
	Echinopanax elatus				*Cotoneaster melanocarpa*	*Spiraea chamaedryfolia*
	Acer barbinerve		*Cornus macrophylla*			
	Euonymus pauciflorus		*Spiraea pubescens*			
					Moneses uniflora	*Moneses uniflora*
					Ramishia secunda	*Ramischia secunda*
Maianthemum bifolium	*Solidago virgaurea* var. *dahurica*	*Maianthemum bifolium*		*Poa annua*	*Poa nemoralis*	
					Milium effusum	
	Maianthemum dilatatum	*Carex lanceolata*	*Carex lanceolata*	*Carex creba*	*Trisetum spicatum*	
		Lathyrus humilis	*Lathyrus humilis*			
Oxalis acetosella	*Oxalis acetosella*	*Dendranthemum erubescens*	*Dendranthemum erubescens*	*Deyeuxia* sp.	*Helictotrichon pubescens*	*Lathyrus gmelinii*
	Clintonia udensis					*Asperula odorata*
Mitella nuda	*Streptopus koreanus*	*Clematis macropetala*	*Aster ageratoides*	*Polygonatum sibiricum*	*Aegopodium alpestre*	*Carex pediformis*
Athyrium brevifrons	*Dryopteris austriaca*	*Aquilegia yabeana*	*Pedicularis musciola*	*P. kansuensis*	*Cicerbita azurea*	*Calamagrostis* sp.
Smilacina dahurica	*Cacalia kamtschaticum*	*Equisetum pratense*				*Poa sibirica*
Aegopodium alpestre		*Phlomis umbrosa*	*Phlomis umbrosa*	*Polygonum viviparam*	*Stellaria songorica*	*Deyeuxia* sp.
	Carex gotoi	*Veratrum nigrum*	*Aquilegia viridiflora*			*Galium boreale*
Deyeuxia angustifolia					*Geranium collinum*	*Goodyera repens*
		Galium bungei	*Galium bungei*	*Pedicularis* sp.	*Lathyrus gmelinii*	*Vicia crassa*
Pyrola incarnata		*Convallaria keiskei*	*Allium sacculiferum*	*Thermopsis alpina*		
Filipendula palmata	*Gallium*				*Dryopteris*	

Equisetum sylvaticum	*kamtschaticum*				*filix-mas*	
	Angelica grosseserrata					
	Fragaria orientalis	*Pyrola rotundifolia*	*Caltha palustris*			
	Thalictrum tuberiferum		*Vicia unijuga*			
	Moss 80 species					
Hylocomium splendens	*Pleurozium schreberi*	*Mnium* sp.		*Pseudostereodon*	*Aulacomnium* sp.	*Mosses*
Aulacomnium palustre var. *imbricatum*	*Hylocomium splendens*	*Entodon orthocarpus*		*Procerrimus*	*Hylocomium proliferum*	*Brachythecium albicans*
Helodium lanatum	*Rhytidiadelphus* spp.			*Abietinella abietina*		*Drepanocladus vernicosus*
Sphagnum squarrosum	*Ptilium crista-castrens*				*Thuidium abietinum*	
Pleuroziopsis ruthenica	*Pleuriziopsis ruthenica*			*Bryum erthro-carpum*		*Hylocomium proliferum*
Climacium dendroides	*Climacium* spp.			*Timmia mega-politana* etc.	*Dicranum scoparium*	*Pleurozium schreberi*
Mnium sp.	*Mnium marginatum*				*Brachythecium populeum*	
Dicranum undulatum						
Usnea longissima						
				Bryum	*Hypnum revolutum*	

Table 4. Life-form spectra of spruce forests

	Species % 1	Coverage % 1	Species % 2
Phanerophytes	23.24	68.59	19.28
Chamaephytes	4.39	0.36	9.64
Hemicryptophytes	39.55	16.70	42.17
Geophytes	26.37	13.06	21.69
Therophytes	3.28	1.18	1.20
Liana	2.15	0.02	–
Moss			6.02

1. *Picea jezoensis* forest 2. *Picea meyeri* forest

of plant species in the spruce forest are small herbs. Mosses are also abundant in the spruce forests where they cover the ground and tree trunks.

The life forms and biomass of two spruce forests

The *Picea jezoensis* forest is distributed in the mixed coniferous and broad-leaved forest region, while the *Picea meyeri* type forest is distributed in the deciduous broad-leaved forest region. Although these two forest types are found in different regions, they show striking similarities in composition. Hemicryptophytes are the most abundant plants in these forest types, but the cumulative cover value of all hemicryptophytes is less than that of phanerophytes. Phanerophyte species number is less than that of both hemicryptophytes and geophytes, but phanerophytes, by definition, make up the overstory (Table 4; Chen Ling-Zhi 1963, 1964, 1987).

Spruce seedlings and saplings are abundant in spruce forest understories. This is consistent with the known stability of the subalpine spruce forests.

Few studies of *Picea* forest biomass and production have been conducted in northern China. In these studies biomass is computed using measurements of tree diameter, height, density, and age. In order to make comparisons between forests, it would be necessary to find two stands with similar tree diameters and densities.

At the time of the study, the age of the *Picea jezoensis* forest studied in the Changbaishan Mountains was 150–200 years, with 1432 trees/ha (Table 5). The mean diameter at breast height was 29 cm. Biomass of this *Picea jezoensis* forest was 286.8 t $\cdot$ ha.$^{-1}$ including 242.6 t $\cdot$ ha.$^{-1}$ (84%) of above-ground biomass and 44.2 t $\cdot$ ha.$^{-1}$ of below-ground biomass.

Primary production of the tree layer reached 10 $\cdot$ 6 t $\cdot$ ha.$^{-1}$a.$^{-1}$ (Li Wen-ha et al. 1981). For comparison, two *Picea meyeri* forests were studied, one on the sands of Inner Mongolia (Chen Bing-hao & Chen Chu-ying 1980) and

Table 5. Biomass of *Picea* forest

(a) *Picea jeozensis* forest on Changbaishan Mt. (N41°–42°45′, E 127°33′–128°16′) at 1100–1700 m above sea level

Biomass t/ha						Primary production t/ha · a		
	Leaves	Branches	Trunks	Roots	Total	Aboveground part	Underground part	Total
Tree layer	17.47	35.66	183.46	27.94	264.53	9.67	0.93	10.6
Understory	1.28	4.53	–	15.32	21.13	1.19	0.51	1.70
Herbs	0.21	–	–	0.94	1.15	0.21	0.94	1.15
					286.81			13.45

1432 trees/ha, mean diameter at breast height = 29 cm · age = 150–200 yr.

(b) *Picea meyeri* forest on sand dunes in Inner Mongolia (N43°31′, E117°13′) at 1300 m above sea level

Biomass t/ha						Primary production t/ha · a				
	Leaves	Branches	Trunks	Roots	Total	Leaves	Branches	Trunks	Roots	Total
Tree layer	7.83	16.87	45.14	27.41	97.23	0.09	0.18	0.46	0.28	1.01
Understory		5.60		0.52	6.12					
Herbs		5.51			5.51					
					108.88					

765 trees/ha, mean diameter at breast height = 21 cm · age = 55–130 yr.

the other in the Lu Liang Mountains. The *P. meyeri* forest of Inner Mongolia had a mean dbh of 21 cm while the stand in the Lu Liang Mountains had a mean dbh of 22.6 cm. The stand densities of the two *P. meyeri* locations were 765 and 848 trees/ha respectively. The age of these spruce forests was 55–130 years. The biomass of the *P. meyeri* forest on sands was 108.9 t · $ha.^{-1}$ including 81 t · $ha.^{-1}$ (74%) of aboveground biomass and 27.9 t · $ha.^{-1}$ (26%) belowground biomass. The primary productivity of the canopy layer was 1.0 t · $ha.^{-1}$· $a.^{-1}$ (Chen Bing-hao 1980, Xu Wen-duo 1983). Although densities, mean dbh's, and ages of these two *P. meyeri* forest stands were different from the *P. jezoensis* forest, we can still conclude that biomass and primary production of the typical *Picea jezoensis* forest is higher than that of the typical *P. meyeri* forest.

Conclusion

Spruce is an abundant species and the many and varied spruce forests form stable forest community-types as well as a valuable resource in northern China and in subalpine areas elsewhere in the country. The different spruce forest types have different geographical and elevational distributions. Spruce forests can regenerate quite well under natural circumstances. Presently, some spruce forests are still subjected to a clear-cutting management system, which results in the destruction of seedlings and saplings. Often after a clear-cut, *Larix* will be artificially reforested at a site. If a site is left alone, however, *Betula* will intercede. Spruce can recover in a *Larix* or *Betula* forest under natural conditions if there is a seed source nearby. The most promising plan of management involves selective cutting of spruce, which allows the resource to be utilized continuously.

Acknowledgements

I thank Kathleen Baker-Brosh and Robert Peet for editorial and linguistic assistance, Miss Li Pu-ping for kind help to prepare the figures.

References

Editorial Committee of Vegetation of China 1980. Vegetation of China. Science Press, Beijing.

Chen Bing-hao & Chen Chu-ying 1980. The preliminary study on the biomass and production of *Picea koraiensis* forest on sand. Forest Science 4:269–277.

Chen Ling-zhi. 1963. The preliminary study on the structure of *Picea jeozensis* forest on the westsouthern slope of Changbaishan Mt. Acta Phytoecologica et Geobotanica Sinica 1(1–2):6–80.

Chen Ling-zhi 1964. The structure of major plant communities of various vertical zones on the northern slope of Changbaishan Mt. Acta Phytoecologica et Geobotanica Sinica 2(2):207–225.

Chen Ling-zhi 1987. The vegetation of Pang Quanggou Nature Reserve in Shagxi Province (unpublished).

Li Wen-ha, Deng Kum-mei & Li Fei 1981. Study on biomass and primary production of the main ecosystems in Changbai Mt. Research of Forest Ecosystem II.

Xu Wen-duo 1983. *Picea meyeri* and its spruce forest in Nei Mongool Sands. Acta Phytoecologia et Geobotanica Sinica 7(1):1–7.

20. Boreal forest phytogeocoenoses of Hokkaido Island, Japan[1]

SATORU KOJIMA
Department of Biosphere Science, Faculty of Science, Toyama University, Toyama 930, Japan

Abstract

Vegetation and environmental characteristics of coniferous forests of Hokkaido Island, Japan, are discussed. The forests developing in mid-elevations of the island exhibit some vegetational and environmental affinities to the boreal forests in the mid- to high latitudes of the northern hemisphere in the following respects: they developed under the cool, humid Dfc and cooler Dfb climate (Köppen); the forest canopy consists of boreal conifers such as *Abies* and *Picea*, hence showing the same physiognomy; and the vegetation incorporates a fair number of circumpolar or circumboreal species and conforms basically to the vegetation characteristics of Vaccinio-Piceetea. Based on this, the forests are considered to represent boreal phytogeocoenoses of the Japanese Archipelago. On the other hand, the forests are somewhat unique in the following ways: a complete lack of *Pinus* and *Larix*, which are important elements of the boreal forest; predominant occurrence of *Sasa* plants; common presence of lianas; and a lack of well developed moss communities in mesic habitats, which are quite common elsewhere, but in the island mesic habitats are represented by *Sasa* communities instead. These unique features can be attributed to the relatively mild, more humid climate of the island and by the presence of *Sasa* plants, though *Sasa* itself is probably promoted by the climate. Ecological characters of three conifers representing the forests are discussed in relation to soil pH. *Picea glehnii* was found most tolerable to acidic and low nutrient soils, whereas *Picea jezoensis* requires the least acidic and most fertile soils and *Abies sachalinensis* falls somehow between the other two conifers.

Key words: boreal forest; coniferous forest; Hokkaido, Japan; soil characteristics; Vaccinio-Piceetea

[1] This study is partially supported by a Grant-in-aid for Scientific Research of Monbusho, Japan (Grant No. 60540422).

E.O. Box et al. (eds.), Vegetation Science in Forestry, 367–389.

Photo 1. An extensive development of coniferous forest in the central region of Hokkaido Island. Such a forest consists mainly of *Picea glehnii*, *P. jezoensis*, and *Abies sachalinensis*.

Introduction

In central Hokkaido, Japan, where mountains are concentrated, well developed coniferous forests occur extensively (Photo 1). These forests consist mainly and characteristically of such conifers as *Abies sachalinensis*, *Picea glehnii*, and *P. jezoensis* (Photo 2). The forests exhibit some vegetational and environmental affinities to the boreal forests of North America and Eurasia (i.e. to continental boreal forests) and, therefore, they may be regarded representing boreal phytogeocoenoses of the Japanese Archipelago. A phytogeocoenosis is a vegetation-soil complex which constitutes a major part of a terrestrial ecosystem.

In Japan these forests have been customarily treated as subarctic (Nakano 1942, Tatewaki 1958, Yamanaka 1979) or subalpine (Horikawa 1970, Yoshioka 1974, Ishizuka 1974). Phytosociologically they are basically similar to the Vaccinio-Piceetea Br.-Bl. 1939, although they also show some unique features. Suzuki (1960) designated the coniferous forests Vaccinio-Piceetea japonicum. Miyawaki (1967, 1989) called the areas represented by the forests the Vaccinio-Piceetea Region. Kojima (1975, 1979a), introducing the concept of biogeoclimatology (Krajina 1965, 1969), designated these areas as the montane *Abies sachalinensis* biogeoclimatic zone (hereafter called the *Abies*

Photo 2. A well developed and representative stand of the Japanese boreal forest. Dominant trees are mostly *Picea jezoensis* accompanied by some deciduous trees such as *Sorbus commixta*, *Acer japonicum*, and *Acer ukurunduense*.

sachalinensis zone), which is basically comparable to the oroboreal zone of Hämet-Ahti et al. (1974). The present article attempts to describe vegetational and environmental characteristics of the coniferous forests of Hokkaido Island and to compare them with those of the continental boreal forests.

Geography of Hokkaido Island

Hokkaido (78,500 km^2) is the northernmost island of the Japanese Archipelago (Fig. 1), lying between 41°30′ and 45°30′ N and between 130°30′ and 145°45′ E. Elevation ranges from sea level up to 2290 m. The physiography is essentially mountainous. Most high mountains are concentrated toward the center of the island, where the *Abies sachalinensis* zone occurs.

General environmental characteristics of the *Abies sachalinensis* zone

General environmental characteristics of the *Abies sachalinensis* zone are summarized in Table 1. This zone, which is comparable to an altitudinal belt

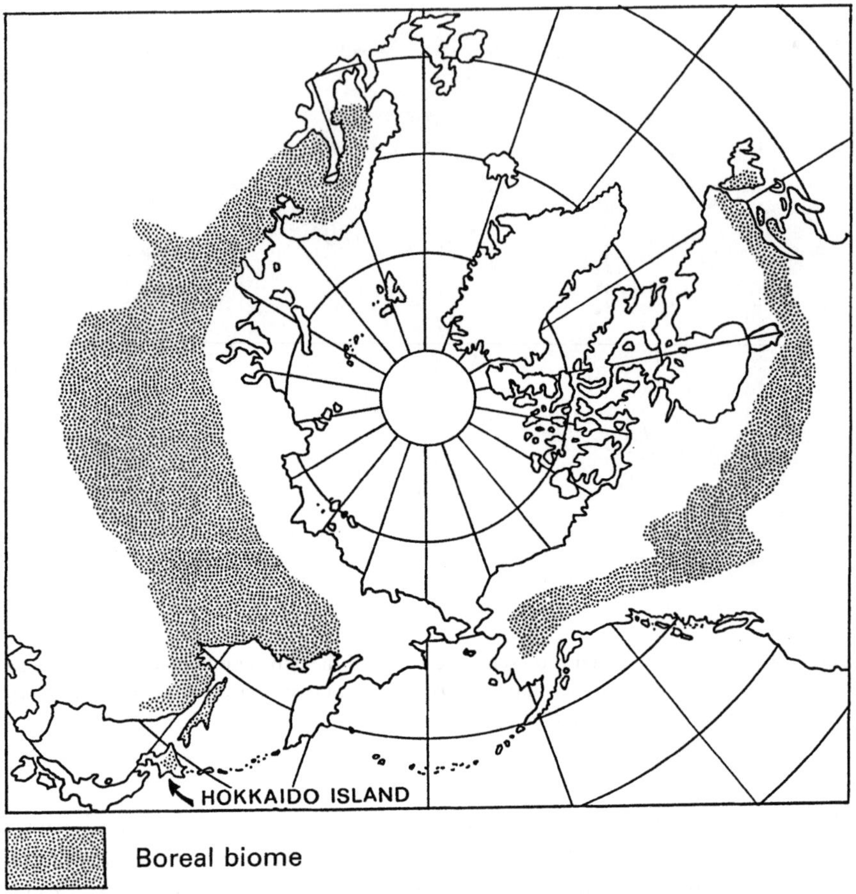

Fig. 1. Geographical range of boreal biome of the world and the location of Hokkaido Island.

characterized by *Abies sachalinensis*, occurs at mid-elevations on the island, from ca. 500 to 1500 m, though varying to some degree from location to location. The climate associated with this zone is cool and humid, which may be classified as Köppen's (1931) Dfb type. Mean annual temperature ranges from −1 to 4°C and mean annual total precipitation from 900 to 1500 mm. Precipitation is adequate throughout the year though usually September and October are the wettest months. The amount of precipitation tends to increase westward in the island. Figure 2 illustrates the climatic range of the *Abies sachalinensis* zone projected on Whittaker's scheme for the climatic ranges of major biomes. There is no weather station which properly represents the climate of the zone. The record at Akanko Weather Station (Table 2) may approximate the climatic characteristics of the zone but is located at the lower (warmer) boundary of the zone. From the climatic characteristics,

Table 1. General environment of the montane *Abies sachalinensis* biogeoclimatic zone (from Kojima 1979)

Elevational ranges of the zone:	
northern Hokkaido	ca. 400 to 1,000 m asl.
central	ca. 700 to 1,500 m asl.
eastern	ca. 100 to 700 m asl.
southern	ca. 500 to 1,100 m asl.
southwestern	ca. 800 to 1,300 m asl.
General physiography:	Mountainous, various aspects and steepness
Climatic characteristics:	
mean annual temperature	−1 to 4 °C
mean monthly temperature	
of the warmest month	16 to 20 °C
of the coldest month	−10 to −20 °C
number of months with	
mean monthly temp. 10 °C	2 to 3
mean monthly temp. 0 °C	5
annual precipitation	900 to 1,400 mm
annual potential	
evapotranspiration	350 to 550 mm
annual soil water surplus	600 to 1100 mm
climate type (Köppen/Trewartha)	Dfb – Dfc
Major pedogenic processes:	Moderate braunification Weak podzolization Mor and moder formation Gleization Paludification Frequent volcanic ash deposits
Major soil type:	Dark brown forest soil and brown forest soil (*cf.* Dystric Cambisol)
Representative higher syntaxonomic unit:	Abietalia sachalinensis (Kojima 1975, 1979)
Zonal phytogeocoenosis:	*Picea jezoensis-Abies sachalinensis – Sasa senanensis* phytogeocoenosis with brown forest soil

the *Abies sachalinensis* zone may be placed in the hemiboreal OC sector of Tuhkanen's (1984) climatic-phytogeographical system.

The geology of the mountains of central Hokkaido varies greatly, from volcanic to sedimentary rocks of various ages, as well as metamorphic rocks (Geological Survey of Japan 1978). Due to intense volcanic activity in the island, deposits of volcanic ash are very common (Hokkaido Kazanbai Meimei Iinkai 1972) and serve as an important soil parent material in the region.

Soils are, in general, strongly acidic and poor in basic cations. They are mostly the brown forest soils of the Japanese soil classification system (Forest Soil Division 1976), which are comparable to Dystric Cambisols of the FAO

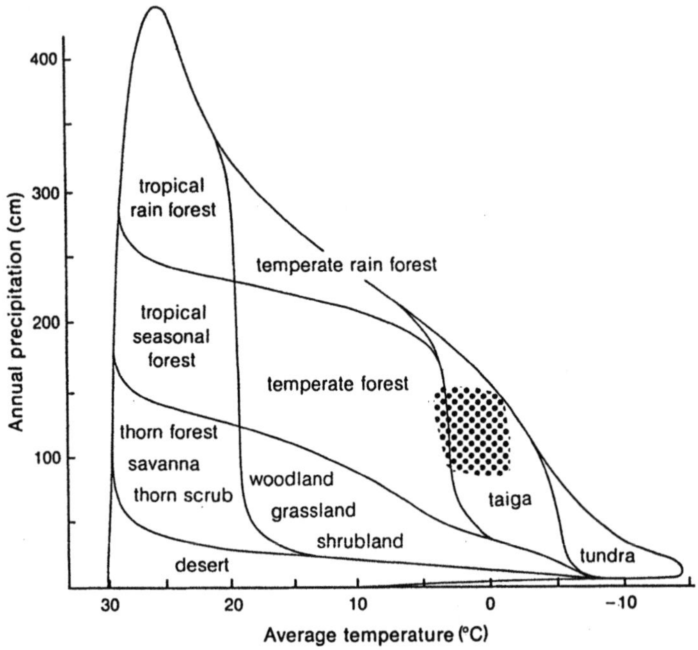

Fig. 2. Climatic range of the Japanese boreal forest as projected in Whittaker's diagram. The dotted pattern indicates the range.

Table 2. Climatic characteristics at Akanko Weather Station (43°25′ N, 144°06′ E, 430 m asl.)

	Mean temperature (°C)	Mean precipitation (mm)
Jan	−11.1	75
Feb	−10.7	15
Mar	−5.8	72
Apr	2.5	87
May	8.7	124
Jun	12.8	137
Jul	17.6	114
Aug	18.6	126
Sep	13.4	180
Oct	7.1	184
Nov	0.1	105
Dec	−7.5	64
Year	3.8	1324

system (FAO-UNESCO 1974), Umbrept or Dystrochrept of the U.S. system (Soil Survey Staff 1960), or Orthic Dystric Brunisol of the Canadian system (CSSC 1978). In general, the soils are deep and horizons are well developed. The A horizon is thick and dark. Litter is relatively well decomposed. Most humus associated with the coniferous forests is identified as mormoder (Klinka et al. 1981). Despite the cool, humid climate and preponderant occurrence of conifers, podzols are rare. Soils are frequently interlaminated with volcanic ash. Buried soils overlain by ash are common.

General vegetation characteristics

The forest vegetation is characterized by the presence of such species as: *Abies sachalinensis*, *Picea glehnii*, *P. jezoensis*, *Sorbus commixta*, *Rhododendron brachycarpum*, *Vaccinium hirtum*, *V. ovalifolium*, *V. smallii*, *Menziesia pentandra*, *Cornus canadensis*, *Oxalis acetosella*, *Ilex rugosa*, *Maianthemum dilatatum*, *Anemone debilis*, *Waldsteinia ternata*, *Rumohra mutica*, *Streptopus streptopoides*, *Dryopteris austriaca*, *Lycopodium serratum*, *L. obscurum*, *Galium kamtschaticum*, *Carex sachalinensis*, *Sasa senanensis*, *Actinidia kolomikta*, and *Hydrangea petiolaris*. Though their occurrence is rather sporadic, the following species are also closely associated with the coniferous forests: *Lycopodium annotinum*, *Pyrola secunda*, *Maianthemum bifolium*, *Listera cordata*, *Clintonia udensis*, and *Coptis trifolia*.

The forest canopy, which is normally closed, consists primarily of the three conifers listed earlier. *Betula ermanii* frequently accompanies the conifers, especially at higher elevations. At lower elevations where the *Abies sachalinensis* zone meets the lowland nemoral zone, deciduous trees such as *Quercus mongolica* var. *grosseserrata*, *Tilia japonica*, *Ulmus laciniata*, and *Magnolia obovata* are mixed with the conifers. The presence of *Sasa* plants under the forest canopy is one of the unique features of the Japanese boreal forest (Photo 3). *Sasa* is a dwarf bamboo which commonly forms extremely dense thickets overwhelming other plants. The species composition of the understory vegetation varies greatly according to habitat conditions. In general, the understory vegetation is more sensitive to habitat conditions than is the tree layer.

Forest types of the Japanese boreal coniferous forest

Based on the floristic structure mainly of the ground vegetation and on environmental characteristics mainly of the soil chemical characteristics, the coniferous forests may be classified into forest types. A forest type, as here defined, is a basic unit of phytogeocoenoses. It is comparable to plant associations of Krajina (1960), which are the smallest unit discriminable vegetationally as well as environmentally.

Indeed, for the Japanese boreal forests, five forest types were distinguished (Kojima 1991), based on relevés representing various phytogeoco-

Photo 3. Sasa plants commonly and characteristically dominate in the understory vegetation. This photograph shows a *Picea glehnii – Sasa chartacea* stand in eastern Hokkaido.

enoses (Kojima 1979b, 1980, 1983, unpubl.). They were: 1. the moss type, 2. the *Sasa kurilensis* type, 3. the *Sasa senanensis* type, 4. *Carex sachalinensis* type, and 5. *Dryopteris crassirhizoma* type (Table 3). These forest types, however, only represent those of well drained circum-mesic habitats. There will be many more types if wetland phytogeocoenoses are included. Following are the brief descriptions of the vegetation and soil characteristics of the five forest types.

1. The moss type is differentiated and characterized by the presence of *Ledum palustre* var. *diversipilosum* and *Vaccinium vitis-idaea*. Other species such as *Linnaea borealis*, *Listera cordata*, *Rubus pedatus*, and *Ephippianthus schmidtii* are nearly exclusively associated with this type, though their coverage is rather insignificant. The forest canopy consists exclusively of *Picea glehnii*. The shrub layer is well developed. It is a characteristic feature that Ericaceous shrubs such as *Menziesia pentandra*, *Ledum palustre* var. *diversipilosum*, *Rhododendron brachycarpum*, *Vaccinium hirtum*, *V. ovalifolium*, and *V. smallii* dominate in the layer. Development of the herb layer is moderate. The moss layer develops well to form a complete carpet-like moss cover. Occurrence of this forest type is generally associated with lithic habitats such as piles of large boulders and blocks, rock slides, colluvial talus, and bedrocks. Soils are shallow, with lithic contact

Table 3. A condensed summary table of vegetation structure, modified from Kojima 1991. The Roman numerals represent constancy class and the rectangular boxes indicate the differential species for the forest types

	Forest type*	M	Sk	Ss	C	D
	Number of relevés	11	10	12	11	6
	Species					
Characteristic species of the Japanese boreal forest						
Tree layer	*Abies sachalinensis*	III	V	V	V	IV
	Picea glehnii	V	III	III	III	I
	Picea jezoensis	–	III	III	V	IV
	Sorbus commixta	II	III	III	–	
Shrub and herb layers	*Abies sachalinensis*	V	V	V	V	V
	Acer ukurunduense	II	IV	IV	V	III
	Sorbus commixta	III	III	IV	III	II
	Rhododendron brachycarpum	III	I	III	I	II
	Cornus canadensis	IV	IV	III	V	IV
	Oxalis acetosella	II	IV	V	V	V
	Ilex rugosa	III	III	I	I	I
	Maianthemum dilatatum	II	III	III	V	I
	Anemone debilis	II	II	II	III	I
	Waldsteinia ternata	II	I	–	II	III
	Menziesia pentandra	IV	V	IV	III	–
	Picea glehnii	V	III	III	II	–
	Vaccinium ovalifolium	III	III	I	I	–
	Vaccinium smallii	II	V	II	II	–
	Vaccinium hirtum	V	III	II	II	–
	Rumohra mutica	II	III	IV	III	–
	Streptopus streptopoides	I	III	II	II	
	Picea jezoensis	–	III	III	III	IV
	Dryopteris austriaca	–	III	IV	III	III
	Lycopodium obscurum	–	II	III	III	III
	Galium kamtschaticum	–	II	II	III	III
	Lycopodium serratum	–	III	IV	IV	IV
	Actinidia kolomikta	–	II	V	V	V
	Hydrangea petiolaris	–	III	IV	I	III
Differential species of the forest types						
	Ledum palustre var. *diversipilosum*	III	–	I	–	–
	Vaccinium vitis-idaea	III	–	–	–	–
	Sasa kurilensis	–	V	–	–	–
	Viburnum furcatum	–	III	I	I	–
	Skimmia japonica var. *intermedia*	–	III	I	–	II
	Maianthemum bifolium	II	III	I	I	–
	Sasa senanensis	–	I	V	IV	III
	Carex sachalinensis	–	II	V	V	V
	Dryopteris amurensis	I	I	–	III	I
	Lastrea phegopteris	–	–	II	I	I
	Osmunda asiatica	–	–	I	III	–
	Actaea erythrocarpa	–	–	–	III	II
	Trillium tschonoskii	–	I	I	III	III
	Tiarella polyfolia	–	–	II	III	III
	Dryopteris crassirhizoma	–	–	III	IV	V
	Circaea alpina	–	–	I	III	V

Table 3. Continued

Forest type* Number of relevés Species	M 11	Sk 10	Ss 12	C 11	D 6
Pachysandra terminalis	–	–	–	I	IV
Viola selkirkii	–	–	I	II	III
Rubus pseudo-japonicus	–	–	II	II	III
Cacalia auriculata	–	I	–	II	III
Clematis ochotensis	–	–	–	I	III
Cimicifuga simplex	–	–	I	I	III
Taxus cuspidata	II	–	–	II	III
Acer japonicum	–	–	II	I	III
Kalopanax septemlobus	–	–	II	I	III
Tilia japonica	–	–	I	I	III
Rubus idaeus	–	I	I	I	III
Adoxa moschatellina	–	–	–	–	III
Adenocaulon himalaicum	–	–	–	–	III
Phryma leptostachya	–	–	–	–	III
Vitis coignetiae	–	–	–	–	III
Average number of vascular species per relevé	15.9	24.4	25.8	32.0	40.0
Average total volume of standing trees (m^3/ha)	464.4	627.4	664.1	867.9	654.7

* M: moss type, Sk: *Sasa kurilensis* type, Ss: *Sasa senanensis* type, C: *Carex sachalinensis* type: D: *Dryopteris crassirhizoma* type.

at a depth of 30–50 cm; soils may rarely be deeper as in pockets between large boulders. In this forest type, plants with a shallow root system have an advantage. In fact, most of the species, including *Picea glehnii*, characteristically have shallow rooting systems. Because of the almost completely ombrotrophic conditions, the soils are extremely acid and poor in basic cations. Soil pH is 3.2 and 4.1, on average, for the organic and mineral soils respectively. Both figures are the lowest among the five forest types. Base saturation, too, is the lowest of the five forest types, 11% and 3% on average for organic and mineral soils respectively. Reflecting the low base status of the soils and lithic conditions, species diversity is also very low in this forest type. The average number of vascular species/relevé is 15.9, which is the lowest number among the five forest types. For the same reason the total volume of standing trees is very small, only 464 m^3/ha on average, which is also the smallest figure of the five types.

2. The *Sasa kurilensis* type is differentiated by the presence of *Sasa kurilensis*, *Viburnum furcatum*, *Skimmia japonica* var. *intermedia*, and *Maianthemum bifolium*. Other species such as *Weigela middendorffiana*, *Rhododendron albrechtii*, *Clintonia udensis*, and *Coptis trifolia* are more or less associated with this type. The forest canopy involves all three conifers,

but *Picea glehnii* tends to dominate. The *Sasa* layer, exclusively of *Sasa kurilensis*, develops very well and often exceeds 2 m in height. The shrub and herb layers are weakly developed, being overwhelmed by the *Sasa* plants. As a general tendency, this forest type develops better at higher elevations and/or at locations where a large amount of snow accumulates. The geographical range of *Sasa kurilensis* is known to be associated with regions of heavy snow (Suzuki 1978, Yoshino 1978, Toyooka et al. 1983). Due to the insulation effect of the snow, in most instances the ground is not frozen, which permits the establishment of some temperate species as long as they can tolerate the long duration of the snow cover. Such species include *Skimmia japonica* var. *intermedia* (Uemura et al. 1986) and *Viburnum furcatum*. Soils are generally deep and horizons well developed. Soil base status is intermediate, and average soil pH is 4.4 for organic and 4.6 for mineral soils. Average base saturation is 25% for organic and 9% for mineral soils. Tree growth is intermediate, and the average total volume of standing trees is 627 m^3/ha.

3. The *Sasa senanensis* type is differentiated only by *Sasa senanensis*. The species has a substantially higher dominancy in this forest type though it may occur widely in other forest types. On the other hand, this forest type includes most of the characteristic species of the Japanese boreal forest. This suggests that the *Sasa senanensis* type represents the typical or zonal vegetation of the Japanese boreal forest. Tatewaki (1958) considered the *Sasa* type of forest to represent the climax potential. The forest canopy generally involves all three conifers in more or less even proportion. The shrub layer develops moderately, and the herb and moss layers develops weakly. This type is the most extensive in the *Abies sachalinensis* zone covering a wide range of habitats. Soils are deep and horizons well developed. The soil base status is intermediate. Average soil pH is 4.3 for organic and 4.8 for mineral soils. Base saturation is also intermediate, 20% for organic and 6% for mineral soils on average. Tree growth is intermediate, with total volume of standing trees is about 664 m^3/ha.
4. The *Carex sachalinensis* type is differentiated by the presence of *Carex sachalinensis*, *Dryopteris amurensis*, *Lastrea phegopteris*, *Osmunda asiatica*, *Actaea erythrocarpa*, *Trillium tschonoskii*, and *Tiarella polyfolia*. Species diversity of this type is fairly high at 32.0 vascular species/relevé on average, which is the second highest value among the five forest types after the *Dryopteris crassirhizoma* type. This higher diversity is mainly contributed by shrub and herbaceous species. The forest canopy consists of a mixture of the three conifers, of which *Picea jezoensis* is generally slightly more important than the others. The *Sasa* layer, exclusively *Sasa senanensis*, develops moderately, which permits better development of

the shrub and herb layers. This forest type usually develops on lower slopes and in valley bottoms with gentle topography. It is fairly extensive throughout the *Abies sachalinesis* zone. Soils are deep and well developed. Soil base status is intermediate to slightly higher, and average soil pH is 4.3 for organic and 4.6 for mineral soils. Average base saturation is 25% for organic and 14% for mineral soils. Tree growth is excellent because of the deep, moderately well drained soils, and gentle topography. The average total volume of standing trees can reach 868 m^3/ha, the highest figure among the five forest types.

5. The *Dryopteris crassirhizoma* type is differentiated by the presence of *Dryopteris crassirhizoma*, *Circaea alpina*, *Pachysandra terminalis*, *Viola selkirkii*, *Rubus pseudo-japonicus*, *Cacalia auriculata*, *Clematis ochotensis*, *Cimicifuga simplex*, *Adoxa moschatellina*, *Adenocaulon himalaicum*, *Phryma leptostachya*, *Rubus idaeus*, *Vitis coignetiae*, *Taxus cuspidata*, *Acer japonicum*, *Tilia japonica*, *Kalopanax septemlobus*, *Fraxinus mandshurica*, and *Ulmus laciniata*. This forest type shows the highest average species diversity among the five forest types, reaching 40.0 spp./relevé. This forest type also includes a fair number of species which are essentially elements of the nemoral vegetation of lower elevations, including most of the species listed above plus *Circium kamtschaticum*, *Laportea bulbifera*, *Osmorhiza aristata*, *Mitella nuda*, *Cacalia hastata* var. *orientalis*, *Urtica platyphylla*, and *Cardamine leucantha*. This forest type usually develops at slope bases and in valley bottoms or ravines along small streams, where seepage provides continuously moist conditions. Soils are generally coarse-textured, consisting mainly of fluvial material and rarely also of colluvial deposits. The horizons may not necessarily be well developed. The soil base status is very high, and average soil pH is 5.4 for organic and 5.2 for mineral soils. Base saturation is also very high, 40% for organic and 38% for mineral soils. All these figures are indeed the highest among the five forest types. It is suspected that soils with a high base status may permit the establishment of nemoral elements even in the *Abies sachalinensis* zone where the climate may not necessarily be favorable for them. Tree growth is rather intermediate, the average total volume of standing trees being only 655 m^3/ha.

Differentiation of the forest types

The above sequence of the forest types appeared to reflect primarily soil base status (Kojima 1991). The moss type develops in the most acidic, least fertile habitats, whereas the *Dryopteris crassirhizoma* type occurs in habitats of the highest base status. At the same time, the above sequence of the forest types seems also to reflect a topographic gradient and moisture avail-

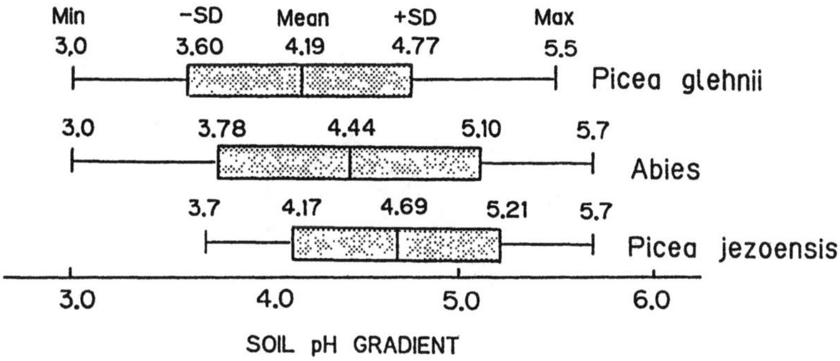

Fig. 3. Ecological distribution of the three conifers along a soil pH gradient.

ability. The moss type usually occurs on upper slopes and/or rocky habitats where soils are shallow and excessively drained. The *Dryopteris crassirhizoma* type, on the other hand, develops on lower slopes and slope bases, where some seepage effects are expected. The other types occur between these extremes.

Under a humid climate, as in the Japanese boreal forest, soil water deficits normally do not occur. The moisture regime is, therefore, not a very critical factor in vegetation differentiation. Under a humid climate, soils are continuously being leached and basic cations removed, resulting in a depauperization and acidification of the soils. This soil impoverishment is more intense in well drained and/or other water-shedding habitats. In water-receiving habitats, such as slope bases and seepage sites, soils are enriched with basic cations which have been transported by seepage water. The above sequence of forest types is in fact a consequence of this differentiation mechanism.

Ecological amplitude of the conifers along a soil-pH gradient

Soil pH is an indirect measure of soil fertility based on the availability of basic cations (Kojima 1987). An attempt was made to examine the ecological amplitude of the three conifer species along a soil-pH gradient and to determine their ecological characteristics. For this, 78 relevés selected from studies previously made (Kojima 1979b, 1980, 1983, and unpublished data). From them, the pH values and cover values of the three conifer species, expressed in Krajina's (1933) species significance classes, were transcribed. The pH for each relevé was calculated as the average of the L-H and A horizons, in which most of the roots were concentrated.

The soil pH of the 78 relevés ranged from 3.0 to 5.7, with an average of 4.44 and standard deviation of 0.655. Figure 3 shows the overall ecological ranges of the three conifers along the soil-pH gradient. The relevés were

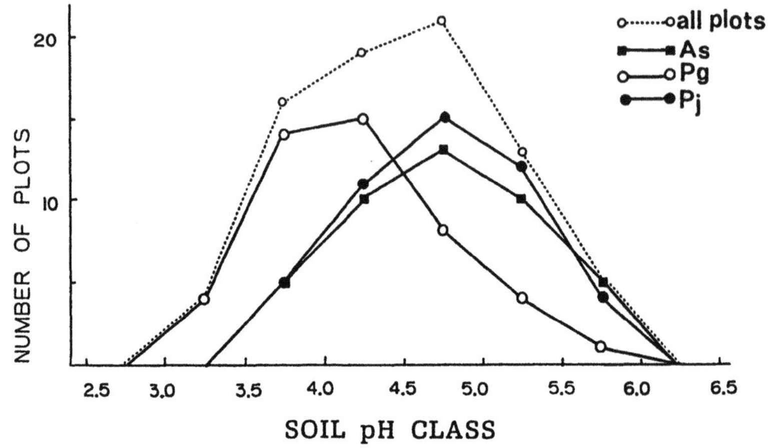

Fig. 4. Frequency distributions of the three conifers in relation to soil pH classes.

grouped into six classes of soil-pH, at intervals of 0.5 pH units. The frequency distributions of soil pH and of the three conifers, based on the six pH classes, are shown in Figure 4. A test of statistical goodness of fit was applied, which showed that the distribution of soil-pH fits a normal distribution fairly well (Chi-square value 2.039, d.f. = 5). This suggests that the more general soil types associated with the coniferous forests are normally distributed.

The conifers, however, show quite a different distribution pattern (Fig. 4). *Picea glehnii* occurs mostly in the pH class 4.0–4.4, with a long tail to the right in the figure. A Chi-square test revealed that its distribution does not fit the normal distribution of the relevés in general. *Picea jezoensis* and *Abies sachalinensis* do not occur at soil pH less than 3.4, and their modes occur 1 pH class higher than that of *Picea glehnii*. This pattern is shown more clearly when coverage values of the tree species are used (Fig. 5). *Picea glehnii* is strongly dominant at lower pH, while the other two conifers have higher coverage at higher soil pH. A shift in species dominance takes place around a pH value of 4.5, which is in fact near the overall mean of soil pH for the boreal sites.

These analyses suggest an antagonistic relationship between *Picea glehnii* and the other two conifer species. This antagonism is corroborated by correlation coefficients among the coverage value of the three conifers and soil pH in the 78 relevés (Table 4). Occurrence of *Picea glehnii* shows negative correlations with both *Abies sachalinensis* (−0.409) and *Picea jezoensis* (−0.687), and is also negatively correlated with soil pH (−0.637). On the other hand, both *Abies sachalinensis* and *Picea jezoensis* are positively

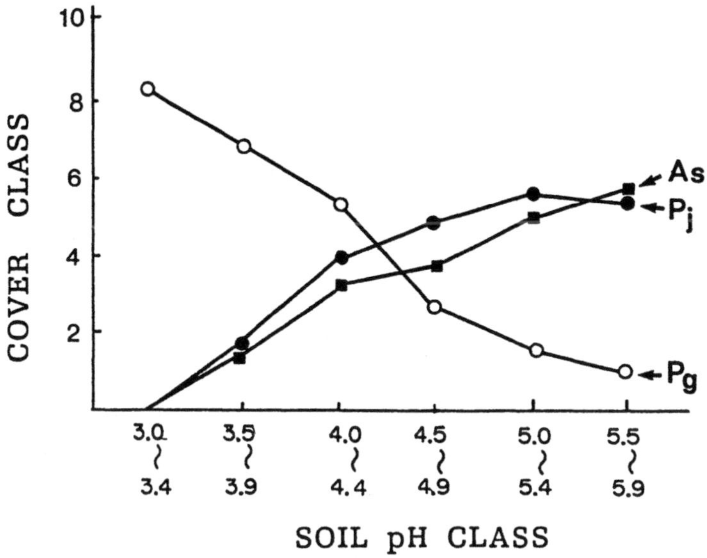

Fig. 5. Average cover classes of the three conifers in relation to soil pH classes. The coverage is expressed in Domin-Krajina's species significance classes (Krajina 1933).

Table 4. Correlation coefficients of soil pH and the three coniferous species. The calculations were made based on the soil pH (average of the L-H and A horizons) and each species' coverage in the tree layer expressed in Domin-Krajina's (Krajina 1933) species significance class for 78 relevés

	Abies	*Picea gl.*	*Picea jz.*
pH of LH & A horizons	0.453*	−0.637*	0.497*
Abies sachalinensis		−0.409*	0.255
Picea glehnii			−0.687*

* $p < 0.01$.

correlated with soil pH. Correlation between the two species is not significant.

These findings suggest that the three conifer species in Hokkaido differ significantly in ecological amplitude along a soil-pH gradient. The frequency as well as coverage of *Picea glehnii* is highest in the pH interval 4.0–4.5, whereas those by the other conifers are highest at soil-pH of 4.5–4.9. The widths of the amplitudes do not differ much. Of the three species, *Picea glehnii* is the one which can grow in the least fertile and most acidic habitats, while *Picea jezoensis* and *Abies sachalinensis* require more fertile habitats.

Some common features of the Japanese and continental boreal forests

The Japanese boreal forest possesses a number of common vegetational and environmental characteristics to those of the continental boreal forests, by which it can also be identified as boreal. First, the climate of the Japanese boreal forest basically corresponds to that of the continental boreal forests. The geographical range of the boreal forest is roughly bounded to the north by the 13°C isotherm for July (or the warmest month) and in the south by the 18°C isotherm, except that where drier conditions prevail the southern boundary lies further north (Larsen 1980). The climate of the Japanese boreal forest basically satisfies this requirement, as shown by the comparison of selected continental and Japanese boreal weather stations in Table 5. Most boreal climates are classified as Köppen's Dfc type, whereas the Japanese boreal region falls along the Dfc-Dfb boundary.

The forest canopy of the Japanese boreal forest consists primarily of *Picea* and *Abies* species, as in other boreal forests. The physiognomic type of the Japanese boreal forests is the same as that of the continental boreal forests and classified as type 1A9c (Mueller-Dombois and Ellenberg 1974). Regional occurrences of the major tree species of the boreal forests are summarized in Table 6.

The vegetation of the Japanese boreal forest includes a certain number of circumpolar or circumboreal elements which are common either throughout the entire boreal region or at least common to the boreal region of Eurasia or North America. Examples include *Lycopodium annotinum*, *Lastrea phegopteris*, *Dryopteris austriaca*, *Pyrola secunda*, *Vaccinium vitis-idaea*, *Ledum palustre*, *Linnaea borealis*, *Oxalis acetosella* (Eur.), *Maianthemum bifolium* (Eur.), *Cornus canadensis* (N.Am.), and *Lycopodium obscurum* (N.Am.). The presence of *Vaccinium* is another common feature which characterizes the vegetation as Vaccinio-Piceetea, though species differ from region to region. Table 7 shows vicariant relationships of *Vaccinium* species in the various boreal regions.

Some unique characteristics of the Japanese boreal forest

Within the broad framework of boreal forests, some features of the Japanese boreal forest are unique. First of all, the Japanese boreal forest occurs under a slightly warmer, more humid climate than that of the continental boreal forests. The mean annual temperature of the Japanese boreal forest is roughly 5 to 6°C higher than that of most continental boreal forests. The total annual precipitation, which exceeds 1000 mm in Japan, is much greater than in other boreal forests. The continentality of the Japanese boreal forests, as measured by Conrad's (1946) index, is also more moderate (Table 7).

Table 5. Climatic characteristics of some selected weather stations representing the boreal forest

Station	Haparanda	Surgut	Yakutsk	Akanko	Watson Lake	Fort McMurray	Amos
Country	Sweden	Russia	Russia	Japan	Canada	Canada	Canada
Latitude	65°50′ N	61°15′ N	62°05′ N	43°25′ N	60° 07′ N	56° 39′ N	48°34′ N
Longitude	24°09′ E	73°30′ E	129°45′ E	144°06′ E	128°49′ W	111°13′ W	78°08′ W
Altitude (m asl)	6	44	103	430	689	369	310
Köppen's climatic type	Dfc	Dfc	Dfc	Dfb	Dfc	Dfc	Dfc
Mean annual temperature (°C)	1.1	−3.1	−10.3	3.8	−3.3	−0.2	0.9
Mean maximum temperature (°C)	15.5	17.5	18.7	18.6	14.9	16.4	16.7
Mean minimum temperature (°C)	−11.6	−22.0	−42.0	−11.1	−26.7	−21.8	−17.6
Number of months MMT* >10°C	3	3	3	4	3	3	4
Number of months MMT* >0°C	6	7	7	4	7	5	5
Accumulated degree-month temperature 0°C	57.5	57.4	61.6	80.7	55.2	69.3	70.3
Mean annual total precipitation (mm)	538	518	236	1324	425	472	865
Amount of precipitation from May to August (mm)	183	249	131	501	181	252	380
Continentality index (after Conrad 1946)	33.5	57.0	94.5	492	61.3	57.1	54.8

* MMT: mean monthly temperature.

Table 6. Vicariant relationships of the major tree species representing the boreal forest regions

	EURASIA west cent east	HOKKAIDO	NORTH AMERICA west cent east
PICEA	P. abies P. obovata P. jezoensis	P. jezoensis P. glehnii	P. glauca P. mariana
ABIES	A. sibirica A. nephrolepis	A. sachalinensis	A. lasiocarpa A. balsamea
PINUS (Diploxylon)	P. sylvestris		P. contorta P. banksiana
LARIX	L. sukaczewii L. daurica L. sibirica		L. laricina
POPULUS (Leuce)	P. tremula	P. tremula	P. tremuloides
BETULA	B. pubescens B. platyphylla	B. platyphylla	B. papyrifera

Table 7. Geographical relations of *Vaccinium* growing in the boreal forest regions

Eurasia	Hokkaido	W. North America	E. North America
V. vitis-idaea	(*V. vitis-idaea*)	*V. vitis-idaea*	*V. vitis-idaea*
V. uliginosum	(*V. uliginosum*)	*V. uliginosum*	*V. uliginosum*
V. myrtillus	*V. hirtum*	*V. myrtilloides*	*V. angustifolium*
	V. ovalifolium		
	V. smallii		

Those in the parentheses indicate rare occurrences.

A complete lack of *Pinus* (of Diploxylon group) and *Larix*, both of which are important elements of the continental boreal forests, is a very characteristic feature of the Japanese boreal forest (Table 8). *Abies* and *Picea* are practically the only conifers in the Japanese boreal forest, though *Taxus cuspidata* occurs sporadically. *Pinus pumila*, which predominates at higher elevations does not really correspond to the boreal but rather to the subarctic or subalpine ecosystems of Japan. *Larix gmelinii* was present on Hokkaido until ca. 15000 years ago, as evidenced by pollen records, but disappeared with the climatic amelioration (Tsukada & Nakamura 1989).

Another very unique feature is a presence of and understory dominance by *Sasa* species (Photo 3). *Sasa* is a very complicated, problematic taxonomic unit. Thirteen species of *Sasa* have been reported in Hokkaido (Ito 1969),

Table 8. Sequences of forest types in relation to habitat conditions in the different boreal regions

Habitat \ Region	W. Eurasia	Hokkaido	W. North America	E. North America
Dry	Lichen type (*Pinus sylvestris – Calluna* – lichens)	Moss type (*Picea glehnii – Pleurozium*)	Lichen type (*Picea mariana – Arctostaphylos* – lichens)	Lichen type (*Pinus banksiana – Empetrum* – lichens)
Mesic	Moss type (*Picea abies – Vaccinium myrtillus*)	Sasa type (*Picea glehnii/ jezoensis – Sasa senanensis*)	Moss type (*Picea glauca/ mariana – Hylocomium/ Pleurozium*)	Moss type (*Picea mariana/ glauca – Pleurozium/ Hylocomium*)
Moist	Herb type (*Picea abies* – herbaceous)	Fern type (*Picea jezoensis – Dryopteris crassirhizoma*)	Equisetum type (*Picea glauca – Equisetum*)	Peat-moss type (*Picea mariana – Ledum – Sphagnum*)

of which the major species (s.l.) covering extensive areas are *Sasa chartacea*, *S. kurilensis*, and *S. senanensis*. These have different ecological requirements and thus segregate according to local conditions, particularly snow accumulation patterns. In brief, *Sasa chartacea* is associated with the least snow accumulation, its distribution coinciding well with an isoline representing maximum snow accumulation less than 50–75 cm. The other two *Sasa* species occur where snow accumulation exceeds 50–75 cm. *Sasa kurilensis* tolerates the deepest snow accumulation, and its geographical distribution is closely associated with the "snow belt" of the Japanese Archipelago (Suzuki 1978). It not only tolerates a great amount of snow but also requires it, in order to protect itself from winter freezing and desiccation. In Hokkaido, *S. kurilensis* occurs widely in mountainous areas of higher elevations where snow accumulations are large. *Sasa senanensis* is intermediate in its snow relationships. It is the most extensively occurring *Sasa* species in Hokkaido, covering more than 50% of all *Sasa*-covered terrain (Toyooka et al. 1983). When *Sasa* dominates in the ground layer, it suppresses establishment of other plants, including trees. The biomass of *Sasa* plants is recorded as ca. 13 t/ha as a very general average, which is roughly 5 to 8% of all above-ground biomass of the boreal forest stands. This value can be as high as 22 t/ha (Toyooka et al. 1983).

Despite the microthermal humid climate and prevalence of conifers which supply acidic litter, podzols are relatively rare in the Japanese boreal region. Most soils are brown forest soils. Horizons are well developed, but in most instances eluviated horizons are not discernible. In this respect, *Sasa* seems to play a significant role (Kojima 1979a). *Sasa* collectively is often a dominant plant in the understory vegetation. Litter of *Sasa* generally has pH as high as 6.3 while that of conifers as low as 2.2 (Uchida 1959). Moder humus, even mull humus, develops in the *Sasa* stands even under the coniferous forests. Shiozaki and Narita (1963) recognized that mor humus in the moss stand became gradually transformed to mull humus along with an invasion of *Sasa senanensis* after windfall. This was also accompanied by a change in soil chemical status such as increase of pH and calcium in the F as well as A horizons. *Sasa* seemingly takes a relatively large amount of basic cations deep from the bottom of the solum and returns them to the surface soils, thus, promoting biogeochemical cyclings in the forest stands. This would decelerate the rate of leaching and consequently the progress of podzolization.

The ubiquitous presence of lianas may be another characteristic of the Japanese boreal forest. Major species include *Actinidia kolomikta*, *Hydrangea petiolaris*, and *Schizophragma hydrangeoides*. Other occasional occurrences include *Rhus ambigua*, *Schisandra chinensis*, and *Vitis coignetiae*.

The occurrence of boreal forest types in relation to habitat moisture conditions, for the various boreal regions, is summarized in Table 8. The Japanese boreal forest clearly shows quite a different pattern. Generally, in the continental boreal forests, dry habitats have well-developed lichen layers, whereas the dry Japanese boreal forest is represented by the moss type. In the continental boreal forests, this moss type usually develops in mesic habitats. In Hokkaido, though, the mesic habitats are occupied by the *Sasa* type. Such a shift of forest types may be attributed to the mild, humid climate of Hokkaido. The abundant precipitation promotes development of mosses instead of lichen layer even in well-drained comparatively xeric habitats. On the other hand, the aggressive occurrence of *Sasa* in mesic habitats suppresses the development of mosses and results in extensive establishment of the *Sasa* type throughout the mesic habitats. For the establishment of the *Sasa*, however, a mild, humid climate with much snow is necessary. If the climate were much drier and colder, as in the continental boreal forests, *Sasa* would not have become established, since *Sasa* cannot tolerate winter desiccation and freezing.

Conclusion

The Japanese boreal forest developing at mid-elevations of Hokkaido is characterized by the presence of *Abies sachalinensis*, *Picea glehnii*, and *P. jezoensis*. Its vegetation structure is basically that of the Vaccinio-Piceetea. It also shows a number of environmental and vegetational features common to the Eurasian and North American boreal forests: 1. a similar cool, humid climate basically of the Dfc type (plus the cool part of the Dfb), 2. a forest canopy consisting almost entirely of *Abies* and *Picea*, thus the same physiognomy as the continental boreal forests; 3. a flora including a fair number of circumpolar or circum-boreal elements; and 4. characteristic importance of *Vaccinium* species.

Nevertheless, the Japanese boreal forest also possesses some unique, distinguishing features: 1) a slightly warmer, more humid climate (mean annual temperature 5 to 6 °C higher than in the continental boreal forests, mean annual precipitation generally over 1000 mm); 2) complete lack of *Pinus* (Diploxylon) and *Larix*, both of which are important elements of other boreal forests; 3) extensive occurrence of extremely dense *Sasa* understories; 4) common occurrence of lianas; and 5. generally high productivity, with total volume of standing trees ca. 600–700 m^3/ha for most forest stands.

One unique feature of the Japanese boreal forest is that it lacks lichen development on well-drained habitats, as is very common elsewhere. Instead, the well-drained habitats are occupied by moss synusiae, which occur on mesic habitats in the continental boreal forests. The prevalence of *Sasa* on mesic sites discourages the establishment of mosses, so the well-drained and/or extremely stony habitats are only the place where the moss type can become established, provided sufficient moisture is available. Such a shift of community types seems to be attributed mainly to climatic differences between the Japanese and continental boreal forests and to the presence of the Sasa understory, promoted by the mild, humid maritime climate of the Japanese boreal forest, which regulate vegetation and soil development to a great extent.

References

Conrad, V. 1946. Usual formulas of continentality and their limits of validity. Trans. Am. Geophysic. Union 27:663–664.

CSSC (Canada Soil Survey Committee) 1978. The Canadian system of soil classification. Can. Dept. Agric. Publ. 1646. Ottawa. 164 pp.

Dahl, E. 1956. Rondane: mountain vegetation in south Norway and its relation to the environment. Norsk. Vidensk.-akademi, Oslo. Matyematisk. Naturvidensk. Klasse Skrift. 1956(3):1–374.

Damman, A. W. 1964. Some forest types of central Newfoundland and their relation to environmental factor. For. Sci. Monogr 8, Soc. Am. Forsters, Washington DC. 62 pp.

FAO-UNESCO 1974. Soil Map of the World, Vol. I, Legend. 59 pp. UNESCO, Paris.

Forest Soil Division. 1976. Classification of forest soil in Japan. Bull. Gov. For. Exp. St. 280:1–28 (in Japanese).

Geological Survey of Japan. 1978. Geological map of Japan (1:1,000,000). 3 maps.

Hämet-Ahti, L., Ahti, T. & Koponen, T. 1974. A scheme of vegetation zones for Japan and adjacent regions. Ann. Bot. Fenn. 11:59–88.

Hokkaido Kazanbai Meimei Iinkai. 1972. *Hokkaido no kazanbai bunpu-zu* (A map of volcanic ash distribution in Hokkaido). Hokkaido Agric. Exp. St. 1 map (in Japanese).

Horikawa, Y. 1970. Natural vegetation. In: Atlas of Japan, Physical, Economic, and Social, pp. 8–9. Inst. Soc. Educ., Inf. Tokyo.

Ishizuka, K. 1974. Mountain vegetation. In: Numata, M. (ed.), The Flora and Vegetation of Japan, pp. 173–210. Kodansha, Tokyo.
Ito, K. 1969. *Sasa*. Sapporo Rin-yu. Nos. 138–144.
Klinka, K., Green, R. N., Trowbridge, R. L. & Lowe, L. E. 1981. Taxonomic classification of humus forms in ecosystems of British Columbia. First Approximation. Victoria: Land Manag. Rep. 8, Ministry Forests, British Columbia. 54 pp.
Kojima, S. 1972. Phytogeocoenoses in the North Klondike River Valley, Yukon Territory. In: Krajina, V. J. (ed.), Progress Report, pp. 29–34. Univ. British Columbia, Vancouver.
Kojima, S. 1975. Biogeoclimatic ecosystem classification of Hokkaido Island, Japan. Paper presented at XIII Pacific Sci. Congr., Vancouver. Abstract p. 103.
Kojima, S. 1979a. Biogeoclimatic zones of Hokkaido Island, Japan. J. Coll. Lib. Arts, Toyama Univ. 12:97–141.
Kojima, S. 1979b. Vegetation and soils in the headwater area of the Tokachi River, Hokkaido, Japan. Sapporo: Hokkaido Electric Power Co. 58 pp. (in Japanese, with English abstract).
Kojima, S. 1980. Vegetation and soils in the headwater area of the Tokachi River, Hokkaido, Japan (II). Sapporo: Hokkaido Electric Power Co. 130 pp. (in Japanese, with English abstract).
Kojima, S. 1983. Phytogeocoenoses of the montane Abies sachalinensis biogeoclimatic zone in Kitami District, Hokkaido, Japan. I. Analysis of coniferous forest phytogeocoenoses. J. Coll. Lib. Arts, Toyama Univ. 16:172–281.
Kojima, S. 1987. An attempt to evaluate fertileness of habitat. In: Takeda, Y. (ed.), Papers on Plant Ecology and Taxonomy to the Memory of Dr. S. Nakanishi, pp. 443–456. Kobe Geobot. Soc. 443–456 (in Japanese, with English synopsis).
Kojima, S. 1991. Classification and ecological characterization of coniferous forest phytogeocoenoses of Hokkaido, Japan. Vegetatio 96:25–42.
Köppen, W. 1936. Das geographische System der Klimate. In: Handbuch der Klimatologie in funf Banden. Köppen, W. & Geiger, R. (eds.), Band I, Teil, C. p 46. Gebruder Borntrager, Berlin.
Krajina, V. J. 1933. Die Pflanzengesellschaften des Mlynica-Tales in den Vysoke Tatra (Hohe Tatra). Bot. Centralb. 50, II:774–957; 51, II:1–224.
Krajina, V. J. 1960. Can we find a common platform for different schools of forest type classification? Silva Fenn. 105:50–55.
Krajina, V. J. 1965. Biogeoclimatic zones and biogeocoenoses of British Columbia. Ecol. W. North America 1:1–17.
Krajina, V. J. 1969. Ecology of forest trees of British Columbia. Ecol. W. North America 2:1–146.
La Roi, G. H. 1967. Ecological studies in the boreal spruce-fir forests of the North American taiga. I. Analysis of the vascular flora. Ecol. Monogr. 37:229–253.
Larsen, J. A. 1980. The boreal ecosystem. Academic Press, New York. 500 pp.
Miyawaki, A. 1967. *Nihon no shokusei* (Vegetation of Japan.) Encyclopedia of science and technology, 3. Vegetation of Japan, pp. 2–9. Gakushuu Kenkyuusha, Tokyo.
Miyawaki, A. 1989. Vaccinio-Piceetea Region. In: Miyawaki, A. (ed.), Vegetation of Japan, Hokkaido, pp. 304–307. Shibundo, Tokyo.
Mueller-Dombois, D. & Ellenberg, H. 1974. Aims and Method of Vegetation Ecology. Appendix B, pp. 466–488. New York: John Wiley & Sons.
Nakano, H. 1942. *Honpou shinrin gunraku no sosei* (Structure of forest communities of Japan.) Bot. Mag. Tokyo 56:186–190 (in Japanese).
Rowe, J. S. 1972. Forest regions of Canada. Dept. Environ., Can. For. Serv. Publ. 1300, Ottawa. 1 map. 172 pp.
Schmithüsen, J. 1976. Atlas zur Biogeographie. Bibliographisches Institut, Mannheim.80 pp.
Shiozaki, M. & Narita, K. 1963. An investigation of the changes in the chemical and physical properties of soils of the cut over area (III). Ann. Rep. Hokkaido Branch, Gov. For. Exper. Sta. 1963:119–125 (in Japanese, with English summary).
Soil Survey Staff. 1960. Soil classification, a comprehensive system, 7th approximation. U.S.D.A., U.S. Gov. Print. Office, Washington, DC. 265 pp.

Sørensen, T. 1948. A method establishing groups of equal amplitude in plant sociology based on similarity of species content and its application to analyses of the vegetation on Danish commons. K. Dansk. Vidensk. Selsk. Biol. Skr. 5:1–16, 34.

Suzuki, S. 1978. Index to Japanese Bambusaceae. Gakken, Tokyo. 384 pp. (in Japanese).

Suzuki, T. 1961. *Nihon no shinrin-tai zenron* (An outline of the forest zones of Japan.) Chiri 9:1036–1043 (in Japanese).

Tatewaki, M. 1958. Forest ecology of the islands of the north Pacific Ocean. J. Fac. Agric. Hokkaido Univ. 50:371–486, 30 plates.

Toyooka, H., Sato, A. & Igarashi, M. 1983. Distribution map of *Sasa* groups in Hokkaido. Hokkaido Br., Forestry & Forest Products Res. Inst. 4 maps. 36 pp.

Tsukada, M. & Nakamura, J. 1989. *Daiyonki-matsu* no *shokuseisi* (Vegetation history of the late Quaternary Period.) In: Miyawaki, A. (ed.), Vegetation of Japan Hokkaido, pp 96–130. Shibundo, Tokyo.

Tuhkanen, S. 1984. A circumboreal system of climatic-phytogeographical regions. Acta Bot. Fenn. 127:1–50.

Uchida, T. 1959. An investigation of the forest humus layers of the needle-leaved forest in Hokkaido. Bull. Gov. For. Exper. Sta. 114:53–205 (in Japanese, with English summary).

Uemura, S., Takeda, Y. & Nakanishi, S. 1986. Behaviour of the main temperate plants in Hokkaido along climatic gradients. Jpn. J. Ecol. 36:141–152 (in Japanese).

Whittaker, R. H. 1970. Communities and Ecosystems. MacMillan Publishing Co., New York. 162 pp.

Yamanaka, T. 1979. *Nihon no shinrin shokusei* (Forest vegetation of Japan.) Tsukiji Shokan, Tokyo. 219 pp. (in Japanese).

Yoshioka, K. 1974. Vegetation map of Japan. In: Numata, M. (ed.), The flora and vegetation of Japan. Kodansha, Tokyo.

Yoshino, M. M. 1978. Altitudinal vegetation belts of Japan with special reference to climatic conditions. Arctic & Alp. Res. 10:449–456.

21. Spatial and temporal patterns of subalpine *Abies* forests in Central Japan

TAKASHI KOHYAMA
Faculty of Education, Kagoshima University, Kagoshima 890, Japan; Present address: Center for Ecological Research, Kyoto University, Shimosakamoto, Otsu 520-01, Japan

Abstract

Two *Abies* species, *A. mariesii* and *A. veitchii*, dominate in the subalpine belt of central Japan. They are recognized as typical climax species, showing similar ecological characteristics, and form either pure *Abies* forests or mixed forests with *Betula ermanii*. A characteristic regeneration pattern called the 'Shimagare' phenomenon (fir waves) is commonly observed in pure *Abies* forests near mountain summits, producing dieback zones in horizontal stripes across mountain slopes. This pattern is maintained by even-aged dense regeneration of *Abies*. Stands with *Betula* hardly show such regeneration. The repetition of abrupt stand-level dieback and subsequent canopy gap formation characterize the structure of *Betula*-mixed stands in various stages. Coexistence of the two *Abies* species is explained in terms of differences in the regeneration process. Stand mosaic patterns are also observed between *Abies* stands and stands of other conifers, *Tsuga diversifolia* and *Picea jezoensis* var. *hondoensis*. Competition-induced exclusion of other species creates such patterns. It is proposed that the mosaic landscape in subalpine coniferous forests results from intra- and inter-specific competition during the regeneration processes.

Key words: dieback, fir waves, regeneration dynamics, subalpine forest

1. Introduction

Natural forests in subalpine belt of the central mountains in Honshu, the main island of Japan, are often dominated by two *Abies* species. The focus of this paper is the dynamic mosaic phenomena in stand distribution in the subalpine *Abies* forests, including the relationship between the two co-

E.O. Box et al. (eds.), Vegetation Science in Forestry, 391–407.

occurring *Abies* species, and their relationship with other tree species. The mosaic phenomena are discussed in relation to the regenerative traits of the *Abies* species.

What we call subalpine coniferous forests appear below the scrub belt dominated by *Pinus pumila* (Pallas) Regel and above the deciduous forests dominated by *Fagus crenata* Blume and *Quercus mongolica* Fischer var. *grosseserrata* (Blume) Rehd. et Wils. The altitudinal range of the subalpine coniferous forests varies from mountain to mountain, but generally lies somewhere in the range 1500–2700 m above sea level. The two *Abies* species, *Abies mariesii* Mast. and *A. veitchii* Lindl., are climax dominants at higher elevations in the subalpine belt. Other common coniferous trees are *Tsuga diversifolia* (Maxim.) Mast. and *Picea jezoensis* (Sieb. et Zucc.) Carr. var. *hondoensis* (Mayer) Rehder. These form climax forests at lower altitude and seral forests at higher altitude, succeeded by climax *Abies* forests. Coniferous forests often contain hardwoods, primarily a birch species, *Betula ermanii* Cham. (See Ishizuka 1974, Kimura 1977, Franklin et al. 1979 for general description of Japanese subalpine coniferous forests.)

2. Stand types in *Abies* forests

The wave regeneration (Sprugel 1976, 1984) or 'Shimagare' (meaning 'stripes of dead-trees' in Japanese) phenomenon (Oshima et al. 1958, Kohyama 1988) is a typical example of a natural stand mosaic. Figure 1 shows the wave regenerating forest on the southwest slope of Mount Shimagare, in the northern Yatsugatake Mountains. This striped pattern of stand-level dieback is maintained by an upward drift (about 1 m per year) of the dieback fronts, due to the prevailing wind, which is succeeded by dense even-aged cohorts of young *Abies* (Kohyama and Fujita 1981). The whole *Abies* forest is maintained dynamically by stand-level dieback as a natural process (*cf.* Mueller-Dombois 1986). The Shimagare phenomenon is not rare in the subalpine *Abies* forests on windward slopes near mountain summits (Oka 1983, Kohyama 1988).

A particular characteristic of subalpine *Abies* forests is that both pure coniferous forests of the two *Abies* species and *Betula*-mixed *Abies* forests occur (Kimura 1977, Kohyama 1984a, 1987, 1988). As Fig. 2 shows, *Betula* trees in mixed forests expand their crowns above the *Abies* trees. Generally, such mixed forests are twice as tall and have several times the biomass, which mainly results from the longer duration of the canopies (Kohyama 1988). The pattern of wave regeneration can be observed only in pure *Abies* forests (Kohyama 1984a, 1988). The north-facing slope of Mount Chausu, just opposite the wave-regenerating forest on Mount Shimagare, is shown in Fig.

Fig. 1. A subalpine *Abies* forest on the southwestern slope of Mount Shimagare (alt. 2,395 m) in the northern Yatsugatake Mountains. The horizontal white belts of standing dead trees are conspicuous. July 1977.

Fig. 2. Profile of a *Betula ermanii*-mixed *Abies* stand in Mugikusa Pass in the northern Yatsugatake Mountains. Wide emergent crowns of *Betula* cover the *Abies* trees. The foreground is covered by *Betula* regeneration invading after artificial clear-felling. October 1978.

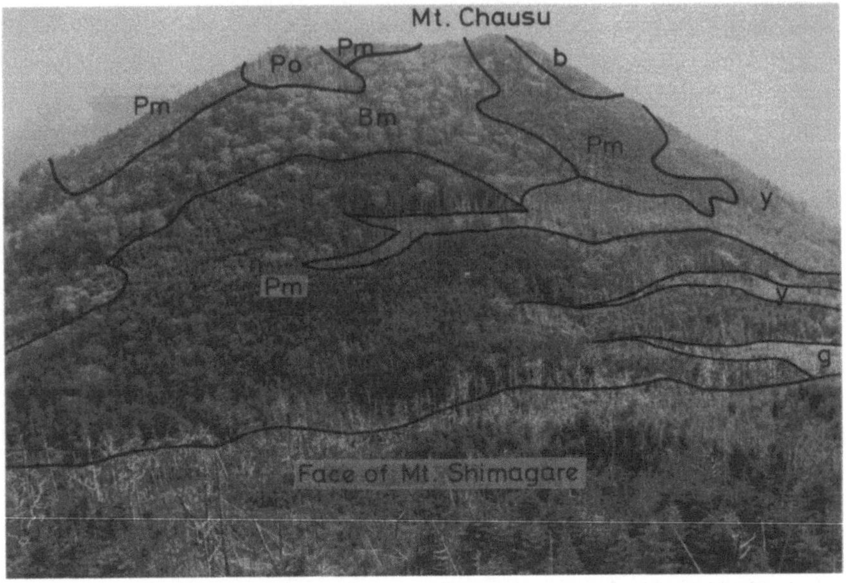

Fig. 3. The northern face of Mount Chausu (alt. 2,383 m) viewed from the summit of Mount Shimagare. The dark crowns are *Abies* trees, the light-colored are *Betula* trees. Pm = mature pure *Abies* stand; Po = over-mature (dieback) pure stand; Bm = mature *Betula*-mixed stand; y = young regeneration; g = grassland; b = bare land. August 1979.

3. The light-colored crowns of *Betula* cover the dark-colored crowns of *Abies* in mixed stands. The stand mosaic composed of pure *Abies* stands and *Betula*-mixed stands is clear. A similar contrast can be observed between pure *Abies balsamea* stands and *Betula papyrifera*-mixed stands in the subalpine forests of New York State (Reiners & Lang 1979, D. G. Sprugel, pers. inform.).

3. Regenerative nature of *Abies*

Here we see the regenerative traits of *Abies*, which cause the mosaic pattern in the landscape. Figure 4 shows the appearance of a 4-year-old *Abies* cohort, originating in the 1974 mast year, along the gradient from a stand edge to a closed-canopy stand. *Abies* seeds germinate successfully on moist moss beds under the dense canopy. They fail to germinate in bright openings, perhaps due to desiccation. The dim stand floor is, however, a harsh environment for seedling survival. Under dim conditions they cannot grow beyond 20 cm in height, even if they survive for 50 years. Such suppressed seedlings have creeping stems. Brighter conditions are required for them to grow upwards (Kohyama 1983).

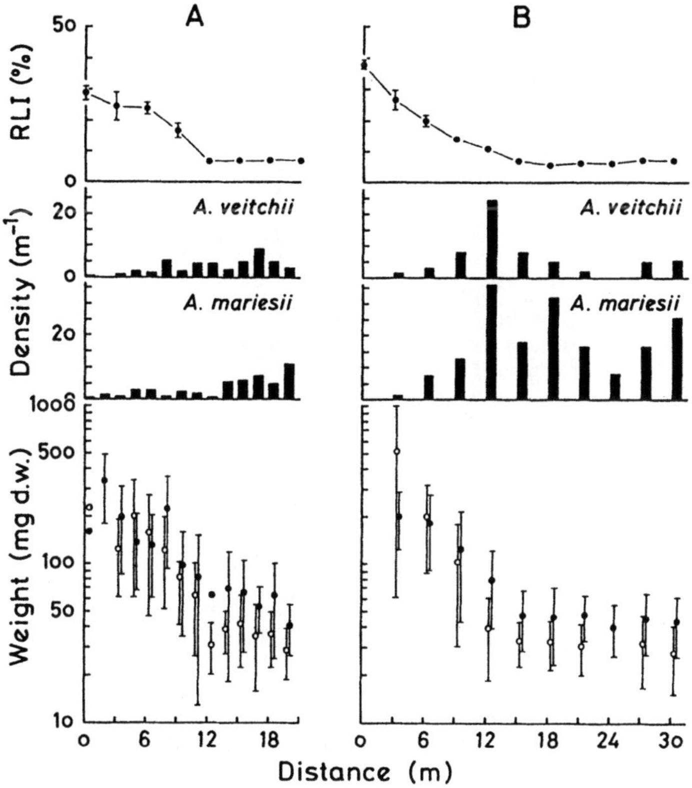

Fig. 4. Occurrence of *Abies* seedlings along two transects from stand edge to closed stand. Top: relative light intensities (mean and standard deviation); middle: density of 4-year-old seedlings of 1974 origin; bottom: mean and standard deviation of seedling dry weight. Open circles = *A. veitchii*, solid circles = *A. mariesii* (Kohyama 1983).

Figure 5 shows suppressed saplings with umbrella-shaped crowns. Such suppressed saplings are often found under a closed canopy, particularly in *Betula*-mixed forests Kohyama 1984a). The growth pattern is quite different from saplings growing on open sites. Suppressed saplings do not grow upward, and their assimilates are allocated to lateral branch shoots (Kohyama 1980). Stem analysis of suppressed saplings shows that they once grew rapidly under better-lit conditions but then were suppressed (Fig. 6). It is clear that suppressed saplings are the group excluded from the canopy due to the competition for light. Two juvenile pools, or advance regeneration, can be thus recognized on the forest floor (Fig. 7); the seedling bank and the sapling bank (Kohyama 1983). Their boundary occurs at 20 cm in height and 5 g in dry weight. Above this size, individuals depend only on assimilates from lateral-branch needles and thus can allocate assimilates exclusively to lateral-

Fig. 5. Suppressed saplings of *Abies mariesii* with umbrella-shaped crowns on the floor of a *Betula*-mixed *Abies* stand in the northern Yatsugatake Mountains. October 1978.

branch shoots under dim conditions. Both banks contribute to regeneration as a reaction to canopy breakage. The relative importance of each bank in regeneration is determined by stand history (related to intensity and duration of suppression) and the simultaneity and/or duration of the canopy dieback.

The regeneration pattern of subalpine *Abies* forest observed in the Yatsugatake Mountains is summarized in Fig. 8. In a pure stand such as the Mount Shimagare forest, the shading effect by the canopy is severe, and suppressed saplings cannot survive until the time of stand dieback (Kohyama & Fujita 1981). In such cases regeneration from a seedling bank takes place (Fig. 8a). The severe competition between canopy trees in a pure stand creates an even-sized, small-crowned tree population. This population is sensitive to physical stress such as prevailing wind and can easily collapse all at once (Kohyama & Fujita 1981: average duration of dying-off is estimated to be 2.6 years). On the contrary, the shading effect is not severe in *Betula*-mixed stands and a sapling bank can survive continuously (Kohyama 1984a). Light conditions are also better for canopy trees of *Abies*, and they have vertically deeper crowns. Thus, stand dieback in *Betula*-mixed stands progresses more

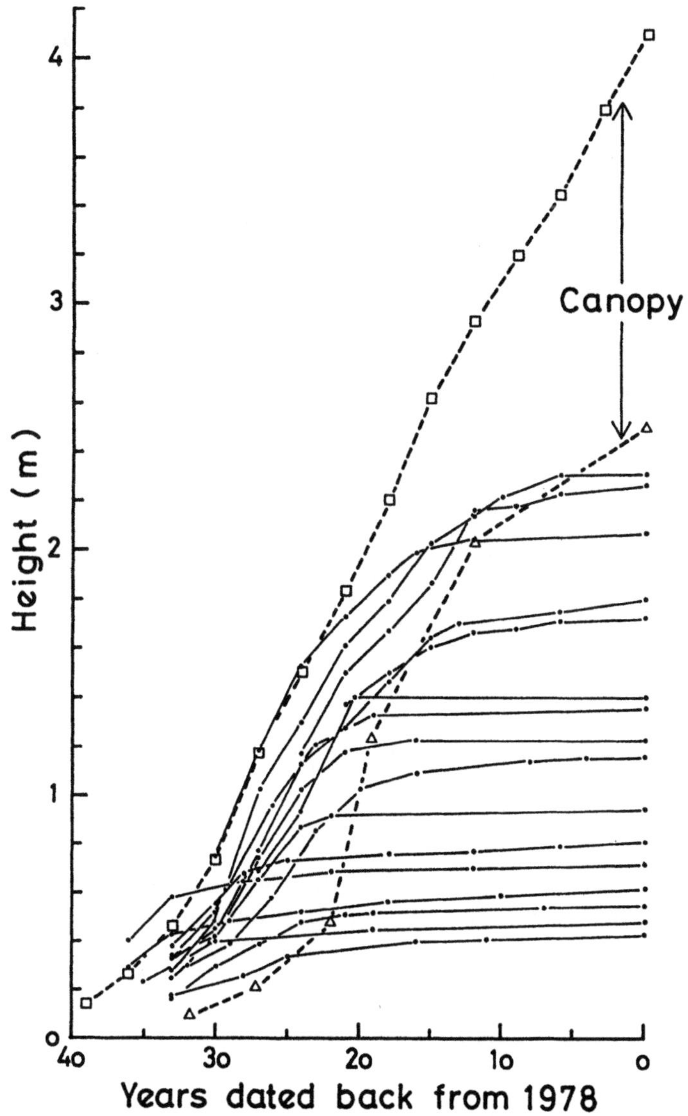

Fig. 6. Time course of height growth of suppressed saplings of *Abies* spp. in a 40-year-old pure *Abies* stand on Mount Shimagare. Canopy change is given by broken lines (Kohyama 1984b).

slowly than in pure *Abies* stands, and regeneration from a sapling bank succeeds the stand dieback (Fig. 8b).

It is interesting that this process of regeneration prevents the invasion of light-demanding *Betula* seedlings. Therefore, the next generation becomes a pure *Abies* stand, which will be replaced by regeneration from a seedling bank, occasionally allowing the invasion of light-demanding *Betula*. *Betula*

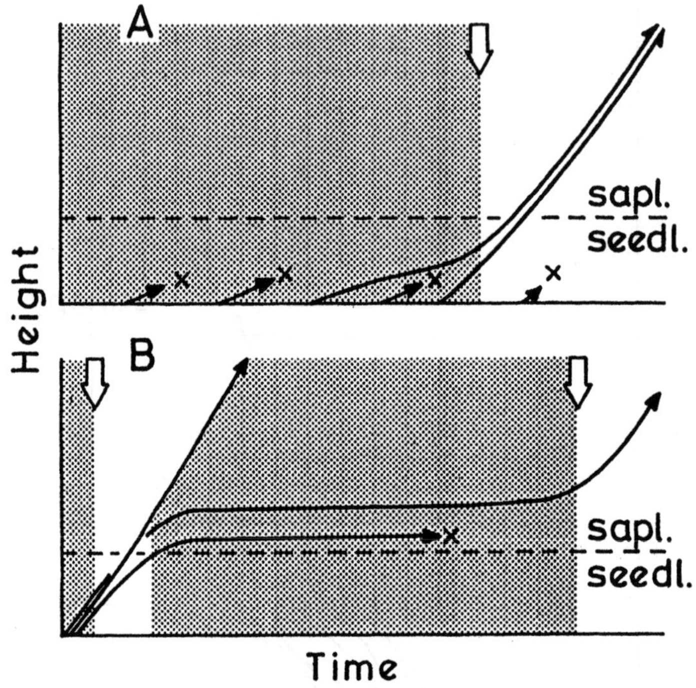

Fig. 7. Two regeneration types of *Abies* in the Yatsugatake Mountains, as expressed by height growth. A: seedling bank type. B: sapling bank type. Open arrows indicate canopy breakage, crosses show individual death (Kohyama 1984b).

regeneration requires a large-scale abrupt stand-level dieback (as seen at the road-side artificial clear-felling in Fig. 2), often caused by typhoons.

4. Coexistence due to regeneration dynamics

The difference in shade-tolerance between the two *Abies* species means that *Abies mariesii* is more persistent in forming a bank of suppressed saplings. By contrast, *Abies veitchii* shows a higher growth rate during regeneration and will overtop *Abies mariesii* when the two species sprout together from a seedling bank (Kohyama 1984a, Kimura et al. 1986). It is the existence of these two types of advanced regeneration that enables the two *Abies* species to coexist, as the model below proves.

The inter-specific competition model with two-regeneration types (Kohyama 1984a), which is based on the idea of undifferentiated multi-site dynamic models (Levin 1976, Whittaker and Levin 1977), is illustrated in Fig. 9. This model assumes that the whole forest K is divided into four types of micro-sites: X and Y the sites occupied by the two species, R_1 and R_2 the sites

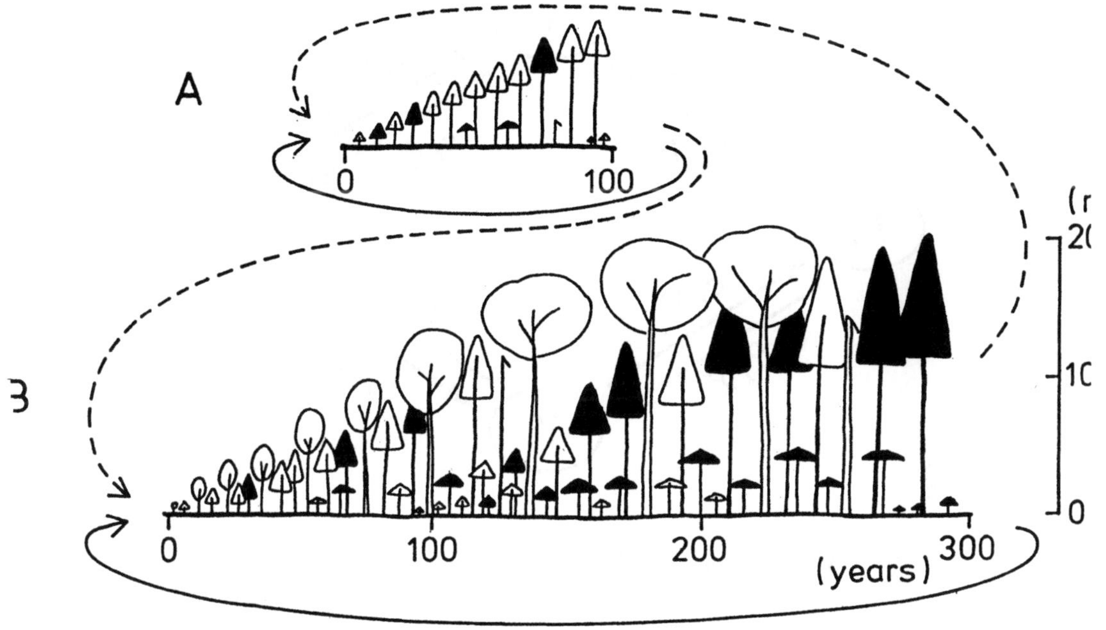

Fig. 8. Regeneration dynamics of subalpine *Abies* forests in the Yatsugatake Mountains. Triangular crowns are *Abies* (open = *A. veitchii*, solid = *A. mariesii*), circular crowns are *Betula ermanii*. A: Shimagare type of regeneration. B: *Betula*-mixed type of regeneration. Redrawn from Kohyama (1988).

open for two regeneration types. Transition rates between categories are given with arrows. If each species has an advantageous regeneration type, i.e. the inequality in Fig. 9 is satisfied, then it is possible for the two species to coexist. Coexistence occurs, however, only within a narrow range of the ratio of the parameters, as is shown later. It is expected from species characteristics that the relatively less tolerant *Abies veitchii* can form monospecific, self-maintaining forests, while the more tolerant *Abies mariesii* is likely to regenerate in *Betula*-mixed forests (Fig. 8). The classic view of succession theory that most shade-tolerant species can finally form self-maintaining forests is no longer valid, and the view of regenerative dynamics is only effective in explaining the species pattern in subalpine *Abies* forests.

5. Geographic gradient

Abies veitchii is dominant on the Pacific side of Japan, where there is less snow, while *Abies mariesii* dominates on the Japan Sea side and in northern Honshu, where there is heavy snow. In some mountains with the heaviest

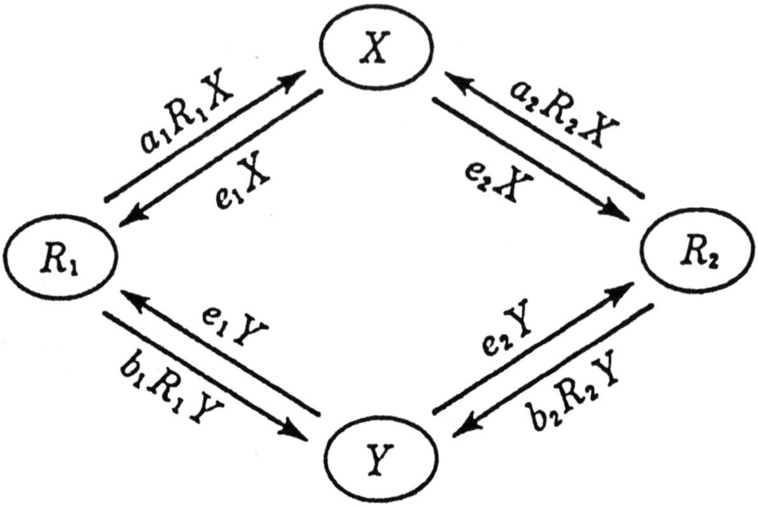

$$K = X + Y + R_1 + R_2$$

necessary condition for coexistence:

$$(a_1 - b_1)(a_2 - b_2) < 0$$

Fig. 9. Two-species dynamic model of Kohyama (1984a), considering two types of regeneration sites. X is the site occupied by canopy of species one; Y that occupied by species two; R_1 is the site open for regeneration type one; R_2 that open for regeneration type 2; e_i is per capita rate of canopy dieback, a_iR_i and b_iR_i are per capita rates of recruitment.

snowfall, no coniferous subalpine forest occurs at all (Sugita 1987). The two subalpine *Abies* species coexist within a limited range along this east-west gradient across Honshu (Kohyama 1984a, 1988).

Wave regeneration with recreation of pure *Abies* stands, i.e. the Shimagare phenomenon, occurs in regions with less snowfall (Oka 1983, Kohyama 1988). In heavy-snow regions, dense pure stands of *Abies mariesii* forest (no *A. veitchii* there) are rarely formed, but they are thinned due to rime-ice damage in winter (Saito 1977). Figure 10 shows a landscape with such a thinned *A. mariesii* forest. Solitary *Abies* trees have living lower branches near the ground because they are protected by winter snow. The competition between *Abies* trees is not strong there, and stand-level dieback rarely occurs. In such thinner forests, regeneration from a sapling bank maintained around canopy trees is most frequent, because the space between canopy trees is covered by dwarf bamboo (Fig. 10), which prevents *Abies* from regenerating (Iizumi 1981). The change in forest structure and regeneration type along

Fig. 10. A thin *Abies mariesii* forest on Mount Hakkoda (alt. 1,585 m), northern Honshu (photo by A. Makita in October 1989). The light-colored area is dwarf bamboo scrub.

the snowfall gradient is summarized in Fig. 11. The geographic distribution of the two *Abies* species can also be understood in terms of regeneration dynamics. Regions with less snow are dominated by *Abies veitchii*, which prefers regeneration from a seedling bank, while regions with heavy snow are dominated by *Abies mariesii*, which prefers regeneration from a sapling bank. (Further, *Abies mariesii* has stronger wood and is more resistant to stress by snow deposits than is *A. veitchii*.)

Let us analyze the effect of the east-west gradient on the abundance of the two *Abies* species, using the above model with two regeneration types (Fig. 9). When the ratio of the formation of the two types of regeneration sites, e_1 versus e_2, changes continuously along the gradient, each species can be potentially distributed along the whole range in the single-species system. However in the two-species system, the two species coexist only within a narrow range (Fig. 12). It is therefore concluded that competitive exclusion plays an important role in the geographic distribution pattern of the two *Abies* species.

6. Stand mosaics with other species

Here we turn to the relation between *Abies* forests and other coniferous subalpine forests. Mosaics of stands at different developmental stages have

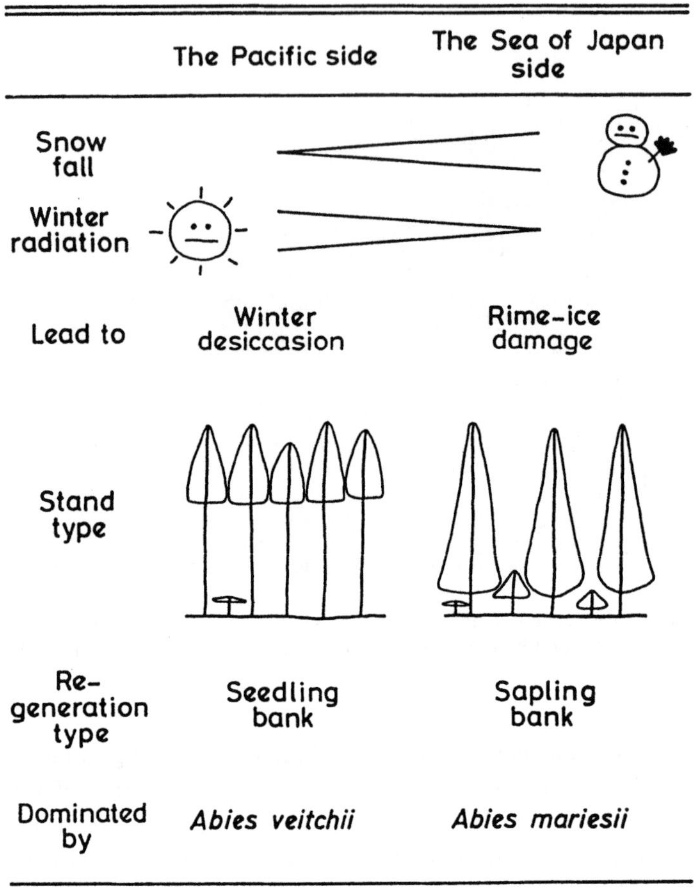

Fig. 11. Difference between the Pacific side and the Sea of Japan side of Honshu, in climatic conditions and in the structure and ecology of subalpine *Abies* forests.

also been reported for *Tsuga diversifolia* forests (Ohsawa 1981, Kanzaki 1984). Further, stand mosaics between *Abies*-dominated forests and *Tsuga*-dominated forests have been observed with clear boundaries (Kimura 1963, Franklin et al. 1979). A vegetation map of the northern Yatsugatake Mountains is shown in Fig. 13a. The profile diagram in Fig. 13b shows such a clear boundary between *Abies* and Tsuga stands. Kimura (1963) discovered that *Tsuga* forests on andesitic lava in the Yatsugatake Mountains are established on shallower, more uneven soils than are *Abies* forests. *Picea jezoensis* occasionally occurs only in a narrow range between *Tsuga* and *Abies* stands, and *Betula ermanii* is admixed only in deep-soiled *Abies* forests.

On the contrary, Ohsawa (1984) and Nakamura (1985) found that *Abies* forests can flourish prior to *Tsuga* forests on the scoria of Mount Fuji. There,

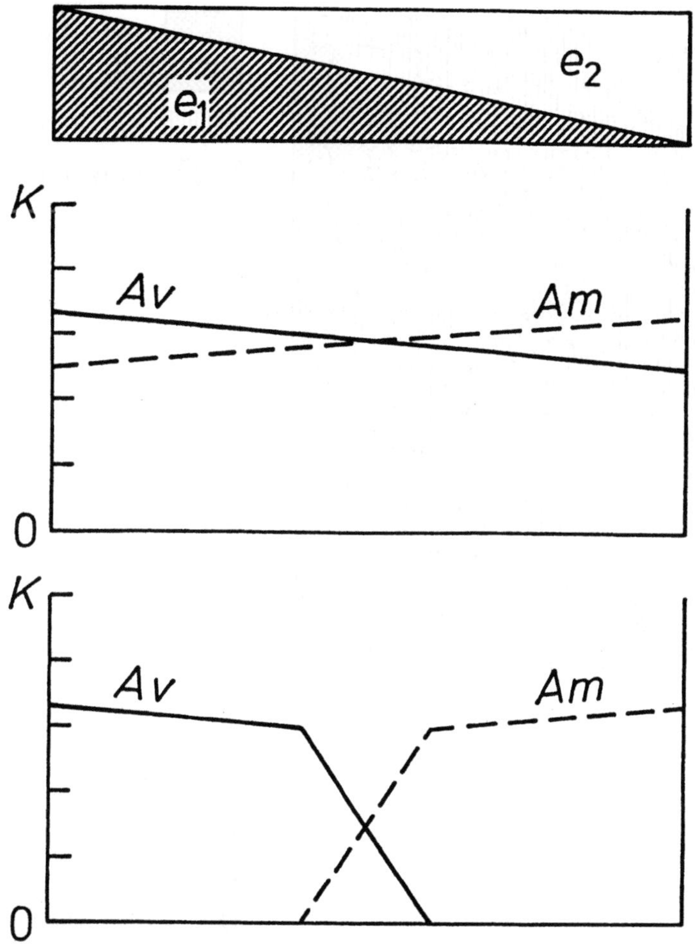

Fig. 12. Dependence of the result of the Kohyama (1984a) model, shown in Fig. 9, on change of the e_1/e_2 ratio, i.e. the ratio of the two regeneration types. e_1 = seedling bank type; e_2 = sapling bank type; *Av* = *Abies veitchii*; *Am* = *Abies mariesii*. Each species can exist if there is no competition (middle), but coexistence is possible only in a limited range when they compete with each other (below). [Calculated for the case that $a_1 : a_2 : (e_1 + e_2) = b_2 : b_1 : (e_1 + e_2) = 3:2:1$.]

Tsuga forests can be then replaced by *Abies* forests as the soil develops. *Abies* seems to require an even soil texture so it is easier to establish on fine scoria than on lava. Therefore, it is neither easy nor meaningful to discuss the position of these species on a simple gradient of primary succession. Rather, the tendency of the dominant species to separate clearly in the landscape is ecologically notable. Kanzaki & Yoda (1986) indicated that *Tsuga* has a longer life span, while *Abies* has more frequent establishment in

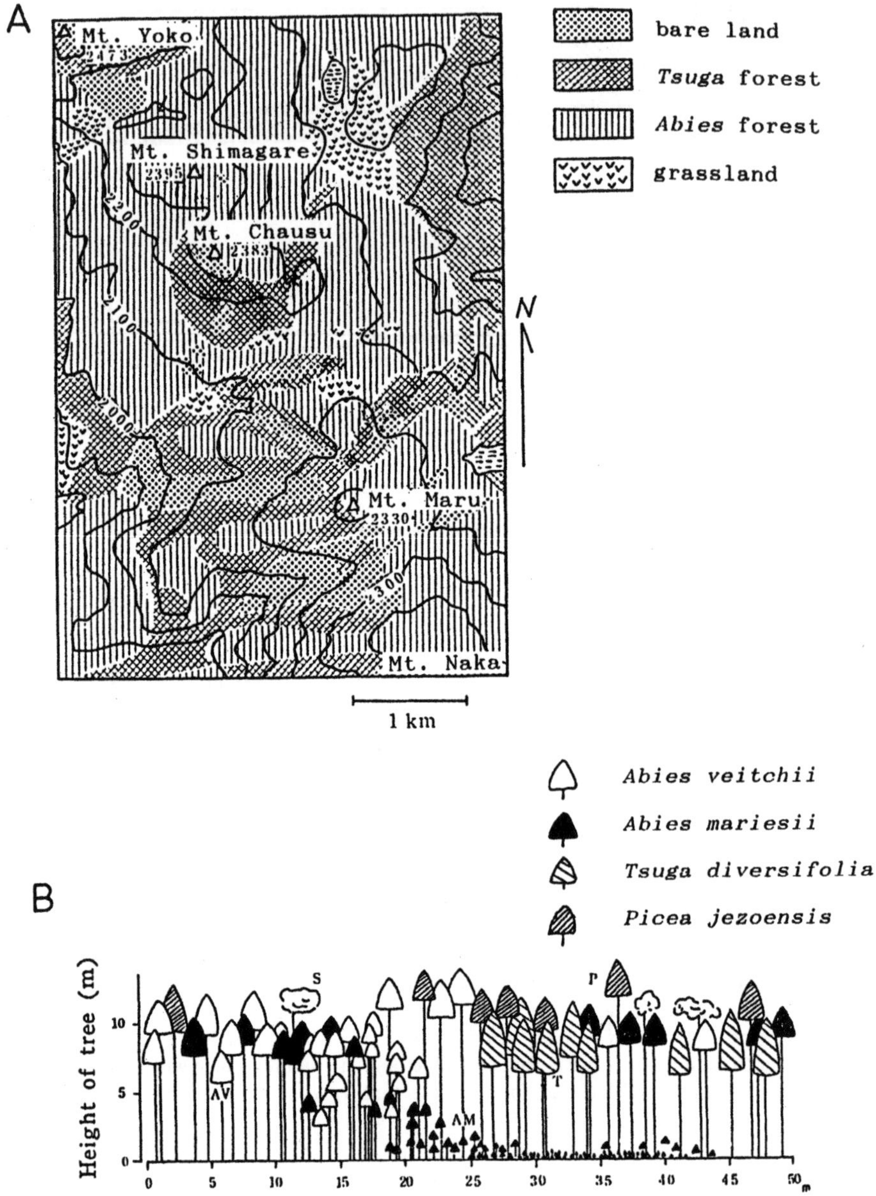

Fig. 13. A: vegetation map in the northern Yatsugatake Mountains. B: forest profile along a transect from *Abies* stand to *Tsuga* stand; an abrupt boundary exists between the two stands. Reproduced from Kimura (1963).

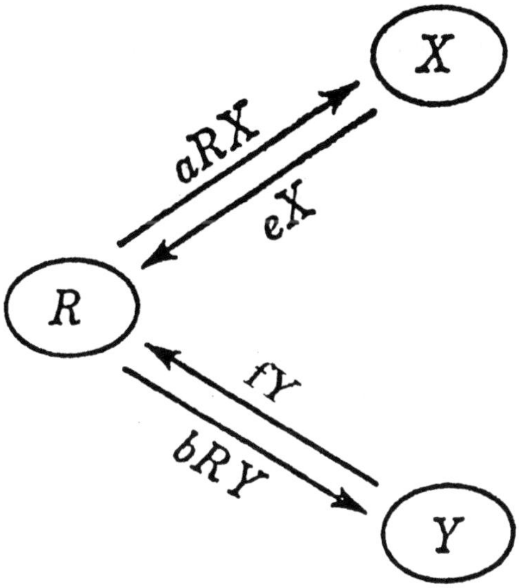

$K = X + Y + R$

demographic difference:

$a > b$ (reproduction for X)
$e > f$ (survival for Y)

Fig. 14. Two-species dynamic model of multi-site mosaic forest with only one type of regeneration site R. X is the site occupied by canopy of species one, Y is occupied by species two. e and f are per capita rates of canopy dieback, and aR and bR are per capita rates of recruitment, for species one and two respectively.

open sites in *Tsuga*-dominated forests. The coexistence between Engelmann spruce (*Picea engelmannii*) and subalpine fir (*Abies lasiocarpa*) in North America has been explained by a similar demographic contrast (Shea 1985, Veblen 1986). Spruce has a longer life span, while fir is more rapid in recruitment. It is doubtful, however, whether such a demographic difference is enough to explain their stable coexistence.

The competition model in Fig. 14 is similar to that of Fig. 9 but has only one regeneration type. The mathematical expression of this model is identical to the Lotka-Volterra competition equation, where the interaction coefficient is unity. Demographic difference can be expressed by inequalities $a > b$ and $e > f$ (X for *Abies* with rapid recruitment; Y for long-lived *Tsuga* or *Picea*). It is well known that coexistence between the two species is not stable in this model, even though the above inequalities are satisfied. We should add more factors, such as spatial clumping of species and temporal fluctuation in reproduction (Chesson & Werner 1981, Shmida & Ellner 1984) which may promote coexistence between *Tsuga* and *Abies*, or between *Picea* and *Abies*. In this context, we must draw attention to the tendency for subalpine coniferous forests in central Japan to occur in mosaics of mono-dominant stands (except for the mixture of the two *Abies* species). This tendency reflects the mechanism of competitive exclusion. In conclusion, inter-specific competition plays an important role in the landscape mosaic of subalpine forests, due to the segregation of dominant species with clear boundaries, as well as in the geographic segregation of the species distributions. Subalpine forests have relatively simple tree compositions, with only a few dominant species. They are, therefore, suitable objects for analyzing the mechanisms of tree community organization. I believe that further analyses of subalpine coniferous forests will present new views of plant communities which overcome the 'individualistic view' of species distribution.

Acknowledgements

I am grateful to Dr. Takehiro Masuzawa and Prof. Elgene Box for giving me the opportunity to present this paper and for helpful suggestions on the manuscript. I thank Mr. Akifumi Makita (Fig. 10), Prof. Makoto Kimura (Fig. 13), the Botanical Society of Japan (Fig. 4), and the Society for the Study of Species Biology (Figs. 6–7) for permission to reproduce a photograph and figures.

References

Chesson, P. L. & Werner, R. R. 1981. Environmental variability promotes coexistence in lottery competitive systems. American Naturalist 117:923–943.

Franklin, J. F., Maeda, T., Oshumi, Y., Matsui, M. & Yagi, H. 1979. Subalpine coniferous forests of central Honshu, Japan. Ecological Monographs 49:311–334.

Iizumi, S. (ed.) 1981. Aomori-todomatsu rin no seitaigakuteki kenkyu (Ecological Studies on the *Abies mariesii* Forest). Mount Hakkoda Botanical Laboratory, Sendai 117 pp. (in Japanese).

Ishizuka, K. 1974. (Mountain zone). In: Numata, M. (ed.) The Flora and Vegetation of Japan, pp. 173–210. Kodansha, Tokyo/Elsevier, Amsterdam.

Kanzaki, M. 1984. Regeneration in subalpine coniferous forests I. Mosaic structure and regeneration process in a *Tsuga diversifolia* forest. Botanical Magazine, Tokyo 97:297–311.

Kanzaki, M. & Yoda, K. 1986. Regeneration in subalpine coniferous forest II. Mortality and the pattern of death of canopy trees. Botanical Magazine, Tokyo 99:37–51.

Kimura, M. 1963. Dynamics of vegetation in relation to soil development in northern Yatsugatake Mountains. Japanese Journal of Botany 18:255–287.

Kimura, M. 1977. (Plant succession in subalpine region.) In: Numata, M. (ed.) pp. 21–30. Succession of Plant Community and Its Mechanisms, Asakura Shoten, Tokyo (in Japanese).

Kimura, M., Kimura, W., Honma, S., Hasuno, T. & Sasaki, T. 1986. Analysis of development of a subalpine *Abies* stand based on the growth-processes of individual trees. Ecological Research 1: 229–248.

Kohyama, T. 1980. Growth pattern of *Abies mariesii* saplings under conditions of open-growth and suppression. Botanical Magazine, Tokyo 93:13–24.

Kohyama, T. 1983. Seedling stage of two subalpine *Abies* species in distinction from sapling stage: a matter-economic analysis. Botanical Magazine, Tokyo 95:49–65.

Kohyama, T. 1984a. Regeneration and coexistence of two *Abies* species dominating subalpine forests in central Japan. Oecologia (Berlin) 62: 156–161.

Kohyama, T. 1948b. Rinboku no demography wo meguru mondaiten (Problems on the demography of forest trees.) Shuseibutugaku-Kenkyu (Kyoto) 8:1–10 (in Japanese).

Kohyama, T. 1987. Kita-Yatsugatake no akozantai shirabiso-ooshirabiso-dakekanba konkorin no dotai (Dynamics of subalpine fir-birch forests in northern Yatsugatake Mountains.) Bulletin of the Botanical Society of Nagano 20:36–81 (in Japanese).

Kohyama, T. 1988. Etiology of "Shimagare" dieback and regeneration in subalpine *Abies* forests of Japan. GeoJournal 17:201–208.

Kohyama, T. & Fujita, N. 1981. Studies on the *Abies* population of Mt. Shimagare I. Survivorship curve. Botanical Magazine, Tokyo 94:55–68.

Levin, S. A. 1976. Population dynamic models in heterogeneous environments. Annual Review of Ecology and Systematics 7:287–310.

Mueller-Dombois, D. 1986. Perspectives for an etiology of stand-level dieback. Annual Review of Ecology and Systematics 17:221–243.

Nakamura, T. 1985. Forest succession in the subalpine region of Mt. Fuji, Japan. Vegetatio 64:15–27.

Ohsawa, M. 1981. A basic unit in forest community dynamics: a case study in the subalpine forest of Japan. BIOTROP Special Publication (Bogor) 13:43–62.

Ohsawa, M. 1984. Differentiation of vegetation zones and species strategies in the subalpine region of Mt. Fuji. Vegetatio 57:15–52.

Oshima, Y., Kimura, M., Iwaki, H. & Kuroiwa, S. 1958. Ecological and physiological studies on the vegetation of Mt. Shimagare I. Preliminary survey of the vegetation of Mt. Shimagare. Botanical Magazine, Tokyo 71:289–301.

Oka, S. 1983. Shimagare gensho no bunpu ni kansuru saikento (Reconsideration on the distribution of the Shimagare phenomenon in Japan.) Journal of Geography (Tokyo) 92:219–234 (in Japanese).

Reiners, W. A. & Lang, G. E. 1979. Vegetational patterns and processes in the balsam fir zone, White Mountains, New Hampshire. Ecology 60:304–317.

Saito, K. 1977. Tohoku Nippon akozantai sin'yojurin no ruikei to bunpu (Types and distribution of subalpine coniferous forest in northeast Japan.) Bulletin of Yamagata University, Natural Science 9:265–293 (in Japanese).

Shea, K. L. 1985. Demographic aspects of coexistence in Engelmann spruce and subalpine fir. American Journal of Botany 72:1823–1833.

Shmida, A. and Ellner, S. 1984. Coexistence of plant species with similar niches. Vegetatio 58:29–55.

Sprugel, D. G. 1976. Dynamic structure of wave-regenerated *Abies balsamea* forests in the north-eastern United States. Journal of Ecology 64:889–911.

Sprugel, D. G. 1984. Density, biomass, productivity, and nutrient cycling changes during stand development in wave-regenerated balsam fir forests. Ecological Monographs 54:165–186.

Sugita, H. 1987. Akozantai sin'yojurin no bunpu jotai to sekisetsu-shin oyobi akozantai-iki no hirosa tono kankei: Jo-etsu sanchi wo chusin to suru chiiki ni tuite (Development of subalpine coniferous forests in relation to the snow accumulation depth and the extent of subalpine area: the case of the Jo-etsu Mountains and vicinity, central Japan.) Japanese Journal of Ecology 37:175–181 (in Japanese).

Veblen, T. T. 1986. Treefalls and the coexistence of conifers in subalpine forests of the Central Rockies. Ecology 67:644–649.

Whittaker, R. H. & Levin, S. A. 1977. The role of mosaic phenomena in natural communities. Theoretical Population Biology 12:117–139.

22. Physiological ecology of the growth characteristics of Japanese mountain birch in Northern Japan: a comparison with Japanese white birch

TAKAYOSHI KOIKE
Forestry and Forest Products Research Institute, Hokkaido Research Center, Sapporo 062, Japan

Abstract

An ecophysiological comparison was conducted of the growth characteristics of Japanese mountain birch (*Betula ermanii*) and Japanese white birch (*Betula platyphylla* var. *japonica*). The optimum temperatures for net photosynthesis and for leaf development of mountain birch were lower than for white birch. Mountain birch completes its shoot development more quickly than white birch. Under field conditions, mountain birch develops continuously from early leaves to late leaves, while white birch expands its late leaves only after the complete expansion of early leaves. The time lag in growth between early and late leaves of white birch disappered with a high temperature treatment. Leaves of mountain birch were injured by exposure to −3°C for 2 hours, while the leaves of white birch could survive −3°C for 8 hours. This pattern was observed throughout the growing period. The distribution and habit of the two birch species is discussed based on those growth characteristics.

Key words: Japanese birch, low-temperature tolerance, optimum temperature, shoot development

Introduction

Japanese mountain birch (*Betula ermanii* Cham.) is a light-demanding species (Watanabe 1971, Koike 1987). It usually appears in mixed coniferous forests because of its longevity which exceeds 250 years (Watanabe 1971, 1979) and invades open spaces or steep slopes where no coniferous species can grow (Ohsawa et al. 1973). In northern Japan, mountain birch occurs widely in the montane and subalpine belts up to 1600 m (Nakano 1968, Ishizuka &

E.O. Box et al. (eds.), Vegetation Science in Forestry, 409–422.

Sugahara 1986, Okitsu 1987). It sometimes forms timberlines or pure stands below the subalpine timberline commonly formed by dwarf pine (*Pinus pumila* Regel) (Watanabe 1971, Okitsu 1987). Mountain birch also grows along seashores in eastern Hokkaido, where summer is cool and humid (Tatewaki 1958, Watanabe 1979). Why does mountain birch grow over such a wide range, especially higher mountain sites and cooler coastal areas? An answer is sought by comparison with a contrasting birch species, Japanese white birch.

Japanese white birch, *Betula platyphylla* Sukatch. var. *japonica* Hara, is a typical early-successional species (Tabata 1966, Koike and Sakagami 1985) and grows well on lowlands and foothills, in both dry and sub-humid soils (Tabata 1966, Sato et al. 1986). Under natural conditions, white birch invades after forest fires and land clearing, forming pure stands. Its lifrspan is 70 to 100 years at most (Watanabe 1971). White birch ranges from lowlands to montane forests, up to 1000 m on Honshu, the central part of Japan (Takahashi 1962, Komizunai 1972) and up to 700 m in central Hokkaido, northern Japan (Nakano 1968).

Seeds of white birch germinate uniformly, and the seedlings usually suffer from cold damage at higher sites. In contrast, the seedlings of mountain birch may avoid frost damage because of the non-uniform germination of the seeds (Komizunai 1972). In European mountain birch, the duration of the growth period is strongly correlated with seedling survival (Kullman 1986). The leaves of mountain birch are damaged by late frost, however, while those of white birch are not (Koike et al. 1983).

The vertical distribution of species is essentially affected by climatic conditions (Woodward 1987, 1989) or how the species can use the climatic conditions (Grace 1987, Körner & Larcher 1988). As Harper et al. (1961) have explained, closely related species can grow at similar sites because they respond in slightly different way to environment. Former studies, however, only described the vegetational characteristics of birch species. There was little research on ecophysiological aspects of the two species (Okitsu 1987).

In this chapter, the physiological ecology of growth characteristics of the two contrasting birch species is examined, in order to clarify their growth habits and distributions. Four aspects of their ecology will be addressed in the following sections: 1) shoot growth of mountain birch in natural environments, 2) comparison between mountain birch and white birch under field conditions, 3) shoot development of the two birch species in regulated environments, and 4) low-temperature tolerance of leaves during the growing period.

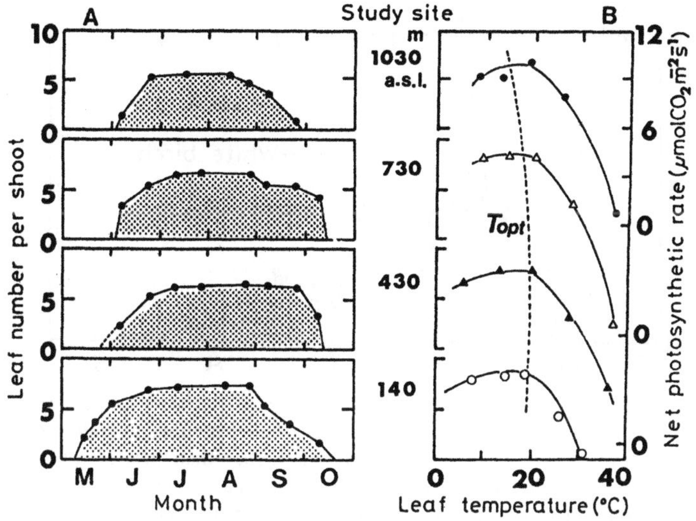

Fig. 1. Seasonal change in the number of leaves per shoot (A) and the optimum temperature for photosynthesis (T_{opt}) (B), for mountain birch grown at different elevations. The number of leaves per shoot and T_{opt} of mountain birch seem to be independent of growth elevations (Koike unpubl.).

Shoot growth of mountain birch in natural environments

Because of the growing period decreases with increasing elevation (Tranquillini 1979), shoot length of European mountain birch, *Betula pubescens* (Sveinbjörnsson 1981) and the diameter growth of evergreen conifers (Benecke & Havranek 1980a) both decrease with increasing elevation. The foliation period of mountain birch also decreases with increasing altitude (Fig. 1A). The number of leaves per current shoot in the middle portion of crown is almost constant, five or six, regardless of elevation. The number of leaf primordia in the middle of the crown of mounatin birch seems, thus, to be fixed and independent of elevation.

The optimum temperature for photosynthesis (T_{opt}) changes markedly with altitude in *Eucalyptus pauciflora* (Slatyer 1978) and evergreen conifers (Benecke & Havranek 1980b), and slightly in herbaceous plants (Körner & Diemer 1987). This shift seems to be an adaptation to an elevational decline of mean temperature (Berry &Björkman 1980). No difference in the maximum photosynthetic rate (P_{max}) or in T_{opt} was observed in mountain birch (Fig. 1B), Moreover, no difference in the P_{max} of European mountain birch grown at different elevation was reported for *in situ* measurements (Sveinbjörnsson 1981). The T_{opt} of mountain birch was around 15°C, independent of growing elevation, which is 5°C lower than that of some trees

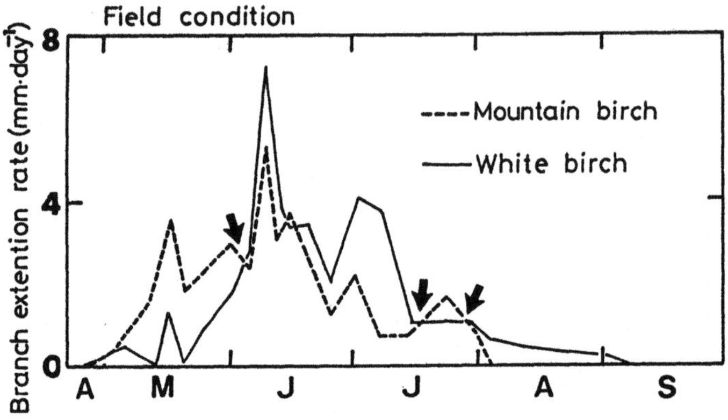

Fig. 2. Seasonal change in the rate of branch extension of mountain birch and white birch. During May the growth rate of mountain birch was higher than that of white birch. Shoot development in mountain birch ceased one month earlier than in white birch. Arrows indicate points at which the faster growth rate of the two species changes (adapted from Koike 1987).

originating in temperate forests (Larcher 1984). Küppers et al. (1987) also reported that the T_{opt} of *Eucalyptus* species from various altitudes was unchanged.

Based on these elevational data, the physiological growth traits of mountain birch might be suitable for low-temperature environments. The young shoots of mountain birch are sometimes killed by late frost, however, while white birch is not damaged by late frost (Koike et al. 1983). Therefore, comparative observations between mountain birch and white birch will be needed in order to know their growth characteristics.

Comparison of shoot growth in mountain birch and white birch under field conditions

The shoot development of *Betula* has been classified as heterophyllous (see Kozlowski & Clausen 1966, Koike & Sakagami 1985). Birch species first flush their early leaves and then expand their late leaves. Under timberline or shade environments, the proportion of early leaves in mountain birch increased (Fujimoto & Miyakawa 1991). The extension rate of shoots in the middle of the crown of mountain birch, during May, was two times higher than that of white birch during May (Fig. 2). From mid-June to mid-July, the shoot extension rate of white birch was twice that of mountain birch. Mountain birch expanded its leaves continuously from early leaves to late leaves. In contrast, the late leaves of white birch were expanded only after

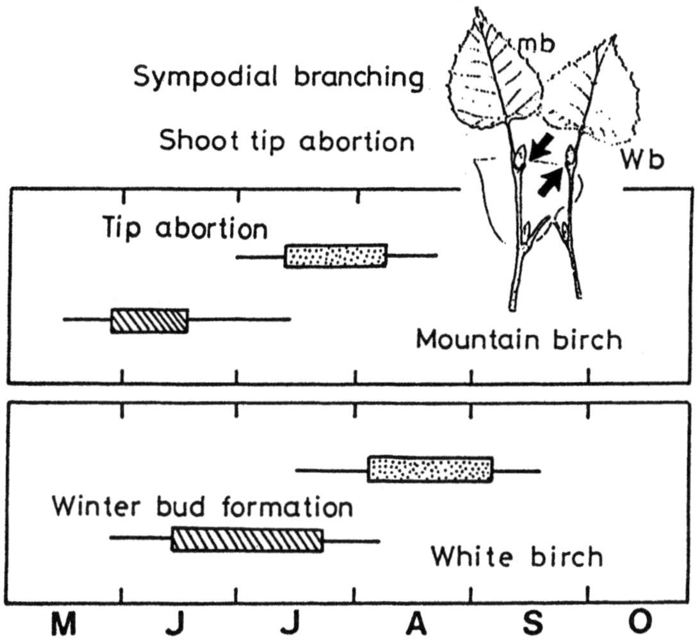

Fig. 3. The time and duration of tip abortion and formation of winter buds of mountain birch (Mb) and white birch (Wb). The horizontal lines and wider bars show the temporal ranges and peaks for the events, respectively. Mountain birch completes shoot development earlier than white birch (adapted from Koike 1987). Birch stops shoot growth with tip abortion because of sympodial branching.

the complete expansion of early leaves (Koike 1987). Shoot growth of mountain birch ceased by late July, whereas that of white birch continued until early September. According to provenance tests, both within and between species of *Betula*, those plants from higher latitude or elevation complete their growth earlier than those from lower latiude or elevation (Sharik & Barnes 1976, Eriksson & Jonsson 1986). Growth cessation of *Betula pendula* is induced by the joint effect of heat sum and photoperiod (Koski & Sievanen 1985), factors likely to be important in mountain and white birch as well.

Winter bud formation in mountain birch starts in late May, 20 days earlier than in white birch (Fig. 3). In both species, winter bud formation occurs first in the lower crown and progresses toward the top. The duration of forming winter buds in mountain birch was half the time as long as that in white birch. Both species stopped their shoot growth with tip abortion because of sympordial branching (Kikuzawa 1983). The occurrence of tip abortion in mountain birch began in late July and was completed by mid-August, while that in white birch started in mid-August and lasted until mid Septem-

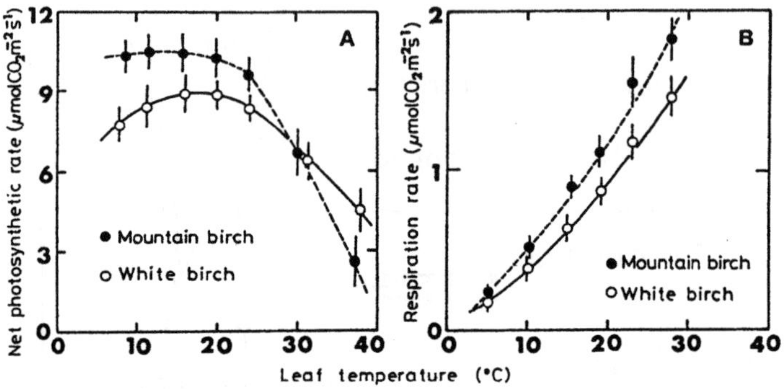

Fig. 4. Temperature dependence of net photosynthesis (A) and respiration (B) of mountain birch and white birch. Net photosynthetic rate was measured under saturated photon flux density (PFD) of 680 μmol $\cdot$ $m^{-2} \cdot s^{-1}$ (Koike 1987).

ber (Fig. 3). Also, early leaves of mountain birch developed exponentially at 1.5°C lower daily mean temperature than for white birch (Koike 1987). The temperature required for developing 50% of all leaves per shoot was 1.3°C lower in mountain birch than in white birch at 800 m (Yanagisawa 1954).

The optimum temperature for photosynthesis (T_{opt} was 15°C for mountain birch and 19°C for white birch (Fig. 4A). Above T_{opt}, net photosynthesis in mountain birch was depressed more than in white birch. White birch also showed a depression of net photosynthetic rate below T_{opt}. No significant seasonal change in T_{opt} can be observed for either species (Koike 1987). Below 28°C, the net photosynthetic rate per unit leaf area was higher in mountain birch than in white birch. One variety of European mountain birch forming timberlines shows higher photosynthetic rates than in ordinary Euporean mountain birch (Hällgren et al. 1982). The temperature dependence of photosynthetic rates of some Australian species seems closely related to their native distributions (Hill et al. 1988). Temperature dependant photosynthesis may characterize the specific pattern of distribution of trees.

At the same temperature, the leaf respiration rate of mountain birch was higher than that of white birch (Fig. 4B). Körner & Larcher (1988) showed that plants acclimated at lower temperature have higher mitochondrial respiration than plants acclimated at higher temperature. Plants which are acclimated to low temperatures seem to utilize changing temperatures efficiently for energy production. Dark respiration rates may show a compensatory trend in response to low temperatures. These results may imply that mountain birch can develop its shoots at lower temperature than can white birch.

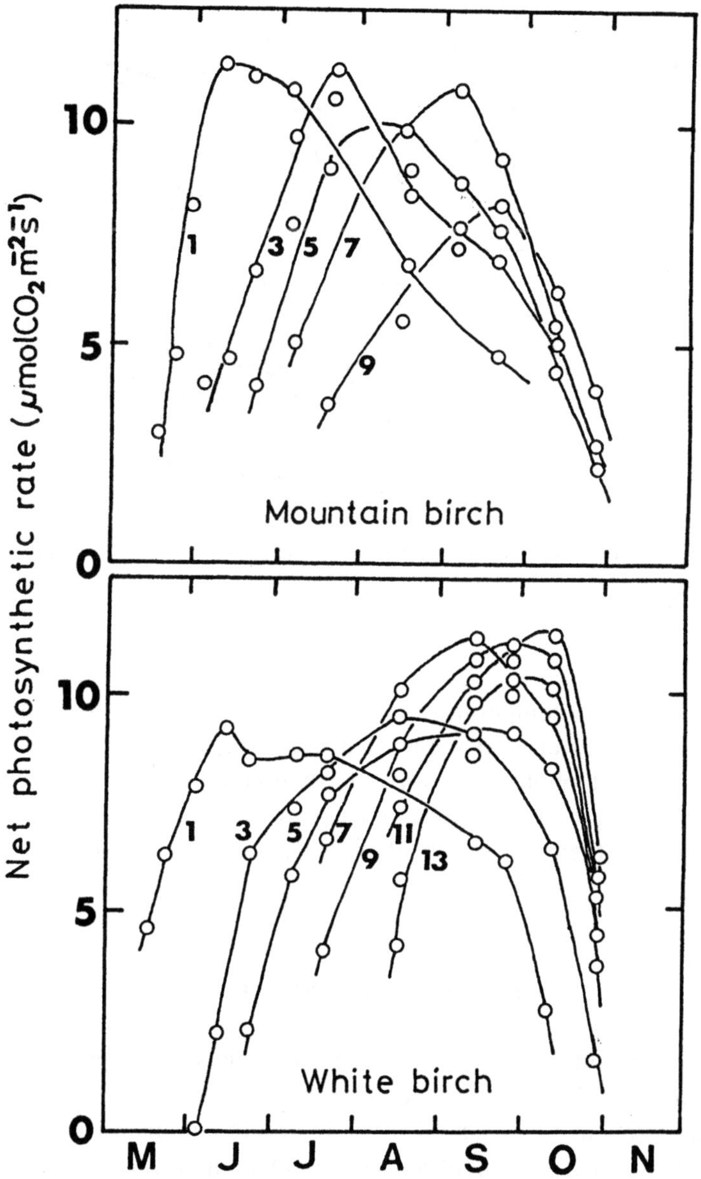

Fig. 5. Time course of net photosynthetic rate by individual leaves. Leaves are identified by leaf order counted from the base of the shoot. Net photosynthesis was measured under PFD of 680 μmol · $m^{-2} \cdot s^{-1}$ and optimum temperatures (Koike, unpubl.).

In mountain birch, the peak photosynthetic rates of individual leaves (Pi) occurred earlier in the growing season, in June and July, whereas the Pi of white birch was not attained latter in the growing season (Fig. 5). For leaves beyond the ninth on the corrent shoot, Pi in white birch declined sharply

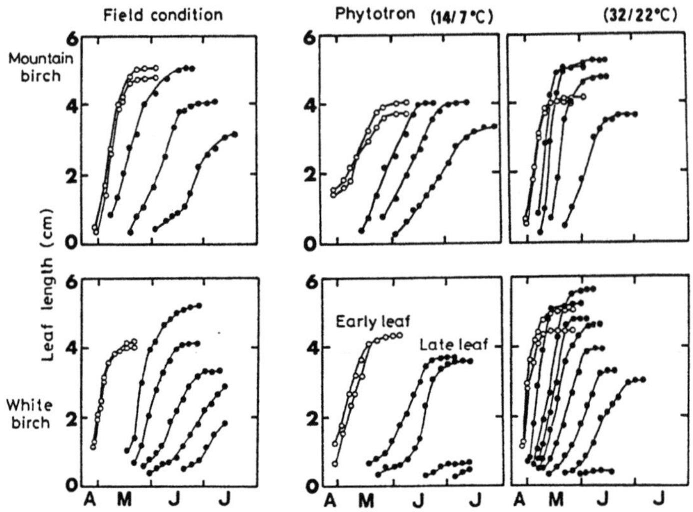

Fig. 6. Time course of seedling leaf growth of mountain birch and white birch under field and regulated conditions. Under high-temperature treatment (day/night: 32/22°C), white birch developed leaves and shoots continuously (Koike et al. 1990).

when the minimum temperature fell below 5°C (Koike & Sakagami 1985). Based on these patterns, it seems that mountain birch completes its shoot development and photosynthesiszes most efficiently early in the growing period. In contrast, white birch first expands its early leaves early in the growing period, then produces its late leaves until September.

Shoot development under controlled environments

Early leaves produce energy for the development of late leaves (Kozlowski & Clausen 1966). Why is there any time lag between the expansion of early leaves and late leaves, especailly in white birch? An answer to this question was sought through treatment with different temperatures in regulated environment. Under field conditions, mountain birch develops continuously from early leaves to late leaves while white birch expands late leaves only after the complete expansion of early leaves (Fig. 6). This time lag between early and late leaves in white birch disappered, however, under treatment with higher temperatures. Under low-temperature treatment, the time lag was increased, even in mountain birch. This time lag between early and late leaves in white birch could be an adaptaion to avoid late frost.

The number of leaves per shoot of mountain birch was five or six, regard-

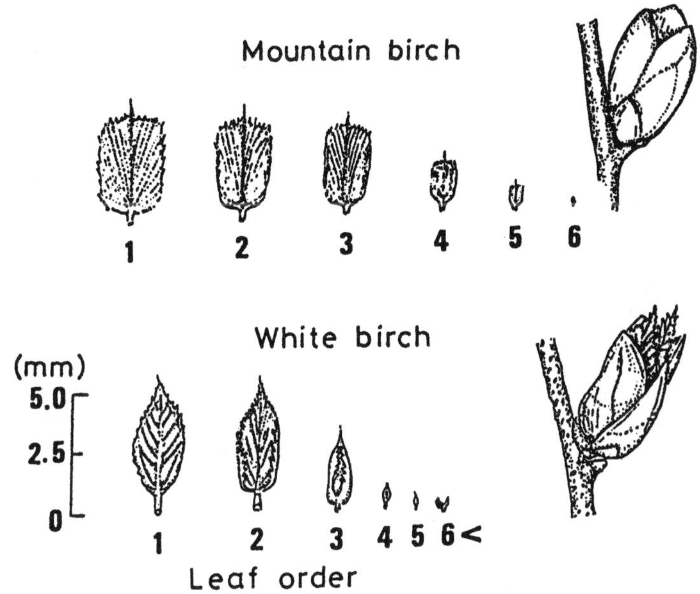

Fig. 7. Leaf primordia of winter buds of mountain birch and white birch. The first and second leaf primordia become early leaves and the rest becomes late leaves (Koike unpubl.).

less of temperature (Fig. 6). In contrast, the number of leaves per shoot of white birch increased with increasing treatment temperatures. The size of the leaf primordia for early leaves was larger than those of late leaves, for both birch species (Fig. 7). In mountain birch, the size of the third and fourth leaves was always intermediate between the early and late leaves. Based on the measurements in natural environments (Koike 1987) and under regulated conditions, it seems that mountain birch completes its shoot system before early frost, while white birch avoids damage by late frost and efficiently uses higher temperatures for expanding its shoots as long as possible, until the coming of low temperatures.

The optimum temperature (T_{opt}) of mountain-birch seedlings treated with both low and high temperature was almost the same, around 18°C (Fig. 8). This stability of T_{opt} was also observed in saplings from different elevations (Fig. 1B). In contrast, T_{opt} of white-birch seedlings shifted from 18°C in the low-temperature treatment to 27°C in the high-temperature treatment (Fig. 8). Mountain birch thus seems suitable for growing in cooler environments. Why, however, does mountain birch suffer from late frost in spring while white birch is not affected (Koike et al. 1983)?

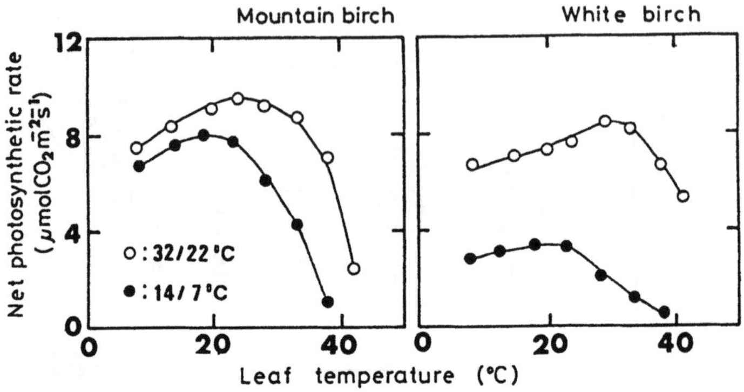

Fig. 8. Temperature dependence of net photosynthetic rate of birch seedlings treated with high and low temperature. Seedlings were cultivated under regulated conditions from before bud opening. Early leaves were used in order to assure an uniform leaf age of around 55 days when photosynthesis was measured (Koike unpubl.).

Low temperature-tolerance of leaves during the growing season

Frost plays a decisive role in the global distribution of plant species (see Sakai and Larcher 1987, Woodward 1987), as well as in the elevational distribution of closely related species (Davidson & Reid 1985). Timing is perhaps the most important aspect in the frost tolerance of species (Körner & Larcher 1988). In summer, 80% of the leaves of mountain birch were killed by a low temperature of −3°C for 4 hours, whereas 90% of the leaves of white birch survived (Fig. 9). This relationship was observed throughout the growing season (Koike at al. 1990). In late summer, however, the shoots of mountain birch were more resistant against low temperatures than those of white birch (Takatoi et al. 1963).

Foliar resistance to frost does not always coincide with the natural distribution pattern, as in Tasmanian tall trees (Reid and Hill 1990). Actually the leaves of mountain birch suffered more from low temperature than those of white birch, but mountain birch could recover from damage by late frost (min. temp. −2°C for 6 hours) by opening its winter buds and adventitious buds by early June (Koike et al. 1983). This means that shoots and buds of mountain birch can recover from damage caused by frost as late as early May.

Concluding remarks

Mountain birch generally occurs at higher elevation than white birch. Every growth parameter of mountain birch, including the optimum temperatures

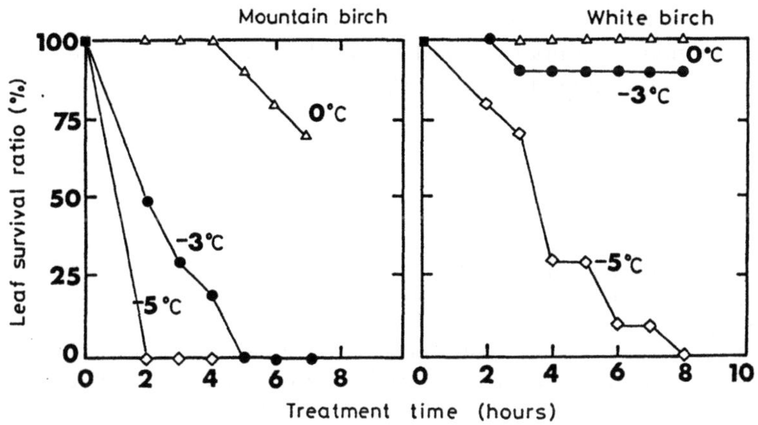

Fig. 9. Leaf survival ratio of birch under various combinations of temperature and treatment time during the growing period. The criteria for leaf survival were determined the water absorption of shoots after the low-temperature treatments (Koike et al. 1983).

for photosynthesis and for leaf development as well as the earlier completion of shoot growth, is suitable for growing in a cooler climate than that of white birch (Koike 1987). The leaves of mountain birch are damaged by late frost, however. while white birch avoids frost damage. What is the meaning of this contradiction?

According to the geographical distributions described by Hämet-Ahti & Ahti (1969), Watanabe (1979), and Okitsu (1987), mountain birch also occurs along seashores or lakeshores with cool and humid summer climate. In contrast, white birch (including the main species *Betula platyphylla*) can grow further north, at the edge of Siberia, where summer is quite warm (Budyko 1971). Based on these facts, it appers that mountain birch can grow under cooler climatic conditions, due to its higher capacity for low-temperature utilization and avoidance of early frost by means of quickly developing shoots. In contrast, white birch seems to develop its shoots rapidly by using higher temperatures and expands its shoots until late summer. Once shoot development in white birch is complete, the leaves are resistant to low-temperature damage.

Acknowledgements

I thank T. Masuzawa, Ch, Körner, H. Turner and Y. Sakagami for their proper guidance, B. Sveinbjörnsson, R. Häsler and K. Takahashi for invaluble comments, and Y. Fujimura and M. Koike for their aid during field measurements, and E. O. Box and R. K. Peet for editorial and linguistic assistance. Financial support in part by the Japanese Ministry of Agr. For. and Fish. is acknowledged (BRP 93-IV-B-2).

References

Benecke, U. & Havranek, W.M. 1980a. Phenological growth characteristics of trees with increasing altitude, Craigieburn range, New Zealand. New Zeal. For. Serv. Tech. Paper 70:155–173.

Benecke, U. & Havranek, W. M. 1980b. Gas-exchange of trees at altitudes up to timberline, Craigieburn range, New Zealand. New Zeal. For. Serv. Tech. Paper 70:195–212.

Berry, J. & Björkman, O. 1977. Photosynthetic response and adaptation to temperature in higher plants. Ann. Rev. Plant Physiol. 31:491–543.

Budyko, M. I. 1971. Climate and life. (in Russian; translated into Japanese by Z. Uchijima and S. Iwakiri 1973). Univ. of Tokyo Press, Tokyo.

Davidson, N. J. & Reid, J. B. 1985. Frost as a factor influencing the growth and distribution of subalpine Eucalypts. Aust. J. Bot. 33:657–667.

Eriksson, G. & Jonsson, A. 1986. A review of the genetics of *Betula*. Scand. J. For. Res. 1:421–434.

Fujimoto, S. & Miyakawa, M. 1991. Growth characteristics of *Betula ermanii* in paticular references to response pattern at timber lines. J. Fac. Agr., Hokkaido Univ. 65:219–228.

Grace, J. 1987. Climatic tolerance and the distribution of plants. New Phytol. 106 Suppl.:113–130.

Hällgren, J.-E., Sundbom, E. & Strand, M. 1982. Photosynthetic responses to low temperature in *Betula pubescens* and *Betula tortuosa*. Physiol. Plant 54:275–282.

Harper, J. N., McNaughon, I. H. & Sagar, G. R. 1961. The evolution and ecology of closely lelated species living in the same area. Evolution 15:209–227.

Hämet-Ahti, L. & Ahti, T. 1969. The homologies of the Fennoscandian mountain and costal birch forests in Eurasia ancl northl America. Vegetation 19:208–219.

Hill, R. S., Read, J. & Busby, J. R. 1988. The temperature-dependence of photosynthesis of some Australian temperate rainforest trees and its biogeographical significance. J. Biogeography 15:431–449.

Ishizuka, M. & Sugahara, S. 1986. Conmposition and structure of natural mixed forests in central Hokkaido I. Composition differences and species characteristics by elevation and from disturbances. J. Jpn. For. Soc. 68:79–86.

Kikuzawa, K. 1983. Leaf survival of woody plants in deciduous broad-leaved forests. 1. Tall trees. Can. J. Bot. 61:2133–2139.

Koike, T., Fujimura, U. & Sakagami, Y. 1983. The frost damage in a *Betula ermanii* plantation caused by a late frost in early May. Trans. Jpn. J. For. Soc, Hokkaido Br. 32:107–109 (in Japanese).

Koike, T. & Sakagami, Y, 1985. Comparison of the photosynthetic responses to temperature and light of *Betula maximowicziana* and *Betula platyphylla* var. *japonica*. Can. J. For. Res. 15:631–635.

Koike, T. 1987. The growth characteristics in Japanese mountain birch (*Betula ermanii*) and white birch (*Betula platyphylla* var. *japonica*), and their distribution in the northern part of Japan. In Fujimori, T. & Kimura, M. (eds.), Human Impacts and Management of Mountain Forests, pp. 189–200. FFPRI Ibaraki, Japan.

Koike, T., Fujimura, Y., Takahashi, K. & Sakagami, Y. 1990. The growth characteristics of *Betula ermanii* in Hokkaido. Trans. J. Jpn. For. Soc. 101:435–436 (in Japanese).

Komizunai, M. 1972. Some experiments on the germination of *Betula ermanii* and *B. platyphylla* var. *japonica*. Jpn. J. Ecol. 22:19–23 (in Japanese).

Körner, Ch. and Diemer, M. 1987. *In situ* photosynthetic responses to light, temperature and carbon dioxide in herbaceous plants from low and high altitude. Funct. Ecol. 1:179–194.

Körner, Ch. & Larcher, W. 1988. Plant life in cold climates. In: Long, S. F. & Woodward, F. I. (eds.), Plants and Temperature, pp. 25–57. The Comp. Biol. Ltd., Cambridge, U.K.

Koski, V. and Selkainaho, J. 1982. Experiments on the joint effect of heat sum and photoperiod on seedlings of *Betula pendula*. Comm. Inst. Fol. Fenn. 105:34.

Koski, V. and Sievanen, R. 1985. Timing of growth cessation in relation to the variations in the growing season. In: Tigerstedt, P. M. A., Puttonen, P. & Koski, V. (eds.), Crop Physiology of Forest Trees, pp. 167–194. Univ. of Helsinki, Dept. of Plant Breed, Helsinki, Finland.

Kozlowski, T. T. and Clausen, J. J. 1966. Shoot growth characteristics of heterophyllous woody plants. Can. J. Bot. 44:827–843.

Kulluman, L. 1986. Demography of *Betula pubescens* spp. tortuosa sown in contrasting habitats close to the birch tree limit in Central Sweden. Vegetation 65:13–20.

Küppers, M., Swan, A. G., Tompkins, D., Gabriel, W. C. L., Küppers, B. I. L. & Linder, S. 1987. A field portable system for the measurement of gas exchange of leaves under natural and controlled conditions: examples with field-grown *Eucalyptus pauciflora* Sieb. ex Spreng. spp. *pauciflora*, *E. behariana F. Muell.* and *Pinus radiata* R. Don. Plant Cell Environ. 10:425–435.

Larcher, W. 1984. Ökologie der Pflanzen, 4 ed. Eugen Ulmer, Stuttgart: UTB 232. Verlag. 403 pp.

Nakano, M. 1968. Regeneration of birch species. Hakadate Japan.: Hakodate Regional Forest Office. 92 pp (in Japanese).

Read, J. & Hill, R. S. 1988. Comparative responses to temparature of the major canopy species of Tasmanian cool temperate rainforest and their ecological significance. I. Folia forest resistance. Aust. J. Bot. 36:131–113.

Read, J. & Hill, R. S. 1990. Comparative responses to temparature of the major canopy species of Tasmanian cool temperate rainforest and their ecological significance. II. Net photosynthesis and climate analysis. Aust. J. Bot. 38:185–205.

Ohsawa, M., Tezuka, T. and Numata, M. 1973. Altitudinal distribution of forest vegetation on Mt. Poroshiri, the Hidaka range, Hokhaido, Japan. Mem. Nat. Sci. Mus. 6:183–198 (in Japanese).

Okitsu, S. 1987. The *Betula ermanii* zone. In: Ito, K. (ed.), Vegetation of Hokkaido, pp. 168–193. Hokkaido Univ. Press, Sapporo, Japan (in Japanese).

Sakai, A. & Larcher, W. 1987. Frost Survival of Plants. Ecol. Stud. 62, Springer Verlag, Berlin/Heidelberg/New York/Tokyo, 321 pp.

Sato, A., Ishizuka, M., Sugawara, S. & Toyooka, H. 1986. Changes in the structure and the above-ground biomass of a young *Betula platyphylla* var. *japonica* stand on a peat soil. J. Jpn. For. Soc. 68: 205–207 (in Japanese).

Sharik, T. L. & Barnes, B. V. 1976. Phenology of shoot growth among diverse populations of yellow birch (*Betula alleghaniensis*) and sweet birch (*B.lenta*). Can. J. Bot. 54:2122–2129.

Slatyer, R. O. 1982. Photosynthetic adaptation in altitudinal populations of the Australian snow-gum, *Eucalyptus pauciflora*. In: Waring, R. H. (ed.), Carbon Uptake and Allocation in Subalpine Ecosystems as a Key to Management, pp. 31–37. Carvallis, U.S.A.: Forest Res. Lab. OSU.

Sveinbjörnsson, B. 1981. Bioclimate and its effect on the carbon dioxide flux of mountain birch (*Betula pubescens* Ehrh.) at its altitudinal tree-line in the Tornetrask area, Northern Sweden. Nordicana 47:111–122.

Takahashi, Ke. 1962. Studies on vertical distribution of the forest in Middle Honshu. Bull. Gov. Forest Exp. Sta. 142:1–171.

Tabata, H. 1966. A contribution to the biology of Japanese birches. Mem. Col. Sci. Kyoto Univ. Ser. B 32:239–271.

Takatoi, I., Watanabe, T. & Kamata, U. 1963. The frost hardiness of birch and alder trees in the autumn, their water contents, refractive indicies, and the concentration of their cortical cell saps. Bull. Gov. For. Exp. St. Hokkaido. S38:100–107 (in Japanese).

Tatewaki, M. 1958. Forest ecology of the islands of the North Pasific Ocean. J. Fac. Agr., Hokkaido Univ. 50:371–186.

Tranquillini, W. 1979. Physiological Ecology of the Alpine Timberline. Ecol. Stud. 31, Springer-Verlag, Berlin/Heidelberg/New York, 137 pp.

Watanabe, S. 1971. The alpine plants on the Hidaka Range – Notes on the subalpine *Betula ermanii* forest. Kitano-Yamanami 1:68–72 (in Japanese).

Watanabe, S. 1979. The subarctic summer green forest zone in the northeastern Asia. Bull. Yokohama Phytosoc. Japan 16:101–111.

Woodward, F. I. 1987. Climate and Plant Distribution. Cambridge University Press, Cambridge U.K. 174 pp.

Woodward, F. I. 1989. The impact of environmetal stresses on ecosystems. In: Jones, H. G., Flowers, T. J. & Jones, M. B. (eds.), Plants under Stress, pp. 11–29.

Yanagisawa, T. 1954. Some phenological observations of the forest trees at various altitudes of Mt. Muine in Hokkaido. Bull. Gov. For. Exp. St. 70: 71–92 (in Japanese).

23. Successional process on Mt. Fuji: from alpine herbaceous perennials to dwarf scrub

TAKEHIRO MASUZAWA
Department of Biology, Faculty of Science, Shizuoka University, 836 Ohya, Shizuoka 422 Japan

Abstract

The successional process from herbaceous perennials to woody scrub was investigated in the alpine belt of Mt. Fuji. The study site is located at the timberline of the southeast slope, which is mainly covered by various types of *Polygonum cuspidatum* patches. The structure of these patches is related to successional stage at alpine timberline. Shoot height, shoot density, biomass above ground and soil nutrients were measured in some patches. Shoot density and biomass are low in the center of large patches and other species can grow in this "dead center". These various patch sizes play a significant role in the successional process in the alpine belt of Mt. Fuji.

Key words: doughnut phenomenon, Mt. Fuji, *Polygonum cuspidatum*, timberline

Introduction

On Mt. Fuji, many vegetation studies have been conducted, from the forest at the base to the timberline (e.g. Hayata 1911, 1929, Tohyama 1965a,b, 1968, Ito 1972, Miyawaki & Hamada 1973). The plant communities near timberline of Mt. Fuji, including those on the southeast slope, are not in a stable climax stage but rather still in a stage of advancing succession (Hayata 1911, Tohyama 1968, Masuzawa 1985). The area from the vicinity of the Hoei Crater to the lower southeast slope is in this successional state, and the otherwise bareland is covered only by a very open community of perennial herbaceous plants (Masuzawa 1985). In this situation, the herbaceous perennial community is playing an important role in that succession proceeds from the herbaceous community to a woody community. That is, in almost all cases, the succession from alpine grassland to forest is started by the invasion

E.O. Box et al. (eds.), Vegetation Science in Forestry, 423–431.

Fig. 1. The study area, located near Mt. Hoei-Zan on Mt. Fuji. The timberline occurs at 2400 to 2500 m above sea level (white arrow).

by woody plants into the patches of *P. cuspidatum*, *P. weyrichii* and *Carex doenitzii*. Therefore, it is assumed that these *Polygonum* patches provide important sites for the establishment of the woody plants.

The purpose of this paper is to discuss the advance of successional processes promoted by these patches, based on a pattern of structural change as indicated by data on patch size, population, and biomass.

Study area

Mt. Fuji (3776 m) is the highest peak in Japan and is located near the southern coast of the central part of the country. Most of the slope is covered with various types of vegetation up to a height of about 2500 m, except for apart of the southeast slope where the vegetation is recovering from damage by the latest volcanic eruption, from a parasitic crater Hoei-Zan, in 1707.

In general, the timberline occurs on Mt. Fuji at 2400 to 2500 m above sea level. The study area (35°21′N, 138°45′E) is situated at the timberline on the southeast slope, near Hoei-Zan (Fig. 1). Previous meteorological investigation of this area has determined that the mean air temperature in summer varies between about 10°C and 15°C, and that the maximum air temperature

in summer is 22°C; the minimum air temperature in winter is −23°C (Masuzawa *et al.* 1982, Masuzawa 1983, Masuzawa *et al.* 1988, 1989).

The precipitation is high throughout the year, especially during the summer growing season. Relative humidity is high from May to October, especially from June to September, with the mean over 80% because of frequent foggy conditions in the afternoon. Annual precipitation on the foothills of Mt. Fuji ranges from 1300 to 1600 mm. The soil of the study site has rather low fertility (Masuzawa 1985).

Vegetation

The timberline at the study site is composed of four deciduous tree species, *Alnus maximowiczii*, *Betula ermanii*, *Salix reinii* and *Larix kaempferi* (Masuzawa 1985, Sakio & Masusawa 1988). There is no dwarf-pine belt (*Pinus pumila*) in the vertical sequence of vegetational belts, as found on many high mountains in Japan (Hayata 1911, 1929, Tohyama 1968). Instead of a *Pinus pumila* community, *Larix kaempferi* reaches the timberline (Table 1). The alpine semi-desert above the timberline is thinly covered with herbaceous perennials: *Polygonum weyrichii*, *Artemisia pedunculosa*, *Carex doenitzii*, *Polygonum cuspidatum*, *Campanula punctata* and *Arabis serrata*. Patches of *P. weyrichii* and *P. cuspidatum* are dominant near the timberline, and *A. serrata* extends farthest above timberline.

Structure of patches

The study area is at the second of three Hoei-zan craters on the southeast slope of Mt. Fuji, at 2400 m above sea level. *P. cuspidatum* and *P. weyrichii* dominate on this slope, and seedlings of woody plants are only sparsely present. A quadrat of 2500 m^2 (50 m × 50 m) was set out here, and representative large, medium and small vegetative patches were selected within this area. Vegetation structure (height of shoots, number of shoots, density, biomass) was investigated in each of these patches. The belt-transect method was adopted for each patch, extending from the upper side to the lower side of each patch.

P. weyrichii forms patches by vegetative propagation, and 65 patches were present within the quadrat investigated (Fig. 2). Among them, patches having areas of 0–5 m^2 occupied about 72% of the area, the largest patch reaching 65 m^2. The total patch area was 370 m^2 and the coverage was about 15%. The number of shoots, density of shoots, biomass above ground and dry weight per shoot are shown in Table 1. There was a high correlation

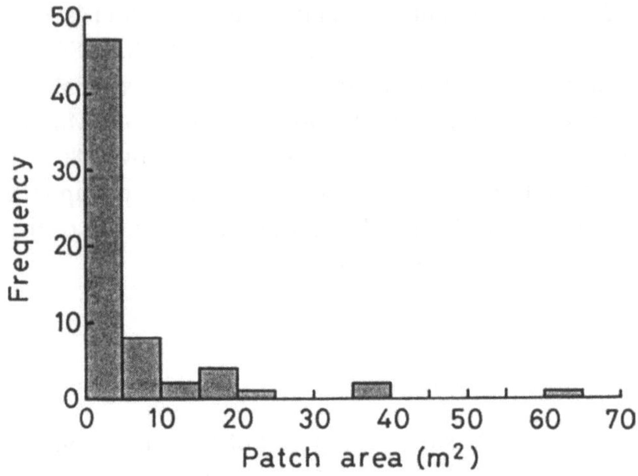

Fig. 2. Frequency distribution of patch area of *P. cuspidatum* in the sample plot (50 m × 50 m) at the end of the growing season of 1987).

Table 1. Structure of *P. cuspidatum* patches; S, Small; M, Medium; L, Large patch

	Patches of *Polygonum cuspidatum*		
	Small	Medium	Large
Area (m^2)	1.44	5.6	31
Number of shoots per patch	164	708	1727
Shoot density (number/m^2)	113.9	126.4	55.7
Biomass above ground, per patch (g)	123.1	967.7	3445.5
Biomass (g/m^2)	85.5	127.7	111.6
Dry weight per shoot (g)	0.75	1.37	2.00

between the measure D^2H and dry weight of a shoot in all of the large, medium and small patches (D = diameteter, H = height). Thus the biomass of each patch could be estimated from this correlation. Biomass above ground in the large, medium and small patches was respectively about 3400 g, 970 g and 120 g. The number of shoots were 164 in small, 708 in medium and 1727 in large patches. On the other hand, average shoot density and average biomass per square meter showed highest values (126.4 g and 127.7 g respectively) in medium patches.

The dry weight per shoot increased from small to large patches, reaching 2.0 g per shoot. The variation of shoot height in each patch measured along the belt transect is shown in Fig. 3. Shoot height in small patches increased from the periphery to the central part but in medium patches showed a

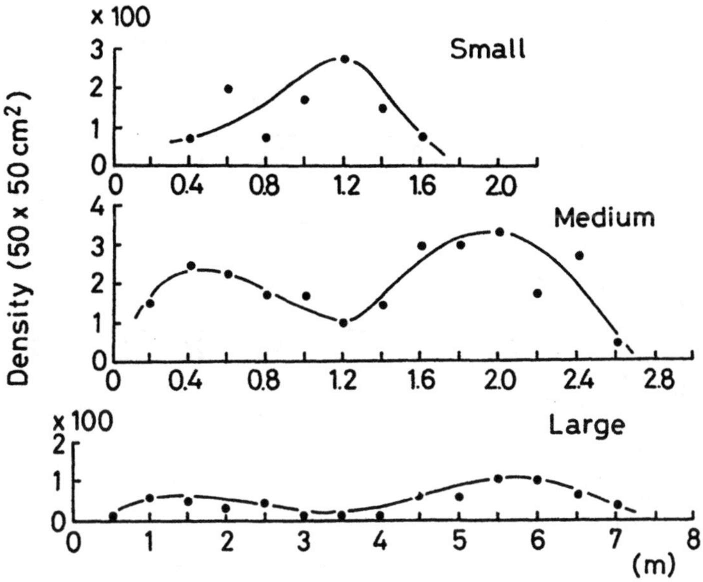

Fig. 3. Belt transect data of plant height in small, medium and large patches. The belts were oriented from the upper side to the lower side of each patch, on an erasy slope. S, Small; M, Medium; L, Large patch.

plateau structure. In large patches, average shoot height was 40 cm at the periphery and about 10 cm in the central part.

In the center the shoots of *P. weyrichii* were competitive with those of other species, especially *Saussurea triptera*. *P. weyrichii* gradually expands new shoots radially from the central part to the periphery. When *P. weyrichii* grows into a large patch, the biomass at the center decreases and other species invade the central part. All individuals invading the large patch were perennial and included *Saussurea triptera*, *Arabis serrata* and *Artemisia pedunculosa*. Figure 4 shows the variation of shoot density (50 cm × 50 cm) in each patch. Shoot density in small patches increased from the periphery to the center. In medium patches, shoot density in the center was lower, about 100 per 2500 cm^2, although the height of the shoots did not change. This means that, in spite of the unchanged height, shoot density has already decreased in patches of medium size. In large patches, there was a similar tendency. Average shoot density was lower in large patches, about 100 at the periphery but only about five in the center. In the central part, most shoots of *P. cuspidatum* had been replaced by other invading species.

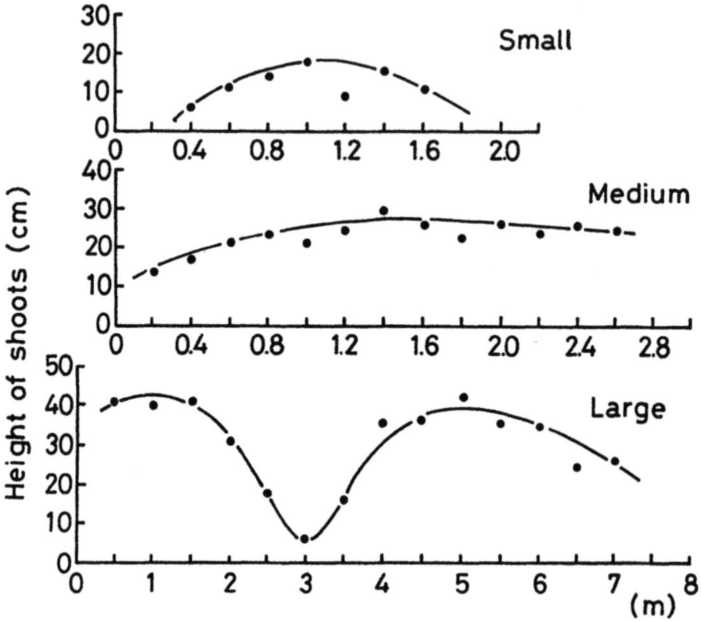

Fig. 4. Belt transect data of shoot density in small, medium and large patches. Density is shown by number of shoots par 2500 cm^2 (50 cm × 50 cm).S, Small; M, Medium; L, Large patch.

Table 2. Nitrogen concentration and ignition loss in the soil in central and peripheral parts of a medium patch at 0 cm, 15 cm, 30 cm depth from ground surface

	Central part		Peripheral part	
Depth (cm)	Nitrogen (‰)	Ignition loss (%)	Nitrogen (‰)	Ignition loss (%)
0	0.59	2.98	0.19	0.98
15	0.13	1.03	0.05	0.46
30	0.18	1.04	0.05	0.68

Soil conditions

Soil nutrients changed with patch size. The nitrogen content and ignition loss at various depths were compared between the periphery and center of medium patches (Table 2). The soil is older in the center than at the periphery of patches. The nitrogen content near the center was higher than that at the periphery for each depth. In the central part, nitrogen was 0.59‰ at the ground surface, 0.13‰ at 15 cm depth and 0.18‰ at 30 cm depth.

Ignition loss at the ground surface also differed between the center and the periphery, but the values at 15 cm and 30 cm depth were similar. In other

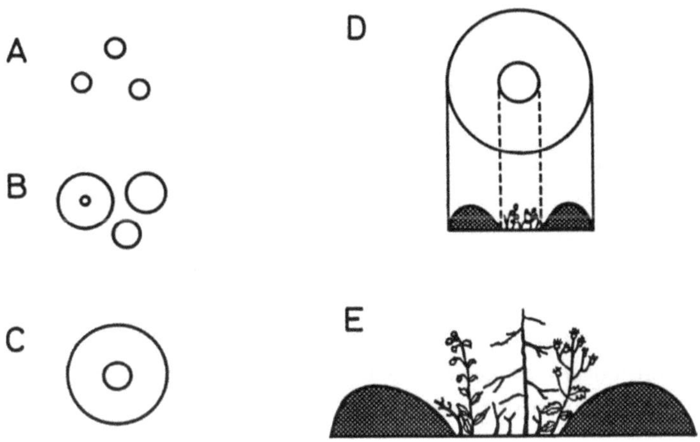

Fig. 5. The process of "doughnut" formation at timberline on Mt. Fuji. A, Small patches of *P. cuspidatum*; B, Medium patch with small dead center; C, Large patch with dead center; D, Top view and cross section of a large patch; E; Cross section of a large patch invaded by other plants; *Saussurea triptera*, *Artemisia pedunculosa*, *Polygonum weyrichii*, and *Larix kaempferi*.

words, litter decomposition in the central part advanced more rapidly than that at the periphery, providing better soil conditions inside the patch.

Doughnut phenomenon

The process of particularization in shrubs and other bush-like plants, which produces a dead center and outward growth (doughnut phenomenon), is shown in Fig. 5. After the seed germination of *P. cuspidatum*, the seedling forms a small patch and increases the number of shoots. This small patch expands radially as the shoots grow and increase in number. Thereafter, the shoots grow to a similar height at the periphery and in the central part of a patch. In a medium patch, the shoot density decreased in the central part, though the height of shoots was similar throughout.

The respiration by rhizomes of *P. cuspidatum* was measured at the periphery and the dead center in large patches. The respiration rate of rhizomes at the center of *P. cuspidatum* patches was measured in order to find the relationship between rhizome activity and the low biomass of shoots in the center. Respiration values of about 1.3 mg CO_2 g^{-1} hr^{-1} in the central part and about 0.5 mg CO_2 g^{-1} hr^{-1} at the periphery of a large patch were obtained in August. There was a similar pattern of respiration in September. Judging from the respiratory activity of rhizomes in August, individuals showed higher respiration at the periphery than in the central part of patches.

In a large patch in September, the respiration of rhizomes in the periphery was also higher than that in the central part.

Patch growth of herbaceous perennials was studied for *Phragmites communis* (Buttery and Lambert 1965, Haslam 1969, 1970, 1971), *Phragmites australis* (Kudo and Ito 1988) and *Pteridium aquilinum* (Watt 1940, 1974). However there have been no ecophysiological studies on the patch structure according to patch size, especially on the activity of rhizomes and on the soil development in patches. Based on the above results, it appears that the activity of the central part of each patch is lower than that of the peripheral part. This is considered to be because the underground organs near the center are crowded and aged, causing withering from center outward, and resulting in the doughnut phenomenon.

Acknowledgements

The author is greatly indebted to Mr. H. Kudoh, Mr. T. Hayashi, Miss N. Nakamura and Miss S. Suzuki of Shizuoka University for their kind assistance. Thanks are also due to the members of the Shizuoka Regional Forestry Office.

References

Buttery, B. R. & Lambert, J. M. 1965. Competition between *Glyceria maxima* and *Phragmites communis* in the region of Surlingham Broad: I. The competition mechanism. J. Ecol. 43:163–182.

Haslam, S. M. 1969. Stem types of *Phragmites communis*. Trin. Ann. Bot. 33:127–131.

Haslam, S. M. 1970. Variation of population type in *Phragmites australis*. Trin. Ann. Bot. 34:147–158.

Haslam, S. M. 1971. The developments and establishment of young plants of *Phragmites communis*. Trin. Ann. Bot. 35:1059–1072.

Hayata, B. 1911. The vegetation of Mt. Fuji, pp. 1–125. Maruzen, Tokyo.

Hayata, B. 1929. Succession in the vegetation of Mt. Fuji. Act. For. Fenn. 34:1–28.

Ito, E. 1972. Silviculture of Mt. Fuji Shizuoka Pref. Dept. of Agr. Forestry 1–119 (in Japanese).

Kudo, G. & Ito, K. 1988. Rhizome development of *Phragmites australis* in a reed community. Ecol. Res. 3:239–252.

Masuzawa, T., Takeyama, K. & Sakio, H. 1982. An ecological study of microclimate at timberline on Mt. Fuji: I. Measurement of diurnal change in air and soil temperature in spring. Rep. Fac. Sci., Shizuoka Univ. 16:93–101.

Masuzawa, T. 1983. An ecological study of microclimate at timberline on Mt. Fuji: II. Air temperature, relative humidity and light condition. Rep. Fac. Sci., Shizuoka Univ. 17:91–99.

Masuzawa, T. 1985. Ecological studies on the timberline of Mt. Fuji I. Structure of plant community and soil development on the timberline. Bot. Mag. Tokyo 98:15–28.

Masuzawa, T. Nakamura, N., Suzuki, S. & Wada, K. 1989. Seasonal change in the air temperature and the warmth index over three years at the timberline on Mt. Fuji. Rep. Fac. Sci., Shizuoka Univ. 24:89–96.

Miyawaki, A. & Hamada, T. 1973. Ein kurzes vegetationskundliches Gutachten der Umgebung des Berges Fuji. Sci. Rep. Mt. Fuji Numazu City 35–63 (in Japanese).

Sakio, H. & Masuzawa, T. 1988. Ecological studies on the timberline of Mt. Fuji II. Primary productivity of *Alnus maximowiczii* dwarf forest. Bot. Mag. Tokyo 100: 349–363.

Tohyama, M. 1965a. Cool temperate montane forest of Mt. Fuji. Agr., Hokkaido Univ. 5:111–124 (in Japanese).

Tohyama, M. 1965b. Forest vegetation on the lava flows of Mt. Fuji. J. Fac.Agr., Hokkaido Univ. 5:125–137 (in Japanese).

Tohyama, M. 1968. The alpine vegetation of Mt. Fuji. J. Fac. Agr., Hokkaido Univ., 55: 459–467.

Watt, A. S. 1940. Contributions to the ecology of bracken (*Pteridium aquilinum*). I.The rhizome. New Phytol. 39: 401–422.

Watt, A.S. 1974.Contributions to the ecology of bracken (*Pteridium aquilinum*). IV. The structure of the community. New Phytol. 46:97–121.

24. A study of the population dynamics of *Picea asperata* Mast

JIANG HONG
Institute of Botany, Academia Sinica, Beijing, China

Abstract

Picea asperata Mast (spruce, dragon pine) is the dominant canopy species in the forests of southwestern China. Its distribution ranges from the upper reaches of the Minjiang River and the large and small branches of the Jinchuan River of Sichuan Province, to the Beilongjiang and Taohe Rivers of Gansu Province. This study was undertaken to better understand the population dynamics and life cycle of spruce because of its prominence and economic importance. Ecologically, spruce is a significant successional species and is considered vital for watershed conservation in some areas, especially in the upper reaches of the Yangzhi River.

Key words: life table, southwestern China, spruce

Introduction

To study the population dynamics of spruce, I used a life table approach. A life table is a systematic account of the number of individuals surviving to each age or stage in the life cycle of a species, in nature or in an experimental situation. Life tables can aid us in determining the different factors causing the death of individuals in a population and even help ascertain the key factor. Life tables allow us to estimate the mortality trends of individuals in the next generation. As a result, life tables are valuable for production management decisions.

E.O. Box et al. (eds.), Vegetation Science in Forestry, 433–439.

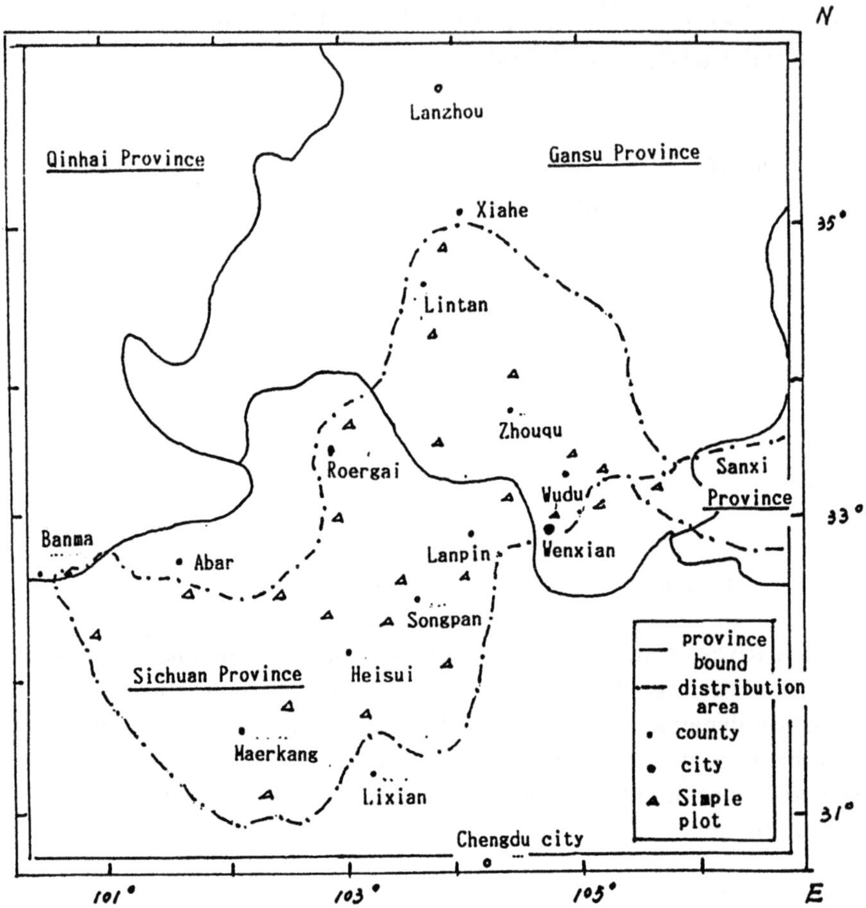

Fig. 1. Spruce (*Picea asperata*) distribution area in China.

Methods

Study area

The study area includes counties in three of the southwestern provinces of China: in Sichuan Province, Songpan, Lanpin, Roergai, Maerkang, Lixian, Heisui, and Abar; in Gansu Province, Diebu, Zhouchu, Zuoli, Lintan, Xiahai, Xiahe, Wudu, and Wenxian; and Banma County in Qinghai Province. The study area lies within an approximately 32–35° N latitude and 101–105° E longitude (Fig. 1). The elevation of the study area varies from 2400 to 3500 m. The study area can be divided into two physiographically different distribution districts: the mountain gorge area and the plateau hill area. The annual mean temperature of the area is 5.8°C, the all-time high being 37°C and the all-time low being − 35°C. Mean annual rainfall is 500 mm, mean annual relative humidity is 60%, and the frost-free period is 100 days. The soils are predominantly mountain brown forest soils with a pH range of 4.5–7.5 and a humus horizon thickness ranging from 1–36 cm. The forests in the study include pure spruce forests and mixed spruce-fir forests in which spruce is the dominant species.

Field methods

This study is based on 328 plots, each 40 × 30 m. We measured environmental factors, forests composition, numbers of trees in each plot, and canopy height of some trees as well as distance to each surrounding tree. In 14 plots, we measured the height, diameter, and age for all trees and seedlings.

A generalized life table for *Picea asperata*

One way to compile a life table is to follow the fate of the individuals in a cohort from birth until the last member of the cohort dies. This procedure produces a 'dynamic' life table. The 'static' life table, which estimates age-specific death risks from the age structure of a population at one moment in time, is used in situations where it is not practical to follow a cohort for an extended period. Our objective was to construct a static life table for spruce. Due to the fact that the study area has extremely varied topography and community composition, we first constructed 130 static life tables corresponding to the different forest types and environmental conditions. We then combined the data to form a general life table that averages the 130 individual life tables (Table 1).

Some of the most prominent characteristics in the life table are:

1. The seedling pool is large; 56,197 individuals/hectare. Mortality in the seedling pool is about 99.4% per year.
2. The number of young trees, those less than 5-years-old, is significantly smaller than the number in the seedling pool, the number remaining at five years-of-age being about 315 individuals/hectare.
3. The probability that a tree will enter the canopy is related to the heterogeneity of the habitat. The times of highest relative mortality appear to be from 2–5 years of age and from 30–40 years of age. The probability that a seedling will enter the canopy increases with formation of gaps in the canopy, and the likelihood that a seedling will enter the canopy is greatly reduced if the canopy is well developed and dense.
4. The spruce population survivorship is consistent with a Deevey III curve (Fig. 2).

Table 1. The static life table of spruce population.

Age (years)	Survi-vorship	No. dying during interval	Average mortality		No. surviving to years x			Kill power
x	lx	dx	qx	ex	ax	Logax	Loglx	kx
1	1000.0000	994.3947	994.3947	0.5	56197	10.937	6.908	5.184
5	5.6053	2.1353	380.9524	2.4	315	5.753	1.724	0.480
15	3.4699	1.3880	400.0000	2.6	195	5.273	1.244	0.511
30	2.0820	1.3880	666.6666	3.0	117	4.762	0.733	1.099
50	0.6940	0.2135	307.6923	6.9	39	3.664	−0.365	0.368
70	0.4805	0.0000	0.0000	8.8	27	3.296	−0.733	0.000
90	0.4805	0.0534	111.1112	7.8	27	3.296	−0.733	0.118
110	0.4271	0.0178	41.6667	7.7	24	3.178	−0.851	0.043
130	0.4093	0.0356	86.9565	7.0	23	3.135	−0.893	0.091
150	0.3737	0.0712	190.4762	6.6	21	3.045	−0.984	0.211
170	0.3025	0.0000	0.0000	7.0	17	2.833	−1.196	0.000
190	0.3025	0.0356	117.6470	6.0	17	2.833	−1.196	0.125
210	0.2669	0.0178	66.6667	5.8	15	2.708	−1.321	0.069
230	0.2491	0.0000	0.0000	5.1	14	2.639	−1.390	0.000
250	0.2491	0.0356	142.8571	4.1	14	2.639	−1.390	0.154
270	0.2135	0.0178	83.3334	3.8	12	2.485	−1.544	0.087
290	0.1957	0.0356	181.8182	3.0	11	2.398	−1.631	0.201
310	0.1602	0.0534	333.3334	2.6	9	2.197	−1.832	0.405
330	0.1068	0.0178	166.6666	2.7	6	1.792	−2.237	−0.154
350	0.1246	0.0534	428.5715	1.4	7	1.946	−2.083	0.560
370	0.0712	0.0356	500.0000	1.0	4	1.386	−2.643	0.693
390	0.0356	0.0000	0.0000	0.5	2	0.693	−3.336	0.693

Results and discussion

The life table revealed two major age-stages in which mortality of spruce is especially high relative to other age-stages. One of the stages is from the young seedling to the seedling stage (<5 years). The other stage exhibiting high mortality is from the seedling stage to the young tree stage. The two positions in the life cycle of the spruce population are called "environment sifter" in Fig. 3.

Our other studies suggest that mortality during these stages is controlled by different factors. The major cause of mortality during the seedling stage, drought, is particularly important because spruce seedlings have short roots which cannot absorb adequate water as soil surface moisture decreases. We observed young seedlings in the moss and litter to be sparse and root length of naturally regenerating spruce seedlings to be short (Table 2). The average height of a one-year-old spruce seedling is 2.5–2.9 cm and the root length is 4–4.5 cm. The potential for drought stress is evident when this average root measurement is compared to the average thickness of the litter and moss layer, generally about 4.8–5.8 cm. These data suggest that roots of spruce

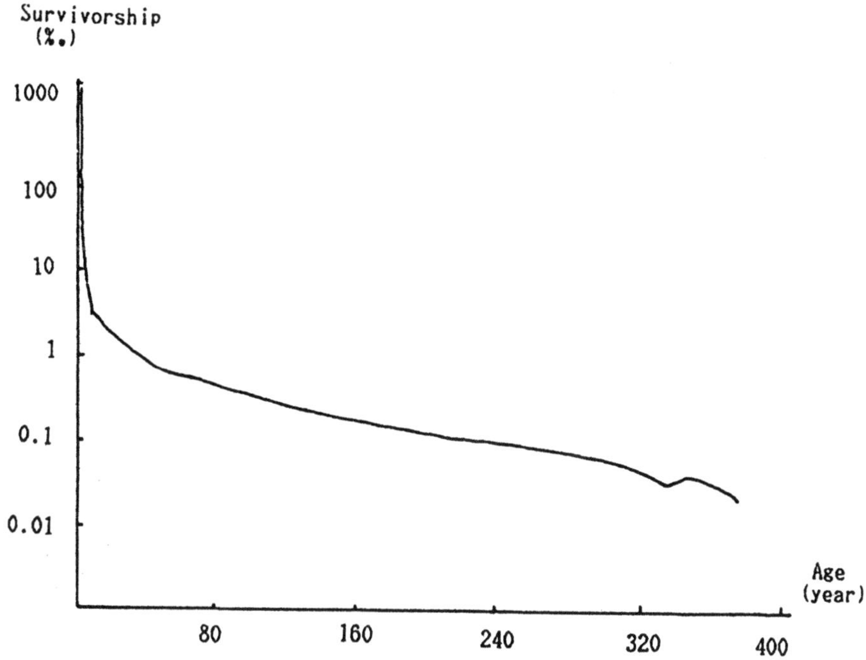

Fig. 2. The spruce population survivorship curve.

seedlings cannot reach the mineral soil, and thus lack of water is the probable reason for the sparsity of seedlings in moss and litter.

Another factor restricting spruce regeneration is the subalpine climate of the study areas. Spruce seeds commonly germinate during the month of June, although smaller numbers germinate during May, July, August, and September. The spruce seedling must fully establish primary xylem during the first growing season in order to pass from the young-seedling stage to the seedling stage. The process of xylegenesis takes approximately 45–60 days after germination occurs, with the results that those seedlings germinating late in the season usually fail to complete xylegenesis.

The major cause of mortality for spruce between the seedling and young tree stages is lack of light. During this stage of growth, the seedling must survive the low light environment under the canopy. Our earlier results showed that a spruce can tolerate only 20–30 years of suppression under a closed canopy (see Fig. 3). For development to continue beyond 20–30 years, an increase in light is required. There is intense competition for light among the young trees, so most individuals are lost to this self-thinning process. Those young trees that manage to survive this stage enter the canopy. Although considerable competition still exists after entry into the canopy, this

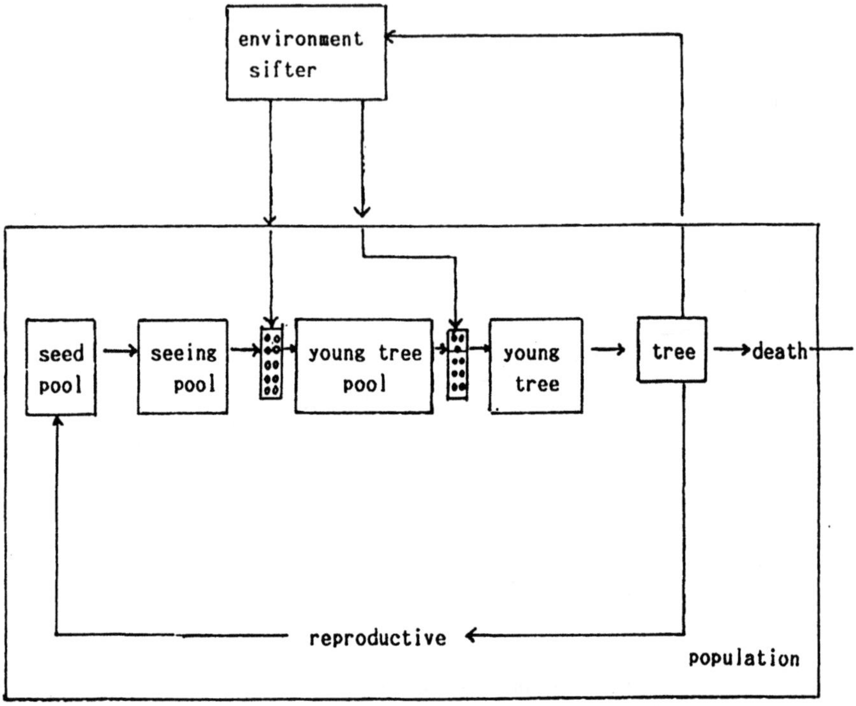

Fig. 3. The variate proceed and control factor of spruce population size.

Table 2. The seedling root development state of spruce

Seedling Age	Thickness of litter cm	Seedling height cm	Root diameter cm	Root length cm	Root wide cm	Root number
1	4.8	2.9	1.0	4.5	2.0	2
2	5.0	8.2	2.2	5.9	2.7	4
3	5.2	10.2	2.5	6.3	3.5	4
4	5.8	11.6	3.2	8.2	4.1	5
5	4.8	15.6	4.0	15.0	6.0	6

Table 3. The self-thinning of spruce young tree population.

Age type	5	10	15	20	25
Live trees	458	292	333	125	125
Standing dead trees	1700	1530	820	400	225
Sum	2158	1822	1153	525	350

primarily affects tree size and shape and is little threat to the tree's existence. Our results are consistent with those of Spurr (1973) who reported that light is often an important factor in spruce regeneration and that spruce rarely can regenerate under its own canopy.

Conclusions

The age-stages of highest relative mortality for spruce trees in the forests of southwestern China are the young-seedling to seedling stage, and the seedling to young tree stage. The factor causing greatest mortality during the young-seedling to seedling stage is drought stress, while the major factor responsible for mortality from the seedling to young tree stage is lack of light. Nonetheless, we cannot exclude the possibility that other environmental factors may be responsible for mortality during these age-stages.

Acknowledgements

I thank Kathleen Baker-Brosh and Robert K. Peet for editorial and linguistic assistance.

References

Harper, J. L. 1977. Plant Population Biology. Academic Press, New York.
Dirzo, R. 1984. Perspectives on Plant Population Ecology. Sinauer Associates Inc., Sunderland, Mass.
Daubenmire, R. 1981. Plant Community. Academic Press, New York.
Silvertown, J. W. 1982. Introduction to Plant Population Ecology. Longman, London.
Spurr, S. H. 1973. Forest Ecology, second edition. Ronald Press, New York.

25. Alpine vegetation in a global perspective

GEORG GRABHERR
Department of Vegetation Ecology and Conservation Biology, Institute of Plant Physiology, University of Vienna, Althanstr. 14, A-1090 Vienna, Austria

Abstract

The ecological position of the nemoral high mountain systems of Europe is discussed in a global context by focusing on dissimilarities with other high mountain systems, as opposed to the more typical approach of reporting similarities.

According to Walter's (1983) zonal classification, the nemoral high mountain systems of Europe (i.e. Alps, Carpathians, Pyrenees, Dinarids) are unique in their location at the intersection of the nemoral and mediterranean zones. Mountains at the same latitude in East Asia, North America, and the Southern Hemisphere all lack this interzonal character. As a consequence, the zonal vegetation types of the area, such as European alpine grasslands, have to be considered as endemic. Even structural characters of the vegetation, like the combination of sedges with masses of lichens, seem to have no true equivalents in other mountains systems of the world. As a consequence, the term "alpine" is best used only in a loose way; classification approaches based on altitudinal belts and environments should be preferred.

Key words: alpine vegetation, Alps, high mountain ecosystems, life zones, plant communities, vegetation structure

Introduction

The term "alpine" is commonly used for high mountain environments above timberline, such as for the mountains of easternmost Asia (see Ishizuka 1974, Miyawaki & Nakamura 1988, Holzner & Hübl 1988, Zhu & Rowe 1987). However, the word derives from and applies specifically to areas of high mountain climate and vegetation of the Alps, which have served as a model

E.O. Box et al. (eds.), Vegetation Science in Forestry, 441–451.

for vegetation studies since the beginnings of geobotany. This dual usage of the term has led to considerable confusion. To clarify the meaning of this term, it seems appropriate to list the special features of the Alps, and to seek dissimilarities with other mountain systems of the world. The goal of this exercise is not to replace the term "alpine", but to clarify the position of the Alps within the high mountain environments of the world.

The position of the Alps in relation to Walter's life zones

Walter's (1983) zonal classification of world vegetation is especially useful for comparing high mountain environments. In his scheme zonobiomes are defined as large, homogeneous regions with a specific climate and, accordingly, vegetation characterized by more or less well defined transition zones. The major mountain systems of the world are listed in Table 1 and classified according to which of the 9 principal zonobiomes occur on them.

The large mountain systems of Europe (i.e. the Alps including the Carpathians, the Pyrenees, and some mountains of the Balkan Peninsula) can be considered as interzonal mountains. Their geographical position between the nemoral zonobiome of Central Europe and the mediterranean zonobiome of Southern Europe (Fig. 1) makes them unique compared with the other mountain systems of the world. No other mountain system is situated between these two distinct life zones. Beside the fact that mediterranean environments are completely absent in East Asia, most other mountain ranges at similar latitudes, like the Japanese Alps, are island mountains subject to oceanic climate. Also the third nemoral mountain system, the Appalachians, is surrounded by life zones different from the mediterranean one. In the southern hemisphere, the mountains of New Zealand and some parts of the Patagonian Andes have to be considered, but these mountain systems are also predominantly oceanic in character and do not have links to mediterranean environments.

Another distinctive characteristic of the vegetation of the European nemoral mountains is that it was more severely modified by the ice age climate than was the vegetation of East Asia or eastern United States. Most of the Arcto-Tertiary flora of Europe was eliminated, and re-establishment of the vegetation was predominantly from southern sources which might have amplified the interzonal character of the Alps and the other European nemoral mountains.

The present environmental character of the Alps is as distinctive as its floristic origins. In contrast to the mediterranean mountains, there is no summer drought in the Central European mountains (Fig. 1), and compared with the boreal mountains of the north, the absence of permafrost (with a

Table 1. Classification of the high mountain systems of the earth according to Walter's (1983) life zone system (zonobiomes). Large mountain systems are almost interzonal or multizonal, solitary mountains like tropical volcanoes are intrazonal. The interzonal position of the Alps between the nemoral and mediterranean life zone has no true equivalent in other continents, and even the Central European mountains are different

I	– Zonobiome of the equatorial humid diurnal climate with evergreen tropical rain forest
II	– Zonobiome of the humido-arid tropical summer-rain region with deciduous forests
III	– Zonobiome of the subtropical arid climate with deserts
IV	– Zonobiome of the winter rain region with an arido-humid climate and sclerophyllic woodlands (mediterranean)
V	– Zonobiome of the warm-temperate humid climate
VI	– Zonobiome of the temperate-nemoral climate
VII	– Zonobiome of the arid-temperature climate
VIII	– Zonobiome of the cold-temperate boreal climate
IX	– Zonobiome of the arctic tundra climate

	max. Altitude	1	2	3	4	5	6	7	8	9	
EUROPE – ASIA MINOR											
Lebanon	3086 m			0	0						interzonal
Mts. of Kurdistan	35607 m				0			0			interzonal
Taurus Mountains	3916 m				0			0			interzonal
Pindus Mts. (Olymp)	2911 m				0						intrazonal
Taygetos	2404 m				0						intrazonal
Mts. of Crete	2456 m				0						intrazonal
Mt. Etna	3323 m				0						intrazonal
Teide (Tenerife)	3718 m				0						intrazonal
Mts. of Corsica	2710 m				0						intrazonal
Sierra Nevada	3481 m				0						intrazonal
Mts. of the Castile Region	2592 m				0						intrazonal
Cantabrian Mts.	2615 m				0	0					interzonal
Pyrenees	3404 m				0	0	0				interzonal
Appennino	2914 m				0		0				multizonal
Alps	4807 m				0		0				interzonal
Dinaric Alps	2751 m				0		0				multizonal
Rhodope Mountains	2925 m				0		0				interzonal
Pontic Mountains	3937 m					0		0			interzonal
Mount Elbrus	5604 m					0		0			interzonal
Caucasus Mountains	5633 m					0	0	0			interzonal
Carpathians	2655 m						0	0			interzonal
High Mts. of Scotland	1343 m						0		0		multizonal
Scandinavian Mountains	2469 m						0		0	0	multizonal
Ural Mountains	1894 m							0	0	0	multizonal
High Mts. of Armenia	5165 m							0			intrazonal
High Mts. of Iceland	2119 m								0	0	multizonal
High Mts. of Greenland	3700 m									0	intrazonal
AUSTRALIA – OCEANIA											
High Mts. of Hawaii	4214 m	0									intrazonal
High Mts. of New Guinea	5029 m	0									intrazonal
Australian Alps	2280 m					0					intrazonal
New Zealand Alps	3764 m					0	0				interzonal
Ruapehu (New Zealand)	2797 m					0	0				interzonal

Table 1. Continued

	max. Altitude	1	2	3	4	5	6	7	8	9	
ANTARCTIC MOUNTAINS											
South Georgia	2804 m									0	intrazonal
High Mts. of the Antarctica	5140 m									0	intrazonal
ASIA											
High Mountains (mainly volcanos) of the Malayan archipelago											intrazonal
Kerinjti (Sumatra)	3805 m	0									
Kinabalu (Borneo)	4101 m	0									
Slamat (Java)	3428 m	0									
Lokilaloiki (Celebes)	3311 m	0									
Tsukao (Taiwan)	3930 m	0	0								interzonal
Himalaya Mountains	8848 m		0					0			intrazonal
Karakorum	8611 m		0					0			interzonal
High Mts. of SW-China	7590 m		0		0			0			multizonal
High Mts. of SE-Persia	4042 m			0							intrazonal
High Mts. of Yemen	3760 m			0							intrazonal
Zagros (Persia)	4547 m			0				0			interzonal
Hindu Kush	7690 m			0				0			interzonal
High Mts. of Baluchistan	3519 m				0			0			multizonal
Qin Ling Mts. (China)	4107 m					0	0				interzonal
Fujisan	3776 m					0					intrazonal
Chang-bai-shan (China)	2744 m						0				intrazonal
Japanese Alps	3192 m						0				intrazonal
Pamir	7495 m							0			intrazonal
Kun-lun	7723 m							0			intrazonal
Nan-shan	6345 m							0			intrazonal
Tien-shan	7439 m							0			intrazonal
Altai	4506 m							0			intrazonal
Sajan Mountains	3491 m							0	0		interzonal
Changai Mountains	3905 m							0	0		interzonal
Stanowoj Mountains	2999 m								0		intrazonal
Verchoyansk Mountains	2959 m								0		intrazonal
Tschersker Mountains	3147 m								0		intrazonal
Kamchatka Volcanos	4750 m								0	0	multizonal
Kolyma Mountains	1962 m								0	0	multizonal
Korjaken Mountains	2562 m								0	0	multizonal
Anadyr Mountains	1843 m									0	intrazonal
Byrranga Mountains	1146 m									0	intrazonal

few exceptions) and the obligatory presence of a snow cover during winter are the most striking differences. The nemoral mountain climate favors mesophytic plants which need not be especially frost resistant. Sakai & Larcher (1987) presented figures for frost tolerance and found surprising sensitivity in species which are typical of snow-protected habitats.

Table 1. Continued

AMERICA	max. Altitude	1	2	3	4	5	6	7	8	9	
Sierra Central (DR)	3175 m	0									itnrazonal
Middle American Volcanos											intrazonal
Tajumucu	4210 m	0									
Popocatepetl	5451 m		0								
Citlaltepetl	5700 m		0								
South American Andes	6958 m	0	0	0	0	0	0	0		0	multizonal
Sierra Madre	3559 m		0								intrazonal
Sierra Nevada	4418 m				0			0			interzonal
Cascade Mountains	4392 m				0		0	0	0		multizonal
White Mountains	3997 m							0			intrazonal
Appalachian Mountains	2037 m					0	0		0		multizonal
Wasatch Mountains	3710 m							0			intrazonal
Rocky Mountains	6193 m							0	0		multizonal
Mackenzie Mountains	2972 m								0		intrazonal
Brooks Range	2816 m								0	0	interzonal
AFRICA											
Cameroon Mountains	4070 m	0									intrazonal
East African Mountains											intrazonal
Ruwenzori	5109 m	0									
Mt. Elgon	4321 m	0									
Mt. Kenya	5190 m		0								
Mt. Kilimanjaro	5895 m		0								
Mt. Meru	4525 m		0								
Virunga Volcanos	4507 m		0								intrazonal
Drakensberg	3462 m		0								intrazonal
Ethiopian Highlands	4620 m		0	0							multizonal
Djebel Marra	3003 m			0							interzonal
Tibesti	4165 m			0	0						intrazonal
Ahaggar	2070 m				0						
High Atlas											
Mts. of the Cape Region											

Vegetation types

The uniqueness of the European high mountains becomes clear in any effort to classify the widespread alpine grasslands of the Alps into phytosociological units, or to examine their distribution. Three major vegetation types have been distinguished by floristic differences since the earliest phytosociological publications near the beginning of this century (Rübel 1912, Braun 1913, Braun-Blanquet & Jenny 1926, Braun-Blanquet 1948; summaries in Ellenberg 1988, Ozenda 1985). Grasslands on acid soils are combined in the class *Caricetea curvulae* Br.-Bl. 1948 and those on limestones in the class *Seslerietea variae* Oberdorfer 1978. The third class, which was originally included in the

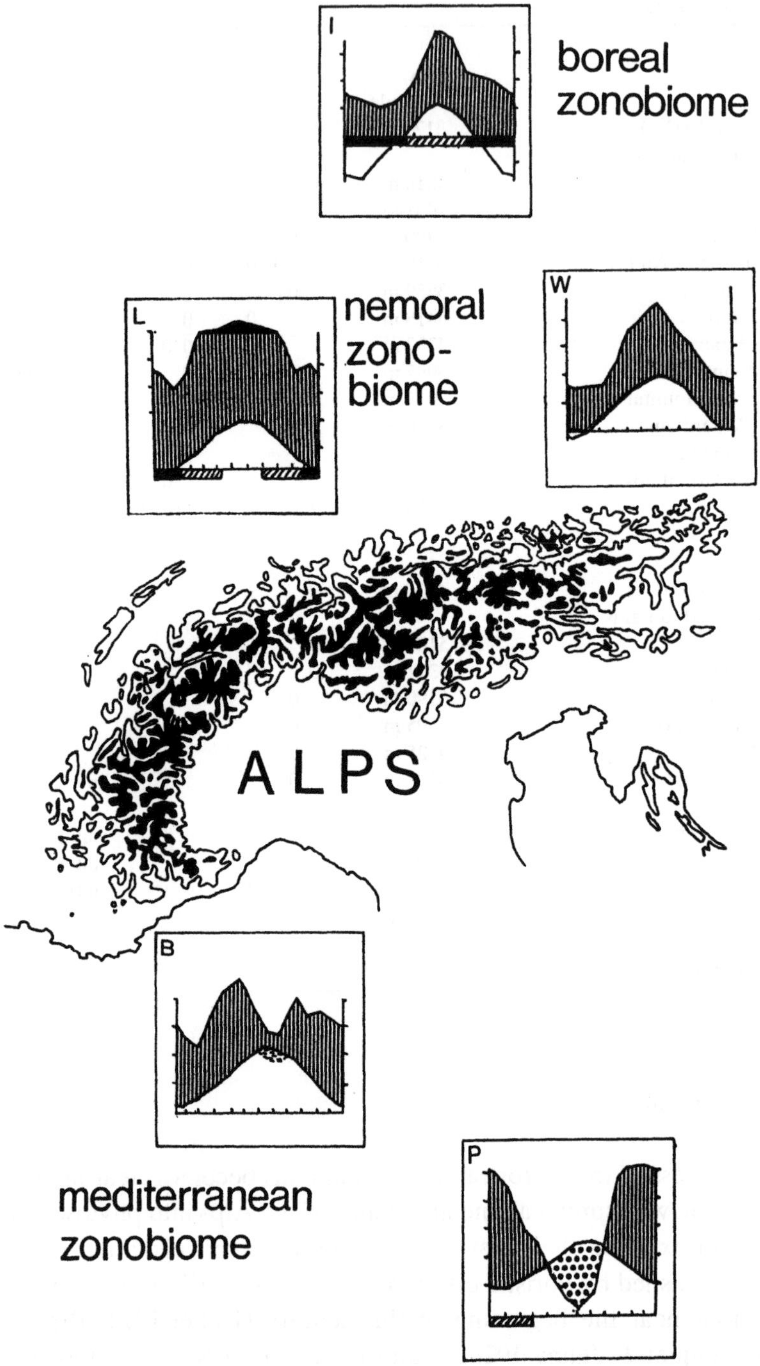
I
boreal zonobiome
L
nemoral zono-biome
W
ALPS
B
mediterranean zonobiome
P

limestone grasslands, was separated by Ohba (1974) as the *Carici rupestris-Kobresietea bellardi* and consists of grassland types of windswept ridges usually free of snow and therefore exposed to extreme winter temperatures.

Interestingly the *Kobresietea* show a world wide distribution (Fig. 2) and can be divided into zonal and azonal communities (Komarkova 1981). Their center of diversity occurs in the highlands of Central Asia where they are exposed to very dry, cold winters. Not surprisingly, in the Alps these communities occur on exposed ridges where the zonal sedge communities cannot survive. They behave, in fact, as a typical azonal vegetation type and should be placed in an ecological classification of alpine vegetation (of the Alps) under that category (Table 2). The Japanese high mountain grasslands, which are restricted to small areas (Miyawaki & Nakamura 1988), differ in some respects from the Alpine ones, though *Carex rupestris* and *Kobresia bellardi* are characteristic components. These species grow together with taller grass species (e.g. *Calamagrostis nana*) and are tremendously effected by soil frost action (Holzner & Hübl 1988). However, like the Alpine *Kobresietea*, the Japanese communities are azonal derivates of the main types of Central Asia, the Arctic, and the mountain tundra of the southern Rocky Mountains (Komarkova 1981).

In contrast, the zonal communities, dominated by sedges and grasses, and occupying huge areas in the alpine belt of the Alps, seem to have no equivalent in other mountains of the world. The most important species, such as *Carex curvula*, *Carex sempervirens*, *Carex firma*, and a series of *Festuca* and *Sesleria* species, are European endemics and may even be considered as endemic supraspecific taxa (Pignatti & Pignatti 1975). Probably they survived the glacial periods in southern refugia, or even in some parts of the Alps which remained free of glacial ice.

Fig. 1. Map of the Alps and climate diagrams (after Walter & Lieth) which show the position of the Alps between the nemoral and mediterranean life zone. Areas above 1000 m are indicated by the closed line, and areas above 2000 m are shaded black. Climate diagrams: abscissa – month of the year, left ordinate – mean monthly temperature (10 °C – intervals), right ordinate – mean monthly precipitation (20 mm – intervals); where the curve for precipitation drops below the temperature curve (dotted area; see P=Palermo), drought periods are indicated (for details see Walter 1983). Lindau (L) in the north has a typical suboceanic nemoral climate, the climate of Wiener Neustadt (W) tends to be more continental, that of Brescia (B) in the south shows some typical mediterranean characters, like a summer minimum for precipitation. The diagrams of Ivalo (I: Northern Finland) and Palermo (Southern Italy) are inserted for comparison.

Fig. 2. The world distribution of *Kobresia* grasslands. The highlands and high mountain systems (7, 8) of Central Asia are the center of diversity where *Kobresia* grasslands form the zonal vegetation. *Kobresia* communities are also widespread in the Arctic. In the marginal areas (3, 4, 6) *Kobresia* grasslands are restricted to special, azonal habitats; the Caucasus mountains are in an intermediate position. The dark-shaded areas indicate the distribution of alpine sedge grasslands (i.e. the classes *Caricetea curvulae* and *Seslerietea variae*) which are endemic to Europe (modified after Ohba 1974).

Vegetation structure and life schedules

All the dominant grasses and sedges of the grasslands of the Alps are long-lived perennials that form large, clonal populations (Grabherr 1989). Their competitive ability in alpine environments is comparatively high and the associated forb and grass species do not contribute much to the total biomass

Table 2. Habitats and vegetation classes above timberline in the Alps. Habitats with dominance of one particular environmental factor are considered as azonal. On wind-swept ridges, exposure to extreme low temperatures is the determining factor; rock crevices offer only limited space, which also precludes larger competitors; plants on screes are exposed to mechanical stress; long-lasting snow restricts the length of the vegetation period; high nutrient levels on habitats with accumulation of detritus favors tall, rapidly growing species

ZONAL VEGETATION
Dwarf shrub heath (natural or anthropogeneous)
CETRARIO-LOISELEURIETEA
Limestone grasslands
SESLERIETEA VARIAE
Grasslands on acid soils
CARICETEA CURVULAE
AZONAL VEGETATION
Grasslands on wind swept ridges
CARICI RUPESTRIS - KOBRESIETEA BELLARDI
Vegetation under recurrent disturbance
THLASPIETEA ROTUNDIFOLIAE
Vegetation on rocks (space limited)
ASPLENIETEA TRICHOMANIS
Vegetation under surplus of water and snow
SCHEUCHZERIO - CARICETEA FUSCAE
SALICTEA HERBACEAE
Vegetation under surplus of nutrients
MULGEDIO - ACONITETEA

(Grabherr et al. 1977, Klug-Pümpel 1982, Grabherr 1989). In addition, lichens often fill the gaps between the small tussocks. For example, in the *Carex curvula* grassland, lichens contribute more than 60% of the total above-ground biomass (Grabherr et al. 1977). In typical arctic tundras of the North sedges codominate with mosses, and are restricted on wet sites.

Azonal habitats

The special character of the European high mountain vegetation is most evident in its climax vegetation types. Like other mountains systems, the Alps, Pyrenees, and Carpathians show a high variety of special habitats that can be classified according to the most important habitat factors (Table 2). Their phytogeographical relations differ from type to type. Rock communities contain many endemics; vegetation types maintained by recurrent disturbance are composed of widespread species; wetlands show a striking concentration of species of boreal-arctic origin; tall forb communities of nutrient-rich sites are dominated by species that belong to widespread genera, but the particular species are often endemic. Tall forb communities are especially prominent in the snow-rich Japanese Alps (Miyawaki & Nakamura 1988, Holzner & Hübel 1988), which may indicate that the dissimilarities between the different nemoral mountain systems may not be as pronounced with respect to azonal vegetation.

Conclusion

The alpine vegetation of the Alps and other mountains of Central Europe has several unique characteristics. It has to be considered as a comparatively restricted type of high mountain vegetation. This has created and will continue to create confusion when these characters are used to describe high mountain vegetation in other parts of the world. Alternatives and concepts with an adapted terminology for each life zone might be useful, as was demonstrated for the mediterranean mountains by Ozenda (1990). The term "alpine" should only be applied in an informal way for treeless high mountain vegetation. Derivates like "subalpine", "upper alpine", "afroalpine" and others are likely to be misunderstood and should be avoided for mountains others than nemoral or warm temperate ones.

References

Braun, J. 1913. Die Vegetationsverhaltnisse der Schneestufe in den Rätisch-Lepontischen Alpen. Neue Denkschr. Schweiz. Naturf. Ges., Zürich 48:1–347.

Braun-Blanquet, J. 1948. Übersicht der Pflanzengesellschaften Rätiens. Vegetatio 1:29–41.

Braun-Blanquet, J. & Jenny, H. 1926. Vegetations-Entwicklung und Bodenbildung in der alpinen Stufe der Zentralalpen. Denkschr. Schweiz. Naturforsch. Ges. 63:183–349.

Ellenberg, H. 1988. Vegetation Ecology of Central Europe. Cambridge University Press, New York. 731 pp.

Grabherr, G. 1989. On community structure in high alpine grasslands. Vegetatio 83:223–227.

Grabherr, G., Mähr, E. & Reisigl, H. 1977. Nettoprimärproduktion in einem Krummseggenrasen (*Caricetum curvulae*) der Ötztaler Alpen, Tirol. Oecologia Plantarum 13:227–251.

Holzner, W. & Hübl, E. 1988. Vergleich zwischen Flora und Vegetation der subalpin-alpinen Stufe in den japanischen Alpen und in den Alpen Europas. Veröff. Geobot. Inst. ETH, Stiftung Rübel, Zürich 98:299–329.

Ishizuka, K. 1974. Mountain vegetation. In: Numata, M. (ed.) The Flora and Vegetation of Japan. Elsevier, Amsterdam.

Klug-Pümpel, B. 1982. Effects of microrelief on species distribution and phytomass variations in a *Caricetum curvulae* stand. Vegetatio 48:249–254.

Komarkova, V. 1981. Holarctic alpine and arctic vegetation: Circumpolar relationships and floristic-sociological, high-level units. In: Syntaxonomie, pp. 451–476. Dr. W. Junk, The Hague.

Miyawaki, A. & Nakamura, Y. 1988. Überblick über die japanische Flora in der nemoralen und borealen Zone. Veröff. Geobot. Inst. ETH, Stiftung Rübel, Zürich 98:100–128.

Ohba, T. 1974. Vergleichende Studien über die alpine Vegetation Japans. 1. *Carici rupestris-Kobresietea bellardii*. Phytocoenologia 1:339–401.

Ozenda, P. 1985. La végétation de la chaine alpine dans léespace montagnard européen. Masson, Paris, 331 pp.
Pignatti, E. & Pignatti, S. 1975. Syntaxonomy of the *Sesleria varia*-grasslands of the Calcareous Alps. Vegetatio 30:5–14.
Rübel, E. 1912. Pflanzengeographische Monographie des Berninagebietes. Bot. Jahrb., Leipzig 47:1–646.
Sakai, A. & Larcher, W. 1987. Frost Survival of Plants. Springer, New York.
Walter, H. 1983. Vegetation of the Earth and Ecological Systems of the Geo-biosphere. Springer, New York.
Zhu, T.-C. & Rowe, J. S. 1987. A comparison of alpine tundra floras: N.E. China and N.W. North America. Linzer biol. Beitr. 19(1):29–41.

26. Towards a regional forest classification for the humid tropics of Asia

PETER S. ASHTON
Arnold Arboretum, Harvard University, Cambridge, Massachusetts, USA

Abstract

The classification of the forest types of India by H. Champion is put into regional context, modified and extended into the outline for a proposed new classification for the zonal upland primary forest types of the lowlands of the Asian humid tropics.

Key words: Asia, forest classification, lowland humid tropics, regional types, zonal forest

Introduction

It would be fair to say that knowledge of the forest types of lowland tropical Asia is greater than that of the Neotropics but not as complete as that of Africa, where the integrative role of A.E.T.F.A.T. (1959) has set an example that no other continent has so far succeeded in emulating. This collaboration resulted in continental vegetation maps of unprecidented uniformity and precision. Two pioneering efforts do exist in Asia, though, which provide a basis at least for reviewing what we know and drawing up a schematic classification as a basis for discussion and refinement. These are the landmark and remarkably detailed preliminary survey of the Forest Types of India and Burma (Champion 1935), and its revision for modern India alone with S. K. Seth (1968); and the magnificent mapping program of the Institut de la Carte Internationale du Tapis Végétal, the University of Toulouse, and the French Institute at Pondicherry. Both rest on a combination of a wealth of personal experience on the ground and a rather comprehensive though inevitably regionally patchy literature, equally essential for success. The comprehensive forest working plans which have been and remain a tradition of India, and formerly Burma, were a particularly important source of ground information.

E.O. Box et al. (eds.), Vegetation Science in Forestry, 453–464.

Table 1. H. G. Champoin's classification of tropical lowland forests in South Asia and Burma (excluding Sri Lanka)

Wet evergreen forests	Dense tall forest, entirely evergreen or nearly so.
Semi-evergreen forests	As above, but dominants include some deciduous species.
Moist deciduous forests	Dominants mainly deciduous but subdominants and lower storeys largely evergreen. Top canopy rarely dense, and even, but over 25 m.
Littoral and swamp forests	Mainly evergreen, or varying density and height, but always associated with wetness.
Dry deciduous forests	Entirely deciduous or nearly so. Top canopy rather light and rarely over 25 m high, usually 8–20 m.
Thorn forests	Deciduous with low thorny trees and xerophytes predominating. Canopy more or less broken. Height under 10 m.
Dry evergreen forests	Hard leaved evergreen trees predominate with some deciduous emergents, often dense, but usually under 20 m high.

Champion and Seth's classification (Table 1) gives primary emphasis to physiognomy including leaf phenology, and structure including life forms and morphological characters, with less emphasis on species composition and dynamics. This was done explicitly in recognition of the influence of historical biogeography on the flora, and of the confounding effects, and often local specificity of successional vegetation, widespread though it was even then. Their approach indeed provided a regional frame. But the downplaying of dynamic aspects led to a static rigidity which has made it inadequate in an era of actively managed, not to say degraded forests. The subordination of floristic information, and even of dominant species, to structure and physiognomy has obscured those major regional differentiations which are related to geology and soil, rather than to climate and climate-correlated factors such as fire.

The Pondicherry group have consciously attempted to address both these shortcomings. Their method depends, first, on a classification of climate based on criteria found, by empirical experience in many regions of the world, to be the most correlated with vegetation: mean temperature of the coldest month, annual rainfall, and length and number of dry seasons. Second, it depends on comprehensive ground truth, extrapolated by means of aerial photographs, which has permitted use of dominant species as a major basis for their vegetation classification. In effect, soil-correlated variation on a geographical scale is thereby incorporated, though the necessary soil survey data to define the interrelationship between vegetation and soil do not yet exist. Comprehensive ground truth has also provided the means to understand more comprehensively the dynamic relationships between different vegetations and to identify seres more rigorously. The Pondicherry group have now extended their work from India, including the Himalayas and Cambodia, to Indonesia, where they have published 'bioclimate' maps for Sumatra and Borneo, and a vegetation map of Sumatra.

The principal shortcomings of current regional classifications

Vegetation classification has to be a continuing process, a method of successive approximations dependent on continuing flow of new information concerning natural plant communities, and also appraisal of the currently rapid rate of man-induced vegetational change. It is the first of these which principally concerns us here. Particularly, detailed understanding of the relationship between natural vegetation and soils, although perhaps more studied in Asia than in the tropics elsewhere, remains at best patchy and preliminary. Nevertheless, a broad pattern for zonal lowland forests has emerged sufficiently to provide our principal conclusions. This cannot be said for the montane forests or forests on extreme sites where local historical biogeographic and habitat variation mandates a higher level of ground documentation than in the lowland hills, and whose frequent inaccessibility has left vast areas, particulary of Upper Burma and Laos, virtually unknown; nor of the alluvial forests which have manifested much local variation but which have extensively disappeared.

The principal deficiencies in current classification concern the wet evergreen forests of the aseasonal and shortly seasonal lowlands: the zonal forests on yellow-red ultisol soils. I have called these mixed forests on account of the absence of simple dominant species in the emergent canopy (Ashton 1967). Champion's classification included no examples of these 'true' wet evergreen forests because they occur in aseasonal climates, in Southwest Sri Lanka and in the area formerly known to biogeographers as Malaysia (e.g. van Steenis 1950), and now called Malesia to distinguish it from the political entity. Not only do Champion's wet evergreen types bear major differences from the wet evergreen forests of the aseasonal wet lowlands, some of which have been alluded to by Whitmore (1984, see also Ashton 1967), but the relationship between his so-called evergreen and semi-evergreen forests has been obscured by his consideration of the influence of rainfall seasonality on canopy deciduousness to the virtual exclusion of edaphic influences.

The Toulouse-Pondicherry group have also not yet provided a satisfactory classification of mixed wet evergreen forests, particularly in aseasonal climates. This is again because differences in rainfall seasonality and amount fail to correlate with vegetational change in regions where precipitation exceeds evaporation on the average every month of the year; and also because forest types are difficult to classify where single species dominance is absent and the relationship between structure and floristic composition complex.

It is within the wet evergreen forests, then, that we must first search for new approaches to a regional classification.

Towards a new regional classification of lowland zonal forests

The historical-biogeographic barriers of Wallace's line, the Bashi Channel south of Taiwan, and the Torres Strait are already well known (Whitmore 1981, Merrill 1909, 1923, van Steenis 1950, Walker 1972). Although Merrill (1909, 1923) suggested that Wallace's Line might better be placed east of the Philippines, the family dominance of Dipterocarpaceae in the mixed wet evergreen forests of lowland peninsular Malaysia, Sumatra, Borneo and the Philippines so clearly distinguishes them from those further east that, vegetationally, there is no question that Wallace (1860) was right. As is also the case at the Torres Straits, though, there is no such clear dividing line between Bali and Lombok in the characteristics of the seasonal deciduous and semi-evergreen forests, so that a gradual transition appears to occur, in what is left of these forests, from seasonal central and east Java through to northern Australia. But these eastern Indonesian forests will not concern us here.

A remarkably similar, though certainly less prominent historical biogeographic boundary in lowland evergreen forests occurs west of the India-Burma border ranges, along a line which runs north across the Brahmaputra valley between Darrang and Lakhimpur to the Himalayan foothills. This line marks the western limit of the seasonal wet climates of the Indo-Burmese region, climates that do not occur again until the Western Ghats of peninsular India. That the biogeographic boundary may be along the eastern edge of the intervening dry corridor is supported by the boundary that occurs there in the distribution of the dominant species that characterize the fire-climax dry deciduous dipterocarp forest, a forest formation of more seasonal climates which is to be further discussed. *Shorea robusta*, the sal, which is the dominant and often only canopy species in much of this extensive forest type in eastern India and Bangladesh reaches east to Meghalaya and Tripura, and up the Brahmaputra to the Sibsagar forests in Lakhimpur district. Immediately to the east of the frontier ranges, the Dry Deciduous Dipterocarp forests are co-dominated or locally dominated by one or more of a group of some six fire-resistant dipterocarps in the genera *Dipterocarpus* and *Shorea*. The geographical distance between the two types barely exceeds 150 km.

Research into the nature of this biogeographical boundary has been impeded by the international frontier, but there is a wealth of opportunity here for graduate student research in India and Burma.

The range of climates, at least in terms of the length of the dry season, experienced by the lowland evergreen and semi-evergreen forests of the Western Ghats and of Indo-Burma are the same, yet the great majority of the more abundant canopy tree species and of some genera, such as *Cullenia* and *Durio*, are confined to one or the other region. One notable exception is the widely abundant *Mesua ferrea*, which ranges from southern India to Peninsular Malaysia, and also to northern Sumatra where it may be introduced. I do not know of one species which is sufficiently ubiquitous, yet confined to the wet seasonal lowland evergreen forests, which could be used to define them as a whole.

The other major classes of zonal forest can be explained by ecological factors.

Whitmore, 1984, has reviewed biogeographic evidence for the major biogeographic boundary, which he terms the Kangar-Pattani Line, which occurs in northwest peninsular Malaysia across to southeast peninsular Thailand. This may rather exactly correlate with a change from a climate (or soil water conditions) which is effectively aseasonal to the south of the Line, to one with a short annual period of lower atmospheric moisture or soil water stress to the north. The exact climatic relationship must await more detailed vegetation mapping, coupled with transplant and ecophysiological experiments on juveniles of critical species. Among the many changes that appear to occur across the Line in the composition and function of the forests are, to the north, a massive drop in species richness, particularly occasioned by virtual elimination of the series of ecologically sympatric congeneric species so characteristic of western Malesian forests; more frequent, annual flowering of canopy trees, although flowering intensity varies between years; the dampening or absence of mast fruiting; and the death of nearly all seedlings of climax species within one year of germination in most years, during their first dry season. The case for these additional changes currently rests on my own field observations. Once again, there are plenty of opportunities for graduate research here, but students must get there without delay, because the recent cessation of guerilla activity has been followed by a log-rush.

On the basis of their composition, I have termed the western Malesian zonal lowland evergreen forests Mixed Dipterocarp forests, and those abutting to the north and throughout continental Asia Seasonal Wet Evergreen Dipterocarp forests.

Political history has also clouded the very real differences which exist between the wet evergreen forests of the Western Ghats and those of southwest Sri Lanka. Champion did not include the latter in his classification. The differences between these South Asian types broadly mirror the differences north and south of the Kangar-Pattani line, but here the two are separated by a broad belt of strongly seasonal climate in central and northern Sri Lanka, as well as by the Palk Strait. There is only a narrow zone of wet seasonal climate in lowland Sri Lanka, mainly confined to the northern foothills. The seasonal wet evergreen lowland element is poorly represented there. But the wet zone of Sri Lanka is an area of less than 10,000 km^2, of which about one third is montane. It has apparently only had intermittent connection with the rain forests of the Far East (Ashton & Gunatilleke 1987). It is therefore not surprising that the total indigenous angiosperm flora is somewhat less than 2500 species but that over 60% of these in the primary forests are endemic. The differences in flowering phenology and seedling survivorship between Lankan and Ghat forests are also likely to parallel those south and north of the Kungar-Pattani Line, but Lankan forests have about one fifth the richness of western Malesian forests (e.g. see Ashton 1984). This is in part because there are far fewer ecologically sympatric series of congenerics in Sri Lankan than in western Malesian lowland forests. These Lankan forests represent a distinct, albeit depauperate outlier of the Mixed Dipterocarp forest nevertheless, characterized by abundance of the endemic and speciose canopy dipterocarp taxon *Shorea*, section *Doona*. The western Malesian Mixed Dipterocarp forests are well distinguished by the red meranti field group of *Shorea*, comprising sections Mutica, Brachypterae, Rubellae, Ovales, and Pachycarpae.

The seasonal boundary of the seasonal evergreen forests also requires review. Champion, besides acknowledging historic-biogeographic differences east and west of the Brahmaputra, recognized two ecological bases for division of his evergreen and semi-evergreen lowland forests: Rainfall and latitude and thus, by implication, minimum monthly temperature. He used presence of deciduousness in the canopy as the principal criterion for distinguishing between his tropical wet evergreen and tropical semi-evergreen types; but he admitted that the latter comprise a heterogeneous group, many of which could be regarded as phasic communities between true evergreen and moist forests with a fully deciduous canopy, termed by him mixed deciduous forests. Frequent additional differences between evergreen and semi-evergreen forests include lower stature, a tendency towards greater dominance of the most abundant canopy species and an impoverishment, in both species and numbers, of the emergent canopy in the latter. He acknowledged that the type occurs both in drier and more seasonal climates, that is with less than 2500 mm mean annual rainfall and/or 5–6 dry months, and also on shallow or freely draining soils prone to water stress in wetter, less seasonal climates. Reading his accounts, one finds no obvious floristic difference between the evergreen and semi-evergreen types.

Current knowledge of Asian forests suggests that there are two causes for the unsatisfactory nature of this division. First, patterns of variation in forest structure and leaf phenology do not go hand in hand. An obvious example of this is the Dry Evergreen forests of southeastern Peninsular India and northern and eastern Sri Lanka, which occur on freely draining soils in a climate with low but rather well distributed rainfall. Here, all but a few scattered semi-evergreen trees, such as *Chloroxylon swietenia*, are evergreen, yet the forest is short and of generally even canopy.

Second, canopy deciduousness, though broadly correlated with rainfall seasonality, is also generally more prevalent in disturbed and successional forests, and on comparatively low nutrient soils. Many species such as *Tectona grandis* (teak), which are successional towards the wet end of their range, become mature-phase components towards the dry. Perhaps all deciduous species of the everwet lowlands, such as *Pterocymbium*, *Salmalia* and *Octomeles*, are successional. Canopy deciduousness can therefore often be indicative of disturbance rather than of a distinct forest type.

The issue of soil nutrients is of more particular interest in our present context, though, because, as will be explained, it may provide us with a basis for a more workable and consistent forest classification.

The only detailed comparisons of edaphic and vegetational variation in tropical forests have been carried out in Brunei and Sarawak, northwest Borneo (Ashton 1964, in prep., Austin et al. 1972, Ashton & Hall 1992). This research demonstrated relative autonomy of species composition and stand (in particular, canopy) structure there. Stand stature does indeed correlate with soil depth and presumably with short-term water storage capacity (see also Brunig 1970). The density of emergents appears to be influenced by periodic catastrophe, though, particularly soil surface stability and exposure to gusts of wind, or occasional intense drought. Species composition within one climatic zone, though modified by natural disturbance regimes as well as man, is principally correlated at community level with nutrient-related soil characteristics. In Borneo, a major floristic division occurs between forests on udult ultisols, generally relatively rich in clay and with relatively high surface pH and mineral soil nutrient concentrations; and on humic ultisols, characteristically with a surface root-matted horizon of raw humus of variable thickness but also with generally lower pH (3.5–4.1), mineral soil clay content and nutrient concentrations. Forests on udult soils appear to be of more uniform species composition than those on humults, where there is evidence that composition is more closely tied to soil nutrient concentrations.

Throughout western Malesia, widespread species characteristic of humult ultisols appear to include *Dryobalanops aromatica* and *Shorea curtisii*, those of udults *Shorea parvifolia*, *Shorea leprosula* and, in the Philippines where lowland humult soils are rare, *Shorea polysperma*. Clearly, though, the relationship between soil chemistry and water economy is more complex than implied here. *Shorea curtisii*, for instance, occurs in Borneo on lowland humult ultisols but in Peninsular Malaysia follows inland ridges and also occurs on low hills near the coast. On the ridges, it appears to be closely associated with the distribution of soil surface raw humus, but this is not apparent in the second habitat. Both habitats are prone to soil water stress, as are many humult soils in Borneo, which are generally sandy. The distribution of *Shorea curtisii* is paralleled by a number of other species. Clearly, more research in other parts of the region outside Borneo is mandated.

Humult ultisols occur in the coastal hills and narrow ridges of Sri Lanka's wet zone, where *Shorea worthingtonii* (Section *Doona*) is a characteristic species; and they extend beyond the everwet lowlands to the margins of Champion's Tropical Lowland Wet Evergreen forests. In Indo-Burma these bear a strictly evergreen forest characterized by, among others, *Dipterocarpus costatus*, *Dipterocarpus dyeri*, *Shorea hypochra*, and *Heritiera javanica*. Some species of these soils, such as *Hopea ferrea* and *Vatica cinerea* also occur on limestone at the climatically wet end of their range. These forests tend to be somewhat shorter in stature than those on udults in the same region which include such species as *Parashorea stellata*, *Shorea assamica*, *Dipterocarpus kerrii*, *Amesiodendron chinense*, *Pterospermum* spp. and deciduous successional species such as *Bombax* spp. and *Tetrameles nudiflora*. The *Dipterocarpus costatus-Shorea hypochra* type is best developed on the cretaceous sandstone of the Cardamome and Kamchay mountains of southern Cambodia but recurs in the Tarutao islands of Peninsular Thailand and in the Andamans. In an interesting parallel with *Shorea curtisii*, *Dipterocarpus costatus* and some of its associates also occur on humus-rich soils on high ridges south into peninsular Malaysia and at mid-elevations in northern Thailand and the Arakan Yuma north to Chittagong.

The tropical wet evergreen lowland forests of peninsular India appear to be confined to udult soils. They have their own characteristic species, which include *Dipterocarpus indicus*, *Hopea parviflora*, *Calophyllum wightianum*, *Cullenia excelsa*, and also the widespread deciduous successional species *Salmalia* (*Bombax*) *malabarica* and *Tetrameles nudiflora*.

Along the Western Ghats of peninsular India, the seasonal evergreen forest becomes gradually impoverished and shorter northwards, and no distinct transitions are apparent. This is not so in Indo-Burma. There, the Seasonal Mixed Wet Evergreen Dipterocarp forests essentially occur throughout the range of climates in which commercial rubber plantations can be grown, that is from the Kangar-Pattani line north to Chumphon in Thailand, then along the west peninsular coast through Burma and northwest along the monsoonward slopes of the Arakan ranges into the Chittagong Hills tracts of Bangladesh, also from Cochinchina northwards into the coastal flanks of the Annamite range approximately to Quang Tri and in the southern flanks of the Khorat plateau from Khao Yai eastwards into southern Laos.

North of these forests in Indo-Burma, Champion rightly indicates that there are other zonal evergreen forest formations. These, it now appears, comprise at least two distinct types on a regional scale. One occurs in more seasonal climates, with up to five months dry season, and in even drier climates as a more or less narrow riverine gallery forest. This type is well known in Thailand as Dry Evergreen forest. The type typically occurs on yellow to rust-red oxisols and is characterized by *Dipterocarpus turbinatus*, a species which occurs from southern Vietnam and Laos to the Patkoi Hills of Upper Burma, the southern flanks of Meghalaya, Tripura and Mizoram, thus coinciding with the range of the type. This type may include varying numbers of deciduous species in the canopy, including *Lagerstroemia*, *Afzelia* and other legumes, taxa which transgress the ecotone into the moist deciduous forests into which these forests grade on drier soils and more seasonal climates. Edaphic types do not appear to be easily recognizable, perhaps because the sandy more acid soils of these climates are so dry that they support fire climax, deciduous vegetation instead. The forests do vary considerably in composition, particularly in the presence and degree of dominance of leading canopy species and the composition of the understorey. The local abundance of bamboo appears to indicate periodic penetration of fire.

Another regional type was recognized as northern tropical wet evergreen forest, by Champion (1935) and Champion & Seth (1968), who nevertheless also included the Cachar evergreen forests, which I consider to be Dry Evergreen Dipterocarp forest, and two transitional types in this category. These northern Seasonal Wet Evergreen Dipterocarp forests, as I propose to name them, mark the latitudinal margin of lowland tropical evergreen forest in east Asia. They are remarkably rich in species, and maintain a high level of endemism of trees as well as shrubs and herbs. They vary geographically in composition, with a major local center of endemism in southern Guanxi and northern Tonkin. Species which characterize these forests include *Dipterocarpus retusus*, *Shorea assamica* ssp. *assamica* towards the west, *Parashorea chinensis* and *Dracontomelum duperreanum* towards the east. These forests are rich in shrubs and small trees, including submontane elements such as *Camellia* and *Strobilanthes*, and herbs.

Northern Seasonal Wet Evergreen Dipterocarp forests occur in climates of moderate to high mean annual rainfall with a minimum of about 2000 mm, but in which the dry season is relatively short and confined to the cool season, which is characterized by clear days and abundant dew formation at night. These forests extend over the moist foothills from Lakhimpur on the Brahmaputra eastwards, through Arunachal Pradash and upper Burma to the Chiang

Rai region of northernmost Thailand, Upper Laos, Southern Yunnan and Guangxi, and Vietnam north of Vinh. Now they are perhaps the most endangered of all tropical evergreen forest types.

At the northern margins of this forest, in south China, clearing and grazing leads to invasion by a grassland flora rich in temperate elements, including agrostoid grasses, *Viola*, *Corydalis*, *Ranunculus* and *Delphinium*, associated with cool-season ground frosts. Whereas the northern limits of the Dry Evergreen Dipterocarp forests are determined by seasonal drought, those of the Northern Seasonal Evergreen Dipterocarp forest appear to be determined by minimum winter temperature.

The drier forest climates of tropical Asia are occupied by a diversity of deciduous forests, in which fire, if not necessarily annual, is a major ecological factor. The interface between evergreen, including semi-evergreen, forests and the deciduous forests has for centuries been determined by man-induced fire rather than annual water stress, as fire exclusion experiments long ago demonstrated (Slade 1895, Milroy 1930). In effect, deciduous forests at the moist end of their present climatic range are invariably successional stages, deflected from proceeding to an evergreen, fire sensitive mature phase.

The ubiquity of fire and other disturbance partially obscures the relationship between deciduous forest types and soil. Here once again Champion, placing physiognomic and phenological characteristics above floristic, made his major division between moist and dry types of deciduous forests, and also differentiated the tropical thorn forests of the desert fringe (Table 1).

Within his two deciduous types, Champion recognized a group of northern communities in most of which the dipterocarp *Shorea robusta* is an important and often dominant canopy component; and a group of southern communities in which teak, *Tectona grandis*, is often dominant. In both groups, however, mixed deciduous community types are also included. This categorization cannot be sustained on a regional scale. First, the mixed deciduous forests of southern and northern India, and indeed those to the east into Vietnam, possess a floristic and indeed physiognomic continuity which defies classification on regional scale, although species richness and endemism increase towards the east. Second, teak, the major component of the southern forests, occurs within the latitudinal range of the northern deciduous forests of Champion in Burma and northwestern Thailand. *Shorea robusta*, sal, does not occur in Indo-Burma. The northern boundary of peninsular Indian teak and the southern boundary of sal almost exactly coincide in the hills of central India (Champion and Seth 1968). Further south they overlap, but the two species rarely co-occur in the same stand. Though there are certainly exceptions, particularly in the moist deciduous forest regions where a long history of fire and other disturbance has played a major part in maintaining these forests, teak is associated with a distinct floristic element which does not occur in sal forests, and *vice versa*. Teak and its associates prevail on

Table 2. Zonal lowland forests of tropical Asia. Suggestions for a regional classification

1. Lowland evergreen forests (fire sensitive)
   - 1.1. Aseasonal evergreen forests (mast fruiting)
     - 1.1.1. Mixed dipterocarp forests
       - 1.1.1.1. Meranti forests (Sunda shelf, Philippines)
         1. *Dryobalanops aromatica* – *Shorea curtisii* forest (humult ultisols)
         2. *S. parvifolia* – *S. polysperma* forest (udult ultisols)
       - 1.1.1.2. Dun (*Shorea*, section *Doona*) forests (Sri Lanka)
     - 1.1.2. Mixed Sahul forests (Moluccas, New Guinea)
   - 1.2. Seasonal evergreen dipterocarp forests (annual fruiting)
     - 1.2.1. Seasonal mixed wet evergreen dipterocarp forests (1–3 month dry season)
       - 1.2.1.1. Southern mixed wet evergreen forests
         1. *Dipterocarpus costatus* forest (humult ultisol)
         2. *Parashorea stellata* forest (udult ultisol)
       - 1.2.1.2. Northern mixed wet evergreen forests (*S. assamica*, *D. retusus*)
       - 1.2.1.3. Pen. Indian forests (*D. indicus*, *Cullenia excelsa*)
     - 1.2.2. Dry evergreen forests (4–6 month dry months)
       - 1.2.2.1. Eastern dry evergreen forests (*Dipterocarpus turbinatus*)
       - 1.2.2.2. Pen. Indian/Lankan dry evergreen forests (*Manilkara hexandra*, *Memecylon edule*)
2. Tropical deciduous forests (fire tolerant)
   - 2.1. *Dalbergia*/*Lagerstroemia*/teak forests (high nutrient, mesic)
     - 2.1.1. Mixed deciduous forests
     - 2.1.2. Teak forests
   - 2.2. Deciduous dipterocarp forests (low nutrient)
     - 2.2.1. In-Kanyin (*D. obtusifolius*/*S. obtusa* forests (Indo-Burma)
     - 2.2.2. Sal (*S. robusta*) forests (India)

clay-rich soils which can be broadly described as mesic, although they may become extremely dry seasonally at the dry end of the climatic range. Sal and its associates occur on acid, sandy or lateritic soils. East of the India-Burma frontier ranges, as indicated earlier, sal is replaced by a group of Indo-Burmese endemic dipterocarps, including *Dipterocarpus tuberculatus*, *D. oblongifolius*, *D. intricatus*, *Shorea obtusa* which is closely related to sal, and *S. siamensis*. But the floristic associates remain largely the same, so that these Dry Deciduous Dipterocarp forests are clearly a historical-biogeographic analog of the *Shorea robusta* forests of India. Notable is the general paucity of bamboos in these forests, except for members of the dwarf bamboo genus *Arundinella*. Though the tall bamboo *Dendrocalamus strictus* occurs widely in sal as well as teak forests, its congener *D. hamiltonii* is characteristic of teak and associated mixed moist deciduous forest in Indo-Burma. Indeed, at the moist end of their range, teak and associated mixed deciduous forests are often distinguished, where not degraded by frequent fire, by their tall bamboo subcanopy, whereas the Dry Deciduous Dipterocarp forests generally bear a grassy understorey, with scattered shrubs. Rigorous regional analysis, once again, is needed, but it would appear that the major discontinuity is between what could be called the Dry Deciduous Dipterocarp and associated forests, and the Mixed Deciduous and associated teak dominated forests. Within each, I hypothesize, a floristic continuum will be found from the moist to the dry end of their wide and parallel climatic ranges.

In summary, it is suggested that classifications which emphasize physiognomy and phenology over floristics do not provide a sufficient basis for the more detailed, silvical research required for the intensive management that is now mandated for the restricted natural forests which remain. Rather, a classification based on floristic criteria is needed and will, it is suggested, yield a more ecologically informative base and more consistent discontinuities in the field. A tentative beginning, for testing and modification, is provided in Table 2.

References

Ashton, P. S. 1964. Ecological studies in the mixed dipterocarp forests of Brunei State. Oxford Forestry Memoire:25.

Ashton, P. S. 1967. Climate versus soils in the classification of Southeast Asia tropical lowland vegetation. J. Ecol. 55:67.

Ashton, P. S. 1984. Biosystematics of tropical forest plants: A problem of rare species. In: Grant, W. F. (ed.) Plant Biosystematics, pp. 947–518. Academic Press, Toronto.

Ashton, P. S. in prep. Comparative ecology of mixed dipterocarp forests in northwest Borneo. II. Patterns of floristic variation.

Ashton, P. S. & Gunatilleke, C. V. S. 1987. New light on the plant geography of Ceylon. I. Historical plant geography. J. Biogeogr. 14:249–285.

Ashton, P. S. & Hall, P. 1992. Comparisons of structure among mixed dipterocarp forests of north-western Borneo. J. Ecol. 80:459–481.

Association pour l'études taxonomiques de la Flore d'Afrique Tropicale (AETFAT) 1959. Vegetation map of Africa south of the tropic of Cancer. Explanatory notes by R. W. J. Keay. Oxford.

Austin, M. P., Ashton, P. S. & Grieg-Smith, P. 1972. The application of quantitative methods to vegetation survey. III. A reexamination of rain forest data from Brunei. J. Ecol. 60:305–324.

Brunig, E. F. 1970. Stand structure, physiognomy and environmental factors in some lowland forests in Sarawak. Trop. Ecol. 11:26–43.

Champion, H. G. 1935. A preliminary survey of the forest types of India and Burma. Indian Forest Records, n.s., Silviculture 1:1.

Champion, H. G. & Seth, S. K. 1968. A revised survey of the forest types of India. Manager of publications, Delhi.

Merrill, E. D. 1909. The Malayan Australasian and Polynesian elements in the Philippine flora. Ann. Jard. Bot. Btzg. 2.3:277–306.

Merrill, E. D. 1923. An enumeration of Philippine flowering plants. Philip Bur. Science Publ. 18. 4:5–124.

Milroy, A. J. W. 1930. The relation between sal forests and fire. Indian For. 56:442.

Slade, H. 1895. Too much fire protection in Burma. Indian Forester 22:172.

Steenis, C. G. G. J. van. 1950. The delimitation of Malaysia and its main plant geographical divisions. Flora Malesiana 1:1. 1xx–xxv.

Walker, D. 1972. Bridge and barrier: The natural and cultural history of Torres Strait. Research School of Pacific Studies, Australian National University. Publication B1/3, Canberra.

Wallace, A. R. 1860. The zoological geography of the Malay archipelago. J. Linn. Soc., London 4:172–184.

Whitmore, T. C. 1981. Wallace's line and plate tectonics. Clarendon Press, Oxford.

Whitmore, T. C. 1984. Tropical rain forests of the Far East. Clarendon Press, Oxford.

27. Structure and dynamics of dry dipterocarp forest, Sakaerat, northeastern Thailand

PONGSAK SAHUNALU & PRICHA DHANMANONDA
Dept. of Silviculture, Faculty of Forestry, Kasetsart University, Bangkok 10900, *Thailand*

Abstract

Investigations of the stand characteristics and changes in mortality and growth of tree species in the typical deciduous dry dipterocarp (broad-leaved) forest occurring in mainland Southeast Asia, the most extensively distributed forest type of Thailand, were carried out in four stands of 1 ha square in Sakaerat Environmental Research Station, northeastern Thailand. The four stands show sporadic, discontinuous crown cover with large canopy openings as well as stratification of the whole stand into-two layers, with some parts showing three layers. Some quantitative stand parameters vary from stand to stand, but all stands contain relatively few species. Association types of the four stands, based on the highest importance value index (IVI) of the two main species, were recognized as (1) *Shorea floribunda-Quercus kerrii*, (2) *Shorea obtusa-Shorea siamensis*, (3) *Shorea obtusa-Pterocarpus macrocarpus* and (4) *Shorea siamensis-Shorea floribunda*. The distributional pattern of all individuals seems random, while tree species having the highest importance value index (IVI) suggest a contagious distribution. Changes in number of individuals, death and recruitment (to DBH $\geqslant$ 4.5 cm) were followed for four years, together with their mean annual diameter increment, relative growth rate and basal area cover for the stands and for the selected tree species. These stand parameters changed differently depending on species composition, stand density and behavior of the dominant species.

Key words: dry dipterocarp forest, dynamics, northeastern Thailand, Sakaerat, structure

Introduction

Ecological studies of tropical forest vegetation have mostly concentrated on that occurring in humid climates, partly due to the attractive characteristics

E.O. Box et al. (eds.), Vegetation Science in Forestry, 465–494.

of the forest structure, species diversity and high productivity compared to the temperate forest vegetation. Among these studies ecological characteristics in the static form are numerous, but there are few reports on the dynamic aspects. In particular, studies on forest types which occur in dry climates are almost unavailable, although their importance at local and regional levels of the tropics is obvious. Dry dipterocarp forest or dry deciduous dipterocarp forest (Royal Forest Department 1962) is a deciduous broad-leaved forest community type occurring on relatively dry sites and is mainly composed of trees belonging to the family Dipterocarpaceae. It occurs widely in mainland Southeast Asia, i.e. in Vietnam, Laos, Cambodia, Burma, Thailand and a small area of northeast India. In Thailand it is found in the north, northeast and central regions, where the annual rainfall is 1000–1500 mm with a 5–6 month dry period, which is the approximate period of leaflessness (with frequent ground fire). This forest type covers approximately 45% of the total forested area of Thailand (Neal 1967). As it occurs naturally at altitudes less than 1000 m above sea level and on shallow, poor soils unsuitable for agricultural use, it is extensively utilized as a source of wood, fuelwood, charcoal, feeding ground for domesticated animals and for other minor forest products, as well as protective forest in some watershed areas, wildlife sanctuaries and some recreation sites. As the population pressure on land use intensified in recent years, most of the disturbed forest areas in Thailand were released for agricultural purposes; the dry dipterocarp forest, especially that remaining on unproductive soils, would potentially be managed as a source of forest products. Although management of other forest types in the country has been practiced and knowledge of their structure and dynamics is relatively large, knowledge of the dry dipterocarp forest has not been well documented. Structural analysis of dry dipterocarp forest has been done to some extent, e.g. by Ogawa et al. (1961), Kutintara (1975) and Bunyavejchewin (1983), but dynamics of growth, death, recruitment and so on have not yet been fully studied. This paper deals with this forest type at a particular site considered to be typical for the deciduous broad-leaved forest vegetation of the relatively xeric habitat of the tropical forest condition. Forest structure and dynamics are studied in order to clarify stand characteristics and changes in tree populations as well as some aspects of stand and individual tree growth. For this part of the studies, emphasis is only on the canopy tree layer. The present report represents part of the long-term study of forest vegetation in the Sakaerat Environmental Research Station.

Study site and methods

The study was carried out at the Sakaerat Environmental Research Station, which is located on the southeast fringe of the Korat Plateau (lat 14°31′ N,

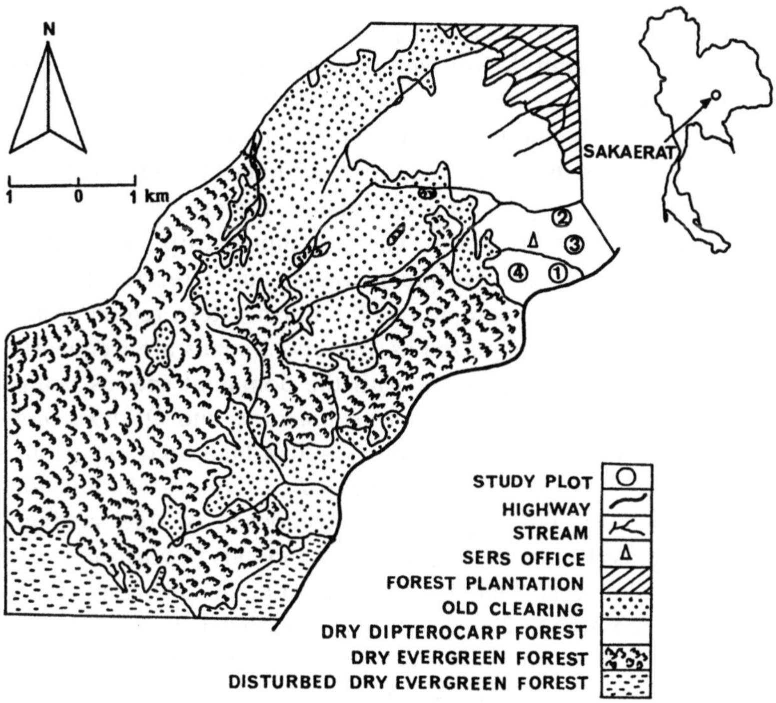

Fig. 1. Vegetation map of the Sakaerat Environmental Research Station (SERS) and locations of the study plots.

long 101°55′ E) in northeastern Thailand (Fig. 1). The area encompasses 81 km^2 of a reserved forest with two dominant forest types, dry evergreen forest and dry dipterocarp forest. The area has also been designated a biosphere reserve, as it has suffered little disturbance compared to much of the reserved forest area of the country and has been well protected since the establishment of the station in 1967. The topography of the site is mountainous, with slopes ranging from 10 to 30%. It is underlain by sandstone formed in the Triassic to Cretaceous periods, classified as the Korat geological group (Moormann & Rojanasoonthorn 1972). The altitude of the area is 200–800 m above sea level. The soil is generally poor, characterized by rock outcrops and dominated by laterite. The soil texture is mostly sandy loam and clay loam. The climate is classified as a "Tropical savanna type" according to Köppen (Griffith 1978), with a bimodal rainfall pattern, each period lasting for 2–4 months. The total precipitation is 1100–1300 mm per year. The summer dry period lasts from March to mid-May and the cool dry season follows from November to February. The daily mean temperature is 29°C during the hottest month and 18.5°C during the coldest month, with an annual mean temperature of 26°C. Ground fire sometimes occurs during the dry season, especially in the dry dipterocarp forest area, but has not been so severe ever since the area was protected with an extensive network of firelines laid out around the dry dipterocarp forest area of the station. Vegetation types of the area are dry evergreen forest (29.5 km^2), dry dipterocarp forest (12.2 km^2), bamboo forest (14 km^2), grassland, and an agricultural area (37.9 km^2). The dry evergreen forest occupies the southwestern portion, while the dry dipterocarp forest is on the northeastern portion of the reserve area (Wacharakitti et al. 1980). The dry dipterocarp forest of this area occupies less than the dry evergreen forest, but it represents the general feature of the forest type common to the northeastern region. Although its stratification is usually recognized as 3 layers, 2 layers are frequently found. The upper layer is 21–35 m high, with the dominant species being *Shorea obtusa*, *Shorea siamensis*, *Dipterocarpus intricatus* and *D. tuberculatus*. The middle layer, if present, is commonly 11–20 m high, and important species include *Quercus kerrii*, *Gardenia sootepensis*, *G. obtusifolia* and *Randia tomentosa*. The lower layer is generally composed of tree saplings, dwarf bamboo-like grass (*Arundinaria pusilla*) and cycads (*Cycas siamensis*).

Four study plots,each 1 ha square, were set up in 1984 in a little disturbed and generally flat area at 290–365 m above sea level. Each plot was divided into 100, 10 m × 10 m quadrats, and all trees with diameter at breast height (DBH) equal to or more than 4.5 cm were tagged with aluminium number tags just below a painted ring marking DBH for repeat measurement. Tree censuses were generally carried out at interval of 1 year, with the last measurement taken in 1988. Stem positions and crown projection area were mapped, species composition was checked, and dead trees and those recruited to reach the fixed DBH limit of 4.5 cm were identified periodically. Total height of trees (H) and height to the lowest living branch (H_B) were also measured at the beginning of the study. Stand profiles were drawn for each randomly selected transect belt of 10 m × 100 m, some of which were combined to form 20 m × 100 m belts.

Structural characteristics of the four plots were investigated to obtain their basic stand information. The importance value index (IVI), which is the sum of the percentages of relative density, relative frequency and relative dominance (Curtis & McIntosh 1951), was calculated for each component species of the stands. The relative density and relative frequency were determined from all 10 m × 10 m quadrats, and the relative dominance was computed from the total basal area at breast height relative to the sum of basal area of all species, with each fraction expressed in a percentage with a sum of 300. The spatial pattern of all living trees (DBH ≥ 4.5 cm) and of the selected tree species having the first five ranks of IVI values were studied by using the Morisita I_δ *Index* (Morisita 1959); defined as:

$$I_\delta = q \frac{\sum_{i=1}^{q} n_i(n_i - 1)}{N(N-1)}$$

where q is the number of quadrats, n_i is the number of individuals in the ith quadrat, and N is the total number of individuals in q quadrats. The index has a value of 1 when the distribution is random, <1 when uniform, and >1 when clumped. The index was applied by changing the side length of each quadrat size from 10 m up to 100 m within 1 ha plots. In this case the number of quadrats varied from 1 to 100 contiguous quadrats.

Information on species diversity was computed using the Shanon-Wiener index H^* (MacArthur & MacArthur 1961), given by:

$$H^* = -\sum_{i=1}^{n} p_i \log_2 p_i$$

where n is the total number of species, and p_i is the proportion of species i to the total species. In this study, those species which could not be identified by botanical name though the vernacular name was known, was treated as one species.

The crown-depth diagram of Ogawa et al. (1965) was applied to verify the stratification of the study stands. Since the variation in tree height and crown cover were common in the sparsely dipterocarp forest, the whole stand was partitioned into five segments of 20 m × 100 m to observe the vertical arrangement of the stands together with the whole stand analysis. Other methods of stratifying the canopy structure were also attempted (Dhanmanonda unpubl.) and found to be useful, such as those of Hozumi (1975) and Yamakura (1987).

Growth increment of all living individuals (DBH $\geq$ 4.5 cm) and the five selected tree species having the highest values of IVI in each plot were studied in two ways, absolute and relative growth rates of DBH, defined as follows:

$$\Delta D = \frac{D_{t_1} - D_{t_0}}{t_1 - t_0}$$

where ΔD (absolute growth rate) is the mean annual increment of DBH, D_{t_1} is the DBH in 1988, D_{t_0} is the DBH in 1984 and the period between t_1 and t_0 is therefore 4 years; and

$$\mathrm{RGRD} = \frac{\ln D_{t_1} - \ln D_{t_0}}{t_1 - t_0}$$

where RGRD is the relative growth rate of DBH and ln is the natural logarithm.

The mortality rate of trees (λ) was computed as:

$$\lambda = \frac{\ln N_{t_0} - \ln N_{t_1}}{t_1 - t_0}$$

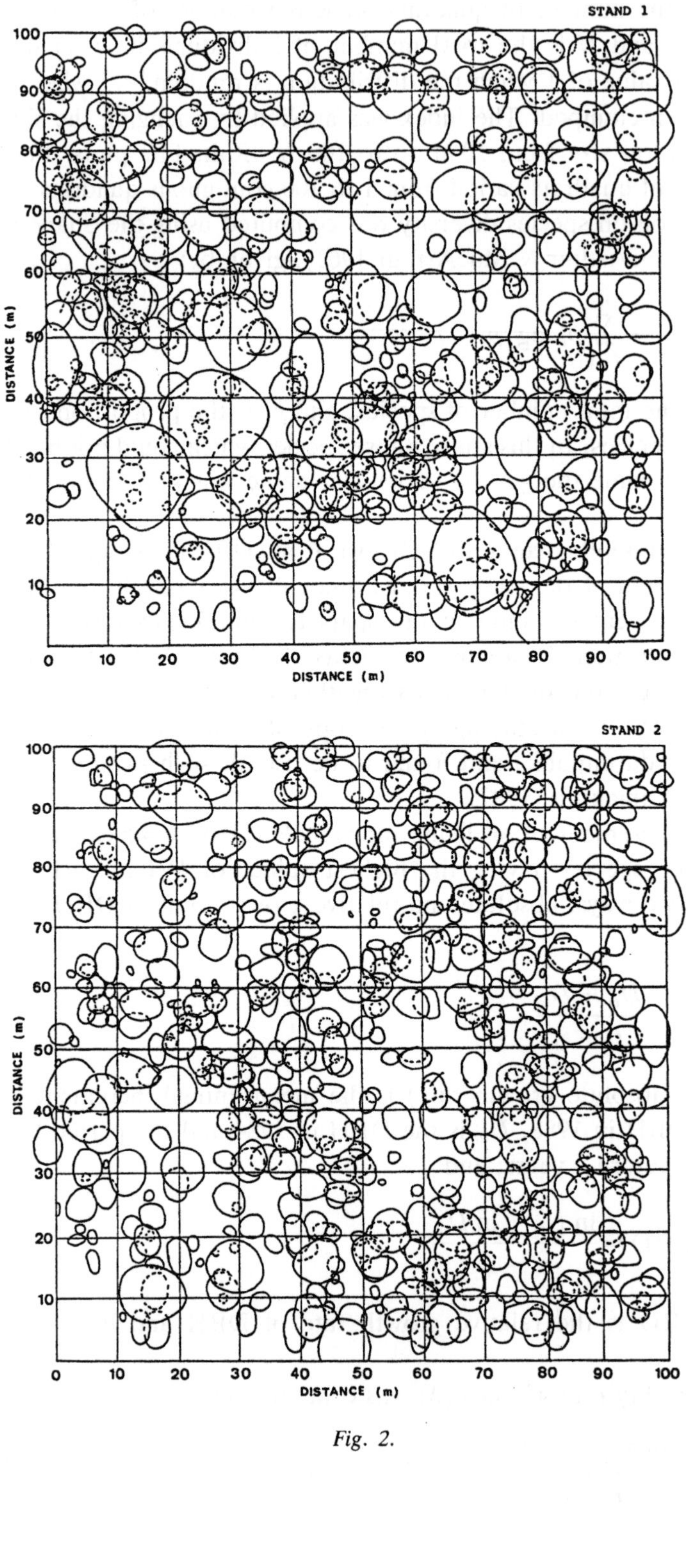

Fig. 2.

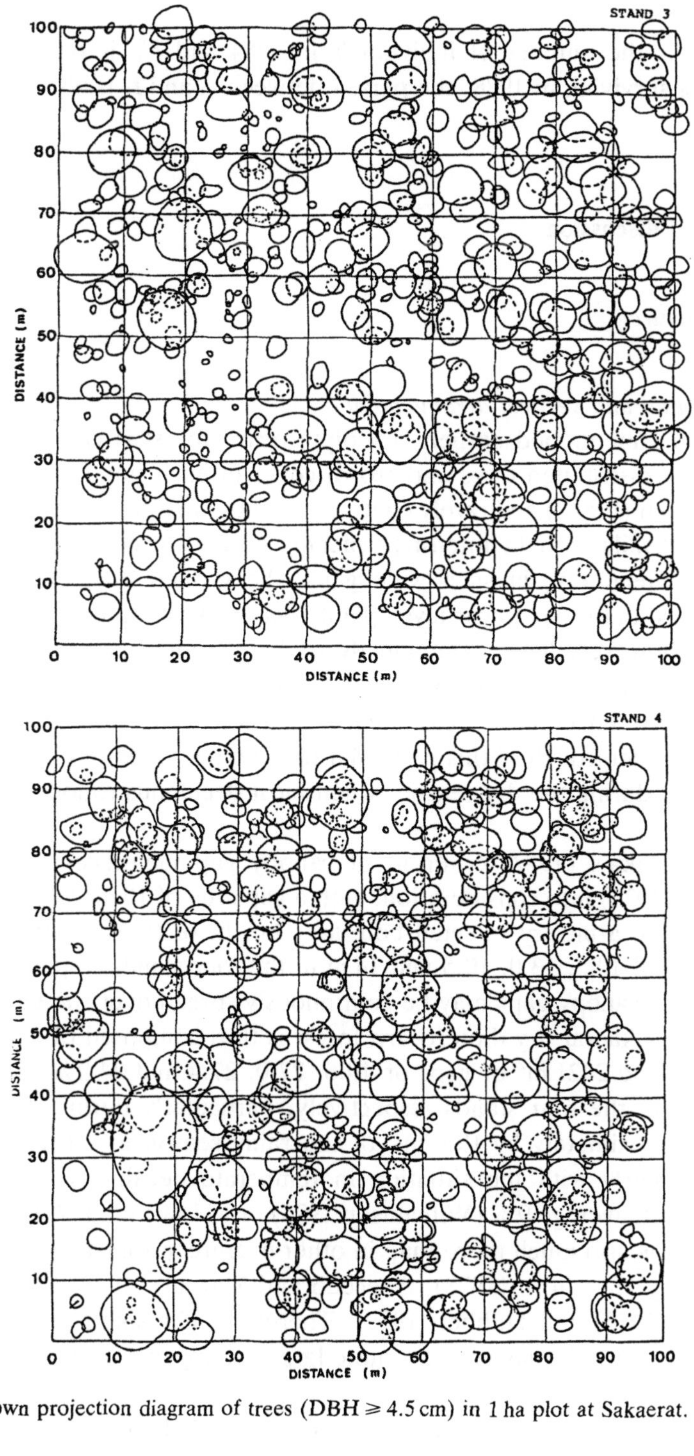

Fig. 2. Crown projection diagram of trees (DBH ≥ 4.5 cm) in 1 ha plot at Sakaerat.

where N_{t0} is the number of individuals in 1984 and N_{t1} is the number of individuals surviving in 1988; the period between t_0 and t_1 is also 4 years.

Results and discussion

Stand structure

1. *Horizontal and vertical arrangement*
As has been mentioned earlier, dry dipterocarp forest is characterized by the sparsely distributed trees and crown cover of each stand which is more or less broken (Fig. 2) and rarely continuous, thus creating numerous canopy openings. Available light intensity reaching ground level is most probably abundant. Generally, in such wide gaps, the ground area may be filled by seedlings and other lower-layer plant species. Dry dipterocarp forest at Sakaerat, however, has a common dwarf bamboo-like grass, *Arundinaria pusilla* (Graminae), together with many seedlings of the taller tree species and the scattered cycad, *Cycas siamensis*. These constitute a special and important undergrowth of this particular forest type. Openings may be more pronounced during the dry season, when most deciduous canopy trees shed their leaves.

Taking Naka's (1982) criterion of canopy gap identification following Whitmore (1975) into consideration, with a gap area of 10 m^2 as a minimum, as well as the arbitrary selected boundary height of trees occurring in the canopy gap of 10.5, 13.0, 10.5 and 14 m in the four stands respectively, one sees that the canopy gaps may be continuously spread over the entire stand areas. The Sakaerat dry dipterocarp forest is thus composed of about 1965, 2728, 2740 and 2421 m^2 of gaps or 19.6, 27.0, 27.4 and 24.2% of the total stand area in the four stands respectively, particularly during the full-leaf season; approximately by the end of the rainy season in the region. This contrasts clearly with the situation in temperate deciduous broad-leaved forests, which average 9% (Williamson 1975) or as low as 3.2–2.4% (Runkle 1982). This is also much more than in other evergreen forests, such as those described by Hartshorn (1978), Avecedo and Marquis (1978), Poor (1968), Whitmore (1975) and Hubbel (1979). Therefore, if a single overstory tree dies by any cause, it may create a quite large canopy gap in the dry dipterocarp forest.

Tree layers (DBH $\geqslant$ 4.5 cm in this study) tended to be averaged as a simple single monolayer, somewhat resembling an ordinary tree plantation. A portion of a 10 m $\times$ 100 m transect randomly selected from the 1 ha square plot in order to provide the stand profile (Fig. 3) shows a vertical arrangement which tended to be more or less 2–3 layers of the canopy tree. The broken crown cover and little overlapping in the crown, from the dominant canopy to the successive subordinate tree species, are clearly observed. Analysis of stratification using stand profile diagrams adopted from Richards (1952) was found to be insufficient to describe the more quantitative aspects of stratification, particularly in a narrow strip of forest (i.e. the present 10 m × 100 m transect). Based on Ogawa's method of crown-depth diagram analysis (Ogawa et al. 1965), applied to each segment of a long five-belt transect (20 m × 100 m) of each 1 ha plot, together with the whole stand (1 ha area),the relationships between total tree height (H) and height to the lowest living branch (H_B) were studied. An example of this analysis is shown in Fig. 4. This type of analysis for classifying the stratification of these stands was also attempted (Dhanmanonda unpubl.), which included use of the numerical methods of Hozumi (1975) and Yamakura (1987). All these methods proved to be useful but, at this moment, Ogawa's method was found to be convenient and sufficient, and needs none of the complex further calculations and graphical interpretations, particularly for the dry dipterocarp forest studied here, which has a rather simple stratification. Generally, the vertical arrangement of tree layers is not complex, but stratification varies between stands and within similar stands of different locations. As a whole, dry dipterocarp forest at Sakaerat shows mostly two layers. First and successive tree layers varied with different compositions of important trees, and their associations of this forest type, while on other sites three clearly segregated strata may be apparent, i.e.; the dry dipterocarp forest in Namphrom basin (Sahunalu et al. 1979).

Generally, the top tree layers are above 10.5, 13.0, 10.5 and 14.0 m high in the four stands respectively, while the second tree layers are lower and a third layer is found only in some parts of stands 2 and 4, below 10 m high. Such stratum separation is judged from three principle criteria: the minimum point of the vertical crown curve together with the abrupt downward change in height curve and the step-like distribution of tree grouping on the H–H_B diagram (Ogawa et al. 1965). The most important is the degree of crown overlapping, which has been taken into consideration. In this case, trees with dominant crown, not touched by any neighbors (open circle in Fig. 4) and partly overlapping (circle with dot), completely shaded or suppressed under the dominant and co-dominant tree layers (black circle), were thought to be arrayed in response (and well adapted) to available light conditions. The present study has shown that, in the four stands, approximately 45, 47, 50 and 54% of the canopy trees do not overlap in stands 4, 2, 1 and 3 respectively, suggesting 100% light interception by this tree layer. The numbers of tree species belonging to this layer are 16, 15, 17, and 18 species, with total

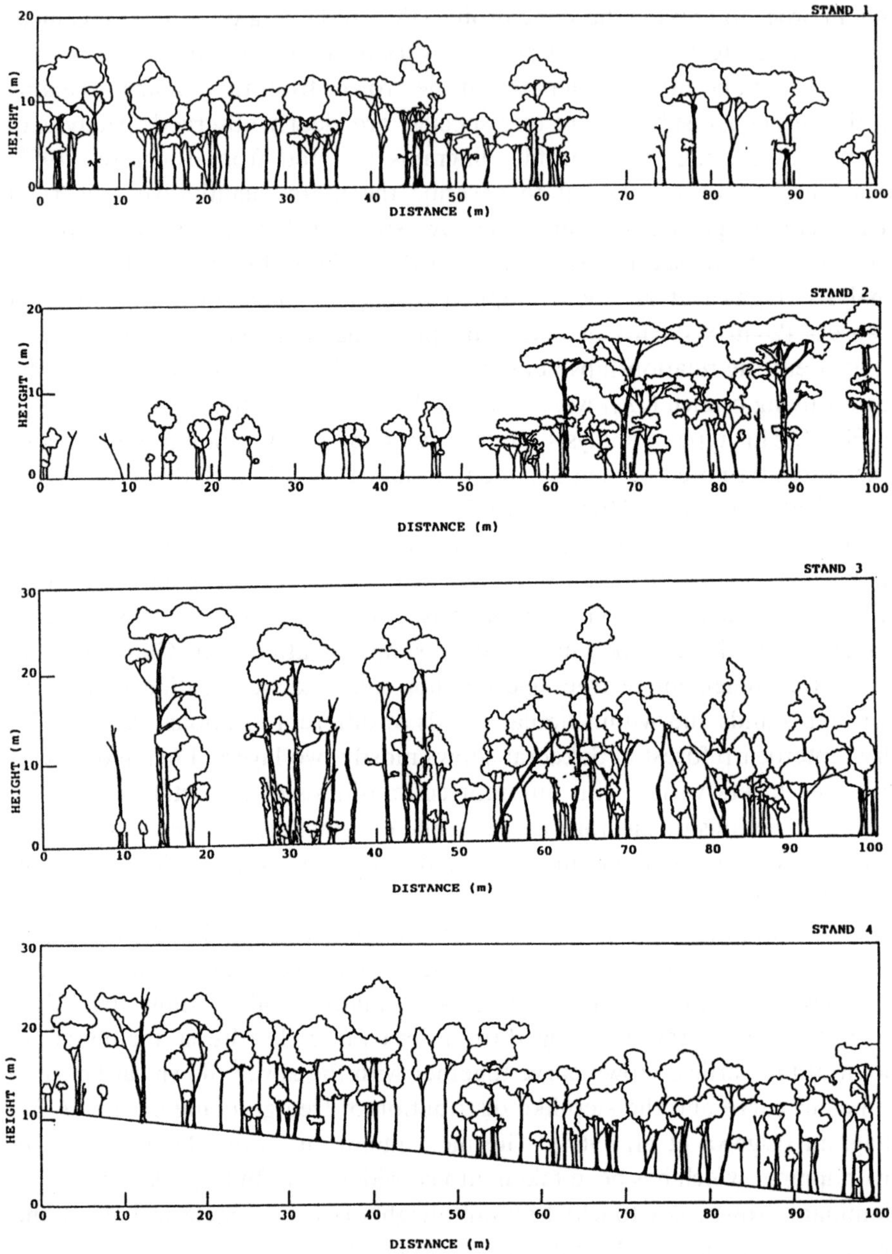

Fig. 3. Forest profile (trees with DBH $\geq$ 4.5 cm) along randomly selected belt transects of $10 \times 100\,m^2$ in each stand at Sakaerat.

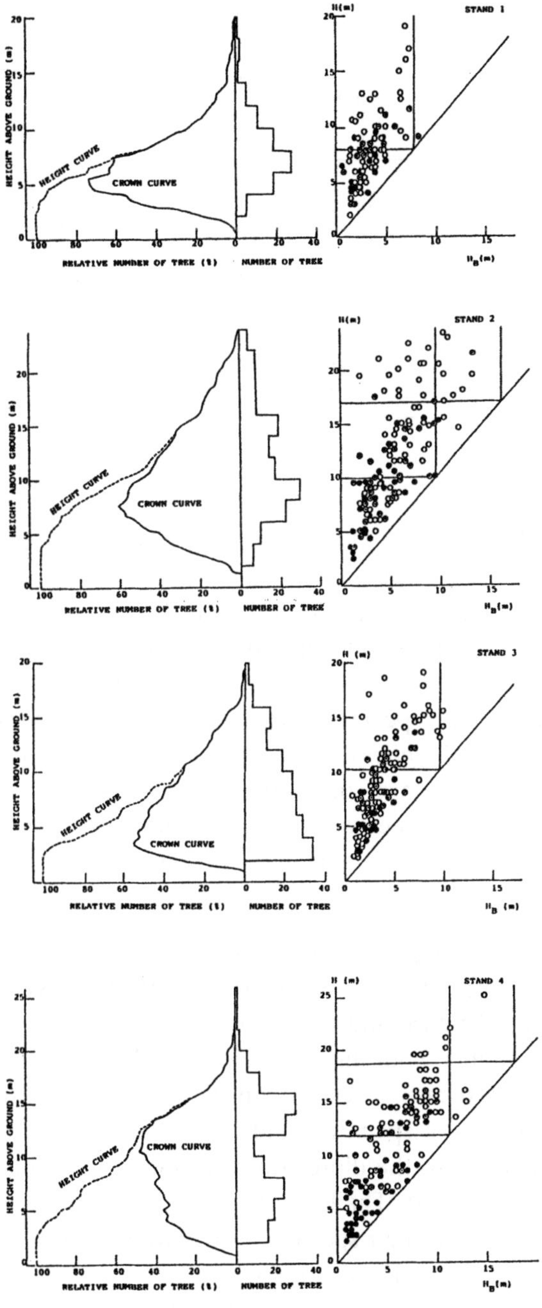

*Fig. 4.*An example of crown-depth diagram analysis and relationship between tree height (H) and height of the lowest living branch (H_B), for trees (DBH $\geqslant$ 4.5 cm) in an arbitrarily selected segment ($20 \times 100\,m^2$) of each (1 ha) stand. Open circle: non-overlapping crown, circle with dot: partly overlapping crown, solid circle: completely shaded crown. Relative number of trees represents the number of trees having height and crown height in each class in proportion to the total trees in the stand. For full analysis see Ogawa et al. (1965).

Table 1. Stand characteristics of the four dry dipterocarp forest stands studied at Sakaerat,Thailand

	Stand			
	1	2	3	4
Plot size (ha)	1	1	1	1
Altitude (msl)	284	360	362	365
Aspect	N	N	S	N
Slope		gentle		
Stand density (trees/ha)	554	667	733	701
No. of species	33	32	28	33
Species diversity (*H**)	3.97	2.86	2.42	3.52
Total basal area (m^2/ha)	14.41	19.28	14.21	18.09
Average height (m)	7.95	12.08	7.84	10.25

*H** = Shanon–Weiner index.

basal area at DBH of 8.66, 13.95, 7.91 and 9.77 m^2/ha respectively. These veteran tree species are *Dipterocarpus intricatus*, *Shorea obtusa*, *S. obtusa* and *S. siamensis* in the respective stands. The partially overlapping crowns are certainly exposed to full light but may occasionally be interfered with by neighboring trees (23, 24, 28 and 35% of trees in stands 1, 3, 4 and 2 respectively). These tree species are *Quercus kerrii*, *S. floribunda*, *Pterocarpus macrocarpus*, *Mangifera caloneura*, *Xylia xylocarpa*, *Sindora siamensis* and others. There are some less important species which maintain their status as suppressed trees with the remainder serving as ground cover; some represent the development stage of the canopy layer regeneration.

2. *Structural composition and its importance*

Among the four stands, the density of trees (DBH $\geqslant$ 4.5 cm) varied from 554 to 733 trees per hectare and the number of species from 28 to 33, with a few unknown climber and shrub species (Table 1). The occurrence of climbers in dry dipterocarp forest is considered to be unusual. Both tree density and abundance of species found in this forest type at Sakaerat are not directly comparable to the dipterocarp savanna forest studied by Ogawa et al. (1961) in northwestern Thailand. The features of these forest types are quite similar, however, in their sporadic distribution of trees and their fewer species compared to other closed forests.

Using the same measure of tree density (only trees bigger than 4.5 cm DBH), one finds the Sakaerat dry dipterocarp forest to be dense but more abundant in species than the Ping Kong dipterocarp savanna forest in northern Thailand and the Namphrom dry dipterocarp forest in northeastern Thailand (*cf.* Ogawa et al. 1965,Sahunalu et al. 1979). The relatively low species diversity in stand 3 (highest stand density) resulted from fewer species than in the other three stands; its lower basal-area cover is due to the abundance of small trees in the lower two size classes (Fig. 5), although the same proportion of moderately sized trees occurs among other stands (as well as the absence of large trees). The large basal-area cover in stand 2 and stand 4 is obviously due to the presence of a few large trees in the large size classes (Fig. 5).

The diameter distribution of trees in the four stands followed well the negative exponential decrease pattern from the lowest size class (4.5–9.5 cm) to the largest size class (around 54.5–64.5 cm), suggesting that the inverse J-shape distribution pattern covers much of the tree population. The larger size classes beyond that range are exceptionally variable, since large trees are sporadic and involve less common species. The knowledge of size-class distribution in relation to age distribution is not known for this forest type because it is impossible to determine the age of trees directly, due to the well known lack of annual rings in most trees. Therefore, the regeneration status could not be reliably suggested from this size-class distribution alone, as it is usually assummed for temperate forests.

Species compositions of the four study stands together with their importance value indices (IVI) are shown in Table 2, in which it is found that 15 species are common to the four stands, 13 species are only in a single stand, 6 species are only in two stands, and 13 species are only in three stands. It can be seen that the most important species and their associations vary between stands, and that the stands may be recognized as four different association types, based on the IVI values of the two most abundant species, namely *Shorea floribunda-Quercus kerrii*, *Shorea obtusa-Shorea siamensis*, *Shorea obtusa-Pterocarpus macrocarpus* and *Shorea siamensis-Shorea floribunda* for stands 1, 2, 3 and 4 respectively. Kutintara (1975) recognized 6 association types of dry dipterocarp forest in northwestern Thailand, while Ogawa et al. (1961) primarily divided this forest type into 3 broad association types and Sukwong (1974) added another type.

The classifications of Kutintara (1975) and Sukwong (1974) are remarkably different from the present study since they included the dry dipterocarp forest on a relatively high-altitude site, where local pines (*Pinus kesiya* and *P. merkusii*) may be found associated with the Dipterocarp species. The Sakaerat dry dipterocarp forest site is on a rather low, moderately dry hill with shallow soil, where pines or even other common species of dry dipterocarp forest (such as *Dipterocarpus tuberculatus* and *D. obtusifolius*) are not frequently found. The Sakaerat stands are dominated in particular by *Shorea* (Dipterocarpaceae), and the whole forest community may belong to association type 1 (*Shorea obtusa–Shorea siamensis*) in the classification of Ogawa et al. (1961). The association type may be different from stand to stand within one site in this area, however, and a species which predominates on

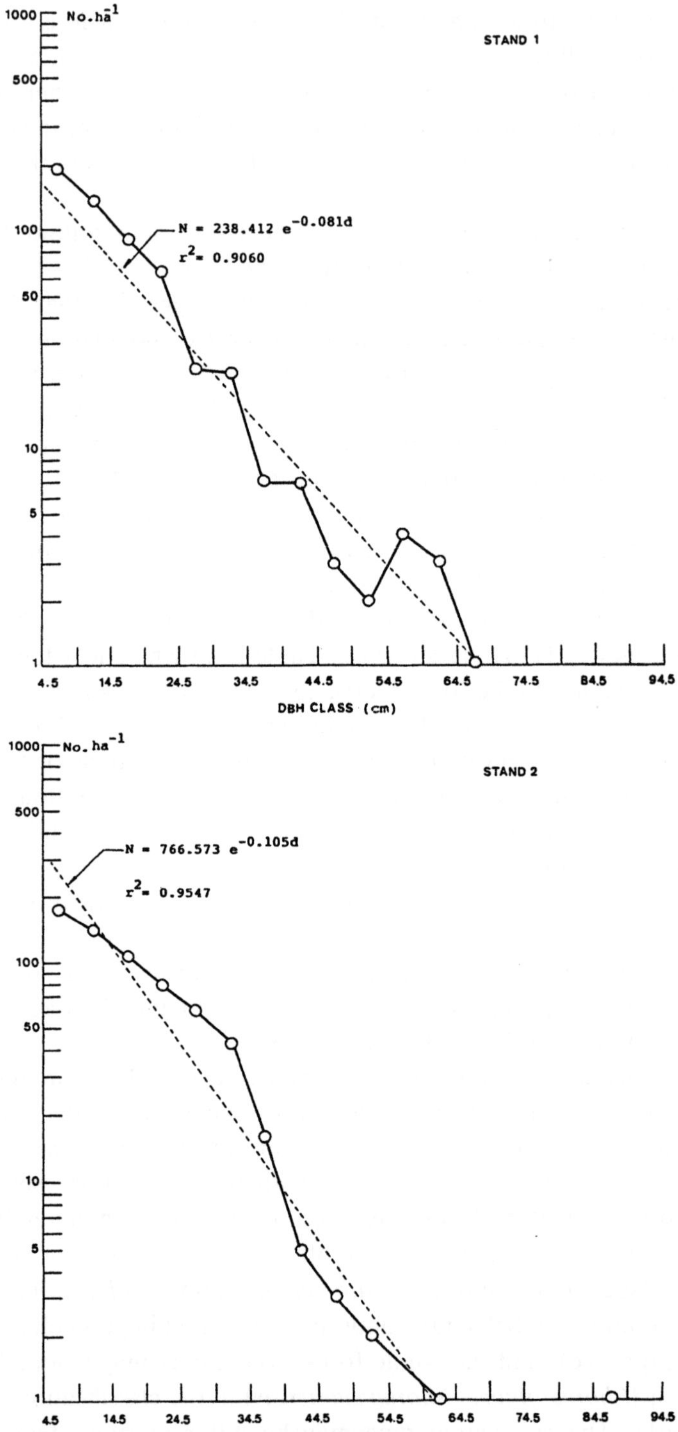
No.ha^{-1}
STAND 1
N = 238.412 e$^{-0.081d}$
r^2= 0.9060
1000
500
100
50
10
5
1
4.5
14.5
24.5
34.5
44.5
54.5
64.5
74.5
84.5
94.5
DBH CLASS (cm)
No. ha^{-1}
STAND 2
N = 766.573 e$^{-0.105d}$
r^2= 0.9547
DBH CLASS (cm)

Fig. 5.

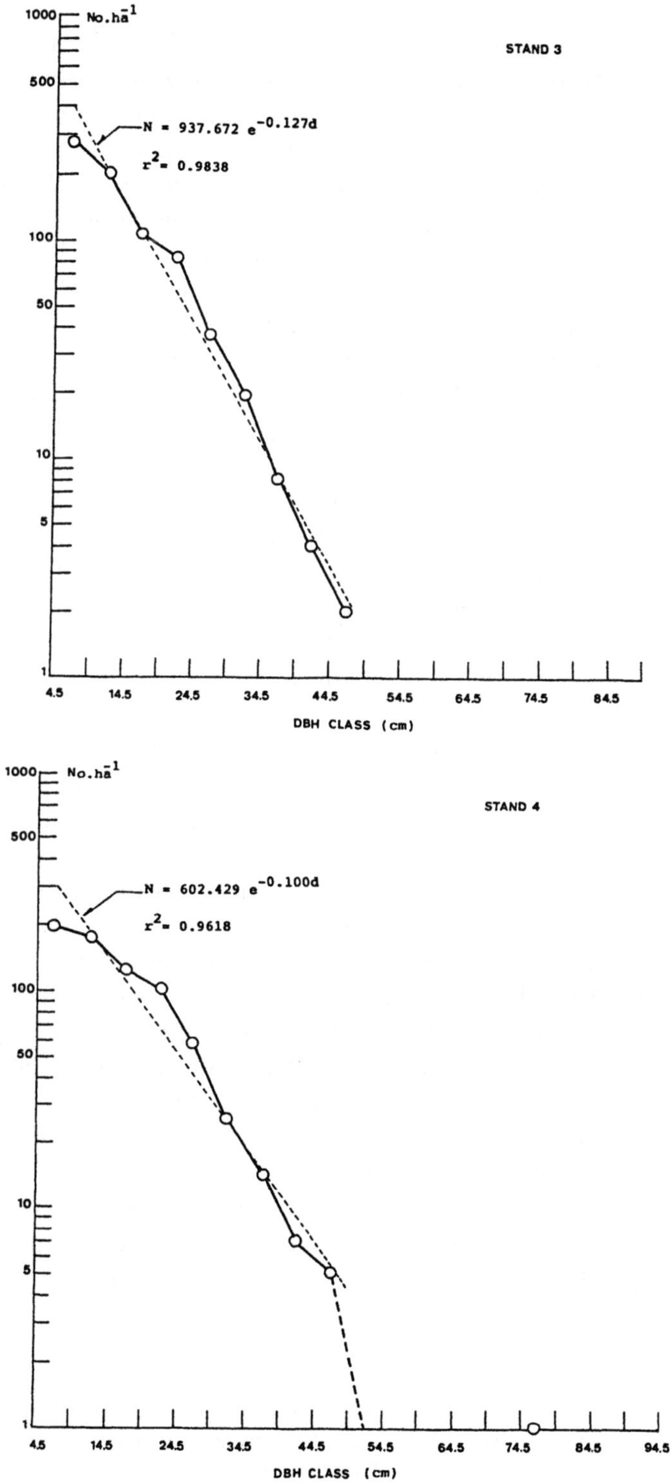
No.ha^{-1}
STAND 3
N = 937.672 e$^{-0.127d}$
r^2= 0.9838
1000
500
100
50
10
5
1
4.5
14.5
24.5
34.5
44.5
54.5
64.5
74.5
84.5
DBH CLASS (cm)
No.ha^{-1}
STAND 4
N = 602.429 e$^{-0.100d}$
r^2= 0.9618
1000
500
100
50
10
5
1
4.5
14.5
24.5
34.5
44.5
54.5
64.5
74.5
84.5
94.5
DBH CLASS (cm)

Fig. 5.

one site may become a sub-dominant species on another, such as *S. siamensis* and *S. floribunda*. It is remarkable that *Pterocarpus macrocarpus*, an important commercial tree belonging to the Leguminosae family, is not only a sub-dominant species of *S. obtusa* (stand 3) but also one of the five important species in every stand (Table 2). Although *S. siamensis*, *S. floribunda*, *Q. kerrii*, *D. intricatus* and *Sindora siamensis* are common species in every stand, they attain high performance in one or two stands only, while *Mangifera caloneura* and *Morinda coreia* appeared to be among the five leading species only in stand 2 and stand 3 respectively (Table 2).

Factors determining the distribution of the important species of dry dipterocarp forest were discussed by Kutintara (1975) and Bunyavejchewin (1983), but only two of these species (*S. obtusa* and *S. siamensis*) will be dealt with here. Kutintara (1975) found that, of several topographic and edaphic factors, the percentage of rock and slope together with the high content of *P* and *K* in soils had a positive correlation with the distribution of *S. siamensis*, suggesting that this tree species will occur only on a sandy soil of rocky type and on mesic to extremely xeric sites with a long dry period. Bunyavejchewin (1983) found that the distribution of *S. siamensis* had positive correlations with Ca and Mg content but was negatively correlated with elevation, *K* and soil particle density.

With regard to the distribution of *S. obtusa*, Kutintara (1975) found a positive correlation with percentage of silt in the upper root zone, and with rock content, slope, and *P* and *K* of the upper soil layer. He concluded that *S. obtusa* can succeed over a wide range of environmental conditions ranging from extremely xeric with shallow, rocky, sandy soil to deep, well-drained sandy clay in mesic to slightly hydric habitats. However, Bunyavejchewin (1983) suggested that the distribution of *S. obtusa* has a positive correlation with particle density of soils but a negative correlation with elevation, organic matter, *K* content and slope. Obviously, species composition of dry dipterocarp forest is controlled by several complexes of environmmental factors, and the factors controlling the variation of species distribution in the Sakaerat dry dipterocarp forest, on the same site and under the same climatic regime, still remain to be investigated further.

The importance value index of each species is depicted following its magnitude as shown in Fig. 6. Among the four stands studied, stands 1, 3 and 4 are characterized by a lognormal distribution, while stand 2 strongly suggests

Fig. 5. Distribution of trees (DBH $\geqslant$ 4.5 cm) of Sakaerat dry dipterocarp forest. Broken thin line shows the negative exponential decrease of the number of trees as determined by the model: $N = Ae^{-\beta d}$, where N = number of trees per hectare, d = mid-point of DBH class (cm), A *and* β = constants of the relationship between N and d, r^2 = coefficient of determination.

Table 2. Species importance value indices (IVI, %). Absence of species in a stand is indicated by a dash (–)

Species	Stand			
	1	2	3	4
Albizzia odoratissima	1.97	4.65	5.18	3.72
Antidesma laurifolium	1.40	–	0.47	1.62
Antiaria toxicaria	2.01	–	–	–
Antocephalus chinensis	–	–	6.72	–
Aporusa villosa	4.70	0.44	–	2.88
Artocarpus lakoocha	–	–	0.47	–
Bauhinia sp.	0.69	4.24	–	1.43
Bauhinia pottsii	–	–	3.54	–
Bombax insigne	0.01	–	–	–
Canarium kerrii	–	–	–	2.36
Careya sphaerica	2.69	0.47	4.32	1.61
Cratoxylon annamense	–	–	–	2.36
Dalbergia assamica	–	–	2.57	–
Dalbergia cultrata	9.22	1.96	–	9.46
Dalbergia oliveri	–	–	2.88	2.16
Dillenia obovata	13.41	–	1.00	–
Diospyros ehretioides	3.66	1.02	2.84	1.59
Diospyros kerrii	–	1.00	–	–
Diospyros mollis	1.32	–	2.72	0.87
Diptreocarpus intricatus	36.27	6.40	20.93	9.49
Dipterocarpus tuberculatus	1.96	–	–	–
Eugenia cumini	0.06	0.49	4.26	0.81
Gardenia sootepensis	0.46	4.47	2.49	6.78
Irvingia malayana	1.99	2.52	1.14	1.61
Kydia calycina	0.67	0.64	–	–
Lannea coromandelica	2.93	1.10	–	–
Lithocarpus polystachyus	0.93	–	4.55	–
Mangifera caloneura	5.71	14.48	1.97	–
Mitragyna brunonis	14.81	4.78	–	8.05
Morinda coreia	4.57	7.24	9.44	10.12
Nauclea brunnea	0.45	–	–	1.57
Nauclea orientalis	–	0.90	–	0.41
Parinari anamense	–	0.64	2.23	2.14
Phyllanthus emblica	0.48	0.45	0.57	1.21
Prima pyramidata	–	–	–	0.45
Pterocarpus macrocarpus	36.72	28.14	41.20	20.97
Quercus kerrii	42.58	7.88	2.28	23.56
Sindora maritima	–	–	–	14.35
Sindora siamensis	7.98	4.70	11.88	–
Shorea floribunda	46.16	4.03	7.54	72.78
Shorea obtusa	22.84	123.84	147.82	–
Shorea siamensis	5.73	49.71	–	77.66
Stereospermum neuranthum	–	–	–	0.60
Terminalia chebula	0.93	0.46	–	0.40
Vitex canescense	–	0.45	–	–
Vitex peduncularis	7.91	1.35	1.02	3.54
Xylia xylocarpa	15.85	14.74	8.93	7.26
Unidentified (climbers)	2.45	5.94	3.01	2.23

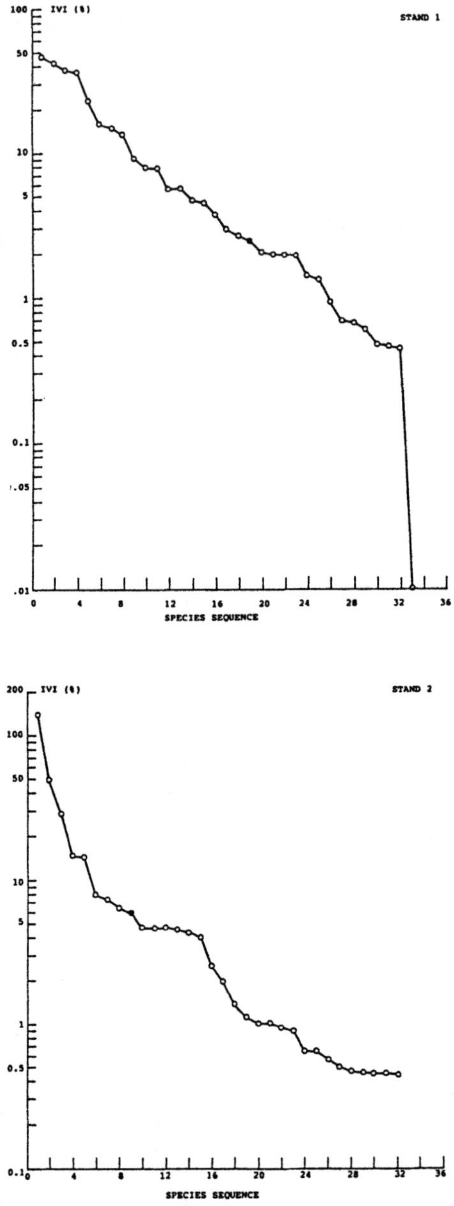

Fig. 6.

a geometric series following Whittaker's (1975) hypothesis. The performance of tree species is clearly dependent on a wide range of habitat factors.

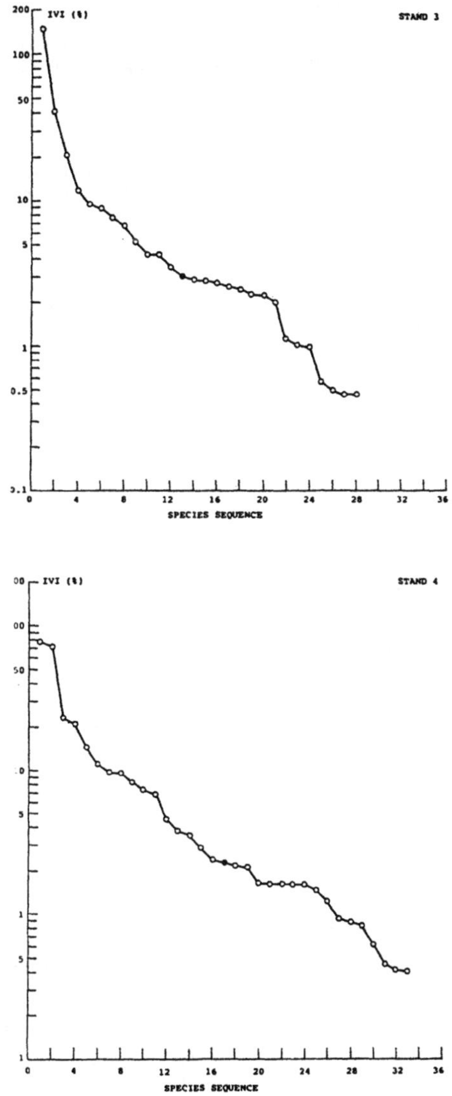

Fig. 6. Species rank arranged according to their importance value indices in Sakaerat dry dipterocarp forest. Solid circle is the IVI of climber.

3. *Spatial distribution pattern of canopy trees*

The spatial arrangement of all individuals of each stand, based on the I_δ index of Morisita (1959) is very similar in that the value of I_δ is near 1, suggesting that trees with DBH7 $\geqslant$ 4.5 cm are randomly distributed. These I_δ values of the five species with the highest IVI value in each stand are shown in Fig. 7. These species show a similar, contagious pattern or occur as clumps regardless of differences in stand density, species composition and species

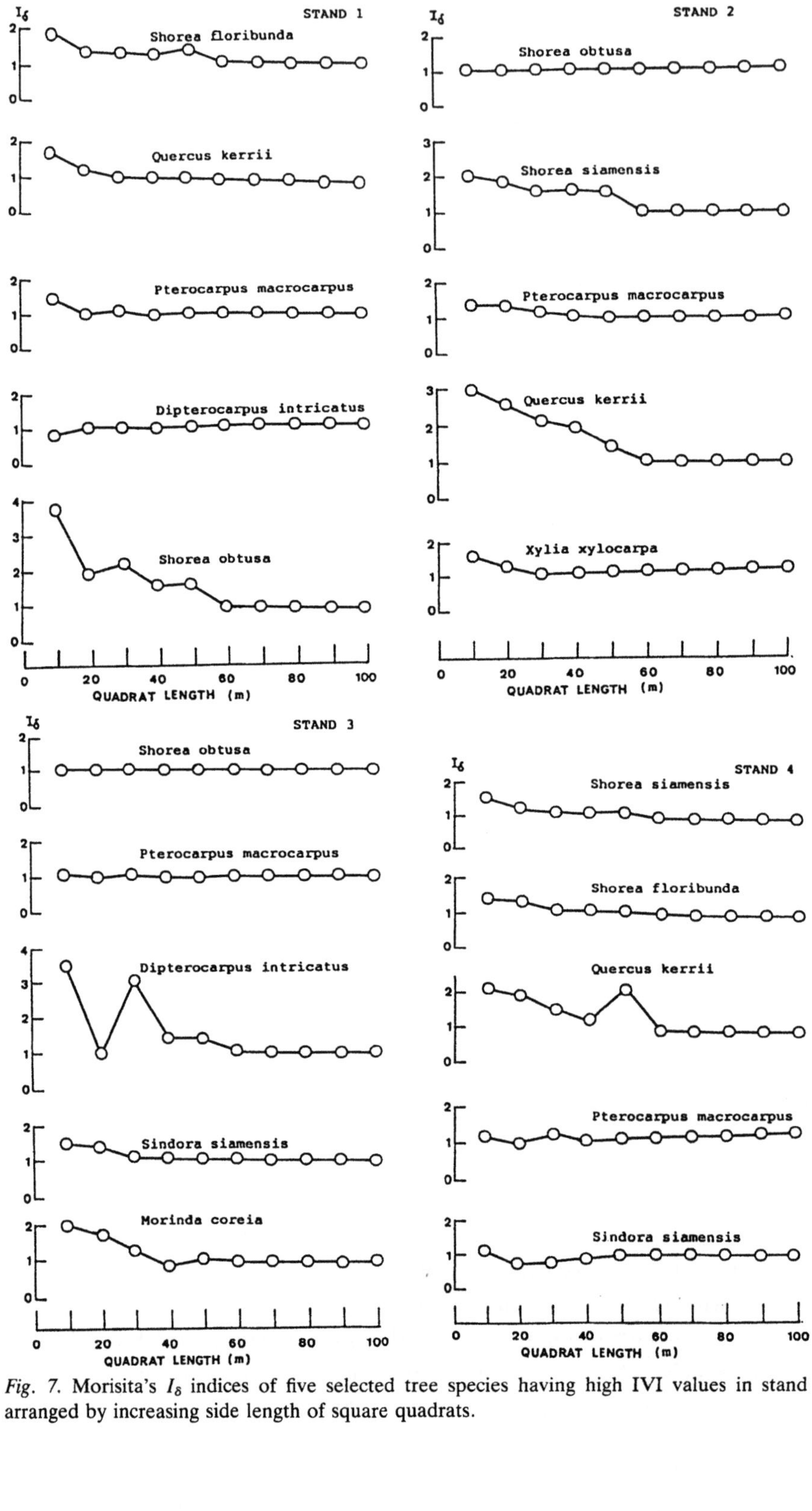

Fig. 7. Morisita's I_δ indices of five selected tree species having high IVI values in stand 1, arranged by increasing side length of square quadrats.

Table 3. Changes in tree individuals (DBH $\geq$ 4.5 cm) over the period 1984–88 in the four stands at Sakaerat. Numbers represent individual trees per hectare. (Figures in parentheses are percentages of the initial number of trees)

	Stand			
	1	2	3	4
No. of individuals, 1984	554	667	733	701
No.of trees dying (1984–88)	28	30	62	41
	(5.05)	(4.50)	(8.46)	(5.85)
No.of recruits reaching DBH $\geq$ 4.5 cm	29	29	176	31
	(5.23)	(4.35)	(24.01)	(4.42)
No.of individuals, 1988	555	666	847	691
Net change (4 yrs)	+1	−1	+144	−10
	(+0.88)	(−1.05)	(+15.55)	(−1.43)

diversity of the stands. This pattern is similar to that found for various forest types in Thailand (Ogawa et al. 1961) and some temperate forest types, such as those studied by Williamson (1975). The clumped pattern of tree species may suggest some aspect of species organization or natural regeneration and survival of these species. The spatial distribution of these tree species tended to be in group or patch rather than a random or regular dispersion. Little is known however, about the regeneration process or behavior of the species of dry dipterocarp forest, although it is believed that this forest type may regenerate by seeds and by stem coppice. This aspect is poorly documented and needs to be investigated further.

Stand dynamics

1. *Tree population changes*

During the four-year period 1984–1988, the number of dead trees and of trees recruited to DBH $\geq$ 4.5 cm were examined. These records enabled computation of net changes in numbers of tree populations of each stand, as shown in Table 3. Stands 1, 2 and 4 show very small gains and losses resulting from an almost identical number of dead and recruited trees. Stand 3, however, shows remarkable gain and loss, dead trees being 8.5% of the initial number (only 4.5–5.9% in the other stands). The net gain (15.6%) of the tree population in stand 3 is attributable to the large number of tree recruits, 24% of the initial tree number (only 4–5% in the other stands). It is noted that recruits outnumbered dead trees by almost a factor of three in stand 3, the largest gain in number of trees in all stands.

The mortality rates (λ) in the four stands (Table 4) were in the range 0.012–0.022 yr^{-1} and were highest in stand 3. These mortality rates were not significantly different in stands 1, 2 and 4, but the larger loss (−1.43%) in

Table 4. Mortality rates of trees with DBH ≥ 4.5 cm in the four stands

	Stand			
	1	2	3	4
No. of individuals, 1984	554	667	733	701
No.of dead trees (1984–88)	28	30	62	41
Mortality rate (no. per year)	0.013	0.012	0.022	0.015

stand 4 (Table 3) indicates the poorest regeneration and most instability in stand maintenance, as compared to other stands.

The number of trees dying during the four-year period is shown in Fig. 8 and Table 5. It is evident that the smaller size classes and the more abundant species of each stand are more susceptible to death than others, with the exception of *S. floribunda* (the dominant species of stand 1, of which only a single tree died during the period). Tree death may be due to several factors. In the present study, tree competition is considered to be a major cause. This is particularly true for stands 3 and 4, where tree density was higher than in the other two stands.

Fire is a principal factor controlling the dry dipterocarp forest and is believed to be the most important factor for maintaining the forest structure and species composition. Sabhasri et al. (1968) recognized the dry dipterocarp forest as a pyric climax forest because, if it is well protected from fire, tree species belonging to more mesic forest types will invade the dry dipterocarp forest. Although the Sakaerat reserve forest has been well protected from fire for several years since the establishment of the research station, unexpected ground fires sometimes occur during the dry season and may affect tree mortality, especially of small trees. Other factors such as drought, strong wind, pests, diseases and grazing are considered to be minor causes or may be disregarded, since most tree component of this forst type are relatively tolerant of these factors.

The numbers of trees recruited to the size class DBH ≥ 4.5 cm during the four-year period are listed in Table 6. The principal component species of each stand were observed, with 12 species for stands 1, 2 and 4, and 22 species for stand 3. The numbers of individuals of each species recruited varied between stands, but the numbers of individuals of the leading dominant species are still large compared to other species in stands 2, 3 and 4, suggesting a rather stable maintenance of species composition of the stands. In stand 1 (Table 6), though, the recruitment of *P. macrocarpus* (third highest IVI in the stand, see Table 2) seems to be superior to *Q. kerrii*, the second dominant species, which showed a relatively larger number of dead individuals (Table 5). The maintenance of species composition of the stand

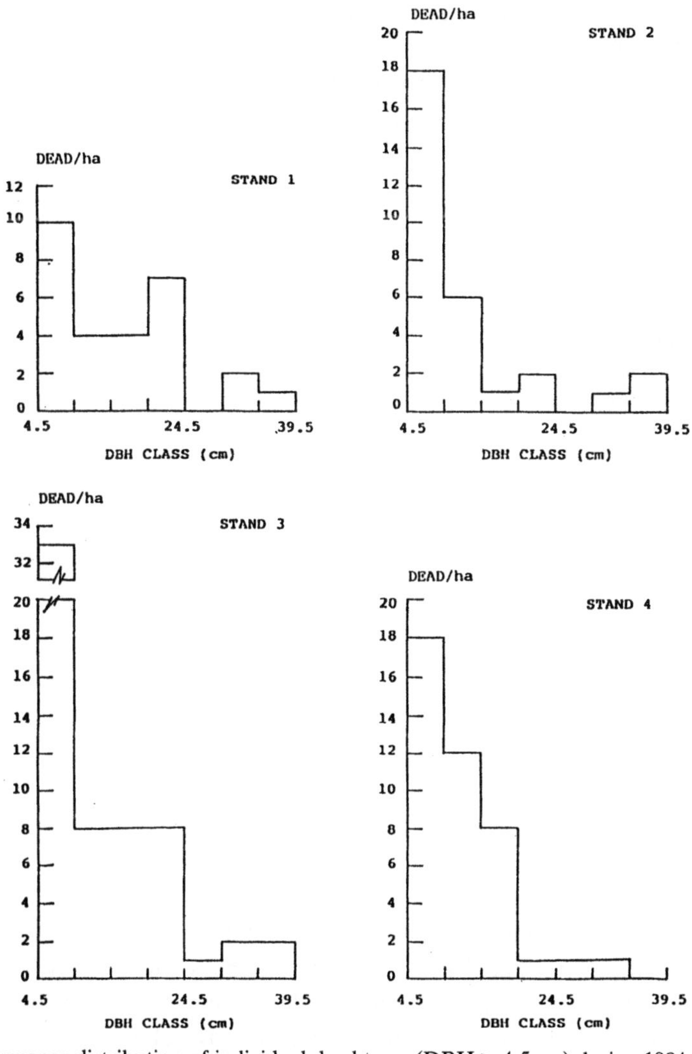

Fig. 8. Frequency distribution of individual dead trees (DBH ⩾ 4.5 cm) during 1984–88, by size class.

may be slightly changed if this replacement trend continues over the long term. This tendency may be more marked because *P. macrocarpus* occurs widely and is an important species component in all four stands.

2. *Tree coverage changes*

Change in basal area at breast height of the four stands was computed from the standing trees, dead trees and new recruits, as shown in Table 7. In contrast to the net changes in number of trees, the basal area increased in

Table 5. Numbers of trees dying in each stand during 1984–88. The absence of species in a stand from the beginning is indicated by a dash (–)

Species	No. of individuals			
	Stand			
	1	2	3	4
Albizzia odoratissima	–	–	1	3
Bauhinia pottsii	–	–	1	–
Bauhinia sp.	–	1	–	–
Dalbergia cultrata	2	1	–	1
Dalbergia oliveri	–	–	–	1
Dillenia obovata	2	–	1	–
Diospyros ehretioides	2	–	1	–
Dipterocarpus intricatus	–	–	1	–
Eugenia cumini	–	–	–	2
Lithocarpus polystachyus	–	–	–	3
Mangifera caloneura	–	–	2	–
Mitragyna brunonis	1	–	–	1
Morinda coreia	–	–	–	1
Nauclea brunnea	1	–	–	1
Pterocarpus macrocarpus	4	1	3	3
Quercus kerrii	10	–	1	5
Shorea floribunda	1	–	3	7
Shorea obtusa	1	15	46	
Shorea siamensis	–	11	–	6
Vitex peduncularis	1	–	–	1
Xylia xylocarpa	2	1	1	2
Unidentified (climbers)	–	–	1	1
Total	28	30	62	41

all four stands, with some variation between stands. The large gain in stand 4, 7.6% of the initial basal area, is larger than in the other three stands, although the number of new recruits was low. This is because the surviving trees had more rapid growth than in other stands. The percentage gain in basal area in stand 3 was 5% similar to stands 1 and 2, but the net gain was only 0.71 m^2/ha in spite of the large gain in the number of trees. This was due to higher mortality (7.2%) and slower growth of the surviving trees, as compared to the other three stands.

3. *Growth of stands and selected important tree species*

The mean annual increment (Table 8) of diameter at breast height (DBH) of the surviving trees and the relative growth rate were not significantly different in stands 1 and 2. The largest increment and relative growth rate were found in stand 4 and the lowest in stand 3. Growth of trees in the dry dipterocarp forest studied is not comparable to those of other forest types of the region or to those of humid climates. This forest type is considered

Table 6. Numbers of trees recruited (to DBH ⩾ 4.5 cm) during 1984–88 in the four stands of dry dipterocarp forest, Sakaerat, Thailand

Species	Stand			
	1	2	3	4
Albizzia odoratissima	1	–	4	2
Aporusa villosa	1	–	3	–
Artocarpus lakoocha	–	–	2	1
Bauhinia sp.	–	–	4	–
Bombax insigne	–	–	1	–
Canarium kerrii	–	–	3	–
Careya sphaerica	–	–	2	–
Cratoxylon anamense	–	–	11	1
Croton oblongifolius	–	–	1	1
Diospyros ehretioides	–	–	1	1
Gardenia sootepensis	–	1	–	3
Kydia calycina	–	1	–	–
Lannea coromendelica	–	3	–	2
Lithocarpus polystachyus	–	3	–	2
Mangifera caloneura	1	–	–	–
Mitragyna brunonis	1	2	1	3
Morinda coreia	3	1	1	–
Nauclea brunnea	2	1	1	–
Nauclea orientalis	–	1	–	–
Plachonella cambodiana	–	–	1	–
Pterocarpus macrocarpus	11	2	8	3
Quercus kerrii	1	1	–	5
Shorea obtusa	3	10	85	–
Shorea floribunda	–	–	4	3
Shorea siamensis	–	1	–	5
Sindora siamensis	–	–	5	1
Terminalia chebula	1	–	–	1
Vitex peduncularis	–	1	–	–
Xylia kerrii	–	–	11	1
Xylia xylocarpa	3	5	11	2
Unidentified (climbers)	1	1	1	1
Total	29	29	176	31

to be the most xeric type, where favorable conditions for growth are rarely found and the growth habit of the component species may exhibit different patterns. It is generally believed that the growth of this forest type is the slowest. This forest type normally occurs on dry sites with shallow soils, as often associated with rock outcrops, with 4–5 months dry season and where soil water deficit is expected to be frequent. Moreover, the deciduous character of the trees, with a relatively long leafless period, is another factor responsible for the slow growth of trees in dry dipterocarp forest.

Variations in growth of different component species of each stand are also pronounced. Five species with high IVI were selected from each stand, and

Table 7. Changes in basal area of trees with DBH ⩾ 4.5 cm. (Figures in parentheses are percentages of the initial basal area)

	Stand			
	1	2	3	4
Basal area in 1984 (m^2/ha)	14.41	19.28	14.21	18.09
Basal area of dead trees (m^2/ha)	0.70	0.51	1.02	0.56
(4.86)	(2.65)	(7.18)	(3.10)	
Basal area of recruits reaching DBH ⩾ 4.5 cm (m^2//ha)	0.70	0.08	0.43	0.08
	(0.50)	(0.41)	(3.03)	(0.44)
Basal area in 1988 (m^2/ha)	14.53	19.88	13.90	18.90
Net change (4 yrs)	+0.83	+1.11	+0.71	+ 1.37
	(+5.76)	(+5.76)	(+5.00)	(+ 7.57)

Table 8. Absolute growth rate (ΔD) and relative growth rate (RGRD) of diameter at breast height in the four stands at Sakaerat, Thailand

	Stand			
	1	2	3	4
No. of individuals, 1988	526	637	671	660
ΔD (mm/yr)	1.42	1.43	0.82	1.64
RGRD × 10^{-3} (per year)	11.77	11.60	8.66	17.60

For the calculation of ΔD and RGRD see text.

their mean annual increment and relative growth rate of DBH were compared, as shown in Table 9. There were large variations between species of the same stand and between the association types of the four stands. Both growth measures showed a similar trend. It appears that the leading dominant species of a stand does not show better growth than other associated species. Strikingly rapid growth of *P. macrocarpus* in every stand was clearly observed. Also, the growth of *S. siamensis*, the leading dominant species of stand 4 (considered to be the most xeric stand), was relatively rapid as compared to other species common to this forest type. The rapid growth of the whole of stand 4 appears to be the result of the better growth of these five major component species: in other stands rapid overall growth was due to the rapid growth of *P. macrocarpus*. The slower growth of stand 3 was partly attributed to the relatively slow growth of the major component species, which may have been affected by severe tree competition (stand with the highest tree density) but perhaps also by the slightly slow growth rate of *S. obtusa*, which appeared to be the most abundant species in the stand and showed similar growth rates in other stands.

Table 9. Mean relative (RGRD) and absolute (ΔD) growth rates for the five most important species of each stand of dry dipterocarp forest at Sakaerat (Thailand)

Species	IVI ranks				Mean RGRD$\times 10^{-3}$ (yr^{-1})				Mean ΔD ($mm\ yr^{-1}$)			
Stand:	1	2	3	4	1	2	3	4	1	2	3	4
Shorea obtusa	5	1	1	–	11.02	5.45	6.77	–	1.59	0.42	0.49	–
Shorea siamensis	–	2	–	1	1	14.78	–	16.20	–	2.02	–	2.39
Shorea floribunda	1	–	–	2	7.53	–	–	3.30	0.92	–	–	0.54
Quercus kerrii	2	–	–	3	7.79	–	–	10.80	1.19	–	–	1.23
Pterocarpus macrocarpus	3	3	2	4	28.63	23.91	15.74	29.50	3.31	2.55	2.12	2.97
Xylia xylocarpa	–	4	–	–	–	19.78	–	–	–	1.66	–	–
Dipterocarpus intricatus	4	–	3	–	8.14	–	0.50	–	1.40	–	0.03	–
Mangifera caloneura	–	5	–	–	–	8.56	–	–	–	2.38	–	–
Sindora siamensis	–	–	4	5	–	–	7.79	11.30	–	–	0.63	1.73
Morinda coreia	–	–	5	–	–	–	3.38	–	–	–	0.13	–
Total stand					11.77	11.60	8.66	17.60	1.42	1.43	0.82	1.64

Conclusion

The structure of the Sakaerat dry dipterocarp forest is generally characterized by two tree layers in each of the four stands studied, with tree height ranging from 10.5 to 14.0 m for the top layer. Some randomly selected areas within these stands, however, may be observed to have three layers. More than 45% of the trees are isolated, with crowns unshaded by neighboring trees and coverage ranging from about 8 to 10 m^2 per hectare (total basal area at breast height). Crowns partly shaded by one or more adjacent tree crowns constitute 23 to 35% of the total, the rest are totally shaded by higher crown trees.

The species composition of this forest type, particularly in the Sakaerat area, is dominated by *Shorea floribunda*, *Shorea obtusa* and *Shorea siamensis*, though the associated species are different from stand to stand. Judging from their importance value index (IVI), the four stands studied may be recognized as four association types: *Shorea floribunda–Quercus kerrii*, *Shorea obtusa–-Shorea siamensis*, *Shorea obtusa–Pterocarpus macrocarpus* and *Shorea siamensis–Shorea floribunda*. Each association type has a characteristic, inversed J-shaped tree size distribution, though some exceptions may be noted for the most isolated, uncommon large trees. Instead of direct relationships of species abundance and habitat factors, species importance-species rank curves are depicted, which clearly suggest a lognormal distribution for three stands, with one stand intermediate between the lognormal and geometric distributions. The species importance-species rank curves suggest that most trees and stands have a wide range of habitat requirements in common.

The spatial arrangement of the tree layer in the stands suggests that trees are distributed randomly but that the relatively more important species tend to occur in groups (though sparsely distributed).

Changes in tree populations during the four-year period are found to be different in the four stands, resulting from tree death and replacement by new recruits. Two stands gained 0.2 to 16.0% and two stands lost 1.1 to 1.4% of the initial number of individuals. All stands show similar patterns of tree mortality rates but differ in species death and recruitment. The most remarkable tree death and recruitment involve *Pterocarpus macrocarpus* and *Shorea obtusa*, which are relatively common species in all four stands. In spite of the large gain and loss of individuals, all stands gained in basal area coverage, about 5 to 7% of the initial basal area. The slight gain in basal area coverage was due to the slow growth rates of the important species. The remarkable increase in basal area of the stands is traceable to the rapid growth rate of *Pterocarpus macrocarpus*, one of the most important species in general, and to the growth rates of other principal species, *Shorea obtusa*, *S. floribunda* and *S. siamensis*. The growth rates differ between species and between stands, however, depending on successful establishment and survival of the component species of the stands.

Acknowledgements

We express our thanks to the National Research Council of Thailand (NRCT) for financial support for this study; to the Center for Southeast Asian Studies, Kyoto University (CSEAS) for providing facilities for preparation of this manuscript; and to Dr. I. Yamada of the CSEAS for his hospitality to one of us (PS) during his stay in Kyoto. Field assistance by our graduate students during 1984–1988 is also acknowledged.

References

Avecedo, M. & Marquis, R. 1978. A study of the light gaps of tropical rain forest at Llarona, Peninsula de Osa. Org. Trop. St. 78-3:290–303.

Bunyavejchewin, S. 1983. Canopy structure of the dry dipterocarp forest of Thailand. Thai Forest Bull. (Botany) No 14. Forest Herbarium, Royal Forest Dept., Bangkok. 93 pp.

Curtis, J. T. & McIntosh, R. P. 1951. An upland forest continuum in the prairie forest border region of Wisconsin. Ecology 32:476–496.

Griffiths, J. F. 1978. Applied climatology: an introduction, 2nd ed. Oxford University Press. 136 pp.

Hartshorn, G. S. 1978. Tree falls and tropical forest dynamics. In: Tomlinson, P. B. & Zimmerman, M. H. (eds.) Tropical Trees as Living Systems, pp. 617–638. Cambridge University Press, London.

Hozumi, K. 1975. Studies on the frequency distribution of the weight of individual trees in a forest stand. V. The M-w diagram for various types of forest stands. Jap. J. Ecol. 25:123–131.

Hubbel, S. P. 1979. Tree dispersion, abundance and diversity in a tropical dry forest. Science 203:1299–1309.

Kutintara, U. 1975. Structure of the dry dipterocarp forest. Ph.D. Dissertation. Colorado State Univ., Fort Collins. 242 pp.

MacArthur, R. H. & MacArthur, J. W. 1961. On bird species diversity. Ecol. 42:549–598.

Moormann, F. R. & Rojanasoonthorn, S. 1972. The soils of the Kingdom of Thailand: Explanatory text of the general map. Dept. of Land Development, Bangkok.

Morisita, M. 1959. Measuring of the dispersion of individuals and analysis of the distributional patterns. Mem. Fac. Sci. Kyushu Univ., Ser. E. (Biol.) 2:215–235.

Naka, K. 1982. Community dynamics of evergreen broadleaf forest in southern Japan. I. Wind-damaged trees and canopy gaps in an evergreen oak forest. Bot. Mag. Tokyo 95:385–399.

Neal, D. G. 1967. Statistical description of the forest of Thailand. Military Research and Development Center, Bangkok. 346 pp.

Ogawa, H., Yoda, K. & Kira, T. 1961. A preliminary survey on the vegetation of Thailand. Nature and Life in Southeast Asia 1:21–157.

Ogawa, H., Yoda, K., Kira, T., Ogino, K., Shidei, T., Ratanawongse, D. & Apasutaya, C. 1965. Comparative ecological studies on three main types of forest vegetation in Thailand. I. Structure and floristic composition. Nature and Life in Southeast Asia 4: 13–48.

Poor, R. A. A. 1968. Studies in Malaysian rain forest. I. The forest on triassic sediment in Jenka Forest Reserves. J. Ecol. 56: 143–196.

Richards, P. W. 1952. The Tropical Rain Forest. An Ecological Study. Cambridge Univ. Press, London. 450 pp.

Royal Forest Department 1962. Types of forest of Thailand. Report No. 44, 12 pp.

Runkle, J. R. 1982. Patterns of disturbance in some old-growth mesic forests of eastern north America. Ecology 63:1533–1546.

Sabhasri, S., Boonnittee, A., Khemnark, C. & Aksornkoae, S. 1968. Structure and floristic composition of forest vegetation at Sakaerat, Pak Thong Chai, Nakorn Ratchasima: I. Variation of floristic composition along transect through dry evergreen and dry dipterocarp forest. ASRCT (App. Sci. Res. Corp. Thailand).

Sahunalu, P., Jamroenpruksa, M., Dhanmanonda, P., Suwannapinan, W. & Prachaiyo, B. 1979. Comparative structural characteristics of 3 forest types at Namphrom basin, Chaiyaphoom province. For. Res. Bull. No. 63. Kasetsart Univ., Bangkok.

Sukwong, S. 1974. Deciduous forest ecosystem in Thailand. A paper presented at the UNESCO seminar on deciduous forest ecosystems, 25–30 November 1974. Sakaerat Experiment Station, Nakornratchasima, Thailand. 16 pp.

Wacharakitti, S., Intrachantra, P. & Mungkorndin, S. 1980. Natural resources and land use studies of SERS. For. Res. Bull. No. 69. Kasetsart Univ., Bangkok.

Whittaker, R. H. 1975. Communities and Ecosystems, 2nd ed. Macmillan Publ. Co., New York. 385 pp.

Whitmore, T. C. 1975. Tropical Rain Forests of the Far East. Clarendon Press, Oxford. 675 pp.

Williamson, G. B. 1975. Pattern and seral composition in an old-growth beech-maple forest. Ecology 56(3):727–731.

Yamakura, T. 1987. An empirical approach to the analysis of forest stratification. I. Proposed graphical method derived by using an empirical distribution function. Bot. Mag. Tokyo 100:109–128.

28. Mosaic structure and tree growth pattern in a monodominant tropical seasonal evergreen forest in Thailand

MAMORU KANZAKI[1], KYOJI YODA[1] & PRICHA DHANMANONDA[2]
[1]*Faculty of Science, Osaka City University, Osaka 558, Japan*; [2]*Faculty of Forestry, Kasetsart University, Bangkok 10903, Thailand*

Abstract

The mosaic structure, species composition and growth pattern of trees were examined in a tropical seasonal evergreen forest in Thailand. The gap, building, and mature phases covered 5.9, 20.2, and 73.9% of the total forest, respectively, and the gap formation rate was estimated as $0.005\ \text{ha} \cdot \text{ha}^{-1} \cdot \text{yr}^{-1}$ based on 4 years of observation. The maximum size of gaps in the forest was $410\ \text{m}^2$. The main component species of the forest were *Hopea ferrea*, accounting for 47.4% of the total basal area, and several sub-canopy species. Two types of canopy species were found: fast-growing species with bell-shaped size structure (corresponding to late-secondary species or late-seral species) and slow-growing species with L-shaped size structure. The dominant canopy species, *Hopea ferrea*, belongs to the latter category. The turnover time of the forest (the mean duration between succeeding gap formations) was estimated as 199 years from the lifespan of *Hopea ferrea*, and 200 years from the gap formation rate. The monodominance of the forest is considered to be maintained by the high shade tolerance of *Hopea ferrea* and the small gap size in the forest.

Key words: diameter growth, gap, *Hopea ferrea*, lifespan, patch dynamics, turnover time

Introduction

Forests can be recognized as mosaics involving patches of various stages in a regeneration cycle (Watt 1947). The horizontal classification of a forest into gap, building and mature phases is a convenient method to describe the forest structure (Whitmore 1975). Mosaic maps of forests and temporal changes in the mosaic pattern yield important information because the species

E.O. Box et al. (eds.), Vegetation Science in Forestry, 495–513.

composition, species richness and stability of forests are closely related to the coarseness of the mosaic pattern and to the frequency of the disturbances that form gaps in the forest (Pickett & Thompson 1978, Connell 1978, Hart 1990). One purpose of this paper is to assess three descriptors which indicate the dynamic features of forests, in a tropical seasonal evergreen forest in Thailand: the gap formation rate, turnover time (White & Pickett 1985), and gap size distribution.

Forest dynamics ultimately depend on the growth of constituent trees of the forest. In tropical forests, where the growth of trees cannot be evaluated from annual rings, tree growth is estimated from repeated measurement of tree size (Lieberman et al. 1985). The growth of trees in an old-growth forest, however, probably varies over the developmental phases of the forest, i.e. gap, building, and mature phases, and the analysis of growth must be done with respect to the forest regeneration cycle. The growth pattern of trees also varies among species, and analyses for every component species are required (Lieberman et al. 1985). The abundance of component species in tropical forests makes such analyses difficult, and some categorization of species into ecological groups is necessary (Swaine & Whitmore 1988, Lieberman et al. 1985, Lieberman & Lieberman 1987, Swaine et al. 1987). The second purpose of the paper is to examine the diameter growth pattern of trees by phases and by species groups.

Study site

The study site is at the Sakaerat Environmental Research Station, Nakorn Ratchasima province, northeast Thailand, about 180 km northeast of Bangkok (lat. 14°30′ N, long. 101°56′ E). The station was established in 1967 and consists of 29.5 km^2 of seasonal evergreen forest and 12.2 km^2 of drought-deciduous forest (Wacharakitti et al. 1980). In Thailand, these forests are called dry evergreen forest and dry dipterocarp forest, respectively (Royal Forest Department 1962).

The site is in the tropical monsoon climate zone. Mean annual temperature at the station is 26.2°C. Annual rainfall is 1240 mm, and monthly rainfall is less than 50 mm during the dry season, from December to February. The site is located on a cuesta-like table mountain which is composed of sandstone classified into the Phra Wihan formation of the Khorat group (Methikul & Silpalit 1968). Red-yellow podzolic soil covers the site and is more than 60 cm deep (Bos & Thunduan 1968).

Two types of seasonal evergeen forest are observed in the area. One type is dominated by *Hopea ferrea*, the other by *Shorea henryana*. A study plot was set up in the most strictly preserved *Hopea ferrea* forest. The species composition is poorer and monodominance pattern is clearer in the *Hopea ferrea* forest than in the *Shorea henryana* forest (Bunyavejchewin, 1986). The canopy of the forest attains 30 m to 40 m in height, and three vertical strata, over 30 m, 30 m to 19 m, and 17 m to 5 m, are found, in addition to a sapling stratum (Yamakura 1987). The geometric mean of relative illuminance values at ground level is 1.2% in the rainy season and 1.7% in the dry season (Yoda et al. 1983). Significant depression of the soil respiration rate is found during dry season (Yoda & Nishioka 1982), suggesting that the stress of low humidity decreases biological activity in the dry season.

Methods

Mosaic map

A belt plot, 50 m wide and 526 m long (2.63 ha), was set up along a small trail in the forest, using a traverse compass (Pocket Compass LS-25, Ushikata) and measuring tapes. The forest was divided into three developmental phases according to canopy height, as follows:

Gap:		canopy height < 10 m
Building:	$10 \text{ m} \leq$	canopy height < 20 m
Mature:	$20 \text{ m} \leq$	canopy height

Boundaries between the phases were checked in the forest, and a mosaic map was prepared in 1985.

Tree census

All living trees more than 30 cm DBH in the belt plot were mapped, and diameter at 1.3 m height (DBH) was measured in 1987 and 1989. For a detailed tree census, three quadrats ($900 \text{ m}^2 \times 2$ and $400 \text{ m}^2 \times 1$) were set up in the mature phase, four quadrats ($225 \text{ m}^2 \times 3$ nad $100 \text{ m}^2 \times 1$) in the building phase, and three quadrats ($20 \text{ m}^2 \times 2$ and $30 \text{ m}^2 \times 1$) in the gap phase. Height and DBH were measured for trees over 3 cm DBH in the mature phase, over 1 cm DBH in the building phase, and over 0.5 m height in the gap phase. A tree census in the 10 quadrats was conducted in 1985 and was repeated in 1987 and 1989. Enumerated trees were examined by the local botanist, and local names were recorded in the field. Specimens were collected from all trees except common species that could be identified easily. These specimens were identified by Dr. Thawatchai Santisuk of the Royal Forest Department of Thailand. Nomenclature basically followed Smitinand (1980).

Table 1. Area percentage of the three developmental phases in a 2.63 ha plot and the rate of gap formation

	Gap	Building		Mature	Total
Area in 1985 (%)	5.9	20.2		73.9	100.0
Area in 1989 (%)	7.9	19.9		72.2	100.0
Gap formation rate (ha ha^{-1} yr^{-1})					
in each phases		0.0044		0.0053	
in non-gap area			0.0051		
in whole area					0.0050

Phases are defined by canopy height: <10 m for gaps, 10–20 m for the building phase, and ≧20 m for the mature phase. Gap formation rates were calculated from changes in the mosaic pattern over 4 years.

All dead trees over 20 cm DBH, in the belt plot, were mapped in 1985, and trees that died after 1986 were added to the map, in 1987 and in 1989. We recorded their DBH and whether they contribute to canopy gap formation at the time of the mapping.

Results and discussion

Mosaic structure

The forest consisted of 5.9% gap, 20.2% building phase, and 73.9% mature phase in 1985 (Table 1); the gap percentage increased to 7.9% in 1989. The patches of the three phases were well mixed and constituted a fine mosaic pattern (Fig. 1). The gap fraction in the total area of tropical forests varies from 4% to 25%, according to a review by Brokaw (1985). The fraction in this forest was relatively small, even considering the differences in the definition of gaps among different authors.

The maximum gap size was 410 m^2, but smaller gaps were more frequently observed in the plot (Table 2), as reported in various forest types (Brokaw 1982, Runkle 1982, Nakashizuka 1984). Small gap size suggests the absence of major disturbances in this forest, such as windstorms and fires. From the increasing gap area over 4 years, the annual gap formation rate was estimated as 0.005 ha · ha^{-1} · yr^{-1} in the whole forest, or 0.0051 ha · ha^{-1} · yr^{-1} in the non-gap area (area of building + mature phases, Table 1). The gap formation rate in the area of the building phase was 0.0044 ha · ha^{-1} · yr^{-1}, and was comparable to that in the mature area (Table 1). The gaps in the building area, however, were formed by canopy trees falling into the building-phase area from the mature-phase area. The gap formation rate in this forest (based

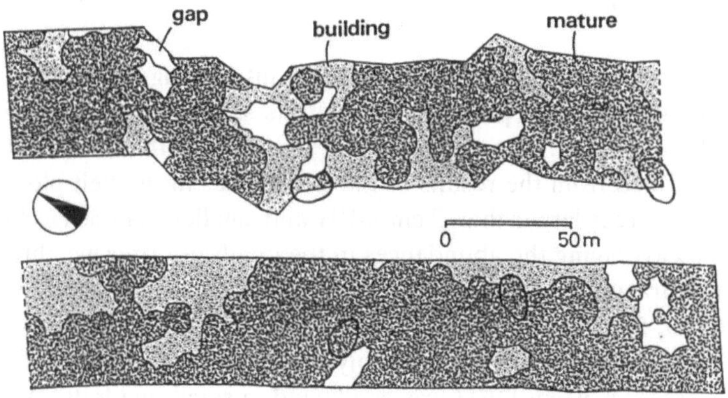

Fig. 1. Mosaic structure of gap, building, and mature-phase patches in a 2.63 ha belt plot (50 m wide and 526 m long) in a tropical seasonal evergreen forest in Thailand. Phases were defined by tree height as follows: ≧20 m for mature phase, 10–20 m for building phase, and <10 m for gaps. The map was prepared in 1985. The area enclosed by thick lines represents gaps formed during the study period, from 1985 to 1989. All of the recent gaps were in contact with older gaps or building-phase patches.

Table 2. Gap area distribution in the 2.63 ha plot, in 1989. A gap was counted when more than half of it was included in the 2.63 ha plot

gap size (m^2)	Number of gaps	Area percent
25–125	8	21.3
125–225	6	34.5
225–325	2	17.1
325–425	2	27.1
Total	18	100.0

on the whole forest area) was small but was within the range of rates reported from various tropical forests, 0.005 to 0.02 $ha \cdot ha^{-1} \cdot yr^{-1}$ (Runkle 1985, Brokaw 1985).

Four new gaps were formed from 1985 to 1989 in the belt plot, and all of them were in contact with older gaps or building-phase patches (Fig. 1). Thus, the edges of mature-phase patches facing building-phase patches or gaps were considered to be the most fragile part of the forest, characterized by a high probability of gap formation. The same phenomenon was reported by several other studies (Runkle 1985, Lawton & Putz 1988).

Species composition

Species composition in 1989 was obtained by integrating the results from the tree censuses in the belt plot and in quadrats set in the gap, building, and mature-phase patches. The abundance of trees larger than 30 cm DBH was directly obtained from the results of the tree census in the belt plot, and the abundance of trees larger than 3 cm DBH and smaller than 30 cm DBH was obtained by averaging the abundances in the quadrats, after weighting based on the area fractions of the gap, building and mature phases in the forest.

In 2.63 ha, 52 tree species were observed (excluding unidentified species), but the number of species was probably underestimated because the census for trees less than 30 cm DBH was conducted in small quadrats. Among the 52 species, 12 species were canopy species, with maximum DBH at least 50 cm, and 28 species were sub-canopy species (10 cm ≤ maximum DBH ≦ 50 cm) (Table 3).

A single canopy species, *Hopea ferrea*, accounted for 47.4% of the total basal area. The other canopy and sub-canopy species, such as *Hydnocarpus ilicifolius*, *Lagerstroemia duperreana*, *Shorea henryana*, and *Irvingia malayana*, followed with less than 6% relative basal area each.

Size structure and species abundance

The frequency distribution of DBH values was prepared for the trees (DBH ≧ 3 cm) in the belt plot, for each species, by integrating the results from the tree censuses in the belt plot and in the 10 quadrats. We classified the species into two categories, species with L-shaped size structure and species with bell-shaped size structure. The L-shaped group was represented by *Hopea ferrea* and bell-shaped by *Shorea henryana* (Table 3). The categorization of the species reflects the shade-tolerance of the young trees of each species.

The tree species with L-shaped size structure had more individuals in both the canopy and sub-canopy species groups (Fig. 2). In other words, rare species are characterized by a lack of small trees, as pointed out by Hubbell (1979). Among canopy species, *Hopea ferrea* had a huge number of trees compared to the other canopy species. Thus the monodominance pattern was clear in the canopy species group. Conversely, the abundance of the sub-canopy species with L-shaped size structure, such as *Hydnocarpus ilicifolius*, *Memecylon ovatum*, *Memecylon geddesianum*, *Walsura trichostemon*, and *Aglaia pirifera*, was much more evenly distributed. The species composition in each of the three phases, obtained from the tree censuses in the 10 quadrats, is shown in Table 4. *Hopea ferrea*, and the five sub-canopy species just mentioned above, had high abundance ratio (the ratio of the fraction of

Table 3. Main species in the 2.63 ha plot (DBH $\geqq$ 3.0 cm). Only the 20 species with the greatest basal area are listed, along with two minor canopy species

Species	Basal area		No. of trees		Max. DBH	Size str.
	(m^2/ha)	Ratio to total	(n/ha)	Ratio to total		
Hopea ferrea (Dipterocarpaceae)	14.29	0.474	264.0	0.144	C	L
Hydnocarpus ilicifolius (Flacourtiaceae)	1.58	0.052	193.0	0.106	SC	L
Lagerstroemia duperreana (Lythraceae)	1.47	0.048	40.8	0.022	C	L
Shorea henryana (Dipterocarpaceae)	1.37	0.045	6.0	0.003	C	B
Irvingia malayana (Ixonanthaceae)	1.30	0.043	1.1	0.000	C	B
Mammea siamensis (Guttiferae)	1.10	0.036	39.1	0.021	SC	L
Memecylon ovatum (Memecylaceae)	0.91	0.030	202.5	0.111	SC	L
Walsura trichostemon (Meliaceae)	0.90	0.029	204.9	0.112	SC	L
Eugenia sp.1 (Myrtaceae)	0.85	0.028	16.9	0.009	C	L
Memecylon geddesianum (Memecylaceae)	0.83	0.027	91.5	0.050	SC	L
Aglaia pirifera (Maliaceae)	0.56	0.018	143.6	0.078	SC	L
Parkia sumatrana (Mimosaceae)	0.51	0.016	1.9	0.001	C	B
Litsea s. (Lauraceae)	0.43	0.014	31.2	0.017	SC	L
Dialium cochinchinense (Caesalpiniaceae)	0.33	0.011	13.0	0.007	C	L
Canthium sp.1 (Rubiaceae)	0.33	0.011	96.7	0.053	SC	L
Ficus sp. (Moraceae)	0.32	0.010	0.3	0.000	C	B
Dipterocarpus costatus (Dipterocarpaceae)	0.25	0.008	0.3	0.000	C	B
Melodorum fruticosum (Annonaceae)	0.24	0.008	48.2	0.026	SC	L
Stereospermum fimbriatum (Bignoniaceae)	0.22	0.007	0.3	0.000	C	B
Canthium sp.2 (Rubiaceae)	0.17	0.005	15.3	0.008	SC	B
Other minor canopy species						
Terminalia nigrovenulosa (Combretaceae)	0.11	0.003	7.4	0.004	C	L
Peltophorum pterocarpum (Caesalpiniaceae)	0.10	0.003	0.3	0.002	C	B
TOTAL 58 species	30.16	1.000	1821.9	1.000		

The total number of tree species found in the plot was 52, excluding liana and unidentified species. The species were categorized into 3 groups: canopy species (C: max. DBH $\geqq$ 50 cm), sub-canopy species (SC: 50 cm > max. DBH $\geqq$ 10 cm), and understory species (10 cm > max. DBH). The size structure of the population was categorized into 2 groups, L-shaped (L) and bell-shaped (B), basedon the DBH distribution pattern of each species. Species identification was performed by Dr. Thawatchai Santisuk. Nomenclature followed Smitinand (1980).

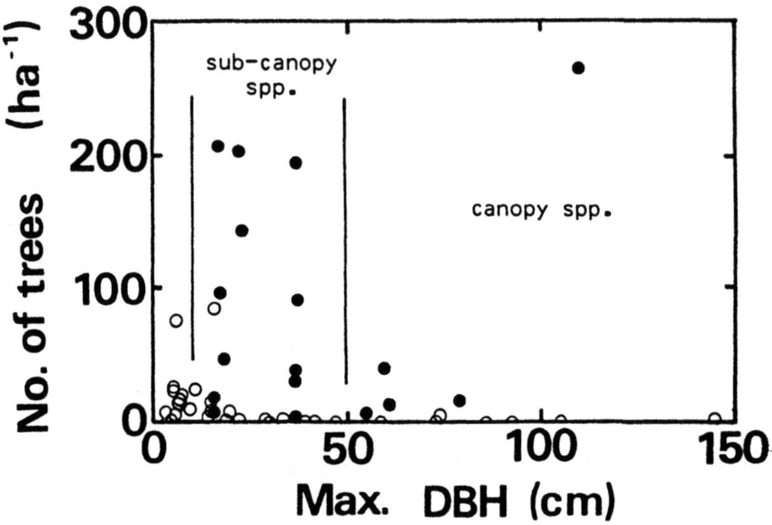

Fig. 2. Relationship between abundance and maximum size of trees in each species. The number of trees per species is plotted against the maximum DBH of the species in the 2.63 ha plot. For canopy species (maximum DBH $\geqq$ 50 cm) and sub-canopy species (10 cm $\leqq$ max. DBH $<$ 50 cm), population size structures were categorized as L-shaped (closed circles) and bell-shaped (open circles). *Hopea ferrea* (max. DBH = 109.4 cm, no. of tree = 264 ha^{-1}) dominated among the canopy species.

stems to the total stems) in all phases. These values of relative species abundance indicate that *Hopea ferrea* vigorously regenerates in gaps of the forest. Since the gap trees of *Hopea ferrea* and shade-tolerant sub-canopy species frequently have scars caused by fallen gap-maker trees, most individuals seem to have been established under a closed canopy before the gaps were formed.

Diameter growth of trees

A relationship between the initial DBH of trees and diameter growth for the following two years was obtained from two census intervals, from 1985 to 1987 and from 1987 to 1989, by phase and by species or species group. Figure 3 shows this relationship for *Hopea ferrea* in the mature phase. Since the number of sample trees with large DBH was small, the horizontal axis (DBH) is log-transformed. Bars indicate the mean value of diameter growth for each size class. Even the observed data had a wide range of variation around the means, the mean growth rate peaked around 20 cm to 50 cm DBH and decreased in both directions. The reduced growth rate for smaller trees is

Table 4. Canopy and sub-canopy species composition in quadrats in the three developmental phases

Species	Ratio of the number of stems to total		
	Mature	Building	Gap
Canopy species (max. DBH ≥ 50 cm)			
L-shaped size structure			
Hopea ferrea	0.173	0.086	0.263
Dialium cochinchinense	0.002	0.009	0.012
Eugenia sp.1	0.014	0.004	0.008
Lagerstroemia duperreana	0.006	0.013	0.004
Bell-shaped size structure			
Parkia sumatrana	0.002	0.003	0
Ficus sp.	0.002	0	0
Terminalia nigrovenulosa	0.002	0.003	0
Sub-canopy species (10 cm ≦ max. DBH < 50 cm)			
L-shaped size structure			
Aglaia pirifera	0.079	0.195	0.142
Memecylon ovatum	0.126	0.114	0.089
Walsura trichostemon	0.146	0.028	0.055
Memecylon geddesianum	0.040	0.054	0.036
Hydnocarpus ilicifolius	0.111	0.080	0.033
Canthium sp.1	0.064	0.024	0.029
Melodorum fruticosum	0.007	0.003	0.019
Linociera sp.3	0.008	0.010	0.004
Microcos paniculata	0.002	0.006	0.004
Litsea sp.	0.027	0	0.002
Mammea siamensis	0.025	0.009	0
Diospyros sp.2	0	0.003	0
Bell-shaped size structure			
Canthium sp.2	0.007	0.018	0.002
Adenanthera pavonina	0.005	0.003	0.002
Drypetes sp.	0.004	0.001	0
Siphonodon celastrineus	0.002	0	0

See Table 1 for the defintion of phases and size-structure types.

probably caused by shading by taller trees, while the reduced growth rate of large trees may be caused by senescence.

In the following analysis, the relationships between DBH and mean growth rate were compared among species groups. Since the growth rate of sub-canopy species did not differ between the L- and bell-shaped size-structure groups, these two species groups were lumped together for the sub-canopy species. In mature-phase patches (top panel of Fig. 4), the canopy species with bell-shaped size structure grew rapidly in the diameter classes as high as 100 cm but grow much more slowly in smaller size classes. This pattern suggests that these species are potentially rapid-growing but shade-

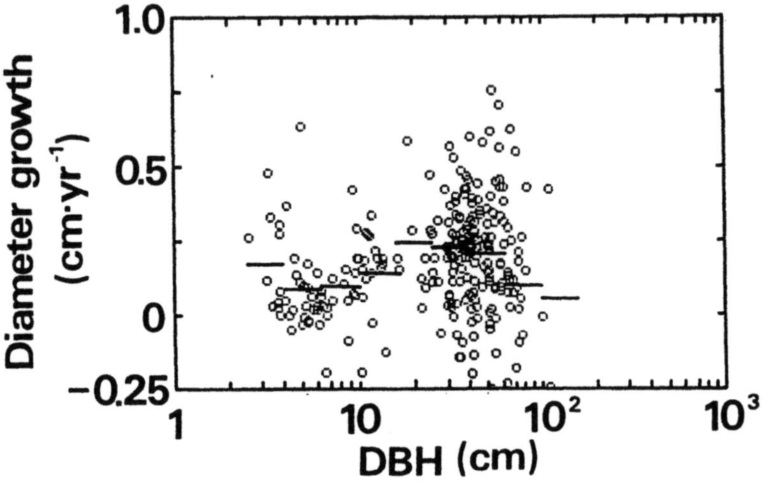

Fig. 3. Relationship between the diameter growth rate and DBH for *Hopea ferrea* in mature-phase patches. Bars indicate the mean diameter growth in each size class. The data are based on DBH measurements for all enumerated trees in the belt plot and the ten quadrats.

intolerant species. The growth rate of sub-canopy species, on the other hand, decreases with increasing tree size above a DBH value of about 20–30 cm.

In the building phase (bottom panel of Fig. 4), the growth rates of *Hopea ferrea* and the sub-canopy species both increase with increasing tree size, with *Hopea ferrea*'s growth rate higher than that of the sub-canopy species in size classes over 10 cm DBH. The growth rate of canopy species other than *Hopea ferrea* was the slowest in every size class. The canopy species with bell-shaped size structure, such as *Shorea henryana* and *Parkia sumatrana*, grew rapidly under non-shaded conditions and were shade-intolerant. Since these species germinated even under closed canopy, they cannot be categorized as pioneer but rather as late-secondary species (Whitmore 1982) or late-seral species (Swaine & Whitmore 1988).

Age estimation for canopy trees

The time needed for trees to grow to a certain size was frequently determined based on the mean growth rates of trees in successive size classes. Small trees overtopped by canopy trees, however, showed very slow growth rates (Fig. 4) and had high mortality rates (Fig. 5) in each phase. In order to estimate the age of trees that survive to mature size, the growth rate of the overtopped, shaded trees should be neglected (Ashton 1981, Swaine et al. 1987). In this paper we assume that trees in the largest size class in existing

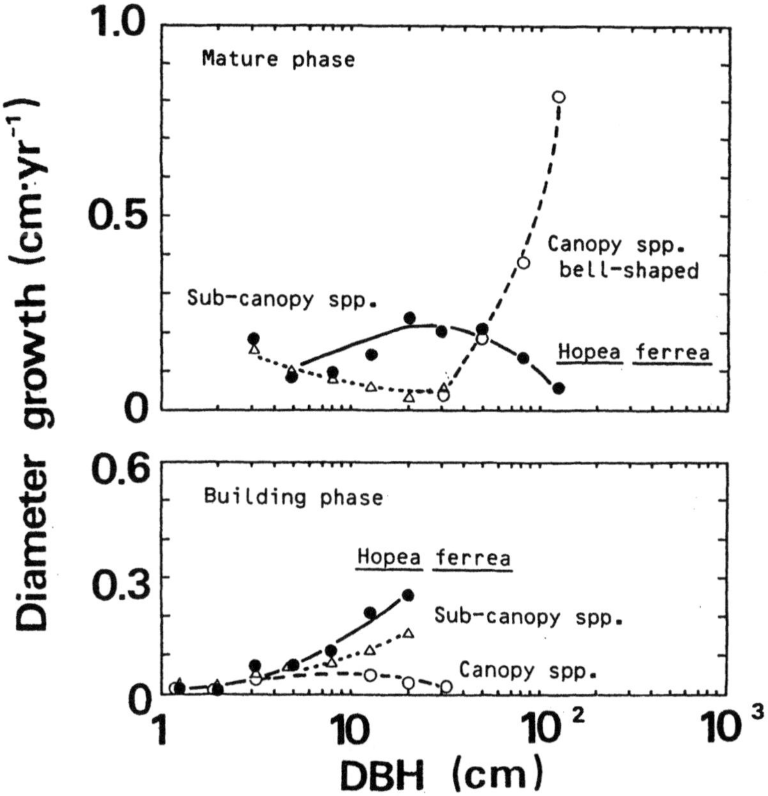

Fig. 4. Relationships between average diameter growth rate and DBH by phase and by species group. In mature phase (top panel), species are categorized as canopy species with bell-shaped size structure (open circles), *Hopea ferrea* (solid circles), and sub-canopy species (triangles). In building phase (bottom panel), species are categorized as canopy species other than *Hopea ferrea* (open circles), *Hopea ferrea* (solid circles), and sub-canopy species (triangles).

gap and building phases will survive and reach the canopy of mature-phase areas.

The mean diameter growth rates of *Hopea ferrea*, by size class and by phase, are shown in Fig. 6. The highest growth rate in the gap and building phases and the growth rates of trees over 25 cm DBH in the mature phase is traced by the thick descending line. This line represents the size-growth rate relation of non-shaded *Hopea ferrea* trees which occupy the upper part of the canopy throughout their life-cycle and have a high probability of occurring in the canopy in the mature phase. The relation between DBH (D) in cm and diameter growth rate of non-shaded *Hopea ferrea* (R) in $cm \cdot yr^{-1}$ was described by the linear model,

$$R = -aD + b,$$

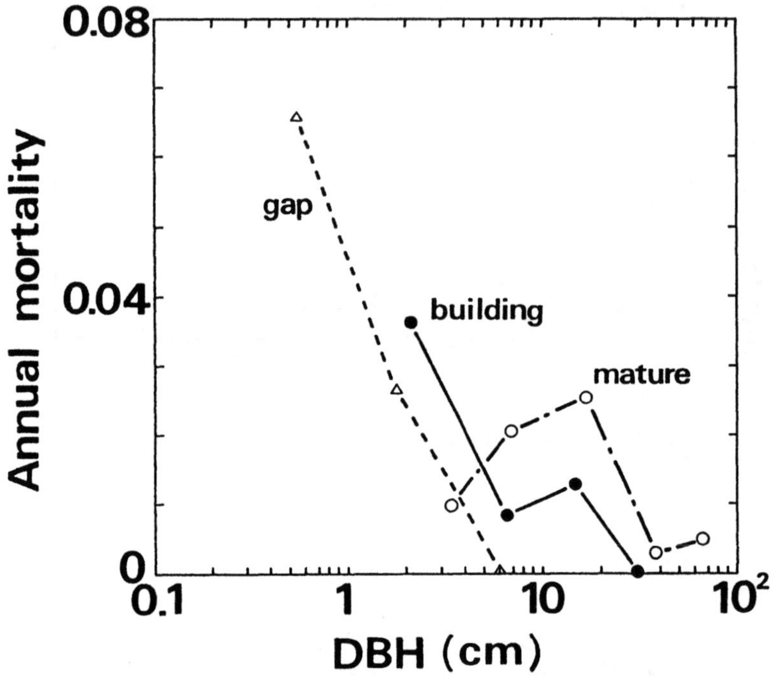

Fig. 5. Relationship between DBH and annual mortality of trees, by developmental phase. The mortalities are based on data from tree censuses in the belt plot (for trees over 30 cm DBH in 1987 and 1989) and in the 10 quadrats (in 1985, 1987, and 1989). Annual mortality is defined as $1 - \exp\{(\ln N_1 - \ln N_2)/2\}$, where N_1 and N_2 are the numbers of trees at the initial census and at the census after 2 yrs, respectively.

where $a = 0.002661$ and $b = 0.353$. Replacing R by dD/dt, a differential equation was obtained,

$$\frac{dD}{dt} = -aD + b,$$

in which t is time in years and $D = 0$ at $t = 0$ as an initial condition (i.e. omitting negative or zero growth, so $-aD + b > 0$). This differential equation has the general solution

$$\ln(-D + b/a) = -at + C,$$

where C is an integral constant. Consequently,

$$D = b/a - e^{C}e^{-at}. \qquad (1)$$

From the initial condition and equation (1),

$$e^{C} = b/a.$$

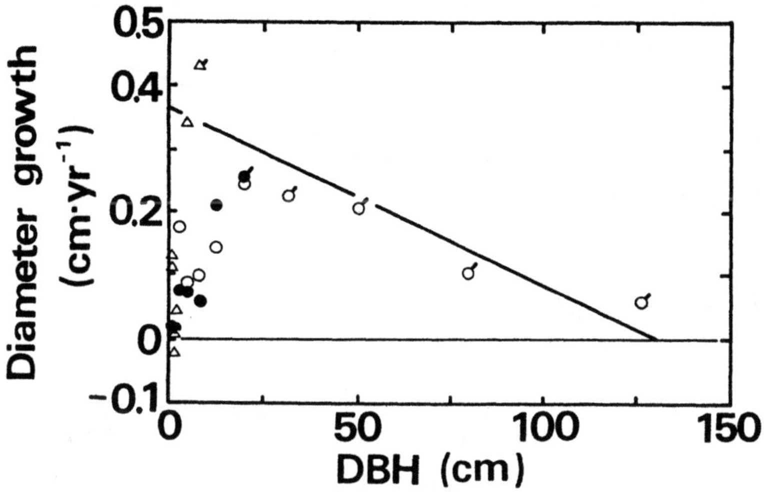

Fig. 6. Diameter growth pattern of *Hopea ferrea* in the three phases: gap (triangles), building (closed circles), and mature (open circles). The descending line represents the highest growth rate over all size-classes, and was obtained by linear regression for data marked by tails: $R = -0.002661D + 0.353$, where R is diameter growth rate in cm yr^{-1} and D is DBH in cm($r^2 = 0.8014$).

Thus, the growth curve of non-shaded *Hopea ferrea* can be written as

$$D = 132.6\,(1 - e^{-0.002661\,t}). \qquad (2)$$

From this equation, the time required for trees to reach a certain DBH is expressed as,

$$t = -375.8 \ln(1 - D/132.57). \qquad (3)$$

The diameter growth curve obtained is asymptotic (Fig. 7). It differs from the linear growth pattern of *Parashorea malaanonan* reported by Brown (1919) and from that of tropical trees in Ghana as reported by Swaine et al. (1987), who reconstructed growth curves of trees based on mean growth rates for successive diameter classes, excluding data from overtopped slow-growing trees. The decreasing annual ring width on stem disks is frequently observed in temperate-zone trees, though, so an asymptotic growth curve seems to be realistic.

Estimation of turnover time from life-span of Hopea ferrea

Since the forest is dominated by a single dominant canopy species, *Hopea ferrea*, the turnover time of the forest, i.e. the mean time interval from gap formation to the next gap formation at the same place, is considered to be

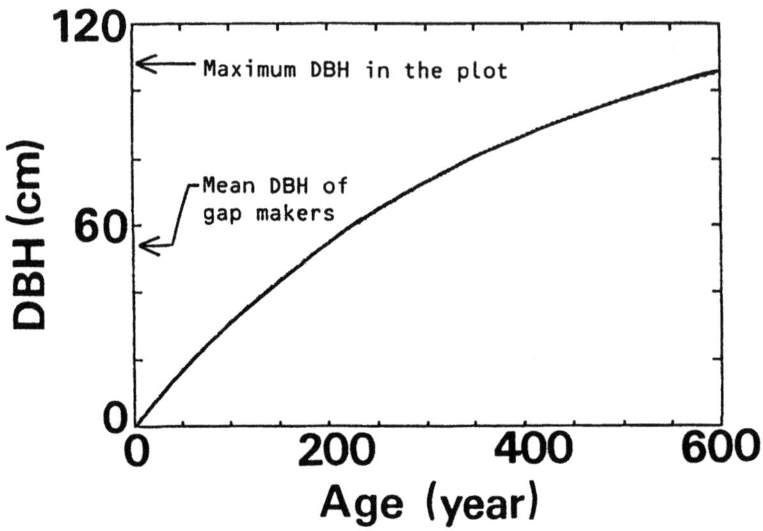

Fig. 7. Growth curve for non-shaded *Hopea ferrea* obtained from the size-growth relation shown in Fig. 4. The growth model is expressed by an asymptotic function, $D = 132.6 \{1 - \exp(-0.002661t)\}$, where D is DBH and t is time in years. The observed maximum DBH of *Hopea ferrea* in the plot was 109.4 cm, and the mean DBH of trees dying and causing gap was 54.6 cm in the plot.

almost equal to *Hopea ferrea*'s mean lifespan. For simplification we assume that (1) gaps are filled by a cohort of *Hopea ferrea* of 1.3 m height at the time of gap formation, (2) the diameter growth follows the growth curve shown in equation (2), and (3) non-shaded *Hopea ferrea* trees die when they reach 54.6 cm DBH, the mean DBH of trees dying and causing gaps (Fig. 8). Non-shaded *H. ferrea* trees would reach 54.6 cm DBH at 199 years, based on equation (3), and the turnover time would be 199 years.

Next we estimated the duration of the gap, building and mature phases in the regeneration cycle of the forest. The criteria for the three phases, 10 m and 20 m tree height, corresponds to 7.8 cm and 21.1 cm in DBH, based on the allometric relation between DBH and tree height(H) (Ogawa et al. 1965) of non-shaded *Hopea ferrea* trees,

$$\frac{1}{H} = \frac{1}{1.62D} + \frac{1}{48.4} \quad [\text{m, cm}],$$

obtained by non-linear least-squares methods (Nagano and Kira 1978). Based on the growth model (equation 3), a gap will enter the building phase (trees over 7.8 cm DBH) at 23 years after gap formation and then reach the mature phase (trees over 21.1 cm DBH) at 65 years after gap formation. So, the

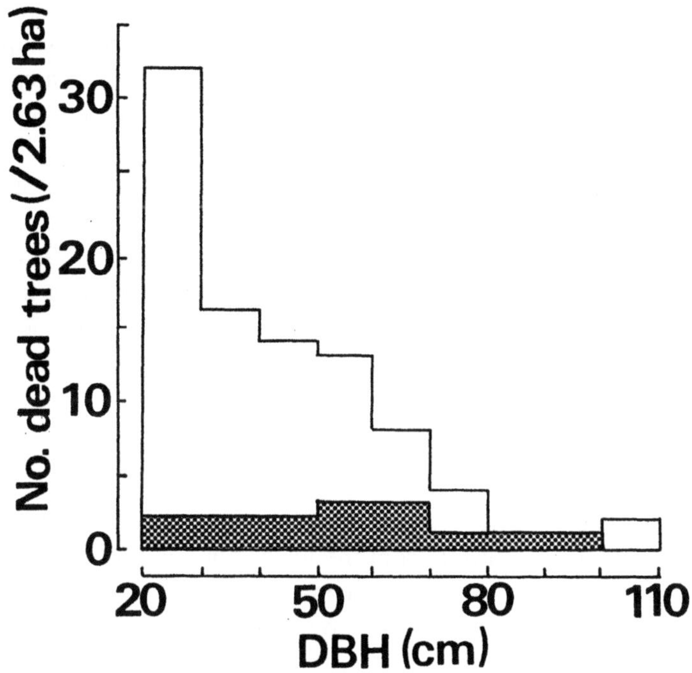

Fig. 8. Size distribution of dead trees found in the 2.63 ha plot, in 1989. Trees dying and causing gaps (gap makers) are indicated by the dotted area. The mean DBH of gap makers was 54.6 cm.

duration of the gap, building and mature phases was estimated to be 23 years, 42 years, and 134 years, respectively.

Estimation of turnover time from the mosaic pattern

The reciprocal of the gap formation rate for the whole forest (0.005 $ha \cdot ha^{-1} \cdot yr^{-1}$) indicates the turnover time of the forest. This would be 200 years, which is almost equal to the turnover time estimated from the lifespan of *Hopea ferrea*.

Another method to estimate the turnover time is available from the gap fraction of total forest area. If we assume that the gap fraction (G) is constant through time, and that gaps change to non-gap areas at 23 years after gap formation, as estimated above, then the turnover time (T) is expressed as

$$T = (1/G)\,23\ [\text{year}, \text{ha} \cdot \text{ha}^{-1}]. \qquad (4)$$

Substituting the gap area fraction in 1985 (0.059) and in 1989 (0.079), equation (4) yields turnover times of 386 years and 288 years, respectively. These values are considerably larger than the estimates based on the lifespan of

Hopea ferrea and on the gap formation rate. Since the gap fraction was not constant even over the 4 years, but gradually increased, the assumption of a constant gap fraction is invalid. Therefore, the first two estimates of turnover time seem to be more reliable than the estimate based on the gap fraction.

The observed turnover time, about 200 years, is longer than the figures reported from various other tropical forests, 60 to 135 years (Brokaw 1985) and is closer to estimates in temperate forests (Nakashizuka 1984, Runkle 1985). The longer turnover time probably results from the long lifespan of the dominant species, *Hopea ferrea*, and the scarcity of short-lived, fast-growing species.

Monospecific dominance in the forest

The number of species in the forest studied was obviously lower than that in tropical rainforests, resulting from the lower floristic diversity of much of Thailand, where there is a clear dry season (Whitmore 1975). The monodominance of *Hopea ferrea* in the forest, however, seems to be closely related to the shade tolerance of *Hopea ferrea*.

Saplings of *Hopea ferrea* and of the shade-tolerant sub-canopy species cover the forest floor densely preceding gap formation, and these saplings cover the gaps at high densities after gap formation (Table 4). These saplings probably impede the establishment of more light-demanding species, such as *Shorea henryana* and *Parkia sumatrana*, in the gaps. Furthermore, small gaps in this forest are also likely to favor the successful recruitment of *Hopea ferrea*. The minimum gap size required for the successful establishment of the light-demanding pioneer species is considered to be 400 m^2 to 1000 m^2 (Whitmore 1982, Brokaw 1985), and gaps in this forest were too small for successful establishment of light-demanding short lived species, such as *Mallotus* sp. (which accounted for 0.43% of the trees in gaps). The new gaps formed over the 4 years studied were located at the edge of mature-phase areas, in contact with old gaps or building-phase patches. Such a spatial arrangement of gap formation could also favor the recruitment of *Hopea ferrea*. Saplings of *Hopea ferrea* located at the edge can probably avoid the risk of death by sudden increase of radiation at the gap opening by acclimating to the intermediate light conditions at the edges.

Monodominant and mixed forests

Hart (1989) compared the forest structure and characteristics of dominant species in monodominance forest and mixed forest in humid tropical Zaire and concluded that the dominant species of the monodominant forest, *Gilbertiodendron dewevrei*, was shade-tolerant.

At the Sakaerat Environmental Research Station, species-richer forest dominated by *Shorea henryana* occurs next to *Hopea ferrea* forest. Bunyavejchewin (1986) reported that an average of 57 tree and shrub species was found in a 1 ha plot in the *Hopea ferrea* forest, and that *H. ferrea* accounted for about 50% of the total basal area, corresponding with the results of this study. On the other hand, he found an average of 75 species in a 1 ha plot in the *Shorea henryana* forest, and that *S. henryana* accounted for about 25% of the total basal area. The absence of a shade-tolerant dominant seemed to increase the species richness of the *Shorea henryana* forest. Reviews by Connell (1978) and Hart (1990) suggested that monodominant forests composed of shade-tolerant canopy species are considered to be a more stable forest type in tropical rainforests, and that species-rich rainforests tend to converge to monodominant forests of shade-tolerant species if major disturbances do not occur. This hypothesis clearly explains the successional relations in both forest types. This hypothesis, however, should be checked by direct observation of the disturbance regimes and population dynamics of the component species in both the monodominant and mixed forests.

Acknowledgements

We would like to express our sincere thanks to Dr. Thawatchai Santisuk for his kind support in the identification of species. Thanks are also due to Mr. Katsuhiko Kimura, Miss Duangdao Faidang, the staff of the Research and Training in Reafforestation Station, and the staff of the Sakaerat Environmental Research Station for their kind cooperation in the field research work. The study was financially supported by the International Scientific Research Program (No. 59041056, No. 62041088, and No. 63041110) and Grant-in-Aid for Co-operative Research (A) (No. 04304003) of the Ministry of Education, Science and Culture (Monbusho) in Japan.

References

Ashton, P. S. 1981. The need for information regarding tree age and growth in tropical forests. In: Bormann, F. H. & Berlyn, G. (eds.) Age and Growth Rate of Tropical Trees: New Directions for Research. Yale University School of Forestry and Environmental Studies Bulletin No. 94.

Bos, F. & Thunduan V. 1968. Detailed Soil Survey of the North-eastern Portion of ASRCT Sakaerat Experimental Station (Amphoe Pak Thong Chai, Changwat Nakhon Ratchasima). ASRCT Report No. 1, ASRCT, Bangkok. 22 pp.

Brokaw, N. V. L. 1982. Treefalls: frequency, timing, and consequences. In: Leigh, E. G., Rand, A. T. & Windsor, D. M. (eds.) The Ecology of a Tropical Forest: Seasonal Rhythms and Long-term Changes, pp. 101–108. Smithonian Institution, Washington, D. C.

Brokaw, N. V. L. 1985. Treefalls, regrowth, and community structure in tropical forests. In: Pickett, S. T. A. & White P. S. (eds.) The Ecology of Natural Disturbance and Patch Dynamics, pp. 53–69. Academic Press, Florida.

Brown, W. H. 1919. Vegetation of Philippine Mountains. Manila. Cited from: Richards, P. W., 1952. The Tropical Rain Forest. Cambridge University Press, Cambridge.

Bunyavejchewin, S. 1986. Ecological studies of tropical semi-evergreen rain forest at Sakaerat, Nakhon Ratchasima, northeast Thailand, I. Vegetation Patterns. Natural History Bulletin of the Siam Society 34:35–57.

Connell, J. H. 1978. Diversity in tropical rain forests and coral reefs. Science 199:1302–1309.

Hart, T. B. 1990. Monospecific dominance in tropical rain forests. Trends in Ecology and Evolution 5:6–11.

Hart, T. B., Hart, J. A. & Murphy, P. G. 1989. Monodominant and species-rich forests of the humid tropics: causes for their co-occurrence. The American Naturalist 133:613–633.

Hubbell, S. P. 1979. Tree dispersion, abundance, and diversity in a tropical dry forest. Science 203:1299–1309.

Lawton, R. O. and Putz, F. E. 1988. Natural disturbance and gap-phase regeneration in a wind-exposed tropical cloud forest. Ecology 69: 764–777.

Lieberman, D., Lieberman, M., Hartshorn, G. and Peralta, R. 1985. Growth rates and age-size relationships of tropical wet forest trees in Costa Rica. Journal of Tropical Ecology 1:97–109.

Lieberman, D. & Lieberman, M. 1987. Forest tree growth and dynamics at La Selva, Costa Rica (1969–1982). Journal of Tropical Ecology 3:347–358.

Methikul, A. & Silpalit, M. 1968. Reconnaissance geological survey of the ASRCT Sakaerat Experimental Station (Amphoe Pak Thong Chai, Changwat Nakhon Ratchasima). ASRCT Report No. 4, 9 pp., ASRCT, Bangkok.

Nagano, M. & Kira, T. 1978. Primary production. In: Kira, T., Ono, Y. & Hosokawa, T. (eds.) JIBP Synthesis Vol. 18:69–82. University of Tokyo Press, Tokyo.

Nakashizuka, T. 1984. Regeneration process of climax beech (*Fagus crenata*) forests. IV. gap formation. Japanese Journal of Ecology, 34:75–85.

Ogawa, H., Yoda, K., Ogino, K. & Kira, T. 1965. Comparative ecological studies on three main types of forest vegetation in Thailand. II. Plant biomass. Nature and Life in Southeast Asia IV:49–80.

Pickett, S. T. A. & Thompson, J. N. 1978. Patch dynamics and the design of nature reserves. Biological Conservation 13:27–37.

Royal Forest Department 1962. Types of Forest of Thailand. Royal Forest Department, Ministry of Agriculture, Bangkok. 12 pp.

Runkle, J. R. 1982. Patterns of disturbance in some old-growth mesic forests of eastern North America. Ecology 63:1533–1546.

Runkle, J. R. 1985. Disturbance regimes in temperate forests. In: Pickett, S. T. A. & White, P. S. (eds.) The Ecology of Natural Disturbance and Patch Dynamics, pp.17–33. Academic Press, Florida.

Smitinand, T. 1980. Thai Plant Names. 379 pp.

Swaine, M. D., Hall, J. B. & Alexander, I. J. 1987. Tree population dynamics at Kade, Ghana (1968–1982). Journal of Tropical Ecology 3:331–345.

Swaine, M. D. & Whitmore, T. C. 1988. On the definition of ecological species groups in tropical rain forests. Vegetatio 75:81–86.

Wacharakitti, S., Intrachanda, P. & Mungkorndin, S. 1980. Natural resources and land use studies of Sakaerat Environmental Research Station. Forest Research Bulletin, 69; Faculty of Forestry, Kasetsart University, Bangkok.

Watt, P. S. 1947. Pattern and process in the plant community. Journal of Ecology 35:1–22.

White, P. S. & Pickett, S. T. A. 1985. Natural disturbance and patch dynamics: an introduction.

In: S. T. A. Pickett & White, P. S. (eds.), The Ecology of Natural Disturbance and Patch Dynamics, pp. 3–13. Academic Press, Florida.
Whitmore, T. C. 1975. Tropical Rain Forests of the Far East. Clarendon Press, Oxford.
Whitmore, T. C. 1982. On pattern and process in forests. In: Newman, E. I. (ed.) The Plant Community as a Working Mechanism, pp. 45–59. Blackwell, Oxford.
Yamakura, T. 1987. An empirical approach to the analysis of forest stratification. I. Proposed graphical method derived by using an empirical distribution function. Botanical Magazine Tokyo 100:109–128.
Yoda, K. & Nishioka, M. 1982. Soil respiration in dry and wet seasons in a tropical dry-evergreen forest in Sakaerat, NE Thailand. Japanese Journal of Ecology 32:539–541.
Yoda, K., Nishioka, M. & Dhanmanonda, P. 1983. Vertical and horizontal distribution of relative illuminance in the dry and wet seasons in a tropical dry-evergreen forest in Sakaerat, NE Thailand. Japanese Journal of Ecology 33:97–100.

29. Growth dynamics of some broad-leaved tree species in central Thailand

KOICHI KAMO[1], BOPIT KIATVUTTINON[2] & BUNYARIT PURIYAKORN[2]
[1]*Kansai Research Center, Forestry and Forest Products Research Institute, Momoyama, Fushimi, Kyoto 612, Japan*; [2]*Royal Forest Department, Bangkhen, Bangkok 10900, Thailand*

Abstract

The seasonal diameter increment and annual growth of *Eucalyptus camaldulensis*, *Acacia auriculiformis* and *Azadirachta indica* var. *siamensis* in west-central Thailand were examined in relation to the seasonal and inter-annual rainfall pattern. The monthly values of diameter increment tended to coincide not only among the sample trees of each species but also among the three species studied, responding to the seasonal rainfall pattern. The three species commenced diameter increment at the beginning of the rainy season and continued more or less in relation to fluctuation of rainfall throughout the rainy season, ceasing in the middle of the dry season. Monthly diameter increment had a close correlation with the amount of rainfall during that and the preceding month. Based on this relation, the growth performance in relation to rainfall was analyzed for the three species resulting that *Azadirachta indica* var. *siamensis* grows more efficiently for a given amount of rainfall than *Eucalyptus camaldulensis* and *Acacia auriculiformis*, when soil is not dry; in dry conditions, *Eucalyptus camaldulensis* and *Acacia auriculiformis* are superior to *Azadirachta indica* var. *siamensis* in the growth for a given amount of rainfall. The annual stem-volume increment fluctuated in the stands of three species. The variation in annual rainfall seems to be related to the variation in annual stem volume increment.

Key words: biomass production, dendrometer, diameter increment, drought resistance, fast-growing tree species, phenology, tropical monsoon climate

Introduction

Climate in the tropics is in most cases seasonal to more or less extent in terms of rainfall distribution. This is conspicuous in the tropical monsoon

E.O. Box et al. (eds.), Vegetation Science in Forestry, 515–528.

climate, which prevails in the main plantation areas in Thailand. In these areas various broad-leaved tree species, including *Eucalyptus* and *Acacia*, have been used in reforestation, but the phenology of these species has not yet known. Study of the phenology of these tree species will give us any basic information on suitable schedules for silvicultural operations as well as growth characteristics of the tree species in relation to their environment. Phenology, including seasonal growth of some tree species, has been studied in tropical forest by various authors (Dawkins 1956, Boaler 1963, Alvim 1964, Lowe 1968, Hopkins 1970, Daubenmire 1972, Hazlett 1987, Okimori 1987). Quantitative studies on the relation between tree growth and meteorological events, however, have not been conducted much so far.

In this study, the seasonal and annual growth of *Eucalyptus camaldulensis*, *Acacia auriculiformis* and *Azadirachta indica* var. *siamensis* Valenton was investigated in relation to seasonal and annual rainfall for two years in order to elucidate the growth characteristics of these species in the tropical monsoon climate. *E. camaldulensis* is the main plantation tree species besides *Tectona grandis* and has been planted widely for its fast growth in arid and denuded areas in Thailand. *A. auriculiformis* has been planted in some cutover areas. Both species originated in Australia and have often been planted for re-afforestation in tropical monsoon areas of Southeast Asia. *A. indica* var. *siamensis* is found naturally in tropical seasonal forests in Southeast Asia, but is not often planted, in spite of the usefulness of its wood for construction, furniture or firewood. *E. camaldulensis*, *A. auriculiformis* and *A. indica* var. *siamensis* are all evergreen, although *A. indica* var. *siamensis* in research stands reduces its leaf biomass greatly for two to three weeks in January to February.

Study site and method

The broad-leaved tree stands studied are at the Ratchaburi Forest Experiment Station located in Ratchaburi province in west-central Thailand. The plantations were established on denuded areas with sparse vegetation cover. The soil at the research site is classified as Oxisol (low-humic latosol) with sandy loam. Dry dipterocarp forests, which once covered this area, were heavily exploited for a long time as a source of firewood or charcoal for the pottery industry in this area, which is one of the largest in Thailand. The climate of Ratchaburi is classified as tropical monsoon, with a pronounced dry season from November to April. The mean annual temperature and mean annual rainfall for the last ten years were 29.2°C and 983 mm, respectively. The T/P ratio ((mean annual temperature/mean annual precipitation) * 100) (Murphy 1986) in Ratchaburi was 3.0, while this ratio was 1.6–

2.8 in plantation areas of north and northeastern Thailand, 1.3–2.4 in those of southern Thailand and 2.8–2.9 in those of west-central Thailand (calculating the data from Viriyabuncha & Pitpreecha 1989), indicating that Ratchaburi is drier than the other most plantation areas in Thailand.

A sample plot, 400 m^2 in size, was set up in each stand of *E. camaldulensis*, *Ac. auriculiformis* and *Az. indica* var. *siamensis*. The height and diameter at breast height of all standing trees in the sample plot were measured in November 1987 and 1988, and in December 1989. At the beginning of the study, in November 1987, around 10 sample trees of various size outside each plot were cut and measured for the dry weight in order to estimate the biomass of the stands studied. The biomass of each biomass component, as well as the stem volume, was calculated by allometric methods.

In November 1987, an aluminum dendrometer band (Liming 1957) was installed on the stems at breast height of roughly ten sample trees for each species, chosen from the medium to larger trees in each stand. The smaller trees were not sampled, since the growth of smaller trees may be affected strongly by the trees in their vicinity and since their seasonal growth may be too small to measure. The range of diameter at breast height of sample trees was 7.4–10.9 cm for *E. camaldulensis*, 6.0–9.9 cm for *Ac. auriculiformis* and 6.6–10.9 cm for *Az. indica* var. *siamensis*. Vernier scales on the dendrometers were read twice a month for two years. The dendrometer data during the initial two months were omitted in order to avoid the errors due to increment variation possibly caused by differences in the fit of the dendrometer band on the stem and difference in spring tension.

Results

Growth and biomass of study stands

The growth and biomass of the study stands varied to some extent among the three species studied (Table 1). The stand age was 3.3 years in *E. camaldulensis*, 4.3 years in *Az. indica* var. *siamensis* and 6.3 years in *Ac. auriculiformis*. The stand density varied from 3625 trees/ha for *E. camaldulensis* to 5275 trees/ha for *Az. indica* var. *siamensis*. Average diameter and leaf biomass were almost the same among the three species, while the average height was the greatest for *E. camaldulensis* and smallest for *Az. indica* var. *siamensis*. The above-ground biomass and stand stem volume were larger in *Ac. auriculiformis* than for the other two species. These values show that the three species stand grow smaller among the tropical plantations studied so far (Kamo 1990). This might be due mainly to drier soil conditions in the research site.

Table 1. Dimensions and above-ground biomass of the study stands (1987)

	Tree species		
	E. camaldulensis	*Ac. auriculiformis*	*Az. indica* var. *siamensis*
Stand age (yrs.)	3.3	6.3	4.3
Avg. height (m)	9.9	8.5	6.0
Avg. dbh (cm)	5.8	5.8	5.6
No. trees (n/ha)	3625	5275	4725
Leaf mass (ton/ha)	2.5	2.9	2.4
Above-ground mass (ton/ha)	35.4	51.7	29.8
Stand stem volume (m^3/ha)	59.4	84.0	55.4

Seasonal pattern of diameter increment and rainfall

Variation of the monthly diameter increment among the sample trees was examined for each species studied over a period of two years. The correlation coefficient in almost all combinations of sample trees of the same species was high and significant at the 1% level (Table 2). This suggests that the diameter increment tends to fluctuate monthly in roughly the same way for individual trees of a given species. This result allows use of the mean sample value of the diameter increment to express the diameter increment of each species.

The monthly patterns of rainfall and diameter increment of the three species in 1988–1989 at Ratchaburi are shown in Figs. 1, 2 and 3. Monthly rainfall varied seasonally from 0 mm to 300 mm over the two years. The monthly diameter increment for each species seems to vary in accordance with the monthly rainfall with a bit discrepancy between them. All three species began the diameter increment immediately after the increased rainfall in May, at the beginning of the rainy season. Diameter increased throughout the rainy season from May to October except in June and/or July, a short period of reduced rainfall. In November, at the beginning of the dry season, diameters were still increasing, although rainfall decreased greatly. Diameter growth slowed greatly in December and January, and ceased in February through April, in the latter half of the dry season, in both 1988 and 1989.

The pattern of monthly diameter increment of the three species was different in 1988 and 1989 (Figs. 1, 2, 3), being much less in May and June of 1989 than in May and June of 1988. This followed the rainfall during that time which was much less in 1989 (115 mm) than in 1988 (389 mm).

These patterns of seasonal diameter increment were nearly identical among the three species. The significantly high correlation coefficients for diameter-increment regressions between species (Table 3) suggest that these species grow with almost the same seasonal pattern, at least at Ratchaburi.

Table 2. Correlation coefficient for mean monthly diameter increment between samples trees

Tree No.	1	2	3	4	5	6	7	8	9	10
E. camaldulensis										
2	0.863**									
3	0.554**	0.738**								
4	0.932**	0.946**	0.698**							
5	0.875**	0.851**	0.654**	0.880**						
6	0.902**	0913**	0.770**	0.955**	0.866**					
7	0.802**	0.849**	0.899**	0.948**	0.802**	0.940**				
8	0.848**	0.857**	0.783**	0.899**	0.800**	0.862**	0.851**			
9	0.620**	0.886**	0.607**	0.745**	0.755**	0.726**	0.630**	0.698**		
Ac. auriculifomis										
2	0.706**									
3	0.561**	0.909**								
4	0.506*	0.815**	0.766**							
5	0.306	0.666**	0.449*	0.606**						
6	0.787**	0.810**	0.789**	0.828**	0.529**					
7	0.575**	0.909**	0.698**	0.834**	0.569**	0.772**				
8	0.854**	0.905**	0.750**	0.770**	0.576**	0.878**	0.744**			
9	0.569**	0.887**	0.839**	0.866**	0.627**	0.725**	0.736**	0.877**		
10	0.553**	0.849**	0.743**	0.735**	0.616**	0.759**	0.670**	0.737**	0.735**	
11	0.516*	0.736**	0.766**	0.868**	0.524*	0.842**	0.872**	0.794**	0.657**	0.594*
Az. indica var. *siamensis*										
2	0.903**									
3	0.886**	0.851**								
4	0.677**	0.783**	0.557**							
5	0.729**	0.810**	0.781**	0.829**						
6	0.899**	0.905**	0.844**	0.565**	0.596**					
7	0.839**	0.829**	0.881**	0.555**	0.869**	0.830**				
8	0.608**	0.664**	0.720**	0.773**	0.813**	0.481*	0.642**			
9	0.767**	0.857**	0.799**	0.723**	0.766**	0.704**	0.767**	0.867**		

*Significant at the 5% level.
**Significant at the 1% level.

The relationship between monthly diameter increment and rainfall was also examined by regression analyses. A positive linear relationship seems to exist between the monthly diameter increment and the monthly rainfall, but with some extreme values (see Fig. 4). Figure 5 shows the relationship between the monthly diameter increment and the total amount of rainfall during that and the preceding month. In this graph, the extreme values in Fig. 4 disappeared, and the relationship was improved. The correlation coefficients relating monthly diameter increment and rainfall over the various period of time are shown in Fig. 6. The correlation was highest (significant at 1% level) for rainfall during the preceding plus current months but was high also for rainfall in the preceding one month alone. Correlations were

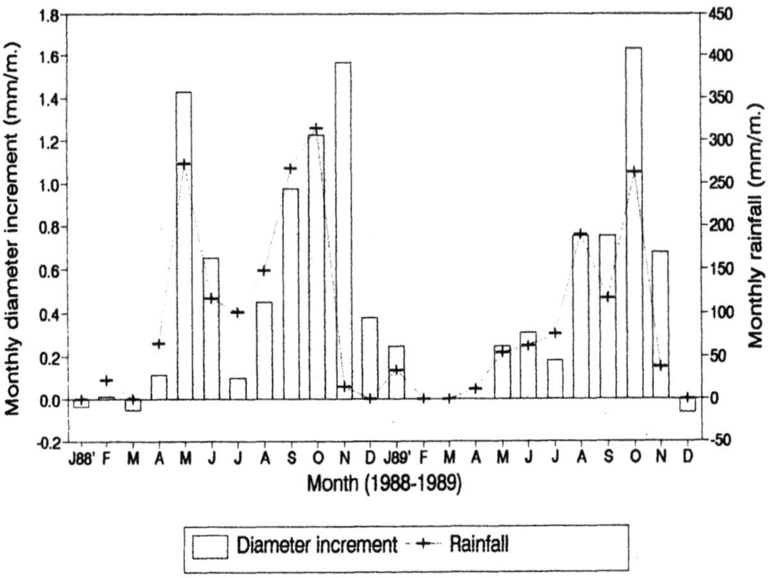

Fig. 1. Monthly variation of rainfall and mean diameter increment of *E. camaldulensis*.

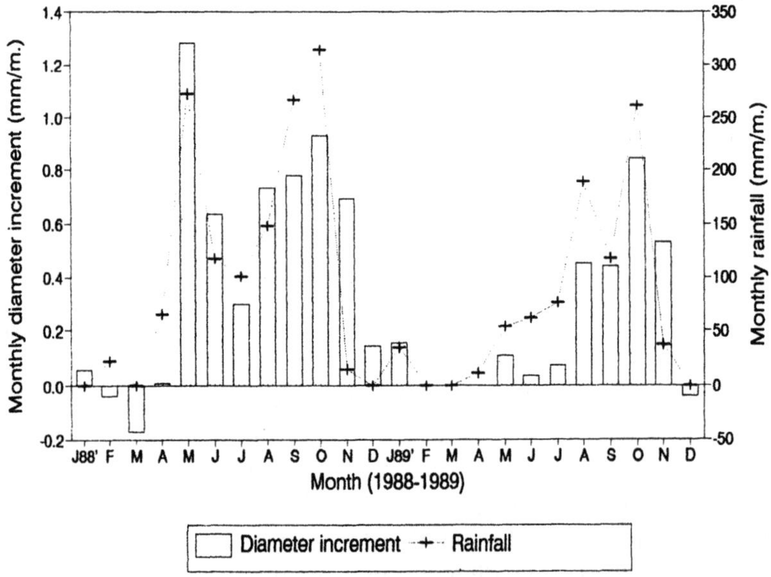

Fig. 2. Monthly variation of rainfall and mean diameter increment of *Ac. auriculiformis*.

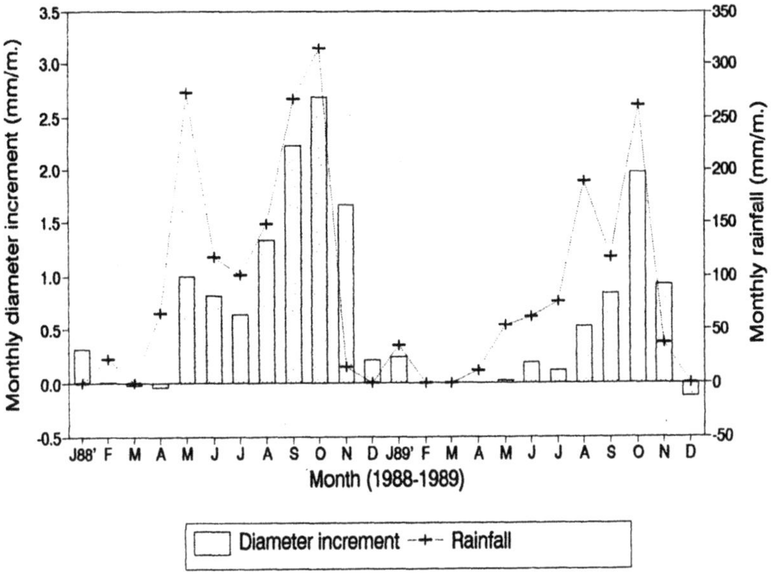

Fig. 3. Monthly variation of rainfall and mean diameter increment of *Az. indica* var. *siamensis.*

Table 3. Correlation coefficient for monthly diameter increment between *E. camaldulensis*, *Ac. auriculiformis* and *Az. inidica* var. *siamensis*

Species	*E. camaldulensis*	*Ac. auriculiformis*
E. camaldulensis	–	–
Ac. auriculiformis	0.895*	–
Az. indica var. *siamensis*	0.836*	0.848*

*Significant at the 1% level.

quite low for rainfall in the two or three months preceding measurement of diameter increment.

The best regression models for the relationship between monthly diameter increment (D) and the amount of rainfall during the preceding and current months (R) were as follows:

$$D = -0.0052 + 0.0027R \ (r = 0.8442) \text{ for } \textit{E. camaldulensis} \quad (1)$$

$$D = -0.0038 + 0.0020R \ (r = 0.8880) \text{ for } \textit{Ac. auriculiformis} \quad (2)$$

$$D = -0.1287 + 0.0043R \ (r = 0.9071) \text{ for } \textit{Az. indica} \text{ var. } \textit{siamensis} \quad (3)$$

Using equations (1), (2) and (3), D/R ratios were calculated. These D/R ratios represent the efficiency of utilizing rainfall in increasing the diameter of the tree species. If this value becomes higher, the tree is assumed to

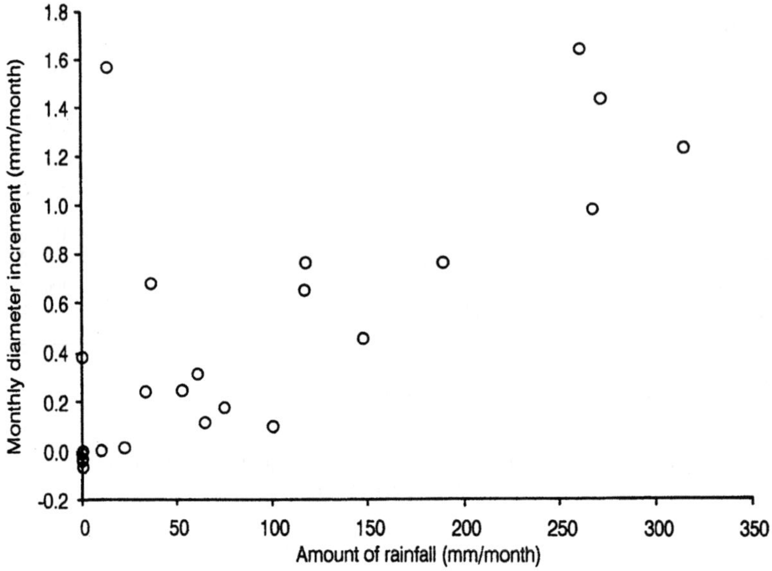

Fig. 4. Relationship between mean monthly diameter increment and monthly rainfall (*Az. indica* var. *siamensis*).

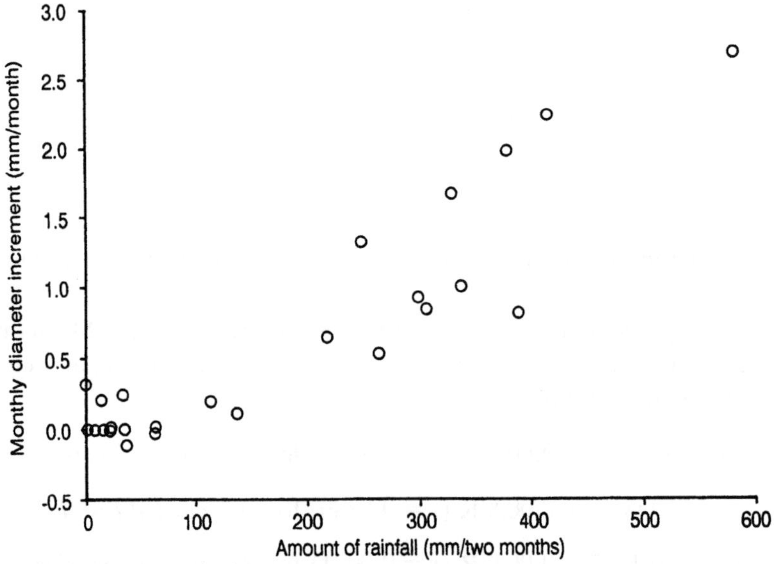

Fig. 5. Relationship between mean monthly diameter increment and amount of rainfall during the month diameter increment was measured and the preceding month (*Az. indica* var. *siamensis*).

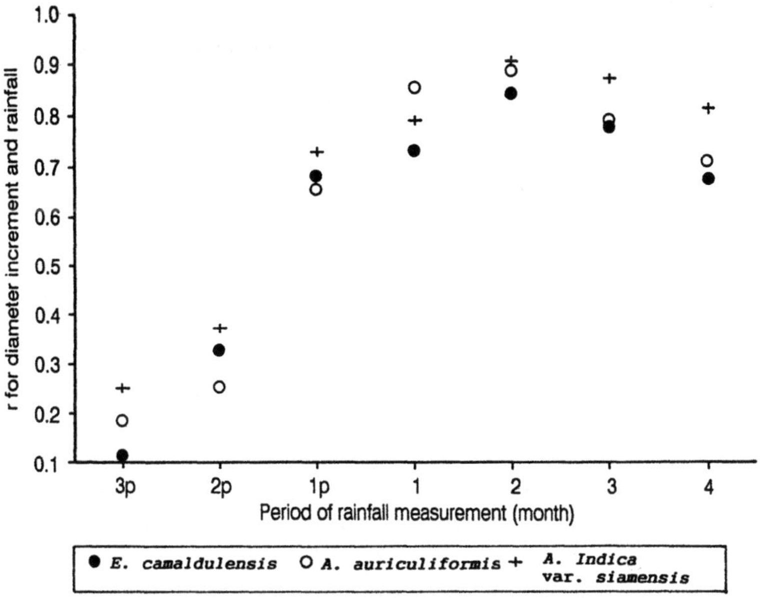

3P: three preceding month
2P: two preceding month
1P: one preceding month
1 : current month (month of diameter increment)
2 : current month + 1P
3 : current month + 1P + 2P
4 : current month + 1P + 2P + 3P

Fig. 6. Correlation coefficient for mean monthly diameter increment and amount of rainfall during varying period.

increase its diameter more for a given amount of rainfall. These D/R ratios are plotted against two-month rainfall amount in Fig. 7. For *E. camaldulensis* and *Ac. auriculiformis*, D/R ratios was almost constant over the range from 100 mm/two months to 600 mm/two months, decreasing a bit for less than around 100 mm/two months. On the other hand, the D/R ratio of *Az. indica* var. *siamensis* decreased gradually as rainfall decreased from 600 mm to around 200 mm/two months and thereafter decreased rapidly. The D/R ratio was highest in *Az. indica* var. *siamensis*, followed by *E. camaldulensis* and then *Ac. auriculiformis* (for rainfall more than 100 mm/two months). For rainfall around 50 mm/two months, the R/D ratio was highest for *E. camaldulensis* and smallest for *Az. indica* var. *siamensis*. From equations (1), (2) and (3), the approximate threshold amount of rainfall for diameter increment could be estimated as about 30 mm/two months for *Az. indica* var. *siamensis* and much less than 30 mm/two months for *E. camaldulensis* and *Ac. auriculiformis*.

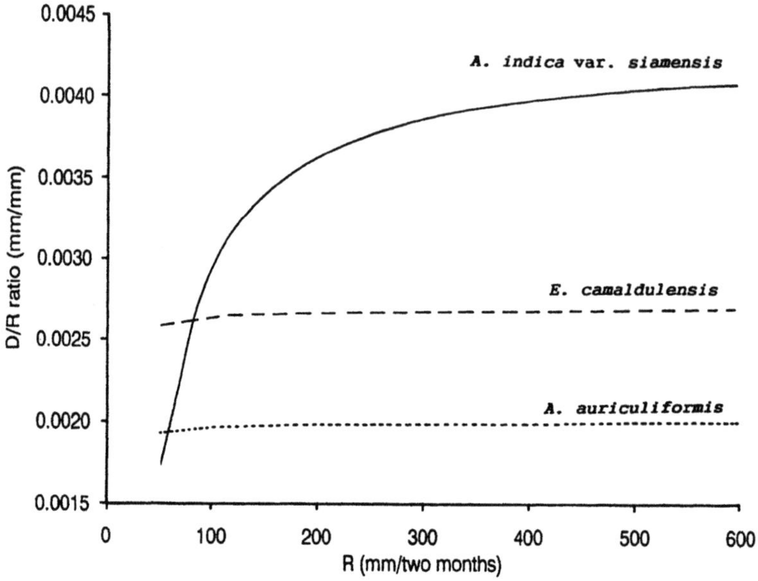

Fig. 7. Relationship of monthly diameter increment per unit of rainfall (D/R ratio) to amount of rainfall during the period of preceding and current month (R).

Annual variation in stem volume increment

The annual stem volume increment in 1988 and 1989 were also examined for stands of the three species (see Fig. 8). The increment was much smaller in 1989 than in 1988 for *Az. indica* var. *siamensis* (by 64%) and *E. camaldulensis* (by 60%), while the increment of *Ac. auriculiformis* in 1989 was smaller than that in 1988 by only 9%. Annual rainfall was much less in 1989 (841 mm) than in 1988 (1322 mm) as well, suggesting there was some relation between annual growth of the three species studied and annual rainfall.

Discussion

Based on monthly average value of rainfall and diameter increment in the humid tropics with no dry season, Whitmore (1974) reported that there was no correlation between rainfall and stem diameter increment. On the other hand, in the dry tropics with a distinct dry season, it was suggested that there was some relationship between tree growth and rainfall seasonality (e.g. Lowe 1968, Daubenmire 1972). How rainfall influences the seasonal growth of trees in tropical monsoon areas, however, has not yet been known quantitatively.

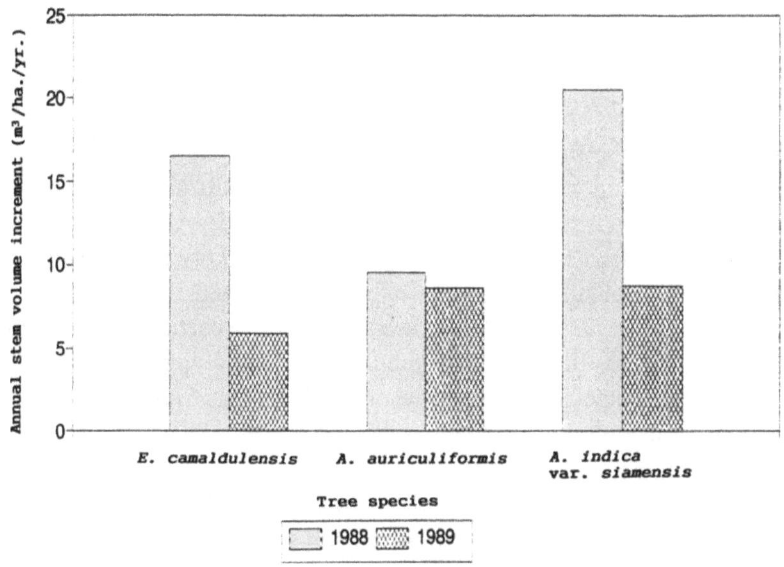

Fig. 8. Annual stem volume increment in 1988 and 1989.

In Ratchaburi with a distinct dry season, the seasonal diameter increment was found to respond to the seasonal rainfall pattern in the *E. camaldulensis*, *Ac. auriculiformis* and *Az. indica* var. *siamensis* stands. These species commenced diameter growth at the beginning of the rainy season, continued their growth, depending on the amount of rainfall, throughout the rainy season, and ceased in the middle of the dry season. This pattern of diameter increment seems to be almost the same not only among individual trees of each species but also among the three species. In the humid tropics of Indonesia without a pronounced dry season, it has been reported that individual trees of the same species show different seasonal diameter increment (Okimori 1987). In seasonal forests in Nigeria, with a less severe dry season than at Ratchaburi, the seasonal cambial activity of individual tree within same species varied, although each species had its own seasonal growth pattern (Hopkins 1962, 1970). At Ratchaburi which is under the drier climate in Thailand, severe seasonal water stress through the fluctuation of rainfall may strongly affect seasonal diameter increment, possibly resulting in the synchronization of diameter increment among individual trees within the same species, and even among different species with different growth characteristics (see Growth and biomass) and different origins. The soil at the research site is poor in terms of water holding capacity (Yoshioka, pers. comm.). This may intensify the seasonal water stress on tree growth at Ratchaburi.

At the beginning of the rainy season, the three species at Ratchaburi increased their diameter immediately after the first rain. Such an immediate response to improved soil moisture was also observed in other studies. The growth of *Pterocarpus angolensis* D.C. at Kabungu, Tanganyika began almost exactly at the time the first rains fell (Boaler 1963). In controlled cacao trees, watering under strong water stress brought about an abrupt increase in stem diameter increment in less than 24 hours (Alvim & Alvim 1978). These observations show that trees under water stress are likely to respond quickly to improved soil moisture conditions by increasing their diameter.

During the rainy season, a drop in rainfall for a certain period brought about a reduction in diameter increment in the three species studied. This suggests that these species studied are possibly on the fringe of water stress at Ratchaburi throughout the year.

The correlation study of diameter increment and monthly rainfall at Ratchaburi suggests that the rainfall of both the current and preceding months affect the monthly diameter increment of each species. The rainfall for more than the past two months by itself does not seem to influence strongly the monthly diameter increment. This implies that available moisture in the soil, stored at most for two months, is the main source of water for monthly diameter increment of the three species at Ratchaburi.

Judging from the monthly pattern of diameter increment, cessation of diameter growth for the three species is from February through April at Ratchaburi. In forestry, this means that silviculture operations related to tree growth, including thinning, pruning, etc., should be conducted in this period.

The rainfall threshold for cessation of tree growth in the tropics is said to be 50–60 mm/month (Kira 1983). The results obtained here show that the three species at Ratchaburi could still grow at a much lower level of rainfall than 50–60 mm/month, suggesting that these three species might be more drought resistant. Comparing the growth performance of the three species, *Az. indica* var. *siamensis* tended to increase its diameter more efficiently than the other two species at rainfall above 100 mm/two months. Below 200 mm/two months, the diameter increment for a given amount of rainfall decreased abruptly and became smallest at rainfall below 50 mm/two months among the species studied. The threshold amount of rainfall for cessation of growth was estimated to be larger in *Az. indica* var. *siamensis* than *E. camaldulensis* and *Ac. auriculiformis*. These mean that *A. indica* var. *siamensis* has greater ability than *E. camaldulensis* and *Ac. auriculiformis* to utilize soil moisture to increase diameter when the soil is not dry; in dry conditions, its capacity to endure drought is assumed to be lower than that of the other two species. The diameter increment for a given amount of rainfall did not vary much for *E. camaldulensis* and *Ac. auriculiformis* over the whole range of rainfall. This indicates that the ability to utilize soil moisture to increase diameter could not change much under natural rainfall levels. This trait, along with the low rainfall threshold for cessation of growth in these species, could be responsible for the strong drought resistance of these species. Comparing growth performance, one finds the diameter increment for a given amount of rainfall to be higher in *E. camaldulensis* than in *Ac. auriculiformis*. This may partly explain the fact that *E. camaldulensis* has been planted more widely than *Ac. auriculiformis* in arid and denuded areas in Thailand.

The close correlation between seasonal diameter increment and seasonal rainfall suggests that annual growth might be affected by the amount of annual rainfall as well. At Ratchaburi, stand stem volume increment of *E. camaldulensis* and *Az. indica* var. *siamensis* fluctuated greatly between 1988 and 1989 in accordance with fluctuation of annual rainfall. Therefore, the fluctuation in stem volume increment was possibly caused by fluctuation in the annual rainfall of the corresponding years. In tropical monsoon areas in the Philippines, some fast-growing tree species (*Gmelina arborea*, *Anthocephlus chinensis*, *Ac. auriculiformis*) showed annual variation in stem volume increment of two to six times for three years (Kamo et al. 1989). Murphy (1986) indicated that annual variation both in amount of rainfall and in intensity and timing of wet and dry periods were often considerable in tropical areas. The coefficient of variation for annual rainfall is often in the range of 30% in tropical areas, while that is approximately 15% in humid temperate regions. It could be inferred, therefore, that large annual variations of growth of some tree species in tropical monsoon areas might possibly have relation to the fluctuation in annual rainfall. This suggests that for making production assessments of some tree species in tropical monsoon areas, the large variation of growth between years possibly due to the variation of annual rainfall should be taken into consideration.

Acknowledgements

This study was conducted as a research activity in the Research and Training in Re-afforestation Project in Thailand, financed by the Japan International Cooperation Agency. We are grateful to Dr. R. Kato, chief advisor of the Research and Training in Re-afforestation Project, Mr. B. Buntawee, Mr. V. Sumantakul, and Mr. P. Kuerkool of Royal Forest Department of Thailand for their kind support; Mr. C. Viriyabunchae for his assistance in field work; Dr. S. Oohata of Kyoto University for his advice in preparing the manuscript; and Dr. P. Sahunalu of Kasetsart University for his reading of the manuscript and valuable comments.

References

Alvim, P. de T. 1964. Tree growth periodicity in tropical climates. In: Zimmermann, M. H. (ed.) Formation of Wood in Forest Trees, pp. 479–495. Academic Press, New York.

Alvim, P. de T. & Alvim, R. 1978. Relation of climate to growth periodicity in tropical trees. In: Tomlinson, P. B. & Zimmermann, M. H. (eds.) Tropical Trees as Living Systems, pp. 445–464. Cambridge University Press, London.

Boaler, S. B. 1963. The annual cycle of stem girth increment in trees of *Pterocarpus angolensis* D.C., At Kabungu, Tanganyika. S. B. Boaler. Commonw. For. Rev. 42:232–236.

Daubenmire, R. 1972. Phenology and other characteristics of tropical semi-deciduous forest in north-western Costa Rica. J. Ecol. 60:147–170.

Dawkins, H. C. 1956. Rapid detection of aberrant girth increment. Emp. For. Rev. 35:448–454.

Hazlett, D. L. 1987. Seasonal cambial activity for *Pentaclethra*, *Goelthalsia*, and *Carapa* trees in a Costa Rican lowland forest. Biotropica 19:357–360.

Hopkins, B. 1962. Vegetation of the Olokemeji Forest Reserve, Nigeria. 1. General features and the research sites. J. Ecol. 50:559–598.

Hopkins, B. 1970. Vegetation of the Olokemeji Forest Reserve, Nigeria. 6. The plants on the forest site with special reference to their seasonal growth. J. Ecol. 58:765–793.

Kamo, K., Ishizuka, M. & Ohsmi, K. 1989. (Growth analysis of fast-growing species and Benguet pine stand. Study on regeneration and afforestation in tropical forest in the Philippines.) Bulletin of Tropical Agriculture Research Center 65:65–79 (in Japanese).

Kamo, K. 1990. (Productivity of fast-growing species in Thailand.) Tropical Forestry 19:26–34 (in Japanese).

Kira, T. 1983. (Ecology of Tropical Forest.) Jinbunshoin, Tokyo. 251 pp. (in Japanese).

Liming, F. G. 1957. Homemade dendrometers. J. Forestry 55:575–577.

Lowe, R. G. 1968. Periodicity of tropical rain forest tree *Triplochiton scleroxylon* K. schum. Commonw. For. Rev. 47:150–163.

Murphy, P. G. 1986. Ecology of tropical dry forest Ann. Rev. Ecol. Syst. 17:67–88.

Okimori, Y. 1987. (Phenological study on growth of tropical trees.) Ph. D. Thesis, Kyoto University (in Japanese).

Viriyabuncha, C. & Pitpreecha, K. 1989. Climatic data [1978/1987] and diagram of stations under silvicultural research sub-division, Division of silviculture, Royal forest department. Technical paper no. 23, Forest ecology section, Silvicultural research sub-division, Division of silviculture, Royal forest department, Thailand.

Whitmore, T. C. 1974. Change with time and the role of cyclones in tropical rain forest on Kolombangara, Solomon Islands. Institute paper no. 46, Commonwealth Forestry Institute University of Oxford.

30. Stratification of several peat swamp forest types in Brunei Darussalam

ISAMU YAMADA
Center for Southeast Asian Studies, Kyoto University, Kyoto 606, Japan

Abstract

The stratification of tropical peat swamp forests in Brunei was studied in several different forest types by measuring tree height, height to the lowest living branch, and crown area. The crown mass of a plot was stratified into one meter units, and the total crown area at each meter in height was calculated. The most distinct two-layer strata were found in Alan bunga and Alan batu forest types, both of which are dominated by *Shorea albida*. With increasing accumulation of peat, the stratification becomes simpler. Secondary forests damaged by insects or fire showed a single canopy layer at a lower height. The successional direction of peat swamp forest is discussed, from mixed peat swamp to Alan bunga. Alan batu is suggested to be a later stage of Alan bunga.

Key words: Brunei, peat swamp forest, *Shorea albida*, stratification, succession

Introduction

Stratification, one of the major structural features of tropical rainforest, has been studied by many tropical ecologists since Davis & Richards (1933) presented their profile diagram. Approaches such as the production-ecological approach (Monsi & Saeki 1953, Kira 1978) and the numerical approach (Hozumi 1975, Yamakura 1987, 1988, Yamakura et al. 1989) have also provided insights into forest structure. Bourgeron (1983) and Brunig (1983) have reviewed many attempts to explain stratification, and Pompa et al. (1988) have recently studied Mexican tropical rainforests and compared several patterns of stratification. Yamada (1990) reported the changing pattern

E.O. Box et al. (eds.), Vegetation Science in Forestry, 529–544.

of vertical stratification along an altitudinal gradient in the montane forest of Java.

Tropical rainforests in Brunei have been studied by Browne (1955), Anderson (1961b), Ashton (1964) and Brunig (1974). Although it is a small country, Brunei contains various types of forest, including mangrove, peat swamp, freshwater swamp, kerangas, mixed dipterocarp forest and montane forest. In the peat swamp forest, six forest seres have been studied by Anderson (1961b) and, recently, by Yamada (1987). To commemorate the independence of Brunei Darussalam from the United Kingdom, the first international cooperation project was launched in 1984 to establish a forest research center. I was dispatched to Brunei from 1984 to 1986, where I carried out a basic forest ecological study of various forest types. This report discusses the stratification of peat swamp forest based on the method used in Yamada (1990).

Study sites and methods of study

The study site was chosen to cover several peat swamp forest types in the Badas peat swamp area along the western part of Sungai Liang in the Belait river basin. Major peat swamp forests such as mixed peat swamp forest, Alan batu forest, Alan bunga forest, and Padang alan forest occur in succession from the sea to the interior. Damaged secondary forest types also occur, such as Kapor paya and Ulat bulu. The depth of peat varies from several centimeters at the shallowest point to over 10 m at the center of the swamp. Plots measuring 50 m × 50 m to 50 m × 100 m were selected at typical points, after surveying the surrounding forest. All the trees over 10 cm DBH were identified, and DBH, total height, and height to the lowest living branch were measured. The locations of trees and a crown projection diagram were drawn. This report will deal only with stratification, not with species composition.

Result and discussion

Alan batu forest

Alan batu is a type of *Shorea albida* (Alan) forest containing the tallest and oldest Alan, which can be found at the periphery of peat swamps. The highest trees reach 60 m and have high buttresses and widespread root systems under the buttresses, which constitute a higher pseudo-floor above the real peat-forest floor. The trees are old, many have dropped branches, and many dead stems are scattered on the floor.

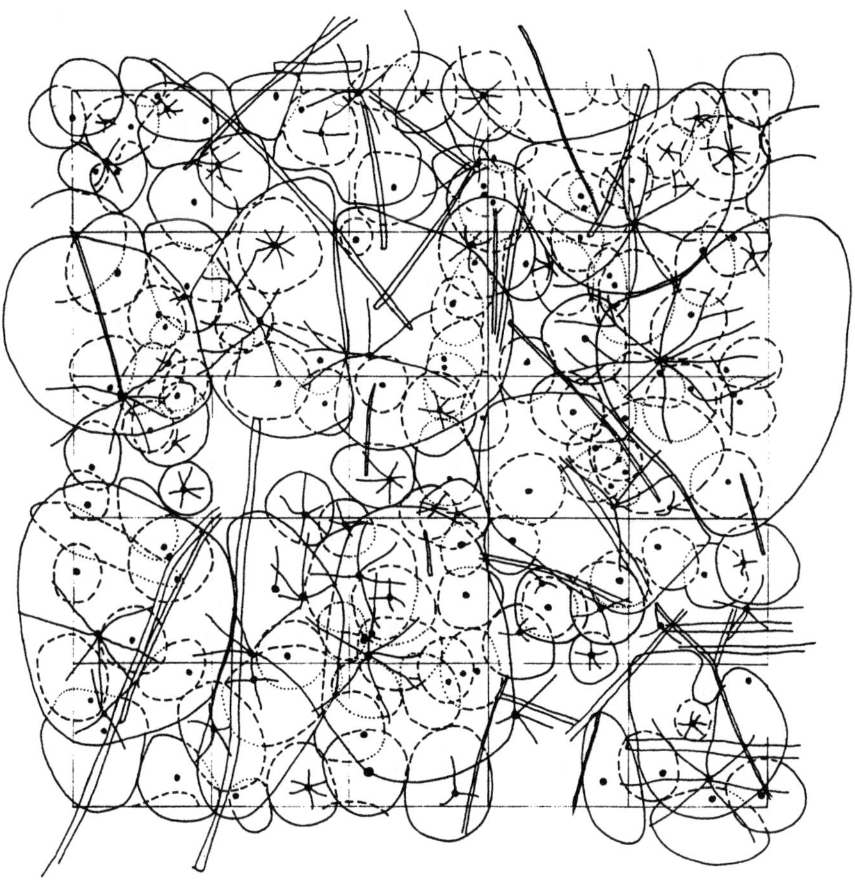

Fig. 1. Crown projection of Alan batu forest. Note the large crowns of *Shorea albida* and many fallen trees.

As shown in Fig. 1, the distribution of species is not complex, and the large Alan have a very wide crown area. Compared with Alan, other species are smaller. The very typical stratification is indicated in Fig. 2. The layer of Alan is far higher than that of other species. A layer just below Alan occurs is almost empty. The layer below that is very dense, and part of it comes up to occupy the empty layer above.

2. *Alan bunga forest*

Further inland from the Alan batu forest, the pure Alan forest type which Anderson (1961b) called Alan consociation begins to appear. This is the richest *Shorea albida* forest, of which more than 80% of the volume is occupied by this one species. The Alan measures nearly 50 m in height and

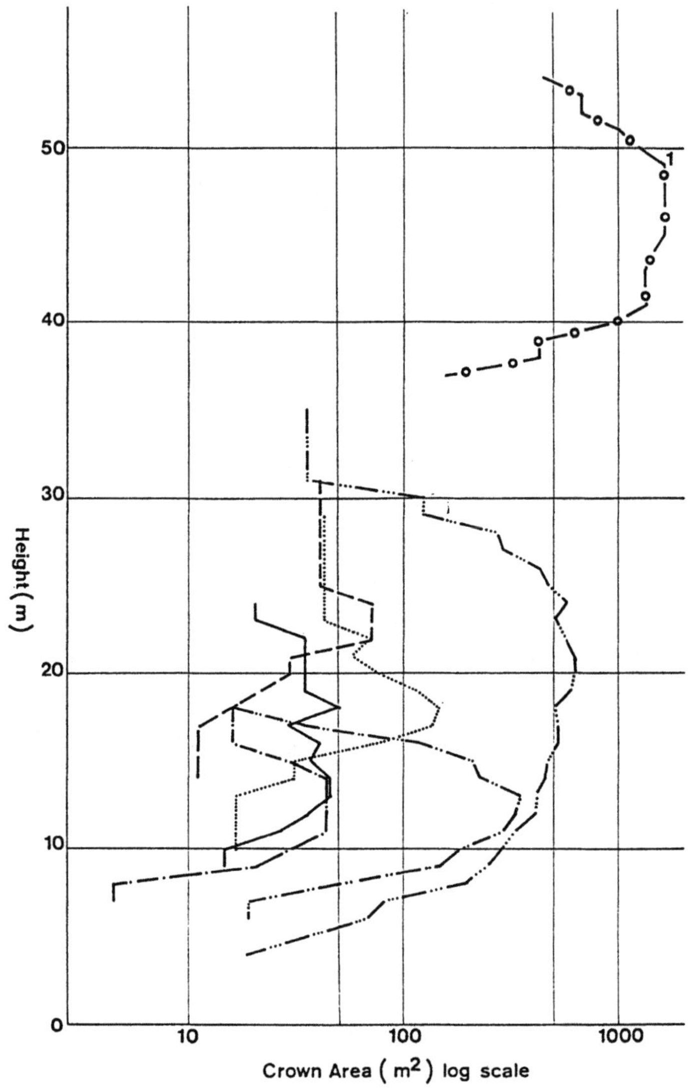

Fig. 2. Stratification of Alan batu forest. *Shorea albida* (1) is dominant in the uppermost layer. There is a wide empty zone between *Shorea albida* and other species.

80–90 cm in DBH, and forms a mono-layer canopy. From the air, the greyish-green crown can be seen to be gathered together to make a very flat canopy. From the forest floor, individual crowns can be seen to be slightly separated from those of neighboring trees.

As shown in Fig. 3, the distribution of Alan with crowns of similar size is very distinctive, and almost no other trees are seen in this layer. Smaller trees can be found much lower. The forest floor is covered by an impen-

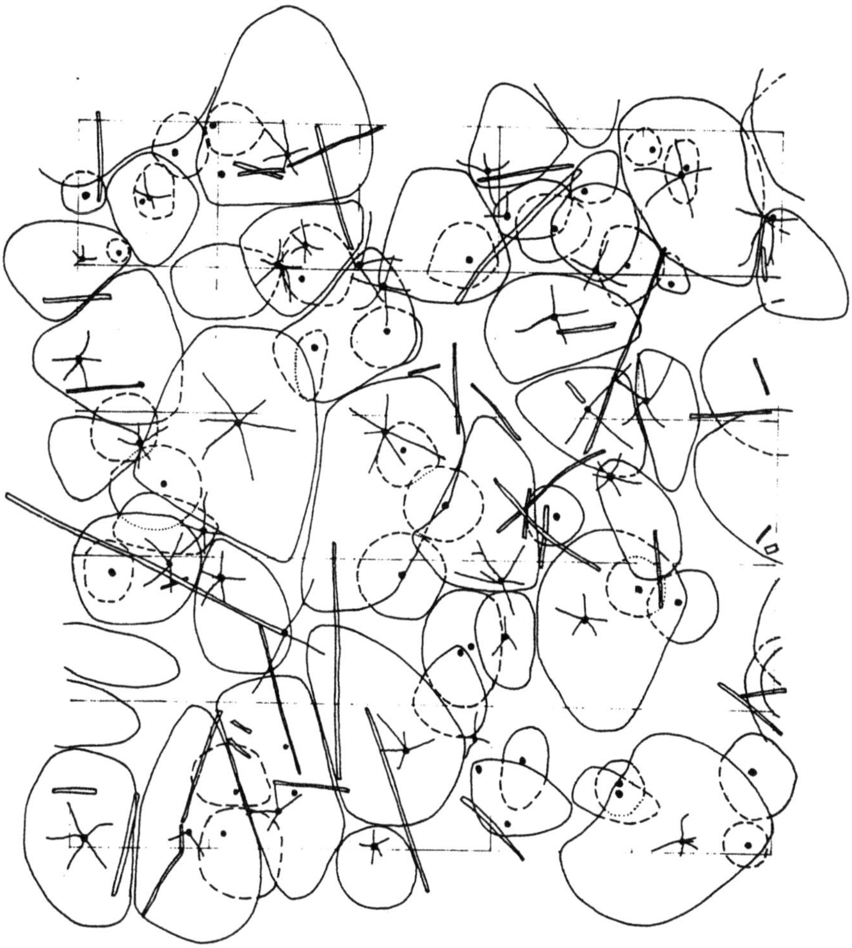

Fig. 3. Crown projection of Alan bunga forest. The similar size of the crowns is notable. *Shorea albida* is dominant, with 80% of the total number of trees.

etrable, vigorous growth of pandanus (*Pandanus andersonii*). The stratification consists of two distinct layers (Fig. 4): the upper layer is Alan only, the lower consists of several species. Compared with the Alan batu forest, the total height is lower but the total crown area in the highest layer is much larger.

3. *Padang alan forest*

Inland from the Alan bunga forest, the tree height gradually diminishes and the species composition becomes simpler. In the middle of the Badas peat

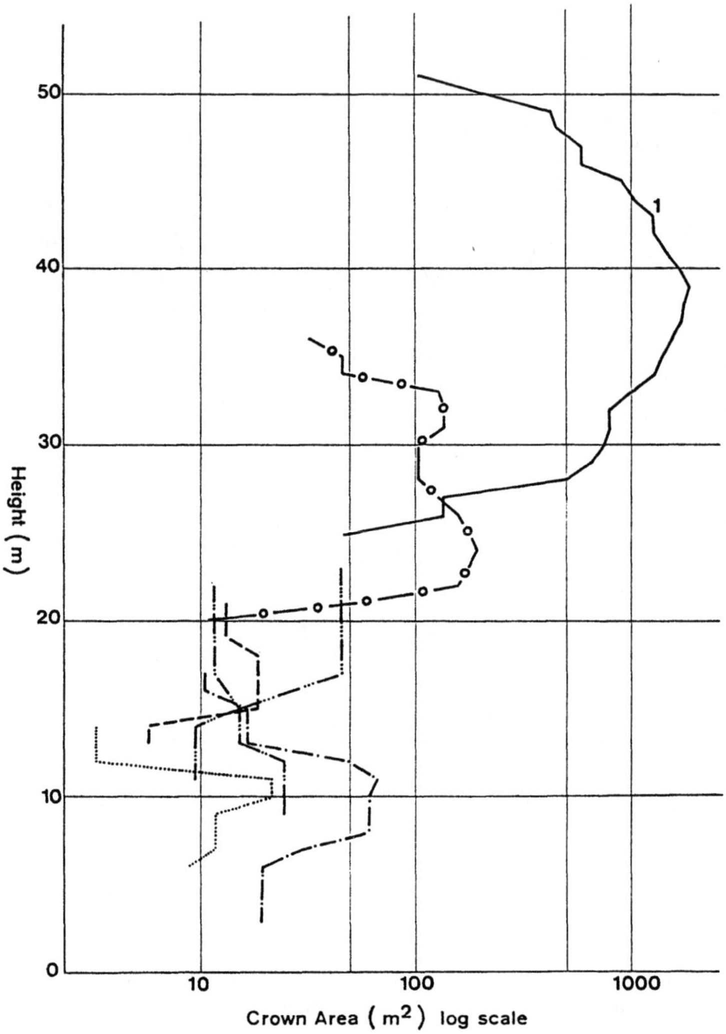

Fig. 4. Stratification of Alan bunga forest. Crown area of *Shorea albida* (1) is very large compared with that of other species.

swamp area, the peat is more than 10 m deep and the tree size is smaller. This type of forest is called Padang alan forest.

Figure 5 shows the pattern of tree distribution, in which the crown area is evenly distributed regardless of species. All the species look very similar in the field. The stratification differs from the former two types (Fig. 6). The total height is lower and the layer of Alan virtually overlaps with that of other species.

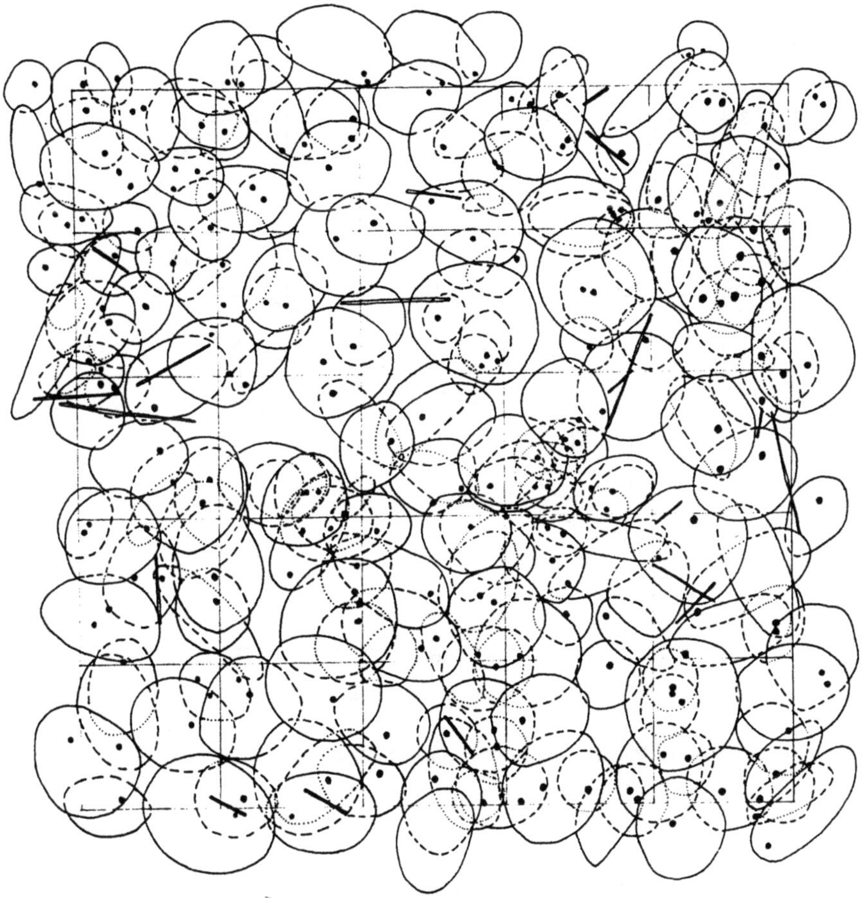

Fig. 5. Crown projection of Padang alan forest. Crown area is evenly distributed regardless of species.

4. *Mixed peat swamp forest*

Mixed peat swamp forest occurs along the Belait river, where the peat is shallowest. It lies about 500 m from the bank of the Belait river and about 3 km inland from the river mouth. This forest has a similar physiognomy to that of the mixed dipterocarp forest. Many species of various size are found, and the forest floor is not difficult to walk through. The largest tree is Kapor paya (*Dryobalanops rappa*), which attains 50 m in height and has large buttresses. The crown projection diagram (Fig. 7) for this type shows the distribution of trees of various sizes, with none of them particularly dominant.

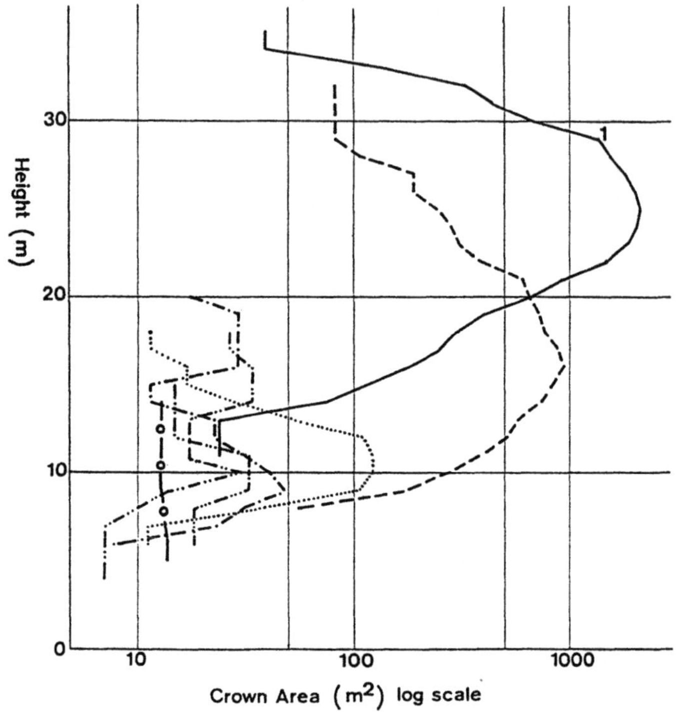

Fig. 6. Stratification of Padang alan forest. *Shorea albida* (1) and other species overlap. The height is become much lower.

The stratification (Fig. 8) is also less distinct than that of other types, and the gradual change of species by layer is notable.

5. *Ulat bulu forest*

Ulat bulu is the name of an insect larva which attacks *Shorea albida* and kills it on a large scale. Areas of such damage can be seen from the air, appearing like craters in the otherwise continuous canopy of Alan forest (Anderson 1961a, Lee 1979).

Damaged forest is found along the oil survey path into padang alan forest. As shown in Fig. 9, the tree size is very similar in each species, of which none is dominant. The forest floor is very dark, and no small seedlings were found. At the first stage of damage, climbers grow vigorously; after 20 years the forest becomes a typical secondary forest with many trees of the same size. The stratification pattern shown in Fig. 10 is very simple, with many species together in the same layer. This stratification type is quite different from the Alan forest types.

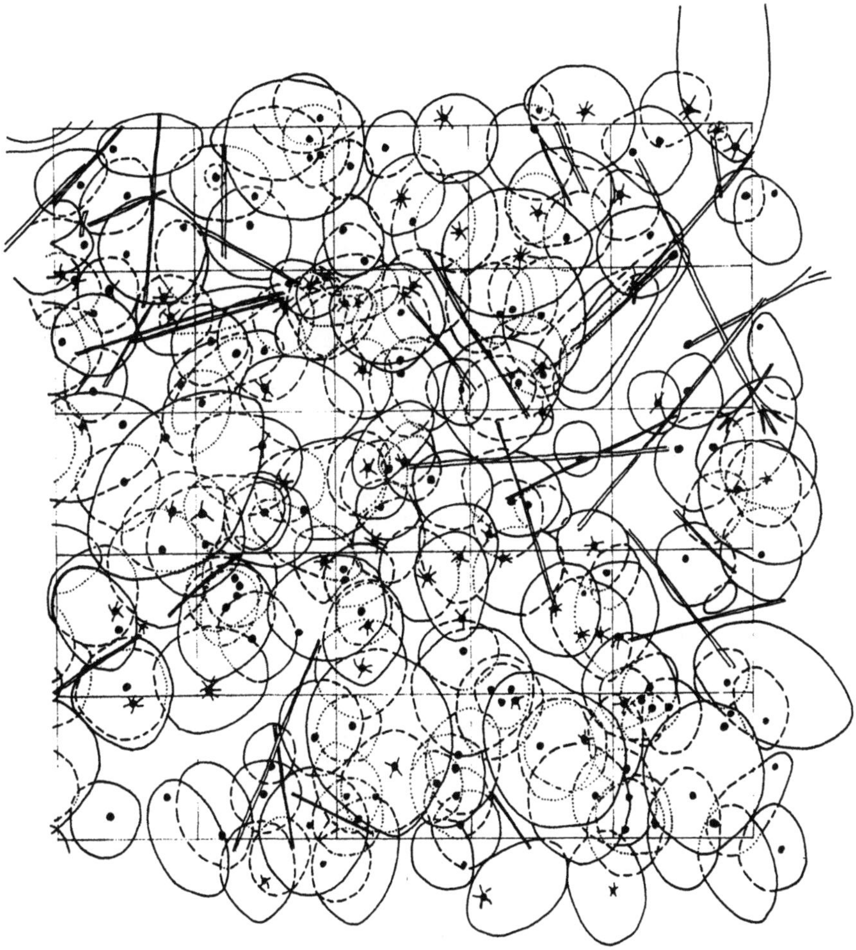

Fig. 7. Crown projection of mixed peat swamp forest. The variety in size crown is characteristic of this forest type.

6. *Kapor paya forest*

At the fringe of the peat swamp, near the sandy seashore, Kapor paya (*Dryobalanops rappa*) has become dominant after the frequent burning of fields. The trees are smaller than other peat swamp forest trees. The conical crown of this species is a typical sight from the road near Seria.

As seen is the crown projection diagram for this type (Fig. 11), it is notable that many trees have similar sized crowns. The stratification is also very simple, as seen in Fig. 12, which shows one large layer of Kapor paya.

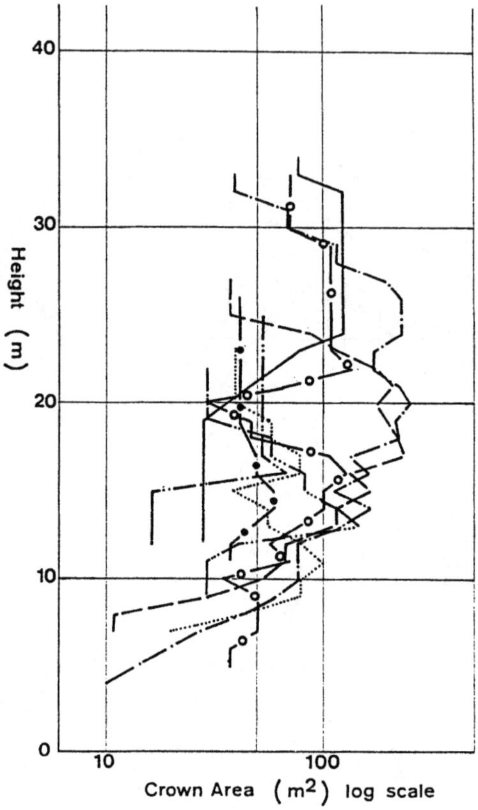

Fig. 8. Stratification of mixed peat swamp forest. No dominant species are seen, and the crowns of many trees occur together in the same layer.

Discussion

Having described the stratification of each forest type in Badas peat swamp, I would like to consider the direction of succession of peat swamp forest types.

The first stage must be the mixed peat swamp forest. At the site which had shallower peat and was near the river bank, the conditions allow many different species to grow. The physiognomy is similar to that of mixed dipterocarp forest, which means that various species can grow harmoniously under the more-or-less favorable conditions. With increasing peat accumulation, conditions become more acidic and fewer species can survive. Alan (*Shorea albida*) must be the most adaptable species, and pure Alan forest (Alan bunga) is seen in those areas of peat which are evidently the most favorable for Alan.

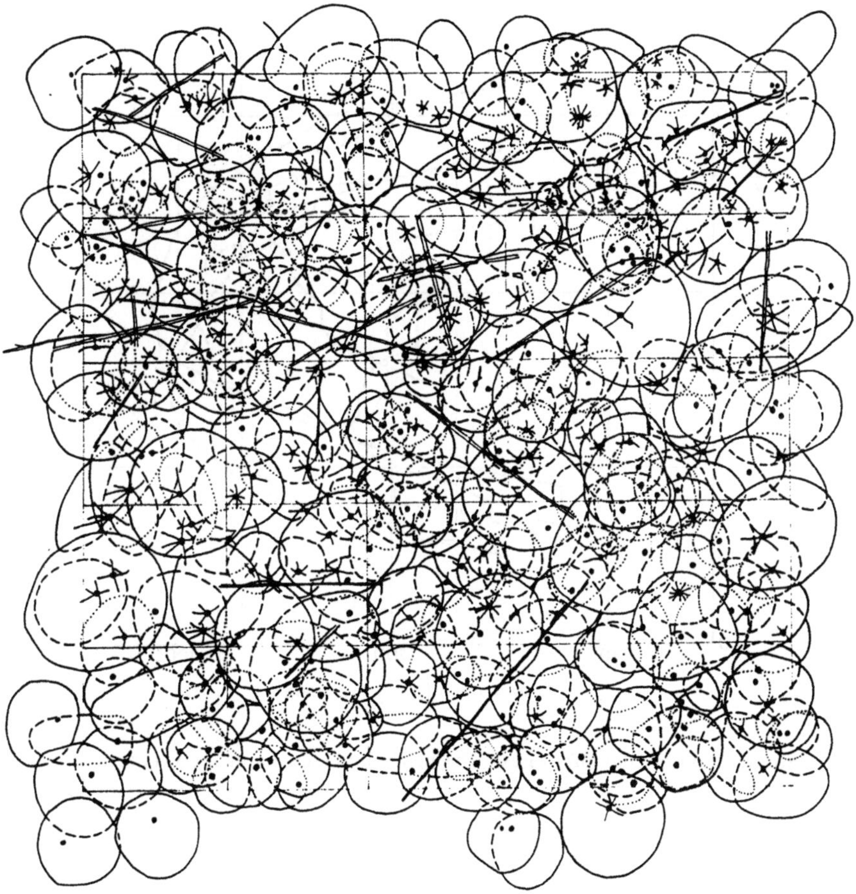

Fig. 9. Crown projection of Ulat bulu forest. Trees of similar crown area are seen all over the plot. The density is very high.

Alan has several characteristics which might cause the establishment of pure forest. First, Alan grows very fast on open sites such as logging areas and roadsides. It has a high light requirement in order to grow to climax forest, and seedlings are seldom found under the canopy. Second, Alan has large buttresses with spreading root systems. This morphological character allows it to occupy a wide space on the peat forest floor. Fast and wide penetration of a root system into the peat is the most important factor in the dominance of a peat swamp area. The third characteristic is gregarious flowering. Species belonging to the family Dipterocarpaceae and some other tropical families generally flower gregariously at irregular intervals of 2–7 years. The cause of this phenomenon is still under discussion, but it is agreed at least that certain climatic fluctuations trigger the flowering activity. Alan

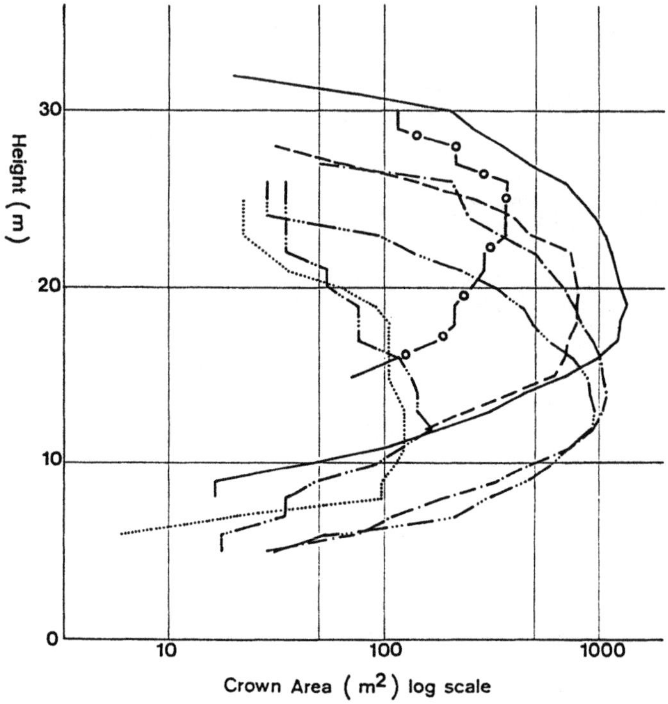

Fig. 10. Stratification of Ulat bulu forest. The height and crown area are very similar in many species.

is said to flower every 25 years, but general observations made by loggers in the forests suggest that the interval may actually be shorter. Fortunately, at the beginning of 1986, gregarious flowering occurred in the Badas area, where 90% of Alan batu and 80% of Alan bunga trees flowered and most subsequently fruited. Together with the root system, gregarious flowering is a key factor in the establishment of a pure forest in peat swamp. Two possibilities can be advanced concerning the relationship between Alan bunga and Alan batu. One is that Alan bunga was established first and, with time, as trees fell one by one, the remaining Alan grew very large, forming the Alan batu forest seen today. The other is that Alan batu originally involved only a small number of trees, which grew large with time. In this case, no successional relationship exists between Alan bunga and Alan batu.

The former possibility seems more realistic. From an initial mixed peat swamp forest, many species would not survive as the peat became deeper, and eventually the most adaptable species, Alan, would establish a pure stand, mainly because of the three characters discussed. This pure forest might remain a long time, but eventually individual trees would die, and the

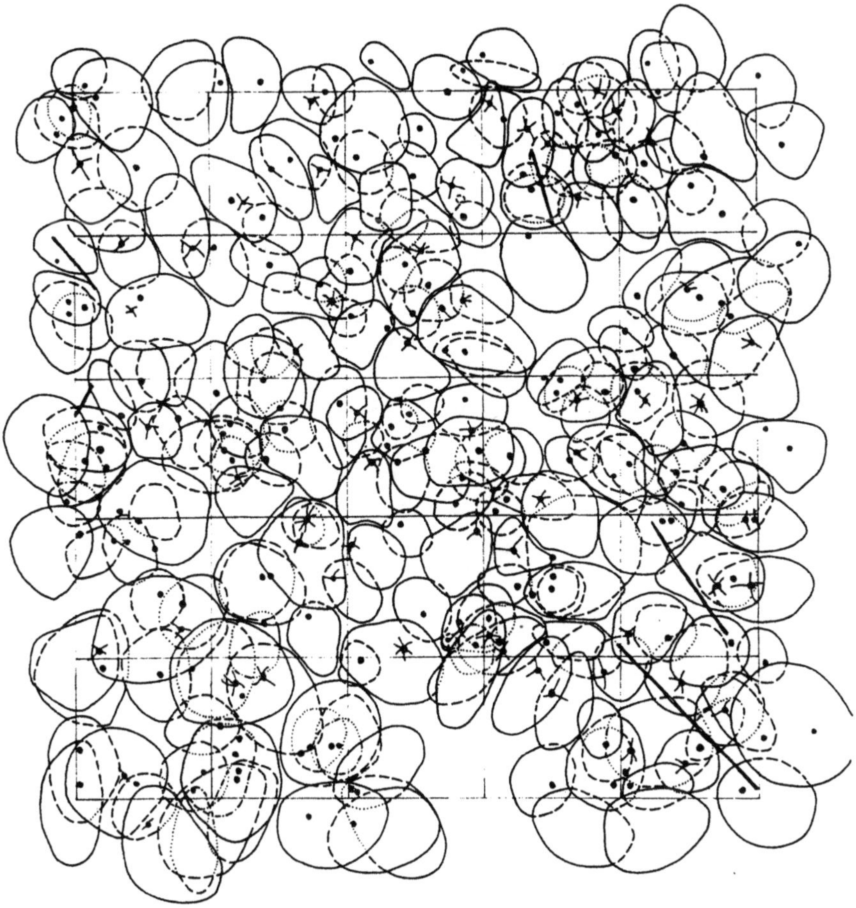

Fig. 11. Crown projection of Kapor paya forest. The similar crown size is notable, many of which are crowns of *Dryobalanops rappa*.

remaining trees would grow as large as those now found in the Alan batu forest. With further accumulation of peat, conditions would become less suitable for Alan, its size would decrease, and other species able to survive the conditions would emerge. The result is Padang alan. This is the direction of the primary succession. From time to time, disturbance by fire or pests would occur. Ulat bulu is a form of insect attack in which Alan is killed on a large scale. Secondary succession then starts, with many climbers followed by the pioneer secondary species. If gregarious flowering occurs in timely fashion, Alan seedlings from the surrounding Alan forest can survive. Much of the secondary stand is homogeneous, and many trees are of such similar age that they are easily damaged by the disturbance. The fortuitous timing of the disturbance and gregarious flowering may be the key to the re-estab-

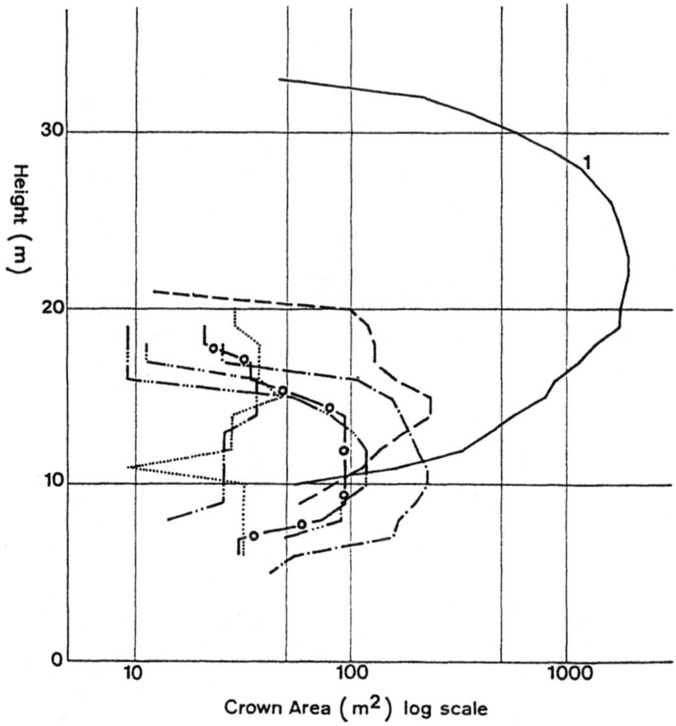

Fig. 12. Stratification of Kapor paya forest. The crown area of *Dryobalanops rappa* (1) is far larger and higher than that of other species.

lishment of Alan, but so far no good timing evidence has been found. In the mixed peat swamp forest, fire kills many species and only Kapor paya can survive in frequently burned areas. All these trends are summarized in Fig 13.

From the standpoint of stratification, the most complex stratification is the mixed peat swamp; the Alan forest is much simpler. In the Alan forest, Alan bunga is the most distinctive type, which forms a pure Alan layer at around 40 to 50 m high. Almost no tree layers were found below the Alan layer. With the accumulation of peat, the alan layer becomes higher and the number of Alan trees decreased, so many other species could invade the empty spaces. On sites with deeper peat, the trees become smaller and the stratification becomes simpler. Disturbed forest results in simple monolayer stratification.

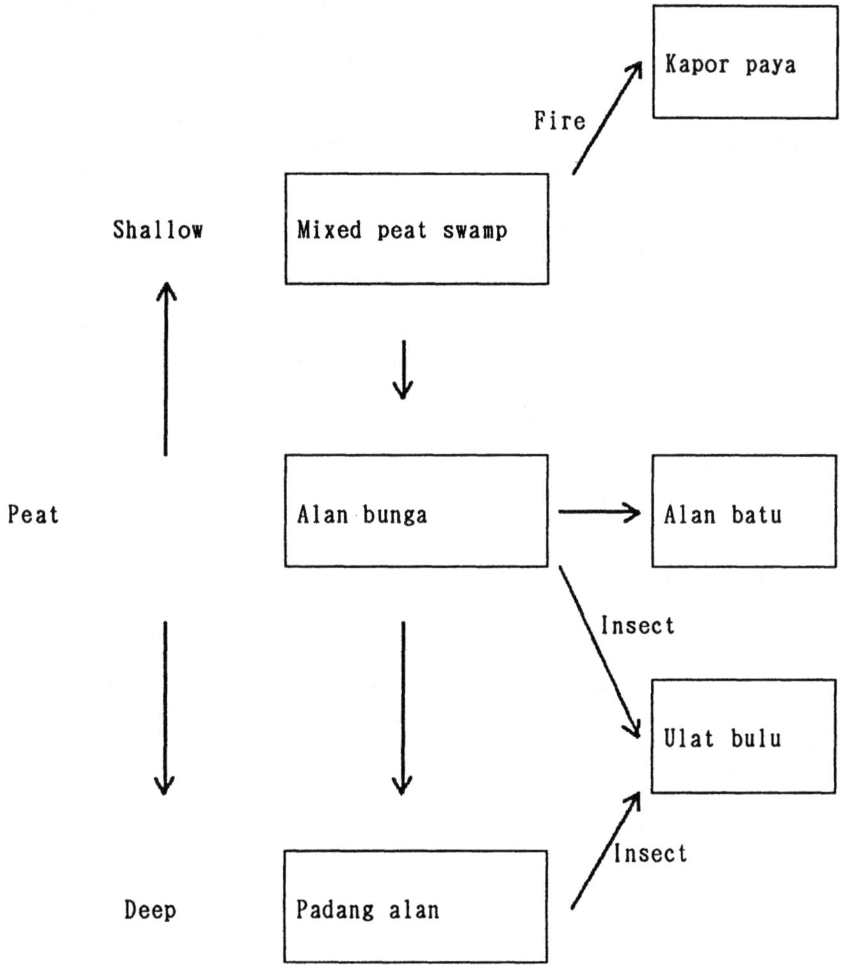

Fig. 13. Direction of succession of peat swamp forest in Badas, Brunei.

Acknowledgement

I would like to thank Hj. Mo. Yassin B. Amp. Salleh, Director of Forestry Department for his administrative cooperation and Mr. Niga bin Nangkat for his great effort in the field survey.

References

Anderson, J. A. R. 1961a. The destruction of *Shorea albida* forest by an unidentified insect. Emp. For. Rev. 40:19–29.

Anderson, J. A. R. 1961b. The ecology and forest types of the peat swamp forests of Sarawak and Brunei in relation to their silviculture. Ph.D. thesis. Edinburgh University.

Ashton, P. S. 1964. Ecological studies in the mixed dipterocarp forests of Brunei State. Oxford. For. Mem. 25.

Bourgeron, P. S. 1983. Spatial aspects of vegetation structure. In: Golley, F. B. (ed.), Tropical Rain Forest Ecosystems, pp. 29–45. Elsevier, Amsterdam.

Browne, F. G. 1955. Forest trees of Sarawak and Brunei and their products. Government Printer, Sarawak, Kuching.

Brünig, E. F. 1974. Ecological studies in the Kerangas forests of Sarawak and Brunei. Government Printer, Kuching. 237 pp.

Brünig, E. F. 1983. Vegetation structure and growth. In: Golley, F.B. (ed.), Tropical Rain Forest Ecosystems, pp. 49–75. Elsevier, Amsterdam.

Davis, T. A. W. & Richards, P. W. 1933. The vegetation of Moraballi Creek, British Guiana: an ecological study of a limited area of tropical rain forest. Part I. J. Ecol. 21:350–384.

Hozumi, K. 1975. Studies on the frequency distribution of the weight of individual trees in a forest stand. V. the M-w diagram for various types of forest stands. Jap. J. Ecol. 2:123–131.

Kira, T. 1978. Community architecture and organic matter dynamics in tropical lowland forests of Southeast Asia, with special reference to Pasoh Forest, West Malaysia. In: Tomlinson, P. B. & Zimmermann, M. H. (eds.), Tropical trees as living systems, pp. 561–590. Cambridge University Press, Cambridge.

Lee. H. S. 1979. Natural regeneration and reforestation in the peat swamp forests of Sarawak. Tropical Agriculture Research series 12:51–60. Tropical Agriculture Research Center. Ibaragi, Japan.

Monsi, M. & Saeki, T. 1953. Uber den Lichtfactor in den Pflanzengesellschaften und seine Bedeutung für die Stoffproduktion. Jap. J. Bot. 14:22–52.

Pompa, J., Bongers, F. & Meave del Castillo, J. 1988. Patterns in the vertical structure of the tropical lowland rain forest of Los Tuxtlas, Mexico. Vegetatio 74:84–91.

Yamada, I. 1987. Report on the forest research in Negara Brunei Darussalam form 1984 through 1986. Japan International Cooperation Agency.

Yamada, I. 1990. The changing pattern of vertical stratification along an altitudinal gradient of the forests of Mt. Pangrango, West Java. In: Baas, P. et al. (eds.), The Plant Diversity of Malesia, pp. 177–191. Kluwer Academic Publishers, the Netherlands.

Yamakura, T. 1987. An empirical approach to the analysis of forest stratification. I. Proposed graphical method derived by using an empirical distribution function. Bot. Mag. Tokyo 100:109–128.

Yamakura, T. 1988. An empirical approach to the analysis of forest stratification. II. Quasi-1/2 power law of tree height in stratified forest communities. Bot. Mag. Tokyo 101:153–162

Yamakura, T. P. Sahunal & Karyono. 1989. A preliminary study of changes in forest stratification along environmental gradients in Southeast Asia. Ecological Research 4:199–116.

31. Dynamics of aboveground big woody organs in tropical rain forest ecosystems

TSUYOSHI YONEDA
Department of Arts and Sciences, Faculty of Education, Osaka Kyoiku University, Kashiwara 582, Japan

Abstract

The dynamics of aboveground big woody plant parts over 10 cm diameter was studied in a lowland rainforest at Pasoh, West Malaysia, and in a foothill rainforest at Pinang Pinang, West Sumatra. Standing biomass of live and dead wood and the rates of growth, death and decomposition were observed. Based on these values, the ecosystem dynamics was analyzed, both in terms of total values and for each diameter class.

The frequency distribution of stem and branch diameter was determined or estimated in these forests for three different classes:
1. live woody organs,
2. freshly supplied woody litter, and
3. older woody litter which had accumulated on the ground.

The estimated living woody biomass in these forests was nearly equal in all diameter ranges, suggesting some similarity in architecture of the forest stands. Mean longevity and mean turn-over time of woody material 20 cm in diameter were estimated at 80 years and 7 years in Pinang Pinang.

The annual decay rate of woody litter was estimated by a logistic equation for dry weight loss, considering effects of bulk density and diameter. Estimated rates in the two forest ecosystems were approximately equivalent to the death rates obtained from observations in five 2-ha plots for 2 years and two 1-ha plots for 4–5 years. Standing biomass of live and dead woody materials in a foothill rainforest were approximately equal to those in a lowland rainforest, whereas the turnover rates were only 70% of the rates in the latter forest.

Key words: big woody organs, decomposition rate, foothill dipterocarp forest, lowland rainforest, organic matter dynamics, turnover time

E.O. Box et al. (eds.), Vegetation Science in Forestry, 545–556.

Introduction

This paper summarizes the dynamics of aboveground big woody organs, stems and branches, in two tropical rainforest ecosystems. One is a lowland dipterocarp rain forest in the Pasoh Forest Reserve, West Malaysia, and the other is a foothill dipterocarp forest on the side of Mt. Gadut, Pinang Pinang, West Sumatra, Indonesia.

Synthetic ecosystem studies had been carried out in Pasoh Forest (2°58′ N, 102°18′ E) within the framework of the IBP (Kira 1987). The study described here was part of this program (*cf.* Yoneda et al. 1977). The altitude at Pasoh is about 130 m and the mean annual rainfall about 1900 mm. The study area was densely covered by undisturbed forests of the *Shorea-Dipterocarpus* type (Soepadmo 1973).

Pinang Pinang Forest is situated on a ridge extending to the west from the summit of Mt. Gadut (1895 m), about 17 km northeast by east of Padang (00°53′ S, 100°21′ E). The altitude ranges from 490 m to 520 m. Annual rainfall is extremely high, usually over 7000 mm. *Swintonia schwenchii* was dominant in the emergent tree layer. The species composition of the canopy layer is diverse and consists of many families. Intensive ecological and floristic studies had been carried out in this area as a Sumatra Nature Study (SNS) project from 1980 to 1986. The study reported here was also launched within this framework (Yoneda et al. 1990).

The aims of this study was to reveal the dynamics of large woody biomass (stems, branches, trunks, etc.) in tropical rain forest ecosystems, in terms of the stock of live and dead wood and the rates of growth, death and decomposition. I analyzed the ecosystem dynamics in terms of total values and for different diameter classes.

Standing amounts

Biomass of stems and branches

The total length of aboveground woody organs of a given diameter (d) in the forest was estimated using a method (Yoneda 1982) based on the pipe-model theory (Shinozaki et al. 1964). I defined the length based on the frequency distribution of living stems and branches of given diameters ($\Phi_1(d)$), with standard dimensions of [km cm^{-1} ha^{-1}].

The frequency distribution of wood diameters in the forests shows almost linearly decreasing wood frequency with increasing wood diameter when plotted on log-log coordinates, for stems and branches less than about 10 cm,

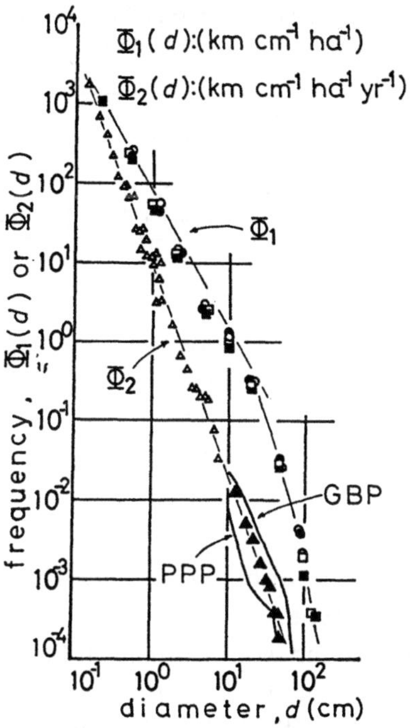

Fig. 1. Frequency distribution of diameters [$\Phi_1(d)$] of all stems and branches in various forest stands and that [$\Phi_2(d)$] of newly supplied woody litter. The straight line represents Eq. (2). [$\Phi_1(d)$]: ○, Pasoh; □, Pinang Pinang; ■, Gajabuih; ●, Airsirah. [$\Phi_2(d)$]: ▲, Pasoh; △, Minamata. Smooth curves: Pinang Pinang (PPP) and Gajabuih (GBP).

and a steeper decrease for larger diameters (Fig. 1). The diagram also shows the corresponding relationships in two additional tropical rainforests in the Mt. Gadut area. One is a foothill dipterocarp forest at Gajabuih, 1.5 km from Pinang Pinang. The stand structure and elevation are similar to those of Pinang Pinang. The other is a hill oak forest at Airsirah, 1130 m in altitude, about 5 km distance from Pinang Pinang. The main canopy was dominated by Fagaceae (*Quercus*, *Lithocarpus*, *Castanopsis*) and was closed at a height of 25–35 m, without emergent trees. The frequency of stem and branch diameters of different sizes in these four tropical rainforests was nearly equal for all diameter classes. The results suggest that these forests are similar in their architecture especially branching. Moreover, the frequency of stem and branch diameters of different sizes in mature evergreen and deciduous broad-leaved forests (Yoneda 1982) and young regenerating coppice forests (Yoneda unpubl.) in temperate western Japan was also close to that in mature tropical rainforsts, for diameters less than a few centimeters. This suggests that the surface structure of their closed canopies, with small

Table 1. Accumulation and rates of supply of big woody litter bigger than 10 cm diameter in a lowland rainforest at Pasoh and a foothill rain forest at Pinang Pinang. Amounts in Pasoh Forest are mean values of five 2-ha plots, including standing dead trees. ρ shows mean bulk density at each arbitrary grade of decay

Arbitrary grade of	Pasoh			Pinang Pinang	
decay	ρ [g cm^{-3}]	Accumulation [ton ha^{-1}]	Supply rate [ton ha^{-1} yr^{-1}]	ρ [g cm^{-3}]	Accumulation [ton ha^{-1}]
0	0.58	2.9	5.8	0.69	0.3
1	0.56	3.1	1.3	0.58	0.3
2	0.54	7.4	0.6	0.50	5.7
3	0.46	14.2	0.9	0.41	16.8
4	0.48	11.9	0.7	0.28	14.1
5	0.39	9.4	0.2	0.15	1.8
Total	–	49.0	9.3	–	39.0

branches and twigs densely distributed, is somewhat similar despite differences in maturity or forest type.

The biomass of aboveground woody organs in Pasoh Forest was estimated at 419 d.m. (dry matter) ton ha^{-1} by the dimensions analysis based on allometric relationships (Kira 1987). The distribution of biomass over stems and branches of different diameters in these four tropical rainforests, calculated with $\Phi_1(d)$, showed a convex curve with a maximum value at 20 cm diameter. The surface area of stems and branches, however, decreased exponentially with increasing diameter. The biomass of big woody organs larger than 10 cm diameter accounted for 88% of the total biomass in Pasoh Forest and was estimated at 369 d.m. ton ha^{-1}. The biomass in Pinang Pinang Forest was estimated at 408 d.m. ton ha^{-1}, based on the $\Phi_1(d)$ distribution (Yoneda et al. 1990).

Necromass of woody litter

All dead trunks and branches over 10 cm diameter were treated as big woody litter. The length and diameter at both side ends of woody litter on the ground were measured *in situ*. For standing dead trees only the *dbh* was measured, while tree height was estimated visually. Woody litter was classified into six arbitrary categories according to their degree of decay (Yoda & Kira 1982). A census of the necromass of standing dead trees was carried out only in Pasoh Forest.

The necromass of big woody litter in Pasoh Forest was estimated at 49 d.m. ton ha^{-1} on an average within the five 2-ha plots (Table 1). The variance (C.V. = 19%) among the plots was relatively small in comparison

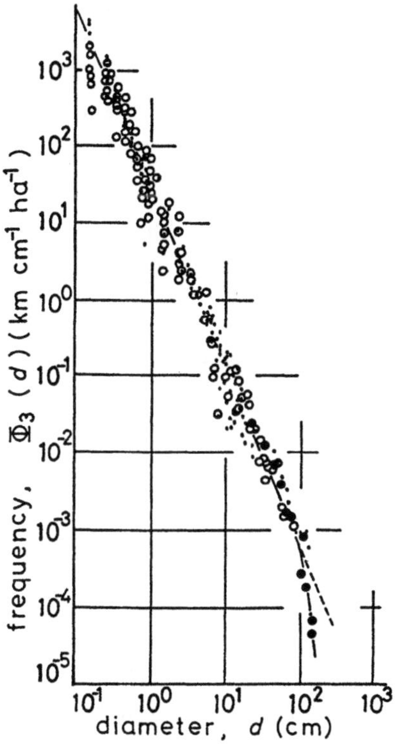

Fig. 2. Frequency distribution of diameter [$\Phi_3(d)$] in accumulated woody litter on the forest floor. The straight line represents Eq. (1). ●, Pason; ○, Pinang Pinang and Airsirah; •, Several forests in Japan.

with the variance (C.V. = 75%) of supply rate of big woody litter. The accumulation on the ground was 35 d.m. ton ha^{-1}, or 71% of the total amount of big woody litter, including standing dead trees. The necromass of big woody litter on the ground at Pinang Pinang was 39 d.m. ton ha^{-1} and accounted for 82% of the total litter on the ground including leaf and other kinds of litter (Yoneda et al. 1990).

Proctor et al. (1983) reported that the necromass of woody litter, including standing dead trees, at four lowland rainforests in Mt. Mulu National Park, Sarawak, ranged from 25 to 64 d.m. ton ha^{-1}. Klinge (1973) estimated the necromass on the ground at a rainforest at Manaus (Brazil) to be 26 d.m. ton ha^{-1}. Yoneda (1986) hypothesized that necromass tends to correspond to forest maturity, irrespective of forest type.

Figure 2 shows the frequency distribution ($\Phi_3(d)$) of dead woody organs on the ground in forests of the Pasoh and Mt. Gadut areas. The frequency of dead stems and branches smaller than 10 cm diameter at Mt. Gadut area is also shown. The relationships between $\Phi_3(d)$ and diameter (d) in the forests, on log-log coordinates, could be approximated by a single straight line with a negative slope, for diameters smaller than about 100 cm. The frequency at Pasoh, though, tended to be slightly higher than at Mt. Gadut. The following empirical equation, obtained in Japan (Yoneda 1982), could be applied to this relationship:

$$\Phi_3(d) = 31.6d^{-2.4} \qquad [\text{km cm}^{-1}\,\text{ha}^{-1}, \text{cm}] \qquad (1)$$

The results show that the maximum diameter of dead woody litter on the ground is a good measure for estimating the total necromass volume, while the weight has to be estimated based on the bulk density.

Turnover rate

Supply rate of woody litter

In Pasoh Forest, the supply rate of woody litter was directly observed for 1.8 years by estimating increments of big woody litter in the five 2-ha plots (Yoda & Kira 1982). Woody litter increments in a foothill dipterocarp forest were estimated from *dbh* census data, obtained six times in two 1-ha plot for 4 years at Pinang Pinang and 5 years at Gajabuih (Yoneda et al. 1990). The death rate of woody organs of a given diameter was estimated from observations of dead trees during these study periods, using the same method as for living woody biomass (as described above). I defined the death rate in a foothill rainforest as the supply rate of woody litter. Hence, the rate would be underestimated since branches of smaller diameter (falling without tree death) are not included.

The supply rate of woody litter in Pasoh Forest varied greatly over the five plots, with maxima ranging from 3.1 to 20.5 d.m. ton $\text{ha}^{-1}\,\text{yr}^{-1}$. The mean value was 9.3 d.m. ton $\text{ha}^{-1}\,\text{yr}^{-1}$. Standing dead trees accounted for 60% of woody litter supply. Of total amount, 62% belonged to decay grade 0, i.e. fresh woody litter with withered leaves still remaining on the branches (Table 1). In the case of accumulated dead wood, 72% belonged to more advanced decompositional stages, mainly grades 3–5 (boles without big branches). The woody litter supply rates at Pinang Pinang and Gajabuih were estimated at 3.8 d.m. ton $\text{ha}^{-1}\,\text{yr}^{-1}$ and 11.9 d.m. ton $\text{ha}^{-1}\,\text{yr}^{-1}$ respectively, and the mean rate was 7.9 d.m. ton $\text{ha}^{-1}\,\text{yr}^{-1}$.

The frequency distribution of diameters ($\Phi_2(d)$: [km $\text{cm}^{-1}\,\text{ha}^{-1}\,\text{yr}^{-1}$]) in newly fallen woody litter at Pasoh could be approximated by an empirial formula obtained from the relationship in an evergreen oak (*Castanopsis* and *Quercus*) forest at Minamata, Japan (Fig. 1). This equation is

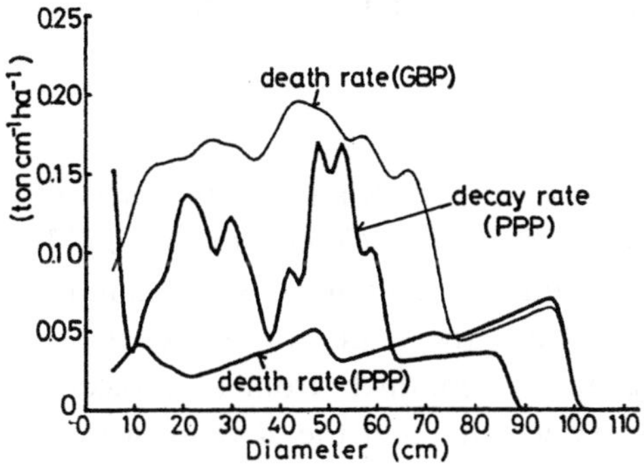

Fig. 3. Relationships between wood diameter and dry-weight loss rates of woody litter on the forest floor at Pinang Pinang (PPP) and death rates of woody organs at Pinang Pinang and Gajabuih (GBP). Decay rate was calculated using Eq. (5). Values were approximated by moving averages (from Yoneda et al. 1990).

$$\Phi_2(d) = 8.0d^{-2.6} \qquad [\mathrm{km\ cm^{-1}\ ha^{-1}\ yr^{-1}, cm}]. \tag{2}$$

The frequency of newly supplied woody litter at Pinang Pinang and Gajabuih were clustered around the regression line. The death rate estimated on dry weight basis, however, showed large difference between the two forests (Fig. 3).

Decay rate

I estimated the decay rate in the tropical rainforests from the rates of CO_2 evolution (R) of woody litter and dry weight loss (r) of one on the ground. Yoneda (1975) showed that relationships between both rates, R and r, and bulk density (ρ) could be approximated by straight lines with negative slopes. This means that the decay rate of woody litter gradually increases as decomposition progresses. The surface of woody litter is well protected by bark tissue with a high resistance to decay. The apparent volume could be assumed to be constant throughout the process of decay, so the following logistic equation was derived to represent the time course of dry weight (w) decrease based on the linear relationshps (Yoneda 1975, 1985, 1986):

$$w/w_0 = 1/\{ID + (1 - ID)\exp(\beta t)\} \tag{3}$$

where w_0 the initial dry weight, ID the index of durability, β the coefficient of weight decrease and t is time. The ID value of woody litter could be obtained approximately by the equation $ID = \rho_0 d^{0.2}$ (Yoneda 1985), increasing as diameter increases (ρ_0 = initial bulk density). Eq. (3) could express fairly well the observed time course of dry-weight decrease of woody litter on the ground in humid forests (Yoneda 1986). The coefficient β is calculated from the equation

$$\beta = 1/t \ln\{(w_0/w - ID)/(1 - ID)\} \tag{4}$$

A linear relationship between R and ρ was observed in Pasoh Forest and the β was estimated at 0.79 yr^{-1} (Yoneda et al. 1977). The rate of weight decrease r was measured with cylindrical wood samples of *Diospyros* left on the ground for three years; the value of β was estimated at 1.10 yr^{-1} by Eq. (4). Disc-shaped samples, 4 cm thick, from a Lauraceous tree showed a more rapid decrease in dry weight, and the β value was 2.20 yr^{-1} twice that of *Diospyros*. The differences among the estimated β values may depend on differences in assessment of the activities of termites. Abe (1979, 1980) revealed the important role of termites in decomposition of woody litter in Pasoh Forest. The observed CO_2 evolution rate of woody litter could hardly assess the activity of termites owing to escape often from wood samples before CO_2 measurement and should be regarded as the rate of mineralization carried out mainly by microbes. Dry-weight loss experiments assess the total weight loss caused not only by mineralization but also by humification, fragmentation, leaching and so on. Experiments using cylindrical samples may be expected to estimate the rate on the ground under natural conditions more faithfully than the method using the disc-shaped samples because cylinders are closer to the natural shape of dead stems and branches.

The β value at Pinang Pinang was estimated at 0.74 yr^{-1}, based on a dry-weight loss experiment using cylindrical wood samples at Airsirah. This value is only 67% of that (1.1 yr^{-1}) at Pasoh but 1.6–2.1 times that (0.36–0.46 yr^{-1}) at Minamata, Japan (Yoneda & Kirita 1978).

Collins (1983) revealed that the activity of termites is higher in dry regions than in a typical rainforest, being suppressed in rainforests under perhumid conditions. Yoneda (1991) assessed the activity of termites in several forests under different humidity conditions in Thailand and showed that activity in dry dipterocarp forests, a kind of savanna forest, was apparently higher than in the rainforests. Accurate assessment of termite activity would be extremely important for analysis of material cycling in these ecosystems.

Annual dry-weight loss of woody litter on the forest floor was estimated using the following equation, obtained by differentiating the logistic equation Eq. (3) with respect to time (Yoneda 1986):

$$Dw(d) = Aw(d) \times \beta \times (1 - ID) \tag{5}$$

where $Dw(d)$ and $Aw(d)$ denote annual rate of dry weight loss and the necromass of woody litter classified by decay class (Grades 0–5) and diameter class. Figure 3 showed the $Dw(d)$ distribution by diameter at Pinang Pinang; $Dw(d)$ varied greatly over different diameters but did not exceed the supply rates of woody litter at Pinang Pinang and Gajabuih. The total decay rates at Pasoh and Pinang Pinang were 10.3 d.m. ton $ha^{-1}yr^{-1}$ and 7.5 d.m. ton $ha^{-1} yr^{-1}$, respectively. For standing dead trees, β was defined as half of that of the woody litter on the ground, based on the results at the evergreen oak forest at Minamata, Japan. The rate was only 1.2 d.m. ton $ha^{-1} yr^{-1}$ at Pasoh because of the low β value, the high bulk density and the large diameters.

Dynamics of woody organs

Frequency dynamics

Assuming that a forest ecosystem is in dynamic equilibrium and the diameter of the woody organs remains constant throughout the process of death and decay, the turnover of woody materials in a forest ecosystem can be analyzed based on the frequency distribution of living organs, newly fallen dead wood and the accumulated woody litter on the ground, based on a turnover theory (Ogawa et al. 1961, Olson 1963). This can be done because the ratio of death rate ($\Phi_2(d)$) to $\Phi_1(d)$ is the average longevity of woody organs, and the ratio of $\Phi_2(d)$ to $\Phi_3(d)$ shows the mean turnover time of woody litter on the ground, at a given diameter.

I tried to estimate the mean longevity of woody organs in a foothill dipterocarp forest using calculated values of $\Phi_1(d)$ and Eq. (2), in order to show the average rate of death at Pinang Pinang and Gajabuih. The estimated longevity of stems and branches was 50, 80 and 100 years at 10, 20 and 30 cm diameter, respectively. It takes 30 years and 20 years on average to increase the diameter of 10 cm and 20 cm woody organs by 10 cm more, respectively. These estimates agree approximately with the estimates of 49 years and 28 years calculated using the annual *dbh* increment each diameter class, based on the *dbh* census for 4 years. Extrapolating the death rate of twigs, 0.2 cm in diameter, using Eq. (2), one gets a longevity estimate of 2.2 years. This suggests that these smaller woody organs are being renewed, almost as fast as leaves. The mean longevity of leaves, at Pasoh and Minamata, was estimated at 1.2 years (Kira 1987) and 1.9 years (Nishioka et al. 1978) respectively, based on the ratio of the annual leaf-fall rate to current leaf biomass.

The mean turnover time ($T_{32}(d)$) of woody litter on the ground is obtained from the ratio of Eqs. (1) and (2):

$$T_{32}(d) = 31.6d^{-2.4}/8.0d^{-2.6} = 4.0d^{0.2} \tag{6}$$

where t is in years and d is cm. The estimated time of woody litter was 6, 7 and 8 years at 10, 20 and 30 cm diameter, and was about 10% of the mean

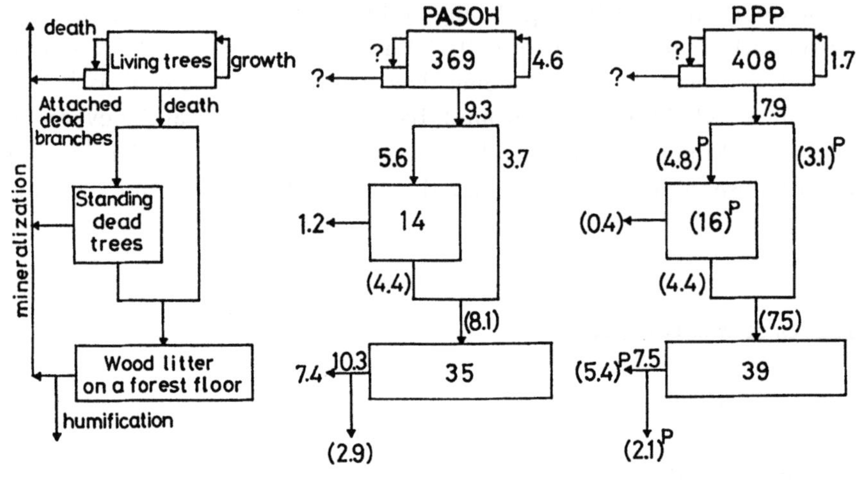

Fig. 4. Flow diagram of aboveground big woody organic matter over 10 cm diameter in a tropical lowland rainforest in West Malaysia (Pasoh) and in a foothill rainforest in West Sumatra (Pinang Pinang; PPP). Figures with and without parentheses are calculated based on dynamic equilibrium in each compartment and on observed values, respectively. Figures with a "p" superscript were estimated using the allocation ratio obtained at Pasoh. Dimensions of storages and fluxes of organic matter are dry matter ton ha^{-1} and dry matter ton ha^{-1} yr^{-1}, respectively. Modified from Yoneda et al. 1990.

longevity. Eq. (6) shows that the turnover time is proportinal to the 0.2 power of diameter. This result agrees with the fact that the index of durability (*ID*) is proportional to the 0.2 power of diameter under the same bulk density conditions.

Organic matter dynamics

Figure 4 shows a flow diagram of aboveground big woody material in a lowland tropical rainforest at Pasoh and a foothill tropical rainforest at Pinang Pinang. As already explained above, the decay rates were estimated by Eq. (5), based on the effects of bulk density and diameter. The growth rate and death rate in the foothill rainforest were averages of the values for Pinang Pinang and Gajabuih because of their large variance.

The annual increment of aboveground woody biomass at Pasoh was estimated at 5.2 d.m. ton ha^{-1} yr^{-1} (Kira 1987); the increment for big woody organs was estimated at 4.6 d.m. ton ha^{-1} yr^{-1} on an assumption that big woody organs accounts for the same proportion, 88%, to total increment as that in the biomass. The annual net production of aboveground big woody organs at Pasoh could then be estimated at 13.9 d.m. ton ha^{-1} yr^{-1} summing the death and growth rates. Sixty percent of annual dead biomass, death rate, was in standing dead trees, but the decay rate was only 1.2 d.m. ton - ha^{-1} yr^{-1}. Thus, 87% of the death rate, 3.7 + 4.4 = 8.1 d.m. ton ha^{-1} yr^{-1}, can be expected to reach the forest floor directly or indirectly through standing dead trees. The annual decay rate of accumulated wood litter on the ground was 10.3 d.m. ton ha^{-1} yr^{-1}, with 72% due to mineralization. If we could assume that each pool is in dynamic equilibrium, then the mean turnover time of big woody organs and big woody litter in this forest ecosystem would be 27 years and 5 years, respectively. The latter value is about ten times greater than that for fine litter, 0.37 years (Kira 1987).

In Pinang Pinang forest the size of each organic matter pool was similar to the corresponding value at Pasoh, but the turnover rate was only about 70% of that at Pasoh. The mean turnover time of big woody organs and woody litter on the ground was estimated at 43 years and 7 years, respectively.

The death rate of big woody organs at Pasoh was approximately equivalent to the fall rate for fine litter, accounting for 32% of the total annual net production, including underground parts (Kira 1987). This suggests that there is a large flux of big woody material at Pinang Pinang which plays an important role in material cycling, as at Pasoh (Yoneda et al. 1977, Kira 1987).

References

Abe, T. 1979. Studies on the distribution and ecological role of termites in a Lowland Rain Forest of West Malaysia. (2) Food and feeding habitats of termites in Pasoh Forest Reserve. Jpn. J. Ecol. 29:121–135.

Abe, T. 1980. Studies on the distribution and ecological role of termites in a Lowland Rain Forest of West Malaysia. (4) The role of termites in the process of wood decomposition in Pasoh Forest Reserve. Rev. Ecol. Biol. Sol. 17:23–40.

Collins, N. M. 1983. Termites population and their role in litter removal in Malaysian forests. In: Sutton, S. L., Whitomore, T. C. & Chdwick, A. C. (eds.), Tropical Rain Forest: Ecology and Management, pp. 311–326. Blackwell Scientific Publications, Oxford.

Kira, T. 1987. Primary production and carbon cycling in a premeval lowland rainforest of Peninsular Malaysia. In: Sethuraj, M. R. & Raghavenda, A.S. (eds.), Tree Crop Physiology, pp. 99–119. Elsevier, Amsterdam.

Klinge, H. 1973. Biomassa y materia orgánica del suelo en el ecosistema de la pluviselva centro-amazonica. Acta Cientifica Venezolana 24:174–181.

Nishioka, M. and Kirita, K. 1978. Litterfall. In. Kira, T. Ono, Y. & Hosokawa, T. (eds.), Primary Production in a Warm-temperature Evergreen Oak Forest of Japan. JIBP Synthesis,Vol. 18, pp. 231–238. University of Tokyo Press, Tokyo.

Ogawa, H., Yoda, K. & Kira, T. 1961. A preliminary survey on the vegetation of Thailand. Nature & Life in SE Asia 1:121–157.

Olson, J.S. 1963. Energy storage and the balance of producers and decomposers in ecological systems. Ecology 44:322–331.

Proctor, J., Anderson, J. M., Fogden, S. C. & Vallack, H.W. 1983. Ecological studies in four

contrasting lowland rain forests in Gunung Mulu National Park, Sarawak. II. Litterfall, litter standing crop and preliminary observations on hervivory. J. Ecol. 71:261–283.
Shinozaki, K., Yoda, K., Hozumi, K. & Kira, T. 1964. A quantitative analysis of plant form – the pipe model theory. I. & II. Jpn. J. Ecol. 14:97–105, 133–139.
Soepadmo, E. 1973. Progress report (1970–1972) on IBP-PT project at Pasoh, Negeri Sembilan, Malaysia. In: Mori, S. & Kira, T. (eds.), Proceedings of the East Asian Regional Seminar for the I.B.P., Kyoto pp. 29–39. Japanese National Committee for IBP, Kyoto.
Yoda, K. and T. Kira 1982. Accumulation of organic matter, carbon, nitrogen and other nutrient elements in the soil of a lowland rainforest at Pasoh, Peninsular Malaysia. Jpn. J. Ecol. 32: 275–291.
Yoneda, T. 1975. Studies on the rate of decay of wood litter on the forest floor. II. Dry weight loss and CO_2 evolution of decaying wood. Jpn. J. Ecol. 25:132–140.
Yoneda, T. 1982. Studies on the rate of decay of wood litter on the forest floor. IV. Turnover of live and dead woody organs in forest ecosystems – An assessment based on the changes in the frequency distribution of their diameter. Jpn. J. Ecol. 32:333–346.
Yoneda, T. 1985. Studies on the rate of decay of wood litter on the forest floor. V. Relation of wood diameter to the rates of dry weight loss and CO_2 evolution of wood litter in evergreen oak forests. Jpn. J. Ecol. 35:157–66.
Yoneda, T. 1986. Decomposition of wood litter in a forest. Jpn. J. Ecol. 36:117–129 (in Japanese with English summary).
Yoneda, T. & Kirita, H. 1978. Fall rate, accumulation and decomposition of wood litter. In: Kira, T., Ono, Y. & Hosokawa, T. (eds.), pp. Primary Production in a Warm-temperate Evergreen Oak Forest of Japan. JIBP Synthesis, Vol. 18, pp. 258–273. University of Tokyo Press, Tokyo.
Yonea, T., Sahunalu, P., Dhanmanonda, P., Tanpival, V. and B. Puriyakorn 1991. Decomposition of woody litter in Thailand. In: Yoda, K. & Sahunalu, P. (eds.), Improvement of Biological Productivity of Tropical Wastelands in Thailand, Report of the Thai-Japan Cooperative Research, pp. 161–183. Dept. Biology, Osaka City University, Osaka.
Yoneda, T., Tamin, R. & Ogino,K. 1990. Dynamics of aboveground big woody organs in a foothill depterocarp forest, West Sumatra, Indonesia. Ecol. Res. 5:111–130.
Yoneda, T., Yoda, K. & Kira, T. 1977. Accumulation and decomposition of big wood litter in Pasoh Forest, West Malaysia. Jpn. J. Ecol. 27:53–60.

32. Vegetation pattern, structure and dynamics in different types of humid tropical evergreen forests of south-east Asia

JUTTA POKER
Universität Hamburg, D-21027 Hamburg, Germany

Abstract

The floristic and architectural structures of forest stands in the tropical climatic zone, especially those in mature phases of dynamic development, are the result of interaction of an extremely great number of biotic, chemical and physical factors. Particularly important are the previous history of site and stand, and the bio-geographical and geological development of fauna, flora and site. Present site conditions and the inherent autecological properties of the tree species determine structural features which change along present edaphic and climatic gradients in a consistent manner. At forest community level on a certain soil-site unit, variation of stand structure in space and time is caused by regeneration and succession, resulting in a shifting mosaic of phasic units with common demographic features. These ecologically fairly uniform units are distinguished by the stage of development of the component trees. An approach toward identifying such units as patches in the mosaic is demonstrated using data sets of permanent sample plots in West Africa and South-East Asia.

Key words: developmental stages, shifting mosaic, spatial distribution, species groups, structure analysis

Introduction

Pattern and structures in space and time in tropical forest ecosystems have received increasingly more attention in recent decades in investigations of forest processes. Vegetation structural analysis refers to stand parameter distributions (diameter, height, basal area, volume), architectural features (canopy and crown construction, physiognomy and pattern of leaf age distri-

E.O. Box et al. (eds.), Vegetation Science in Forestry, 557–571.

bution, height/diameter ratios), pattern of spatial distribution and floristic composition at different scales (area sizes, species, species groups) (Rollet 1978).

Within the tropical climatic zone, general floristic and average architectural structure of mature stands change along edaphic and climatic gradients in a consistent manner. The smaller the forest area in question, that is at forest community level, the more structure analysis is impaired by

- the spatial distribution of developmental stages ('eco-units'): the variation of structural features is linked to the pattern of mosaic patches and is more or less random within small research areas.
- the variation and changes of tree architecture during a tree's life-span: height and crown dimensions even of trees of the same species and diameter class vary so enormously that average values are of limited relevance.

Hypothesis

The process of regeneration, growth, mortality, and decay produces a shifting mosaic-patch pattern with a wide range of unit sizes. It follows that trees within one patch should be of similar development stage; this was already suggested by Baur (1968) in his discussion of the mosaic theory of Aubréville (1938). Until a tree has reached its final stage within the canopy, height increment dominates total growth. If we assume that seedlings and saplings in a newly formed gap start growth at nearly the same time, it may be expected that they also will reach say about 30% of their final height at nearly the same time. In absolute values this may be about 5 m for an understorey species with 15 m maximum height and about 15 m for a neighbouring Dipterocarp with 45 m maximum height. The patch unit in the mosaic should consequently have a demographically common history, and therefore it may be hypothesized that it also possesses common ecological features. Consequently, the analysis of

- mosaic patch sizes and patch distribution
- composition of species and tree dimensions within a patch
- dynamics of and interactions within and between patches

at forest stand level would provide basic information to develop stand growth models for ecosystem research and management purposes. In this paper a method is presented to identify forest mosaic patches from standard vegetation enumeration data sets. As prerequisite, the component species of a stand have to be stratified into species groups to enable comparative analysis of the eco-units.

The research areas

The research areas are located on Hainan Island, South China, in the Philippines, and in Liberia, West Africa. In all three areas permanent sample plots have been established within the framework of long-term research projects. Table 1 summarizes the research areas. In South China, the research area is part of the CERP project, a Chinese-German cooperative ecological research project. Sample plots are located in central Hainan Island in an undisturbed montane rain-forest at an elevation of about 1050 to 1150 m. Up to now, 5.25 ha have been surveyed in a 20 ha research plot.

In the Philippines, the research area consists of 6 plots of one ha each. These were the basic research plots of the Timber Stand Improvement (TSI) project, a cooperation between the Philippine Forest Department and the German Technical Aid Agency. In formerly logged-over areas three plots are located in north and north west Luzon and west Mindanao, and three plots are located in east Mindanao. The sites are very different from each other. The plot altitudes range from 50 to 900 m.

In Liberia, the research areas were established in logged-over forest as part of a timber stand improvement project in cooperation with the German Technical Aid Agency. A continuous 12 ha plot is located in Cavalla Forest; 20 one-ha plots are located throughout a 900 ha area in Grebo National Forest.

The research plots

Within each research area two plots were selected which differed by site and vegetation structure. The distribution of trees per diameter and height class illustrate the differences (Table 2). In South China, the two plots are 100 m apart. The main site difference is a higher portion of steep slopes in plot 2B with a higher level of disturbance through tree fall and, consequently, fewer large trees.

In the Philippines, the site differences are mainly more precipitation and deeper soils in the plots in Eastern Mindanao, where trees are taller. The whole area, but especially the plots in west Luzon and west Mindanao, suffer from disturbance by typhoons.

In Liberia, the main difference between the Grebo and Cavalla sites is that precipitation is lower and past logging has been more intensive in the Cavalla area. The first affects tree height, the second the distribution of trees in the lower diameter classes.

In spite of the differences of sites, species and stand history the stand tables (Table 2) are rather similar; site differences and disturbance affect

Table 1. Location, physical characteristics, stand history, measurement dates and number of measured trees in the permanent sample plots (SP)

	S. China	Phillipines	Liberia
Plot size	SP 2A:1 × 2.75 ha SP 2B:1 × 2.5 ha	Luzon + W Mind.:3 × 1 ha East Mindanao:3 × 1 ha	Cavalla: 1 × 12 ha Grebo: 20 × 1 ha
Elevation	SP 2: about 1050 m 2P 2B: about 1150 m partly moderate, partly steep slopes	Luzon + W M:250–900 m E Mindanao: 50–450 m partly flat, partly moderate slopes	Lowland, party flat, partly gentle slopes
Precipitation	>2500 mm	Luzon + M: 2000–5000 mm E Mindanao: 2500–4000 mm	Cavalla: 1600 mm Grebo: 2000 mm
Stand history	No logging	Selective Logging Luzon + W M: 1960–1968 E Mindanao: 1956–1966	Selective Logging Cavalla: 1969 (int.) Grebo: 1960–1963
Number of measurements	1 (1988/89)	4–10 (L: 1982–1987) (M: 1976–1986)	2 (Cav: 1979/80 + 1986) (Gre: 1978 – 85 + 1988)
Number of trees (>10 cm dbh) no./ha	SP 2A: 1900 SP 2B: 2000 700–800	Luzon + W M: 2000 E Mindanao: 1800 600–700	Cavalla: 6600 Grebo: 7700 400–500

Table 2. Percentages of trees by diameter and height classes, and basal areas, in the research areas in South China, the Philippines and Liberia

Plot sizes (ha)	S China		N + NW Luzon	East	Liberia N.F.	
	2A	2B	W Mindano	Mindano	Cavalla	Grebo
DBH (cm)	2.75	2.5	3 × 1	3 × 1	12	20 × 1
10–19	57	56	75	66	65	58
20–39	29	32	18	23.5	25	29
40–59	9	8	5	6.5	6.5	8
60–79	3	2	1.3	1.5	2.5	3
>80	2.5	1	0.5	1.7	1.5	2.5
Basal area/ha (m^2)	53.5	46.3	26.8	32.7	28.3	28.2
Plot sizes (ha)	S China		N + NW Luzon	East	Liberia N.F.	
	2A	2B	W Mindano	Mindano	Cavalla	Grebo
Height (m)	2.75	2.5	3 × 1	3 × 1	12	20 × 1
0–9	13	16	19	9	11.5	14
10–19	73	76	70	70	59	53
20–29	14	6.5	8	15	21	21
30–39	0.1	1	2.5	4	6	9
>40	–	–	0.05	2.2	2	3.5

mainly the lowest and the highest diameter classes. Similar diameter-class distribution patterns are reported from the 4 ha undisturbed forest in La Selva, Costa Rica (Hartshorn 1978); the 8 ha formerly selectively logged forest in Omo Forest Reserve, Nigeria (Okali & Ola-Adams 1984); and in the 2 ha Kade Forest, Ghana (Swaine et al. 1984). The basal area that a site is able to carry and its distribution over diameter classes seem to follow a set pattern.

Data-base

The vegetation survey followed MAB/IUFRO standard methods (Synott 1979). In South China, about 4000 trees above 10 cm dbh have been enumerated. In the Philippines, each plot was enumerated yearly during observation periods between 4 and 10 years; in total about 10000 trees above 5 cm dbh and 3800 trees above 10 cm dbh were included. Tree height was measured twice; the observation period of height development is from 3 to 5 years. The results of a comparison of stand reactions to silvicultural improvement treatment are published elsewhere (Woell 1989, soon to be published in English). In Liberia, two vegetation surveys have been carried out; tree height was measured during the first survey. In Cavalla Forest, about 6000 trees above 10 cm dbh were included, with an observation period of about

6 years. In Grebo National Forest, about 8000 trees were enumerated; observation periods per one-ha plot vary between 3 and 10 years.

Methods

Classification of species groups

Maximum heights of species were determined by plotting measured top height of those trees per species which showed the largest diameter in the research areas (assuming that these trees will not grow much in height). To minimize errors due to chance or problems in species identification, these maximum dimensions were checked against records in the literature (Voorhoeve 1965, Salvosa 1963, Ashton 1988, van Steenis 1954, 1958). The species have been sorted by decreasing maximum heights. For example, in the Philippines (Fig. 1) the tallest species is Red Lauan (*Shorea negrosensis* Foxw.). In Luzon and West Mindanao, 294 trees occur and the heights of trees above 100 cm dbh are considered. In East Mindanao with a total of 314 Red Lauan trees, the diameter limit had to be reduced to 70 cm since large diameters are rare (the maximum diameter measured was about 2 m). Height ranges are 25–35 m and 25–55 m, respectively. The crowns of all these mature trees are fully exposed to light, in crown-position class 5 following the classification of Dawkins (1958). Regardless of absolute height, the trees are part of the top canopy. The same procedure was applied to the data from South China and Liberia. Groups of species were seperated according to similar maximum dimensions.

Analysis of species group architecture

The species groups' macro-morphological features were analysed in all research plots to see if this leads to ecologically meaningful results. The correlation coefficients between these architectural features were computed. In order to test for trends in species group architecture, weighted means per diameter and height class for all species groups were calculated.

Estimation of developmental stages

As a first step, only height was selected as a determining parameter. Tree height is influenced by micro-site conditions, light conditions, genetical and external factors. Thus, the success of a tree in reaching its species and site-specific potential height is not guaranteed. Consideration of diameter, crown-position class, height and competitive state is necessary to estimate the

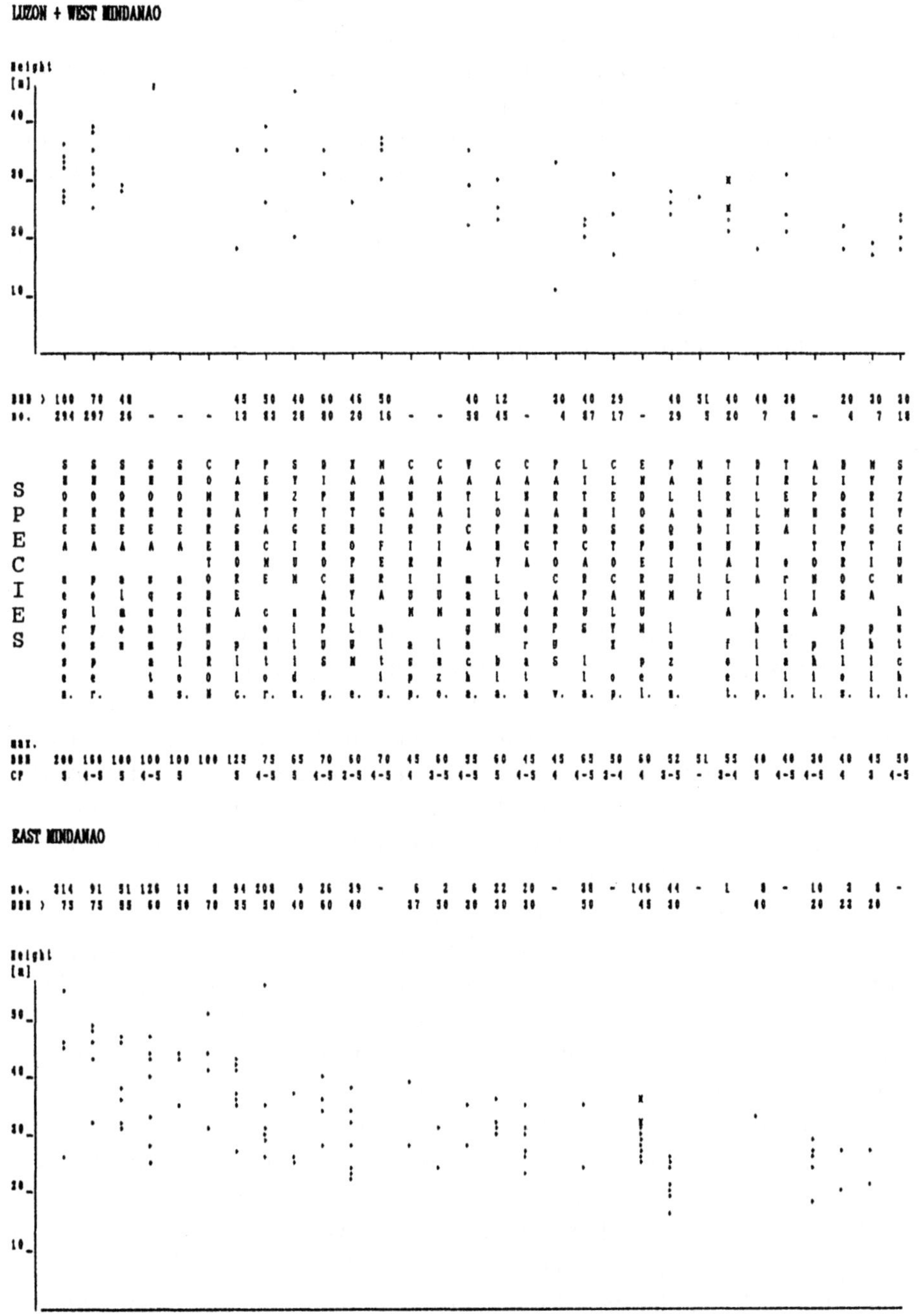

Fig. 1. Measured heights of those trees per species with the largest diameters in the research areas in the Philippines. The 30 tallest species are presented. Species names in the middle have to be read from top to bottom. no.: number of trees >10 cm dbh, DBH >: the height of all trees above the given diameter limit (cm) are presented, max. DBH: maximum measurement of DBH per species, CP: Crown position class(es) of the largest trees per species, Symbol ⊗: height of trees below the given diameter limit.

developmental stage of a tree more reliably. This complex proceeding can only be partly computerized. Thus, in the first test, shortcomings have been taken into account and a time-saving, fully computerized method was chosen: estimation of the developmental stage of each tree is based on a comparison of its actual with its potential height. The potential height is taken from the height limit which determines the species group to which the tree belongs. The developmental stages are expressed in percentage actual/potential height. The percentages were grouped into four classes in 25% steps to consider different growth paces of the trees. The locations of all trees in one sample plot were plotted using different symbols to indicate the developmental stages.

Results

Grouping of species

The species have been classified by height classes according to the average maximum height of mature trees. The height ranges of mature trees can only give a static view into a dynamic system. But in all cases, the height ranges of mature trees are very wide. One reason is that, once a tree has reached the canopy position which satisfies its needs, it accelerates diameter growth at the expense of height growth. Individuals of shade-tolerant species may grow extremely tall compared to the majority of mature trees of this species if they are influenced by tall upper-canopy species. Thus, a forest patch may be dominated by a mature tree of an upper-canopy species with e.g. 25 m height which is not influenced by competitors for light, while another patch within the same stand consists of competing trees about 40 m high where individuals of mid-canopy species have already reached 25 m height. Stratification of the canopy into physio-ecologically meaningful 'layers' is therefore more difficult than is sometimes suggested.

The stratification according to relative height performance of mature individuals of the species in one region seperates mature-upper canopy, mid-canopy and lower-canopy species as well as "true understorey" species. In each stratum, the range of maximum diameters was again classified to seperate species with larger or smaller diameters. The limits for height and diameter, which are the base for the grouping of the species, vary between the research areas (Table 3). These limits result from the comparison of mature dimensions of all species in one area and mark dimensions where significant interruptions within a more or less continuous decrease of mature tree sizes occur. All trees of all species above 10 cm dbh have been stratified into 7 species groups according to their mature stature; thus, the classification

Table 3. Species groups in the two separate plots in each research area, according to maturity status: U1 – trees reach the upper canopy, larger diameters, U2 – trees reach the upper canopy, smaller diameters, M1 – trees reach the mid-canopy, larger diameters, M2 – trees reach the mid-canopy, smaller diameters, L1 – trees reach the lower canopy, larger diameters, L2 – trees reach the lower canopy, smaller diameters, UN – understorey species. no. – number of individuals per area, % = percentages of total number of trees ≥10 cm dbh per area, Ht – height of mature trees (m), DBH – diameter of mature trees (cm), CP – crown position classes of mature trees

Species groups *South China* (A, B = plots 2A and 2B)

	U1	U2	M1	M2	L1	L2	UN
No. A	362	473	456	157	294	61	55
No. B	301	477	494	201	326	84	35
%A	18	26	25	9	16	3	3
%B	16	25	26	10	17	4	2
Ht	>25	>25	20–25	20–25	15–20	15–20	−10–15 m
DBH	>100	60–100	>=60	40–60	>=40	20–40	−15–20 cm

Species groups *Philippines* (L = Luzon + W Mindanao, M = E Mindanao)
(20% of all trees are not identified)

	U1	U2	M1	M2	L1	L2	UN
No. L	841	275	183	46	166	237	37
No. M	741	252	326	32	39	87	28
%L	47	15	10	2.5	9	13	2
%M	49	17	22	2	2.5	6	2
Ht	>35	>35	>25	>25	20–25	15–20	10–15 m
DBH	>100	−80	60–90	30–50	30–50	20–30	10–15 cm
CP	4 + 5	4 + 5	4–5	3–5	3–4 (5)	2–4	1–3 (4)

Species groups *Liberia* (G = Grebo, C = Cavalla)

	U1	U2	M1	M2	L1	L2	UN
No. G	1101	1433	851	1589	1342	1362	31
No. C	1280	1159	778	1302	607	909	5
%G	14	19	11	21	17	18	0.4
%C	21	19	13	22	10	15	0.1
Ht	>40	>40	30–40	25–35	25–30	−20	−10 m
DBH	>100	60–90	60–100	35–50	30–60	20–40	−15–20 cm
CP	4 + 5	3–5	2–4 (5)	2–4 (5)	2–4	1–3	1–2

does not consider the static situation of the vertical stand structure at the moment of the measurement, but the inherent growth dynamic of the species:

- species which constitute the upper canopy
  with larger diameters = U1
  with smaller diameters = U2

- species which grow up into the mid-canopy
  with larger diameters = M1
  with smaller diameters = M2
- species which grow up into the lower canopy
  with larger diameters = S1
  with smaller diameters = S2
- understorey species = UN

Table 3 presents the number of trees ≥10 cm dbh per species group in the three research areas. In the montane area of South China, the number of trees of species which reach the mid-canopy when mature is high in comparison to the other areas. In the Philippines, the Dipterocarps are the reason for the very high proportion of trees of species which reach the upper canopy when mature. In the transition forests of Liberia, the number of trees of species which reach the lower canopy is high.

Within a research area, there are regional differences in the number of trees per species group, mainly due to the abundance of individual species in a certain area. For example, in the Philippine plots, in East Mindanao, Gupas (*Endospermum peltatum* Merr.) (mid-canopy species with larger diameters) is rather abundant but does not occur in West Mindanao or Luzon. Additionally, Dipterocarps have a greater share in the total number of trees. In the plots in the western Philippines species which grow up into the lower canopy are more abundant, such as *Syzygium*, *Aglaia*, a small *Cryptocarya*, *Ficus* and *Memecylon*. This regional difference explains the larger number of trees in the lower height class in the sample plots in the western Philippines in Table 2.

In Liberia, in Grebo National Forest, the number of trees of species with larger diameters which reach the upper canopy when mature is reduced in favour of lower-canopy species with larger diameters (mainly more *Diospyros* and *Memecylon*). This can be contrasted with the Cavalla area, where mainly *Gilbertiodendron*, *Oldfieldia* and *Uapaca* make up the higher proportion of trees of species which reach the upper canopy when mature.

In the South China plots, the distribution of species groups is very similar though many more species occur in plot 2B.

Within a species group in a research area, there are regional shifts of single species. For example, in Liberia about 30% of all trees of upper-canopy species with smaller diameters in Grebo National Forest are *Strombosia glaucescens* Engl., in Cavalla *Chrysophyllum albidum* G. Don. Both species occur in both regions but are either rather abundant or play a minor role. Thus, the ecological function of a species group in a growth region is maintained, but on different sites, different species fulfill this function. Therefore, and because only few species are abundant, a single species is not appropriate to represent a species group.

Table 4. Correlation coefficients of two species groups in the South China research areas. dbh – diameter, ht – height, cl – crown length (depth), cd – crown diameter, h/d – height/diameter ratio.

	Plot 2A			Plot 2B		
	dbh	ht	cl	dbh	ht	cl
U1: reach the upper canopy when mature, larger diameters						
ht	0.72			0.61		
cl	0.67	0.75		0.58	0.79	
cd	0.76	0.74	0.71	0.62	0.68	0.71
h/d	−0.81	−0.5	−0.53	−0.72	−0.18	−0.24
S1: reach the lower canopy when mature, larger diameters						
ht	0.47			0.55		
cl	0.36	0.59		0.43	0.58	
cd	0.34	0.42	0.61	0.43	0.5	0.61
h/d	−0.66	0.16	−0.03	−0.72	0.08	−0.1

Architecture of species groups

The grouping of species according to mature-tree stature allows a more detailed analysis of the architectural structure of a forest stand as a whole. In general, correlation between architectural parameters such as diameter, height, crown dimensions, and height/diameter ratio seldom exceeds a correlation coefficient of .7, which means only 50% of the data fit a regression model. Highest correlation coefficients are obtained for the following relations: crown length as a function of height, crown diameter as a function of height or diameter, and height/diameter ratio as a function of diameter. The correlation coefficients decrease considerably from species which reach the upper canopy when mature to species which reach the lower canopy. They also vary between plots in the same growth region.

The South China data from undisturbed montane rainforest serve as an example. Correlation coefficients for upper and lower-canopy species with bigger diameters when mature are produced in Table 4.

Upper-canopy species with bigger diameters in plot 2A show the highest correlation coefficients of all species groups in all research areas. Still, the distributions of the actual architectural parameters over diameter and height classes vary enormously also in this species group.

Trends in species-group architecture per diameter and height class are: trees of true understorey species below the lower canopy differ in nearly all aspects from the other species groups. Within the same diameter or height class, macro-morphological features change gradually from species which reach the upper canopy when mature to species which reach the lower canopy. Mature crown dimensions of species of lower strata are generally smaller than crown dimensions of immature mid- or upper-canopy species. In South China, there is only a slight trend in height/diameter ratios, while in the Philippine and Liberian data there is a clear decrease of height/diameter ratio per diameter class from upper to lower-canopy species.

Distributions per height class are more indicative because correlation coefficients are higher. Generally, the diameters per height class decrease from species which reach the upper canopy to species which reach the lower canopy when mature. This is not so in the Philippine plots, where small, young dipterocarps are relatively abundant after logging. The bole length/tree height ratio is fairly similar throughout all species groups in all areas. But in the larger height classes, upper and mid-canopy species with larger diameters have a shorter bole length and correspondingly longer crowns than the species with smaller diameters of the same strata. Crown diameter in the lowest height class increases from upper to lower-canopy species. Mature trees have smaller crown diameters than immature trees of higher-canopy species. Species that grow to large diameters have broader crowns than species with smaller diameters. Average values of parameters expressing architectural features differ between species groups in accordance with ecophysiological concepts (e.g. development of crown biomass until maturity) and stand structure concepts (e.g. strategies to achieve canopy closure). Therefore, the species classification scheme is assumed to be suited to group species with common ecological function in whole-stand development.

Distribution of developmental stages

The spatial distribution of mature trees per species group is patchy. As an example, the distribution pattern of mature trees in the South China plot 2A is shown in Fig. 2.

In plot 2A, there seems to be a tendency that upper-canopy species with larger diameters are more grouped spatially than others. This pattern is formed mainly by Podocarpaceae (*Dacrydium pierrei* Hickel, *Podocarpus imbricatus* Bl., *Podocarpus neriifolius* D. Don), which seem to regenerate in small patches like the firs (*Abies*) in temperate mixed beech (*Fagus*) and oak (*Quercus*) forests.

The test of the hypothesis that the mosaic pattern of the rainforest can be identified by developmental stages of the individual trees, using only height as a determining parameter, is treated in Fig. 3. This shows, as an example, the spatial pattern of a one-ha sample plot in Grebo National Forest, Liberia, including all trees above 10 cm dbh. Spatial combination of trees of similar development stage was rather arbitrary, in order to get a visual impression. Due to the diameter limit of 10 cm dbh, the developmental

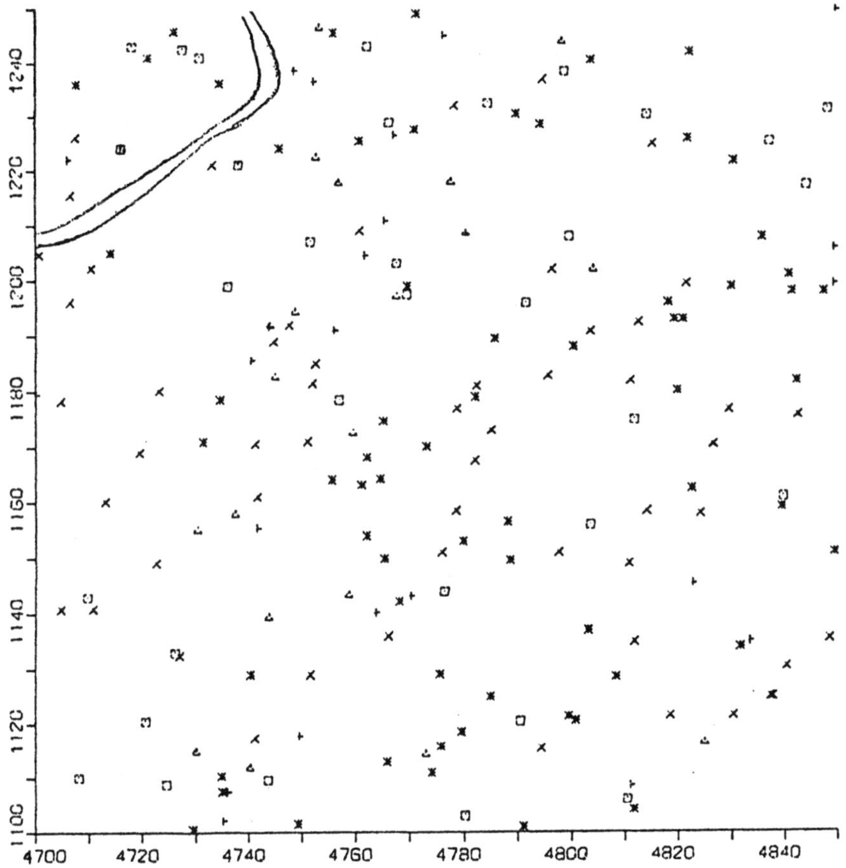

Fig. 2. Location of mature trees in the South China plot 2A, by species group. * species which reach the upper canopy when mature, larger diameters, × species which reach the upper canopy when mature, smaller diameters, □ species which reach the mid-canopy when mature, larger diameters, ▲ species which reach the mid-canopy when mature, larger diameters, + species which reach the lower canopy when mature. Upper left corner: small stream.

class 0–25% does not occur. Further, it is important to mention that species of early successional stages, like *Musanga* sp., are not present in the plot. The emerging pattern of this first trial reveals a mosaic structure which, in spite of all shortcomings of the method used (depending on height only, 'intuitive' spatial combination of trees to separate clusters), supports the hypothesis that developmental stages of single trees are spatially grouped.

Tentative analysis of the mosaic patches indicates that these patches have to be interpreted at different scales. Patches of higher developmental stage (e.g. 75–100%) sometimes include some single trees of lower stages (e.g.

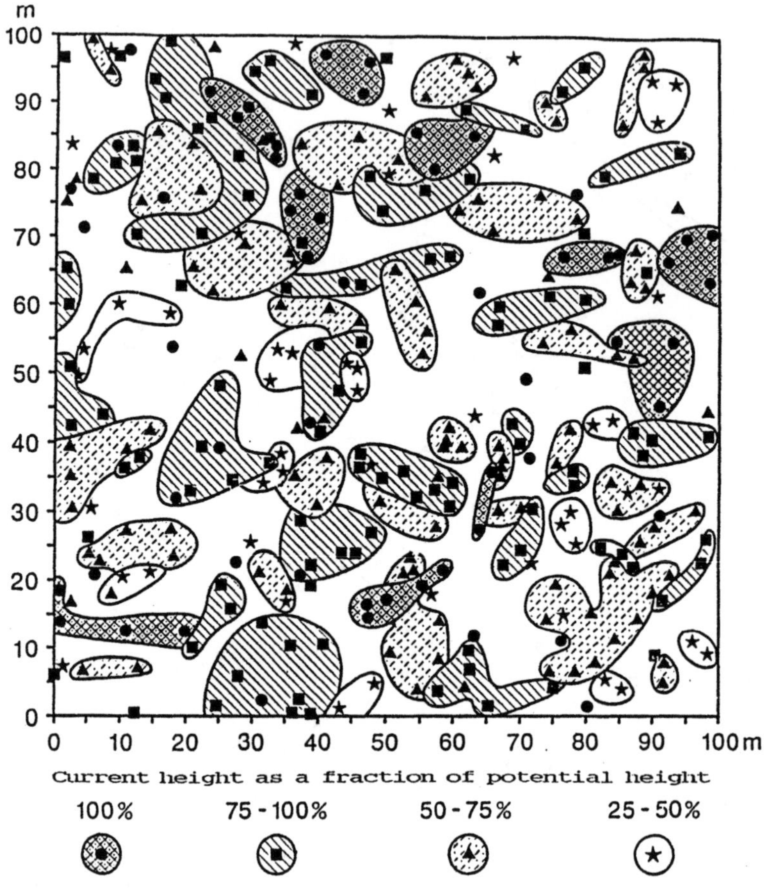

Fig. 3. Tree location in a 1 ha sample plot in Grebo National Forest, Liberia. All trees above 10 cm dbh are divided into 4 classes of maturity according to their height development. Clusters of trees of equal maturity from a mosaic pattern withina matrix of scattered singles. Preliminary results.

25–50%) which indicates that the dynamics of regeneration in this forest stand may be linked to a large extent to very small canopy openings.

Conclusion and outlook

In very different types of humid tropical rainforest the distribution of trees over diameter classes seems to follow a set pattern. The horizontal stand structure is mainly determined by site conditions. Macro-morphological features of different tree species constitute the conditions for survival of the stand. The species may be classified according to their mature height as upper, mid, and lower-canopy species and understorey species, and further subdivided according to mature diameter. Within one species group the species composition varies from site to site. The number of trees per species group is influenced by the successional stage of the stand and the competitive position of the species. Though individual trees of a species group in the same size class show a wide range of variation in their architecture, the average macro-morphological features of species groups also differ.

Regeneration takes place in canopy gaps of varying sizes, resulting in a shifting-mosaic patch pattern of developmental stages. The trees in an eco-unit are of similar developmental stage. Comparison of actual dimensions to potential mature dimensions of a tree on the site in question enables an estimation of the tree's developmental stage. Since trees of similar developmental stage are often spatially grouped, this procedure seems to be a promising method to identify a stand's mosaic structure from standard vegetation enumeration data sets.

Ongoing improvement of the method and analysis of the composition and distribution of the mosaic patches at each measurement date will give more insight into whole-stand development.

References

Ashton, P. S. 1988. Manual of Non-Dipterocarp Trees of Sarawak. Dewan Bahasa dan Pustaka, Kuala Lumpur. 490 pp.

Aubréville, A. 1938. La forêt coloniale: les forêts de l'Afrique occidentale française. Ann. Acad. Sci. colon., Paris 9:1–245.

Baur, G. N. 1968. Ecological basis of rainforest management. Govern. Printer, Sidney. 499 pp.

Dawkins, H. C. 1958. The Management of Natural Tropical High Forest with Special Reference to Uganda. Comm. For. Inst. Paper No. 34, Oxford.

Hartshorn, G. S. 1978. Tree falls and tropical forest dynamics. In: Tomlinson, P. B. & Zimmermann, N. H. (eds.) Tropical trees as living systems, pp. 617–638. Cambridge University Press, London.

Okali, D. U. U. & Ola-Adams, B. A. 1984. Tree population changes in treated rain forests at Omo Forest Reserve, southwestern Nigeria. J. Trop. Ecol. 3, 4: Special Issue:291–314.

Rollet, B. 1978. Organization (Part I, Chap. 5). In: *Tropical forest ecosystems. A state-of-knowledge report*, pp. 112–142. Unesco/UNEP/FAO, Paris.

Salvosa, F. M. 1963. Lexicon of Philippine Trees. FPRI (FOR-PRIDECOM) Bull. 1: 1–136.

Swaine, M. D., Hall, J. B. & Alexander, I. J. 1984. Tree population dynamics at Kade, Ghana (1968–1982). J. Trop. Ecol. 3,4: Special Issue:331–316.

Steenis, C. G. G. J. van (ed.) 1954, 1958. Flora Malesiana, Vols. IV–V. N.V., Djakarta.

Synnott, T. J. 1979: A Manual of Permanent Plot Procedures for Tropical Rainforests. Trop. For. Papers No. 14. C'omm. For. Inst., Oxford. 67 pp.

Voorhoeve, A. G. 1965. Liberian High Forest Trees. Agric. Res. Rep. 625, State Agric. Univ., Wageningen.

Woell, H. 1988. Struktur und Wachstum von kommerziell genutzten Dipterocarpaceen-Mischwaeldern und die Auswirkungen von waldbaulicher Behandlung auf deren Entwicklung, dargestellt am Beispiel von Dauerversuchsflaechen auf den Philippinen. BFH-Mitt. Nr. 161, Wiedebusch, Hamburg. 265 pp.

33. Fine-root amount of young mangrove stands in Trang Province, Thailand

RYUICHI TABUCHI
Hokkaido Research Center, For. & Forest Prod. Res. Inst., Hitsujigaoka, Sapporo 062, Japan

Abstract

The distribution of root density (root amount in unit soil volume) was examined using a boring method in order to estimate the fine root biomass of some young, *Rhizophora apiculata*-dominated natural mangrove stands as well as some 7 to 20 year-old plantations. The biomass of the roots less than 10 mm in diameter, up to 1 m depth, varied from 8 to 25 tons per hectare, the highest being in the 7 year-old stand. The growth process of fine roots with stand development is discussed.

Key words: fine-root biomass, mangrove, southern Thailand, stand development

Introduction

Mangrove forest extends over ca. 2,800 km^2 in Thailand (Aksornkoae 1987) but represents only a small fraction (1%) of the total forest area of the country. Mangrove appears widely along the coast, especially in the south, however, so that 320 km or 95% of the nation's coastline on the Andaman Sea side of the Malay Peninsula is covered by mangrove (Vitayarat 1986). This plays a significant role as a marginal ecosystem and as a fuel-energy resource.

As an ecosystem, mangrove is known to be fragile (Ogino 1989). The accumulation of pyrite originating in seawater may create quite acidic soil, through oxidation if hard desiccation takes place after clearing of a forest stand (Kyuma 1986). Well developed mature mangrove holds a huge amount of root biomass, as much as 500 tons per hectare, more than half of which is fine roots which decay easily (Komiyama et al. 1987; Komiyama et al. 1988; Tabuchi et al. 1983). Such great fine-root biomass may bring anaerobic

E.O. Box et al. (eds.), Vegetation Science in Forestry, 573–583.

conditions with high respiration, if rapid decomposition and poor drainage take place together.

The ongoing yield system of mangrove in Thailand is by clear-felling alternating 40 m-wide strips on a 30 year rotation cycle (Nalampoon 1982). Because of the clear-felling system, studies on carbon dynamics in mangrove soil are urged for the prediction of stand productivity. The fine-root biomass, therefore, is important for the organic carbon budget, but only few data on fine-root amount of mature forests are available. The present paper reports the fine root amounts examined in some young mangrove plantations as well as in immature natural forest.

This study was carried out as a part of a Thai-Japan (Thai Royal Forest Department, and Japan International Cooperation Agency) joint project titled "Research and Training in Re-Afforestation Project (2nd Phase)".

Site and methods

Location

Plantations of *Rhizophora apiculata* (*Ra*) around the Trang River estuary (aged 7, 10, 15 and 20 years old in 1987) as well as *Ra*-dominated natural stands, were selected for root-biomass study. The sites are near Kantang city (7°20′ N, 99°30′ E), Trang Province, southern Thailand. Of these sites, the 15 and 20 year-old stands and the natural stands were located on a small island (Tura islet) (Fig. 1). The rest were located on a nearby island and along the river bank. The mean annual temperature is 28°C. The annual precipitation fluctuates within the range 1500–2000 mm (Srisajjang 1986).

Stand characteristics

Trang is located on the Andaman Sea side of the Malay Peninsula, where the tide level changes twice a day. The sites are innundated twice daily over 2 weeks in a month, even during the dry season, and the forest floor is therefore covered with soft mud.

A permanent plot of ca. $100 \times 100\ m^2$ was established in the natural stand on Tura islet and was divided into a hundred $10 \times 10\ m^2$ subplots. Trunk diameter (D) at 0.3 m height above the root collar was measured for *Rhizophora* spp. larger than 4 cm in diameter; non-*Rhizophora* spp. were measured at 1.3 m height above ground. Of the total of 1,007 trees in the plot, *Ra* was the predominant species (974 trees counted), the remainder were individuals of *R. mucronata* (*Rm*), *Bruguiera cylindrica* (*Bc*), *B. sexangula* (*Bs*) and *Avicennia* spp. (*Av*). The largest tree of *Ra* was 33.4 cm in diameter, but

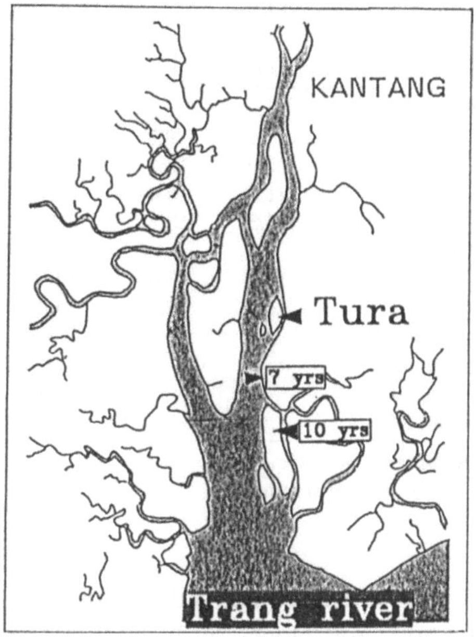

Fig. 1. Brief map of Kantang estuary. Arrows show the location of tura islet (15, 20 yr-old plantations and natural stand), 7 and 10 yr-old plantations.

few big *Av* trees (40 to 100 cm diameter) formed the group of large trees beyond the group of *Ra*. For the root study, four subplots and a site with young trees in a large canopy gap (various basal areas of 5.6 to 65.3 m^2/ha) were chosen in the natural stand plot. For the plantations, the diameter of all trees in the plots sized 150 to 300 m^2 was measured. Some parameters of the stand structures for the sites are listed in Table 1.

Root excavation

A cylindrical soil auger with a 100 cm^2 cross section and 20 cm depth (capacity = 2,000 cm^3) was applied for root excavation. The sampling was done for each 20 cm stratum down to 100 cm depth. Roots were then rinsed and sorted into size classes of 0–2, 2–5, 5–10, 10–20, 20–30, 30–40 and >40 mm in diameter; these were weighed by class. Subsamples were collected to determine the dry-weight ratio. The root density (RD: root dry weight per unit soil capacity, kg/m^3) was subsequently determined, as well as root biomass (RB: kg/m^2), for each root size class. For each site, 25 pits were sampled randomly to determine the average (10 pits for the small gap site).

Another sampling method, the so called monolith method (Komiyama et

Table 1. Parameters of stand structure

Site	Mean D (cm)	Mean H (m)	Tree density (TD: trs/ha)	Basal area (BA: m^2/ha)	Sum of D^2H (10^6 $cm^2 \cdot m$/ha)
Plantations[a]					
7 yrs	5.6	8.3	4,700	18.1	2.17
10	7.0	10.6	5,300	33.5	4.05
15	11.8	16.4	2,470	29.0	6.11
20	14.3	18.2	2,800	46.8	11.53
Natural stand					
All[b]	16.6	20.6	1,012	26.4	8.21
Gap	2.6	5.6	13,600	5.6	0.62
Subplot 1			2,300	65.3	19.45
Subplot 2			1,680	54.3	16.00
Subplot 3			1,810	30.1	7.57
Subplot 4			1,030	25.0	4.14

[a] Stand ages in 1987.
[b] Whole natural stand.

al. 1987) was also applied in order to examine the applicability of the auger method. Soil blocks of 20 × 50 cm^2 surface size and 10 cm thickness were cut out by hand saw down to a depth of 100 cm. Within each, the roots were sampled.

Results and discussion

Stand structure

Frequency curves of *Ra* tree diameter for the planations, natural stand and the gap are compared in Fig. 2. A bell-shaped distribution is obvious for the plantations as well as the gap site; both the mode and the mean shifted towards the larger diameter with age. The diameter range was wider, however, for the natural stand, which still showed a bell-shaped distribution like that of the other young stands. According to a local informant, Tura islet was a non-vegetated small mud flat which appeared only during ebb tide about 70 years ago. Taking these facts into account, one can judge the natural stand on the Tura islet to be a rather young stand, though older than 20 years. The existence of few larger *Av* trees might suggest that the *Av* layer was the remnant of a pioneer stand established on the mud flat and that the *Ra* layer grew up later, sequentially. The bell-shaped distribution with no significant cohort of young trees as successors suggestes that it is in the immature stage.

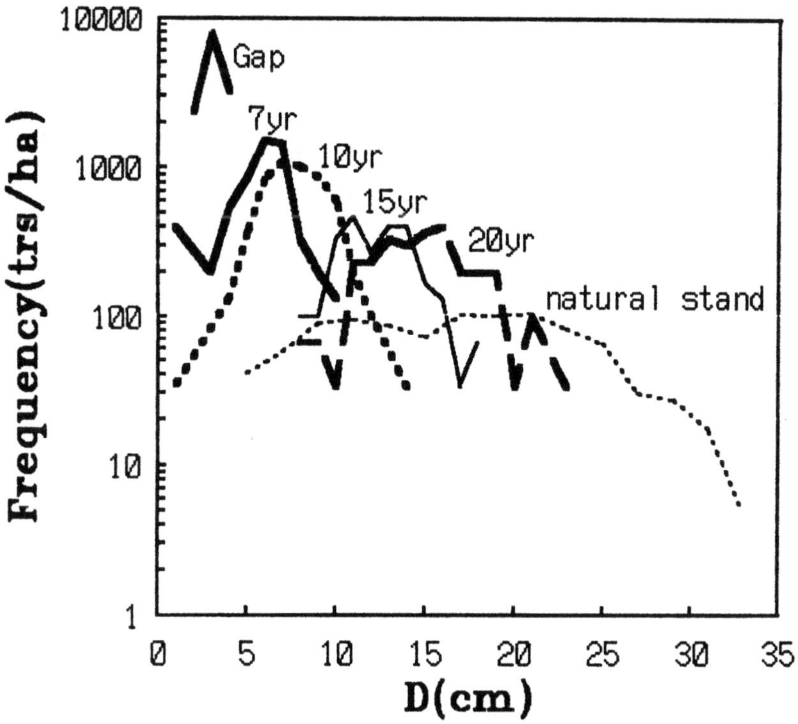

Fig. 2. D frequency curves of stands for *Ra*.

Applicability of auger sampling

The root density of each stratum, derived either by the auger method (RD_A) or by the monolith method (RD_B), was compared on the gap site and the 10 yr-old plantations. The relationship between RD_A and RD_B is shown in Fig. 3. In case that RD_A is nearly equal to RD_B or $RD_A > RD_B$, the auger method was agreed to be reliable enough, since the soil block sampled by the monolith method contains all roots in it. At both sites (see Fig. 3), fine and small roots less than 10 mm in diameter occurred around or above the line $RD_A = RD_B$, but roots larger than 10 mm were seen far below. The auger method was, therefore, applicable for roots smaller than 10 mm in diameter.

Vertical distribution of root density (RD)

The vertical distribution patterns for the density of roots less than 2 mm and from 2 to 5 mm are shown in Fig. 4. The largest root density was observed in the second 20 cm layer (20–40 cm) in some stands, but RD mainly showed

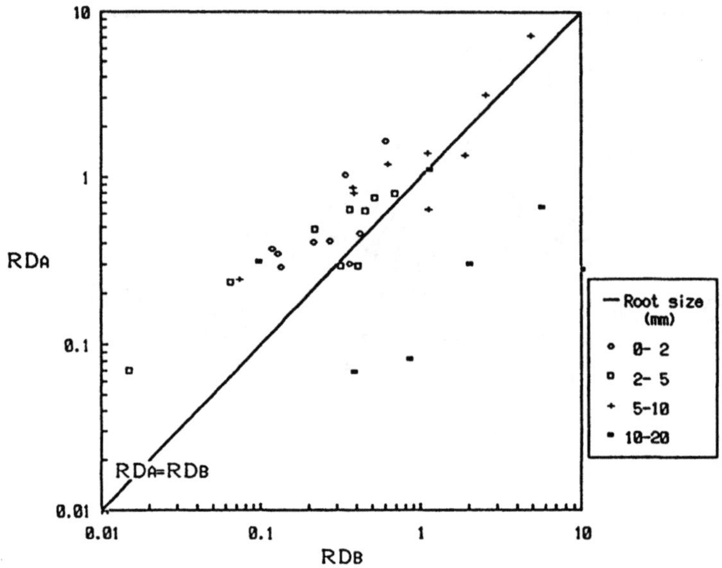

Fig. 3. Relationship between the root density examined with Auger method (RD_A) and with Monolith method (RD_B).

an exponential decrease with depth. The major difference between the plantation sites and some sites in natural stands was that RD decreased more rapidly in plantations than in natural stands. The root size composition in each stratum is presented in Fig. 5 for each site. It is obvious that the more the stand developed, the higher the proportion of large roots. Thus, it is likely that the below-ground rhizosphere extends deeper with stand development, and the smaller roots are replaced by larger roots down to as far as 100 cm depth.

Fine root biomass

The relationship between root biomass (RB: kg/m^2) and basal area (BA: m^2/ha) at each site is shown in Fig. 6 for different root size classes. Fine-root (0–2 mm) biomass varied from 0.135 to 0.672 kg/m^2, being highest at the gap site. Fine-root biomass was observed to decrease with stand development up to around 30 m^2/ha basal area (or about 15 yr-old plantation), then to increase again with stand development. For small roots of 2–5 mm diameter, the root biomass at the gap site fell to very low level, near the smallest for all sites. The same decrease-to-increase trend with area observed for fine roots was also observed for 2–5 mm roots but was less clear. The highest biomass value was in a 7 yr-old plantation (0.679 kg/m^2) and the

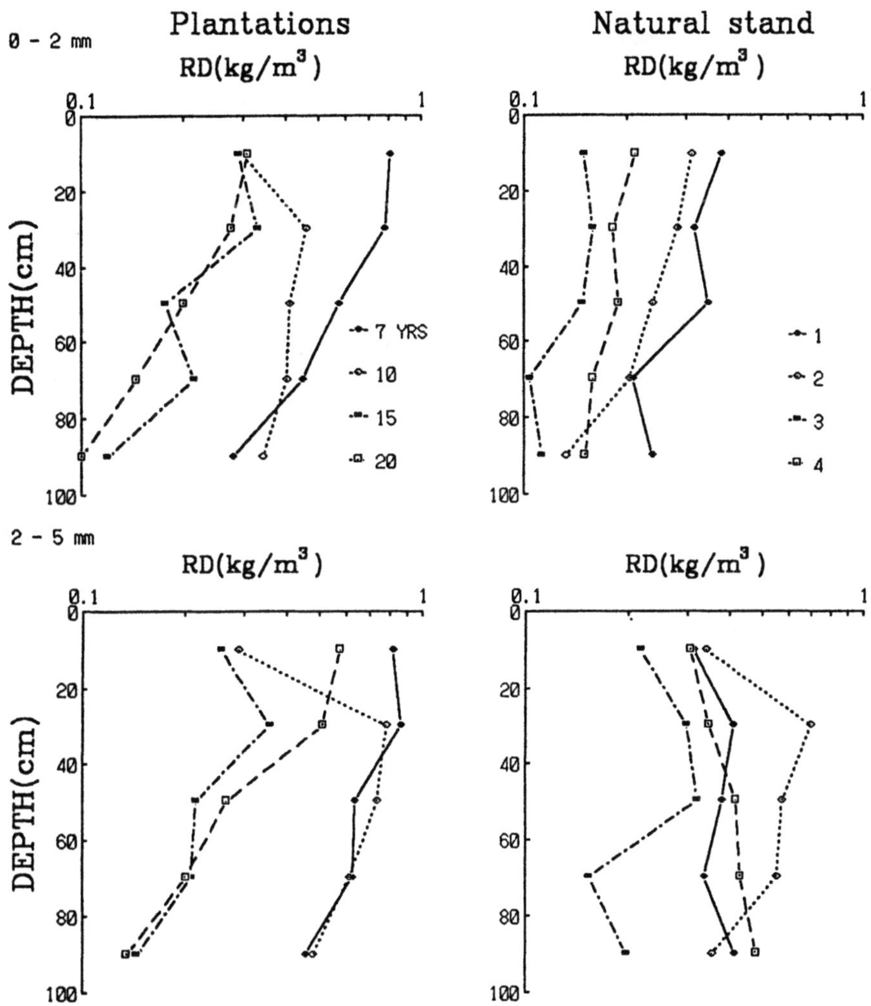

Fig. 4. Vertical distribution of root density. Above: $\phi < 2$ mm. Below: ϕ 2–5 mm. Left: plantations. Right: natural stand. Remark 1, 2, 3, 4 of natural stand corresponds to subplot number 1, 2, 3, 4 in Table 1.

lowest was in a 15 yr-old stand (0.237 kg/m^2). The root biomass for all roots smaller than 10 mm was from 0.75 to 2.47 kg/m^2.

The relationship between fine-root biomass and tree density of a site (TD) is an another indicator of stand development, as shown in Fig. 7. In general, tree density decreases with stand age as long as the stand is not mature enough to produce seedlings. The observed positive correlation suggests that denser stands have more fine-root biomass, down to 100 cm depth. It is therefore assumed that the rhizosphere, which mainly consists of fine and

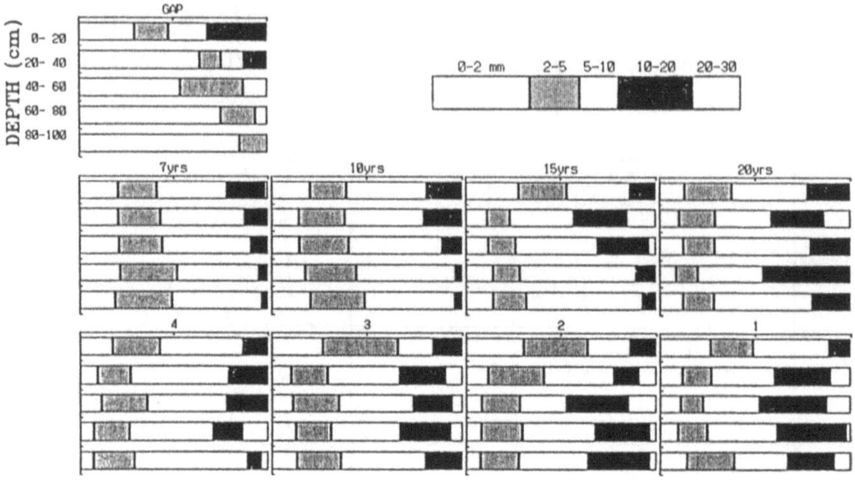

Fig. 5. Root size compositions. 1, 2, 3, 4 of bottom row corresponds subplot number in Table 1.

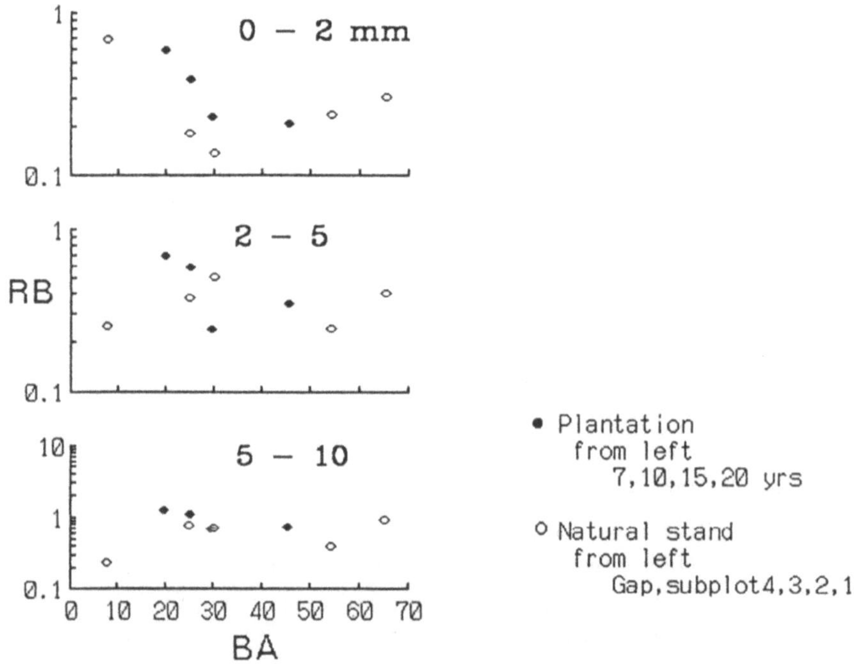

Fig. 6. Relationship between stand basal area (BA: m^2/ha) and root biomass (RB: kg/m^2) by root size.

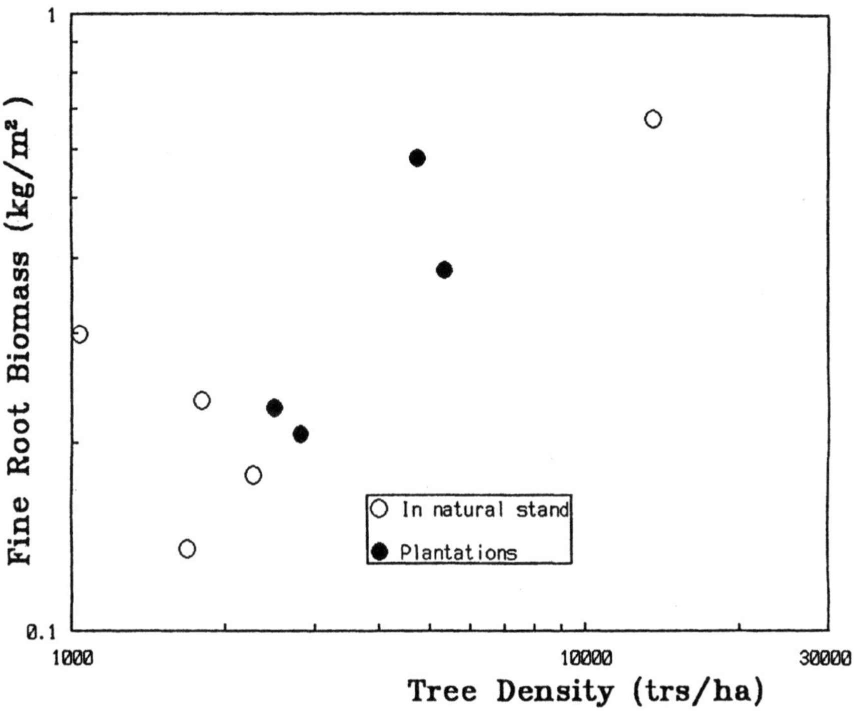

Fig. 7. Relationship between stand density (TD: trees/ha) and fine root biomass ($\phi < 2$ mm).

small roots, accumulates in rather shallow depth while the stand is young and dense, especially at the gap site and the 7 year-old stand. So that those small roots were efficiently sampled. The rather large fine-root biomass in younger stands might, thus, be estimated.

Subsequently, with the development of a stand, the rhizosphere extends deeper and wider, and the proportion of coarse roots becomes greater. Accordingly, the sampling efficiency for fine roots within shallow strata, down to 100 cm depth, may decrease while stands are young. The small fine root biomass was thus estimated on natural stand. Because the network of stilt roots is often well developed in mature stands, however, the mass of fine and small roots growing from the aerial roots reaching the ground would give a huge value for fine-root biomass, as great as some hundred tons per hectare (Komiyama et al. 1987, Komiyama et al. 1988, Tabuchi et al. 1983) as has been estimated for some well developed natural mangrove forests.

The root biomass estimates in this study varied from 1.35 to 6.72 ton/ha for fine roots less than 2 mm in diameter, and between 12.7 and 4.6 ton/ha for roots less than 5 mm. Since the fine-root biomass of terrestrial forests examined by the core sampling method was generally less than 10 ton/ha

(Santantonio et al. 1977), the fine-root biomass of mangrove is not largely different from that of terrestrial forests, at leasst for the young mangrove sites in this study. The difference in maturity between the well developed natural forests mentioned above and those of this study probably makes a great difference in the fine-root biomass. This is reflected in the maximum tree size in the stands as well as in the frequency curve of diameter. For example, a natural mangrove in Ranong Province, southern Thailand, in which there is a huge amount of fine-root biomass (more than 220 tons per hectare) showed a reverse J-shaped frequency curve for diameter, with the largest *Ra* tree being 55 cm in diameter (Tamai et al. 1986). This reverse-J shape suggests that it is in a stable stage and that self-replacement is taking place (Kimmins 1987). The largest D is 33.9 cm, and a bell-shaped D frequency curve is shown for natural stands of this study.

Fine-root biomass as a potential organic matter source, thus, would be kept at a low level over the ongoing rotation period in Southeast Asia, around 40 years or less. Below ground carbon dynamics must be studied further, however, since silvicultural activity in mangrove is still mainly carried out on logged-over sites of original mature natural forest.

References

Aksornkoae, S. 1987. Mangrove resources. In: Arbhabhirama, A. Phantumvanit, D., Elkington, J. & Ingkasuwan, P. (eds.) Thailand: Natural Resources Profile, pp. 144–163. Thailand Development Research Institute, Bangkok.

Kimmins, J. P. 1987. Forest Ecology, 1st ed. Macmillan Publishing Company, New York. 531 pp.

Komiyama, A., Ogino, K., Aksornkoae, S. & Sabhasri, S. 1987. Root biomass of a mangrove forests in southern Thailand. 1. Estimation by the trench method and the zonal structure of root biomass. Journal of Tropical Ecology 3:97–108.

Komiyama, A., Moriya, H., Prawiroatmodjo, S., Toma, T. & Ogino, K. 1988. Primary productivity of mangrove forest. In: Ogino, K. & Chihara, M. (eds.) Biological System of Mangroves: A Report of East Indonesian Mangrove Expedition 1986, pp. 97–117. Ehime University, Matsuyama.

Kyuma, K. 1986. Mangrove accumulates origined soil: Acid sulfate soil. In: Tounan Ajia no Teishicchi (Lowland Swamp of Southeast Asia), pp. 56–79. Tropical Agriculture Research Center, Tokyo (in Japanese).

Nalampoon, A. 1982. Mangrove Forest Management in Thailand. Thai Royal Forest Department, Bangkok. 17 pp.

Ogino, K. 1989. Mangroves of the world: ecological bases for ecosystem rehabilitation. Saishuu to Shiiku (Collecting and Breeding), 51:444–448 (in Japanese).

Santantonio, D., Hermann, R. K. & Overton, W. S. 1977. Root biomass studies in forest ecosystems. Pedobiologia 17:1–31.

Srisajjang, S. 1986. Kantang District. In: Saranukrom watanatham phaktai (Encyclopedia South Thai Culture) 1:113, 10 vols. Srinakharintravirot University, Songkhla. About 300–400 pp. each volume (in Thai).

Tabuchi, R., Ogino, K., Aksornkoae, S. & Sabhasri, S. 1983. Fine root amount of mangrove forest: a preliminary survey. Indian Journal of Plant Science 1:31–40.
Tamai, S., Nakasuga, T., Tabuchi, R. & Ogino, K. 1986. Standing biomass of mangrove forest in southern Thailand. Journal of Japanese Forestry Society 68:384–388.
Vitayarat, P. 1986. Mangrove forest. In Saranukrom watanatham phaktai (Encyclopedia South Thai Culture). 5:2094, 10 vols. Srinakharintravirot University, Songkhla. About 300–400 pp. each volume (in Thai).

34. Structure, litterfall and net primary production in the mangrove forests in East Kalimantan

SUKRISTIJONO SUKARDJO
Indonesian Institute of Sciences, Research and Development Center for Oceanology, Alamat Jl., Pasir Putih I, Ancol Timur, P.O. Box 4801 Jakarta (Java) 11048 Indonesia

Abstract

The composition and productivity of five mangrove associations in Apar nature reserve (128,000 ha) East Kalimantan were studied from August 1983 to August 1984. Measurements were made of the tree composition, litterfall and potential net primary production at high-salinity (33.75 ± 1.75‰) fringing sites and at moderate-salinity riverine (30.25 ± 1.15‰) and inland (28.35 ± 0.85‰) sites. *Rhizophora apiculata* B1. (Importance Value = IV = 151.68) was the dominant tree at fringing sites and had a more even distribution from the shoreline inland and from site to site. *Avicennia officinalis* L. (IV = 168.89) was the dominant tree at the riverine site. *Ceriops tagal* (Perr.) C. B. Robin (IV = 215.66) dominated the inland areas of the riverine site. The fringing mangrove forests had basal area (BA) of 17.72 m^2/ha and 15.18 m^2/ha at two sites and complexity Index (CI) of 19.53 and 10.39 calculated for trees with DBH more than 2 cm. The riverine forests has a BA of 19.47 m^2/ha and a CI of 14.53, while the inland forests had a BA of 24.19 m^2/ha and 37.57 m^2/ha at two sites and a CI of 14.92 and 69.71. Average diameter at breast height (DBH) was greater at the inland sites for each species, but sapling (trees with diameter 2–10 cm) density was more than twice as high at the other plots (256–304/ha) than at the inland sites (112–160/ha). The litterfall varied from 20.50 to 29.35 dry t/ha/year, with the higher values at the inland plots. The estimated potential of net primary production of the mangrove forests during the period of study (August 1983–August 1984) was 40.44 to 43.93 kg C/ha/day at the fringing sites, 42.93 kg C/ha/day at the riverine site, and 44.25 to 45.49 kg C/ha/day at the inland sites. Fluxes estimated litterfall components were 4.71 to 8.20 dry t/ha/year for flowers, fruits and buds; 6.85 to 12.63 dry t/ha/year for leaves; 6.01 to 6.95 dry t/ha/ year for twigs; and 2.31 to 2.96 dry t/ha/year for other detritus.

E.O. Box et al. (eds.), Vegetation Science in Forestry, 585–611.

Key words: Apar nature reserve, East Kalimantan, Indonesia, litterfall, mangrove forest, net primary production

Introduction

In Kalimantan (Indonesian Borneo) mangrove forests occupy a large area as a primary feature of the coastal zone and are estimated to be about 383,450 ha or 9.1% of the total mangrove forest area in Indonesia (Darsidi 1987). The most extensive and luxuriant mangrove forests in Kalimantan are found in East Kalimantan province, with a total area estimated at about 266,800 ha or 69.601% of the total mangrove forest area in Kalimantan. A part of the mangrove forest in East Kalimantan has been designated as a nature reserve, and with the largest area (128,000 ha) in the Apar delta system (Sukardjo 1989a).

Many studies have been conducted on Indonesian mangroves, most concerned with vegetation analysis (e.g Soemodihardjo et al. 1979, 1984, Soerianagara et al. 1987). The productivity and functioning of the mangrove forest have been neglected, except for the Muara Angke-Kapuk mangrove forests (Sukardjo 1989b), and the Segara Anakan and Grajagan areas (Sukardjo in prep.). Comparable data for other Indonesian mangrove forests are badly needed. Very few studies have been done on mangrove forests in East Kalimantan (e.g. Sukardjo 1987, 1988, 1989a).

In designing this study, I relied on the classification of mangrove structure and productivity which has evolved over the past decade from the original description of Lugo and Snedaker (1974), who identified several major mangrove community types based on differences in soil type, nutrient input and hydrology. At present, there are considered to be three main forest types: basin, fringe and riverine (Brown & Lugo 1982, Odum et al. 1982). Each community type has a characteristic structure and range of productivity, litter decomposition and detritus export. Generally, total biomass, tree height, litterfall and decomposition, and freshwater turnover increase from basin to riverine forests (Pool et al. 1975, Brown & Lugo 1982).

This study was undertaken as a necessary prerequisite to the development and implementation of a management plan for the Apar nature reserve. In this paper, I present the most important findings of a one-year research effort that included quantitative descriptions of mangrove associations, the monitoring of litterfall, and the estimation potential net primary production (PN). This constitutes the first part of the ecological study of the mangrove forest of Apar nature reserve; This work was conducted from August 1983 to August 1984 and forms part of a long-term program on comprehensive understanding of mangrove ecology in this region.

The study area

The Apar mangrove forest is located on the southwestern coast of Balikpapan (about 116°10.9′ E long. and 1°56.5′ S lat.). The mangrove forest has developed on a coastal belt with muddy substrate at elevations below 3 m. The 128,000 ha mangrove forest became a nature reserve in 1982 and has been protected from logging activities for more than 50 years (Direktorat Jendral Perlindungan Hutan dan Pengawetan Alam, pers. communication 1989).

Access to the study area from nearby Tanah Grogot, the district capitol, is provided by the Kandilo river. The study area is on a deltaic system and includes some small rivers and shallow basins. The study site is typical of mangrove forests and extends through the three most common plant associations of the mangrove forest: the fringing, the riverine and the basin forests. The mangrove forest is almost completely surrounded by freshwater swamp forests, covering an extensive area of the province (Fig. 1).

Methods

The mangrove forest vegetation was inventoried at five different locations, namely sites I, II, III, IV and V. A 50 m x 50 m plot was established at each location in August 1983. Each plot was divided into 25 subplots of 10 m x 10 m (Fig. 2).

Inventories included species identification and measurement of tree height and diameter at breast height (DBH) of all trees larger than 10 cm in each 10 m × 10 m plot. Saplings (trees with diameter 2–10 cm) were sampled within 5 m × 5 m subplots which were nested within the larger plots (10 m × 10 m). Stem diameter at 30 cm above the highest prop-root for *Rhizophora* or above the ground level for other species was also measured. Plant species were identified using the descriptions made by Ding Hou (1952) (for Rhizophoraceae) and Backer & Bakhuizen van den Brink Jr. (1963–1968). Profiles of 20 m × 50 m (depending on stand density) were drawn to depict the structure of these forest types.

Frequency, density, basal area and Importance Value (IV) were calculated for each mangrove species based on Cox (1967), Curtis & MaCintosh (1951), and Mueller-Dombois & Ellenberg (1974). Complexity Indices (CI) were calculated for trees with DBH more than 2 cm, based on the method of Pool et al. (1977). The CI is the product of the number of species, basal area (m^2/0.1 ha), maximum tree height (m), and number of stems/0.1 ha, multiplied by 10^{-3}, based on 0.1-ha plots (Holdridge et al. 1971).

At each 10 m × 10 m subplot, soil samples up to a depth of 20 cm were collected, and soil salinity during low tide was measured, using a salinometer.

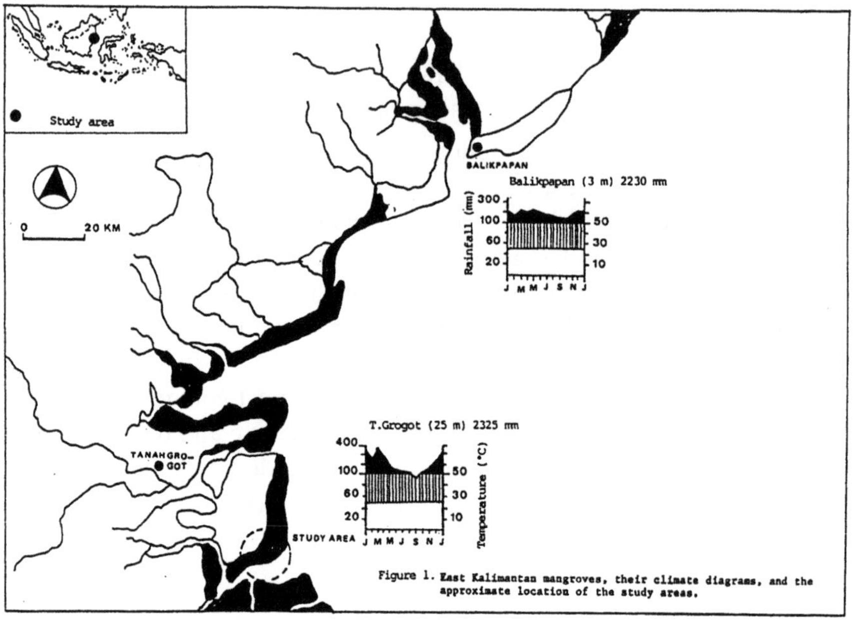

Figure 1. East Kalimantan mangroves, their climate diagrams, and the approximate location of the study areas.

Fig. 1. East Kalimantan mangroves, their climate diagrams, and the approximate location of the study areas.

Soil analyses were performed at the Chemistry section of the Department of Natural Sciences, Bogor Agricultural University, Bogor.

Litter production is usually monitored by collection in litterfall traps (Newbould 1967). Litter traps of 1 m × 1 m × 0.25 m, each with 1 mm nylon mesh base, were suspended diagonally by nylon rope between trees in every other subplot (Fig. 2), at a height of 2 m above the ground, beyond the reach of high tides. Sampling at weekly intervals was carried out from 8 August 1983 to 10 August 1984. A total of 48 collections was made in each trap. The litter in each trap was sorted into its various components (species leaves, fruits + flowers + buds, and twigs). The litter was then oven-dried at 105°C for 5 days and weighed. The total amount of litter in each subplot was converted to daily average amounts, with no correction made for either leaching or other losses.

The litter pool was measured weekly at each of the five sites by using a quadrat of 1 m × 1 m located adjacent to the litter traps (Fig. 2). Sampling sites, number of collections and collection times parallel the collections for litter production. All standing-crop samples were collected, dried, sorted and weighed using the method described above for litterfall.

The turnover of mangrove litter was determined by using the mathematical model $1/k = X_{ss}/L$ where L is litter production, X_{ss} is the steady-state ac-

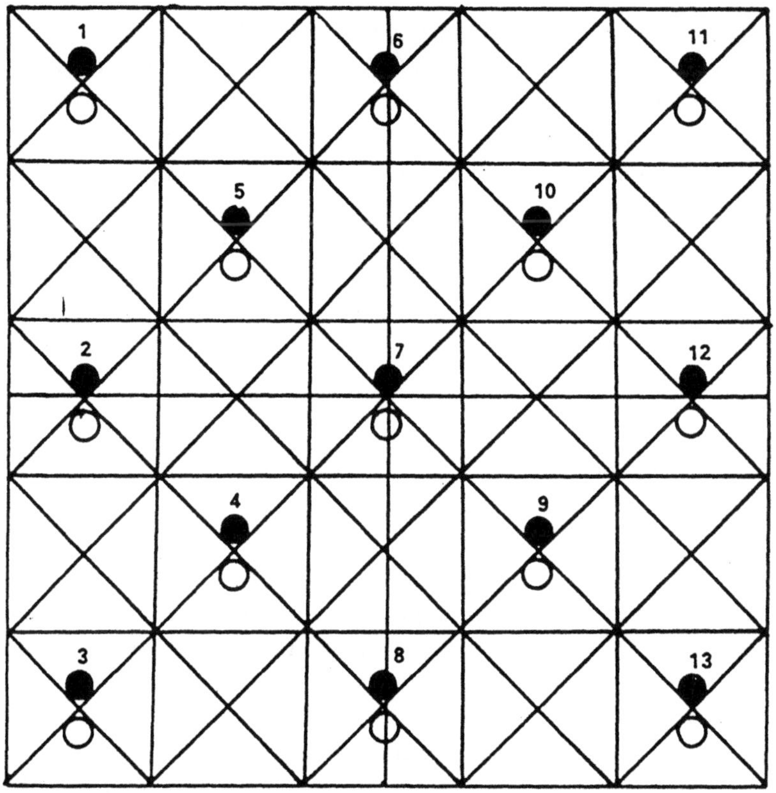

Fig. 2. Plot (0.25 ha) diagram showing the relative positions of litter traps (●) and sampling sites for litter accumulation (○).

cumulation of surface litter on the forest floor (litter pool), and 1/k is the residence time in year (Nye 1961, Olson 1963, Gosz et al. 1973). The equation assumes a steady state (input = output), so the rate of change ($dX/dt = 0$, where X is standing crop and t is time) is zero, and litter production equals litter losses. The mathematical approach of Nye (1961) and Olson (1963) was also used to calculate the time required to attain 95% (3/k) and 99% (5/k) of the steady-state litter accumulation at each site. Fluxes (t/ha/year) of, for instance, dry matter (X) are described by the equation $dX/dt = -kXo\exp(-kt)$, where Xo is the original amount of the component of detritus (t/ha), k is the decay constant (per year), and t is time in years. At $t = 0$, $dX/dt = -kXo$.

The survey method used to estimate productivity was described in detail by Bunt et al. (1979). This procedure is also followed in this study, viz:

1. To estimate the cover percentage of the forest canopy. Basically, it is assumed that light transmission through a well developed mangrove forest

canopy follows the relationship log Io/I = k (Ch1), where Io = incident photosynthetically active radiation (PAR), I = PAR penetrating the forest canopy, and k = 0.31, i.e. a constant derived from calibration with leaves of known total chlorophyll content (Ch1) per unit area.
2. To convert canopy cover into chlorophyll a, the following formula is used: 0.762/0.31 × log Io/I × canopy cover (%), where 0.762 is the ratio of chlorophyll a to total chlorophyll (Bunt et al. 1979).
3. To convert this to a productivity estimate, use the formula: PN = A × chlo. a × 12, where A = 0.57 mg C/mg Chlo., and a is the coefficient of assimilation.

Results and discussion

Floristic composition

Twelve mangrove species and 1 marginal species were recorded in the study plots (Table 1). This is a larger number than at the Banyuasin South Sumatra study plots (Sukardjo and Kartawinata 1979, Sukardjo et al. 1984, Yamada and Sukardjo 1979).

The only mangrove species encountered on at all five study sites was *R. apiculata*. It is the dominant tree and sapling species at sites I and II, with Importance Values (IV) 151.68 and 144.53 for trees, and 176.20 and 113.31

Table 1. Species found in the study plots, with their Importance Values (IV) (as trees, DBH more than 10 cm)

Species	Plot site I	II	III	IV	V
Am = *Avicennia marina*	52.25[a]	54.85[a]	42.90[a]	–	–
Ao = *A. officinalis*	44.95	–	168.89[b]	–	–
Bc = *Bruguiera cylindrica*	17.99	35.52	–	–	31.11
Bg = *B. gymnorrhiza*	–	–	–	–	55.69
Bp = *B. parviflora*	–	12.13	16.22	6.25	–
Bs = *B. sexangula*	–	–	25.74	14.25	8.37
Ct = *Ceriops tagal*	–	–	–	215.66[b]	98.61[b]
Kc = *Kandelia candel*	–	–	10.16	–	–
Ra = *Rhizophora apiculata*	151.68[b]	144.53[b]	36.09	50.24[a]	82.49[a]
Rm = *R. mucronata*	33.10	–	–	–	–
Sa = *Sonneratia alba*	–	52.97	–	–	–
Xm = *Xylocarpus moluccensis*	–	–	–	13.60	18.60
Fm = *Ficus microcarpa*	–	–	–	–	5.13

Importance Value is the sum of the relative frequency, relative density and relative dominance.
[a]Co-dominant species.
[b] Dominant species.

for saplings, respectively. Sukardjo & Kartawinata (1979) and Sukardjo et al. (1984) also described riverine forests in Banyuasin South Sumatra as mangrove forests with *R. apiculata* dominating the study plots, just as I found at Apar nature reserve, East Kalimantan. Moreover, *R. apiculata* occurs throughout the study plots (Table 1). This result supports the hypotheses of van Steenis (1958), de Haan (1931) and Watson (1928) that *R. apiculata* has less stringent habitat requirements and is capable of growing in any of the mangrove zones. On the other hand, only two mangrove species were confined to the shoreline and river-bank areas, viz *R. mucronata* and *Kandelia candel*, respectively. This indicates that they occur specifically in deltaic areas. Thus, both species are only subordinant species of the assemblages. Thus, the identification of mangrove habitats in the Apar deltaic system rests on the fact that mangroves inhabit geomorphic units that have distinct morphology, substrate, salinity, etc. Further, based on the tide data from 1983–1985 (Dinas Hidro-Oseanografi ALRI 1983–1985), the mangrove forests at Apar nature reserve are frequently inundated by tides and are composed of 12 true mangrove species and one marginal species. This characterization and the data presented in Table 1 give some insight into the factors responsible for the variation in zonation patterns (*cf.* Thom et al. 1975); for example, on depositional areas such as the convex banks of meanders or prograding mudflats, where soft mud occurs in the outer zone, *Rhizophora* comprises the outer fringe.

Forest structure

Table 2 summarizes the results of the vegetation analyses using Cox's method (1967) and only three of the mangrove tree species, viz *A. officinalis*, *C. tagal* and *R. apiculata*. These species had the highest importance values (Tables 1 and 2). No single tree species dominated all study sites, nor did any species have the same importance value at all sites. Based on the highest importance values of the tree species (Tables 1 and 2), the following three forest communities can be recognized: 1) *R. apiculata-A. marina* (sites I and II), 2) *A. officinalis-A. marina* (site III), and 3) *C. tagal-R. apiculata* (sites IV and V). At site V there were more trees (700/ha), and most stands had trees with diamater more than 25 cm, resulting in the greatest basal area (37.57 m^2/ha, see Table 3). These stands, however, had few saplings (112 saplings/ha). The higher sapling densities are found in sites I, II and III, reflecting their exposure along the shoreline (Fig. 3). The variation in the structure of the mangrove forest at the Apar nature reserve is attributed to variation in the soil salinity, to the frequency and period of tidal inundation, and to the consequent effect of these factors on soil redox potential through chemical and physical properties of their particular habitats (see Table 4).

Table 2. Stand composition of mangrove forests in the Apar nature reserve, East Kalimantan. (Trees > 10 cm DBH, Saplings = trees with diameter 2–10 cm)

Site type species	Mean Dominant Height (MDH, m)	Trees			Saplings		
		Density (stems/ha)	Basal Area (m^2/ha)	Importance Value (IV)	Density (stems/ha)	Basal Area (m^2/ha)	Importance Value (IV)
Fringing forest:							
Site I.	27.35 ± 3.75	524	15.58		288	2.14	
A. officinalis		84	1.92	44.95	32	0.22	33.15
A. marina		52	2.83	52.28[a]	48	0.36	57.01[a]
B. cylindrica		16	1.70	17.99	16	0.12	17.05
R. apiculata		340	9.30	151.68[b]	176	1.33	176.20[b]
R. mucronata		32	2.83	33.10	16	0.11	16.58
Site II.	30.50 ± 1.50	188	13.36		256	1.82	
A. marina		32	2.65	54.85[a]	64	0.48	72.80[a]
B. cylindrica		16	1.70	35.52	32	0.20	37.78
B. parviflora		4	0.70	12.13	16	0.12	19.98
R. apiculata		120	5.48	144.53[b]	96	0.73	113.31[b]
S. alba		16	2.83	52.97	48	0.29	56.13
Riverine forest:							
Site III	21.90 ± 1.40	264	17.25		304	2.22	
A. marina		40	1.65	42.90[a]	64	0.49	68.12
A. officinalis		144	12.06	168.89[b]	128	0.97	129.54[b]
B. parviflora		12	0.90	16.22	16	0.10	16.01
B. sexangula		24	0.99	25.74	32	0.20	25.82
K. candel		8	0.61	10.16	16	0.11	16.46
R. apiculata		36	0.99	36.09	48	0.35	44.05

Landward site:							
Site IV	35.45 ± 0.77	188	22.99		160	1.29	
B. parviflora		4	0.25	6.25	16	0.11	29.64
B. sexangula		8	0.90	14.25	16	0.11	29.64
C. tagal		144	14.50	215.66[b]	112	0.85	181.44[b]
R. apiculata		24	6.50	50.24[a]	16	0.11	29.64
X. moluccensis		8	0 .75	13.60	16	0.11	29.64
Site V	37.90 ± 1.70	588	36.80		112	0.77	
B. cylindrica		64	0.50	31.11	15	0.10	40.85
B. gymnorrhiza		144	3.15	55.69	17	0.12	42.25
B. sexangula		24	0.19	8.37	14	0.11	41.35
C. tagal		168	17.10	98.61[b]	18	0.14	46.50[b]
F. microcarpa[c]		12	0.09	5.13	13	0.10	40.15
R. apiculata		128	15.40	82.49[a]	19	0.13	49.70[a]
X. moluccensis		48	0.37	18.60	16	0.07	43.20

[a] *Co-dominant species.*
[b] *Dominant species.*
[c] *Marginal species.*
MDH: Mean Dominant Height is the mean high of the five tallest tree per plot.

Table 3. Structural indices of five mangrove associations in the Apar nature reserve

Site (1)	Plot size (ha) (2)	Stems/ha (DBH > 2 cm) (3)	Trees/ha (DBH > 10 cm) (4)	Mean tree height (m) (5)	No. of species with DBH > 2 cm (6)	Basal Area (m^2/ha) (7)	Complexity index[a] (DBH > 2 cm) (8)	Community type (9)
I	0.25	812	526	27.50	5	17.72	19.53	*R. apiculata-A. marina*
II	0.25	444	188	30.50	5	15.18	10.29	*R. apiculata-A. marina*
III	0.25	568	264	21.90	6	19.47	14.53	*A. officinalis-A. marina*
IV	0.25	348	188	35.45	5	24.19	14.92	*C. tagal-R. apiculata*
V	0.25	700	588	37.90	7	37.57	69.71	*C. tagal-R. apiculata*

[a] The product of columns 3, 5, 6 and 7 divided by 10^5.

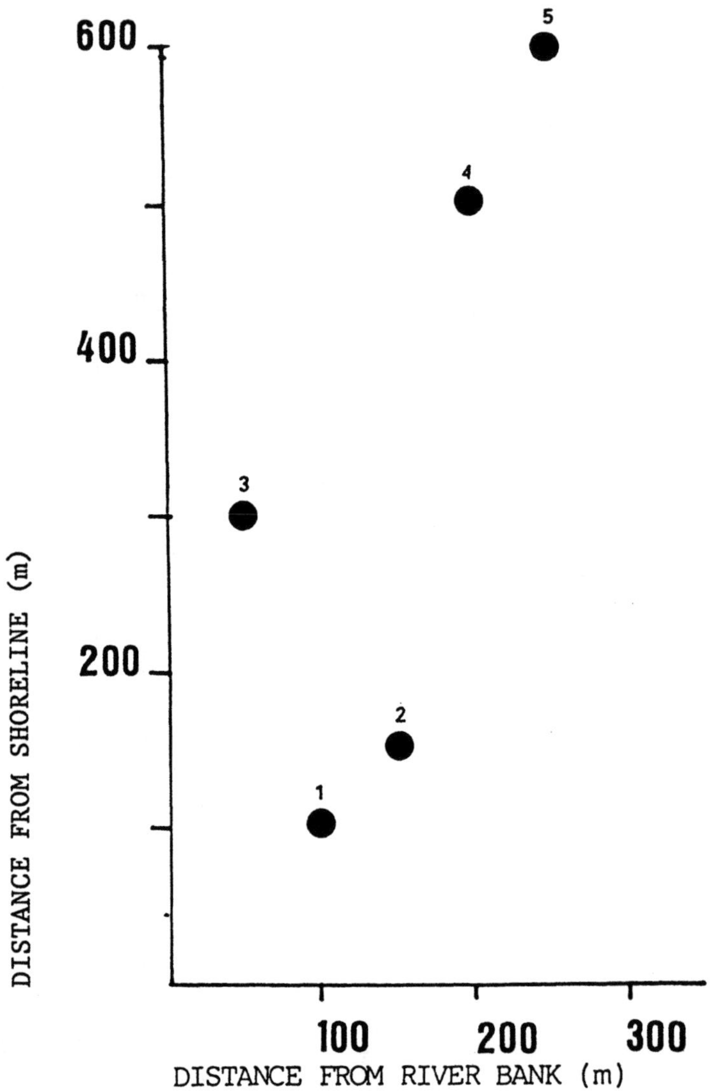

Fig. 3. Distance of plots from river bank or shoreline.

For example, the fringing sites of my study plots (sites I and II) began at low elevation, where the substrate is always inundated by tides twice daily and is exposed during low tides. In these areas the predominant plant associations involved *R. apiculata* (120–340 trees/ha, 96–176 saplings/ha), *A. marina* (32–52 trees/ha, 0–32 saplings/ha), and *Sonneratia alba* (0–16 trees/ha, 0–48 saplings/ha) (Table 2).

Table 4. Soil data from Apar mangrove system. The values shown for P, N and E_h are means and standard deviations from five samples per core at each 10 m × 10 m subplot, to a depth of 20 cm (August 1983 to August 1984)

Site plot	Particle size (%)			Texture class	E_h (mV at 5 cm depth	Soil salinity (‰)	Available P_2O_5 (ppm)	NO_3-N (ppm)	NH_4-N (ppm)
	Sand	Silt	Clay						
I	29.9	39.9	30.2	Clay loam	−162 (30.5)	33.7 (1.75)	33.11 (3.14)	27.19 (1.96)	205.2 (19.9)
II	30.0	39.9	30.1	Clay loam	−160 (22.7)	32.5 (1.15)	33.54 (2.23)	27.43 (1.96)	207.4 (17.5)
III	30.1	39.8	30.1	Clay loam	−158 (27.3)	30.2 (0.95)	34.74 (2.75)	29.06 (1.62)	220.1 (17.1)
IV	31.2	35.9	32.9	Clay loam	−120 (29.5)	28.5 (0.85)	38.12 (2.41)	31.25 (1.65)	236.5 (14.5)
V	31.3	35.5	33.2	Clay loam	−115 (35.5)	28.2 (0.80)	39.72 (3.92)	32.43 (2.35)	245.6 (16.9)

E_h (Redox potential) was measured by immediate insertion of a Pt/SCE combination electrode into the soil. The measured potentials were corrected to E_h (vs. hydrogen electrode reference) by addition of +244 mV to the readings.

No emergent trees are recorded in the mangrove forest at Apar nature reserve. By using Ogawa's method (1965), the mangrove forest at Apar nature reserve can be classified as a unistratal canopy with mean dominant height (MDH = mean height of five tallest trees per plot) ranging from 21.90 to 37.90 m (Tables 2 and 3). Table 3 summarizes the results of the mangrove forest inventories. The variation in 'complexity' between stands is evident. The differences in the complexity index (CI) of these mangrove forest types is due to changes in density and tree diameter within the stands. On the more favorable sites, there were many stems, more stems in the larger diameter classes, and the forest was taller (*cf.* site V). The CI thus varies with plant vigor and may be higher or lower depending upon which diameter classes are being included in the inventory. Based on Table 3, it can be concluded that the mangrove associations at Apar nature reserve would appear to be simple and the contribution of other synusia to structural complexity in the tropical region, including some that appear seasonally, would be very interesting for further study. Depending upon the prevalent soil conditions at each plot site, however, the structural development of the mangrove forest (density, basal area, height etc.) at Apar nature reserve may become proportionally higher or lower. Table 1 illustrates the response of particular tree species to such conditions. This response, coupled with the frequent tidal inundation and/or freshwater flooding, which can be observed in the field, results in a profusion of stems of various diamaters (Fig. 4).

Until recently, researchers have tended to think that community structure in mangrove systems is related to soil salinity gradients (e.g. Lugo and Snedaker 1974). The soil salinity recorded in the study sites, however, is below 40‰. Information summarized by Brown & Lugo (1982) indicates that soil salinities below 50‰ do not significantly affect structure or productivity. The height of mangrove trees did increase considerably (Figs. 3 and 5), though, with distance from the shoreline or river-bank. Also, salinities decrease with distance landward, from 35.50‰ in site I to 27.50‰ in site V. This suggests that substrate at particular sites is an important factor regulating both forest structure and net primary productivity at Apar nature reserve. Hence, the soil salinity gradient is still needed in order to interpret the structural characteristics at Apar mangrove forest. Moreover, in the study plots, the substrate is a uniform clay loam (Table 4). The grain size and sediment type can directly or indirectly influence various aspects of hydrology and soil fertility at Apar mangrove forest. Thus, soil type and texture appear to alter substantially. Depending on the soil type, substantially different soil salinities at Apar mangrove forest can occur at identical heights above a given datum.

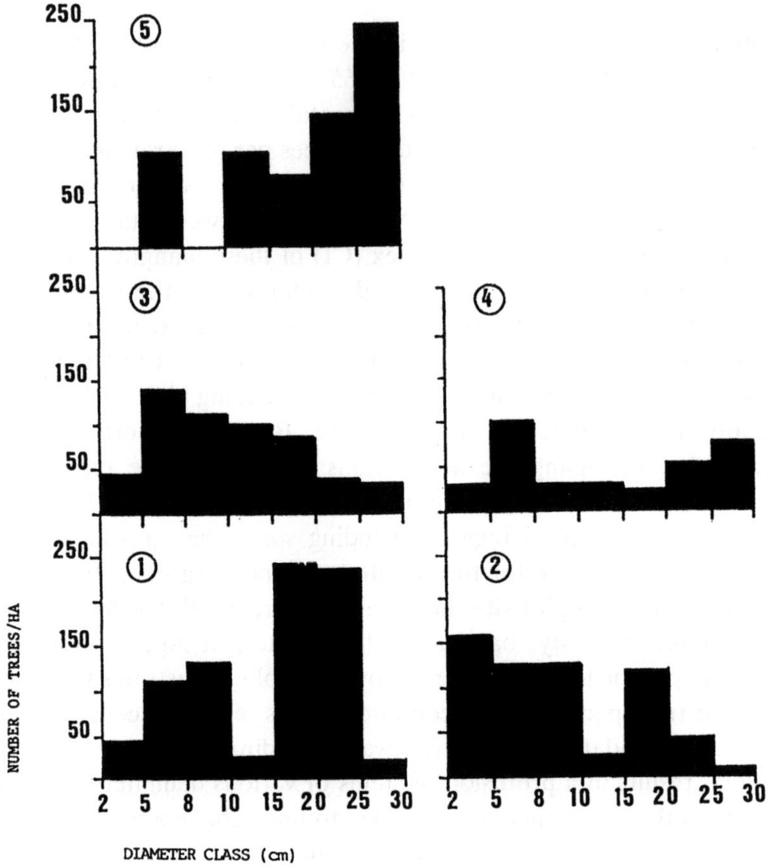

Fig. 4. Number of trees by diameter class at each site.

Net primary production (*PN*)

Average PN per day ranged from 40.44 to 45.49 kg C/ha/day in the five community types (Table 5). There was no apparent difference in PN between forest community types. Comparison with the mangrove forests at Muara Angke-Kapuk, Jakarta (Sukardjo 1989b), Grajagan, East Java and Segara Anakan, Central Java (Sukardjo, in preparation), suggests that the PN in the Apar nature reserve, East Kalimantan was greater as a whole. At Apar nature reserve, the PN at ten 10 m × 10 m subplot sites dominated by *R. apiculata* (sites I and II) ranged from 32.06 to 50.10 kg C/ha/day, and at ten 10 m × 10 m subplot sites dominated by *C. tagal* (sites IV and V) from 39.44 to 51.54 kg C/ha/day. Clearly, environmental conditions at particular sites are more important than forest type.

Regardless of forest type and the location of the particular sites, however,

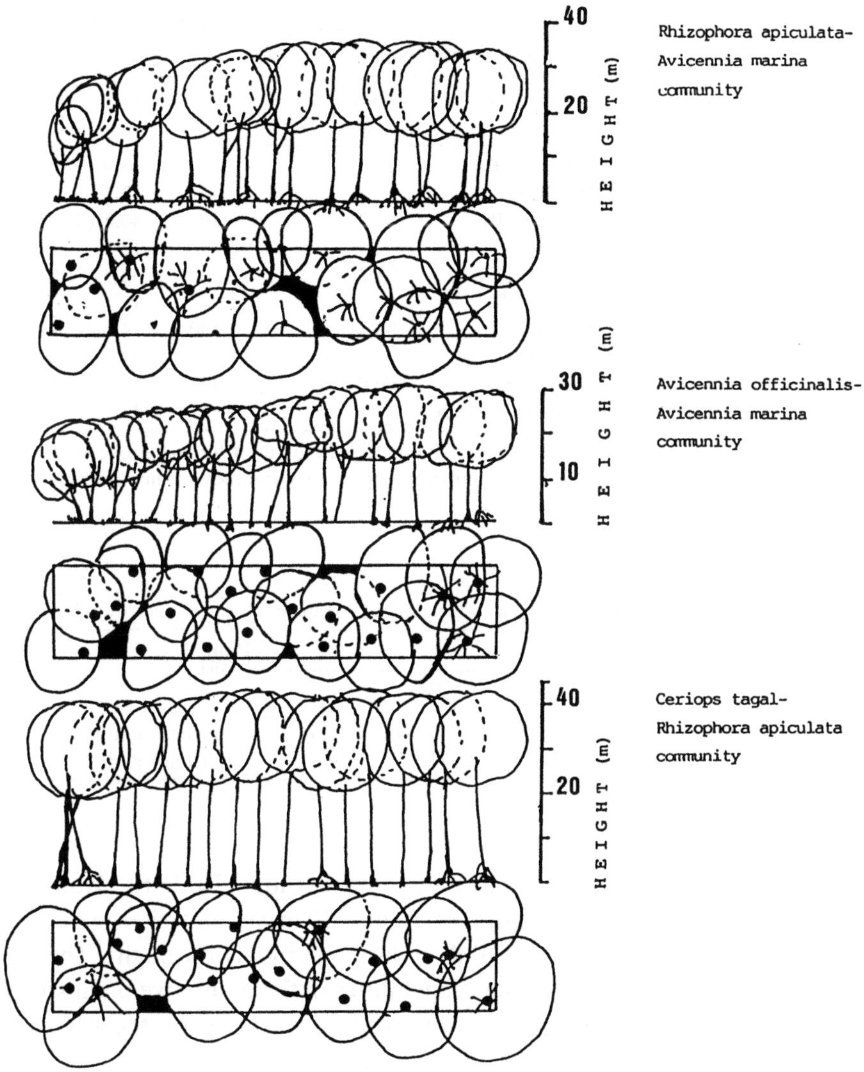

Fig. 5. Structure profiles of the mangrove forest types in the study area (based on the 20 m × 50 m plots).

the highest values of PN tended to be recorded at upstream sites (sites IV and V, 500 to 600 m away from the shoreline) subject to regular freshwater influence. The variability in PN seemed to be most pronounced in the upper intertidal zone most exposed to inundation by sea water. Thus, Tables 5 and 6 suggest that the differences in net primary production at my study sites are influenced mainly by nutrient loading and freshwater turnover.

The PN varied seasonally also, depending upon rainfall. After high rainfall (more than 300 mm) PN rates increased considerably; during periods of

Table 5. Estimated average Potential Net Primary Production (P_N), in kg C/ha/day, for the mangroves in the Apar nature reserve, East Kalimantan

Site	Community type	Tree species	Canopy closure (%)	n	P_N
I	*R. apiculata-* *A. marina*	Am, Ao, Bc Ra, Rm	97.7	198	43.9 (6.17) 12
II	*R. apiculata-* *A. marina*	Am, Bc, Bp Ra, Sa	87.7	197	40.4 (8.38) 12
III	*A. officinalis-* *A. marina*	Am, Ao, Bp Bs, KC, Ra	95.2	201	42.9 (5.35) 12
IV	*C. tagal-* *R. apiculata*	Bp, Bs, Ct Ra, Xm	97.7	190	44.2 (4.09) 12
V	*C. tagal-* *R. apiculata*	Bc, Bg, Bs Ct, Fm, Ra, Xm	99.0	195	45.5 (6.05) 12

For each site, n = number of light readings. The abbreviations used for mangrove species are given in Table 1. Standard error of the mean is shown in parenthesis, and the number of determinations are shown below the mean.

Table 6. Estimated Potential Net Primary Production (P_N) during the wet and drought periods

Period	P_N (in kg C/ha/day)					Rainfall (mm)
Site	I	II	III	IV	V	
March	50.10	48.82	48.28	48.34	51.54	>300 mm/month
September	37.76	32.06	37.58	40.16	39.16	<100 mm/month

drought (less than 100 mm/month) they remained low (Table 6). Moreover, the range of PN values for different 10 m × 10 m subplot sites was wider during periods of abundant rainfall than during periods of drought. This may be due to variation in gas-exchange capabilities in the mangrove plant species in combination with differences in soil properties caused by the variability in soil depth, frequency and period of inundation.

It is interesting, then, to examine the differences in soil factors between the different sites. Detailed studies of soil factors in relation to mangrove growth at Apar nature reserve (see Table 4) have shown the following features (Table 7):

1. Above-ground biomass (trunk volume) was highly significanthly and positively correlated with redox potential (Eh), NH_4-N and NO_3-N. In addition, a significant and negative correlation was found between biomass and soil salinity. All of these soil parameters were averages over depth (0–20 cm) and time (12 months).
2. The ranges of soil parameters recorded at the study sites were: Eh = -115 (±31) to −162 (±45) mV,
   soil salinity = 28.20 (±0.85) to 33.75 (±1.75) ‰,

Table 7. Correlation matrix for the estimated above-ground biomass, redox potential (E_h, mV), available P_2O_5 (ppm), NH_4-N (ppm), NO_3-N (ppm) and soil salinity (‰)

Site	Attribute	Redox potential	Available P_2O_5	NH_4-N	NO_3-N	Soil salinity
I	Biomass	0.983	0.994	0.986	0.968	−0.920
	Redox potential	–	0.973	0.938	0.950	−0.927
	Available P_2O_5		–	0.971	0.964	0.876
	NH_4-N			–	0.949	−0.935
	NO_3-N				–	−0.845*
II	Biomass	0.991	0.989	0.985	0.965	−0.917
	Redox potential	–	0.970	0.935	0.947	−0.918
	Available P_2O_5		–	0.968	0.961	0.873
	NH_4-N			–	0.946	−0.930
	NO_3-N				–	−0.842*
III	Biomass	0.975	0.986	0.978	0.960	−0.913
	Redox potential	–	0.965	0.930	0.942	−0.914
	Available P_2O_5		–	0.963	0.956	0.869
	NH_4-N			–	0.941	−0.925
	NO_3-N				–	−0.838*
IV	Biomass	0.958	0.995	0.990	0.969	−0.929
	Redox potential	–	0.969	0.964	0.909	−0.813*
	AvailableP_2O_5		–	0.988	0.954	0.912
	NH_4-N			–	0.987	−0.936
	NO_3-N				–	−0.975
V	Biomass	0.996	0.955	0.977	0.966	−0.926
	Redox potential	–	0.965	0.964	0.906	−0.810*
	Available P_2O_5		–	0.985	0.951	0.910
	NH_4-N			–	0.984	−0.933
	NO_3-N				–	−0.972

Trunk volume in $m^3/100\,m^2$ is estimated using Rochow's method (1974).
* Significance at $p < 0.05$, Significance at $p < 0.01$.

NH_4-N = 205.21 (±17.91) to 245.60 (±16.90) ppm,
NO_3-N = 27.195 (±1.995) to 32.425 (±2.353) ppm,
available P_2O_5 = 33.115 (±3.145) to 39.724 (±3.923) ppm.

The highly significant positive correlation between biomass and NH_4-N and NO_3-N implies a general nitrogen limitation for the growth of the mangrove species at these study sites. The higher PN rates coincide with favorable growth periods. Further, the data presented in Table 4 strongly suggest that the Apar soils are generally phosphorus-deficient, with most samples having soil P (available P_2O_5) values in the order of only 29.97 to 43.647 ppm. Low P (available P_2O_5) levels are evident in *R. apiculata-A. marina* (sites I and II), and *A. officinalis-A. marina* (site III) forest communities, with predominantly tidally influenced soils. On the other hand, the *C. tagal-R. apiculata* forest communities (sites IV and V) had P levels higher than those found in sites I, II and III. This is attributed to the fact that the forest area frequently received freshwater flooding from upstream, so the river is important in

Table 8. Major litter components at each site for August 1983 to August 1984

Site	Total litter (dry t/ha/year	Litter components (dry t/ha/year)			Community type*
		Leaves	Fruits + Flowers + Buds	Twigs	
I	21.10	7.78	7.25	6.07	*R. apiculata-A. marina*
II	20.50	7.56	7.04	5.90	*R. apiculata-A. marina*
III	28.82	10.62	9.90	8.29	*A. officinalis-A. marina*
IV	25.92	9.56	8.90	7.46	*C. tagal-R. apiculata*
V	29.35	10.82	9.00	9.53	*C. tagal-R. apiculata*

* Community type based on the dominant and co-dominant species.

supplying P. Moreover, the Eh values indicate intensely reducing soil conditions. These results were consistent with the higher sand content (29.90–31.30 %) in these soils (Table 4). Interestingly, my observation indicates that crab bioturbation is a common process that results in soil turnover, aeration and ground water/soil water mixing.

Soil NH_4-N and NO_3-N levels were also generally low at all sites, compared with concentrations at the mangrove forests in Cimanuk delta complex, West Java, Banyuasin South Sumatra and Muara Angke-Kapuk Jakarta (Sukardjo 1982, 1989b, Sukardjo & Kartawinata 1979, Sukardjo et al. 1984). These results indicate the general nitrogen limitation at the Apar mangrove forests. Hence, the overall trend of productivity at the Apar mangrove forests appears to be linked primarily with both nitrogen limitation and the very low phosphorus availability of the soils. Many workers have shown that increased nutrient availability, especially of P, is related to higher productivity in mangroves (Onuf et al. 1977, Snedaker & Brown 1981, Brown & Lugo 1982, Boto et al. 1984).

Litterfall and litter accumulation

The results of litterfall and litter accumulation measurements are summarized in Tables 8 and 9, and in Figs. 6, 7 and 8. Figure 6 shows the seasonal pattern of litterfall for five mangrove associations, and Figs. 7 and 8 the combined pattern of leaf fall and litter accumulation.

The *C. tagal-R. apiculata* forest communities (sites IV and V) exhibited the highest litter production (25.92 to 29.35 dry t/ha/year), followed by the riverine forest (site III: *A. officinalis-A. marina*, 28.82 dry t/ha/year) and the fringing forest (sites I and II: *R. apiculata-A. marina*, 20.50 to 21.10 dry t/ha/year) (Table 8). Total litterfall for the entire year, as well as the component totals in each category, were highest at site V. These values are higher than values reported for fringing and riverine mangrove forests in Florida (Pool et al. 1975, Brown & Lugo 1982, Twilley 1982), in Mexico (Day Jr.

Table 9. Comparison of site characteristics with litter production and accumulation at the five sites (August 1983 to August 1984)

Site	Distance to shoreline/river bank (m)	Density (stems/ha)	Basal area (m^2/ha)	Litter production (dry/ha/year)	Litter accumulation by component (dry t/ha/year)				
					Total	Leaves	Flowers + Fruits + Buds	Twigs	Other*
I	100/100	812	17.72	21.10	48.70	16.10	12.40	13.90	6.30
II	150/150	444	15.18	20.50	54.40	18.20	12.50	15.90	7.80
III	300/50	568	19.47	28.82	50.10	17.30	14.20	13.90	4.70
IV	500/200	348	24.19	25.92	65.70	25.90	19.20	14.40	6.20
V	600/250	700	37.57	29.35	70.50	30.30	17.90	16.70	5.60

* Unidentified materials.

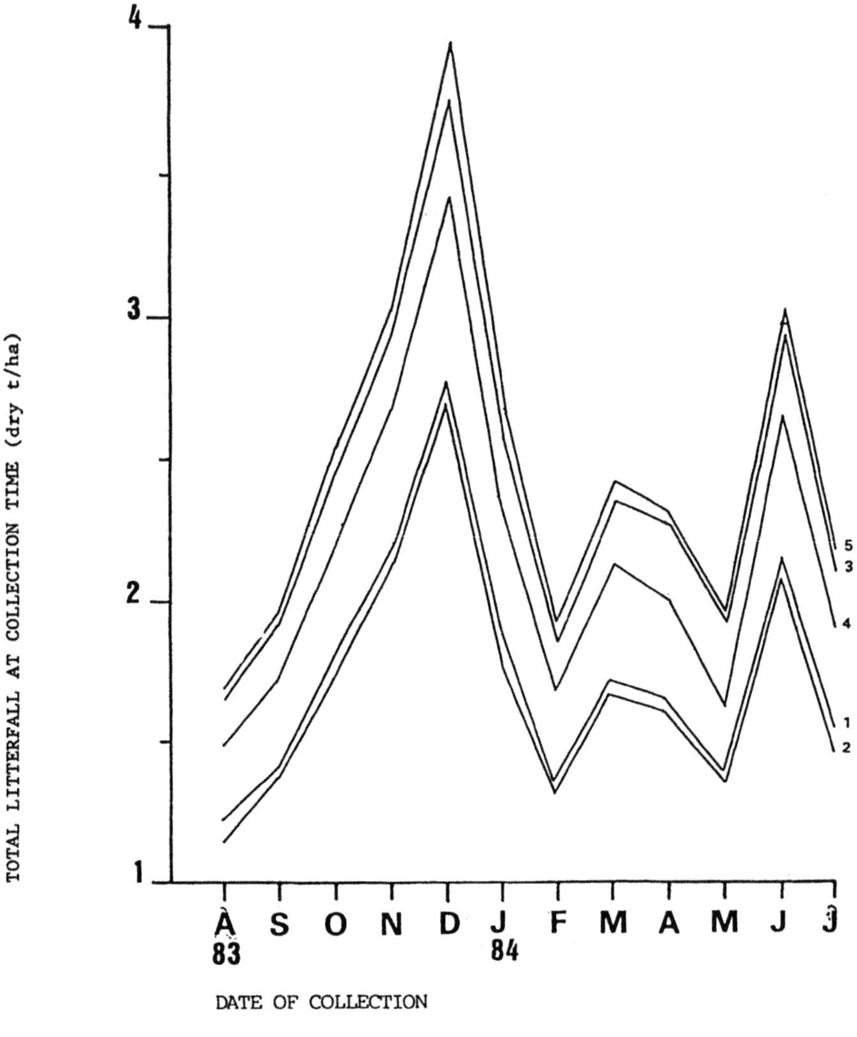

Fig. 6. The monthly litterfall measured at the five sites over the study period 1983–1984.

et al. 1987), in Malaysia (Sasekumar & Loi 1983), in Papua New Guinea (Leach & Burgin 1985), in New Zealand (Woodroffe 1982), in Australia (Bunt 1978, 1982, Goulter and Allaway 1979), in Muara Angke-Kapuk Indonesia (Sukardjo 1989b), and in Saudi Arabia (Saifullah et al. 1989). Throughout the range of mangrove forests, there seems to be a general trend for peak litterfall to occur during the rainy season. In my study area, the highest litterfall at all five sites occurred during the rainy season, in the months with highest rainfall (Fig. 6). This peak litterfall, at a time when there is abundant water flowing through the area, may help to account for the high export of materials from the forests.

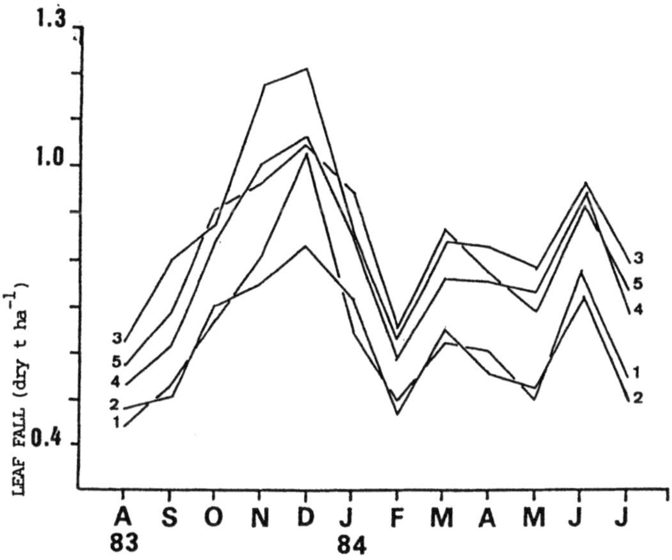

Fig. 7. Leaf litterfall in the five mangrove associations of the Apar Nature Reserve.

Leaffall was the dominant component of litterfall throughout the year at the five sites (Fig. 7, Table 8). There were definite peaks in litterfall. Litterfall increased at the onset of the dry season (August), with *Avicennia* (site III) dropping its leaves one month before the *Ceriops* (sites IV and V) and *Rhizophora* (sites I and II) forest communities (Fig. 7). This suggests that soil salinity plays a role principally in the senescence of the leaves. Figure 9 shows this phenomenon expressed as an inverse relationship between soil salinity and litter production ($r = -0.849$). Thus, the physiography of the Apar study area supports the best developed mangrove forests, as reflected in luxuriance, abundance and diversity (Tables 1, 2 and 3). The fall of flowers, fruits, buds and twigs, and other woody detritus showed strong peaks during the rainy season at all five sites. The smaller fruits, flowers and buds of *A. marina* and *A. officinalis* however, were very numerous; even at sites I and II, *R. apiculata* individuals (120–340 trees/ha, 96–176 saplings/ha) were more abundant (Table 2). Total fruit fall was higher at site III, where the forest is of the riverine type, with *A. officinalis* (144 trees/ha, 128 saplings/ha) and *A. marina* (40 trees/ha, 64 saplings/ha) the dominant and co-dominant species, respectively.

Table 9 shows the variation in litter pools at sites I to V, reflecting differences in the vegetation structure and litter input. In *Ceriops* forest

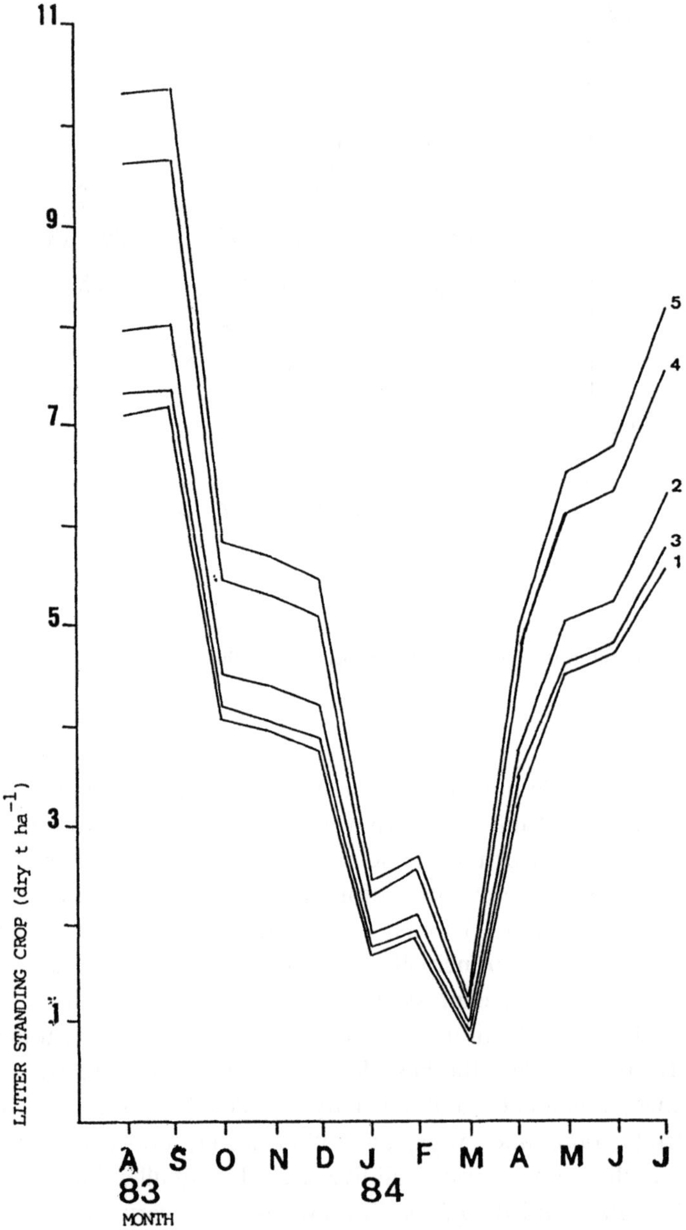

Fig. 8. The monthly litter pool (standing crop) at the five study sites during the period 1983–1984).

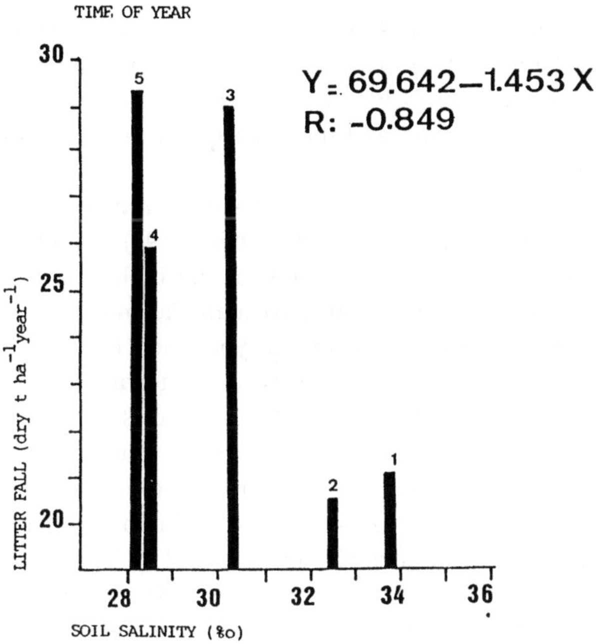

Fig. 9. Annual litterfall rate for the five study sites vs mean soil salinity (‰) (0–20 cm depth).

Table 10. Calculated litter-decay constants (reciprocal of resident time, years). The calculation of the litter decay constant, k, and the times X_{ss} (95%) and X_{ss} (99%) to reach 95% and 99% of steady-state conditions follows Olson (1963)

Site	k	X_{ss} (95%) (years)	X_{ss} (99%) (years)
I	0.43	6.93	11.56
II	0.37	7.94	13.23
III	0.57	5.21	8.69
IV	0.39	7.61	12.68
V	0.41	7.20	12.01

communities (sites IV and V) the litter pools are larger, whereas in the *Rhizophora* forest communities (sites I and II) litter tended to accumulate in prop-root systems, or to be washed out by waves and/or tidal currents. A large pulse of twigs and other woody litter, and a decrease in the loose-litter pool at all five sites occurred during the months with high rainfall. This loss was related to washouts that occurred during the heavy rain.

Table 10 shows the annual turnover rate of total litter. Litter turnover rate was highest in the *Avicennia* forest community (site III). There appeared to be higher rates of litter disappearance during rainy periods than during dry periods, when loose litter tended to accumulate. Generally, the annual

"samphire" communities of *Arthocnemum* spp., meadows of *Sporobulus virginatus*, and bare salt flats are much more extensive in tropical Australia. The main reasons are probably the generally lower and more markedly seasonal rainfall in Australia and the consequent poor, irregular supply of river sediments.

The information acquired through this study, in combination with other research in the area (e.g. Sukardjo 1987, 1988, 1989a), is useful in developing plans for management and conservation of the biological resources of the Apar deltaic system. It was possible to map the mangrove forests of the Apar nature reserve area onto airphoto types, and by only limited ground sampling to relate the airphoto patterns to the terrain and vegetation features. I have demonstrated that the mangrove forests at Apar nature reserve are very productive. Many workers have shown that organic detritus from the mangrove forest is an important source of food for many marine organisms, and the mangrove forest area is an important habitat for many commercial fish and crustaceans (Odum & Heald 1972, 1975, Staples 1980a, 1980b, Bell et al. 1984, Robertson & Duke 1987). Apar Bay is the most important commercial fishery area in East Kalimantan (Dinas Perikanan Propinsi Kalimantan Timur, pers. comm. 1984–1989). Thus, the productivity and extent of mangrove forests in the area are directly related to high fishery productivity. The results presented herein also suggest how the forest can be managed to maintain high productivity. Local government should avoid activities which lead to increased stress on mangroves.

The Apar deltaic system in East Kalimantan has a sparse human population and is largely unspoiled; it is readily accessible, however, from Tanah Grogot. If areas of mangrove vegetation are to be set aside for purposes of mangrove study and recreation, the Apar nature reserve area well deserves consideration.

Acknowledgements

I wish very much to thank Prof. Dr. Toru Yano, the Director of the Center for Southeast Asian Studies, Kyoto University, Kyoto Japan for his administrative support during my work here. Prof. Dr. Isamu Yamada provided constructive criticism to the typescript abstract. I thank Miss Hiroko Muguchi for typing the manuscript.

References

Backer, C. A. & Bakhuizen van den Brink, R. C. Jr. 1963–1968. Flora of Java, I-III. Noordhoff, Groningen.

Bell, J. D., Polrand, D. A., Burchmore, J. J. Pease, B. C. & Middleton, M. J. 1984. Structure of a fish community in a temperate tidal mangrove creek in Botany Bay, New South Wales. Austr. J. Mar. Freshwat. Res. 31:137–146.

Boto, K. G., Bunt,J. S. & Wellington, J. T. 1984. Variations in mangrove forest productivity in northern Australia and Papua New Guinea. Estuar. Coast. Shelf Sci. 19:321–329.

Brown, S. & Lugo, A. E. 1982. A comparison of structural and functional characteristics of saltwater and freshwater forested wetlands. In: Gopal, B., Turner, R. E. & Wetzel, R. G. (eds.) Wetland Ecology and Management, pp. 109–130. International Scientific, Jaipur.

Bunt, J. S. 1978. The mangrove of the eastern coast of Cape York Peninsula north of Cooktown. Great Barrier Reef Mar. Park Author Working Paper No. 1:253–269.

Bunt, J. S. 1982. Studies of mangrove litterfall in tropical Australia. In: Clough, B. F. (ed.), Mangrove Ecosystem in Australia, pp. 223–237. Austr. Nat Univ. Press, Camberra.

Bunt, J. S., Boto, K. G. & Boto,G. 1979. A survey method for estimating potential levels of mangrove forest primary production. Marine Biology 52:123–128.

Cox, G. W. 1967. Laboratory manual of general ecology. Vm. C. Brown, Dubuque Iowa.

Curtis, J. T. & Mcintosh, R. P. 1951. An upland forest continuum in the prairie forest border region of Wisconsin. Ecology 32:476–496.

Day Jr, J. W., Corner, W. H., Ley-Lou, F., Day, R. H. & Novarro, A. M. 1987. The productivity and composition of mangrove forest, Laguna de Terminos Mexico. Aquatic Botany 27:267–284.

De Haan, J. H. 1931. De Tjilatjapsche vloedbosschen. Tectona 24:39–76.

Dinas Hidro-Oseanografi ALRI 1983–1985. Daftar pasang surut. Jan Hidros ALRI.

Ding Hou. 1958. Rhizophoraceae. Flora Malesiana 5:429–493.

Gosz, J. R., Liken, G. E. & Bormann, F. H. 1973. Nutrient release from decomposing leaf and branch litter in the Hubbard Brook forest New Hampshire. Ecology Monographs 43:173–191.

Goulter, P. F. E. & Allaway, G. W. 1979. Litterfall and decomposition in a mangrove stand, *Avicennia marina* (Forsh.) Vierh. in Middle Harbour Sydney. Aust. J. Mar. Freshwater Res., 30:541–546.

Holdridge, L. R., Grenke, W. C., Hatheway, W. H., Liang, T. & Tosi, J. A. 1971. Forest environment in tropical life zones. Pergamon Press, N.Y. 747 pp.

Leach, G. J. & Burgin, S. 1985. Litter production and seasonality of mangrove in Papua New Guinea. Aquatic Botany 23:215–224.

Lugo, A. E. & Snedaker, S. C. 1974. The ecology of mangrove. Ann. Rev. Ecol. Syst. 5:39–64.

Mueller-Dombois, D. & Ellenberg, H. 1974. Aims and methods of vegetation analysis. Wiley, New York. 547 pp.

Newbould, P. J. 1967. Methods for estimation the primary production of forest. IBP Handbook No. 2. Blackwell Sci. Publ., Oxford. 62 pp.

Nye, P. H. 1961. Organic matter and nutrient cycles under moist tropical forest. Plant and Soil 13:333–346.

Odum, W. E. & Heald, E. J., 1972. Trophic analyses of an estuarine mangrove community. Bull. Mar. Sci. 22:671–738.

Odum, W. E. & Heald, E. J. 1975. The detritus-based food web of an estuarine mangrove community. In: Cronin, L. E. (ed.), Estuarine Research, pp. 265–286. Acad. Press. Inc., New York.

Odum, W. E., McIvor, C. C. & Smith, T. S., III. 1982. The ecology of the mangroves of South Florida: a community profile. US Fish and Wild-life Service. Office of Biol. Services, Washington DC. FWS/OBS-81124. 154 pp.

Ogawa, H., Joda, K., Kira, T., Ogino, K., Shidei, T., Ratanawongse, D. & Apasutaya, P. 1965. Comparative ecological study on three main types of forest vegetation in Thailand: I. Structure and floristic composition. Nature and Life in Southeast Asia IV:13–48.

Olson, J. S. 1963. Energy storage and the balance of producers and decomposers in ecological system. Ecology 44:322–331.

Ong, J. E., Gong, W. K. & Wong, G. H. 1982. Productivity and nutrient status of litter in a mangrove forest. Biotrop Spec. Publ. No. 17: 33–42.

Onuf, C. P., Teal, J. M. & Valiela, I. 1977. Interaction of nutrient status, plant growth and herbivory in a mangrove ecosystem. Ecology 58:514–526.

Pool. D. J., Lugo, A. E. & Snedaker, S. C. 1975. Litter production in mangrove forests of Southern Florida and Puerto Rico. In: Walsh, G. E., Snedaker, S. C. & Teas, H. J.(eds.) Proc. Intern. Symp. Biol. and Manage. of Mangroves, pp. 213–237. Inst. of Food and Agric. Sci., Univ. of Florida, Gainesville.

Pool, D. J., Snedaker, S. C. & Lugo, A. E. 1977. Structure of mangrove forests in Florida, Mexico and Central America. Biotropica 9:195–212.

Robertson, A. I. & Duke, N. C. 1987. Mangroves as nursery sites: comparisons of the abundance and species composition of fish and crustaceans in mangroves and other nearshore habitats in tropical Australia. Mar. Biol. 96:193–205.

Saeffulah, S. M., A. K. Khafaji and A. S. Mandura 1989. Litter production in a mangrove stand of the Saudi Arabia Red sea coast. Aquatic Botany 36:79–86.

Sasekumar, A. & Loi, J. J. 1983. Litter production in three mangrove forest zones in the Malay Peninsula. Aquatic Botany 17:283–290.

Snedaker S. C. & Brown, M. S. 1981. Water quality and mangrove ecosystem dynamics. EPA, Office of Research and Development, Gulf Breeze, FL. EPA. 80 pp.

Soemodihardjo, S., Nontji, A. & Djamali, A. (eds.) 1979. Prosiding Seminar Ekosistem Hutan Mangrove (Proc. of the Seminar on Mangrove Forest Ecosystem). LIPI Panitia Programm MAB Indonesia, Jakarta 1979. 226 pp. (in Indonesian).

Soemodihardjo, S., Soerianagara, I., Sutisna, M., Kartawinata, K., Supardi, Naamin, N. & Rasjid, H. Al. (eds.) 1984. Prosiding Seminar II Ekosistem Mangrove (Proc. of the Second Seminar on Mangrove Ecosystem). LIPI Panitia Program MAB Indonesia, Jakarta 1984. 369 pp. (In Indonesian).

Soerianagara, I., Adisoemarto, S. Soemodihardjo, S. Hardjowigeno, S., Sudomo, M. & Ongkosongo, O. S. R. (eds.) 1987. Prosiding Seminar III Ekosistem Mangrove (Proc. of the Third Seminar on Mangrove Ecosystem). LIPI Panitia Program MAB Indonesia, Jakarta 1987. 198 pp (in Indonesian).

Staples, D. J. 1980a. Ecology of juvenile and andolescent banana prawn, Penaeus merguiensis in a mangrove estuary and adjacent offshore area of the Gulf of Carpentaria. I. Immigration and settlement of postlarvae. Austr. J. Mar. Freshwat. Res. 31:635–652.

Staples, D. J. 1980b. Ecology of juvenile and andolescent banana prwan, *Penaeus merguiensis* in a mangrove estuary and adjacent offshore area of the Gulf of Carpentaria. II. Emigration, population structure and growth of juvenile. Austr. J. Mar. Freshwat. Res. 31: 653–665.

Steenis, C. G. G. J. van 1958. Ecology (The Introductory part to the monograph of Rhizophoraceae by Ding Hou). Flora Mal. 5:431–441.

Sukardjo, S. 1982. Soils in the mangrove of Cimanuk delta, West Java, Indonesia. Biotrop Spec. Publ. No. 17:113–126.

Sukardjo, S. 1987. Soils in the mangrove swamp forests of the Apar Bay, Tanah Grogot East Kalimantan, Indonesia. Paper presented at the International Peat Soc. Symp. on the Tropical Peat and Peatland for Development, Yogjakarta 15–18 February 1987. 25 pp.

Sukardjo, S. 1988. Community of *Acrostichum aureum* L. in the *Avicennia officinalis* L. mangrove forest in the Muara Pasir, Kandilo river East Kalimantan. Rimba Indonesia XXII (1–2):113–126.

Sukardjo, S. 1989a. The biological resources of the mangrove forest in Apar deltaic system, Tanah Grogot East Kalimantan, Indonesia. In: Paris, D. & Prentice, C. (eds.) Proc. of the Conf. on Waterfowl Conservation in Asia, pp. 18–24. Asian Wetland Bureau Publ. No.52. IWRB, Kuala Lumpur, Malaysia.

Sukardjo, S. 1989b. Litterfall production and turnover in the mangrove forest in Muara Angke-Kapuk, Jakarta. Biotrop Spec. Publ. No. 37:129–144.

Sukardjo, S. & Kartawinata, K. 1979. Mangrove forest of Banyuasin, Musi river estuary, South Sumatra. Biotrop Spec. Publ. No. 10:61–68.

Sukardjo, S., Kartawinata, K. & Yamada, I. 1984. The mangrove forest in Bungin river of Banyuasin district, South Sumatra. In: Soepadmo, E., Rao, A. N. & MaCintosh, D. J. (eds.) Proc. of the Symp. on the Mangrove Environ.: Research and Management, pp. 121–141. The University of Malaya, Kuala Lumpur, Malaysia-Unesco.

Thom, B. G., Wright, L. D. & Colleman, J. M. 1975. Mangrove ecology and deltaic-estuarine geomorphology: Cambridge Gulf, Oral river Western Australia. J. Ecology 63:203–232.

Twilley, R. R. 1982. Litter dynamics and organic carbon exchange in black mangrove, Avicennia germinans basin forests in a Southwest Florida estuary. Ph.D Thesis, Univ. of Florida, Ganesville. 260 pp.

Twilley, R. R., Lugo, A. E. & Patterson-Zucca, C. 1986. Litter production and turnover in basin mangrove forests in Southwest Florida. Ecology 67:670–683.

Watson, J. G. 1928. Mangrove forest of the Malay Peninsula. Malay For. Rec. 6:1–175.

Woodroffe, C. D. 1982. Litter production and decomposition in the New Zealand mangrove, *Avicennia marina* var. *resinifera*. N.Z.J. Mar. Fresh wat Res. 16:179–188.

Woodroffe, C. D. 1985a. Variability in detrital production and tidal flushing in mangrove swamps. In: Bradsley, K. N., Davie, J. D. S. & Woodroffe, C. D. (eds.) Coasts and Tidal Wetland of the Australian monsoon region, pp. 201–212. Australian National University, North Australian Research Unit, Mangrove Monograph No. 1.

Woodroffe, C. D. 1985b. Studies of a mangrove basin, Tuff Crater, New Zealand. III. The flux of organic and inorganic particulate matter. Estuar. Coast. Shelf Sci. 20:447–462.

35. Harmonized ecosystem research methodology for an integrated forest research programme in Southeast Asia

EBERHARD F. BRUENIG & SONJA CSOMOS
University of Hamburg, D-1031 Hamburg 80, Germany

Abstract

The sites and vegetation of tropical forest ecosystems of all levels (from mosaic or "eco-unit" to formation class) are extremely heterogeneous, complex and dynamic. Their biotic and atmospheric (physical and chemical) environments are equally heterogeneous and extremely variable as a result of oscillations and secular changes. Consequently, results of research in one specific ecosystem are unique. Generalization for scientific purposes or application in practice is only feasible if the data base is broadened. Research projects must consequently be integrated into regional and global networks, which requires a harmonization of methods.

First attempts in this direction were made by the International Union of Forest Research Organizations (IUFRO) beginning in 1892, in the field of growth and yield research in forest stands. The International Biological Programme (IBP) harmonized its ecosystem research methods in temperate forests but had only one project in tropical rainforest, at Pasoh Forest Reserve, Malaysia. The International MAB-IUFRO Workshop on Tropical Rainforest Ecosystem Research at Hamburg in 1977 was the next step to evolve further the principles of harmonization of basic tropical forest ecosystem research methods and harmonization of data and information systems.

The objectives and advantages of harmonization of methods are briefly described. Direct data pooling even may require standardization. The presently critical situation of the tropical rainforest and the changing global environment call for concerted efforts in scientific research and monitoring, technological development and forestry and land-use practice. In order to provide reliable scientific backstopping effectively and rapidly, a network approach to harmonized long-term ecological research and to comprehensive monitoring is required.

Key words: ecosystem diversity, ecosystem research, global change, harmonized methodology, networking

E.O. Box et al. (eds.), Vegetation Science in Forestry, 613–620.

Situation

Ecological research in the humid to superhumid, predominantly evergreen tropical forest (tropical rainforest), especially since the Second World War, has convincingly shown that:

- there exists an extremely high level of heterogeneity of soil and vegetation at small scales (single trees, gap sizes of 0.05 to 0.3 ha), medium scales (landscape, landform site-type units) and large scales (regions); this heterogeneity exceeds that known from temperate forests;
- the climatic environment is much more variable and extremely stressful episodic events are much more common and ecologically severe than previously thought, with respect to moisture, temperature, radiation and wind.

A. v. Humboldt (1847) and A. F. W. Schimper (1903) assumed that there are unifying principles in the richness of forms and diversity of their mixtures, i.e. the evenness or unevenness, or degree of dominance, of the contributions of the plant or animal species to the number of individuals or amount of biomass. In particular, unifying principles should govern the relationship between the form of plants and their external conditions. This has been vindicated by extensive evidence from comparative rainforest research in the last two decades. The variation of structural features of trees and forests is closely correlated to ecological gradients of soil conditions and altitude. The following features show this relationship particularly consistently and more or less clearly, considering the great heterogeneity of sites and vegetation in the tropical forest ecosystems (Bruenig 1970, 1973):

- sclerophylly
- leaf sizes
- canopy roughness (aerodynamic properties)
- tree species richness (spp./100 indiv. or spp./ha)
- tree species diversity (evenness of mixture)
- tree species distribution patterns
- tree basal-area stocking
- tree height/stem diameter ratio (except for yet unexplainable lower heights in the Central Amazon-Rio Negro basin)
- tree biomass per hectare (except the lower values due to the less steep regression "biomass on basal area in the Amazon-Rio Negro basin", see Kurz 1982).

The analogous shift of these features with latitude has recently been demonstrated in a comparison of forest plots between Sarawak and Brunei at the equator and South China between 18 and 23° N (Huang 1990). In spite of these consistencies in recent research results, the cause-effect interactions and relationships are poorly if at all understood. This is particularly true with respect to the significance of tree species richness and mixture diversity for such ecosystem features as functionality, self-regulatory capacity, resistence, resilience, elasticity, viability and finally utility for humankind. Almost nothing is known about the nature, conditions and cause-effect interrelationships of organizational structure and the microbial-faunal-floral interaction network and the effect of human activities on them.

These deficiencies are caused and continue to be aggravated in the tropical zone by:
- very low levels of basic ecological research
- intellectual and operational isolation of projects
- very poorly functioning and scattered observation (monitoring) stations
- severe understaffing, underfunding and underdevelopment of existing research stations, excessive dependence on external expertise and support.

Problems

Problems for scientific research in natural tropical forest ecosystems include the facts that:
- the site conditions are very heterogeneous and a wide range of conditions has to be covered;
- vegetation is accordingly heterogeneous, amplified by the effect of gap regeneration cycles and long-term successional trends;
- very long time series of quantitative observations are essential;
- a great variety of different ecosystem parameters must be continuously and consistently measured, while research and monitoring instruments and data processing and evaluation facilities change;
- such research is expensive in money, personnel and expertise;
- sufficient, accurate scientific data are not available when they are needed.

Conclusions

The following conclusions can be drawn with respect to tropical forest ecosystem research design:
- the extremely high spatial heterogeneity and the internal, temporally oscillating pulses require adequate coverage by large research plots, a wide scatter of plots and long-term observation;
- the extremely high temporal variability of climatic and biotic external conditions and of atmospheric inputs of chemical compounds, and the site-specific nature of ecosystem response also require long-term observation at a variety of sites;

- on-going global change adds further complication and uncertainty which amplify the need for long-term observation at a variety of monitoring sites.

We can conclude that the existence of unifying principles makes comparative basic ecological research feasible, but the heterogeneity and variablility of internal and external conditions require strict harmonization of methodology within and between projects as part of regional and global networks. Harmonization of research methods means making design, procedures and data consistent and compatible between projects of a particular programme, without imposing rigid standards which could become an obstacle to creativity, local adaptation and methodological progress. The purpose is to provide the sufficient, reliable data-bases which are required for forest and environmental management.

Targets

To achieve this purpose it is necessary to aim at the following targets:
- to establish an international integrative network of core tropical rainforest ecosystem research sites
- to coordinate between the tropical ecosystem research network and global research and monitoring programmes
- to establish national data and information linkage between basic ecosystem research, applied ecological research, applied experimental trials and practice of forest management and land use.

Approach

The following general approach seems feasible and is recommended:
- design of an efficient, integrated multi-purpose data-base system with standardized data norms for forest ecosystem research which is harmonized with the data sets used in applied research, management and forest and land development;
- design of an integrated hierarchic multi-purpose Geographic Information System incorporating basic forest ecosystem research, applied forestry research, forestry and land management, integrated land-use planning and development, and their continuous monitoring;
- defining internationally accepted and applied criteria for the selection of rainforest ecosystem research sites within networks
- development of a loose-leaf international manual of methods of long-term ecological research in tropical rainforest ecosystems, including prescribed

minimum sets of parameters to be measured and system modelling and data processing procedures.

Harmonization and integration

The need for rigour in ecosystem research

P. Groom in his appreciation of A. F. W. Schimper (1903, page XII) stated that Schimper as the first insisted on the employment of "strict" methods in ecological research, especially on the problem of connection between plant form, structure and physiology at different sites. This requires comparable data in compatible data sets obtained by harmonized research methods. This condition applies even more to ecosystem research projects, if satisfactory synthesis of results within an ecosystem project are to be achieved. The extreme heterogeneity of states and processes, the great organizational complexity and the general dynamic variability at all spatial scales of the tropical rainforest add to the necessity for harmonization between ecosystem research projects in order to reduce the risk of spurious results in isolated projects and to increase the data pool on which the results of research can be based.

National harmonization and integration: the case of Sarawak

In response to the report by a mission of the International Tropical Timber Organization (ITTO) of highly qualified professionals to Sarawak (ITTO 1990), the Sarawak Forest Department developed a Research and Development Programme for Forestry in Sarawak which is exemplary and possibly presently unique in its approach to harmonization of methods and integration of planning, implementation and monitoring between research, forestry practice and general land-use and economic development.

The core of the programme is a unified data base and Geographic Information System (GIS) (Fig. 1). The data from forestry research and practice are harmonized and mutually compatible, thereby bridging the notorious gap between research and practice. The levels represented by basic ecosystem research, applied ecological research, forestry, trials and experimentation, and forestry and general land-use practice are integrated by means of the hierarchic structure of the GIS, which is currently developed. A system-analysis approach is consistently applied to assure recognition and comprehensive coverage of all relevant elements and interactions, at all levels from basic research to policy making. The final aim is to base political and manage-

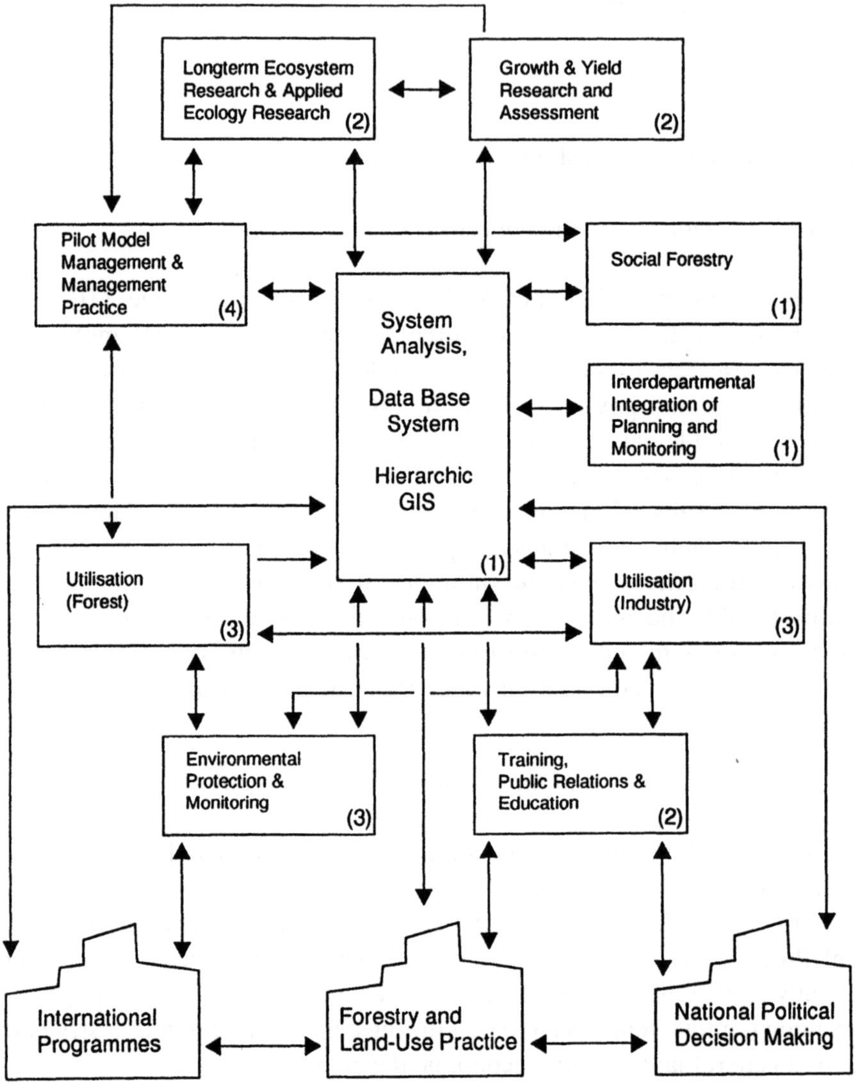

Fig. 1. Structure of an integrated research and development programme in forestry. (Source: Research and Development Programme for Forestry in Sarawak 1990, Forest Department, Kuching, unpublished). In brackets: number of major projects for the period 1991 to 1996. Jagged top-edge: external systems.

ment decisions on a sufficient, reliably integrated and continuously up-dated data base stored in an easily accessible, user-friendly hierarchic GIS.

International linkages

Research and monitoring in the tropical rainforest should be designed to contribute data to related international and regional programmes (Fig. 1), such as:

- IGBP, International Geo-Biosphere Programme
- WCRP, World Climate Research Programme
- TSBF, Tropical Soil Biology and Fertility Programme of IUBS/ICSU
- MAB, Man and Biosphere Programme of UNESCO
- Basic Ecological Research 50 ha-Plot Network in South and Southeast Asia (Manokaran et al. 1990).

This again requires harmonization of the data by standardizing measurement and data processing procedures at national and international levels. INTECOL could play an important role in this endeavour.

References

Adisoemarto, S. & Bruenig, E. F. (eds.) 1979. Transactions of the Second International MAB-IUFRO Workshop on Tropical Rainforest Ecosystems Research. Chair of World Forestry, Hamburg-Reinbek, Special Report No. 2. 295 pp.

Bruenig, E. F. 1970. Stand structure, physiognomy and environmental factors in some lowland forests in Sarawak. Trop. Ecol. 11:26–43.

Bruenig, E. F. 1973. Species richness and stand diversity in relation to site and succession in Sarawak and Brunei. Amazoniana 4:293–320.

Bruenig, E. F. (ed.) 1977/1978. Transactions of the International MAB-IUFRO Workshop on Tropical Rainforest Ecosystems Research. Chapt. G: Special contributions on methodology, pp. 124–264. Chapt. H: An outline for a basic interdisciplinary research programme with emphasis on a common approach to research methodology and measurement techniques, pp. 265–331. Chapt. I: Development of a standard methodology and of principles for the MAB project no. 1, pp. 332–339. Hamburg-Reinbek: Chair of World Forestry. Special Report No. 1. 1st ed. 1977, 2nd ed. 1978. 370 pp.

Bruenig, E. F., Bossel, H., Elpel, K. P., Grossmann, W. D., Schneider, T. W., Wan Zhou-hao & Yu Zuo-yue 1986. Ecological-Socioeconomic System Analysis and Simulation: A Guide for Application of System Analysis to the Conservation, Utilization and Development of Tropical and Subtropical Land Resources in China. Bonn, National MAB-Committee, Mitt. Nr. 24, and German Foundation for International Development (DSE), Feldafing, DOK 1381, SE 78-059086 ox, ISBN 3-926301-01-7, XIII + 388 + VI pp.

Huang, Y. W. 1990. Forest Structure as an Ecological Indicator: A Comparative Analysis of Forest Stands in Sarawak, Brunei and South China. Thesis, Department of Biology, Hamburg University. 139 pp.

ITTO 1990. The Promotion of Sustainable Forest Management: A Case Study in Sarawak, Malaysia. Report to the International Tropical Timber Council. ITTO, Yokohama. 208 pp.

Kurz, W. A. 1982. Biomasse eines amazonischen Feuchtwaldes: Entwicklung von Biomasseregressionen. Mitt. Bundesforsch.anst. Forst-Holzwirtsch. No. 139, III + 98 pp. Hamburg.
Manokaran, N., Lafrankie, J. V., Kochummen, K. M., Quah, E. S., Klahn, J. E., Ashton, P. S. & Hubbel, S. P. 1990. Methodology for the fifty hectare research plot at Pasoh Forest Reserve. Forest Research Institute Malaysia, Kepong, Research Pamphlet No. 104. 69 pp.
Schimper, A. F. W. 1903. Plant-Geography upon a Physiological Basis. Oxford, Clarendon Press, XXX + 839 pp.

Species index

Location index

Subject index

ERRATA

E.O. Box et al. (eds): Vegetation Science in Forestry: Global Perspective Based on Forest Ecosystems of East and Southeast Asia. Handbook of Vegetation Science.
ISBN 0-7923-2679-2.

Table of Contents (pp v–vii)

In addition to minor typographical errors:
Chapter 10, authors, should read: G. Pignatti & S. Pignatti
Chapter 11, title, should read: West and South-East European Deciduous and Deciduous-Conifer Mixed Forests of the Lowlands
Chapter 16, authors, should read: Chen Wei-lie (*only*)
Chapter 19, title, should read: The Spruce Forests of Northern China
Chapter 20, title, should read: Boreal Forest Phytogeocoenoses of Hokkaido Island, Japan
Chapter 26, title, should read: Towards a Regional Forest Classification for the Humid Tropics of Asia
Chapter 28, title, should read: Mosaic Structure and Tree Growth Pattern in a Monodominant Tropical Seasonal Evergreen Forest in Thailand
Chapter 32, title, should read: Vegetation Pattern, Structure and Dynamics in Different Types of Humid Tropical Evergreen Forests of South-East Asia
Chapter 35, authors, should read: E. F. Bruenig & S. Csomos

Chapter 7 (pp 125–140)

Includes errors in captions for Photos 2–12. Correct captions are:
Photo 2 (p. 130): Valdivian laurel forest with epiphytic *Fascicularia bicolor*.
Photo 3 (p. 131): *Dombeyo-Eucryphietum* at Fiordo Comau (42–42.5° S). Near to its southern distribution boundary, *Eucryphia* (flowering) is restricted to a narrow belt above sea-level.
Photo 4 (p. 132): *Eucryphia cordifolia* is often spared from clearing.
Photo 5 (p. 133): Millennial *Fitzroya cupressoides* forest at Alerce Andino National Park.
Photo 6 (p. 134): *Pilgerodendron uvifera* forest with *Drimys winteri* and *Myrtaceae* at Golfo Elefantes (46.5° S).
Photo 7 (p. 134): *Nothofagus betuloides* forest forming the upper altitudinal belt at Queulat National Park (44.5° S).
Photo 8 (p. 135): *Nothofagus betuloides* krummholz at Isla Navarino (55° S).
Photo 9 (p. 135): Old specimen of *Nothofagus obliqua* and *Laurelia sempervirens* are to be found in lowland pastures around 41° S.
Photo 10 (p. 136): The view from SE to NW shows the lakesides of Lago Panguipulli (39° 40′ S) with extensive bamboo thickets (light grey) in place of laurel forest.
Photo 11 (p. 137): Remnants of recently opened laurel forest – tall specimen of *Weinmannia trichosperma* and *Laurelia philippiana* at "Carretera Austral" around 42.5° S.
Photo 12 (p. 138): Soil erosion and landslides following deforestation in the Palena area about 44° S.

MIX
Papier aus verantwortungsvollen Quellen
Paper from responsible sources
FSC® C105338

If you have any concerns about our products,
you can contact us on
ProductSafety@springernature.com

In case Publisher is established outside the EU,
the EU authorized representative is:
Springer Nature Customer Service Center GmbH
Europaplatz 3, 69115 Heidelberg, Germany

Printed by Libri Plureos GmbH
in Hamburg, Germany